Treatise on Heavy-Ion Science

Volume 2
Fusion and Quasi-Fusion Phenomena

Treatise on Heavy-Ion Science

EDITED BY D. ALLAN BROMLEY

Volume 1: Elastic and Quasi-Elastic Phenomena

Volume 2: Fusion and Quasi-Fusion Phenomena

Volume 3: Compound System Phenomena

Volume 4: Extreme Nuclear States

Volume 5: High-Energy Atomic Physics

Volume 6: Astrophysics, Chemistry, and Condensed Matter

Volume 7: Instrumentation and Techniques

Treatise on Heavy-Ion Science

Volume 2
Fusion and Quasi-Fusion Phenomena

EDITED BY

D. ALLAN BROMLEY

Henry Ford II Professor of Physics
Yale University
New Haven, Connecticut

PLENUM PRESS • NEW YORK AND LONDON

Library of Congress Cataloging in Publication Data

Main entry under title:

Treatise on heavy ion science.

Includes bibliographical references and index.

Contents: v. 1. Elastic and quasi-elastic phenomena — v. 2. Fusion and quasi-fusion phenomena.

1. Heavy ions—Collected works. I. Bromley, D. Allan (David Allan), 1926–

QC702.7.H42T74 1984 539.7 84-8384

ISBN 0-306-41572-0 (v. 2)

A Division of Plenum Publishing Corporation

233 Spring Street, New York, N.Y. 10013

Printed in the United States of America

For Pat, Lynn, and David

Preface to Treatise on Heavy-Ion Science

After a long gestation period, heavy-ion physics and chemistry is now, worldwide, the most rapidly growing area of nuclear science, and the concepts, techniques, and instrumentation of this heavy-ion work are finding ever-widening application in other areas of science and technology. Although there remain broad regions at higher energies, with heavier projectiles, and at higher excitations and angular momenta where heavy ions still provide gateways into the totally unknown, intensive studies over the past two decades have provided a sound framework of understanding of many of the salient features of interactions induced by these new heavier projectiles and a basis for coherent planning of future studies.

It thus seemed appropriate, at this point in the history of the field, to pull together in one place and in as coherent a fashion as possible, an overview of what has been accomplished and some enlightened speculation about where we go next. It is my hope that these volumes will provide a definitive reference for those working in this and neighboring fields—both students and professional scientists; beyond that I would also hope that they will make accessible to a much wider audience in other sciences and technologies some of the richness of heavy-ion research, and perhaps help to stimulate the transfer of techniques and concepts that I have already mentioned.

I have been singularly fortunate in planning these volumes in being able to persuade internationally recognized authorities to write in their areas of special expertise and interest, and also fortunate that I have not had to restrict them to any artificial or externally imposed procrustean restrictions on the scope or length of their chapters. I have asked each author to include sufficient background to make the chapter accessible to students and to the nonspecialist, to provide a broad selection of illustrations, and to feel free to extrapolate and to speculate about future directions.

In inviting contributions to these volumes I have made arbitrary decisions concerning both topics and contributors, and I make no claim to completeness. Indeed, a few chapters that I would have liked to include do not appear because of illness or other reasons which prevented their authors from completing them.

I should like to take this opportunity to thank all the authors represented for taking time in already full schedules to share with a wider audience their special experience and expertise in heavy-ion science. As was inevitable from the outset in a multiple-author venture of this scope—with over 65 different authors—my original scheduling and deadlines proved unrealistic. To all those authors who responded to them and produced manuscripts on or before the original deadline—in many cases, I am aware, at substantial personal cost—my most sincere thanks and appreciation. To them, too, however, go my apologies for the delay in bringing their work into print. I have delayed publication for over a year so that I might include a number of very important chapters without which the work would have been obviously incomplete.

Volumes 1–4 of the *Treatise on Heavy-Ion Science* are devoted to aspects of heavy-ion nuclear science, beginning with an overview of the historical development of the science and some of its simpler interactions—elastic and quasi-elastic, fusion and quasi-fusion phenomena—and moving from them to compound system phenomena and to much more complex and less well understood phenomena involving very heavy nuclear species and very high energies. Volume 5 is devoted to high-energy atomic physics, an entirely new field of science first made accessible by the availability of a broad range of heavy-ion beams, and one still very much in its infancy. Volume 6 considers the impact of heavy-ion studies on other sciences including astrophysics, chemistry, surface physics, condensed matter physics, materials science, and heavy-ion-induced fusion power. Volume 7, the concluding volume of this treatise, is devoted to some of the instrumentation peculiar to heavy-ion science and its applications.

Special thanks go to Ellis Rosenberg and Bennett K. Ragnauth of Plenum Press with whom it has been a pleasure to work on these volumes, and to Mary Anne Schulz for all her help in producing them. And I would also acknowledge my indebtedness to the Alexander von Humboldt Stiftung for a Humboldt Award that I was privileged to hold during part of the time these volumes were in preparation.

New Haven, Connecticut D. Allan Bromley

Contributors to Volume 2

John R. Huizenga, Department of Chemistry, and Nuclear Structure Research Laboratory, University of Rochester, Rochester, New York 14627

Ulrich Mosel, Institut für Theoretische Physik, Universität Giessen, 6300 Giessen, West Germany

Andrew M. Sandorfi, Brookhaven National Laboratory, Upton, New York 11973

Wolfgang U. Schröder, Department of Chemistry, and Nuclear Structure Research Laboratory, University of Rochester, Rochester, New York 14627

Preface to Volume 2

As the second volume of this treatise, Volume 2 is devoted to phenomena somewhat more complex than those considered in Volume 1. Mosel considers fusion reactions where a number of open problems remain; the dependence of the fusion cross section on angular momentum and upon the deformation of the participant nuclei are examples. Sandorfi discusses radiative capture of heavy ions; new developments in the instrumentation for such study have had a major impact on the field and beyond delineating fine structure in giant nuclear resonances, attention is currently focussing on higher multipoles at higher excitations and on giant resonances based on excited rather than ground states. Finally, Schröder and Huizenga provide a definitive treatment of current understanding of damped nuclear reactions —an entirely new mechanism intermediate between direct and compound reactions—wherein friction, the last classical variable to be brought into the quantum domain, appears to play an important role.

New Haven, Connecticut — D. Allan Bromley

Contents

3. Damped Nuclear Reactions

Wolfgang U. Schröder and John R. Huizenga

Contents of Volumes 1, 3–7

Volume 1: Elastic and Quasi-Elastic Phenomena

Volume 3: Compound System Phenomena

Volume 4: Extreme Nuclear States

Volume 5: High-Energy Atomic Physics

Volume 6: Astrophysics, Chemistry, and Condensed Matter

Volume 7: Instrumentation and Techniques

1

ULRICH MOSEL is Professor of Physics at the Justus Liebig-Universität in Giessen, West Germany. Receiving his graduate education at the Institute for Theoretical Physics of the University of Frankfurt, and his Dr. Phil. Nat. degree in 1968, he held research associateships at the University of Tennessee (1970–71) and the University of Washington (1971–72) prior to his appointment to this present position. He has held appointments as Visiting Scientist to both the Oak Ridge and Argonne National Laboratories and as Visiting Professor to the Departments of Physics at the University of Washington and the State University of New York, Stony Brook, New York.

Heavy-Ion Fusion Reactions

ULRICH MOSEL

1. Introduction

The fusion of heavy ions has captured the interest of many nuclear physicists over the last few years. This renewed interest was caused, on one hand, simply by the increasing availability of heavy-ion accelerators with suitable energy ranges. A deeper motivation, however, lies in the fact that a heavy-ion fusion reaction is one of the most drastic rearrangements that a nuclear many-body system may undergo.

It also offers the possibility to produce highly excited nuclei. This high excitation energy ends up partly in high thermal excitation, thus making the study of "hot" compound nuclei feasible. The available energy may, however, also be converted into rotational energy. The high angular momenta involved in heavy-ion reactions will then lead to the population of very high spin states. Fusion reactions are thus an ideal tool to study the nucleus at high spins and/or high temperatures. The total excitation energy of the compound nucleus (CN) is given by the sum of bombarding energy and the Q value of the reaction. Its separation into thermal and rotational energy is determined by the angular momentum distribution in the fusion channel.

Finally, it should not be concealed that one of the main motivations for the construction of heavy-ion accelerators was the expectation that once heavy nuclei were given enough energy to overcome the Coulomb barrier, they would fuse to form a heavy and maybe even a superheavy compound nucleus. The obvious failure to produce any of these long-lived nuclei (He 79, Lo 78) and the discovery of other reaction types instead, like deep-

ULRICH MOSEL • Institut für Theoretische Physik, Universität Giessen, 6300 Giessen, West Germany. Work supported by GSI Darmstadt and the Bundesminister für Forschung und Technologie (BMFT).

inelastic and quasifission reactions (see Chapter 3 by Schröder and Huizenga in this volume), has made it necessary to reach a far deeper understanding of heavy-ion-induced fusion reactions than had originally been thought.

In this article a coherent view will be given of the entire field of heavy-ion fusion reactions. After a short discussion of the experimental techniques to determine fusion cross sections and the gross properties of these in Section 2, two possible models will be discussed in Sections 3 and 4. Section 3 will deal with entrance channel effects and their description, mainly in terms of classical trajectory calculations. In Section 4 an alternative explanation of the mechanisms that limit fusion at high energy will be given in terms of yrast line effects.

These analyses are all based on the assumption that all partial waves up to a "critical" one will actually fuse. This is not so in results of time-dependent Hartree–Fock (TDHF) calculations that predict the existence of a fusion "window" such that the lowest partial waves do not fuse. This and the experiments designed to test this prediction are discussed in Section 5.

In Section 6 the effects of deformations on fusion—both dynamic and static ones—are treated. Whereas the main emphasis of this whole chapter lies on the fusion of "lighter" heavy ions, in Section 6.1 also the new phenomena that appear in reactions between heavy nuclei like, e.g., quasifission and the "extra push" for capture, are discussed. These phenomena are directly related to the fact that the dynamical evolution of a fusing heavy nucleus is governed to a large extent by the different shape-degrees of freedom (like necking) and not just by the radial distance between the two interacting nuclei. Finally, a very short discussion of fusion resonances comes at the end, just before a summary and outlook conclude this chapter.

Subbarrier fusion cross sections that are of interest for astrophysical processes are not discussed in this chapter. A good review of this field has recently been given by Rolfs and Trautvetter (Ro 78). The other subjects treated in this chapter have been reviewed before by Lefort (Le 76), Stokstad (St 80d), and Mosel (Mo 78, Mo 80). Furthermore, both fusion and deep-inelastic reactions have been the central topic of a fairly recent conference (Oe 79).

2. Experimental Determinations and General Properties of Fusion Cross Sections

Fusion is defined as the process in which two nuclei amalgamate to form a single, fully equilibrated compound nucleus (CN). This compound nucleus will in general be highly excited and decay by particle emission. The high angular momentum states will preferentially decay by α emission (or fission in the case of heavy nuclei) whereas the low angular momentum states will decay by light particle (p or n) emission. The fusion yield is then given by the sum of the evaporation residue yields plus (in the case of heavy nuclei and high excitations) fission.

The direct experimental methods to determine these fusion cross sections are based on either countertelescope measurements or γ-ray techniques. In the first type of measurements the heavy fragments (i.e., the evaporation residues) resulting from CN formation and subsequent particle evaporation are detected. The nuclear charge Z of the reaction products is identified by using standard $\Delta E - E$ detectors (St 77). On the other hand, in the γ-ray technique (Ko 79a) the γ-ray spectra for the lowest-lying transitions in the heavy residual nuclei are measured. If all important exit channels can be observed, this method gives directly the cross section. Otherwise, for example when the direct ground-state feeding is strong, the absolute cross sections for the production of individual evaporation residues must be deduced through statistical model estimates. The γ-ray method has two limitations: the very high energy γ-rays as well as reactions that feed directly the ground state cannot be observed. On the other hand, the γ-ray technique contains an inherent integration over angle and final state of the evaporation residues and is thus much faster than the particle detection methods.

Both experimental methods to determine the fusion cross section suffer from the difficulty of assigning the observed final nuclei to the class of either evaporation residues (ER) or direct reaction products. This is particularly critical in reactions involving relatively light symmetric or nearly symmetric target–projectile combinations, like, e.g., $^{14}N + ^{12}C$. In such a case determination of the yield of a given nuclide will often not suffice to disentangle fusion and direct reactions. In the given case of $^{14}N + ^{12}C$, for example, the high angular momentum states in ^{26}Al will preferentially deexcite by 3α emission and thus populate states and nuclei that can also be reached by direct reactions. In such a situation, the measured energy and angular distributions have to be exploited as well. These show clearly two contributions, the direct and the ER one (see Figure 2.1) which can be reproduced by statistical model calculations (Go 79a). Therefore, such calculations are also used to deduce the fusion cross section in cases where the distinction between the two processes is not so evident as in this example.

At very high bombarding energies a further complication arises. Wilczyński *et al.* (Si 79a, Si 79b, Wi 80a) have shown that there so-called "incomplete fusion" processes set in, in which only a part of the projectile fuses whereas another (usually smaller) part flies on. Incomplete fusion is thus nothing else than a "massive transfer" reaction. One example discussed by Wilczyński is that of the reaction $^{12}C + ^{160}Gd$, where (^{12}C, α), (^{12}C, 2α), and (^{12}C, 3α) processes are observed. The incomplete fusion reactions are

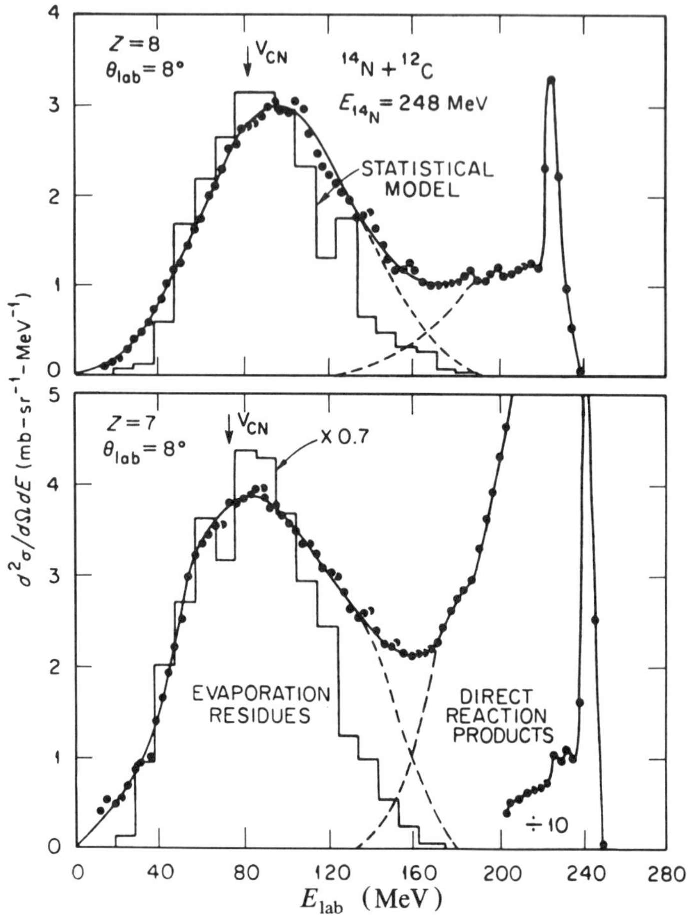

Figure 2.1. Energy spectra obtained in the reaction $^{14}N+^{12}C$ for $Z=7$ and $Z=8$ nuclei in the exit channel. The unfolding of the spectra into direct and compound contribution is indicated by the dashed lines. The histogram shows the predictions of a statistical model calculation (from St 80d).

dominated by the highest partial waves between the fusion and the deep inelastic region. Recently, incomplete fusion has also been observed in reactions between "light" heavy ions (Le 79). The evidence here is based on comparisons of experimental mass yields with those calculated with Hauser–Feshbach codes. It is found (Le 79, Le 81) that at the highest bombarding energies the lighter elements are experimentally considerably more strongly populated than is reproduced by the calculations (see Figure 2.2 for the reaction $^{20}Ne+^{26}Mg$). This is attributed to incomplete fusion where only a part of the projectile actually fuses, the CN formed is lighter than otherwise and thus the ER yield is shifted towards lighter masses. If this effect is taken into account in the statistical model calculations the measured mass distributions can be reproduced (see Figure 2.2). Of course, this analysis again depends on a model and it would, therefore, be very desirable to have direct experimental data, for example by measuring fast light particles in coincidence with the evaporation residues.

The complications that arise at high energies from the presence of incomplete fusion events and from the difficulties of separating the direct from the evaporation residue yield should be kept in mind when experimental cross sections are discussed. Unaffected by these general uncertainties connected with the identification of fusion events, some typical features, however, emerge from the experimental studies that have been performed for many target and projectile combinations.

(1) Fusion cross sections (σ_{fus}) exhibit two different energy regimes. At energies relatively low above the barrier (region I in Figure 2.3) they follow

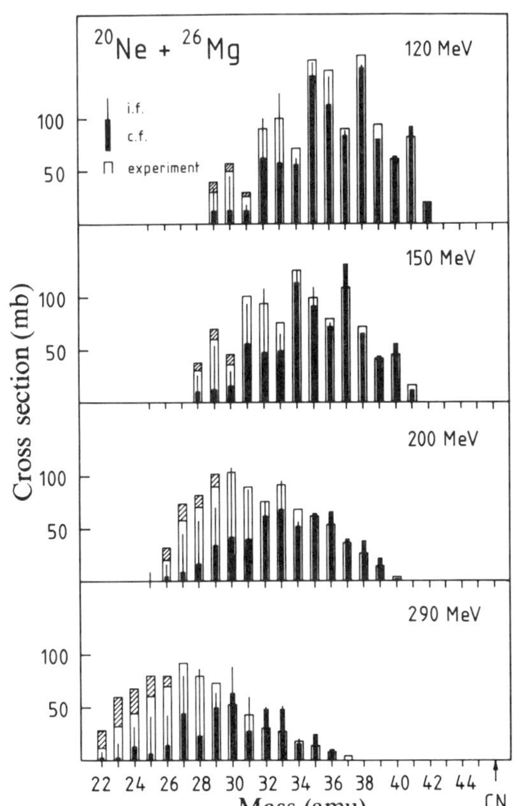

Figure 2.2. Experimental angle-integrated mass yields (vertical rectangular bars) obtained in the reaction $^{20}Ne + {}^{26}Mg$ at three different bombarding energies. The thick lines give predictions of a statistical model calculation assuming that the compound nucleus ^{46}Ti has been formed. The thin lines are results of a statistical model calculation for $^{16}O + {}^{26}Mg$, i.e., incomplete fusion (from Le 81). The shaded areas reflect uncertainties in the experimental cross sections.

the total reaction cross section (σ_R) whereas at the higher energies (region II) they fall clearly below it. This fall is very pronounced for lighter systems for which the fusion cross section exhausts a very large part of the reaction cross section at low energies. There are some indications that at very high energies (region III) fusion cross sections may exhibit even a third regime in which the process is dominated by the fission stability of the compound nucleus (Go 79a, Go 79b).

(2) Fusion cross sections can be very large; they reach maximum values of more than 1 barn in reactions involving relatively light nuclei.

(3) The ratio of fusion to reaction cross section drops strongly with the product of the charges of the two interacting nuclei (Va 79a) (see Figure 2.4). This dependence probably accounts partly for the failure to produce superheavy nuclei in fusion reactions.

(4) Whereas fusion cross sections are in general smooth functions of the energy, the cross sections for light systems exhibit pronounced oscillations (Sp 76a, Sp 76b, Ko 79b).

Qualitatively, fusion can be understood in terms of an interplay between conservative and dissipative forces: The nuclei approach each other

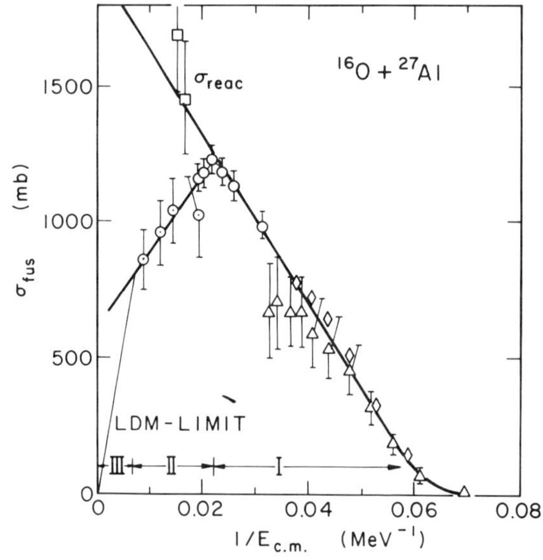

Figure 2.3. Fusion cross section for $^{16}O + ^{27}Al$. The fusion data are from Dauk *et al.* (Da 75, triangles), Eisen *et al.* (Ei 77, diamonds), Back *et al.* (Ba 77b, open circles), and Kozub *et al.* (Ko 75, circles with points). The two reaction cross-section data points (squares) were obtained by Cormier *et al.* (Co 76). The solid line is a fit to the data with the Glas–Mosel formula (Gl 74, Gl 75). The limit predicted by the rotating liquid drop model (LDM) is also indicated. The three energy regimes discussed in Section 2 are indicated at the bottom.

and feel the nucleus–nucleus interaction. Once they come in contact first peripheral reactions like inelastic excitations and single-particle transfer reactions and later on more violent ones will take place and convert kinetic energy into intrinsic excitations (Mo 80). This is what is classically described as friction. If the energy loss is large enough the nuclei will be trapped in the potential well of the nucleus–nucleus potential and kept together there for a time long enough to allow them to resolve their individual structures and merge into a single compound nucleus.

From this picture property (3) listed above can be understood: Since for very heavy systems the Coulomb potential becomes dominant the potential wells become very shallow. Thus the probability for trapping is reduced. This Coulomb effect probably explains the observed scaling of $\sigma_{\text{fus}}/\sigma_R$ with $Z_1 \cdot Z_2$ (Va 79a).

The outstanding problem then is to understand the observed limitation of the fusion cross sections at high energies, i.e., the rather sudden drop of σ_{fus} versus σ_R. There are two possible explanations for the observed drop, one relying on entrance channel and the other one on compound nucleus yrast line limitations. These will be discussed in Sections 3 and 4, respectively.

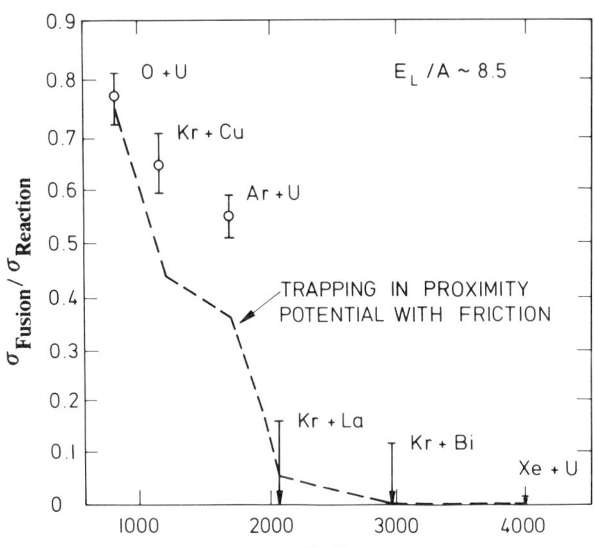

Figure 2.4. Ratio of reaction to fusion cross section for several reactions involving heavy targets. The dashed line is a prediction of the one-dimensional trajectory calculations described in Section 3.2 (from Va 79a).

3. Entrance Channel Effects

3.1. Phenomenological Analyses of Fusion Cross Sections

Entrance channel effects depend on specific properties of the target and projectile through their interaction potential and possibly the dissipative forces. It is now well understood that the low-energy data for fusion, for example up to $E_{\text{c.m.}} \cong 12.5$ MeV, i.e., $1/E_{\text{c.m.}} \cong 0.08\ \text{MeV}^{-1}$, in Figure 3.1 are determined by these properties. The data in this energy range are well described by the classical expression

$$\sigma_{\text{fus}} = \pi R_B^2 \left[1 - \frac{V(R_B)}{E_{\text{c.m.}}} \right] \tag{3.1}$$

The parameters R_B and the potential at R_B, $V(R_B)$, can be determined by fitting this expression to the data. This procedure was first followed by

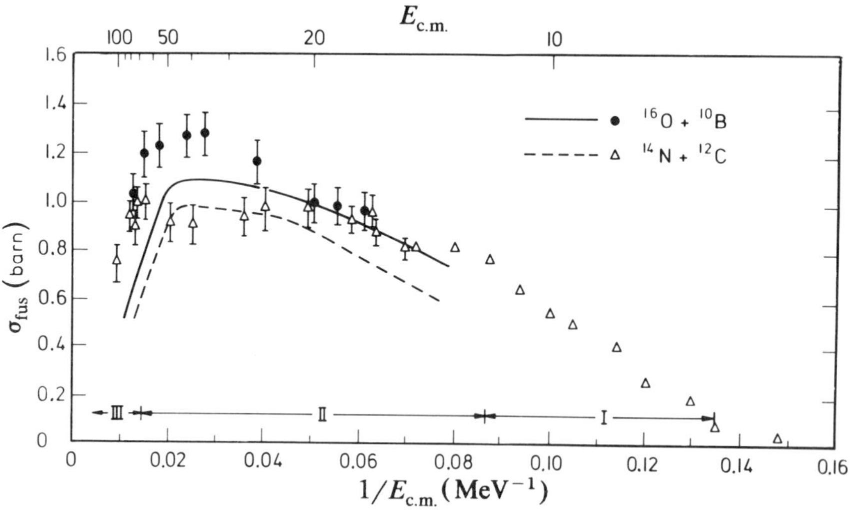

Figure 3.1. Fusion cross sections for $^{16}O + ^{10}B$ and $^{14}N + ^{12}C$, both leading to the same compound nucleus ^{26}Al. The data for $^{14}N + ^{12}C$ are from Gomez Del Campo *et al.* (Go 79a, triangles with error bars) and Kuehner and Almqvist (Ku 64, triangles). The data for $^{16}O + ^{10}B$ (solid circles) are from Gomez Del Campo *et al.* (Go 79b). The two curves give results of one-dimensional trajectory calculations by Vandenbosch in which the radius parameters were specifically adjusted (Va 80). The three energy regimes discussed in the text (section 2) are indicated at the bottom of this figure.

Gutbrod *et al.* (Gu 73). The values for R_B and $V(R_B)$ obtained were

$$R_B = 1.128A^{1/3} + 2.009A^{-1/3} - 1.5A^{-1}\ (\text{fm})$$

$$V(R_B) = \frac{Z_1 Z_2 e^2}{R_B} \tag{3.2}$$

in good agreement with radii and potentials at the interaction barrier as deduced, for example, from optical model analyses. This result implies that all particles that reach the barrier at low energies do indeed get trapped and fuse (neglecting for a moment the comparatively small cross sections for the other reaction channels).

This simple argument can be translated into an argument in angular momentum space. For this purpose one starts with the standard expression for a reaction cross section

$$\sigma_{\text{fus}} = \pi \lambda\!\!\!^{-2} \sum_{l=0}^{\infty} (2l+1) T_l \equiv \pi \lambda\!\!\!^{-2} \sum_l \sigma_l \tag{3.3}$$

where the T_l are the transmission coefficients. Making for simplicity now a sharp cutoff approximation for T_l:

$$T_l = \begin{cases} 1, & l \leq l_0 \\ 0, & l > l_0 \end{cases} \tag{3.4}$$

one obtains

$$\sigma_{\text{fus}} = \pi \lambda\!\!\!^{-2} \sum_{l=0}^{l_0} (2l+1) = \pi \lambda\!\!\!^{-2} (l_0 + 1)^2 \tag{3.5}$$

If l_0, the highest partial wave to fuse, is for example identified with the highest one at energy E which just passes the barrier, one obtains the classical expression of equation (3.1). Replacing the simple sharp cutoff value for T_l by the quantum mechanical penetration probability through a parabolic barrier Wong (Wo 73) has obtained a generalization of equation (3.5) that also includes some barrier penetration effects and thus allows one to determine also the width of the interaction barriers. Potential parameters obtained from such an expression have very recently been compiled by Vaz *et al.* (Va 81a).

A close inspection of the data at low energies, i.e., at large $1/E_{\text{c.m.}}$, shows that the measured cross sections are generally larger than the ones predicted by equation (3.1). This enhancement is due to the quantum mechanical penetrability of the barrier at energies below $V(R_B)$. This is just the effect that is taken into account in Wong's expression (Wo 73) mentioned above.

In the following discussions the maximum angular momentum to fuse, l_0, in energy ranges II and III will be called the "critical" angular momentum and denoted by l_{crit}.

3.2. Trajectory Calculations

The arguments presented above assume tacitly that no energy or angular momentum loss takes place before R_B is reached. Strictly speaking this assumption cannot be correct since peripheral reactions like inelastic excitations and transfer take place due to the long-range tail of the nucleus–nucleus potential. Based on these considerations the procedure described above to extract the potential parameters from the fusion data was criticized by Birkelund and Huizenga (Bi 78, Bi 79).

These authors have performed quite complete and extensive calculations of fusion cross sections in a classical trajectory model. Such calculations were first performed some years ago by Gross *et al.* (Gr 74a, Gr 74b, Gr 78). The newer ones by Birkelund *et al.* (Bi 79), however, have the advantage of starting with realistic nucleus–nucleus potentials as well as a physically motivated and derived treatment of nuclear dissipation that links radial and tangential friction (see below) unambiguously. In all these calculations trajectories of the two nuclei—coupled to the intrinsic excitations—are followed as a function of time. If a trajectory stays trapped in the potential well of the nucleus–nucleus potential such an event is identified with fusion. The highest partial wave that is trapped, l_0, is determined so that the fusion cross section can be obtained from equation (3.5).

The calculations performed by Birkelund *et al.* follow the equations of motion proposed by Randrup (Ra 78a, Ra 78b). In these it is assumed that the two interacting nuclei have one intrinsic degree of freedom each, a rotation, that is described by the angle Θ_1 or Θ_2, respectively. Since in this model intrinsic and relative angular momenta are aligned, the only other two coordinates necessary to describe the motion are the distance r and an angle Θ that describes the orientation of the line connecting the centers of the two nuclei in space. The Lagrangian is then given by

$$\mathscr{L} = \tfrac{1}{2}\mu\dot{r}^2 + \tfrac{1}{2}\mu r^2\dot{\Theta}^2 + \tfrac{1}{5}M_1R_1^2\dot{\Theta}_1^2 + \tfrac{1}{5}M_2R_2^2\dot{\Theta}_2^2 - V(r) \tag{3.6}$$

Here M_1 and M_2 denote the masses of target and projectile, μ their reduced mass, and R_1 and R_2 their radii. The intrinsic moments of inertia are taken to be those of a rigid rotor. The function $V(r)$ describes the nucleus–nucleus potential. In these calculations it is taken to be the proximity potential proposed by Blocki *et al.* (Bl 77).

Energy and angular momentum dissipation is in this model described by the Rayleigh dissipation function

$$\mathscr{F} = F(r)\cdot\left(\dot{r}^2 + \tfrac{1}{2}u_t^2\right) \tag{3.7}$$

where $F(r)$ is the shape function derived by Randrup and u_t is the tangential velocity mismatch in the overlap region (Ra 78a). This mismatch is a function of r, $\dot{\Theta}$, $\dot{\Theta}_1$, and $\dot{\Theta}_2$, as given by Randrup (Ra 78b). The structure of $\mathscr{F}$ shows that it contains both "radial" and "tangential" friction through the $\dot{r}$ and u_t term, respectively. The latter describes the friction occurring at the surface of the two nuclei in a grazing collision. It will thus incite an intrinsic rotation and, therefore, transfer angular momentum out of the relative motion into intrinsic rotation.

The equations of motion can now be obtained from

$$\frac{d}{dt}\left(\frac{\partial \mathscr{L}}{\partial \dot{q}}\right) - \frac{\partial \mathscr{L}}{\partial q_i} = -\frac{\partial \mathscr{F}}{\partial q_i}, \qquad i = 1,\ldots,4 \tag{3.8}$$

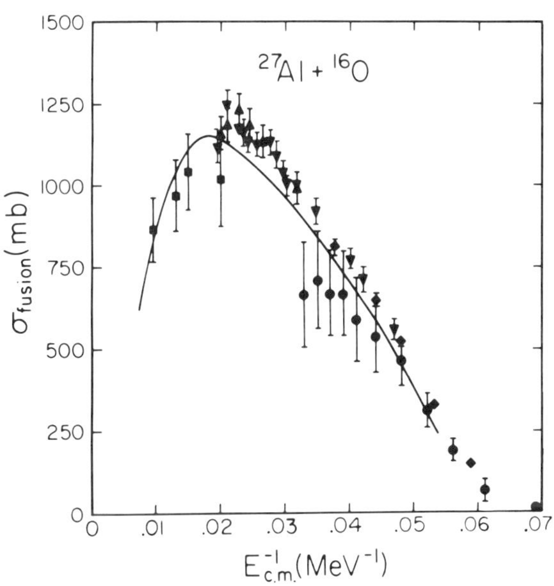

Figure 3.2. Prediction of the fusion cross section for $^{27}Al + ^{16}O$ in the one-dimensional trajectory model discussed in Section 3.2 [from Birkelund *et al.* (Bi 79), the data are from (Da 75, Ei 77, Ba 77b, Ko 75)].

for the four coordinates. They are

$$\mu\ddot{r} = \mu r\dot{\Theta}^2 - \frac{\partial V}{\partial r} - 2F(r)\dot{r}$$

$$\dot{l} = -F(r)u_t r$$

$$\dot{L}_1 = F(r)\frac{\partial u_t}{\partial\dot{\Theta}_1}u_t$$

$$\dot{L}_2 = F(r)\frac{\partial u_t}{\partial\dot{\Theta}_2}u_t \tag{3.9}$$

Here L_1 and L_2 are the two intrinsic angular momenta. These equations of motion are numerically integrated at each energy for different impact parameters. In this way the highest l value is determined for which the trajectory gets trapped. The fusion cross section is then obtained from equation (3.5).

These calculations give a coherent and quite good description of fusion cross sections for many systems throughout the Periodic Table. An example for the type of fit that can be achieved in this model is given in Figure 3.2.

3.3. Fusion Cross Sections at Low Energies

In the low-energy range the calculated fusion cross sections are quite similar to those obtained with the classical formula (3.5), although in this model friction effects take place outside the barrier (Bi 79). This can be understood from the following classical considerations (Gl 78a).

Assume that a particle with initial angular momentum l_0 and c.m. energy E_0 has lost the amounts Δl and ΔE on angular momentum and energy, respectively, by the time it reaches the barrier. Then the classical condition for just reaching the barrier distance R_B reads:

$$(l_0 - \Delta l)(l_0 - \Delta l) = \frac{2\mu R_B^2}{\hbar^2}(E_0 - \Delta E - V_B) \tag{3.10}$$

Here $V_B = V(R_B)$ denotes the barrier height. Inserting this into equation (3.5) gives for the fusion cross section

$$\sigma_{\text{fus}} = \pi R_B^2\left(1 - \frac{V_B + \Delta E}{E_0}\right) + \frac{\pi\hbar^2}{\mu E_0}l_0\Delta l \tag{3.11}$$

(assuming $\Delta l \ll l_0$). One sees that the energy loss decreases the fusion cross section because the particles may not have enough energy left to surpass the barrier and be trapped. On the other hand, the angular momentum dissipation increases the fusion cross section because it lowers the relative angular momentum and thus the centrifugal barrier. Both effects counteract each other.

If now the following relation holds

$$\frac{\Delta E}{\Delta l} = \frac{\hbar^2 l_0}{\mu R_B^2} = \frac{2(E_0 - V_B)}{l_0} \tag{3.12}$$

then equation (3.11) goes over into the classical, friction-free expression

$$\sigma_{\text{fus}} = \pi R_B^2 \left(1 - \frac{V_B}{E_0}\right) \tag{3.13}$$

Thus, equation (3.12) constitutes a *condition* for a situation in which the dissipation "masks" itself as far as its influence on the fusion cross section is concerned.

It is interesting to note that condition (3.12) is identically fulfilled in the calculations outlined above: At energies relatively low above the barrier the highest angular momentum that contributes to fusion is the grazing one. For it the velocity is almost exclusively tangential and thus one has for the energy dissipation rate

$$\dot{E} = -2\mathscr{F} = -F(r)u_t^2 \tag{3.14}$$

On the other hand from equation (3.9) one obtains

$$\dot{l} = -F(r)u_t r \tag{3.15}$$

and thus (at $r = R_B$)

$$\frac{dE}{dl} = \frac{u_t}{r} = \frac{2(E - V_B)}{l} \tag{3.16}$$

with $ru_t = l/\mu$. The equations of motion (3.9) thus fulfill exactly condition (3.12) in the barrier region.

The low-energy data up to the bend in σ_{fus} vs. $1/E_{\text{c.m.}}$ are thus insensitive to the particular type of dissipation. They are instead determined by the conservative potentials alone. Use of a "reasonable" nucleus–nucleus potential (such as the proximity potential) thus ensures agreement with experiment.

A consequence of these considerations is that the radius parameter R_B and the potential height V_B can indeed be extracted from the measured

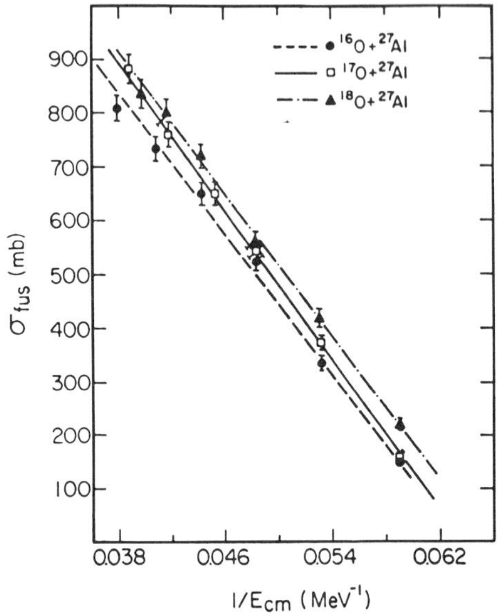

Figure 3.3. Fusion cross sections for the three reactions $^{16,17,18}O+{}^{27}Al$ at low energies. The lines illustrate the $1/E_{c.m.}$ behavior [from Eisen *et al.* (Ei 77)].

low-energy data. This is done best by plotting the fusion cross section σ_{fus} as a function of $1/E_{c.m.}$. The intersect with the $1/E$ axis then gives V_B while R_B is determined by the slope. This is the type of analysis first performed by Gutbrod *et al.* (Gu 73). This method has later been refined by Bass (Ba 77a), who takes the fact into account that the location of the interaction barrier will depend on l. It is, however, not clear if the data are indeed sufficiently accurate to allow such a detailed analysis (Bi 78).

A good example of the sensitivity of this method of extracting barrier heights and radii from the measured fusion cross sections is provided by the results of fusion of ^{16}O, ^{17}O, and ^{18}O with ^{27}Al obtained by Eisen *et al.* (Ei 77). One sees clearly in the plot of σ_{fus} vs. $1/E$ the small differences brought about by the small changes in mass of the oxygen isotopes (Figure 3.3).

3.4. Fusion Cross Sections at High Energies, Critical Distance Models

The bend in σ_{fus} at high energies and its fall below the reaction cross section obviously reflects the fact that not all partial waves that react also fuse. In the so-called critical distance models (Ga 74, Ba 74, Gl 74, Gl 75) this is achieved by postulating the existence of a critical distance R_c which the nuclei have to reach in order to fuse. This corresponds to the use of a particular form factor for the dissipation described above:

$$F(r) = \begin{cases} 0, & r > R_c \\ \infty, & r \leqq R_c \end{cases} \tag{3.17}$$

so that all nuclei that reach R_c get trapped. In this case the fusion cross section can be written down analytically. If penetration through a parabolic barrier is taken into account the result is (Gl 75)

$$\sigma_{\text{fus}} = \frac{\hbar\omega}{2} R_B^2 \frac{1 + \exp[2\pi(E - V_B)/\hbar\omega]}{1 + \exp\{2\pi[E - V_B - (R_c/R_B)^2(E - V_c)]/\hbar\omega\}} \tag{3.18}$$

Here E is the energy in the c.m. system, V_c the value of the potential (nuclear + Coulomb) at R_c, V_B the potential at R_B and $\hbar\omega$ is a measure for the width of the barrier.

Expression (3.18) has been derived under the assumption of l-independent barrier heights and radii. Whereas this is reasonably well fulfilled for heavy nuclei it does not hold for the lighter systems (Ha 80). However, the effects of varying R_B and V_B with l largely cancel each other in the cross-section formula and the cross section as a whole is relatively insensitive to this l dependence (Na 79).

For high energies the expression given above reduces to (Gl 74, Gl 75)

$$\sigma_{\text{fus}} = \pi R_c^2(1 - V_c/E) \tag{3.19}$$

whereas for low energies it goes over into the classical expression

$$\sigma_{\text{fus}} = \pi R_B^2(1 - V_B/E) \tag{3.20}$$

Thus in both energy regimes (I and II) the cross section assumes a linear dependence on $1/E$.

In terms of this model the bend in σ_{fus} is explained by the fact that at low energies all partial waves that surpass the interaction barrier will also reach R_c. At high energies on the other hand the highest partial waves will be kept outside R_c by the centrifugal potential.

In this simple classical model that relies on trapping the nuclei inside the potential well no partial waves can contribute to fusion for which the effective nucleus–nucleus potential (including Coulomb and centrifugal parts) does not exhibit a minimum. For such high energies where these partial waves contribute to the reaction the limiting angular momentum for fusion will become constant and thus σ_{fus} will decrease proportional to $1/E$.

The effects of angular momentum dissipation can be incorporated into this argument and act so as to shift this critical l value to still higher values (Ba 74). The maximum possible angular momentum loss is given by the so-called "sticking limit." If the two nuclei stick together they rotate with a single frequency, i.e., $\dot{\Theta} = \dot{\Theta}_1 = \dot{\Theta}_2$ in equation (3.9). Equating the initial angular momentum

$$L_i = l_i = \mu r^2 \dot{\Theta}_i \tag{3.21}$$

and the final

$$L_f = l_f + L_1 + L_2 = \left(\mu r^2 + \tfrac{2}{5} M R_1^2 + \tfrac{2}{5} M R^2\right)\dot{\Theta}_f \tag{3.22}$$

gives

$$\Delta l = l_i - l_f = l_i \frac{J_s - J_{ns}}{J_s} \tag{3.23}$$

with the sticking moment of inertia

$$J_s = \mu r^2 + \tfrac{2}{5} M_1 R_1^2 + \tfrac{2}{5} M_2 R_2^2 \tag{3.24}$$

and

$$J_{ns} = \mu r^2 \tag{3.25}$$

For a touching distance ($r = R_1 + R_2$) and symmetric systems ($M_1 = M_2$) this yields for the maximum loss of orbital angular momentum

$$\Delta l = \tfrac{2}{7} l_i \tag{3.26}$$

Thus, if the potential for l_i does not exhibit a pocket anymore but that for $l = \frac{5}{7} l_i$ still does, then fusion is still possible due to the angular momentum dissipation effects.

The critical distance model has been widely used to describe fusion cross sections. Although a possible explanation for the existence of a critical distance in terms of the break-up point of the individual shell has been suggested (Mo 80), there is, so far, no compelling evidence for its existence. The model does provide, however, at least a good basis for systematizing fusion data. It should also be noted that the existence of a bend in σ_{fus} vs. $1/E$ was predicted by this model at a time when no complete data covering the full energy range were available for any systems (Gl 74, Gl 75).

The full trajectory calculations with proximity friction performed by Birkelund *et al.* (Bi 79), of course, contain also the limitation at high energies due to the disappearance of the pocket in the nucleus–nucleus potential. The calculated bends in σ_{fus} appear approximately at the right energies and the calculations contain naturally the observed change of $\sigma_{\text{fus}}/\sigma_r$ with $Z_1 \cdot Z_2$ (see Figure 2.4).

In this model the bend is produced not by any critical distance but instead only by the disappearance of the potential minima with increasing l. If no tangential friction were present the bend would be sharp and appear too early; tangential friction smooths this transition. Thus this model really predicts only two basic energy regimes: One for low energies where the barrier is dominant and one for high energies where the effects of the disappearance of the potential pockets with l begin to be felt.

In assessing the physical validity of this particular entrance channel model one has, however, to be aware of the fact that the overwhelming majority of the data described lies in the low-energy domain, governed by the nuclear potential (see discussion above). There a trivial entrance channel effect prevails. Also one has to remember that the $1/E_{\text{c.m.}}$ energy scale favors these low-energy data whereas the crucial part of the comparison concerns the data at high energies.

A closer inspection shows that in the high-energy range the data do indeed show marked deviations from the results of these calculations. An example is given in Figure 3.1, which compares the data for $^{14}N + ^{12}C$ with those of $^{16}O + ^{10}B$ (Go 79b). The observed large difference between these two sets of data cannot be reproduced by the calculations discussed above (Bi 79, Va 80), even though the radius parameters have been adjusted for these particular nuclei by using information on the rms charge radii. Besides not being able to reproduce the difference between the two systems these calculations also give a poor overall fit to the data (Va 80).

As discussed above, only the high-energy data can be used to extract information on the dissipative properties from the fusion cross sections *provided* the conservative potential is known. Unfortunately, at high energies one probes deeper into the overlap region and there the potential becomes less well determined. Also it has been shown (Mo 79a) that even above the bend the fusion cross section is quite insensitive to the particular form factor for the dissipation.

4. Limitations of Fusion Cross Sections by the Yrast Line

4.1. Phenomenological Analyses

As was discussed in the preceding section, the entrance channel models give reasonable overall fits to the fusion data. They show discrepancies, however, when detailed comparisons are performed. It has, for example, long been known that the maximum fusion cross sections for $^{14}N+^{12}C$ and $^{15}N+^{12}C$ differ by about 150 mb (Ha 78, Co 78). Even more striking is the difference in the behavior of the two reactions $^{14}N+^{12}C$ and $^{16}O+^{10}B$ (see Figure 3.1) that both lead to the *same* compound nucleus. These differences cannot be described by the entrance channel models discussed in the preceding section.

It was noted by Harar (Ha 78), however, that the differences in these cross sections seem to be Q-value related. If the yrast line limits the number of partial waves then l_{crit} has to be identified with the yrast spin I and one has the relation

$$E^* = E_{\text{c.m.}} + Q = \frac{I(I+1)}{2J/\hbar^2} \tag{4.1}$$

where J is the moment of inertia of the yrast line. Inserting this expression into equation (3.5) yields

$$\sigma_{\text{fus}} = \frac{\pi J}{\mu}\left(1 + \frac{Q}{E_{\text{c.m.}}}\right) \tag{4.2}$$

Since Q is positive [and $V(R_c)$ negative] this form of the fusion cross section is very similar to that of equation (3.19). Harar (Ha 78) noticed that this model could explain the striking relative differences observed, for example, between $^{14}N+^{12}C$ and $^{15}N+^{12}C$. The absolute cross sections, however, that were obtained from equation (4.2) were too large. To achieve also absolute agreement Harar suggested that a band head energy ΔQ could be added to the rotational energy in equation (4.1), changing equation (4.2) into

$$\sigma_{\text{fus}} = \frac{\pi J}{\mu}\left(1 + \frac{Q - \Delta Q}{E_{\text{c.m.}}}\right) \tag{4.3}$$

That there may indeed be some physical reality in this analysis is illustrated by the striking example of $^{14}N+^{12}C$ on one hand and $^{16}O+^{10}B$ on the other. Whereas the cross sections for these two reactions that lead to the same CN are strikingly different (see Figure 2.3), this difference all but disappears if the data are plotted in the $E^*(l_{\text{crit}})$ representation (Figure 4.1). The strong difference thus seems indeed to be Q-value related. The same observation is true for a number of other combinations of projectiles and targets that lead to the same compound nucleus (Wi 80b, Sa 79).

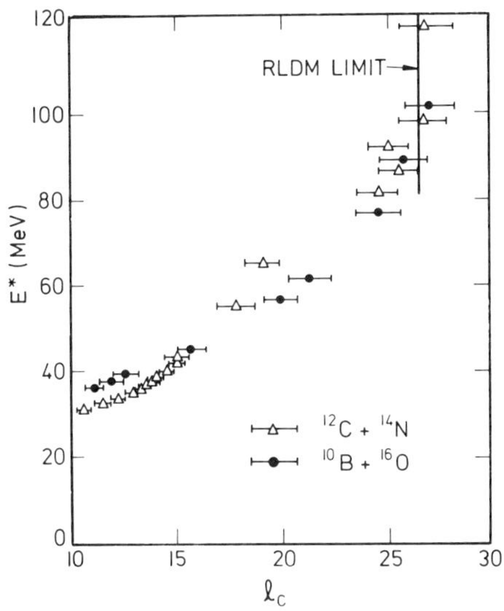

Figure 4.1. Fusion data for $^{12}C + ^{14}N$ and $^{10}B + ^{16}O$ in the $E^*(l_{crit})$ representation. Plotted is the excitation in the compound nucleus ^{26}Al vs. the critical angular momentum (here denoted by l_c). The data are from Gomez Del Campo *et al.* (Go 79a, Go 79b). The rotating liquid drop model (RLDM) limit is also indicated.

4.2. Determination of the CN Yrast Line

It is obvious that the verification of an yrast line effect depends on a knowledge of the yrast line for the compound nuclei in a spin range that corresponds to the region of the bend in σ_{fus} ($I \sim 12$–14 for light compound nuclei in the *sd* shell). Unfortunately, the yrast states in this mass region are experimentally not known beyond $I \approx 10$. This lack of reliable information is due to the low moments of inertia in this mass range. Correspondingly, the 10^+ state in ^{24}Mg (Sz 78) lies high above the threshold for particle emission so that standard γ-spectroscopic methods cannot be applied. It would, therefore, be extremely interesting if particle spectroscopy methods could be extended to a compound nucleus as light as ^{24}Mg. For example, Guerreau has shown that the ratio of multiplicities for α and proton emission can be a sensitive measure for the spin of the emitting nucleus (Gu 80).

Motivated by the lack of reliable experimental information on the location of the yrast line at high spin, calculations have been undertaken to determine the energies of high spin states. Unfortunately, shell model calculations that are very accurate and reliable in the *sd* shell cannot be performed for spins higher than, for example, $I = 12$ in ^{24}Mg (Wa 76). This is the highest spin that can be generated by nucleons in the *sd* shell alone. Higher spins would necessitate the inclusion of intruder states from higher shells. This, however, is not possible due to computer limitations.

In the absence of reliable shell model calculations one has to use less fundamental approaches. The first of these is based on the same methods that are used to explore the structure of high spin states in heavy nuclei. In these calculations the cranking model is used to calculate the energies of yrast states (Gl 78b, Di 80, Ra 81). The Hamiltonian used is that of an anisotropic harmonic oscillator including spin–orbit coupling:

$$\hat{H} = \frac{p^2}{2m} + \frac{m}{2} \sum_{i=1}^{3} \omega_i^2 x_i^2 + C\mathbf{ls} - \omega j_x$$

It is diagonalized in the basis of the cranked harmonic oscillator that can be treated analytically (Gl 79). Details of these calculations as well as an assessment of the reliability of these calculations can be found in a publication by Diebel *et al.* (Di 80). These calculations include all important higher shells by using a sufficiently large and deformed basis. They are thus able to include the effects of intruder shells on the yrast line in contrast to the shell model calculations. On the other hand, they suffer from the fact that the cranking model does not give states with sharp angular momentum. This difficulty can partly be overcome by using the Strutinsky method for rotating nuclei and combinatorial analyses (Di 80). On the basis of such studies the general inaccuracy of the calculations due to these effects were estimated to be $\Delta I \simeq 2$.

Beyond this deficiency the calculations suffer further from the neglect of residual interactions besides those that produce the deformed mean field. In particular pairing-type correlations are completely neglected although it is well known that these influence the moment of inertia significantly. Whereas in earlier calculations the effects of pairing were only estimated (Di 80), a new calculation has now been performed that generates the pair field from a surface–delta interaction and treats both the pairing correlations between like nucleons as well as between protons and neutrons (Mu 81a). The latter is obviously important because light $N = Z$ nuclei show only $T = 0$ pairing and *pn*-pairing correlations survive up to higher spins.

The results of this calculation for ^{24}Mg are illustrated in Figure 4.2, which also shows the data included in the $E^*(l_{\text{crit}})$ representation. The solid curve gives the calculated yrast line which reproduces the experimental yrast line up to the 8^+ state very well but lies still below the empirial "fusion band." However, the pair-correlated yrast line comes quite close to the fusion band in the range $I \approx 12$–14, i.e., just in the region of the bend in σ_{fus}.

In another approach to the yrast lines Thornton *et al.* (Th 80) have recently extracted these from microscopically calculated angular-momentum-dependent level densities. The level densities were obtained in random matrix model calculations using the spectral method of Ayik *et al.* (Ay 74). The extracted yrast lines agree reasonably well with the ones obtained in the cranking approach outlined above. The level densities that are the input to

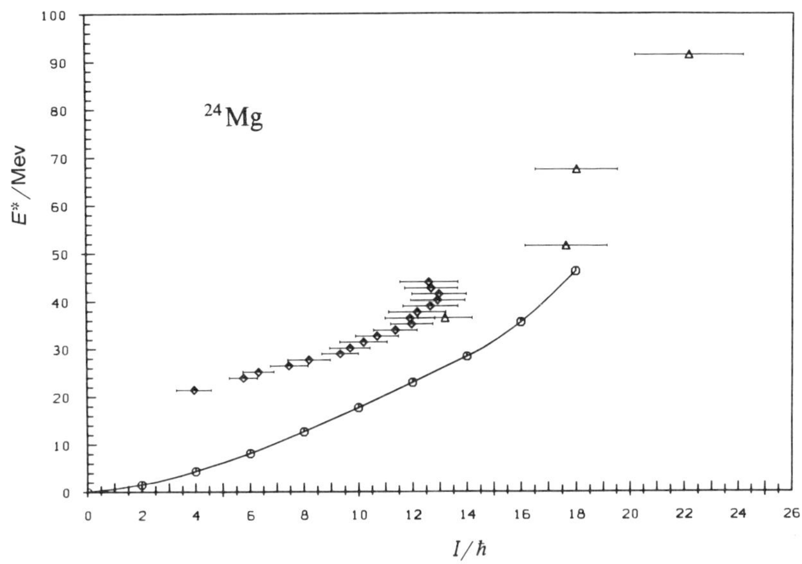

Figure 4.2. Fusion data for $^{12}C + ^{12}C \rightarrow ^{24}Mg$ as obtained by Kovar *et al.* (Ko 80, diamonds) and Namboodiri *et al.* (Na 76, triangles). Also given is the result of an yrast line calculation for ^{24}Mg (solid line and circles) that includes $T = 0$ (pn) pairing (Mu 81). This line reproduces the experimental yrast line for ^{24}Mg up to $I = 8$.

the yrast line calculations depend on average matrix elements of the interaction that have to be determined empirically. This introduces some ambiguity into the calculations and limits their predictive power when states beyond the observed ones are to be predicted. Indeed, the 10^+ state, for example, in ^{24}Mg is calculated to lie about 3–4 MeV below its experimental value (Sz 79). The level densities used by Thornton *et al.* (Th 80) include effects of single-particle orbitals only up the $0f_{7/2}$ state. The true yrast lines at high spin thus probably lie below the ones calculated. Again the empirical fusion bands lie above the calculated yrast line.

4.3. Statistical Yrast Line Effects

One is thus faced with the problem of reconciling two observations with each other: First, that the fusion bands lie above the calculated yrast lines, although they exhibit the bend just where the distance between the two is smallest, and second, that the differences between systems such as $^{14}N + ^{12}C$ and $^{16}O + ^{10}B$ can probably be understood on the basis of Q-value effects. These two observations together suggest that even though the yrast line itself is not reached, the density of available states in the compound nucleus may be too small at the highest partial waves for fusion to occur (Mo 79, Wi 80b, Di 80, Le 80, Va 79b, Va 81b). In this model not the yrast line itself but instead a line somewhat above it, corresponding to some fixed number of intrinsic states and termed the "statistical yrast line," affects the fusion cross section: Vandenbosch (Va 79b, Va 81b) has tried to make this argument more quantitative by demanding that the level density must be high enough so that the states of the compound nucleus do overlap. This leads to a condition $\Gamma/D \approx 1$, where Γ is the width and D the distance between states.

A reliable, quantitative understanding of the fusion cross sections in this model necessitates again a detailed knowledge of the spin-dependent level densities and thus of the yrast lines themselves. Here the main difficulty still persists: Whereas, for example, Lee *et al.* (Le 80) relate the statistical yrast line to a rigid rotor, Kolata in a recent letter (Ko 80a) points out that the bends in the fusion cross sections for light systems are often connected with extrapolations of the empirical ground-state bands (see

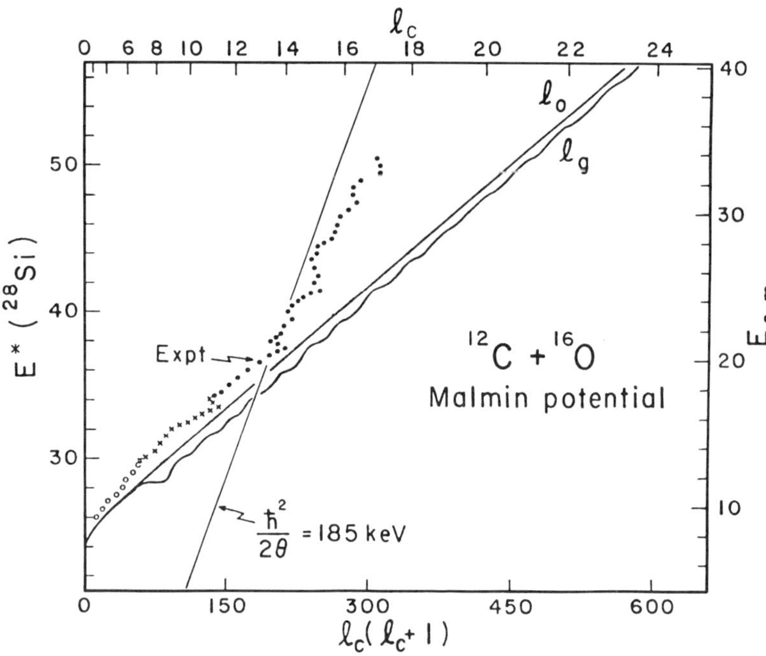

Figure 4.3. Fusion data for $^{12}C+^{16}O$ in the $E^*(l_{crit})$ representation. The two lines denoted l_g and l_0 give the grazing angular momentum and the maximum angular momentum as obtained from the reaction cross section by a sharp cutoff approximation. Also shown is the extrapolation of the experimental yrast line for ^{28}Si. The figure is taken from Kolata (Ko 80a).

Figure 4.3). The moments of inertia of the latter, however, often differ by a factor of 2 from the rigid rotor values.

Another fundamental question still remains unanswered in these models: There is evidence that the spreading width of nuclear states varies weakly with energy. Since this width is the product of a level density ρ and the square of the coupling matrix element V_{if} the weak energy variation implies that $|V_{if}|^2 \sim 1/\rho$ (Ag 77), reflecting the increasing complexity of final and initial states. This same argument may also be valid for fusion. Suppose that there are certain (at present unknown) doorways to fusion. These will generally be mixed with the continuum of CN states so that any one of these latter will carry only a very small part of the total strength. The matrix element V_{fi} will then correspondingly be reduced, thus counteracting the effects of the level density factor.

Applying this same argument to fusion means that the total number of available states may not be the relevant quantity. Instead, the doorway states leading to fusion would have to be identified and only their number should be counted. The nature of these doorway states is at present still unknown.

One final point concerning the possible influence of yrast line or statistical yrast line limitations should be made here. This point is connected with the "experimental" determination of fusion cross sections. These could simply be defined as the sum over all evaporation residue yields. The identification of these yields, however, presents a difficulty in reactions between light reaction partners because of the presence of direct reactions. For example, for $^{12}C + ^{12}C$ the evaporation yields for $Z \leqq 6$ are strongly mixed with the products from inelastic reactions. In this situation statistical Hauser–Feshbach calculations are used to extract the fusion contribution (Go 79, Co 78), as already discussed in Section 2. The "experimental" fusion sections thus contain contributions calculated in a statistical model. Although the energy and angular distributions are in general consistent with the Hauser–Feshbach predictions (Go 79), there remains the logical difficulty if the cross sections thus obtained can be interpreted in terms of an yrast line effect (including a statistical one). The statistical model used to arrive at the "experimental" numbers is, after all, expected to work only in a region of high level densities, i.e., at high *thermal* excitations. On, or close to, the yrast line, however, the excitation energy is present in form of *rotational* energy.

The considerations in this chapter have shown that the differences between fusion cross sections for different systems that lead to the same compound nucleus all but disappear if they are analyzed in the $E^*(l_{\text{crit}})$ representation. The fusion bands obtained in this way obviously separate the strong absorption region (to the left) from regions of weak absorption in the (E^*, l) space of heavy ion interactions. This weak absorption area is first quite small but then—at the point where σ_{fus} bends away from

σ_R—becomes suddenly large and—for fixed energy—extends from l_{crit} to l_{gr}.

The exact cause for this onset of weak absorption is not known yet. Probably, however, the yrast lines of the CN do play a role by limiting the available phase space. If indeed there exist (only a few) specific fusion doorway states then the energy differences of 10–20 MeV between the empirical fusion bands and the calculated yrast lines could still be compatible with a picture of statistical yrast line limitations.

If indeed there are any yrast line effects these should show up on the high angular momentum side of the CN population. The presence of incomplete fusion events may be related to this point: When the two nuclei come into close contact there may not be enough high spin states available for fusion. This allows the direct emission of a fast massive particle to compete favorably with fusion. Through this emission the angular momentum of the system is sufficiently lowered to bring the remaining nucleus–nucleus system into a region of high level density in the compound nucleus.

As was pointed out earlier in this section, the features of fusion cross for reactions leading to the same CN can be understood by invoking a Q-value dependence. It is, however, not possible to give a final answer to the question of entrance channel versus yrast line limitations, because there also seem to exist pairs of reactions that do not fit into this picture: The reactions $^{14}\text{N}+^{10}\text{B}$ and $^{12}\text{C}+^{12}\text{C}$ both leading to ^{24}Mg (Or 82) and the reactions $^{10}\text{B}+^{13}\text{C}$ and $^{11}\text{B}+^{12}\text{C}$ leading to ^{23}Na (Ma 81) show significant deviations in the $E^*(l_{\text{crit}})$ representation. The implications of this discrepancy are at present still unknown.

5. TDHF Descriptions of Fusion

The models that were discussed in the preceding sections all employ macroscopic descriptions of the process. All the relevant degrees of freedom have to be guessed and a number of specific assumptions, such as on friction form factors or moments of inertia, have to be made. Further, the formulations of the two models discussed previously, i.e., the entrance channel and the yrast line models, are mutually exclusive, so that the two mechanisms cannot dynamically compete with each other.

It would, therefore, be highly desirable to have a description of fusion that is free of all these shortcomings and in which the nuclear many-body system is dynamically followed. Such a description is provided by the time-dependent Hartree–Fock (TDHF) theory. Calculations of fusion cross sections in this theory have furthermore been quite stimulating because they

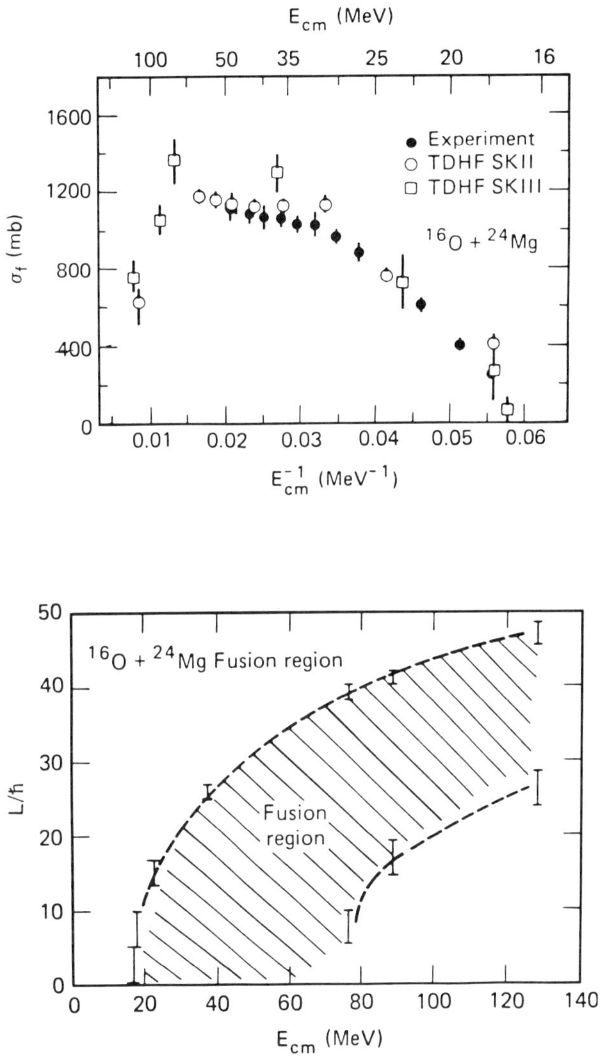

Figure 5.1. Fusion cross section for $^{16}O + ^{24}Mg$. The solid points in the upper part give the experimental values as obtained by Tabor *et al.* (Ta 78). The open symbols give predictions of TDHF calculations with two different versions for the effective nucleon–nucleon force (Kr 81). The lower part of the figure shows the fusion "window," i.e., the hatched area denotes the partial waves that do fuse as a function of center-of-mass energy $E_{c.m.}$ [from Krieger and Davies (Kr 81)].

have predicted the existence of a "fusion window." Whereas all descriptions of fusion cross sections in the preceding sections have assumed that *all* partial waves up to l_{crit} do actually fuse, the TDHF results have cast doubt on this assumption.

Since about 1976 it has become possible to apply the TDHF approximation to the calculation of physical quantities in heavy-ion collisions (Bo 76, Ma 76, Ko 77a). In these studies a wide range of nuclear processes has been observed, including among others fusion. Fusion in these calculations is operationally defined as a process in which a combined nucleus–nucleus system evolves for a sufficiently long period of time without undergoing prompt fission. The calculations are performed for different energies and impact parameters, and the values of l for which fusion (defined in the sense just discussed) occurs are determined (Bo 78, Bo 79, Kr 78, Kr 79). The TDHF calculations do indeed reproduce the empirical fusion cross sections for some systems quite well (see Figure 5.1), whereas in other systems, like $^{40}Ca + ^{40}Ca$, the calculated fusion cross sections depend quite strongly on the force parameters used (Bo 78, Bo 79, Kr 81).

The unexpected outcome of these calculations is that at high bombarding energies the low partial waves do not fuse and that this l window increases with increasing bombarding energy (Bo 79). The fall of the fusion cross section at high energies is then not due to a limitation in l_0 but instead to the opening of this low-l-window. This is illustrated in Figure 5.1. The upper part shows the calculated cross section whereas the lower part of this figure exhibits the fusion window. The shaded area in the lower part gives the range of partial waves that do fuse. It is seen that between 20 and 80 MeV c.m. energy all low partial waves fuse. At about 80 MeV, however, the window opens rapidly and at about 130 MeV, the highest energy calculated, the lowest ≈ 28 partial waves will not fuse. The fusion cross section at this energy is thus predicted to be due to the partial waves between about 28 and 48. Due to the large angular momentum window the fusion cross section starts to drop at about 80 MeV.

In a head-on collision, i.e., at low l values, no rotation can occur and the system must absorb all the available kinetic energy into vibrations. As the energy increases it may not be able to do so without breaking apart. It is thus not surprising that a similar fusion window has also been obtained in a semiclassical treatment of heavy-ion collisions in terms of damped surface modes (Br 78). Alternatively, the energy could also be absorbed into heat, i.e., internal, incoherent excitations, and the presence (or absence) of a fusion window could possibly tell something about this energy absorption mechanism.

The first experimental searches for the low-l cutoff have exploited the fact that the Z and A distributions are sensitive to the distribution of angular momenta in the compound nucleus. The low l values are preferentially connected with single nucleon emission and thus populate the heavier residues. Cutting out the low l values from the l distribution should, therefore, lead to a smaller yield at the upper ends of the Z and A distribution. Up to now all searches for this effect have been negative (Fe 78, Sa 79).

Another type of experiment to search for the l window has been designed to look for the low-energy fragments with low l that do not fuse. Experiments along these lines have very recently been performed by Lazzarini *et al.* (La 81) for the reaction $^{16}O + ^{16}O$. For this particular reaction TDHF calculations predict that at $E_{c.m.} = 34$ MeV the partial waves $l = 0$, 2, 4, and 6 do not fuse. The corresponding events would be connected with processes in which the nuclei just pass through each other and lose about 20 MeV of kinetic energy. The cross section for these events is predicted to be 132 mb. The experiment by Lazzarini *et al.*, however, shows no evidence for the occurrence of such an effect; the total inelastic yield for events with Q values more negative than -15 MeV is 5.9 mb. Thus, no evidence for the occurrence of an l window in the fusion process is seen in this experiment.

The experimental results from the two different methods on the existence of a low-l cutoff in fusion are thus presently all negative. Furthermore, recent one-dimensional calculations extending the TDHF method by including residual interactions in a random matrix model have shown that the fusion window can disappear when the two-body collision terms are taken into account (Gr 81). Three-dimensional calculations are not yet capable of handling these terms but are expected to give similar results. Ultimately, the experiments designed to look for effects of the fusion window may thus tell us more about the deficiencies of the TDHF approximation, such as the neglect of two-body collision terms, than about nuclear fusion.

6. Effects of Deformation on Fusion Cross Sections

6.1. Dynamic Deformations, Fusion of Heavy Nuclei

The considerations in the preceding sections were nearly all restricted to light nuclei. In this mass range structure effects are more likely to show up and also wider energy ranges are accessible here. This restriction to light nuclei was tacitly also made in the theoretical analyses: All the entrance channel models were one-dimensional ones.

In reactions between heavy nuclei the shape polarization effects are expected to be larger than for light reaction partners. A comparison of the configuration of two touching more or less unperturbed nuclei with that at the scission point during a fission process, for light nuclei on one hand and heavy ones on the other, also shows that much wider ranges in deformation space are available for heavy nuclei than for light ones. In other words: The one-dimensional heavy-ion potential *curve* used in the models discussed in the previous sections has to be replaced by a multidimensional potential energy *surface*. Then the potential energy of the nucleus–nucleus system depends not only on the c.m. distance but also on other shape parameters like, e.g., neck-in and elongation degrees of freedom (Bj 72). Heavy-ion-induced reactions in general provide the means to explore the potential energy surface in regions that are otherwise not accessible.

For reactions involving heavy nuclei fusion and fission proceed through quite different regions in deformation space. Whereas the shape of the two nuclei at first contact is approximately still that of the two undisturbed nuclei, e.g., two touching spheres, the nuclear shape at the scission point of heavy nuclei is that of two highly deformed fragments. This can also be seen in Figure 6.1, which shows a potential energy surface for $^{110}Pd+^{110}Pd$ calculated by Nix and Sierk (Ni 77) in the liquid drop model. In the particular shape parametrization used, the point of contact for two spherical ^{110}Pd nuclei lies at a distance $r \approx 1.6$ and an elongation $\sigma \approx 0.71$ (see Figure 6.1). By looking on the potential energy map in this figure it is evident that

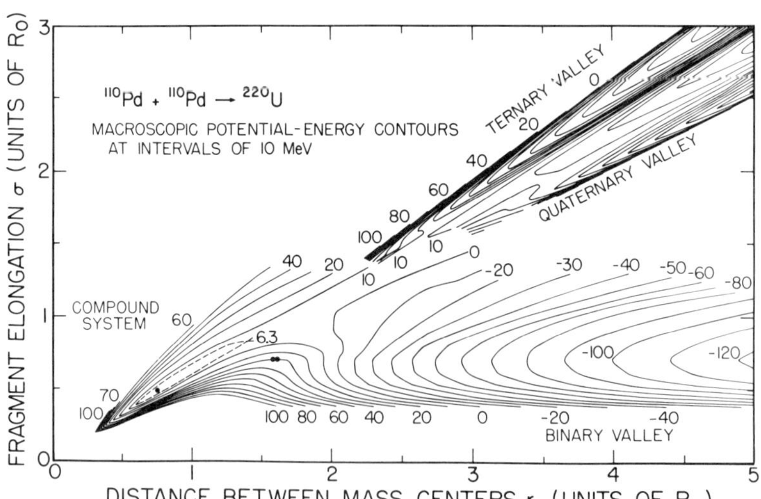

Figure 6.1. Potential energy surface for ^{220}U [Nix and Sierk (Ni 77)] as a function of two shape parameters r and σ. Whereas r directly describes the distance between the centers of mass of the two interacting nuclei, σ is a measure for the deformation of the two. The ground state of ^{220}U is normalized to zero energy; the numbers at the contours give the energies relative to it in MeV. The contact point of two spherical touching ^{110}Pd nuclei lies at $r \approx 1.6$, $\sigma \approx 0.71$.

just reaching this point or one slightly beyond it is not sufficient to guarantee fusion: The fission saddle point lies just inside the contact point. The system after having reached the latter can, therefore, also be injected into the fission path without ever reaching full equilibrium in the shape degrees of freedom. Such events can be identified with deep inelastic (or quasifission) reactions (see the review by Schröder and Huizenga in Chapter 3 of this volume). To reach the spherical compound configuration at $r \approx 0.75$ and $\sigma \approx 0.49$ obviously requires an "extra push."

This expectation is indeed borne out by the result of dynamical calculations performed by Nix and Sierk in 1974 (Ni 74). These calculations, which were based on the liquid drop model and assumed instant clutching at contact, have shown that for example for a central collision of $^{150}Nd + ^{150}Nd$ an energy of at least 50 MeV above the interaction barrier is required to push the system towards the spherical compound configuration (Ni 74).

On the basis of the arguments given above, it is expected that this additional energy is dependent on the mass of the system involved. For compound nuclei lighter than $A_{CN} \approx 220$ the fission saddle point appears at larger distances between the mass centers and corresponds to more elongated shapes than at the contact point. Thus, here the fission saddle point lies "outside" the contact point. The forces connected with an instability of the saddle point against elongation can in this situation not directly inject the dinuclear system into the fission valley. At the same time the energy difference between contact and ground-state configuration becomes larger for lighter nuclei. This increasing energy difference produces stronger forces that pull the system into a shape-equilibrated compound nuclear configuration. With increasing A_{CN} the situation changes, however. The contact configuration moves outside the fission saddle and an "extra push" is needed to reach the ground-state configuration. At the same time the fully equilibrated CN configuration becomes less and less stable with increasing fissility.

When angular momentum is taken into account this argument has to be modified somewhat. Now, for fixed A_{CN} the saddle point moves with l towards smaller deformations. Thus, whereas the saddle lies outward of the contact point for low l values and, therefore, no extra push is needed, for high angular momenta it may lie inside and correspondingly fusion will not occur at the barrier energy. This is illustrated in Figure 6.2 for the same reaction $^{110}Pd + ^{110}Pd \rightarrow ^{220}U$ discussed earlier. This figure shows the dynamical paths for different l values at an energy of 20 MeV above the maximum of the $l = 0$ interaction barrier. Also indicated are the positions of the fission saddle points for the various l's considered. It is seen that $l = 45$ is a critical angular momentum; for all higher l values the dynamical paths stay outside the fission saddle; the system will thus be injected back into the

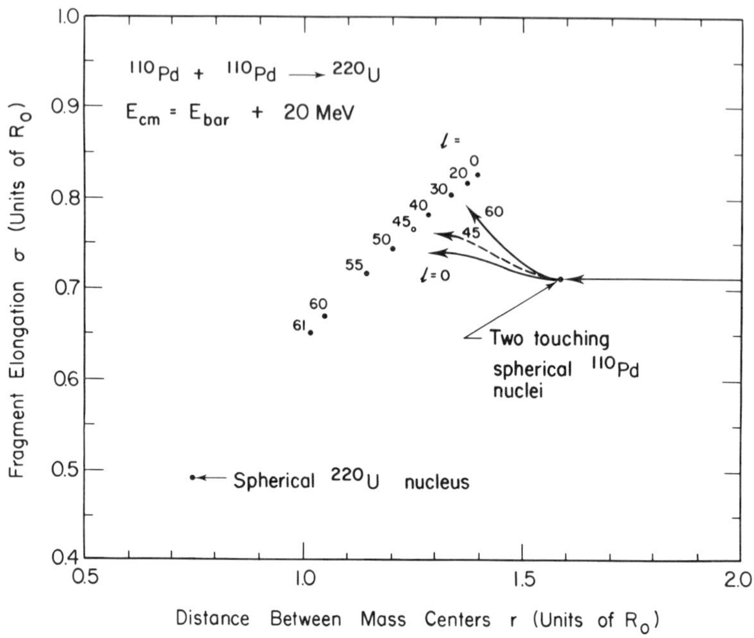

Figure 6.2. The dynamical paths (arrows) for ^{110}Pd + ^{110}Pd on the potential energy surface shown in Figure 6.1. The numbers at the arrows give the angular momenta. The dots denote the location of the fission saddle point for the l-values indicated. The location of the spherical compound nucleus is also shown [Nix and Sierk (Ni 77)].

fission valley, and fusion will not occur. As these calculations contain no energy or angular momentum dissipation beyond that implied by the clutching approximation, the dividing line will in reality probably lie at somewhat lower A_{CN} and l values than given by the calculations of Nix and Sierk (Ni 77).

For asymmetric entrance channels the situation looks slightly more complicated. Following Swiatecki (Sw 81, Sw 82) one can speak here of two saddle points, the conditional one, which occurs when the mass asymmetry of the entrance channel remains frozen, and the unconditional, true fission saddle point. The arguments given just before then apply directly to capture behind the conditional saddle. This, however, is not a sufficient condition for true fusion–fission to occur since in such a case the asymmetry could change and the system could follow a trajectory entirely outside the true, unconditional saddle. Such events are connected with fairly long time scales and should thus *resemble* true fusion–fission events. In particular, in contrast to deep-inelastic reactions which nearly preserve the entrance channel asymmetry, the "capture" events just described would be expected to lead to more symmetric mass distributions resembling true CN fission.

The gross properties of nuclear fission are determined by the fissility parameter $x = Z^2/A$ (Va 73), i.e., the ratio of Coulomb to surface energy, such that nuclei with the same value of x have similar fission properties. Similarly, it is possible to define a parameter that controls the gross properties of fusion reactions. From the work of Bass (Ba 74) one can see that the one-dimensional interaction potential is given by $[s = r - (R_1 + R_2)]$

$$V_l(r) = \frac{Z_1 Z_2 e^2}{r} + \frac{\hbar^2 f^2 l^2}{2\mu r^2} - a_s d \frac{A_1^{1/3} A_2^{1/3}}{R_1 + R_2} e^{-s/d} \tag{6.1}$$

Here the last term represents the attractive surface energy contribution with a_s being the surface energy constant and d a range parameter. The factor f in the second term allows for possible modifications of the simple moment of inertia due to tangential friction. For example, for rolling or sticking motion f assumes the values $5/7$ or $J_{\text{ns}}/(J_{\text{s}} + J_{\text{ns}})$, respectively (see Section 3.4).

Equation (6.1) shows that different nuclear systems with different angular momentum l will all be in equilibrium at contact ($r = R_1 + R_2$) if they agree in the parameter:

$$\left(\frac{Z^2}{A}\right)_{\text{eff}}(l) = \frac{4 Z_1 Z_2}{A_1^{1/3} A_2^{1/3}\left(A_1^{1/3} + A_2^{1/3}\right)} + \frac{4\hbar^2}{m_0 r_0 e^2} f^2 l^2 \frac{A_1 + A_2}{A_1^{4/3} A_2^{4/3}\left(A_1^{1/3} + A_2^{1/3}\right)^2} \tag{6.2}$$

For the first, l-independent term Swiatecki has introduced the name "effective fissility" (Sw 81). The total l-dependent parameter is just the ratio of the sum of Coulomb and centrifugal force to the nuclear proximity force. One expects that systems with the same $(Z^2/A)_{\text{eff}}$ will show a similar dynamical behavior (even though the simple argument given here is based only on a static model).

The dynamic behavior has recently been studied by Swiatecki in a simplified, but physically very transparent model (Sw 81). The description used is basically that of the LDM and includes a very strong one-body dissipation (Ra 78a, Ra 80). Swiatecki finds that the extra push needed is closely described by the expression

$$\Delta E = a \frac{A_1^{1/3} A_2^{1/3}\left(A_1^{1/3} + A_2^{1/3}\right)^2}{A_1 + A_2} \left[\left(\frac{Z^2}{A}\right)_{\text{eff}}(l) - \left(\frac{Z^2}{A}\right)_{\text{eff}}^{\text{thr}}\right]^2 \tag{6.3}$$

where a is a numerical factor and $(Z^2/A)_{\text{eff}}^{\text{thr}}$ a threshold parameter for the effective fissility. The quantity ΔE is in Swiatecki's model the energy needed for capture behind the *conditional* saddle. A true compound nucleus formation could then occur—if at all—only at still higher energies.

The transition from genuine fusion to quasifission with increasing A and the "extra push" seems indeed to be seen in new data on the fusion of very heavy systems with compound nuclear masses in the range $A \approx 234$–272 obtained recently at GSI (Sa 80). In these experiments light targets ranging from ^{26}Mg to ^{64}Ni were bombarded with a ^{208}Pb beam at several bombarding energies between 4.8 and 8 MeV/A. This corresponds to center-of-mass energies from the barrier up to 40% above it in composite systems with total charge between 94 and 110.

The data show that for light targets up to ^{52}Cr the symmetric mass distributions originating in the fission of the amalgamated nuclei are well localized between inelastic events and grow in intensity with increasing incident energy (see Figure 6.3). For the heavier targets the symmetric mass distribution sets in only at energies well above (5%–14%) the Coulomb barrier. The angle- and energy-integrated cross section for all the systems studied is shown in Figure 6.4 as a function of the incident energy. In all the systems the cross section increases steadily with energy but the heavier systems do not show capture until an extra energy above the Coulomb barrier has been supplied.

In these latter systems, furthermore, the cross section exceeds by far the stability limits predicted by the rotating liquid drop model as obtained by Cohen, Plasil, and Swiatecki (Co 74). This is the limit at which the centrifugal stresses overcome the nuclear binding so that the fission barrier reaches zero. The excess of capture cross section beyond the liquid drop limit implies that partial waves contribute to the symmetric yield far in excess of the angular momentum where the fission barrier vanishes. These events are probably those discussed above where the two nuclei overcome the interaction barrier and the conditional saddle but then are injected back into the fission channel. Indeed the data can be well described by equation (6.3) with only two parameters fitted. For example, the threshold parameter turns out to have the value $(Z^2/A)_{\text{eff}}^{\text{thr}} \approx 32.5$ (Sa 81, Bo 82, Sw 82).

Although the data of Sann *et al.* seem to confirm the existence of an extra push, a closer inspection of the data also shows that for the heaviest systems the measured capture cross section at bombarding energies just at the barrier is somewhat larger than predicted by equation (6.3) (see Figure 2 in Sa 81). This small excess yield is in line with the results of Armbruster and his group at GSI, who have studied true fusion reactions between heavy nuclei (Mu 81b, Sc 81). These authors find true fusion at energies directly at the barrier by identifying the ER residues (Sc 81) or the α-decay chains (Mu 81b) of the heavy CN. This result, however, may not be a contradiction to

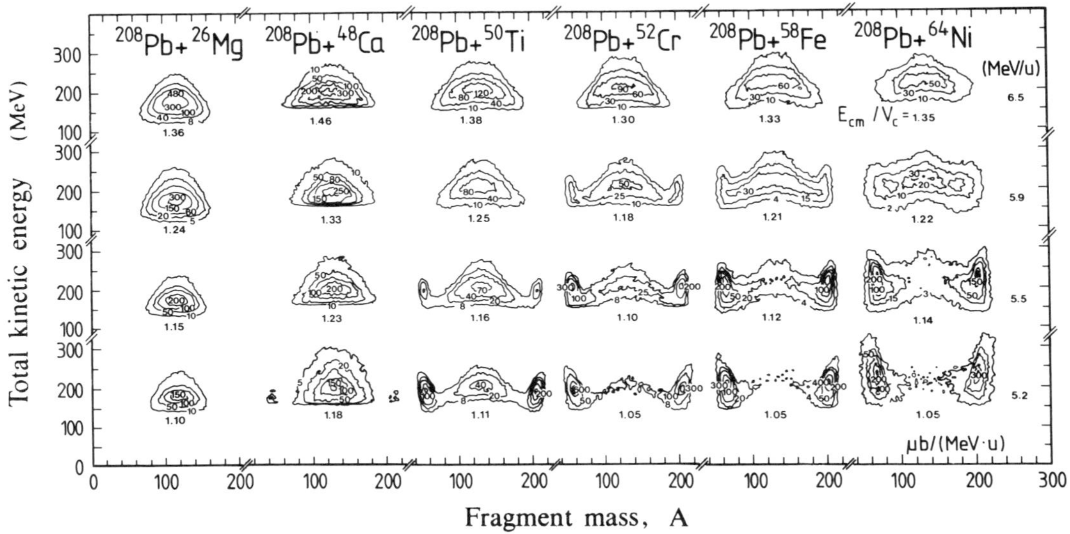

Figure 6.3. Kinetic energy distributions as function of the fragment mass for the reactions indicated. The small numbers below the contour lines give the bombarding energy in units of the Coulomb-barrier height [from Sann *et al.* (Sa 80, Sa 81)].

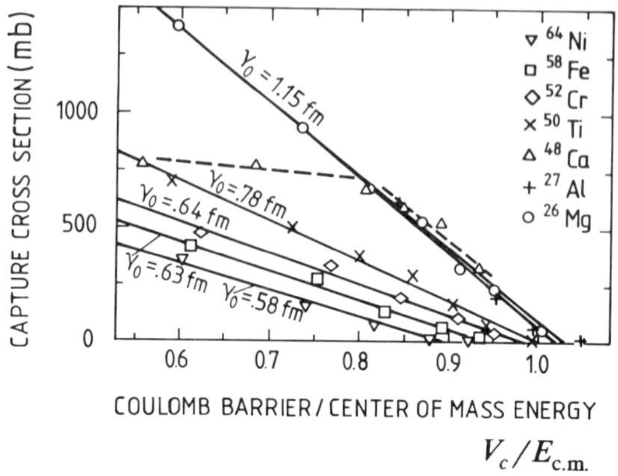

Figure 6.4. Fusion cross sections for reactions between ^{208}Pb and the nuclei indicated in the figure as a function of the Coulomb barrier height divided by the center-of-mass energy. The numbers give the radius parameter r_0 in a fit of the data by equation (3.1) assuming $V(R_B) = V_c$. The lines are drawn to guide the eye (from Sa 81).

the prediction of an extra push as it is probably due to the fact that at low energies, just at the barrier, quantum mechanical barrier penetration effects may become important in just the same way as they show up in lighter systems (see Section 3.1). True CN fusion at higher energies—as predicted by Swiatecki—has so far not been seen. This is probably due to the low survival probability of highly excited heavy compound nuclei.

Furthermore, equation (6.3) shows that the extra push needed—and thus the excitation energy—would be smallest for asymmetric target–projectile combinations. It would thus be very important to verify experimentally if the capture cross sections are indeed similar for systems with the same effective fissility. First experiments along these lines are presently being performed by Gäggeler and Sikkeland (Ga 81). One also has to remember that equation (6.3) has been obtained from very schematic calculations. In particular, the influence of angular momentum has so far been treated only in a very simplified way. It thus remains to be seen if the quantitative dependence of the extra push on l, A, and Z will be confirmed by more detailed calculations.

6.2. Static Deformations

The deformation effects that have been discussed so far are *dynamic* effects: The strongly interacting nuclei get distorted once they come in contact and form highly deformed amalgamated systems. It was discussed that these effects are expected to become important for heavy nuclei.

Another effect of nuclear deformations on the fusion cross sections is that induced by *static* deformations of the interacting nuclei: If one (or both) of the nuclei is deformed one expects an interaction barrier that is strongly dependent on the mutual orientation of the two nuclei. For example, if a deformed prolate nucleus approaches a spherical one the first interaction will occur at larger c.m. distances **r** if the projectile is oriented with its symmetry axis along **r** than in a configuration where it stands vertically on **r** (see Figure 6.5). These orientation effects should thus critically affect in particular the subbarrier fusion cross sections.

If the transmission coefficient through the barrier at a given orientation Θ is denoted by $T_l(\Theta)$ the fusion cross section for this orientation is given

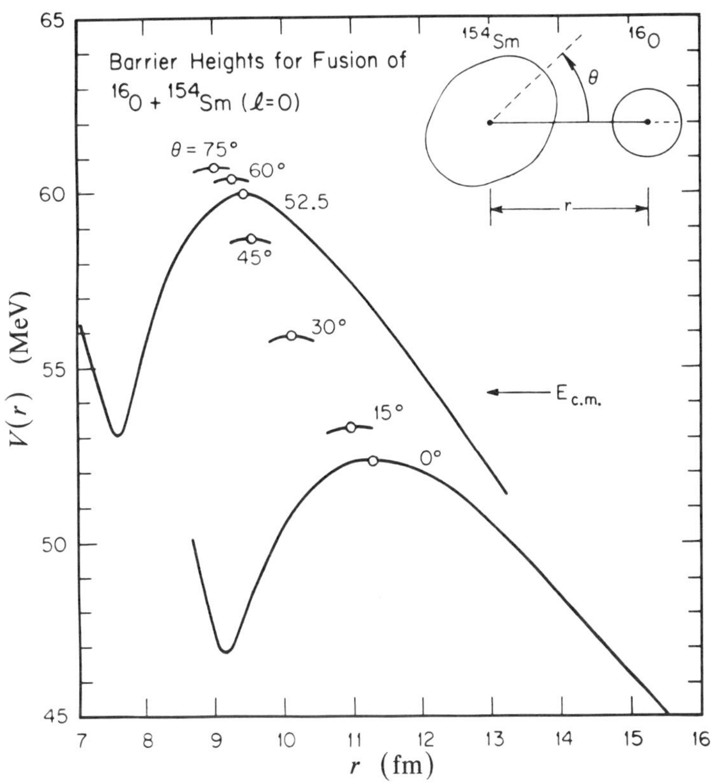

Figure 6.5. Barrier heights for the interaction of $^{16}O + ^{154}Sm$ for $l = 0$ as a function of the angle Θ describing the orientation of the deformed ^{154}Sm. The proximity potential (Bl 77) has been used to calculate the nuclear interaction. Also indicated is the lowest c.m. energy used by Stokstad *et al.* (St 80a) (from St 81).

by

$$\sigma_f(\Theta) = \pi \lambdabar^2 \sum_{l=0}^{\infty} (2l+1) T_l(\Theta)$$

This cross section is plotted in Figure 6.6 (St 81). It is seen to span four orders of magnitude for the example $^{16}O + ^{154}Sm$, depending on the relative orientation.

Since the target is not aligned these cross sections have to be averaged over orientation. The total fusion cross section is then obtained from

$$\sigma_{\text{fus}} = \int_0^{\pi/2} \sigma_f(\Theta) \sin\Theta \, d\Theta$$

Scobel *et al.* (Sc 75) have studied the fusion of ^{35}Cl with the even Ni isotopes whose quadrupole deformations vary widely and even change sign

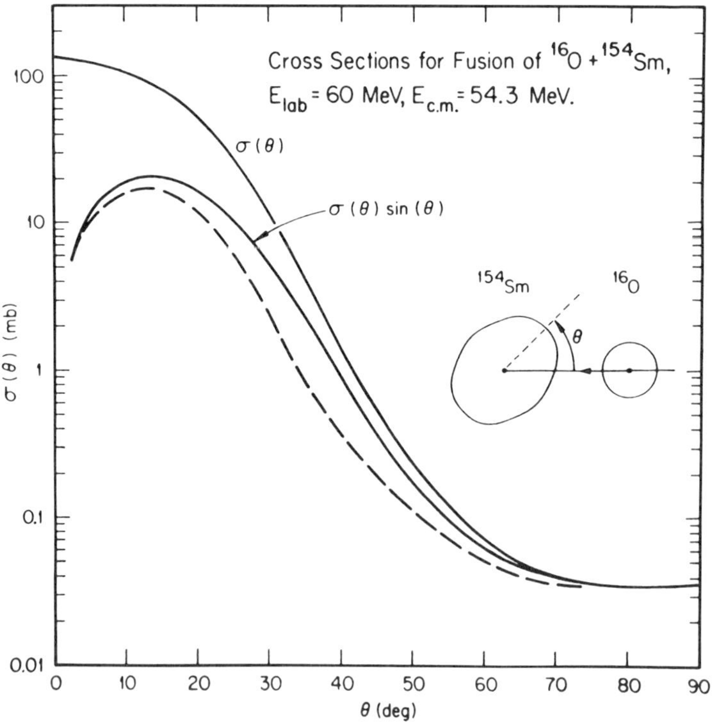

Figure 6.6. Fusion cross sections for $^{16}O + ^{154}Sm$ as a function of orientation (see Figure 6.5), calculated as the total absorption cross section in a complex potential (from St 81). The dashed curve takes a rotation of the target into account.

in going from ^{58}Ni to ^{64}Ni. Those authors found that fits to their data were greatly improved once the deformations were taken into account. More recently Stokstad *et al.* have reported measurements of subbarrier fusion cross sections of ^{16}O (St 80a) and ^{40}Ar (St 80b) with the even Sm isotopes that range from spherical to permanently deformed nuclei. The data for the ^{16}O induced reactions are shown in Figure 6.7. The cross sections for the different isotopes vary far more than would be expected from simple $A^{1/3}$ variations of the radii.

The approach used in the analysis was to fit the fusion cross section for one isotope and use this information to make predictions for the other based on the known changes of the deformation from isotope to isotope. The effects of the static deformations as discussed were found to account correctly for the trends in the data but overestimated the differences in the fusion cross sections for the different isotopes: Besides dynamic effects

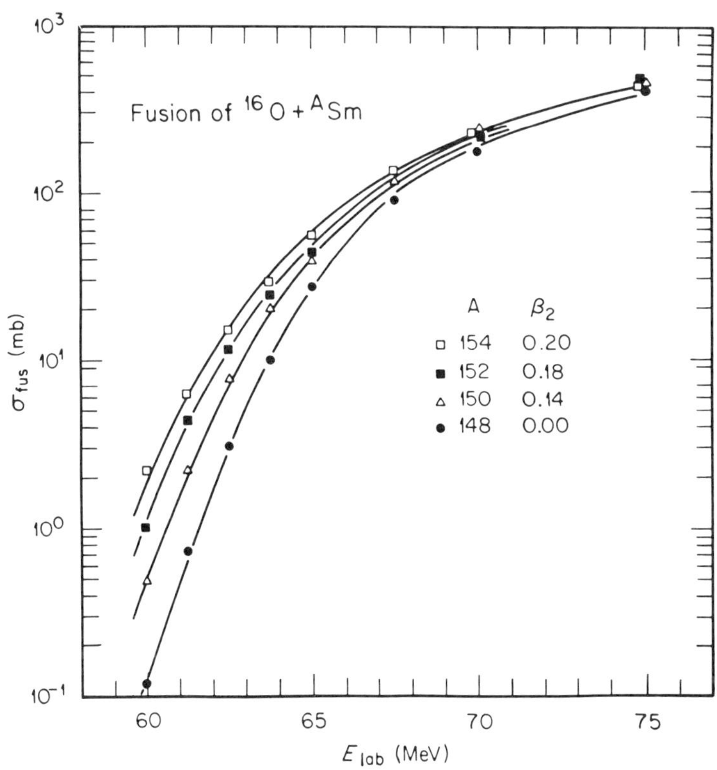

Figure 6.7. Fusion sections for $^{16}O + {}^{A}Sm$ for $A = 148$, 150, 152, and 154. Also given are the deformations of these isotopes. The lines are drawn through the data to guide the eye (from St 80a).

(channel couplings) this discrepancy could be due to the "head-on" approximation made in the calculation of the orientation dependence of the interaction barrier: The higher partial waves are more strongly weighted in the fusion cross section so that peripheral interactions should be taken into account.

The procedure described above to calculate the fusion cross section for deformed nuclei contains an averaging over different orientations. Thus much sensitivity of the cross sections on deformation is lost. Furthermore, the deformation effects are always mixed with effects of the nucleus–nucleus potential; deformation changes and simple potential changes cannot be disentangled from each other. The obvious way out of this difficulty would be to have the nuclei interact in a well-defined spatial orientation, i.e., employ either polarized targets or projectiles. It has been demonstrated that a deformed ^{165}Ho nucleus can be polarized and that this polarization can be maintained during a scattering experiment (Pa 72). However, no fusion data are available for this system.

Alternatively, a group at Heidelberg and Marburg has recently achieved the production and acceleration of aligned (tensor-polarized) ^{7}Li and ^{23}Na beams (St 80c). Both nuclei are deformed in their ground states. So far scattering and fusion data are available for the ^{7}Li beam. The experiments performed exploit the fact that for a spin aligned beam also the spectroscopic quadrupole moments are aligned. Thus, if the spin axis is turned the corresponding mass distributions change their spatial orientation as well. One can, therefore, change the interaction distance between the nuclear surfaces in a well-defined way; for the case of the oblate ^{7}Li, aligned parallel to the scattering plane, for example, there is less overlap between projectile and target and thus less absorption at a given distance than in the unaligned case. Hence, the elastic scattering cross section should be larger and the fusion cross section smaller as compared to the unpolarized situation. This expectation is borne out by the data, as can be seen in Figure 6.8. Here the difference in the fusion cross sections for the reaction $^{7}\text{Li}+{}^{51}\text{V}$ with and without aligned ^{7}Li is shown (Mo 79c, St 80c). The difference amounts to $\sim 22\%$ at $E_{\text{c.m.}} \approx 10$ MeV, which is above the fusion barrier. Similar to the fusion results the effects of the static deformation of ^{7}Li have also been seen in the elastic scattering data (Mo 79c, St 80c). Experiments of this type using the aligned ^{23}Na beam at the Heidelberg EN tandem are presently being prepared.

Altogether these experiments constitute the cleanest source of information on the orientation dependence of nuclear cross sections and thus the nucleus–nucleus interaction. Similar experiments with the aligned ^{23}Na beam promise to be a step forward in studies of deformation effects in nuclear reactions. This is even more so for reaction studies employing polarized targets of well-deformed heavy rare earth nuclei that are presently

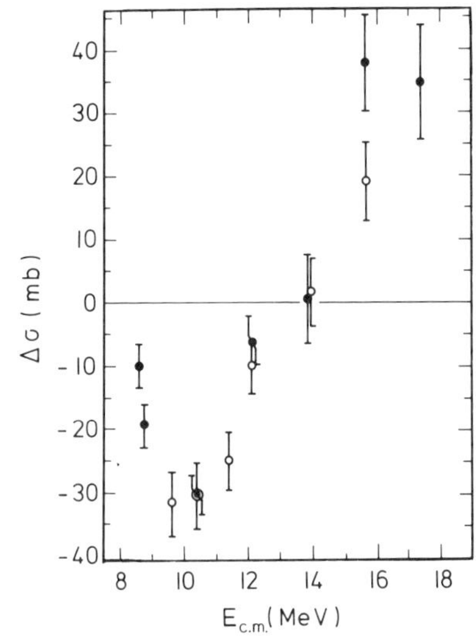

Figure 6.8. Absolute difference between the total reaction cross sections for $^{7}Li+^{51}V$ with aligned and nonaligned ^{7}Li. The solid dots are obtained from an analysis of quasielastic scattering data (St 80c). The oblate ^{7}Li is aligned along the beam axis [from Moebius (Mo 79c)].

being developed by a group at Oak Ridge National Laboratory using laser-pumping techniques (Be 81). These experiments may enable one to measure the mass-deformation of nuclei through subbarrier fusion experiments. At higher energies the static deformation effects will probably be masked by dynamical polarizations, but valuable information could still be obtained about, e.g., the heavy-ion spin–orbit potential.

7. Resonances in Fusion Cross Sections

Besides the observation of large differences in the fusion cross section while going from one nucleus to the neighbouring one as, for example, from $^{14}N+^{12}C$ to $^{15}N+^{12}C$, the observation of oscillatory structure in the energy dependence of the fusion cross sections of p-shell nuclei was one of the most interesting results of fusion studies. The structures have been observed in $^{12}C+^{12}C$, $^{16}O+^{12}C$, and $^{16}O+^{16}O$ (Sp 76a, Sp 76b) (see Figure 7.1). A detailed analysis of fusion data in this mass range has been given by Kovar *et al.* (Ko 79b).

The structure is found to occur primarily in the α-emission channels. Since this is the favored decay mode for high angular momentum states, this could indicate a resonancelike behavior of the highest partial waves. Similar gross structure has also been seen in other inelastic reaction channels, and to some of these gross structures spins have been assigned (Co 77, Ja 79). Using these spin assignments the structures seem to form a rotational band.

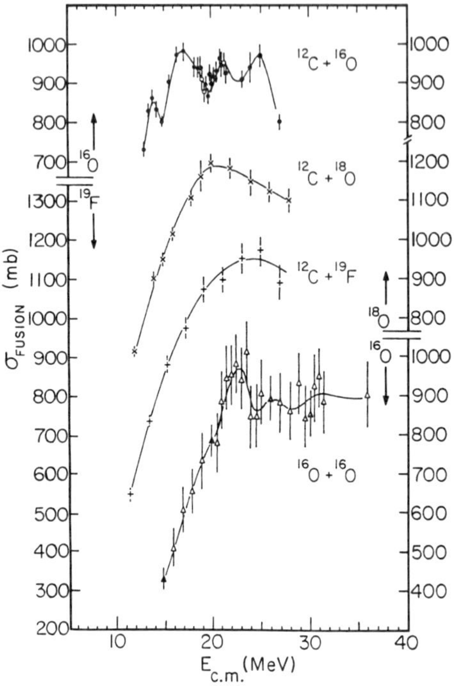

Figure 7.1. Fusion cross sections for O- and C-induced reactions (from Ko 79b).

More recently structure in the fusion yield has also been reported for ^{28}Si + ^{28}Si (Be 79, Di 79).

The first attempts to explain the resonancelike structures in the fusion cross sections were made by simply calculating the total reaction cross sections with optical potentials that fit the elastic scattering data. These potentials are all surface transparent for the reactions listed above. Whereas they give some structure for ^{12}C + ^{12}C and ^{16}O + ^{16}O in the reaction cross section (Ko 78), for ^{12}C + ^{16}O structure is obtained only if the imaginary potential is reduced (Ko 78, Ko 79b). The resonances obtained in such calculations are obviously shape resonances in the heavy-ion potential. The fact that they appear for ^{12}C + ^{12}C and ^{16}O + ^{16}O without modification of the potentials is due to the fact that in these identical systems only even partial waves contribute. Thus the spacing between the resonances is larger, and these can, therefore, be seen better.

However, whereas the bump at $E_{c.m.} = 9$ MeV in the ^{12}C + ^{12}C cross section is reproduced by the optical model calculation and thus is due to a pure shape resonance, it must be noted that the structures in the calculated reaction cross section for ^{12}C + ^{12}C are out of phase with those in the experimental cross sections at the higher energies. This shift has been explained by Vandenbosch (Va 79b) as being due to a limitation in the number of available compound states at high spin. Alternatively, Tanimura and Tazawa (Ta 80a) have explained the shift as an effect of channel coupling that shifts the resonance locations somewhat.

Tanimura (Ta 80b) has also given quite an attractive explanation for the structure in the $^{16}O + ^{16}O$ cross section by assuming that the inelastic 3^- excitation acts as a doorway to all other inelastic excitations. This is mathematically achieved by solving first the coupled channel equations including the 3^- excitation to get the total reaction cross section. The calculation is then repeated with no imaginary potential in the 3^- channel and the inelastic cross section so calculated is identified with the total inelastic cross section.

The result of such a calculation is shown in Figure 7.2. It can be seen that the calculation gives a very good description of the data. This calculation may not be correct in the restriction to only one doorway, as in fusion one expects major polarizations and thus many channel couplings. Nevertheless, it has probably taken into account the right physical process, since one expects also from other model considerations that strong shape vibrations play an important role in the damping of heavy-ion collisions (Br 79).

The discussions above all point to an explanation of the observed structures in the fusion cross section in terms of shape resonances in the

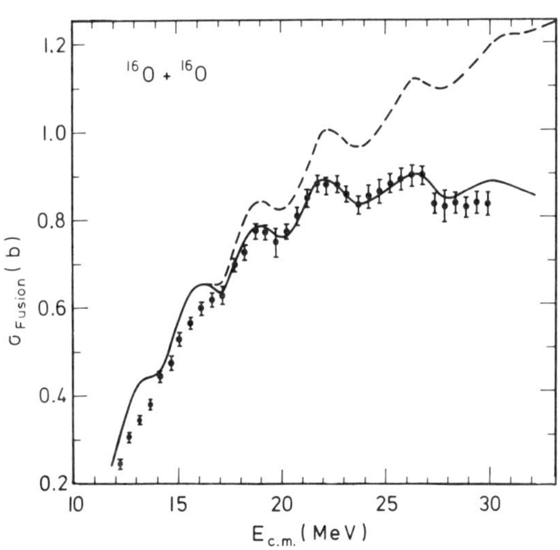

Figure 7.2. Calculated total reaction cross section (dashed curve) and fusion cross section (solid curve) as obtained by Tanimura (Ta 80b) in comparison with experimental values (Ko 77b). The lower part of the figure shows the angular momentum decomposition (from Ta 80b). Note, however, that for $^{16}O + ^{16}O$ several conflicting measurements exist.

nucleus–nucleus potential, possibly shifted by coupled channel (Ta 80a, Ta 80b) effects or level-density considerations (Va 79b). That these shape resonances do indeed survive is due to the surface transparency of the optical potentials and thus to the specific properties of the absorption in these systems (Ha 81). Quite detailed experimental investigations of these points are contained in the work by Kolata *et al.* (Ko 77b, Ko 79a, Ko 79c, Ko 80b).

8. Conclusions and Outlook

It is presently not possible to decide once and for all between entrance channel and yrast line limitations. However, the results on cross sections for different reactions leading into the same compound nucleus seem to support some CN-related effect. If the fusion bands indeed define a statistical yrast line, then information about high spin states in light nuclei may indeed be obtained from the study of fusion reactions. This could be a quite exciting aspect of fusion reactions. The high spin states of light nuclei show all the same physical phenomena as those encountered in heavy nuclei, like, e.g., alignment, backbending, and even a superbackbend (a typical example of these is ^{20}Ne with a band cutoff at $I = 8\hbar$). Since the microscopic structure of the lighter nuclei is much better known than that of the heavy ones, one may have here a unique testing ground for less fundamental models, like, e.g., cranking, that often are applied to the study of heavy nuclei.

It is, therefore, presently the hope that on one side theoretical methods will become available soon that allow one to make more reliable predictions for the yrast lines. On the experimental side it would be useful if particle spectroscopy methods could be improved so that they can reliably be applied to the spectroscopy of high spin states in light nuclei.

Besides this nuclear structure aspect, future research on heavy-ion fusion reactions of light nuclei will probably focus on two aspects. One is the appearance of resonancelike structures in the fusion cross section. These have been observed so far only in reactions involving combinations of ^{12}C and ^{16}O. With the recent observations of resonance structures in elastic and inelastic channels in heavier systems like ^{28}Si + ^{28}Si (Be 79, Di 81) it is a natural challenge to look with increased sensitivity also back into the fusion cross sections for these nuclei. It would be extremely interesting to learn about the possible connection of resonant states of the $(A_P + A_T)$ system with the doorway states to fusion and to identify these latter. This would not only help us considerably to understand the absorptive features of the heavy-ion interaction but would also put all speculations about statistical yrast lines on a safer basis.

The second aspect is the investigation of fusion reactions at high to very high energies. These reactions would lead to the highest excitation energies in the compound nucleus and thus enable one to study nuclear properties in the statistical regime far away from the ground state. Since at the very high energies the cross section for incomplete fusion increases drastically, new experimental techniques, like, e.g., anticoincidence measurements, may become necessary to sort out the true fusion events.

As mentioned in the Introduction, throughout this chapter a slight emphasis has been put on the discussion of reactions involving lighter heavy ions. This is mainly because the data for heavy systems are still quite sparse but also because no quantitative theoretical descriptions exist for these. This is in contrast to the lighter nuclei where a number of models exist and where the data in general show richer features, like, e.g., limitation effects at high bombarding energies, resonance structures, and remarkable differences between adjacent systems with nearly the same masses.

A careful, systematic investigation of fusion reactions between very heavy nuclei has only recently begun. Some of the new developments in this area have been discussed in Section 6.1. There it was shown that finally an order parameter, the effective fissility, seems to emerge that allows one to relate reactions between quite different systems to each other. This concept has undergone very few quantitative checks so far. In particular, with given, fixed mass of the compound nucleus the asymmetry of the entrance channel and the bombarding energy have to be varied in order to see if the capture and fusion cross sections do indeed show the properties expected from the simple models discussed in Section 6.1.

Experiments of this type are of particular importance for an understanding of the reaction mechanism in collisions between very heavy nuclei. Such an understanding is mandatory in the search for superheavy nuclei. The failure to produce these nuclei in nuclear reactions is at present still not understood. Either they are much less stable in their ground states than predicted or the reactions used to produce them have been too violent to allow them to survive. This latter point illustrates why it is important to understand the observation of true fusion of very heavy nuclei at energies just at the barrier. If a superheavy nucleus is to get a reasonable chance for survival (if it exists at all) it has to be produced in the most delicate way possible!

Acknowledgments

I am very much indebted to D. Glas (Giessen) for many years of stimulating, productive collaboration. I also gratefully acknowledge innumerous discussions with R. Vandenbosch (Seattle), D. Kovar, and J. P.

Schiffer (Argonne) on fusion reactions. Finally, the discussion of fusion between very heavy elements has benefited from many helpful comments by P. Armbruster, A. Gobbi, A. Olmi, and H. Sann at GSI, and W. J. Swiatecki (Berkeley).

References

The references for this article were compiled up to the end of 1981.

(Ag 77) D. Agassi, C. M. Ko, and H. Weidenmüller, *Ann. Phys.* (*N.Y.*) **107**:140 (1977).

(Ay 74) S. Ayik and J. N. Ginocchio, *Nucl. Phys.* **A234**:13 (1974).

(Ba 74) R. Bass, *Nucl. Phys.* **A231**:45 (1974).

(Ba 77a) R. Bass, *Phys. Rev. Lett.* **39**:265 (1977).

(Ba 77b) B. B. Back, R. R. Betts, C. Gaarde, J. S. Larsen, E. Michelsen, and T. Kuang-Hsi, *Nucl. Phys.* **A285**:317 (1977).

(Be 79) R. R. Betts, S. B. Dicenzo, and J. F. Petersen, *Phys. Rev. Lett.* **43**:253 (1979).

(Be 81) J. R. Beene, C. E. Bemis, Jr., J. L. C. Ford, Jr., K. A. Erb, and D. Shapira, Oak Ridge National Laboratory, Physics Division Progress Report (1981).

(Bi 78) J. R. Birkelund and J. R. Huizenga, *Phys. Rev. C* **17**:126 (1978).

(Bi 79) J. R. Birkelund, L. E. Tubbs, J. R. Huizenga, J. N. De, and D. Sperber, *Phys. Rep.* **56**:107 (1979).

(Bj 72) S. Bjornholm and W. J. Swiatecki, *Phys. Rep.* **4**:326 (1972).

(Bl 77) J. Blocki, R. Randrup, W. J. Swiatecki, and C. F. Tsang, *Ann. Phys.* (*N.Y.*) **105**:427 (1977).

(Bo 76) P. Bonche, S. E. Koonin, and J. W. Negele, *Phys. Rev. C* **13**:1226 (1976).

(Bo 78) P. Bonche, B. Grammaticos, and S. E. Koonin, *Phys. Rev. C* **17**:1700 (1978).

(Bo 79) P. Bonche, K. T. R. Davies, B. Flanders, H. Flocard, B. Grammaticos, S. E. Koonin, S. J. Krieger, and M. S. Weiss, *Phys. Rev. C* **20**:641 (1979).

(Bo 82) R. Bock, Y. T. Chu, M. Dakowski, A. Gobbi, E. Grosse, A. Olmi, H. Sann, D. Schwalm, U. Lynen, W. Müller, S. Bjornholm, H. Esbensen, W. Wölfi, E. Morenzoni, *Nucl. Phys.* **A388**:334 (1982).

(Br 78) R. A. Broglia, C. H. Dasso, G. Pollarolo, and A. Winther, *Phys. Rev. Lett.* **40**:707 (1978).

(Co 74) S. Cohen, F. Plasil, and W. J. Swiatecki, *Ann. Phys.* (*N.Y.*) **82**:557 (1974).

(Co 76) T. M. Cormier, A. J. Lazzarini, M. A. Neuhausen, A. Sperduto, K. van Bibber, F. Videbaek, G. Young, E. B. Blum, L. Herreid, and W. Thoms, *Phys. Rev. C* **13**:682 (1976).

(Co 77) T. M. Cormier, J. Applegate, G. M. Berkowitz, P. Braun-Munzinger, P. M. Cormier, J. W. Harris, C. M. Jachcinski, L. L. Lee, Jr., J. Barrette, and H. E. Wegner, *Phys. Rev. Lett.* **38**:940 (1977).

(Co 78) M. Conjeaud, S. Gary, S. Harar, and J. P. Wieleczko, *Nucl. Phys.* **A309**:515 (1978).

(Da 75) J. Dauk, K. P. Lieb, and A. M. Kleinfeld, *Nucl. Phys.* **A241**:170 (1975).

(Di 79) S. B. Dicenzo, J. F. Petersen, and R. R. Betts, *Phys. Rev. C* **23**:2561 (1981).

(Di 80) M. Diebel, D. Glas, U. Mosel, and H. Chandra, *Nucl. Phys.* **A333**:253 (1980).

(Ei 77) Y. Eisen, I. Tserruya, Y. Eyal, Z. Fraenkel, and M. Hillman, *Nucl. Phys.* **A291**:459 (1977).

(Fe 78) B. Fernandez, C. Gaarde, J. S. Larsen, S. Pontoppidan, and F. Videbaek, *Nucl. Phys.* **A306**:259 (1978).

(Ga 74) J. Galin, D. Guerreau, M. Lefort, and X. Tarrago, *Phys. Rev. C* **9**:1018 (1974).

(Ga 81) H. Gäggeler and T. Sikkeland, GSI Darmstadt Annual Report (1981).

(Gl 74) D. Glas and U. Mosel, *Phys. Rev. C* **10**:2620 (1974).

(Gl 75) D. Glas and U. Mosel, *Nucl. Phys.* **A237**:429 (1975).

(Gl 78a) D. Glas, Habilitationsschrift, Institut für Theoretische Physik, Universität Giessen, 1978, unpublished.

(Gl 78b) D. Glas and U. Mosel, *Phys. Lett.* **78B**:9 (1978).

(Go 79a) J. Gomez Del Campo, R. G. Stokstad, J. A. Biggerstaff, R. A. Dayras, A. H. Snell, and P. H. Stelson, *Phys. Rev. C* **19**:2170 (1979).

(Go 79b) J. Gomez Del Campo, R. A. Dayras, J. A. Biggerstaff, D. Shapira, A. H. Snell, P. H. Stelson, and R. G. Stokstad, *Phys. Rev. Lett.* **43**:26 (1979).

(Gr 74a) D. H. E. Gross, H. Kalinowski, and J. N. De, Proc. Symp. on Classical and Quantum Aspects of Heavy Ion Collisions, Heidelberg, *Lecture Notes in Physics*, Vol. 33, Springer, Berlin (1974), p. 194.

(Gr 74b) D. H. E. Gross and H. Kalinowski, *Phys. Lett.* **48B**:302 (1974).

(Gr 78) D. H. E. Gross and H. Kalinowski, *Phys. Rep.* **45**:175 (1978).

(Gr 81) P. Grange, J. Richert, G. Wolschin, and H. A. Weidenmüller, *Nucl. Phys.* **A356**:260 (1981).

(Gu 73) H. H. Gutbrod, W. G. Winn, and M. Blann, *Nucl. Phys.* **A213**:267 (1973).

(Gu 80) D. Guerreau and R. Babinet, Proc. Int. Conf. on Nuclear Behaviour at High Angular Momentum, Strasbourg, France (1980), *J. Phys. (Paris)* **C10**:217 (1980).

(Ha 78) S. Harar, in *Nuclear Molecular Phenomena* (N. Cindro, ed.), North-Holland, Amsterdam (1978), p. 329.

(Ha 80) S. Harar and J. P. Wieleczko, Entrance Channel versus Compound Nucleus Limitations in the Fusion Process of $1p$ and $2s-1d$ Shell Nuclei, lecture given at the International School of Poiana Brasov on the Critical Phenomena in Heavy-Ion Collisions, August 1980 (Saclay preprint, 1980).

(Ha 81) F. Haas and Y. Abe, Surface Transparency and Resonant Behavior in Light Heavy Ion Reactions, *Phys. Rev. Lett.* **46**:1667 (1981).

(He 79) G. Herrmann, *Nature* **280**:543 (1979).

(Ja 79) C. M. Jachcinsky, P. Braun-Munzinger, G. M. Berkowitz, R. H. Freifelder, M. Gai, R. L. McGrath, P. Paul, T. Renner, and C. D. Uhlhorn, *Phys. Lett.* **87B**:354 (1979).

(Ko 75) R. L. Kozub, N. H. Lu, J. M. Miller, D. Logan, T. W. Debiak, and L. Kowalski, *Phys. Rev. C* **11**:1497 (1975).

(Ko 77a) S. E. Koonin, K. T. R. Davies, V. Maruhn-Rezwani, H. Feldmeier, S. J. Krieger, and J. W. Negele, *Phys. Rev. C* **15**:1359 (1977).

(Ko 77b) J. J. Kolata, R. C. Fuller, R. M. Freeman, F. Haas, B. Heusch, and A. Gallmann, *Phys. Rev. C* **16**:891 (1977).

(Ko 78) D. G. Kovar, in Nuclear Interactions, *Lecture Notes in Physics*, Vol. 92, Springer, Berlin (1978), p. 329.

(Ko 79a) J. J. Kolata, R. M. Freeman, F. Haas, B. Heusch, and A. Gallmann, *Phys. Rev. C* **19**:408 (1979).

(Ko 79b) D. G. Kovar, D. F. Geesaman, T. H. Braid, Y. Eisen, W. Henning, T. R. Ophel, M. Paul, K. E. Rehm, S. J. Sanders, P. Sperr, J. P. Schiffer, S. L. Tabor, S. Vigdor, B. Zeidman, and F. W. Prosser, Jr., *Phys. Rev. C* **20**:1305 (1979).

(Ko 79c) J. J. Kolata, R. M. Freeman, F. Haas, B. Heusch, and A. Gallmann, *Phys. Rev. C* **19**:2237 (1979).

(Ko 80a) J. J. Kolata, *Phys. Lett.* **95B**:215 (1980).

(Ko 80b) J. J. Kolata, R. M. Freeman, F. Haas, B. Heusch, and A. Gallmann, *Phys. Rev. C* **21**:579 (1980).

(Kr 78) S. J. Krieger and K. T. R. Davies, *Phys. Rev. C* **18**:2567 (1978).

(Kr 79) S. J. Krieger and K. T. R. Davies, *Phys. Rev. C* **20**:167 (1979).

(Kr 81) S. J. Krieger and M. S. Weiss, *Phys. Rev. C* **24**:928 (1981).

(Ku 64) J. A. Kuehner and E. Almqvist, *Phys. Rev.* **134**:B1229 (1964).

(La 81) A. Lazzarini, H. Doubre, K. T. Lesko, V. Metag, A. Seamster, R. Vandenbosch, and W. Merryfield, *Phys. Rev. C* **24**:309 (1981).

(Le 76) M. Lefort, *Rep. Progr. Phys.* **39**:129 (1976).

(Le 79) H. Lehr, W. v. Oertzen, W. Bohne, H. Morgenstern, K. Grabisch, and F. Pühlhofer, Proc. Symp. Deep Inelastic and Fusion Reactions with Heavy Ions, Berlin 1979, *Lecture Notes in Physics*, Vol. 117, Springer, Berlin (1980), p. 354.

(Le 80) S. M. Lee, T. Matsuse, and A. Arima, *Phys. Rev. Lett.* **45**:165 (1980).

(Le 81) H. Lehr, Ph.D. Thesis, Hahn-Meitner-Institut für Kernforschung, Berlin (1981).

(Lo 78) M. A. K. Lohdi, ed., *Proc. Int. Symp. on Superheavy Elements*, Pergamon, New York (1978).

(Ma 76) J. A. Maruhn and R. Y. Cusson, *Nucl. Phys.* **A270**:437 (1976).

(Ma 81) J. F. Mateja, A. D. Frawley, C. C. Dennis, K. Abdo, and K. W. Kemper, *Phys. Rev. Lett.* **47**:311 (1981).

(Mo 78) U. Mosel, Proc. Int. Conference on Nuclear Interactions, Canberra, Australia (1978), *Lecture Notes in Physics*, Vol. 92, Springer, Berlin, p. 185.

(Mo 79a) U. Mosel, *Phys. Rev. C* **19**:2167 (1979).

(Mo 79b) U. Mosel and M. Diebel, Proc. Symp. on Deep-Inelastic and Fusion Reactions with Heavy Ions, Berlin (1979), *Lecture Notes in Physics*, Vol. 117, Springer, Berlin (1980), p. 375.

(Mo 79c) K. H. Moebius, Ph.D. Thesis, Universität Heidelberg (1979), unpublished.

(Mo 80) U. Mosel, in *Heavy Ion Collisions* (R. Bock, ed.), North-Holland, Amsterdam (1980), Vol. 2, p. 275.

(Mu 81a) K. Mühlhans, E. Müller, K. Neergård, and U. Mosel, *Phys. Lett.* **105B**:329 (1981).

(Mu 81b) G. Münzenberg, S. Hofmann, F. P. Heßberger, W. Reisdorf, K. H. Schmidt, J. R. H. Schneider, P. Armbruster, C. C. Sahm, and B. Thuma, *Z. Phys.* **A300**:107 (1981).

(Na 76) M. N. Namboodiri, E. T. Chulick, and J. B. Natowitz, *Nucl. Phys.* **A263**:491 (1976).

(Na 79) K. Nagatani and J. C. Peng, *Phys. Rev. C* **19**:747 (1979).

(Ni 74) J. R. Nix and A. J. Sierk, *Phys. Scr.* **10A**:94 (1974).

(Ni 77) J. R. Nix and A. J. Sierk, *Phys. Rev. C* **15**:2072 (1977).

(Oe 79) W. von Oertzen, ed., Proc. Symp. on Deep-Inelastic and Fusion Reactions with Heavy Ions, Berlin (1979), *Lecture Notes in Physics*, Vol. 117, Springer, Berlin (1980).

(Or 82) M. E. Ortiz, J. Gomez Del Campo, Y. D. Chan, D. E. Disresorio, J. L. C. Ford Jr., D. Shapira, R. G. Stokstad, J. P. F. Sellschop, R. L. Parks, D. Weiser, *Phys. Rev. C* **25**:1436 (1982).

(Pa 72) D. R. Parks, S. L. Tabor, B. B. Triplett, H. T. King, T. R. Fisher, and B. A. Watson, *Phys. Rev. Lett.* **29**:1264 (1972).

(Ra 78a) J. Randrup, *Ann. Phys. (N.Y.)* **112**:356 (1978).

(Ra 78b) J. Randrup, *Nucl. Phys.* **A307**:319 (1978).

(Ra 80) J. Randrup and W. J. Swiatecki, *Ann. Phys. (N.Y.)* **125**:193 (1980).

(Ra 81) I. Ragnarsson, T. Bengtsson, and S. Aberg, Lund Institute of Technology report (1981), Lund-MPh-81/03.

(Ro 78) C. Rolfs and H. P. Trautvetter, *Ann. Rev. Nucl. Sci.* **28**:115 (1978).

(Sa 79) F. Saint-Laurent, M. Conjeaud, S. Harar, J. M. Loiseaux, J. Menet, and J. B. Viano, *Nucl. Phys.* **A237**:517 (1979).

(Sa 80) H. Sann, S. Bjornholm, R. Bock, Y. T. Chu, M. Dakowski, E. Grosse, U. Lynen,

E. Morenzoni, W. Müller, A. Olmi, D. Schwalm, and W. Wolfi, Int. Conf. Nuclear Physics, Berkeley (1980), contributed paper.

(Sa 81) H. Sann, R. Bock, Y. T. Chu, A. Gobbi, A. Olmi, U. Lynen, W. Müller, S. Bjornholm, and H. Esbensen, *Phys. Rev. Lett.* **47**:1248 (1981).

(Sc 81) K. H. Schmidt, P. Armbruster, F. P. Heßberger, G. Münzenberg, W. Reisdorf, C.-C. Sahm, D. Vermeulen, H. G. Clerc, J. Keller, and H. Schulte, *Z. Phys.* **A301**:21 (1981).

(Sc 75) W. Scobel, A. Mignerey, and M. Blann, *Phys. Rev. C* **11**:1701 (1975).

(Si 79a) K. Siwek-Wilczynska, E. H. Du Marche van Voorthuysen, J. van Popta, R. H. Siemssen, and J. Wilczyński, *Phys. Rev. Lett.* **42**:1599 (1979).

(Si 79b) K. Siwek-Wilczynska, E. H. Du Marche van Voorthuysen, J. van Popta, R. H. Siemssen, and J. Wilczyński, *Nucl. Phys.* **A330**:150 (1979).

(Sp 76a) P. Sperr, S. E. Vigdor, Y. Eisen, W. Henning, D. G. Kovar, T. R. Ophel, and B. Zeidman, *Phys. Rev. Lett.* **36**:405 (1976).

(Sp 76b) P. Sperr, T. H. Braid, Y. Eisen, D. G. Kovar, F. W. Prosser, Jr., J. P. Schiffer, S. L. Tabor, and S. E. Vigdor, *Phys. Rev. Lett.* **37**:321 (1976).

(St 77) R. G. Stokstad, D. C. Hensley, and A. H. Snell, *Nucl. Inst. Meth.* **141**:499 (1977).

(St 80a) R. G. Stokstad, Y. Eisen, S. Kaplanis, D. Pelte, U. Smilansky, and I. Tserruya, *Phys. Rev. C* **21**:2427 (1980).

(St 80b) R. G. Stokstad, W. Reisdorf, K. D. Hildenbrand, J. V. Kratz, G. Wirth, R. Lucas, and J. Poitou, *Z. Phys.* **295**:269 (1980).

(St 80c) E. Steffens, Fifth Int. Symp. on Polarization Phenomena in Nuclear Physics, Santa Fe, New Mexico, invited talk, *AIP Conf. Proc.* **69**:1001 (1980).

(St 80d) R. G. Stokstad, Proc. 13th Int. Summer School in Nuclear Physics, Mikolajki, Poland, Sept. 1980, to be published in *Nukleonika*.

(St 81) R. G. Stokstad and E. E. Gross, *Phys. Rev. C* **23**:281 (1981).

(Sz 78) A. Szanto de Toledo, M. Schrader, G. Rosner, E. M. Szanto, and H. V. Klapdor, *Phys. Lett.* **78B**:58 (1978).

(Sw 81) W. J. Swiatecki, Proc. Nobel Symp. on High Spin States, Örenäs, Sweden (1981), *Phys. Scr.* **24**:107 (1981).

(Sw 82) W. J. Swiatecki, *Nucl. Phys.* **A376**:275 (1982).

(Ta 78) S. L. Tabor, D. Geesaman, W. Henning, D. G. Kovar, K. E. Rehm, and F. W. Presser, Jr., *Phys. Rev. C* **17**:2136 (1978).

(Ta 80a) O. Tanimura and T. Tazawa, *Phys. Rep.* **61**:253 (1980).

(Ta 80b) O. Tanimura, *Phys. Lett.* **90B**:204 (1980).

(Th 80) S. T. Thornton, L. C. Dennis, and K. R. Cordell, *Phys. Lett.* **91B**:196 (1980).

(Va 73) R. Vandenbosch and J. R. Huizenga, *Nuclear Fission*, Academic Press, New York (1973).

(Va 79a) R. Vandenbosch, Proc. Symp. on High Spin Phenomena in Nuclei, Argonne National Laboratory (1979), ANL report ANL/PHY-79-4, p. 285.

(Va 79b) R. Vandenbosch, *Phys. Lett* **87B**:183 (1979).

(Va 80) R. Vandenbosch, *Nucl. Phys.* **A339**:167 (1980).

(Va 81a) L. C. Vaz, J. M. Alexander, and G. R. Satchler, *Phys. Rep.* **69**:373 (1981).

(Va 81b) R. Vandenbosch and A. J. Lazzarini, *Phys. Rev. C* **23**:1074 (1981).

(Wa 76) A. Watt, D. Kelvin, and R. R. Whitehead, *Phys. Lett.* **63B**:385 (1976).

(Wi 80a) J. Wilczyński, K. Siwek-Wilczynska, J. van Driel, S. Gonggrijp, D. C. J. M. Hageman, R. V. F. Janssens, J. Lukasiak, and R. H. Siemssen, *Phys. Rev. Lett.* **45**:606 (1980).

(Wi 80b) J. P. Wieleczko, S. Harar, M. Conjeaud, and F. Saint Laurent, *Phys. Lett.* **93B**:35 (1980).

(Wo 73) C. Y. Wong, *Phys. Rev. Lett.* **31**:766 (1973).

2

ANDREW M. SANDORFI is a member of the Physics Department of Brookhaven National Laboratory. He received his undergraduate and graduate education at the University of Toronto in Canada, graduating with the Ph.D. in 1976. Following a brief period at Brookhaven, he joined the High Energy Physics Laboratory staff at Stanford University, where he remained until his return to Brookhaven in 1979. While at Stanford he received the Ernest Rutherford award of the Royal Society of Canada. He is currently the coordinator of LEGS, the Laser-Electron-Gamma-Source at Brookhaven.

Heavy-Ion Radiative Capture*

A. M. SANDORFI

1. Introduction

The radiative capture of light particles ($\leq$ alphas) is a well-established mechanism that has been used for a variety of studies such as the spectroscopy of shell-model states, the decay of giant resonances, and the properties of isobaric analog resonances. In contrast, comparatively little is known about the radiative capture of heavy ions. This is mainly due to the small cross sections encountered. The Coulomb barriers that must be overcome for heavy-ion fusion are much higher. This, together with consistently larger Q values, leads to much higher excitations in the compound nuclei. States at such high energies tend to have much larger total widths, with the result that the branching ratios for radiative decay Γ_γ/Γ tend to be much smaller than in light-ion capture. Consequently, only a few experiments relating to this area of heavy-ion science have been performed. These have been compiled and are listed in Table 1.

1.1. Historical Perspective

Following an initial suggestion[1] that high-energy decay might sometimes compete favorably with particle evaporation following heavy-ion fusion, several searches were made for the radioactive decay of the ground states of nuclei that would be produced if the fused systems γ-decayed

* Supported by the United States Department of Energy under Contract No. DE-AC02-76CH00016.

A. M. SANDORFI • Brookhaven National Laboratory, Upton, New York 11973.

without particle emission.[2-5] These initial experiments, plagued with backgrounds from target impurities, were largely inconclusive. The first unequivocal observation of the radiative capture of heavy ions was reported by W. Feldman and D. W. Heikkinen,[6] who used the first high-resolution anti-coincidence-shielded large-volume NaI(Tl) crystal to detect the high-energy γ rays decaying to the low-lying states in the fused nuclei. These authors surveyed a large range of light heavy-ion projectiles and targets over varying regions of bombarding energy—mostly just above the Coulomb barrier (see Table 1). A pronounced resonance in the $^{7}Li(^{16}O, \gamma_{0+1})$ reaction was

Table 1. Heavy-Ion Capture and Related Measurements[a]

Reaction	Bombarding energy range	Detection technique	Ref.
$^{27}Al(^{16}O,\gamma)^{43}Sc$ $^{31}P(^{16}O,\gamma)^{47}V$	30–38 MeV	Activity and chemical analysis	2
$^{19}F(^{13}C,\gamma)^{32}P$ $^{51}V(^{13}C,\gamma)^{64}Cu$	22 MeV	Activity: $^{32}P(\beta)$, $^{64}Cu(\beta)$	3
$^{31}P(^{14}N,\gamma)^{45}Ti$	27 MeV	Activity: $^{45}Ti(\beta)$	4
$^{56}Fe(e,^{32}P)^{24}Na\,e'$	3 GeV	Activity: $^{32}P(\beta)$	5
$^{7}Li(^{16}O,\gamma)^{23}Na$ $^{11}B(^{12}C,\gamma)^{23}Na$ $^{6}Li(^{16}O,\gamma)^{22}Na$ $^{12}C(^{12}C,\gamma)^{24}Mg$ $^{12}C(^{16}O,\gamma)^{28}Si$ $^{16}O(^{16}O,\gamma)^{32}S$	12.5–30.5 MeV 11.6–18.9 MeV 17.4–24.9 MeV 12.7–23.1 MeV 21.5–36.0 MeV 23.9–41.9 MeV	NaI: High energy γ	6
$^{24}Mg(e,^{12}C)^{12}C\,e'$	26–40 MeV	Damage-track detectors for ^{12}C	7
$^{30}Si(^{14}N,\gamma)^{44}Sc$	40–62 MeV	Activity: γ	8
$^{16}O(^{32}S,\gamma)^{48}Cr$ $^{40}Ca(^{12}C,\gamma)^{52}Fe$	65, 85 MeV 33–48 MeV	Activity: γ Activity: $\beta-\gamma$	9
$^{28}Si(e,^{12}C)^{16}O\,e'$	23–34 MeV	Damage-track detectors for ^{12}C	10
$^{7}Li(^{12}C,\gamma)^{19}F$ $^{9}Be(^{12}C,\gamma)^{21}Ne$ $^{13}C(^{12}C,\gamma)^{25}Mg$ $^{12}C(^{12}C,\gamma)^{24}Mg$ $^{9}Be(^{16}O,\gamma)^{25}Mg$	12.0–27.2 MeV 15.0–18.7 MeV 15.0–23.0 MeV 20.3–33.3 MeV 10.0–33.9 MeV	NaI: High energy γ	50
$^{24}Mg(e,^{12}C)^{12}C\,e'$	21–32 MeV	Damage-track detectors for ^{12}C	11
$^{12}C(^{12}C,\gamma)^{24}Mg$	10–22 MeV	NaI: High energy γ	12

Table 1. Continued

Reaction	Bombarding energy range	Detection technique	Ref.
$^{27}Al(^{16}O, \gamma)^{43}Sc$	50–65 MeV	Activity: γ	13
$^{93}Nb(^{16}O, \gamma)^{109}In$		Activity: γ	
$^{12}C(^{14}N, ^{26}Al)\gamma$	59, 70 MeV	Recoil spectrometer:	14
$^{27}Al(^{14}N, ^{41}Ca)\gamma$	59 MeV	Fused nuclei	
$^{30}Si(^{14}N, ^{44}Sc)\gamma$	59, 70 MeV	detected at $\sim 0°$	
$^{24}Mg(e, ^{16}O\,^{8}Be)e'$	26, 30, 40 MeV	Si detectors: Both fragments detected	15, 16
$^{24}Mg(e, ^{12}C\,^{12}C)e'$			
$^{28}Si(e, ^{16}O\,^{12}C)e'$	26, 30, 40 MeV	Si detectors: Both fragments detected	16
$^{12}C(^{16}O, \gamma)^{28}Si$	16.3–30.1 MeV	NaI: High energy γ	17
$^{24}Mg(\alpha, \alpha'\,^{12}C)^{12}C$	120 MeV	Si detector and ionization counter	18
$^{12}C(^{12}C, \gamma)^{24}Mg$	10–18 MeV	NaI: High energy γ	19
$^{12}C(^{14}N, \gamma)^{26}Al$	12–24 MeV	NaI: High energy γ	28
$^{28}Si(^{12}C, \gamma)^{40}Ca$	16.8–22.0 MeV	NaI: High energy γ	45
$^{24}Mg(^{16}O, \gamma)^{40}Ca$	24.0–27.0 MeV		
$^{198}Pt(^{13}C, \gamma)^{211}Po$	57, 60 MeV	Activity: $^{211}Po(\alpha)$	57
$^{12}C(^{12}C, ^{24}Mg)\gamma$	39.4 MeV	Recoil spectrometer:	63
$^{16}O(^{32}S, ^{48}Cr)\gamma$	85 MeV	Fused nuclei	
$^{14}N(^{30}Si, ^{44}Sc)\gamma$	81.2 MeV	detected at $\sim 0°$	

[a] Measurements are listed roughly in chronological order. (This list was closed in March, 1981.)

observed decaying to the unresolved ground- and first-excited states of ^{23}Na from the region near 25.4 MeV excitation. The resonance, shown in Figure 1, did not appear in any charge-particle decay channels (protons, deuterons, or alphas), and a calculation of compound nucleus formation followed by competitive statistical decay failed to account for an appreciable fraction of the cross section. The resonance was interpreted as a "fission-transition" state of ^{23}Na in which nearly all the excitation energy is tied up as deformation energy. Although the possible existence of such unusual states must have seemed quite exciting, the $^{7}Li(^{16}O, \gamma)^{23}Na$ resonance appeared to be somewhat of an anomaly since none of the other capture reactions investigated showed any evidence of cross sections larger than a few nb/sr and showed no pronounced structure. The lack of interesting structure in several of the heavy-ion capture reactions reported by Feldman and

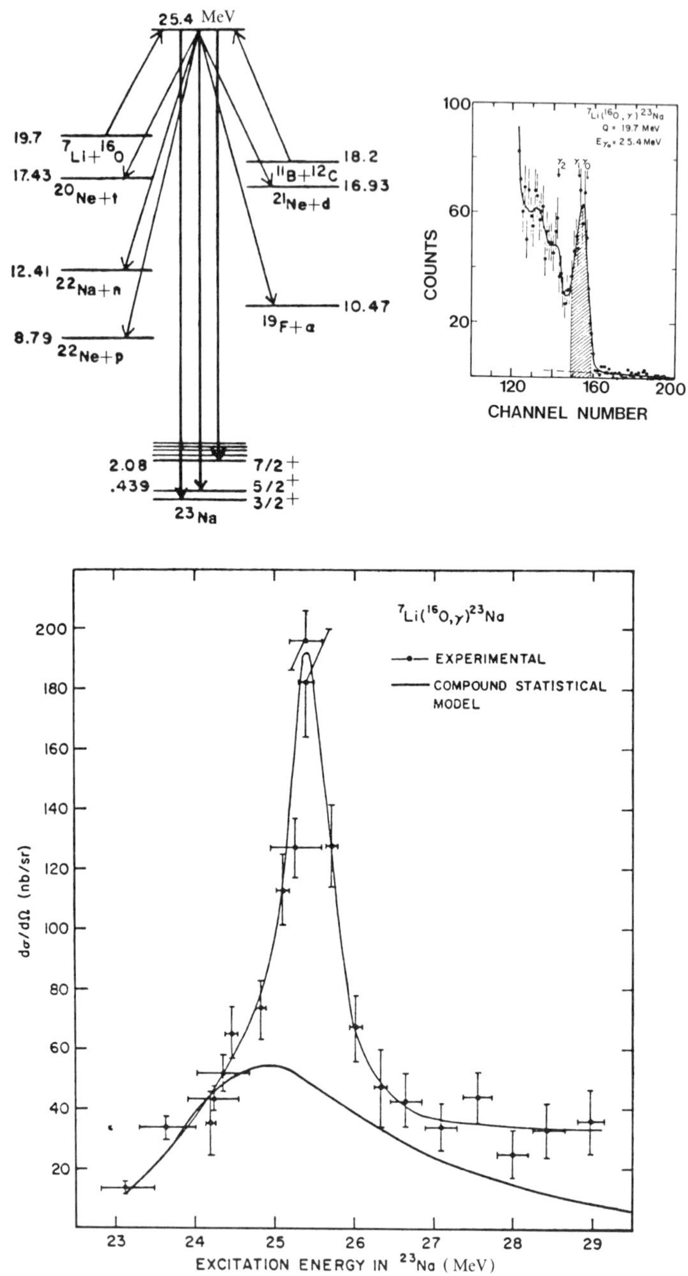

Figure 1. The high-energy portion of a γ-ray spectrum from the $^{7}Li(^{16}O, \gamma)^{23}Na$ reaction is shown at the top of the figure. Below, the excitation function for the summed γ_{0^+1} transitions is compared to a statistical calculation. These data are from Reference 6.

Heikkinen[6] discouraged further detailed measurements by high-energy γ-ray detection. (See, for example, the last paragraph of the review by P. Paul.[20])

Subsequent interest in heavy-ion capture shifted back to the more indirect methods involving the detection of residual activities.[8,9] Thick targets were used to effectively integrate over large ranges of excitation energy and attempts were made to unfold the general behavior of the radiative capture cross section with energy. The results seemed to either conflict or at least to be very different for different projectile–target combinations: the cross section for the $^{40}Ca(^{12}C, \gamma)^{52}Fe$ reaction reported by Imazato[9] decreased sharply with energy, while Zeller *et al.*[8] have claimed that the cross section for $^{30}Si(^{14}N, \gamma)^{44}Sc$ increases even more rapidly than the total reaction cross section.

Renewed interest in the possible existence of isolated heavy-ion capture resonances similar to Figure 1 came from electron-induced fission experiments[7,10,11] (essentially the time-reversed reaction) which indicated that relatively narrow structures in ^{28}Si and ^{24}Mg were decaying into heavy fragments after absorbing a (virtual) photon. In particular, a resonancelike structure near 22 MeV excitation in ^{24}Mg was observed to decay into two ^{12}C nuclei, both in their ground states.[11] The $^{12}C(^{12}C, \gamma_0)^{24}Mg$ radiative capture cross section was subsequently remeasured and several resonances were discovered.[12] Ironically, most of the structure in this reaction does appear between the upper and lower limits of the excitation function examined by Feldman and Heikkinen, but within a 2-MeV interval which they apparently chose not to cover.

The discovery of the $^{12}C(^{12}C, \gamma)^{24}Mg$ resonances has renewed interest in the detection of heavy-ion capture by direct observation of high-energy γ rays. Several other heavy-ion systems have been investigated and these are described in this review.

1.2. General Considerations

The deexcitation process in a radiative capture reaction can follow one of the routes shown schematically in Figure 2. In case (a) the compound nucleus loses energy by the emission of a high-energy γ ray to states below particle-emission threshold (t_p). Subsequent γ decay then leads to the fused nucleus in its ground state. The multipolarity of high-energy γ-ray transitions is very unlikely to be greater than 2.[21] Since the spins of states below t_p are generally low ($J \lesssim 5$), type (a) radiative transitions can be observed only if the spins of the capturing states are relatively low. Such states can be excited only near the Coulomb barrier. As the beam energy is increased above the barrier, the cross section for forming the compound nucleus peaks at increasingly higher spin, and ultimately the cross section for high-energy γ decay to bound levels below t_p must decrease. It is for this reason that all

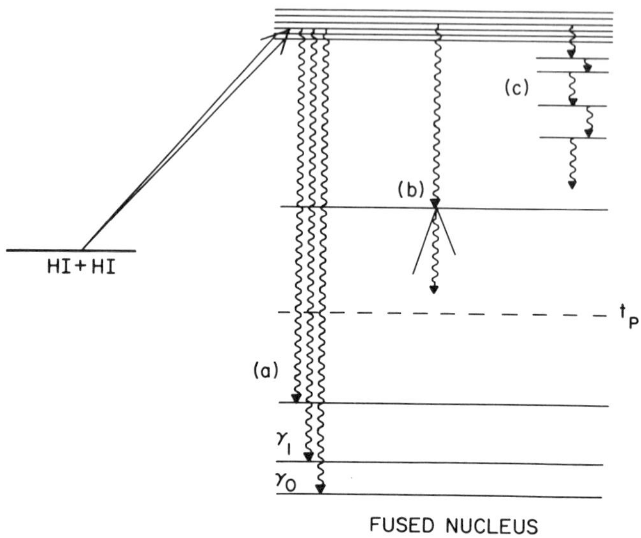

Figure 2. Possible radiative decay routes following heavy-ion capture.

of the capture measurements that study high-energy γ-ray transitions to low-lying states have been concentrated in the vicinity of the Coulomb barrier. These studies are discussed in Section 2.

Type (b) radiative transition (Figure 2) involving high-energy γ decay to states above t_p are of course possible, but become increasingly difficult to separate from the background as the γ-ray energy decreases. The cross section for reaching the ground state of the fused nucleus is further reduced by the γ-ray branching ratio Γ_γ/Γ of the intermediate state, and this is typically 10^{-3} to 10^{-5}. The final nucleus is usually not the fused one and its detection is hard to distinguish from nucleon-evaporation products. A definite signature for such decays would be an enhanced γ transition to a particular intermediate state. Although there are two recent examples of this in high-energy proton[22] and alpha[23] capture, nothing is known about such transitions following heavy-ion fusion. The γ-decay branching ratio of the intermediate state can, however, be much less unfavorable if this state is near the yrast line, and such a scenario is considered in Section 3.

Sequential radiative decays following heavy-ion capture (c in Figure 2) have been speculated to occur over a large range of energies. For capture above the yrast line but below the yrast level + B_n, where B_n is the nucleon binding energy, the compound state might be particle bound by the centrifugal barrier and might, in principle, γ-cascade down to the ground state without particle emission. As illustrated in Figure 3, for a beam energy that reaches an excitation energy E_0^*, such capturing states might be produced in the high-1 tail of the compound-nuclear cross section. This decay mecha-

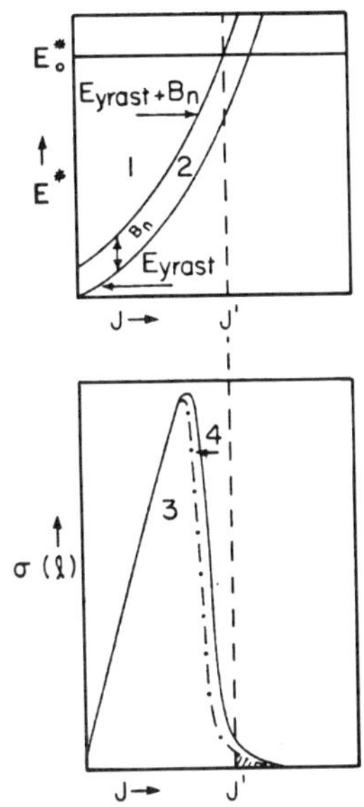

Figure 3. A typical distribution of compound states (region 3) populated in heavy-ion reactions is plotted in the lower part of the figure (Reference 8). Region 4 indicates the relatively small contributions from direct reactions. For initial excitation energies E_0^*, the energy-versus-spin plot indicates how capture through the high-spin tail of the compound nuclear cross section might populate states that are less than the nuclear binding energy B_n away from the yrast line (region 2). Such states might be particle-bound by the centrifugal barrier.

nism is discussed in Section 3, together with the experiments that have attempted to observe it.

In most cases, heavy-ion capture measurements must contend with cross sections that are smaller than those of particle-reaction channels by the ratio Γ_γ / Γ_p of γ-ray-to-particle decay widths. The cross sections that are encountered are thus several orders of magnitude smaller and rather specialized techniques must be employed for their measurement. Wherever such techniques depart from those normally encountered in heavy-ion physics, a short description has been included in the text. Sections 4 and 5 discuss a few aspects of capture measurements that have yet to be fully exploited and that may lead to studies of new phenomena not considered in Sections 2 and 3.

2. High-Energy Gamma Decays to Bound States

The most interesting characteristic of such capture measurements is the presence of resonances that exhibit nonstatistical heavy-ion and γ-ray partial widths. Most of these structures are not correlated either with previously identified features in other heavy-ion reaction channels or with structure in the distribution of electromagnetic multipole strength.

There are three factors that strongly affect the likelihood of observing such resonances. First, since the spins of the final bound states are relatively low, the spins of capture resonances are likely to be relatively low and, as discussed in Section 1.2, such structures will tend to concentrate near the Coulomb barrier and will be greatly diminished at energies above twice the barrier. Second, the average capture yield is indirectly dependent upon the location of giant electromagnetic resonances. The dependence arises simply because the capture cross section is proportional to Γ_γ, the radiative width, and sum rules restrict the total radiative width of a given multipolarity. If nearly all of the strength is contained in a giant resonance (GR) at an energy E_{GR}, then the magnitude of Γ_γ, and hence the capture cross section, cannot help but be greatly reduced at energies significantly different from E_{GR}. However, as discussed below, this is a dependence imposed by sum rules rather than a consequence of structure within giant resonances. Thirdly, the sum-rule restrictions on the total Γ_γ available effectively limit this type of capture measurement to $s-d$ shell nuclei. For heavier projectiles and targets, the Coulomb barrier is significantly higher, while the giant resonances, which change in position with $A^{-1/3}$, are shifted to lower energies. The farther the barrier is from the GR, the more difficult radiative capture is to observe. (The second and third considerations ignore the possibility of high-lying pockets of multipole strength, about which little experimental information is available. The general vicinity of the GRs in $s-d$ shell nuclei is certainly the most likely place to look for radiative capture to bound states, although it should be kept in mind that the GRs that have been observed in light nuclei exhaust significantly less than the sum-rule limits.)

The structures observed in capture have also been studied in the time-reversed reaction of photon-induced (or equivalently electron-induced) fission, and both classes of experiments are discussed in this section.

2.1. Techniques for Studying Fusion Followed by High-Energy γ Decay

The observation of high-energy deexcitation γ rays provided the first definite observation of heavy-ion capture shown in Figure 1. The techniques used in such measurements are similar to those of light-ion capture, and the review by P. Paul[20] describes many of these. Only the problems peculiar to heavy-ion capture are discussed here.

The Brookhaven spectrometer (BNL–MKII) is shown schematically in Figure 4. For reactions near the Coulomb barrier in $s-d$ shell heavy-ion systems the crystal can be positioned fairly close to the target and, since there are often several materials with atomic number Z substantially higher than the target (e.g., Ta, W, etc.), the beam can usually be stopped directly

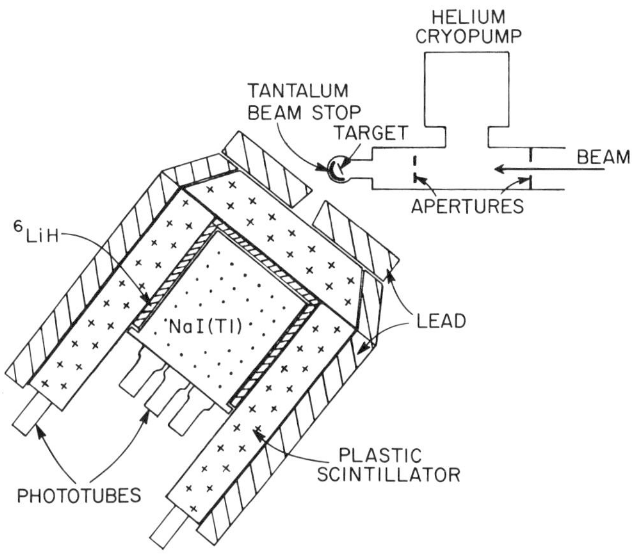

Figure 4. Schematic diagram of the BNL–MKII NaI spectrometer used for heavy-ion capture measurements.

behind the target without significantly increasing the γ-ray background. This allows measurements at forward angles, including 0°.

As shown in Figure 4, the large NaI(Tl) crystal is surrounded by plastic scintillator which is used in anticoincidence to reject events where energy leaves the NaI. This anticoincidence shield not only improves the line shape and resolution of the detector, but also rejects high-energy cosmic ray events with greater than 99% efficiency. The latter is extremely important when low cross sections are being measured.

The time consumption in the measurement of small capture cross sections can be reduced by using high beam currents, and several of the measurements described below utilized average currents of about one particle μA. This introduces three rather unique problems: First, the target thickness can change during the course of a measurement (typically ~ 6 hr in duration), in some cases by as much as 25%. This effect can be taken into account by continuously monitoring the yield of a prominent low-energy γ-ray line. If a separate excitation function is performed on this low-energy line using a comparatively low current beam on a target of known thickness, then the yield of this γ ray can be used to determine the effective target thickness during the course of a high current measurement. For high current experiments above the Coulomb barrier, the average NaI counting rate above the 1-MeV level can easily reach the 100–200-kHz range, and herein lies the second problem. At these rates the plastic anticoincidence shield is counting at well over 1 MHz above threshold (which is usually set as low as possible—typically 50 keV), and there can be an appreciable chance that a valid high-energy event will be rejected owing to an accidental anticoincidence with the shield. Failing that, a valid event can be accidentally rejected by pileup detection electronics (see Reference 20 for typical pulse-processing circuits). As the energy of a heavy-ion beam is increased over the Coulomb barrier, the reaction cross section rises and the average counting rates increase very rapidly. The result is that the rate of accidental rejection, and hence the overall efficiency, can change rapidly with bombarding energy (by as much as 30%)—much more rapidly than in most light-ion-induced reactions. This change in efficiency must be monitored, and a simple method is to passively add the signals from an electronic pulser to the anode signals from the NaI-photomultiplier tubes. The pulser can be set high in energy above the region of interest, and the fraction of these signals that are rejected determines the correction that must be applied to the efficiency. The third problem is one of energy calibration and gain stability. The decays to the ground and first few excited states, when they exist, are usually the highest events in the spectrum and, if the separation of low-lying levels in the fused nucleus is greater than the detector resolution, there is seldom any doubt as to the final state associated with the highest energy transitions. The situation is less clear in cases such as shown in Figure 1a, when a strong transition is observed to levels that cannot be resolved. With a knowledge of the detector line shape for high-energy γ rays, the identity of the final state can be estimated if an accurate calibration is available. The latter is usually obtained from light-ion reactions. For example, the $^{11}B(p,\gamma)^{12}C$ reaction, which populates the giant dipole resonance of ^{12}C, has a substantial cross section over a wide range of proton energies and can provide isolated γ-ray lines which bracket the region of interest. However, heavy-ion reactions generally produce far more γ rays at intermediate energies (5 to 15 MeV) than light-ion reactions, with the result that, for the same count rate in the NaI, the average current flowing down the dynode chains of the photomultiplier tubes is substantially greater. Gain shifts between the "calibrating" light-ion reaction and the capture reaction of interest are inevitable, and the removal of these effects involves considerable effort. Some techniques for gain stabilization are discussed in Reference 20. The gain of the BNL detectors is controlled by a combination of transistor-stabilized photomultiplier bases and an active amplifier with an internal gain determined by the height of an LED signal whose light is injected into each phototube by a system of fiber optics. (This arrangement is described in detail in another publication.[24]) Gain stability at the 0.1% to 0.4% level can be achieved with these techniques.

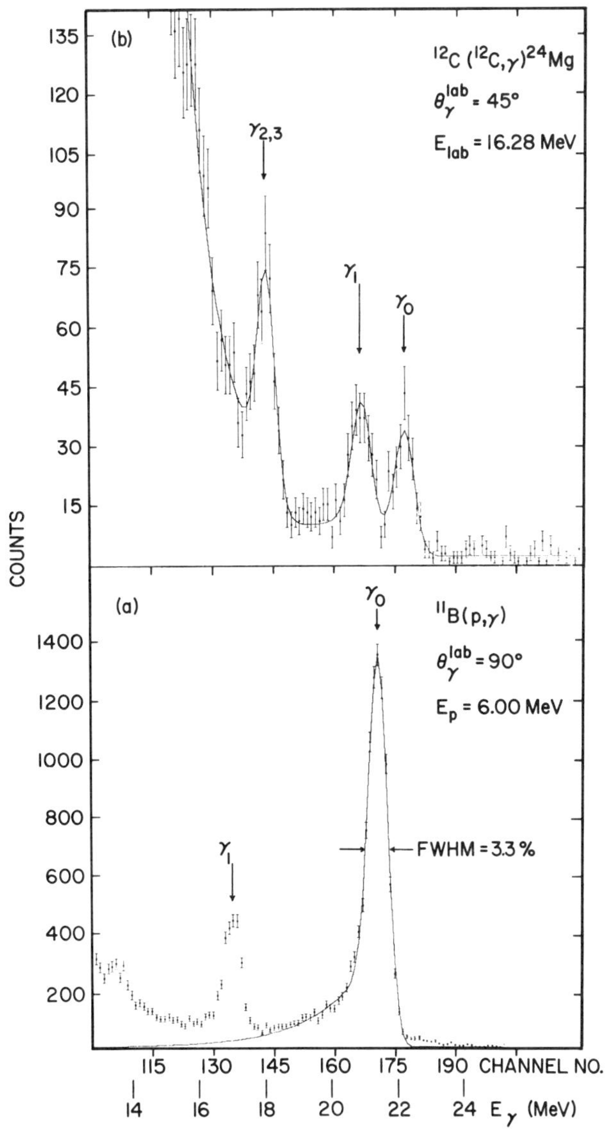

Figure 5. The response of the BNL–MKII detector to isolated high-energy γ rays from the $^{11}B(p,\gamma)^{12}C$ reaction is shown in (a). The solid curve is the intrinsic line shape for this detector, fitted to the γ_0 peak. A typical spectrum of high-energy γ rays from the $^{12}C(^{12}C,\gamma)^{24}Mg$ reaction is shown in (b).

(50-μm) silicon-surface-barrier detectors are mounted behind small ~1.5-kG permanent magnets that sweep away the large fluxes of low-energy Möller-scattered electrons that arise from the interaction of the beam with atomic electrons in the target. The remaining high-energy Mott (elastic) scattered electrons, as well as the fission fragments, are unaffected by these small fields but are far fewer in number. The decay fragments are detected on opposite sides of the electron beam. An event is identified by requiring a coincidence between opposing front detectors while requiring a simultaneous anticoincidence with the second counter of each telescope. Alpha decays, where the α particle penetrates into the back counter of a telescope ($E_\alpha \gtrsim 8$ MeV for 50-μm silicon detectors), can be identified by a coincidence between the front detectors of each telescope and *exactly one* backup counter.

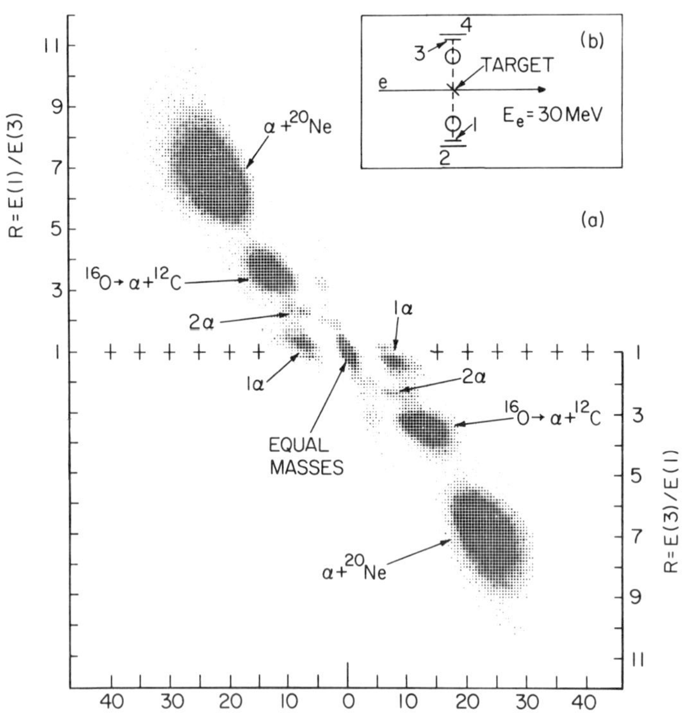

Figure 7. A density plot of the ratio, formed from the energy of the light fragment over that of the heavy fission fragment, is shown here as a function of the time difference between their arrival in opposite counters. The origins of the predominant peaks are indicated. Those labeled "2α" ("1α") are from $^{24}Mg \to ^{16}O + ^{8}Be$ decay where both (only one) α's from the decaying ^{8}Be are detected. The data are from Reference 15.

In the rest frame of the target nucleus, the ratio of the energies of the decay fragments is just the inverse of the ratio of their masses. This is very nearly true in the laboratory, since the velocity of the recoiling target nucleus is small, and can be used as a means of decay-channel identification. The spectrum of this ratio for 30-MeV electrons incident on a partially oxidized ^{24}Mg target is plotted in Figure 7a as a function of the time difference between the arrival of the fragments in opposite detectors.[15] The light fragment from the decay carries more energy and arrives first. With the ratio defined as the greater fragment energy over the lesser ($R \geq 1$ always), the events are naturally sorted into two groups. Labeling the detectors as shown in Figure 7b, events in the upper part of Figure 7a result from the lighter fragment entering detector #1 while the heavier fragment enters detector #3. Events in the lower part of Figure 7a result from the lighter fragment entering detector #3. When both detector systems are at 90° to the incident beam, as in Figure 7, the upper and lower halves of the ratio-vs.-time plot contain identical information. However, for other geometries, for example $\theta_1 = 55°$ and $\theta_3 = 125°$, the R-vs.-T plot provides data at two angles simultaneously.

Events associated with specific two-body-decay channels group together to form peaks in these R-vs.-T plots. Those associated with the symmetric* and asymmetric fission of ^{24}Mg, and with the alpha decay of both ^{24}Mg and ^{16}O, are indicated in Figure 7. (Proton decay is not evident here only because the corresponding recoil nuclei are too low in energy to detect.) An excitation function for one of these decay channels is then contained in the particle-energy spectrum of those events appearing within a single peak. (A more detailed discussion of the techniques involved in these measurements is given in References 15 and 27.)

2.3. Fusion/Fission Resonances in ^{24}Mg

The most extensively studied heavy-ion capture reaction is the fusion of two ^{12}C nuclei leading to low-lying levels in ^{24}Mg. The $\theta = 45°$ excitation functions for the γ_0, γ_1, and $\gamma_{2,3}$ transitions are shown in Figure 8 (the data are from References 12, 19). There are two striking features of these data. First, the yield is localized, especially in the γ_0 channel, extending from 19.5 to about 22.5 MeV excitation. Outside this region the cross section is greatly diminished. Second, within this gross structure, there is considerable finer structure in the form of narrow ($\Gamma \lesssim 0.3$ MeV) resonances, several of which appear correlated in all three γ-decay channels.

* Three-body decays and light contaminants contribute to the "Equal masses" peak of Figure 7. However, the $^{12}C + ^{12}C$ spectrum can be measured by placing a foil in front of one detector to degrade one ^{12}C and thus shift real $^{12}C + ^{12}C$ events to a different position in R versus T (Reference 15).

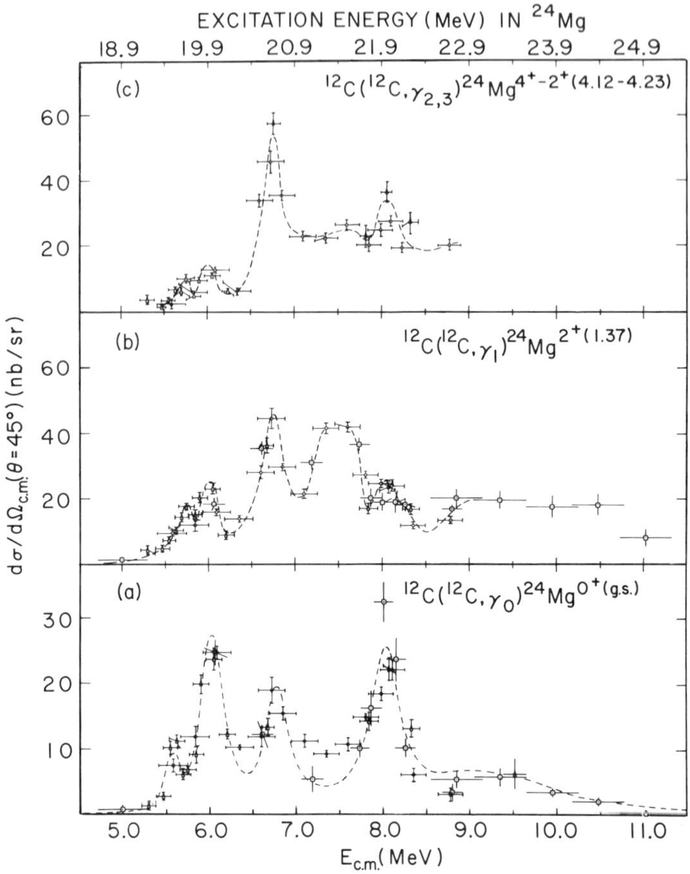

Figure 8. Excitation functions for $^{12}C(^{12}C, \gamma)^{24}Mg$ at $\theta_\gamma = 45°$ are plotted here against $\langle E_{c.m.} \rangle$, the c.m. energy at the center of the target. The corresponding excitation energy scale is given at the top of the figure. Vertical bars indicate statistical uncertainties in the cross section whereas horizontal bars reflect the energy loss in the target. These data are from Reference 19.

For identical bosons in the entrance channel, the J^π of the compound state is limited to $0^+, 2^+, 4^+, \ldots$. The observation of a photon decaying to the 0^+ ground state rules out 0^+ assignments for structures appearing in this channel, and makes multipolarities greater than 2 extremely unlikely.[21] Thus, all of the yield in the $^{12}C(^{12}C, \gamma_0)$ reaction arises from 2^+ states in ^{24}Mg and exhibits a $\sin^2(2\theta\gamma)$ angular dependence. The dashed line through the γ_0 excitation function is the result of a fit to an incoherent sum of Breit–Wigner resonances, and the resonance parameters extracted from this fit are shown in Table 2.[19] The fit falls below the data near 20.3 and 21.3

Table 2. Resonance Parameters Extracted from the $^{12}C(^{12}C, \gamma_0)^{24}Mg$ Excitation Function (Reference 19); Uncertainties are in Parentheses

Centroid			
$E_{c.m.}$ (MeV)	E_x (MeV)	Γ (MeV)	$\Gamma_c\Gamma\gamma/\Gamma^a$ (eV)
5.56(0.05)	19.49	0.130^b	0.011(0.002)
6.01(0.02)	19.94	0.245(0.042)	0.084(0.007)
6.77(0.03)	20.70	0.250(0.050)	0.058(0.007)
8.02(0.02)	21.95	0.342(0.082)	0.090(0.012)
$9.17(0.50)^a$	23.10	$2.0\ (0.6)^a$	$0.17\ (0.03)^a$

^{a}The fifth Lorentzian serves as a parameterization of the background under the four obvious resonances in the γ_0 yield. Different parametrizations can change the strengths ($\Gamma_c\Gamma\gamma/\Gamma$) of the prominent peaks appreciably, but will have relatively little effect on centroids and widths.
^{b}Constrained to the results of Reference 29.

MeV excitation, which may indicate contributions from additional less prominent structures. The dashed lines through the γ_1 and $\gamma_{2,3}$ excitation functions are merely to guide the eye.

The ^{12}C (ground state) yields from the complimentary electrofission reaction are shown in Figure 9.[15,16] In addition to the excitation of the 2^+ strength observed in $^{12}C(^{12}C, \gamma_0)^{24}Mg$, the $^{24}Mg(e, ^{12}C^{12}C)e'$ reaction can also proceed through 0^+ strength–CO excitation.* The yield of ^{12}C nuclei from the fission of 2^+ states in ^{24}Mg has an angular distribution dominated by $\sin^2(2\theta_c)$—the same as the time-reversed capture reaction—and this vanishes at $\theta_c = 90°$. The 90° cross section results principally from the decay of 0^+ levels and is isotropic. There is a small correction (~ 8%) due to 2^+ excitation via longitudinal photon (C2) exchange. This varies as $[P_2(\cos\theta_c)]^2$ and is nonvanishing at 90°. The 0^+ and 2^+ strengths extracted from these data, assuming no CO–C2 interference, are shown in Figure 10.[16] The $^{12}C(^{12}C, \gamma_0)^{24}Mg$ excitation function is shown as the dash-dotted line in Figure 10b. Three of the four primary resonances of Table 2 appear at about the right level in the electrofission data, although the resolution in the excitation function (~ 0.4 MeV) is not as good as that obtained from the capture measurements. The fourth resonance at 20.70 MeV is greatly reduced in Figure 10b, indicating significant CO–C2 interference near this

* The designations CJ and EJ in electron scattering refer to terms in the interaction corresponding to the exchange of longitudinal ($M = 0\hbar$) and transverse ($M = \pm 1\hbar$) virtual photons, respectively. There is no such thing as a transverse-monopole excitation, and 0^+ states are populated only through the CO term.

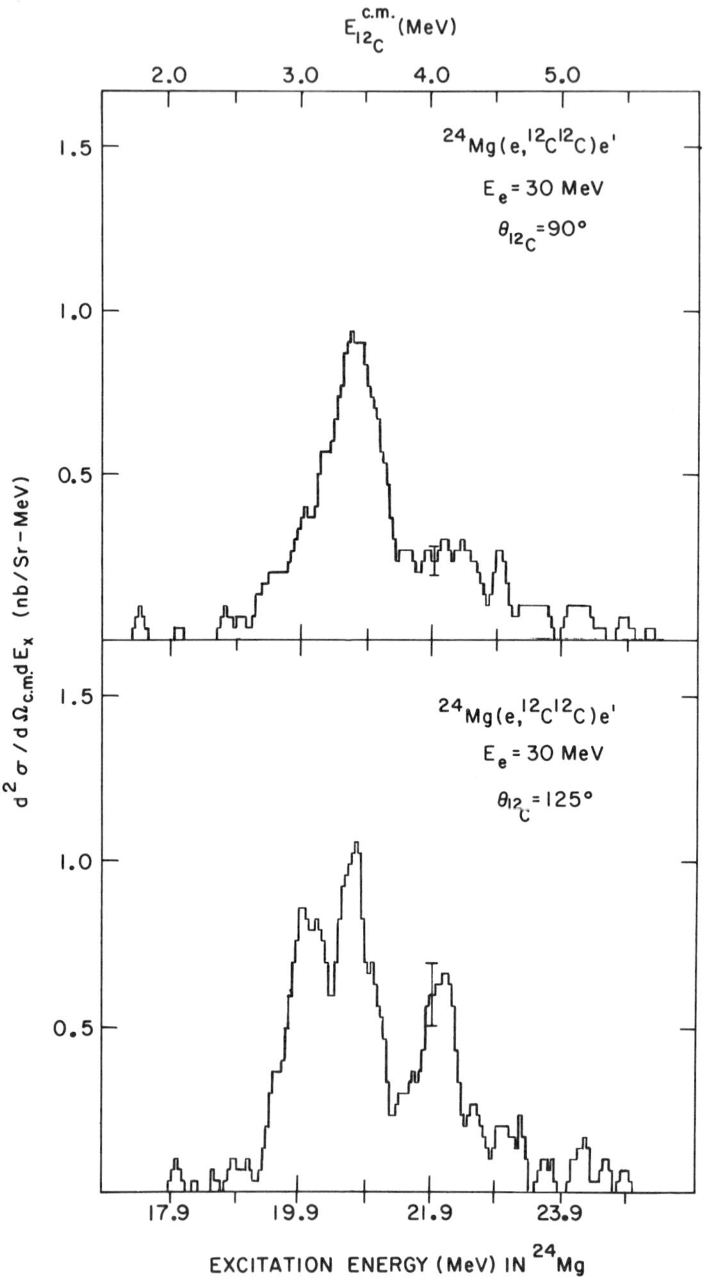

Figure 9. Excitation functions obtained with 30-MeV electrons for the $^{12}C + {}^{12}C$ decay channel of ^{24}Mg are shown here at two angles (Reference 16).

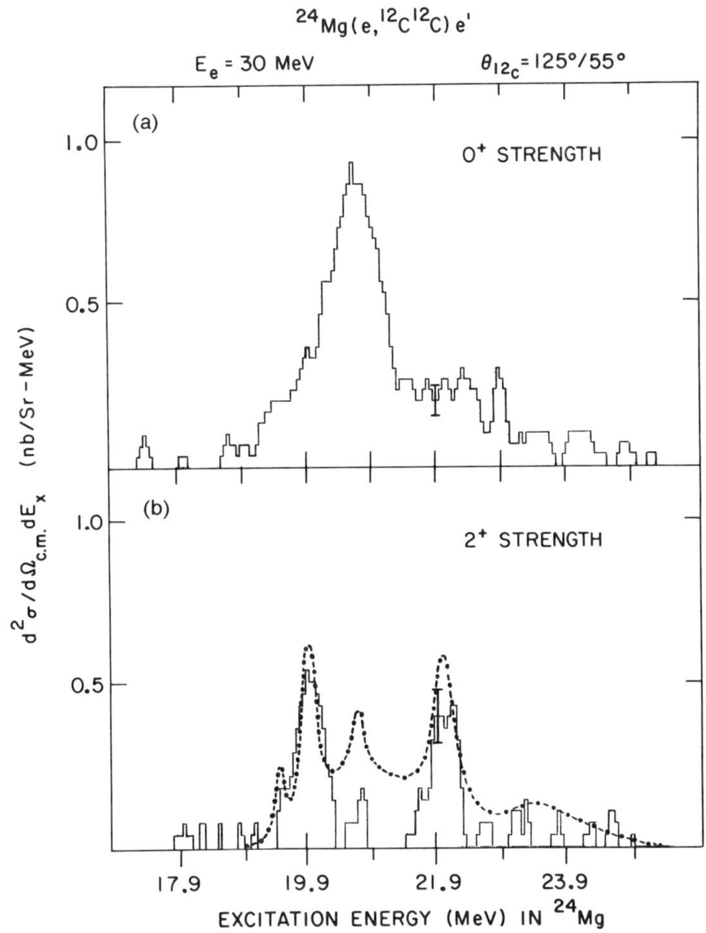

Figure 10. The 0^+ and 2^+ components of the 125°/55° ^{12}C yield of Figure 9 are shown here. The dash-dotted line is the $^{12}C(^{12}C, \gamma_0)^{24}Mg$ excitation function from Figure 8.

energy. This in itself is not surprising since it is very near the peak of the 0^+ strength. However, the presence of strong interference also suggests a structural similarity between the 0^+ and 2^+ states that fission into ground-state ^{12}C nuclei.

The electrofission measurements provide data on the $^{16}O + {}^{8}Be$ channel, which is not accessible through radiative capture. Excitation functions for $^{24}Mg(e, {}^{16}O{}^{8}Be)e'$ are shown in Figure 11 for a sequence of angles.[15] A number of structures are evident. The yield can be decomposed, as indicated on the angle-integrated curve in Figure 11a, into a broad structure and a number of resonances. The broad component (dotted line) is nearly isotropic over the angular range of Figure 11b and could result from CO

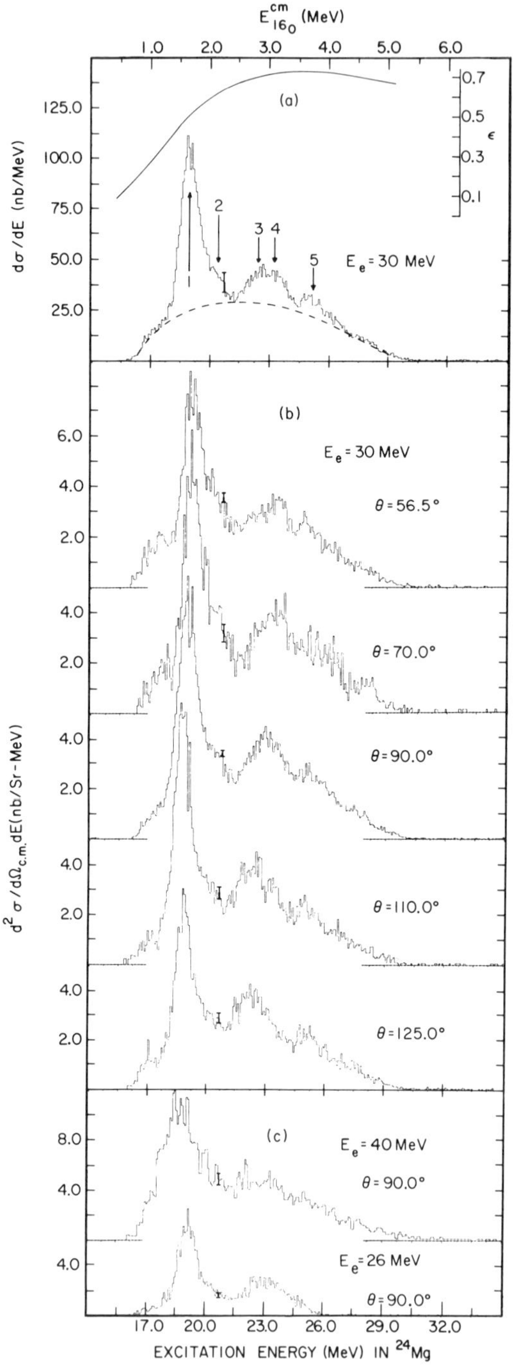

Figure 11. Cross sections for $^{24}Mg(e, {}^{16}O, {}^{8}Be)e'$ are shown in (b) and (c). The angle-integrated spectrum is shown in (a). The data are from Reference 15.

excitations, although other multipoles surely contribute. The angular distribution of the resonance labeled No. 1 at 19.0 MeV indicates strong dipole absorption. Resonances No. 3 and No. 4 are strongly interfering, with the yield of No. 3 peaking at backward angles while that of No. 4 peaks at forward angles. A fit to these distributions is consistent with strong $E2$ and $E1$ excitation at 22.5 and 23.5 MeV, respectively. This 22.5-MeV structure is near the 2^+ 22.0-MeV resonance in the $^{12}C + ^{12}C$ decay channel. At present, it is not possible to determine whether these resonances are uncorrelated or are, in fact, the same structure, shifted in apparent centroid only because of the interference in the $^{16}O + ^{8}Be$ channel. Nonetheless there is, on the whole, little evidence of states in ^{24}Mg that decay into both fission channels, despite the nearly identical Q values and Coulomb barriers for $^{12}C + ^{12}C$ and for $^{16}O + ^{8}Be$ decay. Evidently the structural differences between these channels profoundly affect the yield from fissioning states.

2.4. Comparison between Fusion/Fission in ^{24}Mg and Other Reactions

The study of resonances in light heavy-ion systems has long been a topic of intense interest. This is especially true of the $^{12}C + ^{12}C$ system, which is rich in structure ranging from below the Coulomb barrier up to more than six times the barrier. A large number of resonances have been reported in elastic, inelastic, and a variety of reaction channels (see Figure 14, for example). The previously identified 2^+ $^{12}C + ^{12}C$ resonances above 5.0 MeV c.m. are listed in Table 3. In Figure 12 the $^{12}C(^{12}C, \gamma_0)$ excitation function (Figure 12b, solid curve) is compared with these resonances (Figure 12c and Table 3). The γ_0 peak at 5.6 MeV corresponds to the 2^+ quasi-molecular resonance at 5.63 MeV c.m. However, the peaks at 6.0 and 8.0 MeV c.m. do not correlate with any $^{12}C + ^{12}C$ 2^+ structures. The γ_0 peak at 6.8 MeV c.m. is close to but considerably wider than the 2^+ resonance at 6.64 MeV c.m. ($\Gamma = 0.1$ MeV), so although a contribution to the γ_0 peak from this resonance cannot be ruled out, some significant fraction must be due to a hitherto unreported structure. The main point is that there are

Table 3. Previously Reported 2^+ Resonances in $^{12}C + ^{12}C$ Reactions Above 5.0 MeV c.m. (The Reference Given Is Not Necessarily the Primary Reference.)

$E_{c.m.}$ (MeV)	E_x (MeV)	Γ_{total} (MeV)	Γ_c (MeV)	Γ_c/Γ	Reference number
5.00	18.933	—	—	—	31
5.63	19.563	0.130	0.010	0.08	29
6.26	20.193	0.125	0.016	0.13	32
6.41	20.343	—	—	—	33
6.64	20.573	0.100	0.029	0.29	34

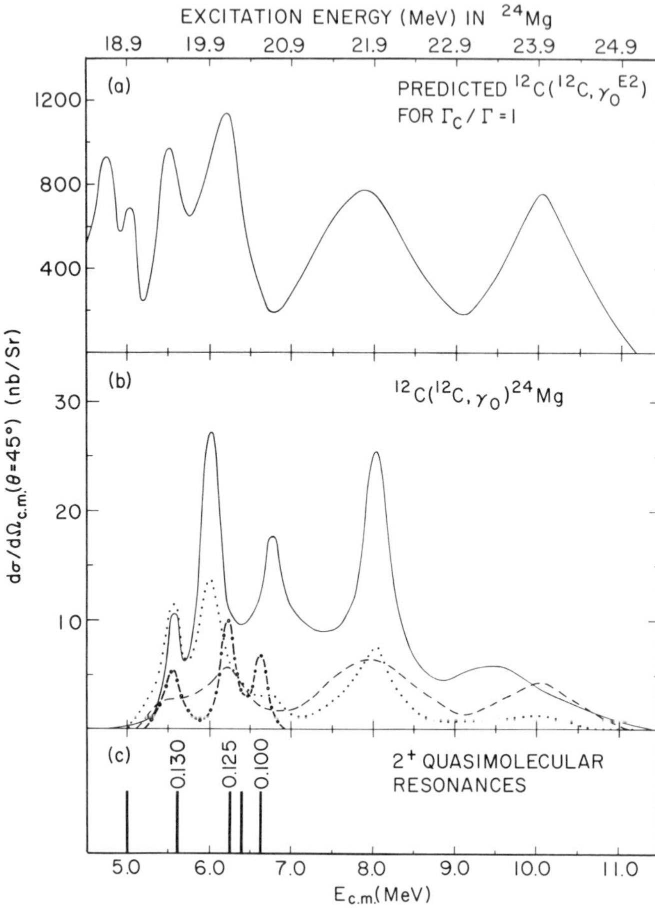

Figure 12. (a) Predicted capture cross sections for $\Gamma_c/\Gamma = 1$. The curve reflects the structure in Γ_γ. (b) The $^{12}C(^{12}C, \gamma_0)^{24}Mg$ excitation function (solid line) and various calculations. The dashed line is the calculation assuming statistical carbon widths. The dotted curve is the prediction assuming resonant carbon widths associated with the resonances of Table 2 and statistical photon widths. The circle-dashed curve is the prediction for the 3 quasimolecular 2^+ states with known ^{12}C widths in Table 3, assuming statistical photon decay. The calculations are taken from Reference 19. (c) The location of 2^+ quasimolecular states (Table 3). Total widths of resonances are indicated in MeV.

striking 2^+ resonances apparent in the γ_0 decay channel that have not been previously identified as 2^+ structures in the $^{12}C + ^{12}C$ system.

The best available data on the distribution of $E2$ strength built on the ^{24}Mg ground state have come from several high-energy $^{24}Mg(\alpha, \alpha')$ experiments,[35,36] which measure the distribution of $E2$ radiative width per MeV, $d\Gamma_\gamma/dE_x$. The integrated photon-interaction cross section can be expressed in terms of the ground-state radiative width Γ_{γ_0} via

$$\int \sigma_\gamma \, dE = \pi^2 \lambda\!\!\!^{-2} \frac{(2J_x + 1)}{(2J_0 + 1)} \Gamma_{\gamma_0} \tag{1a}$$

For $E2$ excitations $J_x = 2$, while the ground-state spin of ^{24}Mg is $J_0 = 0$, and the γ-ray energy is $E_\gamma = \hbar c/\lambda\!\!\!^{-}$. The cross section for the photofission of ^{24}Mg into two ^{12}C nuclei is obtained by multiplying equation (1a) by the branching ratio $\Gamma_{^{12}C}/\Gamma$. The capture cross section can be obtained from this using detailed balance. The $^{12}C(^{12}C, \gamma_0)^{24}Mg$ cross section at an excitation energy E_x in ^{24}Mg is related to $d\Gamma_\gamma/dE_x$ by[19]

$$\frac{d\sigma^{45^\circ}}{d\Omega}(E_x) = 2\left(\frac{25\pi}{64 m_0 c^2}\right) \frac{(\hbar c)^2}{E_x - Q} \frac{\Gamma_c}{\Gamma} \frac{d\Gamma_\gamma}{dE_x} \tag{1b}$$

where the initial factor of 2 arises from the identity of the particles in the incoming channel, and $m_0 c^2$ and Q are the nucleon mass (931.48 MeV) and Q-value (13.93 MeV), respectively. The right-hand side of equation (1) is plotted in Figure 12a under the condition $\Gamma_c/\Gamma = 1$. Thus any structure in the resulting cross section must be due to structure in the photon decay channel. Upon comparing Figure 12a with Figure 12b, solid curve, we see that the 5.6-MeV c.m. resonance seems to be correlated with a peak in the $E2$ strength function and the 8.0-MeV c.m. resonance is observed near the middle of a much broader structure. However, the 6.0- and 6.8-MeV c.m. resonances line up with valleys in the $E2$ distribution. On the whole, the pronounced features of the fusion yields do not directly reflect the structure in Γ_γ.

Similar comparisons applied to the γ_1 and $\gamma_{2,3}$ excitation functions of Figure 8 are somewhat ambiguous. Although the structures evident near 6.0, 6.8, and 8.0 MeV c.m. correlate with 2^+ resonances observed in the γ_0-decay channel, these yields could contain contributions from 0^+ or 4^+ components, and several 4^+ $^{12}C + ^{12}C$ resonances have been reported near these energies.[19] The peak at 5.75 MeV c.m., present in both γ_1 and $\gamma_{2,3}$, and the broad bump at 7.5 MeV c.m. in γ_1 do not correspond in either position or width with any previously identified features of other heavy-ion reactions. Their apparent absence in the γ_0 yield may indicate dominant 4^+ or 0^+ components. There is considerable monopole strength near the peak at 7.5

MeV c.m., 21.4 MeV excitation (see Figure 10a), and these could couple to the first 2^+ of ^{24}Mg via an $E2$ operator.

The 0^+ strength shown in Figure 10a has not been observed in any other reaction. In fact, the giant monopole resonance, which is expected near 21 MeV in ^{24}Mg, has never been identified.

The ^{8}Be + ^{16}O fission resonances of Figure 11 do not correspond to any structures reported in other heavy-ion reactions. The 19.0-MeV peak (No. 1 in Figure 11a) does coincide in spin, position, and width with a peak in the giant dipole resonance (GDR) of ^{24}Mg.[15,30] However, other prominent peaks in the GDR are absent from this decay channel.

2.5. Statistical Model Analysis of ^{24}Mg Resonances

To interpret the unusual features of the fusion/fission reactions discussed above it is essential to know the extent of the contributions to the cross section from the process of compound nucleus formation followed by competitive statistical decay. The most detailed comparisons can be made with the $J^\pi = 2^+ - {}^{24}$Mg symmetric fusion resonances since the ^{12}C(^{12}C, γ)^{24}Mg measurements provide the most extensive data (Figure 8). (A detailed discussion of these calculations is given in Reference 19.)

Following the theory of compound nuclear reactions,[37] the average ^{12}C(^{12}C, γ_0) cross section can be calculated by replacing Γ_c/Γ in equation (1) with $(\overline{\Gamma_c/D})/(\overline{\Gamma/D})$, where Γ_c and Γ are the average carbon and total widths of the underlying 2^+ compound nuclear levels, and D is the average energy spacing of these levels. This calculation assumes that there are a large number of overlapping levels ($\Gamma/D \gg 1$) with random amplitudes and phases and that the decay amplitudes into the entrance and exit channels are uncorrelated. $(\overline{\Gamma/D})$ may be computed according to the Hauser–Feshbach prescription[38]

$$\overline{\Gamma/D} = \frac{1}{2\pi} \sum_{\kappa} T_\kappa \tag{2}$$

where T_κ is the optical model transmission coefficient in channel κ and the sum runs over all open channels. For 2^+ levels in ^{24}Mg $\overline{\Gamma/D}$ ranges from about 3.2 at 19 MeV to 17.5 at 23 MeV excitation. At high energies E_x the density of nuclear levels of spin J (either parity) can be parametrized as[39]

$$\rho(E_x - \delta b, J) = \frac{\sqrt{\pi}\exp\{2[a(E_x - \delta b)]^{1/2}\}}{12a^{1/4}(E_x - \delta b)^{5/4}} \frac{(2J+1)}{2(2\pi)^{1/2}\sigma^3} \times \exp\left(\frac{-(J+1/2)^2}{2\sigma^2}\right) \tag{3}$$

where δ is the average of the neutron and proton binding energies, and b is equal to 2 for even–even nuclei, 1 for odd–even, and 0 for odd–odd. The parameters a and σ have been tabulated by Gilbert and Cameron.[39] Equation (3) determines the average spacing of 2^+ compound-nuclear levels in ^{24}Mg, and combining this with the results from equation (2) implies an average width of about 20 keV for 2^+ states between 19 and 23 MeV in ^{24}Mg. Thus, peaks due to the random coherence in amplitude and phase of strongly overlapping levels cannot be significantly broader than about 50 keV, and since the total widths of the ^{12}C+^{12}C fusion/fission resonances are considerably larger, an Ericson fluctuation phenomenon[40] can be ruled out.

2.5.1. *Statistical ^{12}C Decay from ^{24}Mg*

Under the assumption that a photon excites the nucleus, which then mixes into the compound levels and decays statistically, the carbon strength function $\overline{\Gamma_c/D}$ may be computed as

$$\overline{\Gamma_c/D} = \frac{T_{CC}}{2\pi} \tag{4}$$

where T_{CC} is the ^{12}C+^{12}C transmission coefficient. Changes in the transmission coefficient will affect the gross structure of the cross section, but in this picture finer structure must originate with Γ_γ. Using equations (1), (2), and (4), the $^{12}C(^{12}C, \gamma_0)^{24}Mg$ cross sections from this process have been calculated and are plotted as the dashed line in Figure 12b. Over most of the excitation function, this calculation falls far below the data and accounts for none of the fine structure. However, it does succeed in explaining the drop in cross section below 5.5 MeV c.m. as the result of the Coulomb barrier reduction of T_{CC}. The falloff above 11 MeV c.m. may result from a general decrease in Γ_γ or from the increase in the right-hand side of equation (2) that is expected as the number of open channels increases. Nonetheless, the overall conclusion is that the partial ^{12}C widths for the resonances of Figure 8 are significantly enhanced over statistical widths.

2.5.2. *Statistical γ Decay from* ^{24}Mg

The statistical model can also be used to calculate the decay of a resonance excited in the ^{12}C+^{12}C channel (i.e., a resonance in Γ_c) which mixes into the underlying compound levels, and emits a photon by virtue of the $E2$ strength of these underlying levels. For this the average carbon widths are assumed to have only resonant components,

$$\overline{\Gamma_c/D} = \frac{1}{2\pi} \frac{\Gamma^2}{(E-E_0)^2 + \Gamma^2/4} \frac{\Gamma_c}{\Gamma} \tag{5}$$

The average ^{12}C width of levels in ^{24}Mg has been calculated as a sum of Lorentzians using in equation (5) the positions (E_0) and total widths (Γ) of the resonances in Table 2 assuming for all but the lowest peak, $\Gamma_c/\Gamma = 0.14$ (the average branching ratio for those low spin resonances in this region for which the ^{12}C width has been determined[19]). For the 5.6-MeV c.m. resonance, the measured value of $\Gamma_c/\Gamma = 0.077$ was used.[29] Combining this with equations (1) and (2) gives the dotted curve in Figure 12b. The γ_0 decay of the 5.6-MeV c.m. resonance is completely consistent with statistical mixing into the compound nucleus followed by $E2$ γ decay. The magnitude of the 6.0-MeV resonance predicted by this calculation is about a factor of 2 below the data. This process could account for the 6.0-MeV γ_0 peak if Γ_c/Γ were as large as 0.30. However, this would be grossly inconsistent with elastic-scattering measurements[34] which have assigned $J^\pi = 4^+$, $\Gamma = 0.100$ MeV, and a much smaller carbon width $\Gamma_c/\Gamma = 0.03$ to the only observed peak near this energy (5.98 MeV c.m.). The other peaks in the γ_0 yield are far above the dotted curve. The conclusion from this analysis is that the photon widths of the fusion resonances are, for the most part, significantly enhanced over statistical widths.

Since statistical photon decay causes the 5.6-MeV c.m. resonance to appear in the $^{12}C(^{12}C, \gamma_0)$ excitation function, the other 2^+ quasimolecular resonances of Table 3 would also be expected, at some level. The circle-dashed line of Figure 12b has been calculated using in equation (5) the parameters of those resonances of Table 3 with known ^{12}C widths. Evidently, small contributions due to statistical γ decay from molecular resonances could account for the apparent excess γ_0 yield above the fitted excitation function of Figure 8.

2.5.3. *A New Type of Intermediate Structure*

There are two conclusions to be drawn from the statistical analysis of the preceding sections. First, at least one and possibly all of the previously identified quasimolecular resonances (Table 3) are present in the radiative capture yields, but only at a very low level consistent with a statistical γ decay to ^{24}Mg. Second, the $^{12}C + {}^{12}C$ and the γ_0 decay probabilities of the dominant resonances observed in radiative capture are significantly greater than statistical probabilities for decay from the compound nucleus. The fission/fusion resonances must then reflect correlations between the $E2$ and the $^{12}C + {}^{12}C$ 2^+ strength functions, arising from an unusual form of intermediate structure that is strongly coupled to both the $^{12}C + {}^{12}C$ channel and the low-lying states of ^{24}Mg. This represents the first clear evidence for a close link between states with large $^{12}C + {}^{12}C$ parentage and the *structure* of ^{24}Mg. Furthermore, it is interesting that those states that exhibit this close connection form a new set of resonances which, although undoubtedly present in $^{12}C+^{12}C$ elastic scattering at some level due to their nonstatistical ^{12}C widths, certainly do not correspond to the dominant quasimolecular features seen in $^{12}C+^{12}C$ reactions. That this *new* set of capture resonances has a structure distinct from previously observed resonances may be evident in their γ-decay branches. The 2^+ peaks at 6.0, 6.8, and 8.0 MeV seem to be present in the γ_1 and $\gamma_{2,3}$ excitation functions as well as in γ_0. This is consistent with a close link to the structure of the fused nucleus. The wave functions of the levels in the ground-state rotational band (GSB) of ^{24}Mg are very similar, and thus 2^+ levels that decay strongly to the ground state would also be expected to have significant decay branches to the 2^+ and 4^+ members of the GSB. In contrast, the *old* 5.6-MeV quasimolecular resonance appears only in the γ_0 excitation function. However, the γ decay of this resonance arises from statistical photon emission from an overlapping peak in the distribution of ground-state $E2$ strength. The $E2$ strength built on excited states of ^{24}Mg may be very different, and thus γ decay to the ground state via statistical photon emission does not guarantee a comparable decay rate to other members of the GSB.

Similar considerations apply to the 0^+ strength of Figure 10a which decays by fission. This branch accounts for $(4.2\pm1.2)/\langle\Gamma_{^{12}C}/\Gamma\rangle$ percent of the energy-weighted monopole sum rule S(CO).[15] A recent high-energy $^{24}Mg(\alpha,\alpha')$ experiment has measured an upper limit of 8% of S(CO) in this region of excitation (van der Borg[36]). This implies $\Gamma_{^{12}C}/\Gamma>0.5$, an amazingly large value which again indicates a highly nonstatistical behavior and an unusual form of intermediate structure. Certainly breathing-mode deformations do not easily lend themselves to fission. However, the strong CO–C2 interference evident in Figure 10b—see Section 2.3—may provide the clue since it suggests strong structural similarities. The fissioning 0^+ strength may result, not from a breathing-mode-type density oscillation, but rather from a 0^+ $K=0$ β vibration which is quite akin to quadrupole in deformation.

2.5.4. *Statistical Decay from Mixing in the T = 1 Dipole Resonance*

The cross sections for decay into the $^{8}Be+^{16}O$ channel shown in Figure 11 are about a factor of 10 larger than for symmetric fission into two ^{12}C nuclei. Hauser–Feshbach calculations, similar to those of Figure 12b, for the statistical ^{16}O decay of the $E2$ strength in ^{24}Mg fall a factor of 50 to 100 below the data of Figure 11. However, asymmetric decay following $E1$ absorption is no longer forbidden as it is for symmetric fission, and the total $E1$ photoabsorption cross section in ^{24}Mg is much larger. Most of this $E1$ strength is contained in the $T=1$ giant dipole resonance (GDR), and mixing of the GDR with $T=0$ states in ^{24}Mg can lead to decays into the $T=0$ $^{16}O+^{8}Be$ channel. A prescription for calculating the effect of this mixing has been given by Shikazono and Terasawa.[41] The ^{16}O-decay probability is given by $(\overline{\Gamma_{^{16}O}/D})/(\overline{\Gamma/D})$, with the numerator defined analogously to equation (4). However, the denominator, dominated by proton, neutron, and alpha decay channels with transmission coefficients T_p, T_n, and T_α, respectively, is modified from that in equation (2),

$$\frac{1}{\overline{\Gamma/D}} = \frac{2\pi(1-e^2)e^2R}{(1-e^2+e^2R)\sum\left(\frac{T_p+T_n}{2}\right)+e^2R\sum T_\alpha} + \frac{2\pi e^2(1-e^2R)}{(1+e^2-e^2R)\sum\left(\frac{T_p+T_n}{2}\right)+(1-e^2R)\sum T_\alpha} \tag{6}$$

Here e^2 is the fraction of $T=1$ levels that mix with $T=0$ states, and R is the ratio of the densities of the $T=0$ and $T=1$ levels. The density of the $T=0$ states can be calculated from equation (3) where the energy-dependent terms are shifted by the neutron and proton pairing energies (5.13 MeV in ^{24}Mg),[42] i.e., $\rho_{T=0}=\rho(E_x-5.13)$. For $T=1$ states in a $T_Z=0$ nucleus, the level density can be assumed to be the same as for the adjacent $T=1$, $T_Z=1$ nucleus. The energy-dependent terms in the level density of the $T_Z=0$ nucleus are thus shifted by the energy of the lowest $T=1$ state (9.515 MeV in ^{24}Mg),[43] i.e., $\rho_{T=1}=\rho(E_x-9.515)$.

The GDR is clearly evident[30] in total photoabsorption measurements ^{24}Mg(γ_{TOT}), and a peak in these data at 19.0 MeV is correlated with a similar peak in the $\alpha+^{20}$Ne decay channel.[15,49] A comparison between these

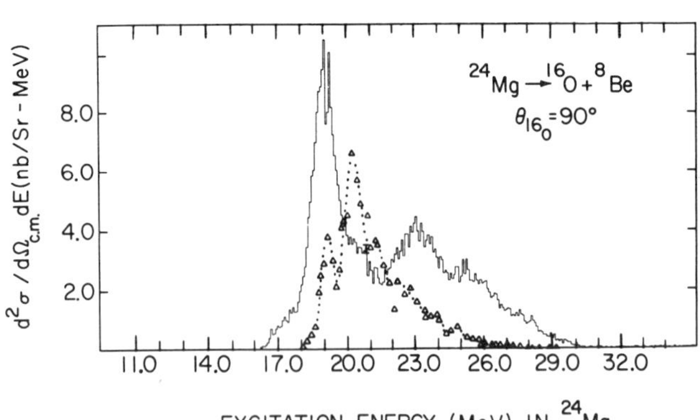

Figure 13. The 90° ^{24}Mg$(e,^{16}$O$,^{8}$Be$)e'$ excitation function is compared to the predictions of statistical decay (triangle points) from the giant dipole resonance. From Reference 15.

measurements gives the branching ratio $\Gamma_{\alpha_0}/\Gamma = 0.0125$ at this energy, and forcing this value from equation (6) determines the isospin-mixing parameter $e^2 = 1.5\%$.

The triangle points of Figure 13 are the results of combining the predictions for statistical ^{16}O decay from equation (6) with the total γ absorption data.[15] The 19.0-MeV resonance appears in asymmetric fission as well as in α decay and may have a significant statistical component. However, purely statistical decay of the GDR also predicts a much stronger peak at 20.4 MeV, and such a peak is all but absent from the data of Figure 13. Furthermore, the predicted decay is confined to a much narrower energy range than is observed. Once again, the fission yields do not simply result from statistical decays of giant multipole strength.

2.6. Fusion/Fission Resonances in ^{28}Si

As discussed in the preceding sections, the capture into and fission from states in ^{24}Mg bear little resemblance to features of other reactions. In particular, the $^{12}C + {}^{12}C$ system exhibits a large number of resonances in a

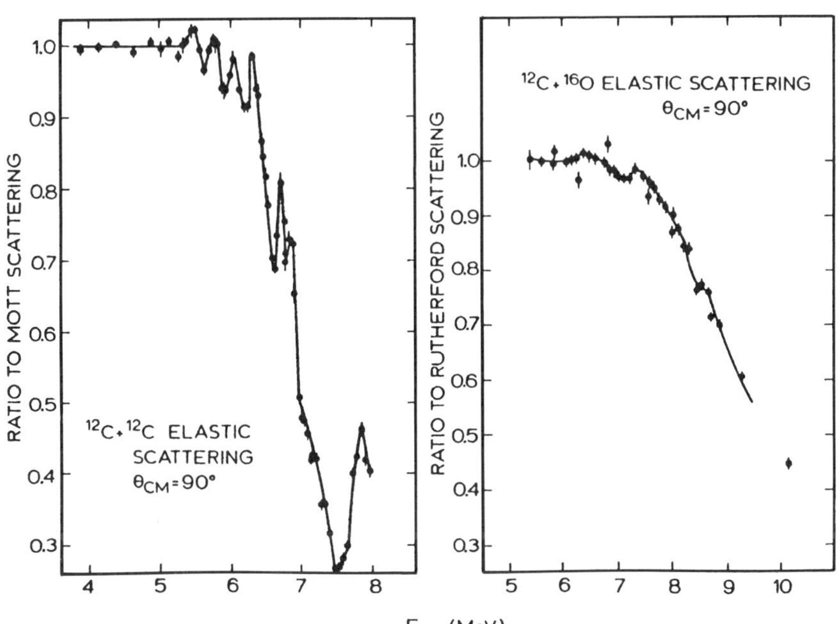

Figure 14. The 90° cross sections for $^{12}C + {}^{12}C$ and for $^{12}C + {}^{16}O$ elastic scattering are shown here. These data are from Reference 46.

variety of channels. The elastic scattering[46] cross section in the vicinity of the Coulomb barrier is shown in Figure 14. Resonances in elastic scattering can appear in radiative capture since the capture cross section is proportional to the elastic width—i.e., Γ_c appears in equation (1). However, the pronounced resonances evident in Figure 14 appear in radiative capture only through statistical mixing into ^{24}Mg levels that possess γ-ray strength, and these cross sections are only in the 5–8-nb/sr range.

The features of $^{12}C+^{16}O$ reactions near the Coulomb barrier are quite different. The elastic cross section[46] shown in Figure 14 displays relatively little structure in this region, although resonances similar to those of $^{12}C+^{12}C$ reactions do appear in other channels and in elastic scattering at much higher energies.[47] Nonetheless, since we have argued that the fusion/fission resonances are not associated with molecular phenomena, structures such as those of Figure 8 might well appear in $^{12}C+^{16}O$ fusion.

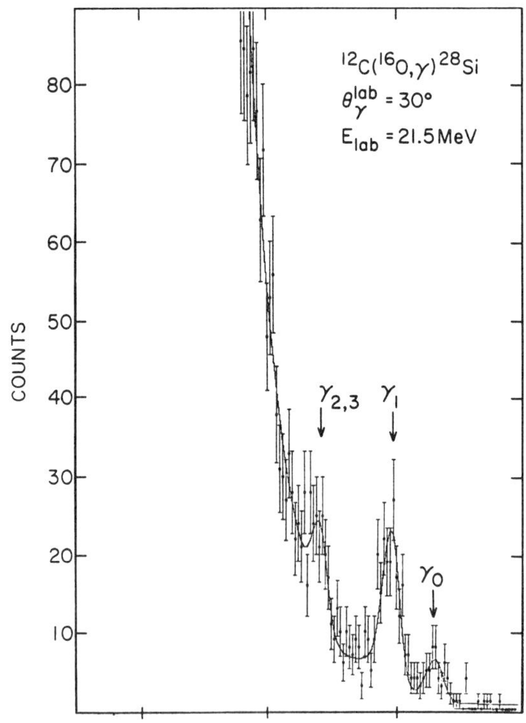

Figure 15. A typical spectrum of high-energy γ rays from the $^{12}C(^{16}O,\gamma)^{28}Si$ reaction is shown here. These data were taken with the BNL–MKII detector (Reference 17).

2.6.1. Structures Observed in Radiative Capture

The $^{12}C(^{16}O, \gamma)^{28}Si$ reaction has recently been studied at Brookhaven.[17] The high-energy portion of a γ-ray spectrum taken at $E_{^{16}O} = 21.5$ MeV is shown in Figure 15 and clearly shows the transitions to the ground and first-excited states of ^{28}Si. The excitation functions for these transitions are shown in Figure 16.[17] There are two striking features of these data. First, there are at least three relatively narrow resonances in the γ_1 yield, which appear with cross sections at about the same level as the resonances of $^{12}C(^{12}C, \gamma)$, despite the increased Q value and Coulomb barrier. Second, the cross sections for γ_0 decay are extremely small.

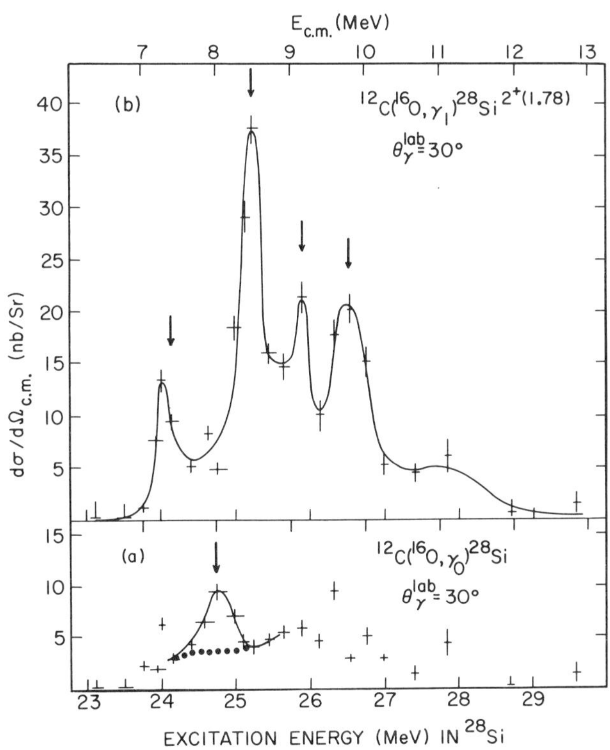

Figure 16. Excitation functions for the $^{12}C(^{16}O, \gamma)^{28}Si$ reaction (from Reference 17). Vertical bars indicate the level of statistical uncertainty, while horizontal bars represent the energy loss in the target. The arrows indicate the energies at which angular distributions have been measured.

Since the ground state of ^{28}Si is oblate[48] while the incoming heavy-ion channel is inherently prolate, the lack of significant γ_0 yield may be viewed as resulting from the inability to connect these two intrinsic shapes with a simple electromagnetic operator. Unfortunately, the same argument can be made for the γ_1 yield since the wave functions of the ground- and first-excited states of ^{28}Si look very much alike.[48] Alternatively, the resonances could be of spin higher than 2^+, 3^- or 4^+, for example. Since the multipolarity of the γ transitions is effectively restricted to 1 or 2,[21] decays to the 1.78 MeV 2^+ first excited state could be observed, while decays to the ground state would be precluded. To test this hypothesis, angular distributions of the γ_1 yield were measured at each of the peaks in the excitation function, and these are shown in Figure 17. Because of the $J^\pi = 2^+$ final state, measurements of $\gamma-\gamma$ correlations would be necessary to uniquely determine resonance spins, and such measurements are extremely difficult owing to the small cross sections involved. Nonetheless, several of the peaks in Figure 16 are quite well isolated and a unique spin and γ-ray multipolarity might dominate each resonance. Each solid curve in Figure 17a is the radiation pattern expected for the transitions indicated in brackets—no other single initial spin or multipolarity is consistent with the data. The 9.8-MeV resonance is likely a 4^+, and so its absence in the decay to the 0^+ ground state is not surprising. However, the distributions for the 7.3-MeV c.m. and 8.5-MeV c.m. resonances suggest $J^\pi = 2^+$ assignments. These could decay to the ^{28}Si ground state. Their absence in γ_0 must reflect some structural variations in these low-lying states of ^{28}Si. Finally, the small γ_0 yield is a mixture of $E1$ and $E2$ components. As can be seen from Figure 17b, these appear not to interfere significantly, since the solid curve is just the incoherent sum of a component of $E1$ decay, given by the dotted curve in Figure 16, and a component of $E2$ decay comprising the peak above it.

An analysis of statistical decay probabilities, analogous to that of Figure 12b, cannot readily be performed for the ^{12}C(^{16}O, γ) reaction. The pronounced resonances occur in the γ_1 excitation function and the multipole strength built on the first excited state of ^{28}Si is unknown. Nevertheless, under the assumption that, on the average, the distribution of $E2$ strength built on the 2_1^+ state is just shifted up by 1.78 MeV from that built on the ground state [the latter has been deduced from ^{28}Si(α, α') measurements[36]], the calculated capture cross sections fall far below the resonances of Figure 16. This is largely due to the increased Q value and Coulomb barrier which pushes the structures to higher excitation energies above the GR in ^{28}Si. The result is that the cross sections one might expect for statistical γ decay of a resonance in the incoming channel, and also for statistical ^{16}O decay from high-lying multipole strength, are quite small. This would again suggest correlations between the elastic and the multipole strength functions arising from an unusual form of intermediate structure.

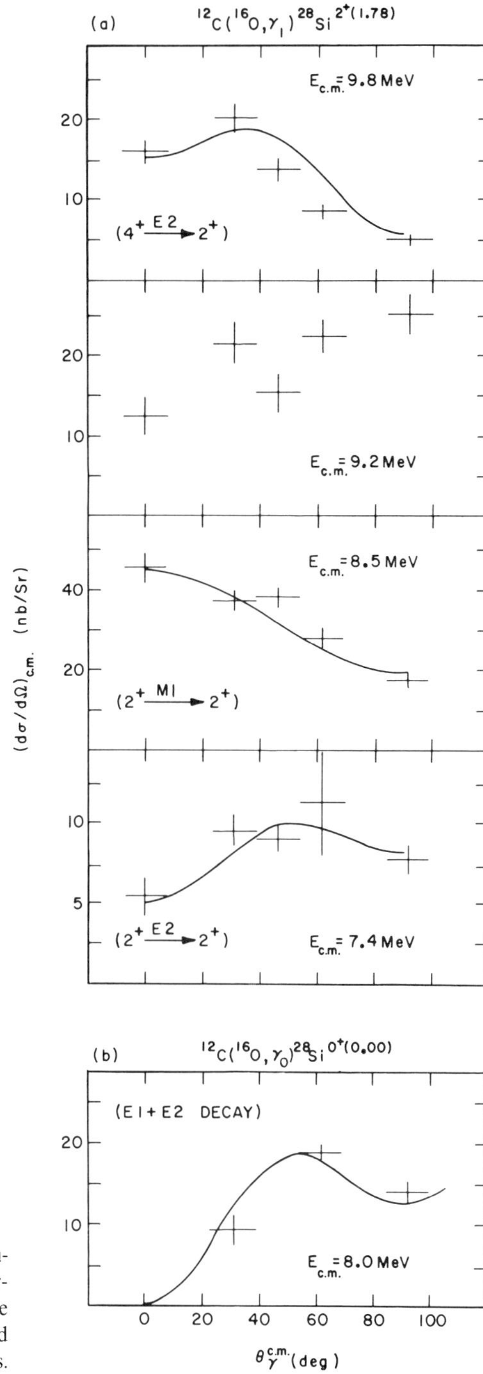

Figure 17. Gamma-ray angular distributions in the $^{12}C(^{16}O, \gamma)$ reaction (Reference 17). The solid curves through the data are the radiation patterns expected for the transitions indicated in brackets.

None of the resonances of Figure 16 have been observed in elastic scattering. The same is true for the dominant $^{12}C(^{12}C, \gamma)$ peaks, but this statement is more significant here since there is far less structure in the elastic excitation function and the presence of a resonance is not easily masked. Since the capture cross sections are proportional to the quantity $\Gamma_\gamma \Gamma_{^{16}O}/\Gamma$, this suggests that the structures seen in radiative capture may well result mainly from resonances in Γ_γ, and as such could be viewed as collective excitations of the low-lying levels in ^{28}Si.

A variety of peaks have been observed in various reaction channels[64] but, generally, these bear little resemblance to the data of Figure 16. One possible correlation is shown in Figure 18, where a peak near 8.5 MeV c.m.

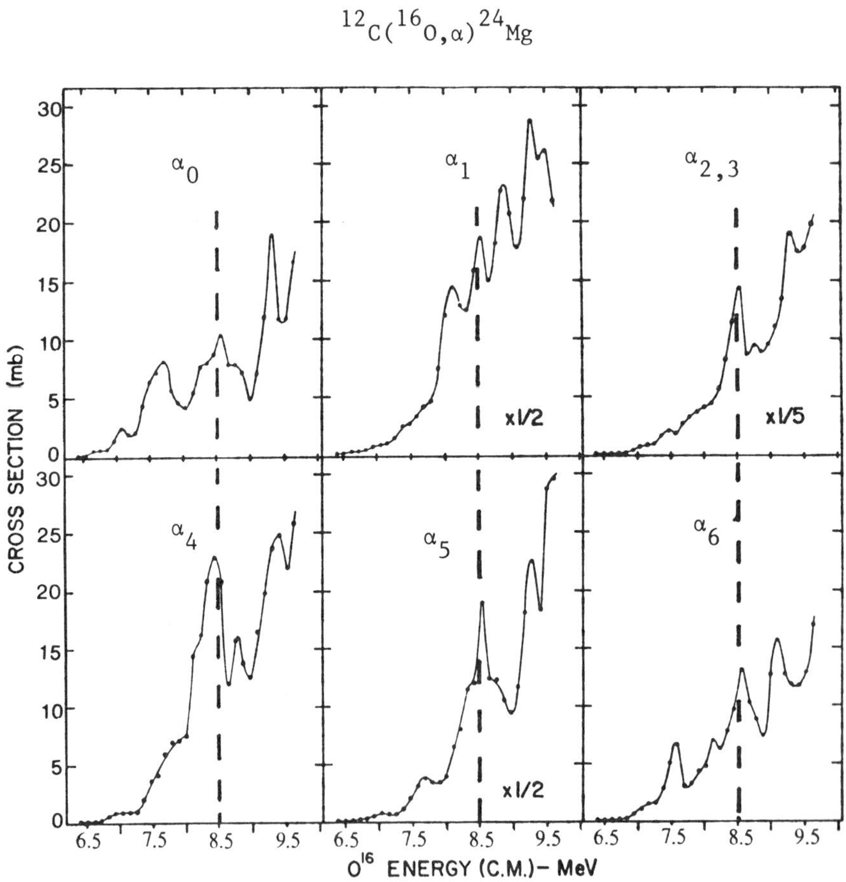

Figure 18. Excitation functions for γ decay from the $^{12}C + ^{16}O$ system. The data are from Reference 49.

persists in several of the $^{12}C(^{16}O, \alpha)^{24}Mg$ decay channels[49] and exhibits a total width, in at least some of the α decay channels, which is similar to the 8.5-MeV c.m. capture resonance. However, the excitation functions in the α- and γ-decay channels are so different that it is difficult to interpret the significance of this correlation.

2.6.2. Structures Observed in Electrofission

The $^{12}C(^{16}O, \gamma_0)$ capture cross sections shown in Figure 16 are very small, and so the electrofission yield to the ground states of ^{12}C and ^{16}O must also be very small. Indeed, the $^{28}Si(e, ^{12}C)$ measurements reported in Reference 10 showed no significant yield to the ground states of ^{12}C and ^{16}O. In contrast, considerable strength was observed to decay into excited states of the fragments. Measurements with detectors that integrated over a range of ^{12}C energies were made at a series of beam energies, and the behavior of the cross section with bombarding energy suggested a structure at about 29 MeV in ^{28}Si with a width of about 2 to 3 MeV.

The angular distribution of the ^{12}C fragments observed in these measurements[10] is shown in Figure 19. As discussed in Section 2.2, electroexcitation in which the final electron energy is near zero is dominated by the exchange of a single virtual photon traveling parallel to the beam direction,

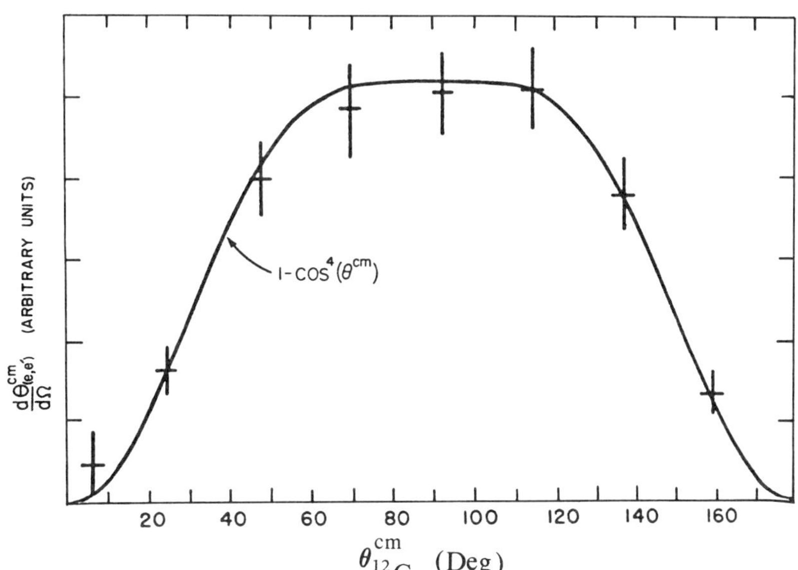

Figure 19. The angular distribution of ^{12}C nuclei from the $^{28}Si(e, ^{12}C)$ reaction (from Reference 10).

and is thus equivalent to photoabsorption. The excitation then aligns the target nucleus with a spin projection of $\pm 1\hbar$ along the beam axis. In studying nuclear deformations leading to fission, it is reasonable to consider cases where the saddle-point configuration is cylindrically symmetric about the fission axis and has a definite value for the projection of the total angular momentum along this direction. If the ^{28}Si electrofission yield is assumed to proceed by $E2$ absorption to states having a definite total angular momentum projection along the scission axis of $K = 2\hbar$, then the resulting ^{12}C angular distribution is

$$W(\theta_{^{12}\mathrm{C}}) \propto |D^2_{21}|^2 + |D^2_{2-1}|^2 = 1 - \cos^4\theta_{^{12}\mathrm{C}} \tag{7}$$

This is just the solid curve through the data in Figure 19.

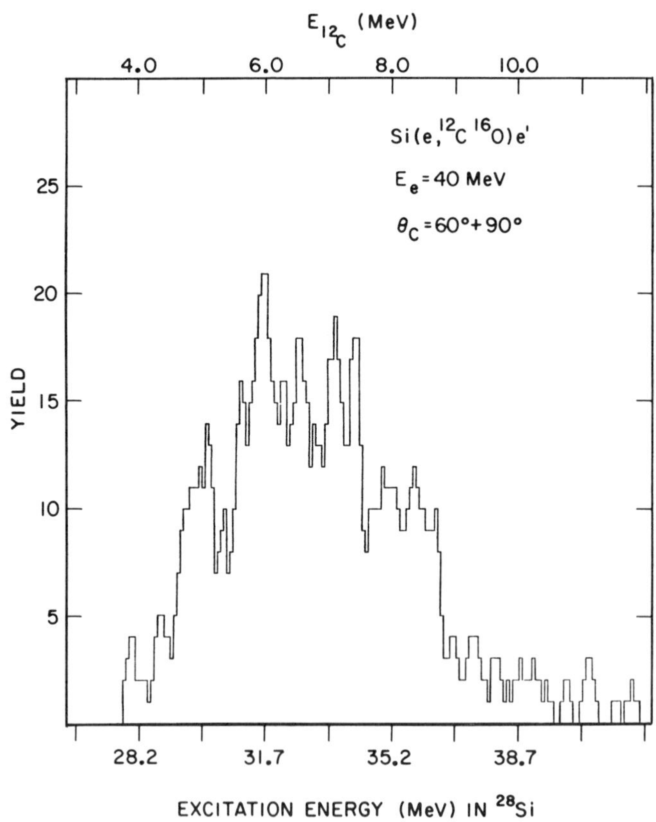

Figure 20. The yield from the $^{28}\mathrm{Si}(e, {}^{12}\mathrm{C}{}^{16}\mathrm{O})e'$ reaction measured with 40-MeV electrons (Reference 16). The excitation energy scale shown at the bottom of the figure assumes the ^{12}C nucleus is in its first excited state.

This decay channel has since been studied in much higher-resolution coincidence measurements using the techniques described in Section 2.2.[16] The energy spectrum of the ^{12}C ions from the $^{28}\mathrm{Si}(e, {}^{12}\mathrm{C}{}^{16}\mathrm{O})e'$ reaction is shown in Figure 20. The excitation energy scale shown at the bottom of the figure assumes the ^{12}C nucleus is in its first excited state (i.e., $Q = -16.75 - 4.44$ MeV). Although $^{12}\mathrm{C} + {}^{16}\mathrm{O}^{(6.1)}$ decays cannot be ruled out, the $K = 2\hbar$ character of the angular distribution suggests that the $^{12}\mathrm{C}^{(4.4)} + {}^{16}\mathrm{O}$ channel provides the more significant contribution. There is considerable structure evident in Figure 20, and again it is concentrated near the Coulomb barrier. The yield disappears for ^{12}C energies in the center of mass below 4 MeV. This corresponds to a $^{12}\mathrm{C} + {}^{16}\mathrm{O}$ center-of-mass energy of 7 MeV, which is just the point below which the capture cross sections of Figure 16 vanish. The centroid of yield in Figure 21 is about 3 MeV higher in excitation than estimated from the earlier $^{28}\mathrm{Si}(e, {}^{12}\mathrm{C})$ singles measurements,[10] although the gross structure is also wider. This may indicate that the earlier measurements were somewhat beam energy limited.

Although angular distributions for the various peaks in Figure 20 are not available, the earlier measurements (Figure 19) indicate that at least the lower-energy components are associated with $K = 2$, 2^+ states. There have been several theoretical attempts to understand fission from such states in ^{28}Si.[51-54] All involve the excitation of states with primarily prolate components. As pointed out in Reference 10, the deformation splitting in prolate nuclei pushes the $K = 2$ components of quadrupole oscillations higher in energy. The corresponding $K = 0$ components may well be below the Coulomb barrier, and this may explain their absence in Figure 19. No $J^\pi = 2^+$ structures have been observed near these excitation energies in other $^{12}\mathrm{C} + {}^{16}\mathrm{O}$ reactions, and the structure of Figure 20 is yet to be fully understood.

2.7. Radiative Capture in Other Heavy-Ion Systems

As indicated in Table 1, there have been several attempts to observe radiative capture in a variety of heavy-ion systems. Most of these earlier studies proved unsuccessful.[6,50] Unfortunately, among these were the $^{12}\mathrm{C}({}^{12}\mathrm{C}, \gamma)$ and $^{12}\mathrm{C}({}^{16}\mathrm{O}, \gamma)$ reactions which later proved rich in interesting structure—Figures 8 and 16, respectively. This was partly due to poor detector resolution and partly to the fact that many of the earlier measurements were performed at energies above the dominant resonances of Figures 8 and 16. The radiative capture cross sections of Figures 1, 8, and 16 begin to rise at an energy given by

$$E_B = \frac{e^2 Z_1 Z_2}{2.0\left(A_1^{1/3} + A_2^{1/3}\right)} \tag{8}$$

where E_B is in MeV, $e^2 = 1.44$ MeV fm, and Z_1, A_1 and Z_2, A_2 are the atomic number and mass of the heavy ions, respectively. The interaction radius of 2.0 $(A_1^{1/3} + A_2^{1/3})$ is much larger than in conventional estimates of the Coulomb barrier, and indicates that the incoming ions begin to fuse at unexpectedly large separations. Furthermore, as discussed in Section 1.2, the rapidness with which the capture cross section falls at energies above E_B is determined by a number of factors that can differ appreciably for different target/projectile combinations. Consequently, many of the systems surveyed in References 6 and 50 may well contain hitherto unobserved capture resonances. At present there is not enough experimental information to predict systematic trends.

The ^{28}Si(^{12}C, γ)^{40}Ca and the ^{24}Mg(^{16}O, γ)^{40}Ca reactions have recently been investigated.[45] In these heavier systems the Coulomb barrier is higher while the electromagnetic multipole strength has shifted to lower energies. The decays to the ground and first few excited levels in ^{40}Ca were observed near the Coulomb barrier of the ^{28}Si + ^{12}C system, although the ^{28}Si(^{12}C, γ_0) and ^{28}Si(^{12}C, γ_1) cross sections were only in the 5-nb/sr range. In contrast, no γ transitions were detected from the ^{24}Mg(^{16}O, γ)^{40}Ca reaction. It is difficult to know if this is the result of structural differences or simply an increased Q value (3 MeV higher). Detailed measurements in these and heavier systems may be worth pursuing, if only to gain a better understanding of the capture phenomena in lighter nuclei.

2.8. Summary

Capture resonances have now been observed near the Coulomb barrier in a number of light heavy-ion systems. They are, for the most part, uncorrelated with structure in other reactions, but tend to appear in the vicinity of giant electromagnetic resonances. In heavier systems where the Coulomb barriers are higher while the giant resonances are lower in excitation, radiative capture has been observed only at greatly reduced levels.

The apparent dependence upon the giant resonances is one that is inflicted by electromagnetic sum rules. This is both a blessing and a curse—a blessing in that the cross sections for capture are proportional to the product of γ-ray and heavy-ion partial widths $\Gamma_\gamma \Gamma_{HI}$, and the giant resonances concentrate the available Γ_γ into a relatively narrow energy range; but a curse in that now one must address the question of the extent to which the collective oscillations that produce the giant resonances are themselves responsible for the features observed in radiative capture. The latter question has been at least partially answered by the calculations described in this section which have demonstrated the nonstatistical character of at least one of, and often both, Γ_γ and Γ_{HI}. These calculations have established correlations between the heavy-ion and the electromagnetic strength functions, the origins of which do not involve giant collective oscillations in any simple way. Large heavy-ion widths of resonances that are connected to the ground and low-lying levels in the compound nucleus by only a simple one-body operator must be associated with unusually large deformations of the nuclear surface. An interpretation of the capture resonances as highly collective excitations of the low-lying compound states is consistent with the anomalously large γ-ray widths observed.

Beyond these simple conclusions, there has yet to be any real theoretical understanding of the resonances observed in heavy-ion capture. Two-centered shell-model and cluster-model calculations are inherently suited to handling the types of excitations discussed above. However, application of such models to the phenomena described in this section has thus far been rather limited.[51-54]

3. Radiative Capture to and along the Yrast Line

We consider here decays of types (b) and (c) in Figure 2, and the extent to which such decays can occur over a range of energies much broader than the vicinity of the Coulomb barrier.

3.1. Mechanisms for Decay without Particle Emission

A typical spin distribution $\sigma(l)$ of the excited compound nuclear levels following heavy-ion fusion is shown in the bottom of Figure 3, and peaks at the grazing angular momentum. Most of the cross section corresponds to complete fusion (region 3 of Figure 3), while at modest energies only a relatively small part results from direct reactions (region 4 of Figure 3). There are two mechanisms by which the compound system can deexcite without particle emission.

The first mechanism has been considered by Zeller *et al.*[8] If, for a given bombarding energy, the spin of the excited compound nucleus is so high that the excitation energy is less than a nucleon binding energy above the yrast line (B_n in Figure 3), then the centrifugal barrier inhibits particle decay and the system has a relatively high probability of cascading down to the ground state along the yrast line. However, the cross sections for producing excited nuclei in this region are contained only in the high spin tail of $\sigma(l)$ and are extremely small (hatched region of Figure 3). Under the assumption that fusion dominates the reaction cross section, the probability for radiative capture is given by

$$\sigma_{\text{capture}} = \sum_{l > J'} \sigma(l) \tag{9}$$

where, as shown in Figure 3, J' corresponds to the intercept of the $(E_{\text{yrast}} + B_n)$ line. The calculation of the dominant features of the $\sigma(l)$ distribution are fairly reliable. However, the evaluation of σ_{capture} involves the very smallest portions of this distribution and is very sensitive to the optical potential describing the incoming channel (see, for example, Figure 6 of Reference 8) and to the deep-inelastic or direct components of the reaction. Furthermore, the angular momentum J' at which the calculation starts is very dependent upon the position of the yrast line. Estimates using equation (9) are likely to be quite unreliable. There are even more troublesome considerations for this mechanism. For sufficiently high bombarding energies J' will be greater than the maximum angular momentum the nucleus can sustain without fissioning.[56] Above this point the notion of an yrast line is meaningless and radiative capture can certainly not occur. For bombarding energies below this limit, the region in which γ decay dominates (shown schematically as region 2 in Figure 3) is not necessarily bounded by the $E_{\text{yrast}} + B_n$ line. It would be if the yrast levels were just members of the ground-state rotational band. If all the excitation energy were contained in the rotational degree of freedom, the binding energy B_n would still have to be overcome before particle decay could occur. However, this situation rarely prevails. At high angular momenta the barriers for particle decay are reduced. This is particularly true for α decay which can carry off many more units of angular momentum than either proton or neutron decay and can thus keep the system in region 1 of Figure 3. A more relevant boundary for region 2 would be the locus of states for which Γ_γ/Γ is a constant, 0.5 for example, and this curve can easily meet the yrast line, often at angular momenta lower than J'. Given these considerations, it is doubtful if this mechanism can make any real contribution to radiative capture.

The second method of reaching the ground state of the fused nucleus is by the decay of moderate-to-high-energy γ rays to levels near the yrast line. This can take place throughout the distribution of $\sigma(l)$ of excited compound states, but must compete with particle emission for the initial decay. Section 2 dealt with the special subset of such high-energy decays where the initial compound states have relatively low spin and where the final near-yrast levels are below particle threshold (t_p in Figure 2). Most of the observed cross sections for such transitions have been concentrated in the vicinity of the Coulomb barrier. The γ decays from higher-spin components of $\sigma(l)$, although difficult to observe directly, can contribute to radiative capture over a much larger energy range. These contributions have been estimated by Branford and Steadman,[13] assuming that the γ decay of the excited compound levels is purely statistical. The various stages of their calculation are depicted in Figure 21 for 55-MeV ^{16}O on ^{27}Al. The primary spin distribution in the compound nucleus ^{43}Sc is fixed by the bombarding energy at an excitation of 48.8 MeV. Fixing the Γ_γ^{E1}, Γ_γ^{M1}, and Γ_γ^{E2} radiative widths for *all* possible transitions at 0.01, 0.10, and 1.00 Weisskopf units, respectively, the calculations of Grover and Gilat[55] were used to predict the statistical populations of states in ^{43}Sc near the yrast line. The dotted region of the $E_x - J$ plane in Figure 21 represents states for which $\Gamma_\gamma/\Gamma \geq 0.5$. The total heavy-ion capture cross sections were then estimated by summing the population probabilities for all such states with $\Gamma_\gamma/\Gamma \geq 0.5$—e.g., the vertically shaded areas in Figure 21.

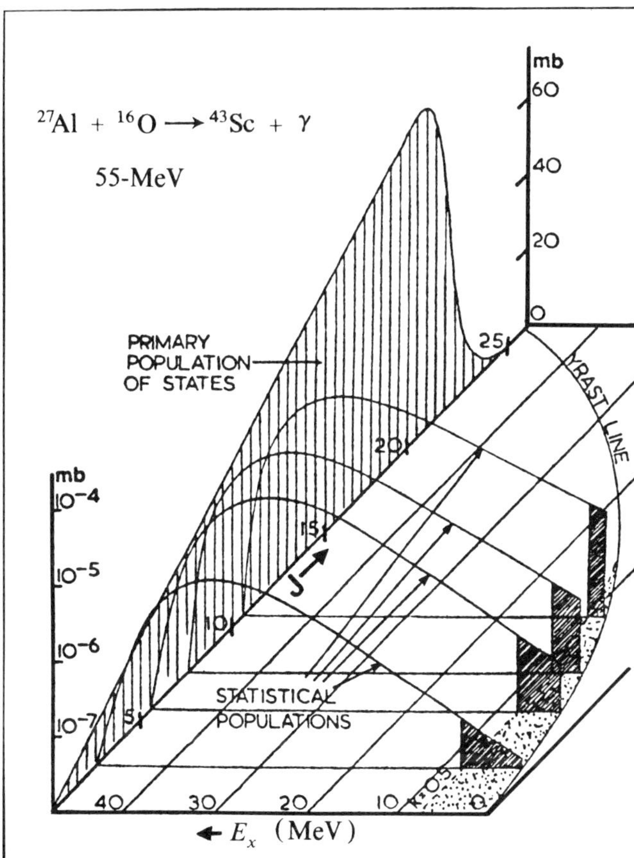

Figure 21. Various stages in a calculation of statistical γ decay of the $^{16}O + ^{27}Al$ system to near-yrast levels in ^{43}Sc are shown here (from Reference 13).

Before comparing such predictions with experimental data, it is useful to keep in mind a shortcoming of these types of calculations. The γ-ray channel has one very important difference from all other decay channels, and that is the limitation imposed on Γ_γ by electromagnetic sum rules. Such limitations are not included in the calculations of Grover and Gilat.[55] Furthermore, most of this γ-ray strength will be concentrated in giant resonances, one for each excited state, and the information available on the location and width of such resonances built on excited states is very sparse. There are only a few recent measurements of GRs built on states up to 15 MeV excitation in a few light nuclei,[23,72] but nothing above this. (Recent attempts to include GRs in a Grover-Gilat type calculation are described in Section 5).

3.2. Activation Measurements

Almost all attempts to detect radiative decays from the $\sigma(l)$ distribution of Figure 3 to the yrast line have utilized reactions where the ground state of the fused nucleus, or a long-lived excited isomeric state, is unstable. The available total cross sections for capture, deduced from measurements of the characteristic radioactivities in the fused nuclei, are listed in Table 4.

Table 4. Activation Measurements of Radiative Capture

Reaction	$\sigma^{\text{capture}}(\mu b)$	Ref.	Background correction[a]
$^{19}F(^{13}C,\gamma)^{32}P$	—	3	✓
$^{27}Al(^{16}O,\gamma)^{43}Sc$	< 0.27	2	✓
$^{27}Al(^{16}O,\gamma)^{43}Sc$	< 0.2	13	
$^{30}Si(^{14}N,\gamma)^{44}Sc$	7.5–22	8	
$^{31}P(^{14}N,\gamma)^{45}Ti$	1 ± 1	4	✓
$^{31}P(^{16}O,\gamma)^{47}V$	<18	2	✓
$^{16}O(^{32}S,\gamma)^{48}Cr$	(7.4)	9	✓
$^{40}Ca(^{12}C,\gamma)^{52}Fe$	—	9	✓
$^{24}Na(^{32}P,\gamma)^{56}Fe$	(64)[b]	5	✓
$^{51}V(^{13}C,\gamma)^{64}Cu$	—	3	✓
$^{93}Nb(^{16}O,\gamma)^{109}In$	0.1	13	
$^{198}Pt(^{13}C,\gamma)^{211}Po$	0.001	57	

[a] A check indicates that a significant correction for target impurities was necessary.
[b] For comparison with other measurements, the ^{56}Fe electrofission cross section of Reference 5 (7 μb at 3 GeV) was converted to an equivalent photofission cross section assuming dipole excitation in a 1-MeV region near the Coulomb barrier at 50 MeV in ^{56}Fe. Detailed balance then gives the cross section for the inverse reaction tabulated here.

Many of the measurements required significant corrections for target impurities, and the uncertainties associated with these corrections may account for the several orders of magnitude variation in $\sigma_{capture}$.

As an example of such experiments, consider the measurements of Branford and Steadman.[13] These authors stopped an ^{16}O beam in an ^{27}Al sample. The ground state of the fused system ^{43}Sc decays with a 3.9-hr half-life, 16% of the time by β^+ emission to the 373-keV level of ^{43}Ca. The signature for a radiative capture event would then be a 373-keV γ ray detected in a Ge(Li) accompanied by two 511-keV γ rays detected 180° to each other in NaI(Tl) crystals. The background-subtracted yield of 373-keV signals, with and without a double 511-keV coincidence, is shown as a function of time after irradiation in Figure 22. There is no positive indication of ^{43}Sc production here. Accordingly, these authors were only able to set an upper limit of 0.2 μb on the average capture cross section for

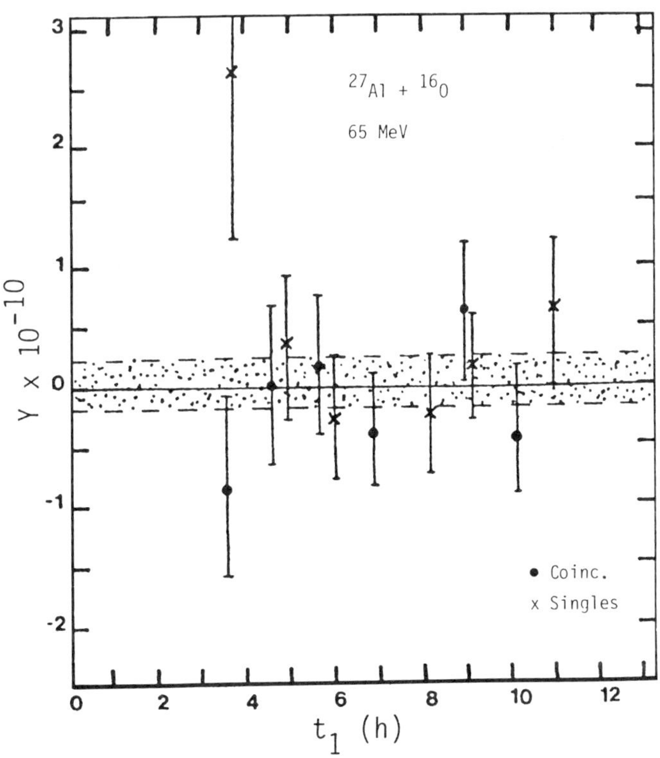

Figure 22. The background-subtracted yield of 373-keV γ rays from the decay of $^{43}Sc(\tau = 3.9$ hr) nuclei formed by the radiative capture of ^{16}O and ^{27}Al is shown here as a function of time after irradiation. Data are from Reference 13.

bombarding energies in the 28.8 to 62.5 MeV range. This number is in agreement with the previous limit set by Coleman *et al.*[2] ($\sigma^{capture} < 0.27\ \mu b$).

The results of the calculations depicted in Figure 21 for statistical γ decay of the $^{27}Al + ^{16}O$ system are compared to this limit in Figure 23. The predictions are still well below the experimental limit, although, as pointed out in the previous section, it is difficult to guess how they would change if the "true" Γ_γ strength functions were known. Nonetheless, the statistical limit is expected to remain below the experimental limit, and is not likely to change significantly for capture in neighboring heavy-ion systems. For this reason, the data of Zeller *et al.*[8] on $^{30}Si(^{14}N, \gamma)^{44}Sc$ are quite surprising. The $^{27}Al(^{16}O, \gamma)^{43}Sc$ and $^{30}Si(^{14}N, \gamma)^{44}Sc$ reactions have nearly identical Q values and entrance-channel Coulomb barriers, and yet the cross sections reported for the latter are a factor of 40 to 100 times larger (7.5 to 22 μb).

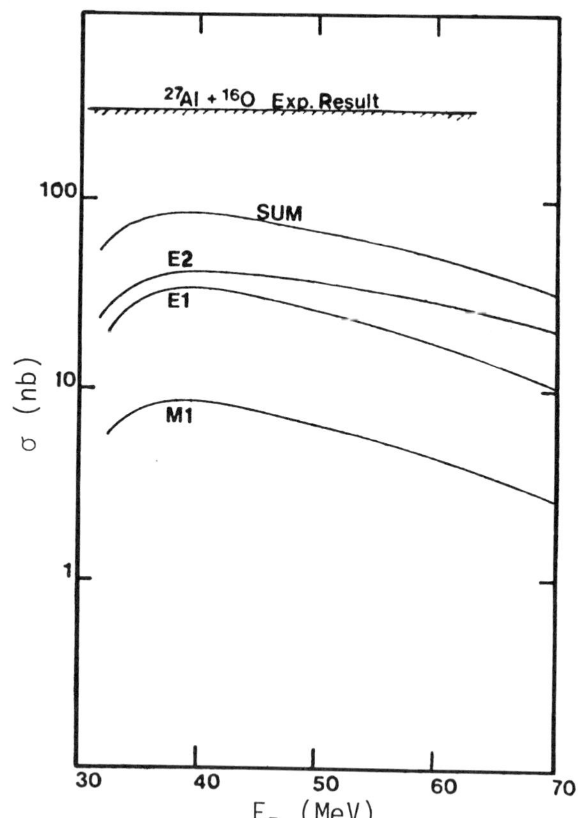

Figure 23. Upper limits for radiative capture in $^{16}O + ^{27}Al$ are compared to the predictions of statistical γ decay depicted in Figure 21.

This either indicates a highly nonstatistical phenomenon or problems with target impurities—traces of phosphorous, for example, can produce ^{44}Sc through the ^{31}P(^{14}N, p)^{44}Sc reaction. Although explanations similar to the latter are preferred in References 13 and 14, nonstatistical behavior cannot be completely ruled out. After all, the contributions to the ^{12}C + ^{12}C capture cross section just from decays to the ground and first three excited levels in ^{24}Mg (Figure 8) is over 2 μb.

Further measurements are necessary to understand the nature of the effect observed by Zeller *et al.*[8] At this point it is not clear if heavy-ion capture has even been observed in activation experiments.

3.3. Direct Detection of Fusion Products

It is possible, albeit more complicated, to directly detect the fused nuclei. In the absence of particle decay, the compound system formed when two heavy ions fuse recoils in the laboratory very near 0° with the velocity of the center of mass. If these nuclei can be separated from the beam particles, they can be distinguished from evaporation residues and reaction products from target contaminants by the monochromaticity of their velocity spectrum. There are several recoil-mass spectrometers, either existing or planned, that could be used for such measurements.[58] A version of this technique has been employed by Viano *et al.*[14] to search for capture in the ^{14}N + ^{12}C, the ^{14}N + ^{30}Si, and the ^{14}N + ^{27}Al systems. Their geometry is shown schematically in Figure 24. The beam particles are separated from the fusion products by an electrostatic deflector. Time-of-flight and energy measurements permit mass identification and determine the velocity. A density plot of time-versus-energy for 70 MeV ^{14}N incident on ^{12}C is shown in Figure 25. The square box indicates where events from heavy-ion capture would be expected.

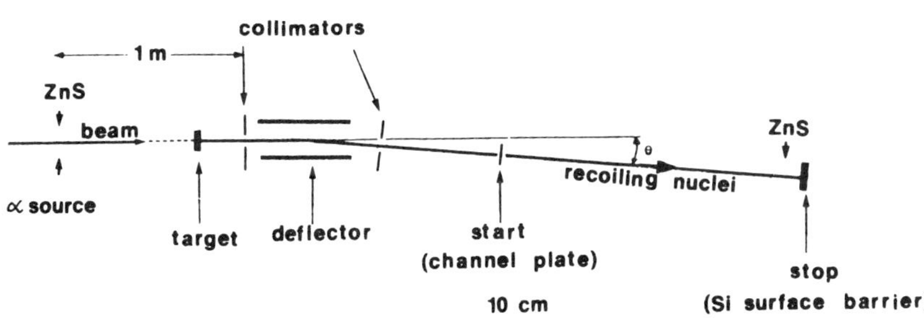

Figure 24. The experimental geometry used by Viano *et al.* (Reference 14) to search for the recoil of fused nuclei produced in radiative capture reactions.

The measurements of Viano *et al.*[14] produced only upper limits on capture cross sections. In particular, the cross sections for $^{30}Si(^{14}N, ^{44}Sc)\gamma$ were found to be less than 25 nb at bombarding energies of 59 and 70 MeV. These limits are again much less than the cross sections reported by Zeller *et al.*[8] for the same reaction. However, this comparison serves to illustrate two problems with this method of recoil detection. First of all activation measurements, like those of Zeller *et al.*, use thick targets that sample the cross section over a large range of excitation energies, while the detection of the recoiling fusion products necessitates the use of thin targets that can sample only a narrow range of energies in ^{44}Sc. The nonstatistical capture phenomena discussed in Section 2 are highly energy dependent, and if such structures occur in the compound nucleus, they can easily be missed in such recoil measurements. Secondly, Viano *et al.* considered only the first mechanism for heavy-ion capture discussed in Section 3.1. Under this mechanism only low-energy γ rays would be emitted as the compound nucleus decayed, and the average deflection of the recoil direction away from 0° would be quite small. Accordingly, the detection system of Reference 14 was designed to limit the deflection angle to 0.35°. However, under the second mechanism of Section 3.1 moderate-to-high-energy γ rays could be emitted as the

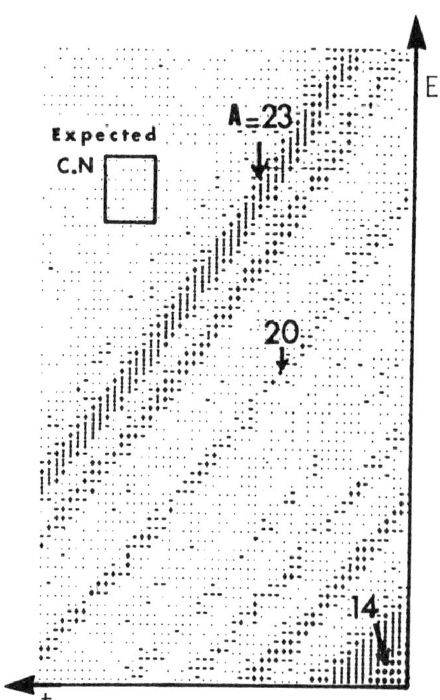

Figure 25. An energy versus time-of-flight plot for the system shown in Figure 24. The box indicates the expected position of radiative capture events. Data are from Reference 14.

nucleus deexcites. Gamma rays greater than 7.5 MeV emitted near 90° would deflect the ^{44}Sc nucleus by more than 0.35° and thus prevent detection. This latter problem has been avoided by Konrad *et al.*,[63] who have used a much larger acceptance (1.72°) recoil spectrometer to study capture into ^{44}Sc. Their upper limit of 1 μb on the ^{14}N(^{30}Si, ^{44}Sc)γ cross section is again much lower than that reported by Zeller *et al.*,[8] and suggests that these activation measurements may be suspect.

The experiments reported by Konrad *et al.* and by Viano *et al.* are inherently restricted to narrow ranges of excitation energy. However, the low backgrounds encountered in these measurements suggest that such techniques could be used to search for capture in many heavy-ion systems, particularly those in which the target has only one naturally occurring isotope so that fusion on a heavier nucleus, followed by neutron decay, cannot masquerade as capture.

An alternate method exists for detecting nuclei produced by complete fusion, although it has yet to be applied to radiative-capture measurements. Ultrasensitive particle identification systems, which use an electrostatic accelerator as one element,[59] have detected nuclei with sensitivities as high as 1 part in 10^{15}. Such a system is shown schematically in Figure 26. Ions from the sample under study are extracted from a sputter source and momentum analyzed. Charge exchange in the terminal of a tandem removes all of the molecular beams that plague conventional mass separators. Such systems could be used first for production of fused nuclei, by bombardment of a thick target with a heavy-ion beam, and then for detection of these

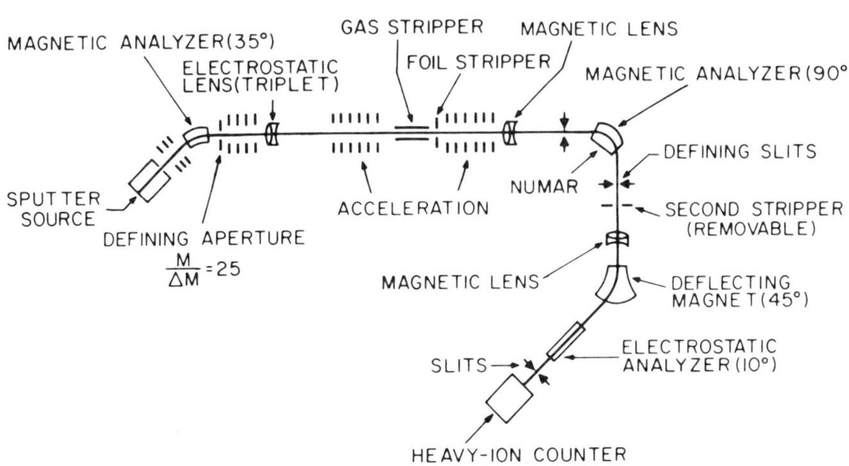

Figure 26. The ultrasensitive particle identification system used by the Toronto–Rochester group (Reference 59).

same complete-fusion products, by placing the activated sample into the ion source and now using the accelerator of Figure 26 as a detector. Such a scenario would be particularly ideal for investigating heavy-ion capture into long-lived nuclei. For example, the $^{12}C(^{14}N,\gamma)^{26}Al$, $^{46}Ti(^{7}Li,\gamma)^{53}Mn$, and $^{122}Sn(^{7}Li,\gamma)^{129}I$ reactions produce nuclei with lifetimes of 7.4×10^{5} years, 3.8×10^{6} years, and 1.6×10^{7} years, respectively. Using these techniques, ^{26}Al and ^{129}I have already been detected[60,61] with sensitivities that indicate that capture cross sections could be measured down to the 10 pb level. Such experiments might well yield the most accurate measurements of the effects discussed in Section 3.1.

3.4. Summary

Considering all previous experiments, there is little known about heavy-ion capture via either of the mechanisms of Section 3.1. Statistical γ decay is likely to contribute the most to capture over large ranges of energy and angular momentum. Unfortunately, estimates of this contribution are, and probably always will be, rather crude owing to the uncertainties in the γ-ray strength functions built upon excited states. (More recent work in this area is discussed in Section 5). If radiative capture cross sections are observed at levels very much larger than those predicted for statistical γ decay, then new phenomena must exist, and these would undoubtedly lead to further investigations.

4. Future Directions

The bulk of this review has dealt with the techniques, aims, and results of the heavy-ion capture experiments that have thus far been undertaken. Interest in these types of measurements will undoubtedly continue. However, these by no means exhaust the potential of heavy-ion radiative capture. I would like to illustrate this point with one hitherto unutilized aspect of such reactions.

The spectrum of high-energy γ rays from the $^{12}C(^{16}O,\gamma)$ reaction decaying to the low-lying states of ^{28}Si is shown in Figure 15, and the energy dependence of these transitions was discussed in Section 2.6. In addition to the γ rays evident here, a much stronger transition has been observed decaying to states near 7 MeV excitation in ^{28}Si.[62] These data, shown in Figure 27, were taken with the BNL–MKII detector, which has an intrinsic resolution of about 3.3% (Figure 5a). Meticulous attention to energy calibration and gain stability are required to pin down the final state in this decay. For at least a few bombarding energies, the final states has been identified as the 6.69-MeV 0^+ in ^{28}Si. Calculations[48] regard this level as the ground state of a prolate shape-isomer in ^{28}Si. The state is not readily populated in light-ion reactions, most of which cause relatively minor distortions of the oblate ground state. However, such states appear enhanced in capture by the inherent prolate character of the incoming heavy-ion channel.

With the exception of the 6.69-MeV level in ^{28}Si, little is known about the existence of shape-isomers in light nuclei. Enhanced γ-ray transitions following heavy-ion capture are excellent signatures for very prolate final states. The $^{16}O(^{7}Li, \gamma)^{23}Na$ reaction is presently under study at Brookhaven. A spectrum of γ rays from this reaction, taken at the peak of the resonance discovered by Feldman and Heikkinen (Figure 1), is shown in Figure 28.[62]

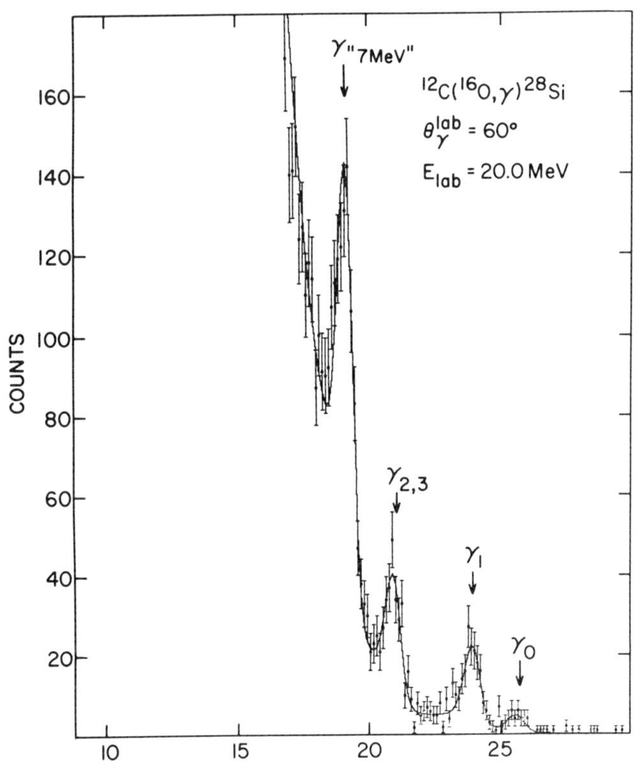

Figure 27. The high-energy region of a γ-ray spectrum observed from the $^{12}C(^{16}O, \gamma)^{28}Si$ reaction (Reference 62).

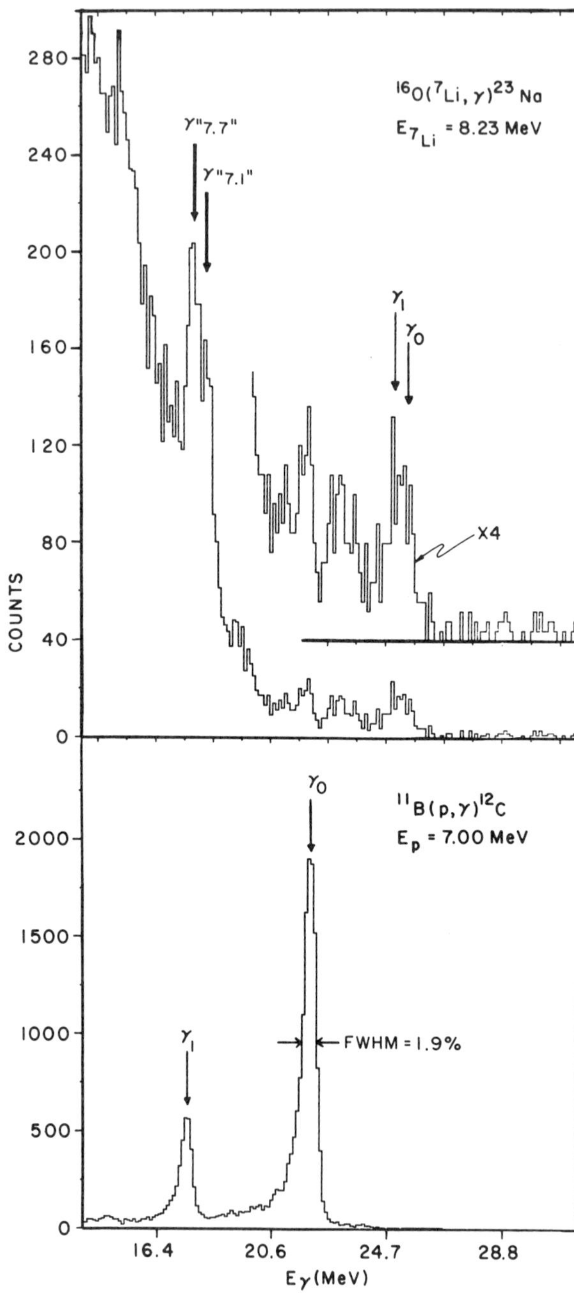

Figure 28. The high-energy portion of the γ-ray spectrum from $^{16}O(^7Li, \gamma)^{23}Na$ measured at the peak of the γ_{0+1} resonance of Figure 1. These data were taken with the BNL–MKIII detector, whose intrinsic response is shown below (Reference 62).

The highest-energy portions of this spectrum are consistent with the earlier data of Figure 1. However, the γ_0 and γ_1 transitions are completely dwarfed by the much stronger decays to levels near 7 MeV excitation in ^{23}Na. These data were taken using the BNL–MKIII detector with substantially improved resolution (2%),[24] and the intrinsic line shape indicates that the peak in Figure 28 results from decays to a doublet of levels at 7.1 and 7.6 MeV in ^{23}Na. These data are highly suggestive of an unusually prolate set of states in ^{23}Na—a shape-isomer—and this is completely unexpected.

During the course of the $^{12}C(^{12}C,\gamma)$ measurements discussed in Sections 2.1 and 2.3, no exceptionally strong transitions were observed to high-lying states of ^{24}Mg. The extent to which shape-isomer-like states occur in other $s-d$ shell nuclei has yet to be investigated, and heavy-ion radiative capture may prove an exceptionally well-suited tool.

5. Review Update

Since this manuscript was completed in the spring of 1981, there have been two important developments that should be pointed out to update this review. In an experiment similar to those discussed in Section 3.3, a recoil spectrometer (a crossed $\vec{E}$ and $\vec{B}$ field velocity selector) was successfully used in coincidence with capture gamma rays to study the $^{12}C(^{16}O,\gamma^{28}Si)$ reaction.[65] The geometry is shown schematically at the top of Figure 29. With the BNL–MKIII detector at 120°, the recoiling ^{28}Si nuclei were focussed through the velocity filter, positioned at $-1.2°$ to compensate for the kick due to the gamma-ray momentum, and entered an $\Delta E-E$ ionization chamber. The large pileup backgrounds present in singles spectra (such as Figure 27) were reduced by up to four orders of magnitude by the coincidence requirement. A coincidence spectrum is shown in the middle of Figure 29, and the gamma-ray yields go almost to zero between many of the peaks. The response of the NaI in the large-solid-angle geometry used for these measurement is shown at the bottom of the figure.

The decay to the 0_3^+ state at 6.69 MeV in ^{28}Si, indicated by the heavy arrow in Figure 29, is clearly the most prominent capture transition. It appears in this coincidence spectrum as a well-isolated peak, although the width of this peak, together with the known line shape of the MKIII detector, requires a small contribution from gamma rays slightly lower in energy, to states at $E_{Si} \sim 7.4$ MeV. This latter component appears as a well defined peak at higher energies where the yield to the 0_3^+ is reduced. As discussed in Section 4, this 0_3^+ has been identified as the bandhead of a $K=0$ prolate shape-isomeric state. The transitions to states at $E_{Si} \sim 7.4$ MeV can be associated with decays to the 2^+ member of this prolate band.

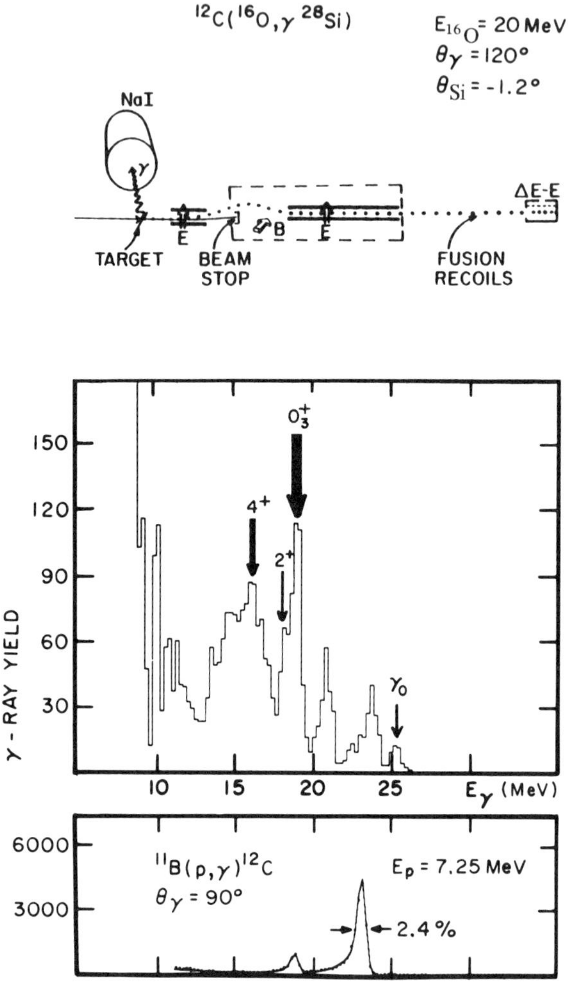

Figure 29. An arrangement of the BNL–MKIII detector and a velocity filter are shown schematically at the top of the figure. A gamma ray plus fusion coincidence spectrum from the $^{12}C(^{16}O, \gamma\,^{28}Si)$ reaction is shown in the center of the figure (Reference 65). The NaI response in the geometry used for this experiment is shown at the bottom.

Transitions to the 4^+ prolate state at $E_{Si} = 9.16$ MeV are also evident in Figure 29, although these too are much stronger at higher energies.

The excitation functions for the decays to the 0^+ and 2^+ prolate states are plotted in Figure 30(a) as a function of the center-of-mass energy. When replotted as a function of gamma-ray energy, which is just the excitation energy above the final state, three of the intermediate structure peaks appear

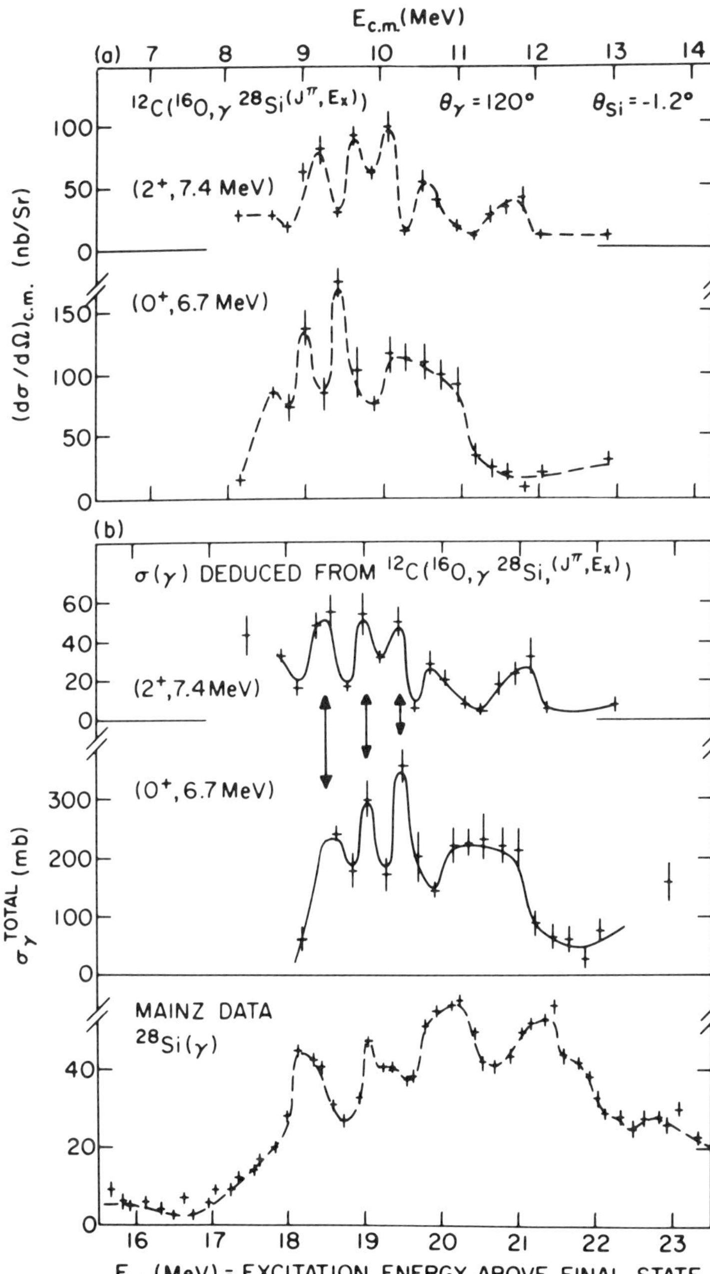

Figure 30. (a) Excitation functions for the $^{12}C + ^{16}O$ capture to the 6.69 MeV 0^+ and the 7.4 MeV 2^+ states of ^{28}Si. Dashed lines are drawn to guide the eye. (b) Total absorption cross sections deduced from (a) under the assumption of statistical fusion widths, and the ground state giant dipole photon strength (Reference 71).

correlated in energy and the full strength distribution bears a strong resemblance to the giant dipole resonance (GDR) seen in photoabsorption. This is evident in Figure 30(b). In contrast to the decays to the ground- and low-lying-states of ^{28}Si, the excitation functions of these transitions are consistent with giant dipole resonances built on the prolate intrinsic shape of ^{28}Si. The observed radiative decays appear enhanced in heavy-ion capture because the entrance channel is inherently prolate. The width of the gamma-ray strength built on the excited prolate states is narrower than the GDR of the oblate ground state, while all models coupling deformation degrees of freedom to the GDR would predict a significantly broader prolate resonance. This is the first time that such a comparison has been possible within a single nucleus. These types of measurements in other nuclear systems would undoubtedly prove most interesting.

Another class of experiments has become quite popular in the last three years. These involve the extraction of statistical gamma-ray spectra produced mostly in heavy-ion fusion reactions.[66] These spectra show an enhancement at a gamma-ray energy near the resonant energy of the GDR built on the ground state of the compound system, and this effect persists over a very broad mass range.[67] As an example, a gamma-ray spectrum from the $^{34}S+^{130}Te$ reaction is histogrammed at the top of Figure 31. The exponential behavior at low energies, extrapolated by the dashed line, is characteristic of statistical photon emission from a nucleus at a constant temperature. The enhancement in the data above the dashed line occurs in a gamma-ray energy region characteristic of the GDR of the compound nucleus ^{164}Er. The shape of the latter in the closest neighboring nucleus for which photoabsorption data exists, namely ^{166}Er, is shown at the bottom of Figure 31.[68] For the most part, heavy-ion fusion spectra, such as shown in Figure 31, can be accounted for by assuming that highly excited states, which are themselves giant dipole resonances built upon lower lying excited states, are statistically populated. The gamma decay from these highly excited levels appears enhanced simply because the gamma-ray strengths associated with these GDRs peak near a common transition energy. The solid curve through the fusion data of Figure 31 was calculated with the statistical-evaporation code CASCADE,[69] modified to use a single Lorentzian shape for the gamma-ray strength distribution built on each state of the residual nuclei populated in the decay cascade. The area (expressed as a fraction of a dipole sum rule S(E1)), centroid energy (E_R), and width (Γ_R) of this excited-state and spin-averaged giant resonance, represented by the Lorentzian in this CASCADE calculation, are similar, though not identical, to the corresponding parameters of the ground state GDR. For the calculation of Figure 31, $S(E1)=100\%$, $E_R=17$ MeV, and $\Gamma_R=13$ MeV.

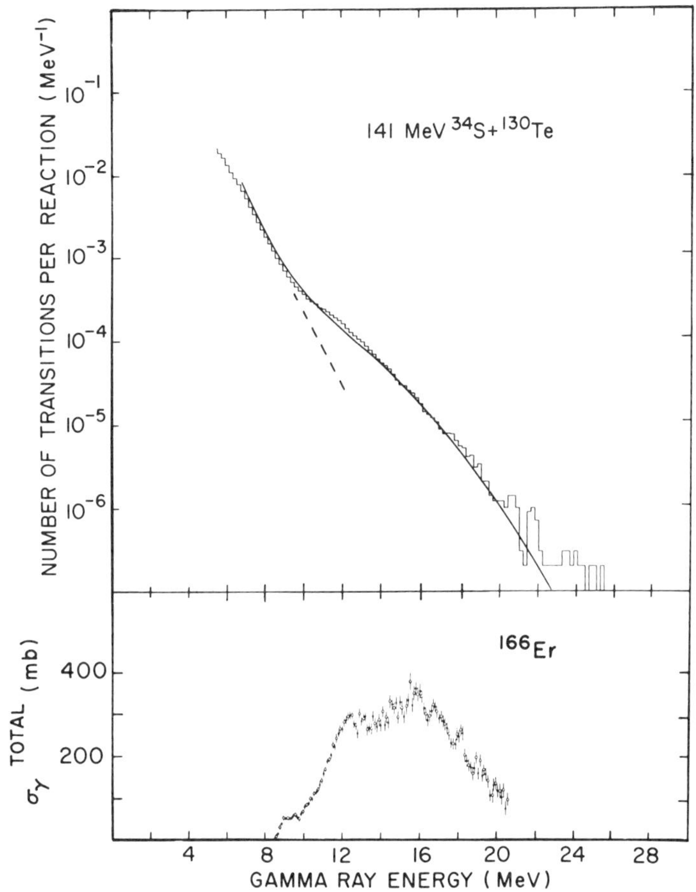

Figure 31. The gamma-ray spectrum from the $^{34}S + ^{130}Te$ reaction is shown at the top of the figure. The vertical scale has been obtained from the measured efficiency and solid angle of the detector, gated against neutrons, assuming an isotropic angular distribution. The solid curve is a statistical calculation of the cross section, assuming a GDR-like gamma-ray strength for each state, divided by the calculated fusion cross section. The ground state GDR in erbium is shown at the bottom (Reference 68).

A recent review of many experiments of this type is given in Reference 70. It is included here because of the role played by radiative capture. If the decay process is indeed statistical, most of the high-energy gamma rays should come from the regions of highest level density, in which case the largest single contribution to the high energy yield should come directly from the compound nucleus. This is indeed inherent in the evaporation calculation of Figure 31. This calculation is shown again in Figure 32, together with the contributions obtained from the various stages of the decay cascade. Above $E_\gamma \cong 11$ MeV, the gamma decay probability increases as the number of evaporated particles decreases, and the total spectrum becomes dominated by radiative capture from the compound nucleus. This decomposition is extremely difficult to verify experimentally but nonetheless is consistent with the available data.[70] The decay mechanism is now very similar to that of Figure 2(b), except that particle evaporation is seen as being much more probable after the initial high-energy transition. The chance of a continued gamma cascade, leaving the compound nucleus in its ground state, is extremely tiny. Nonetheless, the decomposition shown in Figure 32 provides one of the strongest motivations for studying these reactions, since the high-energy parts of such gamma-ray spectra probe the compound nucleus at very high excitation energies (~ 60 MeV for the reaction of Figure 32). However, before this can be exploited, the spin, energy, and deformation dependences of GDRs built on excited states must be understood, and for this a great deal of future effort will be necessary.

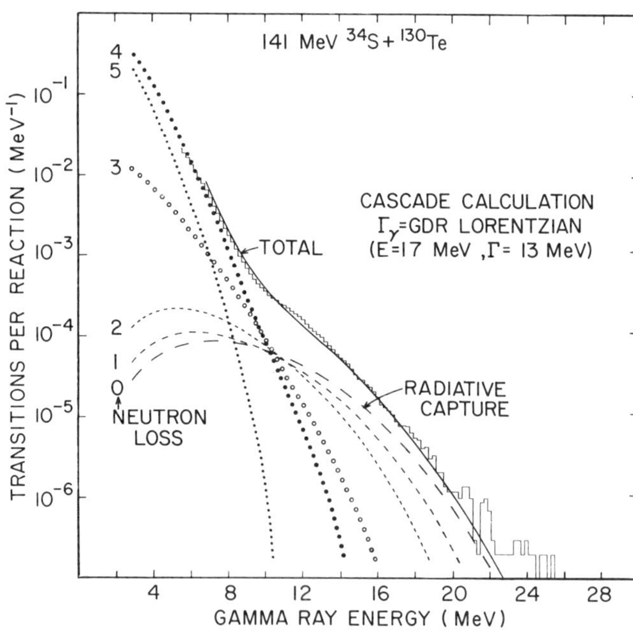

Figure 32. The decomposition of the CASCADE calculation for the $^{34}S + ^{130}Te$ reaction of Figure 31 into the contributions from radiative decays in ^{164}Er, from decays in ^{163}Er,..., from decays in ^{159}Er. The curves are labelled on the left with the number of evaporated neutrons.

Acknowledgments

I am grateful to all those with whom I had the pleasure of collaborating during many of the measurements discussed in this review. I am especially indebted to Alan Nathan for countless hours of discussion over the course of the past years. I would like to thank Dr. S. Skorka for making their data available prior to publication. Finally, I am indebted to Drs. J. Barrette, M. Collins, A. Nathan, and J. Weneser for their critical reading of this manuscript.

References

1. J. J. Pinajian and M. L. Halbert, *Phys. Rev.* **113**:589 (1959).
2. R. F. Coleman, D. N. Herbert, and J. L. Perkin, *Proc. Phys. Soc.* **77**:526 (1961).
3. J. H. Fremlin and P. Reasbeck, *Proc. Phys. Soc.* **82**:110 (1963).
4. B. Linder and A. Zucker, *Phys. Rev.* **127**:1280 (1962).
5. C. B. Fulmer, I. R. Williams, T. H. Handley, G. F. Dell, and L. N. Blumberg, *Phys. Rev. Lett.* **19**:522 (1967).
6. W. Feldman and D. W. Heikkinen, *Nucl. Phys.* **A133**:177 (1969).
7. A. H. Chung, W. T. Diamond, A. E. Litherland, H. L. Pai, and J. Goldemberg, *Phys. Lett.* **53B**:244 (1974).
8. A. F. Zeller, H. S. Plendl, R. H. Davis, M. E. Williams, C. I. Delaune, and R. Holub, *Phys. Rev. C* **13**:661 (1976).
9. J. Imazato, *Z. Phys.* **A277**:117 (1976).
10. A. M. Sandorfi, L. R. Kilius, H. W. Lee, and A. E. Litherland, *Phys. Rev. Lett.* **38**:1463 (1977).
11. A. M. Sandorfi, L. R. Kilius, H. W. Lee, and A. E. Litherland, *Phys. Rev. Lett.* **40**:1248 (1978).
12. A. M. Sandorfi and A. M. Nathan, *Phys. Rev. Lett.* **40**:1252 (1978).
13. D. Branford and S. G. Steadman, *Phys. Rev. C* **18**:2560 (1978).
14. J. B. Viano, A. J. Cole, N. Longequeue, J. J. Lucas, J. Menet, J. C. Saulnier, and J. W. Sunier, *Phys. Rev. C* **20**:551 (1979).
15. A. M. Sandorfi, J. R. Calarco, R. E. Rand, and H. A. Schwettman, *Phys. Rev. Lett.* **45**:1615 (1980).
16. A. M. Sandorfi *et al.*, to be published.
17. A. M. Sandorfi and M. T. Collins, Lecture Notes in Physics **156**:264 (1981).
18. J. Wilczyński, K. van der Borg, H. T. Fortune, J. van Popta, R. H. Siemssen, K. Siwek-Wilczynska, S. Y. van der Werf, and A. van der Woude, *Nucl. Phys.* **A334**:317 (1980).
19. A. M. Nathan, A. M. Sandorfi, and T. J. Bowles, *Phys. Rev. C* **24**:932 (1981).
20. P. Paul, in *Nuclear Spectroscopy and Reactions*, Vol. A, p. 345 (J. Cerney, ed.), Academic Press, New York (1974); M. D. Hasinoff, S. T. Lim, D. F. Measday, and T. J. Mulligan, *Nucl. Inst. Methods* **117**:375 (1974); S. L. Blatt, J. Mahieux, and D. Kohler, *ibid.* **60**:221 (1968).
21. P. M. Endt and C. van der Leun, *At. Data Nucl. Data Tables* **13**:67 (1974).
22. M. A. Kovash, S. L. Blatt, R. N. Boyd, T. R. Donoghue, and H. J. Hausman, *Phys. Rev. Lett.* **42**:700 (1979).

23. A. M. Sandorfi, M. T. Collins, D. J. Millener, A. M. Nathan, and S. F. LeBrun, *Phys. Rev. Lett.* **46**:884 (1981).
24. A. M. Sandorfi and M. T. Collins, *Nucl. Inst. Methods.* (1984) in press.
25. A. M. Sandorfi in *Proc. 3rd Int. Conf. on Clustering Aspects of Nuclear Structure and Nuclear Reactions*, American Inst. of Phys. #47, New York (1978), p. 185.
26. A. E. Litherland, W. T. Diamond, and A. M. Sandorfi, *Nucl. Inst. Methods* **162**:389 (1979).
27. A. M. Sandorfi and J. R. Calarco, to be published.
28. E. Kuhlmann and K. Snover, private communication.
29. E. Almqvist, D. A. Bromley, J. A. Kuehner, and B. Whalen, *Phys. Rev.* **130**:1140 (1963).
30. B. S. Dolbilkin, V. I. Korin, L. E. Lazareva, F. A. Nikolaev, and V. A. Zopevalov, *Nucl. Phys.* **72**:137 (1965).
31. W. Galster, W. Treu, P. Dück, H. Fröhlich, and H. Voit, *Phys. Rev. C* **15**:950 (1977).
32. K. A. Erb, R. R. Betts, S. K. Korotky, M. M. Hines, P. P. Tung, M. W. Sachs, S. J. Willett, and D. A. Bromley, *Phys. Rev. C* **22**:507 (1980).
33. R. Wada, J. Schimizu, and K. Takimoto, *Phys. Rev. Lett.* **38**:1341 (1977).
34. S. K. Korotky, K. A. Erb, S. J. Willett, and D. A. Bromley, *Phys. Rev. C* **20**:1014 (1979).
35. A. Kiss, C. Mayer-Böricke, M. Rogge, P. Turek, and S. Wiktor, *Phys. Rev. Lett.* **37**:1188 (1976); D. H. Youngblood, C. M. Rozsa, J. M. Moss, D. R. Brown, and J. D. Bronson, *Phys. Rev. C* **15**:1644 (1977).
36. F. E. Bertrand, K. van der Borg, A. G. Drentje, M. N. Harakeh, J. van der Plicht, and A. van der Woude, *Phys. Rev. Lett.* **40**:635 (1978); K. van der Borg, Ph.D. Thesis, Groningen, 1979 (unpublished).
37. A. M. Lane, *Nucl. Phys.* **11**:625 (1959).
38. H. Feshbach, C. E. Porter, and V. F. Weisskopf, *Phys. Rev.* **96**:448 (1954).
39. A. Gilbert and A. G. W. Cameron, *Can. J. Phys.* **43**:1446 (1965).
40. T. Ericson, *Adv. Phys.* **9**:425 (1960).
41. N. Shikazono and T. Terasawa, *Nucl. Phys.* **A250**:260 (1975).
42. P. E. Nemirovsky and Yu. V. Adamchuk, *Nucl. Phys.* **39**:551 (1962).
43. P. M. Endt and C. van der Leun, *Nucl. Phys.* **A310**:1 (1978).
44. E. Kuhlmann, E. Ventura, J. R. Calarco, D. G. Mavis, and S. S. Hanna, *Phys. Rev. C* **11**:1525 (1975).
45. A. Sandorfi and M. T. Collins, unpublished.
46. H. Spinka and H. Winkler, *Nucl. Phys.* **A233**:456 (1974).
47. J. Gastebois, in *Nuclear Molecular Phenomena* (ed. N. Cindro), North-Holland, New York (1978), p. 61.
48. S. Das Gupta and M. Harvey, *Nucl. Phys.* **A94**:602 (1967); B. Castel and J. Svenne, *ibid.* **A127**:141 (1969); S. S. M. Wong and G. D. Lougheed, *ibid.* **A295**:289 (1978).
49. D. E. Groce and G. P. Lawrence, *Nucl. Phys.* **67**:277 (1965).
50. J. M. Long, R. E. Peschel, and D. A. Bromley, Contribution to the *Int. Conf. on Nuclear Structure*, Tokyo, 1977, (T. Marumori, ed.), Phys. Soc. Japan (1978), p. 643.
51. K. T. Hecht and D. Braunschweig, *Phys. Rev. C* **17**:396 (1978).
52. M. B. Golin, *Phys. Lett.* **81B**:5 (1979).
53. H. Schultheis and R. Schultheis, *Phys. Lett.* **90B**:185 (1980).
54. H. Schultheis and R. Schultheis, *Phys. Rev. C* **22**:1588 (1980).
55. J. R. Grover and J. Gilat, *Phys. Rev.* **157**:802, 814 (1967).
56. S. Cohen, F. Plasil, and W. J. Swiatecki, *Ann. Phys. (N.Y.)* **82**:557 (1974).
57. D. Branford, F. X. d'Amato, H. W. Sang, Jr., and S. G. Steadman, *Phys. Rev. C* **24**:2339 (1981).
58. H. A. Enge, H. D. Betz, W. W. Buechner, L. Grodzins, W. H. Moore, and E. P. Kanter, *Nucl. Inst. Methods* **97**:449 (1971); H. A. Enge and D. Horn, *ibid.* **145**:271 (1977); Munich Progress Reports 1972, 1974.

59. K. H. Purser, A. E. Litherland, and H. E. Gove, *Nucl. Inst. Methods* **162**:637 (1979).
60. L. R. Kilius, R. P. Beukens, K. H. Chang, H. W. Lee, A. E. Litherland, D. Elmore, R. Ferraro, and H. E. Gove, *Nature* **282**:488 (1979).
61. D. Elmore, H. E. Gove, R. Ferraro, L. R. Kilius, H. W. Lee, K. H. Chang, R. P. Beukens, A. E. Litherland, G. J. Russo, K. H. Purser, M. T. Murrell, and R. C. Finkel, *Nature* **286**:138 (1980).
62. A. M. Sandorfi and M. T. Collins, to be published.
63. D. Evens, K. Rudolph, U. Quade, I. Weidl, G. Löbner, S. Skorka, and P. Konrad, *Nucl. Inst. Methods.* **204**:407 (1983); K. Rudolph, Thesis, Munich 1980 (unpublished); P. Konrad, Thesis, Munich 1980 (unpublished).
64. J. R. Patterson, B. N. Nagorcka, G. D. Symons, and W. M. Zuk, *Nucl. Phys.* **A165**:545 (1971).
65. M. T. Collins, A. M. Sandorfi, D. H. Hoffman, and M. K. Salomaa, *Phys. Rev. Lett.* **49**:1553 (1982).
66. J. O. Newton, B. Herskind, R. M. Diamond, E. L. Dines, J. E. Draper, K. L. Lindenberger, C. Schück, R. Shih, and F. S. Stephens, *Phys. Rev. Lett.* **45**:1383 (1981).
67. A. M. Sandorfi, J. Barrette, M. T. Collins, D. H. Hoffmann, A. J. Kreiner, D. Branford, S. G. Steadman, and J. Wiggins, *Phys. Lett.* **130B**:19 (1983).
68. B. I. Goryachev, Yu. V. Kurnetsov, V. N. Orlin, N. S. Pozhidaeva, and V. G. Shevchenko, *JETP Lett.* **19**:41 (1974).
69. F. Pühlhofer, *Nucl. Phys.* **A280**:267 (1977).
70. J. Barrette and A. M. Sandorfi, Comments *Nucl. Part. Phys.* **A12**:57 (1983).
71. J. Ahrens, H. Borchert, K. H. Czock, H. B. Eppler, H. Gimm, H. Gundrum, M. Kröning, P. Riehn, G. Sita Ram, A. Zieger, and B. Ziegler, *Nucl. Phys.* **A251**:479 (1975).
72. D. H. Dowell, G. Feldman, K. A. Snover, A. M. Sandorfi, and M. T. Collins, *Phys. Rev. Lett.* **50**:1191 (1983).

3

WOLF-UDO SCHRÖDER is Associate Professor of Chemistry at the University of Rochester, Rochester, New York. Educated at the Universities of Darmstadt, Berlin, and Göttingen, he received his Ph.D. from the Technical University of Darmstadt in 1971. From 1968 through 1972, he was a scientific visitor of the European Center for Nuclear Research (CERN) in Geneva, Switzerland, and remained Research Associate at the University of Darmstadt until 1975. From 1975 through 1977, he held a Fellowship from the German Academic Exchange Service (DAAD). He participated in the heavy-ion research program at the University of Rochester, as a Research Associate (until 1979), Senior Research Associate (1979-1980), and a Senior Research Associate and Professor (part-time) of Nuclear Science (until 1983).

JOHN R. HUIZENGA is Tracy H. Harris Professor of Chemistry and Physics and Chairman of the Department of Chemistry at the University of Rochester, Rochester, New York. He received his Ph.D. degree in Physical Chemistry from the University of Illinois in 1949. Huizenga received the E. O. Lawrence Memorial Award in 1966 for research in nuclear fission and the American Chemical Society's Award for Nuclear Applications in Chemistry in 1975. He has served on numerous editorial boards, advisory and visiting committees to universities, government and national laboratories. Recipient of a Fulbright Fellowship in 1954–55 (Amsterdam) and of Guggenheim Fellowships in 1964–65 (Paris) and in 1973–74 (Berkeley, Munich, and Copenhagen), Huizenga is a fellow of the American Physical Society and the American Association for Advancement of Science, and member of the National Academy of Sciences.

Damped Nuclear Reactions

W. U. SCHRÖDER AND J. R. HUIZENGA

1. Introduction

This review focuses on the current status of experimental and theoretical investigations of the mechanisms of reactions between complex nuclei. These processes are characterized by features that are intermediate between those of the comparatively simple quasielastic, few-nucleon transfer reactions, on one hand, and those of the highly complex compound-nucleus reactions, on the other hand. A variety of terms have been associated with these intermediate reaction types, such as quasifission, deep-inelastic scattering, and damped or dissipative collisions. The latter two notations appear to be equally well adapted to the characteristics of the reaction mechanism as it is conceived to date, since they refer to a damping of macroscopic or collective nuclear motion through interactions with intrinsic nucleonic modes.

The growing scientific interest in damped nuclear reactions is attracted by the many opportunities they provide for a study of phenomena occurring in nuclear matter under extreme conditions with respect to shape, intrinsic excitation, spin, mass-to-charge density, etc., and the possibility to control these conditions by selecting appropriate initial and final reaction parameters. Particular goals are to discover the simple modes of nuclear excitation produced in damped nuclear collisions, their foundation on the microscopic processes, and their eventual decay, i.e., the cooperative phenomena and

W. U. SCHRÖDER AND J. R. HUIZENGA • Department of Chemistry, and Nuclear Structure Research Laboratory, University of Rochester, Rochester, New York 14627 relaxation processes occurring within a small quantal system that is initially far from equilibrium. It is inescapable to notice the analogies of these problems with those addressed by other branches of science such as statistical and chemical physics, which invites an interdisciplinary approach to a profound understanding of damped-reaction mechanisms. Damped nuclear reactions, however, not only provide the means to study modes of nuclear motion inaccessible by other methods—they are also amenable to the production of exotic new nuclides far off the beta-stability line.

Damped reactions between complex nuclei were first observed in experiments by Kaufmann and Wolfgang (Kau 59, 61, 61b) studying multinucleon transfer processes induced on a variety of targets by beams of 10 MeV/nucleon ^{14}N and ^{16}O beams. The suspicion of a previously unknown reaction mechanism was based on their observation that the cross sections for final fragmentations did not decrease exponentially with the number of nucleons exchanged between projectile and target nuclei, as had been expected (Bre 56b) from a quantal tunneling process, but exhibited a much weaker dependence. The authors also noted a relatively weak correlation of the cross section with both the binding energy of the exchanged group of nucleons and the Q value of the reaction. Consequently, a grazing-like reaction mechanism was proposed for a broad band of intermediate impact parameters, where the combined Coulomb and centrifugal repulsion is strong enough to overcompensate the nuclear surface attraction. It was envisioned that, in the course of the reaction, the region of contact between the collision partners would be bridged temporarily by a neck of nuclear matter. This neck would become excited by frictional heating and effect a diffusion of nucleons into the main bodies of the interacting nuclei. While some of these early concepts did subsequently not withstand experimental tests, many of the ideas suggested proved to be basically correct. It appears, hence, unfortunate that the work of Kaufmann and Wolfgang received relatively little attention at the time.

It was not until the early 1970s that the importance of the new damped-reaction mechanism was clearly recognized by experimental groups (Gri 70, Gal 70, 75, Art 71, 73, 73b, 73c, Lef 73, 74c, 75b, Wol 74, Vol 74b, 75, Tho 74, Kra 74c, Col 74b, 74c, 75, Gat 74, 75, 75b, Jac 75, Hui 75) and that theoretical concepts were developed (Bon 71, 74, 74b, 75, 75b, Swi 72, Bec 73, Gro 73, 74, 74b, 75b, Wil 73c, Nor 74, 75, Tsa 74, Spe 74, 75, Bro 74, Fin 74, 74b, Gla 74b, 75, Hof 75, Mor 75, Deu 75, 75b, Dav 75) for the interpretation of the unexpected experimental observations. The characteristic features observed for damped reactions at bombarding energies of a few MeV/nucleon above the interaction barrier may be summarized as follows:

1. The reactions are basically binary with only two massive reaction fragments in the exit channel. Subsequently, the excited primary fragments decay via fission and/or the emission of light particles and γ rays.

2. The angular distributions of the reaction fragments change in a characteristic way with the charge product $Z_P \cdot Z_T$ of the projectile–target combination and the bombarding energy. For lighter systems and low bombarding energies, angular distributions are similar to those of a fusion–fission process. With increasing bombarding energy, forward-tending, orbiting-type components develop. Very heavy systems exhibit fragment angular distributions that are focused into a relatively narrow angular range centered close to the grazing angle, for particular bombarding energies. At lower or higher bombarding energies, the angular distributions exhibit backward-tending or forwarding-rising components, respectively. There are definite correlations between the angular-distribution type and the net exchange of mass and charge between the reaction partners, as well as with the amount of kinetic energy dissipated in the reaction. The correlation with the energy loss is the more fundamental, overriding one. Low energy losses correspond to angular distributions characteristic of very fast (10^{-22}–10^{-21} s) processes, while high energy losses are associated with those typical of slower orbiting or fusion–fission-type processes.

3. Fragment mass or charge distributions are bimodal and centered close to masses and charges of projectile and target nuclei. For lighter or asymmetric systems, there is an additional broad component that is centered at symmetric fragmentations. For narrow bins of kinetic energy loss, mass and charge distributions are nearly bimodal Gaussians with widths that increase with increasing energy loss. The additional symmetric component observed for particular systems is associated with the highest possible energy losses. Its width increases with increasing bombarding energy. There are also strong correlations between the net amounts of mass and charge transfer between the reaction partners that develop with increasing energy loss.

4. Large amounts of kinetic energy of radial and orbital motion can be dissipated in a reaction. Final total kinetic energies can be as low as those corresponding to the Coulomb repulsion of highly deformed fission fragments. The partition of the dissipated energy between the final reaction fragments is consistent with an intermediate system that is close to a statistical equilibrium at the instant of scission. Final fragment spins can also be large with a significant alignment, which is strongly dependent on the total kinetic energy loss.

In order to place the damped-reaction process in perspective with other known reaction types, Figure 1 shows an overall classification scheme for nuclear collisions. It is based on the classical concepts of a well-defined impact parameter, a small spreading of the wave packet about an average classical system trajectory, and a shape of the intermediate dinuclear complex that develops with increasing classical interaction time. The neglect of a potential interference or correlation of partial waves with very different angular momenta l implies a localization of different reaction types in l space, an assumption underlying the scheme of Figure 1. Here, different angular momenta are represented by the corresponding distances of closest approach, r, whose relation to an interaction radius R_{Int} determines the reaction characteristics. Experimental systematics suggest that nuclear interactions become important at center separation distances smaller than R_{Int}, corresponding to nuclear surface separations smaller than about 3 fm.

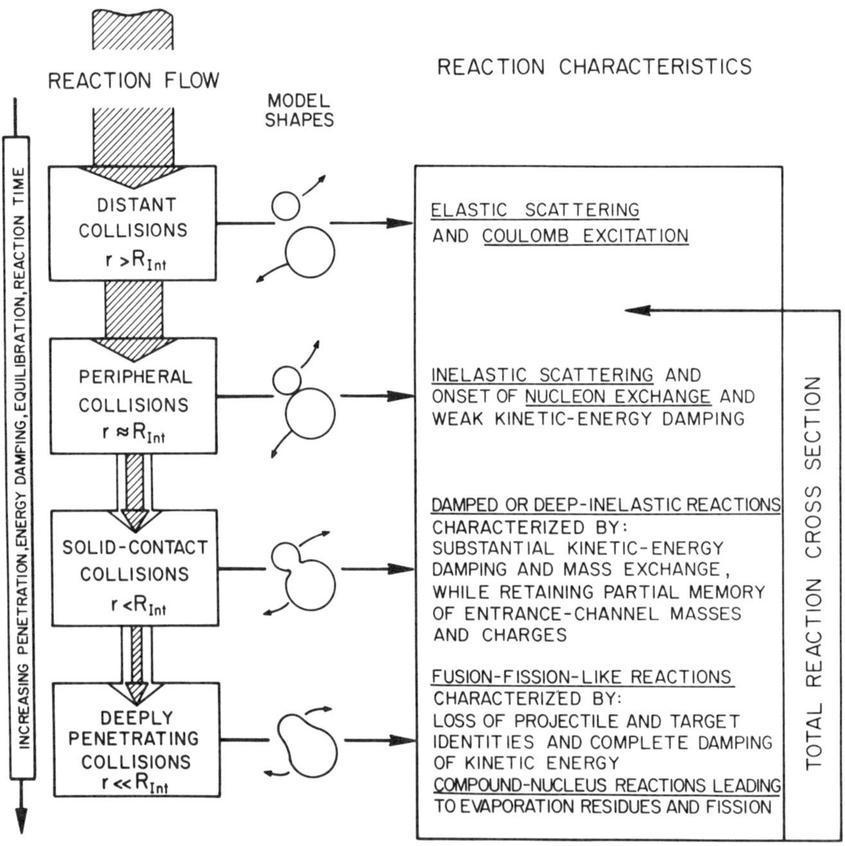

Figure 1. Classification scheme of collisions between complex nuclei based on a classical impact parameter model. The quantity r denotes the distance of closest approach associated with a given impact parameter. Its relation to the interaction radius R_{Int} determines the characteristics of the reaction described on the right. The model shapes apply to the distance of closest approach. The variation in the widths of the hatched vertical arrows illustrates the dominantly peripheral character of collisions between very heavy nuclei. Lighter systems yield deep interpenetrations more readily, as is represented by smaller variations in the widths of the open arrows.

The classification scheme in Figure 1 illustrates the development of the reaction flow, the dinuclear shapes, and the resulting reaction characteristics from top to bottom according to decreasing angular momentum and distance of closest approach, but increasing interpenetration and increasing classical interaction time available to the various relaxation and equilibration processes occurring in the reaction. Distant collisions with $r > R_{\text{Int}}$ do not lead to nuclear reactions. Here, Coulomb excitation and elastic scattering are the only contributions to the total cross section. For somewhat smaller angular momenta associated with r values of the order of the interaction radius, $r \approx R_{\text{Int}}$, inelastic scattering and exchange of a few nucleons are induced in peripheral, grazing-like collisions. Consequently, there is already some loss of kinetic energy of relative motion, i.e., the reaction Q value is slightly negative. For smaller l values with a distance of closest approach significantly smaller than R_{Int}, the interacting nuclei come into solid contact. A window opens between the reaction partners allowing a considerable exchange of mass and damping of the relative kinetic energy. Nevertheless, the constituents of the intermediate dinuclear complex retain their separate identities to a large extent, i.e., carry a memory of the entrance-channel conditions. This memory is lost completely in deeply penetrating fusion–fission-like collisions where the window between the reaction partners becomes so large that a single nucleus with a continuous interior is produced, rather than a dinuclear system. In addition, the kinetic energy of relative motion is completely damped out. A subset of these collision types corresponding to the longest interaction times may eventually produce a compound nucleus characterized by a statistical equilibrium of all degrees of freedom, which may subsequently decay by fission or the evaporation of light particles.

The reaction flow indicated by vertical arrows in Figure 1 is distributed between the various collision types, accounting for the total reaction cross section in rather different ways, depending on the mass of the projectile–target combination and the bombarding energy. For a relatively light system (open arrows) at not too high bombarding energies, a considerable fraction of the reaction cross section is associated with fusion-like collisions and compound-nucleus production. As the bombarding energy is raised, this contribution loses significance, due to an increased centrifugal repulsion. For heavy systems (hatched arrows), the strong Coulomb repulsion prevents deeper interpenetrations such that compound-nucleus formation becomes negligible. For these latter systems, solid-contact collisions result from a broad band of l waves providing reaction products with a continuous range in the extent to which the various degrees of freedom are relaxed. Hence, in the following discussion of the reaction phenomenology, these heavy systems are emphasized.

Over the past years, several intermediate accounts of experimental characteristics and their interpretation have been given (Fle 74b, Gal 76, Mor 76b, Sch 77e, Vol 78, Lef 79b, Gob 80); in addition, there are recent monographs (Nor 80, Hod 78, Bas 80) in which damped nuclear reactions are discussed to some extent. Detailed descriptions of several current theoretical models or concepts are also available (Hof 77b, Gro 78, Has 78, Wei 78, Nör 80c, Mar 80). In addition to a consideration of the most recent developments until 1981, the present review attempts to put the various theoretical approaches to an understanding of damped nuclear reactions in perspective with each other, as conceived from the point of view of an experimentalist. Experimental results will be discussed both with reference to these theoretical concepts and from a more phenomenological view, taking notice of analogies with somewhat better understood nuclear reaction types such as fission. The sheer number of publications on the subject that have appeared over the recent years makes it impossible, however, to give a complete and yet detailed account of these developments and to acknowledge every contribution to the advance of the knowledge about damped nuclear reactions. Hence, it is the aim of this paper to illustrate the main route this advance has taken by a few examples which will be discussed in some detail. The discussion of the experimental material will be interspersed with theoretical analyses that are available only for a limited number of reactions and with a varying degree of diversification. Several theoretical approaches such as time-dependent Hartree–Fock models and hydrodynamical calculations will be referred to only occasionally, as these are dealt with in detail elsewhere. The same applies to elastic scattering, which is discussed here only to the extent that it is significant for the interpretation of damped-reaction features.

The following Section 2 starts out with a review of useful elementary relations between elastic-scattering deflection functions and the reaction cross section. The various determinants of the shape of the differential damped-reaction cross section are illustrated in Section 2.2 employing a particular model. The scope of the subsequent Sections 2.3–2.7 reflects the currently popular view that the many degrees of freedom excited in damped reactions can, at least in principle, reasonably well be subdivided into collective and intrinsic ones. Hence, Sections 2.3, 2.4, and 2.5 deal with classical and quantal approaches to the collective motion and the derivation of static conservative interaction forces. The dynamical response of the nucleonic motion to changing collective coordinates gives rise to reversible and irreversible induced interaction forces which are the subject of Section 2.6. The stochastic aspects of the intrinsic motion associated with transport phenomena in damped reactions are discussed in Section 2.7, while Section 2.8 refers to some applications of transport models and provides some simple illustrations.

Experimental observations are presented in Section 3, along with examples of model interpretations. It also points out some experimental problems that should be considered in the appreciation of data. The section starts with a discussion of the magnitude of the reaction cross section and its dependence on reaction conditions such as the bombarding energy. Proof of the basically binary nature of the damped reaction process is provided in Section 3.2. Sections 3.3–3.6 exhibit the systematics of angular, mass, charge, and final energy distributions of the reaction fragments. The characteristics of direct and sequential multiparticle processes are accounted for by Section 3.7. Section 3.8 emphasizes transitional features of damped reactions for deeply penetrating collisions and explores a potentially new access to an understanding of the damped reaction mechanism. A summary given in Section 4 concludes the presentation.

2. Elements of a Phenomenological Interpretation of Damped Reactions

The ultimate goal of experimental and theoretical investigations of damped reactions between complex nuclei is, of course, to achieve an understanding of the reaction properties in terms of the underlying microscopic mechanisms. To date, however, the development of microscopic quantal reaction theories has not progressed far enough to yield a consistent explanation of the observed reaction features. In this situation, a phenomenological interpretation appears to be useful, where certain reaction patterns are analyzed in terms of macroscopic or microscopic concepts allowing one to study the dependence of model parameters on important reaction conditions. For example, available data suggest that some degrees of freedom pertaining to damped reactions follow closely principles of classical mechanics, while other aspects resemble statistical features familiar from classical diffusion processes. It appears then promising to study damped reactions in terms of conservative and dissipative forces, as well as in terms of transport equations, although this does not justify a neglect of other possible approaches.

Phenomenological models are characterized by a choice of the most significant collective or macroscopic reaction variables associated with the average configuration of the interacting nuclear system, on one hand, and an implicit treatment of intrinsic degrees of freedom by statistical methods, on the other hand. This procedure corresponds to a division of the total Hamiltonian $\mathscr{H}$ into three parts:

$$\mathscr{H} = \mathscr{H}_c + \mathscr{H}_i + \mathscr{H}_{ci} \tag{2.1}$$

The collective part $\mathscr{H}_c$ is often approximated by dynamical models describing the collective motion in terms of classical trajectories, while the intrinsic degrees of freedom corresponding to $\mathscr{H}_i$ are either completely neglected or conceived of as a statistical ensemble in thermodynamic equilibrium. The importance of the the coupling between intrinsic and collective motion represented by $\mathscr{H}_{ci}$ is obvious from the observation of dissipative phenomena in damped reactions. Some of its effects on the collective motion are frequently interpreted in terms of velocity-dependent classical friction forces.

In the following, elements commonly used in phenomenological analyses of damped reactions are reviewed. They include classical and semiclassical descriptions of cross sections, deflection functions, Lagrange–Rayleigh dynamics of relative motion, models of potential energies, and interpretations of dissipative features.

2.1. Elastic Scattering, Deflection Functions, and Cross Sections

The dependences of elastic and reaction cross section on total mass and mass asymmetry of the projectile–target system, on kinetic energy, and on scattering angle are important macroscopic observables yielding information on the nature of the nucleus–nucleus interaction such as the balance between conservative and dissipative forces, their form factors, and the configurations that the interacting system attains during a collision. Fully quantal calculations of elastic scattering and few-nucleon transfer processes have been performed, however, mainly for lighter projectiles. Such calculations are involved, if at all possible, for reactions of medium-weight or heavy systems at center-of-mass bombarding energies $E_{\text{c.m.}}$ of several MeV/nucleon. Here, hundreds of values of angular momentum l may contribute to the reaction cross section associated with scattering waves probing the nuclear interaction force over a considerable fraction of its range.

To some extent, an intuitive picture of the reaction may be obtained from a classical or semiclassical model, where the scattering of two nuclei with frozen configurations is considered in terms of a central interaction potential $V(r)$ depending only on the distance r between their centers. The relative motion is characterized by its de Broglie wavelength

$$\lambda\!\!\!^{-}(r) = k^{-1}(r) = \hbar\{2\mu[E_{\text{c.m.}} - V(r)]\}^{-1/2} \tag{2.1.1}$$

where μ is the reduced mass of the system. A classical approach can be expected (Mes 65) to have some validity, if the potential varies slowly over one wavelength, i.e.,

$$\lambda\!\!\!^{-}|\text{grad}\, V| \ll V \tag{2.1.2}$$

or

$$\Delta L = \hbar ka \gg \hbar|\text{grad}\, V/V| \tag{2.1.2}$$

With a being the range of the interaction force $F = -\text{grad}\, V$, the conditions (2.1.2) require $\lambdabar \ll a$ and a broad band $\Delta L \gg \hbar$ of l waves that sample the force. For typical reactions between heavy nuclei, asymptotically, $\lambdabar := \lambdabar(r \to \infty)$ is of the order of 10^{-2} fm and, thus, small compared to the range $a \lesssim 1$ fm. In addition, ΔL may be of the order of several hundred $\hbar$. While then the asymptotic relative motion is close to its classical limit, the validity of classical approximations has to be evaluated for small separation distances, as well as for other types of collective motion.

The angular distribution of the classical cross section is given by the ring area $d\sigma^{\text{cl}} = 2\pi b\, db$ intersected by trajectories with impact parameter b that are scattered into the solid angle $d\Omega = 2\pi \sin\theta\, d\theta$. It can be written as

$$\frac{d\sigma^{\text{cl}}(\theta, l)}{d\Omega} = \frac{b}{|\sin\theta\, d\theta/db|} = \lambdabar^2 \frac{l+1/2}{|\sin\theta\, d\theta/dl|} \tag{2.1.3}$$

Here, the classical angular momentum $L = p \cdot b$ has been replaced with $\hbar(l+1/2)$, according to the Langer modification (Lan 37). The classical deflection function $\theta(l)$ is characteristic of the interaction potential V and the incident energy $E_{\text{c.m.}}$. The total classical differential cross section becomes

$$\frac{d\sigma^{\text{cl}}(\theta)}{d\Omega} = \frac{\lambdabar^2}{\sin\theta} \sum_l \frac{l+1/2}{|d\theta/dl|} \tag{2.1.4}$$

It contains contributions from elastic and inelastic processes. The total reaction cross section is obtained by subtracting the elastic-scattering amplitude as determined by the corresponding S-matrix elements S_l. Introducing the transmission coefficients $T_l = 1 - |S_l|^2$, differential and total reaction cross sections read

$$\frac{d\sigma_R(\theta)}{d\Omega} = \frac{\lambdabar^2}{|\sin\theta|} \sum_l \frac{l+1/2}{|d\theta/dl|} T_l$$

$$\sigma_R = \pi\lambdabar^2 \sum_l (2l+1) T_l \tag{2.1.5}$$

Hence, the reaction cross section can be evaluated from the elastic-scattering deflection function, provided that information about the transmission coefficients is available.

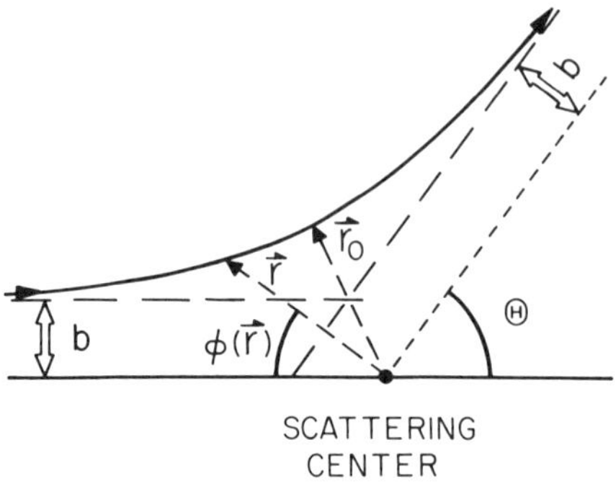

Figure 2. Illustration of the coordinates used in the calculation of the elastic-scattering deflection function associated with trajectories (r, ϕ). The quantity r_0 is the distance of closest approach for a given impact parameter b, and Θ is the associated asymptotic scattering angle.

Employing the coordinate system illustrated in Figure 2, the elastic-scattering deflection function can be derived by considering the radial and tangential kinetic energies

$$E_r = \frac{\mu}{2}\left(\frac{dr}{dt}\right)^2 = E_{\text{c.m.}} - V(r) - \frac{\hbar^2 l(l+1)}{2\mu r^2}$$

$$E_t = \frac{\mu r^2}{2}\left(\frac{d\phi}{dt}\right)^2 = \frac{\hbar^2 l(l+1)}{2\mu r^2} \tag{2.1.6}$$

From these expressions, one derives $d\phi/dr$ and calculates the deflection angle as a function of l:

$$\begin{aligned}\theta(l) &= \pi - 2\int_{r_0}^{\infty} (d\phi/dr)\,dr \\ &= \pi - 2\hbar\left(l + \frac{1}{2}\right)\int_{r_0}^{\infty} \frac{dr}{r^2}\{2\mu[E_{\text{c.m.}} - V_{\text{eff}}(l, r)]\}^{-1/2}\end{aligned} \tag{2.1.7}$$

Here, r_0 is the classical turning point or distance of closest approach defined by $E_r(r_0) = 0$. The effective radial potential is given by $V_{\text{eff}}(l, r) = V(r) + \hbar^2 l(l+1)/2\mu r^2$. The deflection function can also be expressed in terms of a derivative $\theta(l) = 2\,d\delta/dl$ of the phase shift (Sch 78m)

$$\delta = \frac{\pi}{2}\left(l + \frac{1}{2}\right) + \frac{1}{\hbar}(2\mu E_{\text{c.m.}})^{1/2}\left\{\int_{r_0}^{\infty} dr\left[\left(1 - \frac{V_{\text{eff}}(l, r)}{E_{\text{c.m.}}}\right)^{1/2} - 1\right] - r_0\right\} \tag{2.1.8}$$

As a matter of practical concern, one is often interested in the Coulomb deflection angles, where V is replaced by the Coulomb potential $V_{\text{Coul}} = e^2 Z_P Z_T / r$. Z_P and Z_T are the atomic numbers of projectile and target, respectively. For $r > r_0$, one has (see Figure 2)

$$\phi_{\text{Coul}}(l, r) = \int_r^{\infty} dr \left\{ \frac{1}{r^2} \frac{\kappa}{\varepsilon} \left[1 - \left(\frac{1}{\varepsilon} + \frac{\kappa}{\varepsilon r} \right)^2 \right]^{-1/2} \right\}$$

$$= \arcsin\left(\frac{1}{\varepsilon} + \frac{\kappa}{\varepsilon r} \right) - \arcsin\left(\frac{1}{\varepsilon} \right) \tag{2.1.9}$$

Here ε is the eccentricity of the Coulomb orbit:

$$\varepsilon(l) = \left[1 + 2\mu E_{\text{c.m.}} \hbar^2 l(l+1) \left(\mu e^2 Z_P Z_T \right)^{-2} \right]^{1/2} \tag{2.1.10}$$

and κ is given by

$$\kappa(l) = \hbar^2 l(l+1) \left(\mu e^2 Z_P Z_T \right)^{-1} \tag{2.1.11}$$

The asymptotic Coulomb scattering angle $\theta_{\text{Coul}} = \pi - 2\phi_{\text{Coul}}(l, r_0)$ follows from Equation (2.1.9) as

$$\theta_{\text{Coul}}(l) = 2 \arctan[\eta/(l+1/2)] \tag{2.1.12}$$

where $\eta = e^2 Z_P Z_T [\mu/(2\hbar^2 E_{\text{c.m.}})]^{1/2}$ is the Sommerfeld parameter. The distance of closest approach is related to the Coulomb scattering angle by

$$r_0 = \eta \lambda\!\!\!^{-} [1 + \csc(\theta_{\text{Coul}}/2)] \tag{2.1.13}$$

From Equation (2.1.12), one also derives an expression for the Coulomb phase shifts

$$\delta_{\text{Coul}}(l) = \delta_{\text{Coul}}(0) + \sum_{l'=1}^{l} \arctan[\eta/(l'+1/2)] \tag{2.1.14}$$

Obviously, deflection angles calculated according to Equations (2.1.9) and (2.1.12) are reasonable approximations of the elastic-scattering angle only at sufficiently large separation distances r. At smaller distances, the Coulomb interaction deviates from its point-like value (see Section 2.5), but more importantly, the nuclear interaction modifies the deflection angle considerably. The influence of strength and radial form factor of the total interaction potential on the deflection function is illustrated in Figure 3. Here, calculated Coulomb and elastic deflection functions are shown at the

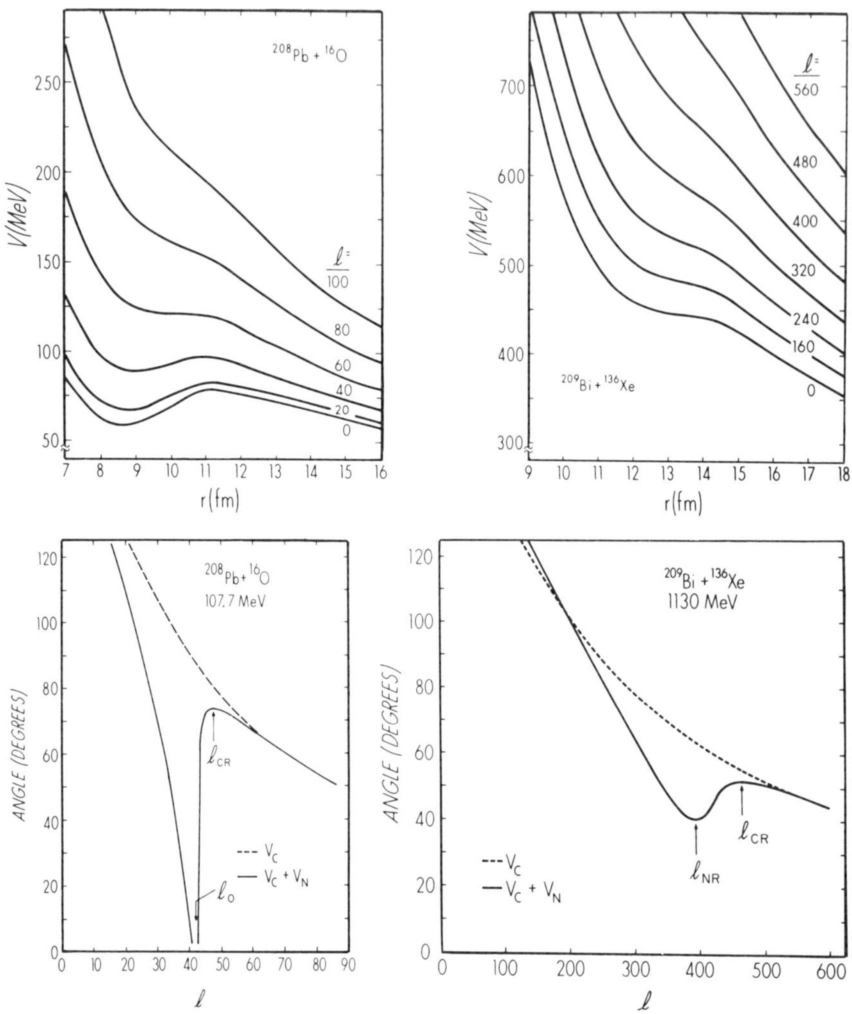

Figure 3. The effective interaction potentials for the systems $^{208}Pb + {}^{16}O$ and $^{209}Bi + {}^{136}Xe$ are shown at the top left and right, respectively, for various angular momenta l. They are sums of Coulomb potential, nuclear proximity potential, and centrifugal potential. At the bottom, the total elastic (solid curves) and the Coulomb (dashed curves) deflection functions are shown for the same systems. The angular momentum associated with the "Coulomb rainbow" is denoted by l_{CR}. The $^{208}Pb + {}^{16}O$ deflection function exhibits an orbiting (spiraling) singularity at $l = l_0$. The deflection function for $^{209}Bi + {}^{136}Xe$ develops an inner (nuclear) rainbow at $l = l_{NR}$.

bottom for the reactions $^{208}Pb + ^{16}O$ and $^{209}Bi + ^{136}Xe$. The effective potentials displayed at the top contain the sum of Coulomb potential for extended nuclei (Bon 76b) and a nuclear proximity potential (Swi 75, Blo 77) to be discussed below. Whereas the potential for the system $^{208}Pb + ^{16}O$ possesses a well-developed minimum, the strong Coulomb potential associated with the system $^{209}Bi + ^{136}Xe$ counteracts the attractive nuclear potential to an extent that the total potential exhibits only a small inflection. The deflection functions follow $\theta_{\mathrm{Coul}}(l)$ for large l. At $l \approx l_{\mathrm{CR}}$, the balance between Coulomb and nuclear forces produces a local maximum in the deflection angle termed a "Coulomb rainbow." For even smaller l values, the $^{209}Bi + ^{136}Xe$ deflection function develops a second rainbow for $l \approx l_{\mathrm{NR}}$, the "nuclear rainbow." For the system $^{208}Pb + ^{16}O$, where the Coulomb repulsion is somewhat weaker, a nuclear rainbow does not appear. Instead, at a certain l value, l_0, the deflection function exhibits an "orbiting" singularity caused by a cancellation of the attractive nuclear force by the repulsive centrifugal and Coulomb forces at a distance r where $E_r = V_{\mathrm{eff}}(l_0, r)$. At rainbow angles, because of $d\theta/dl = 0$, the classical cross section is singular. It is also not a proper approximation in the orbiting region, because here $\lambda\!\!\!^{-}$ becomes very large and $\theta(l)$ attains a logarithmic singularity.

There are other effects that are missed by the classical approach which may be important for an understanding of the elastic-scattering cross section. Apart from fluctuating contributions due to statistical processes, there are the quantal effects of interference of different l waves, barrier penetration, as well as reflection and refraction at real and imaginary potentials, and absorption. Quantum scattering features are included in the fully quantal optical-model calculations, while fluctuative contributions to elastic and reaction cross sections are approached (cf. Section 2.2) with other methods involving semiclassical approximations or phenomenological assumptions.

Semiclassical descriptions of elastic scattering and nuclear reactions have been developed (Mes 65, Ros 62, Bro 72, Bro 72b, Cro 80, Die 73, Fro 78, Gle 75, Kno 74–77, Sch 78m, Koe 75, Lan 74) to various degrees of applicability to heavy-ion reactions and scattering. Gross and Frahn (Gro 76, Fra 76c) have shown that the semiclassical approximation is the leading order of an asymptotic expansion of the scattering amplitude in terms of a large parameter such as the Sommerfeld parameter. It is demonstrated in these studies how effects such as Fresnel and Fraunhofer diffraction arise in different domains of heavy-ion scattering. A detailed discussion of this work would exceed the scope of this chapter. It appears indicated, however, to point out that if the limit of strong absorption is approached in the scattering of relatively heavy ions, comparatively simple models of analysis can be applied. This follows because absorption masks some of the quantal diffraction and refraction patterns.

A rather simple model commonly used to interpret heavy-ion elastic-scattering angular distributions in terms of the reaction cross section has been devised by Blair (Bla 54, Ker 57). Neglecting the effect of the real nuclear potential for grazing collisions, in this model, the differential elastic cross section is written as

$$\frac{d\sigma(\theta)}{d\Omega}=\left|\frac{\lambda\!\!\!^{-}}{2i}\sum_{l}(2l+1)(\mathscr{S}_l-1)P_l(\cos\theta)\right|^2 \qquad (2.1.15)$$

with the "sharp-cutoff" approximation for the reflection coefficient

$$\mathscr{S}_l=|\mathscr{S}_l|\exp[2i\delta(l)]=\begin{cases}0, & l<l_{\max}\\ \exp[2i\delta_{\text{Coul}}(l)], & l\geq l_{\max}\end{cases} \qquad (2.1.16)$$

i.e., assuming total absorption for waves $l<l_{\max}$. P_l in Equation (2.1.15) are the Legendre polynomials. Hence, one obtains

$$\frac{d\sigma(\theta)}{d\Omega}=|f_{\text{Coul}}(\theta)-f_R(\theta)|^2=\left|f_{\text{Coul}}(\theta)-\frac{\lambda\!\!\!^{-}}{2i}\sum_{l=0}^{l_{\max}}(2l+1)\right.$$

$$\left.\times\exp[2i\delta_{\text{Coul}}(l)]\,P_l(\cos\theta)\right|^2 \qquad (2.1.17)$$

Here, f_{Coul} is the Coulomb scattering amplitude leading to the Rutherford cross section

$$\frac{d\sigma_{\text{Ruth}}(\theta)}{d\Omega}=\frac{1}{4}\lambda\!\!\!^{-2}\eta^2\sin^{-4}\frac{\theta}{2} \qquad (2.1.18)$$

$f_R(\theta)$ is the strong-absorption reaction amplitude defined in Equation (2.1.17). The ratio $d\sigma(\theta)/d\sigma_{\text{Ruth}}(\theta)$ given by Equations (2.1.17) and (2.1.18) is an oscillatory function of the cutoff angular momentum $l_{\max}$. However, its average passes through the value $d\sigma/d\sigma_{\text{Ruth}}=1/4$ always at the l value given by Equation (2.1.12), for a range of scattering angles and incident energies. For a given energy or η value, $l_{\max}$ can be inferred from the "quarter-point angle" $\theta_{1/4}$ where $d\sigma(\theta_{1/4})/d\theta_{\text{Ruth}}(\theta_{1/4})=1/4$, according to

$$l_{\max}+\tfrac{1}{2}=\eta\cot\left(\tfrac{1}{2}\theta_{1/4}\right) \qquad (2.1.19)$$

This cutoff angular momentum corresponds to a critical distance of closest approach, the "strong-absorption radius" [cf. Equation (2.1.13)] which is in the present context not distinguished from the "interaction radius" of the Fresnel model,

$$R_{SA} = \eta \lambda\!\!\!^{-} \left[1 + \csc\left(\tfrac{1}{2}\theta_{1/4}\right)\right] \tag{2.1.20}$$

such that l waves with classical turning points $r_0 < R_{\text{SA}}$ are totally absorbed. In terms of this model, the total reaction cross section can be written as

$$\sigma_R = \pi \lambda\!\!\!^{-}{}^2 (l_{\max} + 1)^2 \tag{2.1.21}$$

Since R_{SA} is the classical turning point for the wave with $l = l_{\max}$, it also follows from $E_r(R_{\text{SA}}, l_{\max}) = 0$

$$l_{\max} + \tfrac{1}{2} = \hbar^{-1} R_{\text{SA}} \left\{2\mu\left[E_{\text{c.m.}} - V_{\text{Coul}}(R_{\text{SA}})\right]\right\}^{1/2} \tag{2.1.22}$$

and, hence,

$$\sigma_R = \pi R_{\text{SA}}^2 \left[1 - V_{\text{Coul}}(R_{\text{SA}})/E_{\text{c.m.}}\right] \tag{2.1.23}$$

The conditions of a well-defined radius of the strongly absorbing nuclear interaction potential, together with the "diverging lens" effect of the strong Coulomb repulsion creating a virtual source point of the incident waves, are equivalent to those where classical Fresnel scattering is expected to occur. It has, in fact, been noticed by Frahn and collaborators (Fra 63–74, Ven 63) that the partial-wave expansion [Equation (2.1.15)] in the "sharp-cutoff approximation" [Equation (2.1.16)] reduces to formulas describing classical Fresnel diffraction. In this approach, the ratio of elastic-to-Rutherford cross section is given by

$$\frac{d\sigma(\theta)}{d\sigma_{\text{Ruth}}(\theta)} = \frac{1}{2}\left\{\left[\frac{1}{2} - S(y)\right]^2 + \left[\frac{1}{2} - C(y)\right]^2\right\} \tag{2.1.24}$$

where S and C are the Fresnel integrals of argument

$$y = (2\eta/\pi)^{1/2} \csc\left(\tfrac{1}{2}\theta_{1/4}\right) \sin\left[\tfrac{1}{2}\left(\theta - \theta_{1/4}\right)\right] \tag{2.1.25}$$

For $\theta = \theta_{1/4}$, the integrals vanish, and the ratio of Equation (2.1.24) becomes equal to $1/4$, in agreement with Blair's numerical evaluation. Fresnel diffraction is expected to occur, if $l_{\max} \gg 1$ and $l_{\max} > 1/\sin\theta_{1/4}$, i.e., for strong Coulomb repulsion and moderate energies $E_{\text{c.m.}}$.

The assumption of a sharp boundary between elastic scattering and total absorption has been relaxed in the smooth-cutoff or generalized Fresnel model of Frahn (Fra 78–78c). Adopting a convenient parametrization of the elastic-scattering reflection function over the region in l space contributing to nuclear elastic scattering,

$$|S_l| = \left[1+\exp\left(\frac{l-l_{\max}-1/2}{\Delta}\right)\right]^{-1} \tag{2.1.26}$$

The expression

$$\frac{d\sigma(\theta)}{d\sigma_{\text{Ruth}}(\theta)} = \frac{1}{2}\left\{\left[\frac{1}{2}-S(W)\right]^2+\left[\frac{1}{2}-C(W)\right]^2\right\}F^2 + \begin{cases} 1+[C(W)+S(W)-1]\,F, & \theta<\theta_{1/4} \\ 0, & \theta>\theta_{1/4} \end{cases} \tag{2.1.27}$$

results for the ratio of elastic and Rutherford cross sections. Here,

$$F\big(\Delta(\theta_{1/4}-\theta)\big) = \pi\Delta(\theta_{1/4}-\theta)/\sinh\left[\pi\Delta(\theta_{1/4}-\theta)\right] \tag{2.1.28}$$

is the Fourier transform of the derivative of the elastic reflection function $|S_l|$. The Fresnel integrals $S(W)$ and $C(W)$ are now to be taken for the arguments

$$W = \left[(l_{\max}+1/2)/(\pi\sin\theta_{1/4})\right]^{1/2}|\theta-\theta_{1/4}| \tag{2.1.29}$$

The quarter-point property discussed above is retained in the expression (2.1.27). The angular momentum $l_{\max}$ corresponding to $\theta_{1/4}$ is approximately given by relation (2.1.19). The parameter Δ describing the width of the "l window" contributing to elastic scattering is not specified by the model but may be extracted from a comparison of Equation (2.1.27) to experiments.

Because of the finite width Δ in l space over which the reflection function [cf. Equation (2.1.26)] changes from 0 to 1 both integral elastic and reaction cross sections depend not only on the angular momentum $l_{\max}$ but also on the ratio $\delta = \Delta/(l_{\max}+1/2)$, according to (Fra 78–78c)

$$\sigma = \pi\lambda\!\!\!^{-2}(l_{\max}+1/2)^2(1-2\delta+\delta^2\pi^2/3)$$

$$\sigma_R = \pi\lambda\!\!\!^{-2}(l_{\max}+1/2)^2(1+2\delta+\delta^2\pi^2/3) \tag{2.1.30}$$

The dependence of the predicted function $d\sigma/d\sigma_{\text{Ruth}}$ [cf. Equation (2.1.27)] on the width of the reflection function is illustrated in Figure 4, for the typical reaction ^{165}Ho + ^{56}Fe at $E_{lab} = 462$ MeV studied by Hoover *et al.* (Hoo 81). Here, the quarter-point angle has been fixed at its experimental value. In the sharp-cutoff limit ($\Delta = 0$), the "lit" region $\theta < \theta_{1/4}$ is characterized by a highly oscillatory angular distribution decreasing exponentially in the "shadow" region $\theta > \theta_{1/4}$. The period of the oscillations is given by $\Delta\theta = (4\pi \sin\theta_{1/4})/(l_{max} + 1/2)$. A reaction cross section of $\sigma_R = 2.7$ b results from Equation (2.1.21). As the diffuseness of the reflection function is increased, the oscillations damp out in magnitude and the angular distribution approaches a step function. In this example, the reaction cross section increases to $\sigma_R \approx 5$ b for $\Delta \approx 70$, according to Equation (2.1.30). It should be noted that if the transmission coefficients simply reflect geometrical properties of the interaction potential, there is a unique dependence of the reaction cross section on the bombarding energy (cf. Section 3.1).

While the models discussed above allow a determination of the total reaction cross section σ_R from the variation of the ratio $d\sigma_{exp}/d\sigma_{Ruth}$ of experimental elastic-scattering cross section to the theoretical Rutherford cross section, it is also possible (Woj 77) to deduce σ_R from their difference. If there exists a well-defined angle θ_g separating the domains of pure Rutherford scattering ($\theta < \theta_g$) and total absorption such as assumed in the sharp-cutoff model [cf. Equations (2.1.15)–(2.1.17)], then it follows from

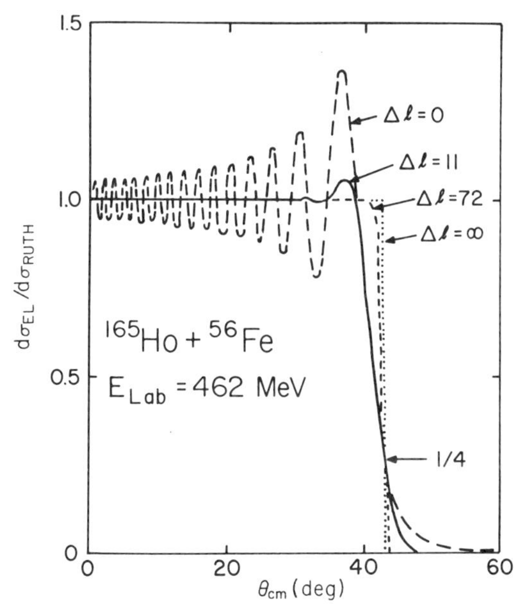

Figure 4. Ratio of the elastic to Rutherford cross section $d\sigma_{EL}/d\sigma_{Ruth}$ for the reaction ^{165}Ho + ^{56}Fe at $E_{Lab} = 462$ MeV as predicted by the Fresnel model, for various diffusenesses Δ ($= \Delta l$) of the reflection function [cf. Equation (2.1.26)]. From (Hoo 81).

Equation (2.1.17) that

$$\frac{d\sigma_R(\theta)}{d\Omega} = \frac{d\sigma_{\text{Ruth}}(\theta)}{d\Omega} - \frac{d\sigma_{\text{exp}}(\theta)}{d\Omega} - 2f^*_{\text{Coul}}(\theta)\cdot f_R(\theta, l_{\text{max}}) \tag{2.1.31}$$

Here, l_{max} is related to θ_g analogously to Equation (2.1.19). Obviously, the expression (2.1.31) may show oscillations. However, if integrated over scattering angle, their effect is small (Woj 77), such that

$$\sigma_R \approx 2\pi \int_{\theta_g}^{\pi} d\Omega \left[\frac{d\sigma_{\text{Ruth}}(\theta)}{d\Omega} - \frac{d\sigma_{\text{exp}}(\theta)}{d\Omega} \right] \tag{2.1.32}$$

gives a reasonable approximation of the total reaction cross section.

The object of the above discussion has been to review methods of practical significance to the determination of the reaction cross section from elastic-scattering data. Of course, a direct measurement of σ_R is also, in principle, possible. However, this would require the simultaneous measurement and identification of all reaction channels, including sequential reactions, an endeavor that may present a formidable task. In addition, in this section classical deflection functions have been considered which prove helpful in understanding gross features of the differential reaction cross section.

2.2. The General Form of the Damped-Reaction Cross Section

In the preceding subsection, it was pointed out how the reaction cross section is related to elastic scattering and the associated deflection function. From such considerations, it is possible to derive the magnitude of the reaction cross section and to understand some of the gross features of its angular distribution such as the angular range where quasielastic events are concentrated. However, because of the strong absorption of those waves with classical turning points inside the characteristic radius R_{SA} [cf. Equation (2.1.20)], such a comparison becomes less meaningful for more intimate reactions. This strong absorption arises from the fact that at incident energies of the colliding system, typically in the range from a few hundred MeV to more than 1 GeV, a tremendous number of reaction channels are open. There has, therefore, been the quest of microscopic reaction theory for a description of damped processes in terms of statistical methods. Several such models have been proposed (Die 76, Hof 76c, Die 77, Str 78d, Aga 78, Aga 79b, Die 80, Lec 81), where the general form of the damped-reaction cross section is derived under various fundamental assumptions. A somewhat detailed discussion of these models will be left to a later section (cf. Section 2.7). In the following, only an outline of the structure of the damped-reaction cross section and its division into an average and a fluctuating part will be given (Str 78d). This will be illustrated in terms of Strutinsky's phenomenological model (Str 64, Str 73, Str 78d, Har 78, Har 80).

If a damped collision is adequately represented by a planar two-body reaction with initial and final channel spins small compared to the angular momentum l, the differential reaction cross section can be written as (Str 78d)

$$\frac{d\sigma_R(\theta)}{d\Omega} = \sum_{\nu} \left| \sum_{l} f_\nu(\theta, l) \right|^2 = \sum_{\nu} \left| \frac{(\lambdabar \lambdabar_\nu)^{1/2}}{2i} \sum_{l} (2l+1) A_l^\nu P_l(\cos\theta) \right|^2 \tag{2.2.1}$$

Here, $A_l^\nu = \eta_l^\nu \exp(2i\delta_l^\nu)$ are the effective reaction amplitudes leading to final channels characterized by sets ν of quantum numbers and the corresponding final de Broglie wave length $\lambdabar_\nu$. If the density $\rho(\nu)$ of these channels is a sufficiently smooth function of ν, it is possible to replace the sum over reaction probabilities $|f_\nu|^2$ by an integral over double-differential average cross sections

$$\left\langle \frac{d^2\sigma_R(\theta)}{d\nu\, d\Omega} \right\rangle = \left\langle \left| \sum_{l} f_\nu \right|^2 \right\rangle \rho(\bar{\nu}) \tag{2.2.2}$$

with

$$\left\langle \left| \sum_{l} f_\nu \right|^2 \right\rangle = \frac{\lambdabar}{4} \sum_{l,l'} (2l+1)(2l'+1) P_l(\cos\theta) P_{l'}(\cos\theta) \langle \lambdabar_\nu A_l^\nu A_{l'}^{\nu *} \rangle$$

Here, the average indicated by the brackets in Equation (2.2.2) is taken over final states with quantum numbers centered around the mean values $\bar{\nu}$, which are identified as macroscopic variables, such as the mean excitation energy or the distance between the collision partners. Writing further the amplitudes as a sum of an average, $\overline{A_l^\nu}$, and a fluctuating part, δA_l^ν, one obtains (Str 78d) for the differential reaction cross section, averaged over macroscopic observables $\bar{\nu}$,

$$\left\langle \frac{d\sigma_R(\theta)}{d\Omega} \right\rangle = \frac{d\sigma_R^c(\theta)}{d\Omega} + \frac{d\sigma_R^{fl}(\theta)}{d\Omega} \tag{2.2.3}$$

This equation contains a sum of a coherent or collective term

$$\frac{d\sigma_R^c(\theta)}{d\Omega} = \left|\sum_l f_l^c(\theta, l)\right|^2 = \left|\frac{(\lambda \bar{\lambda}_\nu)^{1/2}}{2i} \sum_l (2l+1)\bar{\eta}_l \times \exp(2i\bar{\delta}_l) P_l(\cos\theta)\right|^2 \tag{2.2.4}$$

and a fluctuating cross section

$$\frac{d\sigma_R^{fl}(\theta)}{d\Omega} = \frac{\lambda \bar{\lambda}_\nu}{4} \sum_{l,l'} (2l+1)(2l'+1) B(l,l') P_l(\cos\theta) P_{l'}(\cos\theta) \tag{2.2.5}$$

In the above expressions, $\bar{\eta}_l$, $\bar{\delta}_l$ and $\bar{\lambda}_\nu$ are average values of reduced amplitude, phase, and final wavelength, respectively. The correlation matrix in Equation (2.2.4) is defined as $B(l, l') = \langle \delta A_l^\nu \, \delta A_{l'}^{\nu *} \rangle$. Formally, the same results can be derived (Str 80b), if channel spins are nonnegligible. The assumptions about the properties of the S matrix underlying the form of the reaction cross section represented by Equations (2.2.3)–(2.2.5) are not very restrictive and, hence, it is of rather general validity. The coherent part of the cross section contains the quantal effects of interference, diffraction, and absorption. It reflects the quantal fluctuations and dispersions of the macroscopic observables. Incoherent, or statistical fluctuations lead to the appearance of the term $d\sigma_R^{\mathrm{fl}}/d\Omega$ in the reaction cross section.

The relative importance of the two contributions to the reaction cross section can only be evaluated by employing a microscopic model for the reaction amplitudes. Such a model should take explicit account of the possibly strong correlations between channels of different angular momentum, unlike what is usually done in TDHF and classical calculations considering average trajectories for one l value at a time. Neglect of such correlations may result in an underestimation of the reaction cross section and its dispersion. Similar considerations apply, of course, if double-differential cross sections with respect to some macroscopic observables defined in Equations (2.2.2) are calculated.

In order to achieve a comparison of the theoretical reaction cross section of Equations (2.2.3)–(2.2.5) with experiments, at least a phenomenological model has to be adopted for calculating the reaction amplitudes. Similar to the case of elastic scattering, a range of classical and semiclassical approximations may be employed in the evaluation of the reaction cross section. Accepting the hypothesis (Str 64, Str 73) that the damped-reaction cross section is associated with a region of width Δl in angular-momentum space, one may approximate the reduced amplitude by a Gaussian: $\bar{\eta}_l = \eta_0 \exp[(l - l_0)^2/(\Delta l)^2]$ centered at a certain angular momentum l_0. If the width of this "l window" is sufficiently small, i.e., $\Delta l \ll l_0$, a linear approximation

$$\bar{\theta}(l) \approx \bar{\theta}(l_0) + \kappa(l_0)(l - l_0) \tag{2.2.6}$$

of the average deflection function may be valid. Assuming the classical relation $\bar{\theta}(l) = 2\, d\bar{\delta}_l/dl$ to hold for average deflection angle $\bar{\theta}$ and phase shift δ_l, it follows that

$$\bar{\delta}_l = \frac{1}{2}\left[\theta_0(l - l_0) + \frac{\kappa_0}{2}(l - l_0)^2\right] \tag{2.2.7}$$

where $\theta_0 = \bar{\theta}(l_0)$ and $\kappa_0 = \kappa(l_0)$. Evaluation of the coherent part of the differential reaction cross section in the approximation that a summation over l can be replaced by an integration and that the asymptotic form of the Legendre polynomials can be used, yields

$$\frac{d\sigma_R^c(\theta)}{d\Omega} = \lambda\bar{\lambda}_\nu \frac{4 l_0 \eta_0^2}{\left(\xi_q^4 + \xi_q^2 \xi_d^2\right)^{1/2}} \frac{\mathscr{S}(\theta_0, \theta)}{\sin\theta} \tag{2.2.8}$$

The angular dispersion function $\mathscr{S}(\theta_0, \theta)$ is given by

$$\mathscr{S}(\theta_0, \theta) = \exp\left[-\left(\theta_0^2 + \theta^2\right)/\xi^2\right]\left[\cosh\left(\frac{2\theta_0\theta}{\xi^2}\right) + \sin\left(\frac{2l_0 + \theta_0 \xi_d}{\xi}\theta\right)\right] \tag{2.2.9}$$

where the quantity $\xi = (\xi_q^2 + \xi_d^2)^{1/2}$ is the angular dispersion determined by the "quantal dispersion" $\xi_q = \sqrt{2}/\Delta l$ and the "dynamical dispersion" $\xi_d = \kappa_0 \Delta l/\sqrt{2}$. The latter term depends strongly on the form and strength of the nuclear interaction potential determining the slope $\kappa(l_0)$ of the deflection function. In the limit of a strongly varying deflection function, where $\xi_d \gg \xi_q$, the classical form [cf. Equation (2.1.4)] of the cross section for potential scattering is regained.

The expression (2.2.8) for the coherent cross section describes a family of angular distributions dependent mainly on the value of the total dispersion ξ. This is illustrated in Figure 5, where the average angle factor $\bar{\mathscr{S}}(\theta_0, \theta)$ is plotted for various ξ values vs. the ratio θ/θ_0. Here, $\bar{\mathscr{S}}$ differs from $\mathscr{S}$ in that the rapidly oscillating sine term in Equation (2.2.9) has been

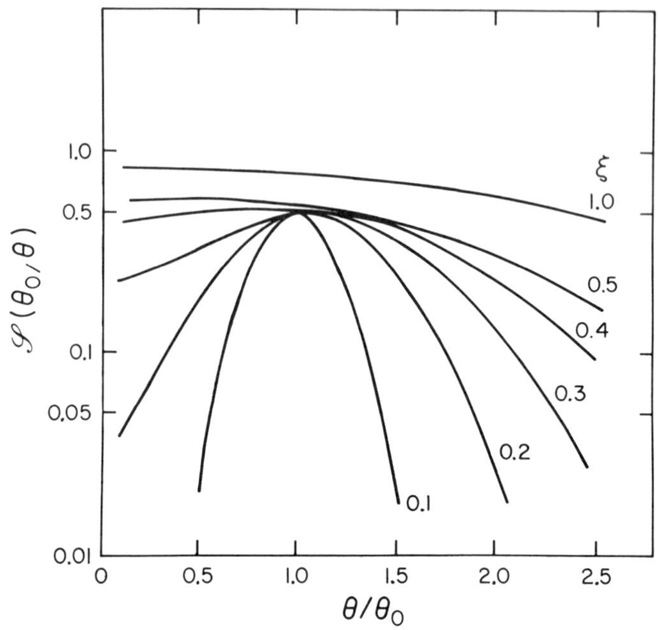

Figure 5. Average shape factor $\mathscr{S}$ of the angular distribution of the coherent reaction cross section, as predicted by the Strutinsky model for various values of the total dispersion ξ, including quantal and dynamical contributions.

dropped. For small dispersions, the angular dependence is described by a Gaussian centered at the mean deflection angle θ_0 corresponding to the centroid, l_0, of the contributing l region. For values $\xi > 0.4$–0.5, the maximum of $\mathscr{S}$ disappears, and a forward-tending, flat distribution results. The characteristic angular width of these distributions is determined by the quantity $\xi/\sqrt{2}$, i.e., in the limit $\xi_q \ll \xi_d$ by the product of the slope κ_0 of the deflection function and the width Δl of the angular-momentum region contributing. In general, however, it is not feasible to interpret an experimental angular distribution in terms of quantal and dynamical dispersions separately, on the basis of this phenomenological model.

The problem of how to interpret angular distributions of the reaction cross section is further aggravated by the existence of the statistically fluctuating contribution $d\sigma_R^{\text{fl}}/d\Omega$ [cf. Equation (2.2.3)], which can be expected to be significant in damped reactions with many open channels. Parametrizing the correlation matrix $B(l, l')$ in terms of a Gaussian form factor such as done above for the reduced amplitude, one obtains (Str 78d) an angular dispersion function of the same structure as that of the function $\mathscr{S}$ [cf. Equation (2.2.9)] corresponding to the coherent part of the cross section. Its dispersion is now determined by the characteristic correlation length associated with B, a typical maximum distance between different values l and l' such that $B(l, l') \neq 0$.

In the above discussion, Strutinsky's phenomenological treatment of the reaction amplitude pertaining to quasielastic and damped reactions has been presented in order to give an example of the nature of processes that can be expected to contribute in these reactions. Coherent and stochastic mechanisms may attain a different relative significance depending on the reaction conditions such as the mass of the interacting system, bombarding energy, or time elapsed in a reaction. Such considerations are not restricted to the distribution of reaction products over reaction angle but are applicable to other macroscopic observables such as the final fragment mass as well.

In most phenomenological approaches, however, some of the above, potentially important aspects of the reaction mechanism are neglected. For example, the coherent part of the cross section is often assumed to be adequately described by its classical limit of potential scattering with the aim of reconstructing the average forces acting between the collision partners. Other models include stochastic or quantal features associated with the transfer of mass. In judging the validity of such models from a comparison with experimental data, one has to be aware of the possibility that several combinations of basic assumptions about the reaction mechanism can all lead to a description of the data of comparable quality. This observation emphasizes the importance of a diversification of reaction models to include as many different aspects as possible.

2.3. Dynamical Models of Collective Motion

As has been discussed in Section 2.1, due to a relatively short de Broglie wave length, the relative motion of two heavy nuclei colliding with incident energies of a few MeV/u is expected to be close to its classical limit, at least for a considerable part of the collision. Furthermore, the experimental observation (cf. Section 3) of mainly two massive fragments emerging from the reaction, on the average relatively intact and close to the initial mass asymmetry, supports the concept of damped reactions as a peripheral contact of two macroscopic, unstructured bodies. It appears then natural to anticipate that the relative motion can, on the average, be well approximated in terms of classical trajectories determined by conservative and dissipative forces acting between the collision partners. This was realized early by Gross *et al.* (Bec 73, Gro 74), who pioneered the application of classical dynamical models to damped reactions. It is, of course, not obvious if the effect of the multiple-scattering processes occurring on the microscopic level when the matter distributions of two complex nuclei in relative motion merge can, in fact, be translated into phenomenological forces providing an acceleration along certain macroscopic coordinates. It is the object of classical dynamical reaction studies to discover whether or not the damped-reaction features can be explained in terms of phenomenological forces or potentials having a universal form described by a few parameters with a smooth dependence on reaction energetics and geometry. Obviously, the nature of such forces, if they exist, will depend on the collective degrees of freedom considered explicitly in the analysis. For example, the interpretation of the scattering of two statically deformed nuclei in terms of a central conservative potential can lead to a potential form factor that is different from the one obtained when the shapes and alignments of the nuclei are explicitly accounted for in the model calculation. Similar arguments can be made with respect to the dissipative model forces.

In order to obtain average, macroscopic deflection functions and cross sections from a model based on the choice of a set $\{q_i\}$ of n collective coordinates, the equations of motion for these coordinates are integrated in general numerically. They are determined by the Lagrange–Rayleigh equations:

$$\left(\frac{d}{dt}\frac{\partial}{\partial \dot{q}_i} - \frac{\partial}{\partial q_i}\right)\mathscr{L} = -\frac{\partial}{\partial \dot{q}_i}\mathscr{F}, \qquad i=1,\ldots,n \tag{2.3.1}$$

Here, $\mathscr{L} = \mathscr{T} - V$ is the Lagrangian, and $\mathscr{F}$ is the Rayleigh dissipation function related to the rate of energy dissipation, dE/dt, where $E = \mathscr{T} + V$. The potential energy $V(q_1,\ldots,q_n)$ is a function of the collective coordinates q_i, whereas the kinetic energy is dependent on the associated velocities $\dot{q}_i$ and the inertia tensor (m_{ij}) according to

$$\mathscr{T} = \frac{1}{2}\sum_{i,j=1}^{n} m_{ij}\dot{q}_i\dot{q}_j \tag{2.3.2}$$

If dissipation of energy is mediated by velocity-proportional friction forces, $F_i = -\partial\mathscr{F}/\partial\dot{q}_i$, the dissipation function is given by

$$\mathscr{F} = \frac{1}{2}\sum_{i,j=1}^{n} c_{ij}\dot{q}_i\dot{q}_j \tag{2.3.3}$$

where (c_{ij}) is the friction tensor. Inertia and friction tensors are, in general, functions of the coordinates q_i. For symmetric tensors, one derives

$$\mathscr{F} = -\frac{1}{2}\frac{dE}{dt} \tag{2.3.4}$$

In the limit of very strong damping of the motion with respect to a certain degree of freedom q_0, the kinetic-energy terms in Equation (2.3.1) are negligible and the corresponding equations of ("overdamped") motion can be derived from the balance $\partial\mathscr{L}/\partial q_0 = \partial\mathscr{F}/\partial\dot{q}_0$ between dissipative and conservative forces.

Classical dynamical calculations have been performed for damped reactions and nuclear fusion by several groups (Gro 74, Gro 74b, Tsa 74, Bon 74b, Spe 74, De 76, Deu 75, Deu 77b, Gro 78, Wil 76, Bir 78b, 79, Sch 78g, 79f, 80, Bec 78b, Blo 80b, Ban 80b, Swi 81) employing various models for the configuration of the interacting system. Among the simplest reaction geometries is the one (Bir 78b) illustrated in Figure 6 pertaining to the scattering of two rigid spherical nuclei induced by the central conservative Coulomb and nuclear interaction potentials V_{Coul} and V_N, respectively. Here, four coordinates are used to describe a collision: the distance r between the mass centers of projectile and target, their relative orientation angles θ_P and θ_T, respectively, and the orientation θ of the total system. The

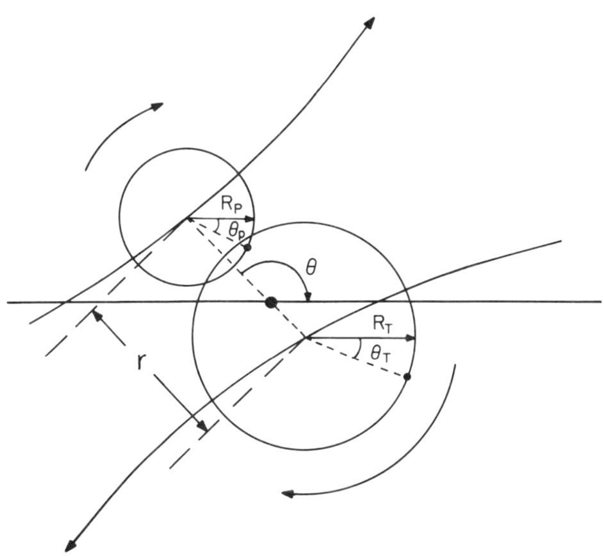

Figure 6. Reaction variables for the scattering of spherical projectile (P) and target (T) nuclei.

Lagrangian can be expressed as

$$\mathscr{L} = \tfrac{1}{2}\mu\left[\dot{r}^2 + (r\dot{\theta})^2\right] + \tfrac{1}{2}\mathscr{I}_P\dot{\theta}_P^2 + \tfrac{1}{2}\mathscr{I}_T\dot{\theta}_T^2 - V_{\text{Coul}} - V_N \tag{2.3.5}$$

In Equation (2.3.5)

$$\mu = \frac{M_P M_T}{M_P + M_T}$$

$$\mathscr{I}_i = \frac{2}{5} M_i R_i^2, \qquad i = P, T \tag{2.3.6}$$

denote the reduced mass μ of the system and the moments of inertia, $\mathscr{I}_i$, of the projectile and target with masses M_P and M_T, respectively. The dissipation function may be approximated in terms of a friction coefficient $c_r(r)$ associated with the radial motion, and the coefficient $c_t(r)$ describing the slowing-down of the relative sliding motion of the two nuclear surfaces. Then

$$\mathscr{F} = \frac{1}{2}\left\{c_r\dot{r}^2 + c_t\left(\frac{r}{R_P + R_T}\right)^2\left[R_P(\dot{\theta}_P - \dot{\theta}) + R_T(\dot{\theta}_T - \dot{\theta})\right]^2\right\} \tag{2.3.7}$$

From the Lagrange–Rayleigh Equations (2.3.1), one obtains the equations of motion

$$\mu\ddot{r} - \mu r\dot{\theta}^2 + c_r\dot{r} + \frac{\partial V_{\text{Coul}}}{\partial r} + \frac{\partial V_N}{\partial r} = 0$$

$$\mu r^2\ddot{\theta} + \dot{I}_P + \dot{I}_T = 0$$

$$\dot{I}_i = \frac{d}{dt}\mathscr{I}_i\dot{\theta}_i = -c_t\left[\frac{r}{R_P + R_T}\right]^2 R_i(\dot{\theta}_i - \dot{\theta}), \qquad i = P, T \tag{2.3.8}$$

The second of these equations describes the total angular momentum balance holding because of the absence of an external torque. I_P and I_T are the intrinsic spins of projectile and target, respectively, and $l = \mu r^2\dot{\theta}$ is the orbital angular momentum.

In the model described above, dissipation of angular momentum ceases when $R_P(\dot{\theta}_P - \dot{\theta}) = R_T(\dot{\theta} - \dot{\theta}_T)$, i.e., when the two nuclei roll on each other's surfaces. From this condition on the angular velocity and the balance of the tangential forces acting on the surfaces of the two nuclear spheres, it follows that the final angular momentum in the rolling limit is reduced to

$$l_r = (5/7)l \tag{2.3.9}$$

from an initial value l, regardless of the masses and radii of the two nuclei. The missing angular momentum $l - l_r$ has been transformed into intrinsic spin I_i of the nuclei:

$$I_i = \frac{2}{7} \frac{R_i}{R_P + R_T} l, \qquad i = P, T \tag{2.3.10}$$

If, in addition to the tangential sliding friction associated with the friction coefficient c_t [cf. Equation (2.3.7)], there is also rolling friction (Tsa 74, Wil 76, Ban 80), angular momentum can be dissipated further, until $\dot{\theta}_P = \dot{\theta}_T = \dot{\theta}$. This "sticking situation" is characterized by a rigid rotation of a dinuclear system around its center of gravity with final orbital angular momentum l_s and intrinsic spins I_i:

$$l_s = \frac{\mathscr{I}}{\mathscr{I} + \mathscr{I}_P + \mathscr{I}_T} l$$

$$I_i = \frac{\mathscr{I}_i}{\mathscr{I} + \mathscr{I}_P + \mathscr{I}_T} l, \qquad i = P, T \tag{2.3.11}$$

where $\mathscr{I} = \mu r^2$ is the moment of inertia for the orbital motion of two nuclei with reduced mass μ, and $\mathscr{I}_P$ and $\mathscr{I}_T$ are the moments of inertia of

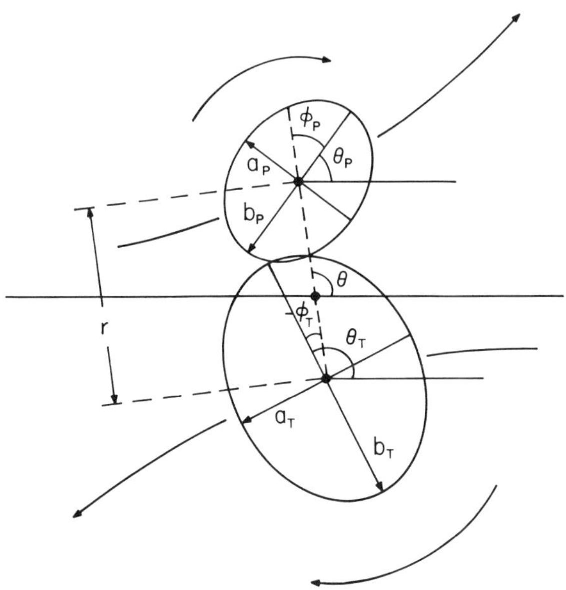

Figure 7. Reaction variables for the scattering of axially symmetric, spheroidal projectile (P) and target (T) nuclei.

projectile and target nuclei, respectively, as defined in Equation (2.3.6). For a symmetric projectile–target system, rolling and sticking limits are realized simultaneously ($l_r = l_s$).

The experimental observation of final fragment kinetic energies much lower than estimated from the Coulomb repulsion of two spherical nuclei clearly indicates the presence of large nuclear deformations in the exit channel, at least under certain reaction conditions. Hence, it appears necessary to include deformation degrees of freedom in classical dynamical calculations. Such models have, indeed, been applied (Deu 76, Deu 77, Swi 79, Swi 80, Sch 80, Sch 81b, Gro 80) in a number of cases. These calculations differ in the choice of coordinates associated with deformations, as exemplified below.

In the model of Deubler and Dietrich (Deu 76, Deu 77), the shapes of the interacting nuclei are approximated by axially symmetric spheroids. The coordinates of the scattering problem are defined in Figure 7. The symmetry axes of projectile and target nuclei are of the length $2b_P$ and $2b_T$, respectively. The perpendicular axes have the lengths $2a_P$ and $2a_T$. The amount of spheroidal deformation is then indicated by the coordinates $q_i = a_i/b_i$. With the set of variables shown in Figure 7, kinetic and potential energies can be written in the form (Deu 76, Deu 77)

$$\mathscr{T} = \tfrac{1}{2}\mu\left[\dot{r}^2 + (r\dot{\theta})^2\right] + \tfrac{1}{2}\mathscr{I}_P\dot{\theta}_P^2 + \tfrac{1}{2}\mathscr{I}_T\dot{\theta}_T^2 + \tfrac{1}{2}B_P\dot{q}_P^2 + \tfrac{1}{2}B_T\dot{q}_T^2$$

$$V = V_{\text{Coul}}(r,\theta_P,\theta_T,q_P,q_T) + V_N(r,\theta_P,\theta_T,q_P,q_T) + V_P(q_P) + V_T(q_T) \tag{2.3.12}$$

Here, B_P and B_T are the inertia parameters associated with projectile and target vibrations, V_P and V_T are the corresponding potential energies. The moments of intertia can be approximated by those of rigid spheroids:

$$\mathscr{I}_i(q_i) = \tfrac{1}{5}M_iR_i^2q_i^{2/3}(1+q_i^{-2}), \qquad i = P, T \tag{2.3.13}$$

Obviously, the inertia parameters for nuclear vibrations have to be calculated in a microscopic model. In a hydrodynamical calculation assuming irrotational flow, Deubler *et al.* (Deu 76, Deu 77) obtained

$$B_i(q_i) = \tfrac{1}{3}B_i(1)q_i^{-4/3}(1+2q_i^{-2}) \tag{2.3.14}$$

The quantities $B_i(1)$ and $V_i(q_i)$ can, in principle, be deduced from experimental information about energies and lifetimes of vibrational nuclear states, provided they do not change significantly with the large amounts of excitation energy acquired in damped reactions.

One then arrives at the following set of equations of motion (Deu 77):

$$\mu\ddot{r} - \mu r\dot{\theta}^2 + c_r\dot{r} + \frac{\partial V_{\text{Coul}}}{\partial r} + \frac{\partial V_N}{\partial r} = 0$$

$$\mu r^2\ddot{\theta} + 2\mu r\dot{r}\dot{\theta} + (\mathscr{I}_P + \mathscr{I}_T)\ddot{\theta} + \sum_{i=P,T}\left[\mathscr{I}_i\ddot{\theta}_i + \frac{\partial}{\partial q_i}\mathscr{I}_i q_i(\dot{\theta} + \dot{\phi})\right] = 0$$

$$\mathscr{I}_i(\ddot{\theta} + \ddot{\phi}_i) + \frac{\partial}{\partial q_i}\mathscr{I}_i\dot{q}_i(\dot{\theta} + \dot{\phi}_i) + c_t\dot{\phi}_i + \frac{\partial V_{\text{Coul}}}{\partial \phi_i} + \frac{\partial V_N}{\partial \phi_i} = 0$$

$$B_i\ddot{q}_i + \frac{1}{2}\frac{\partial}{\partial q_i}B_i\dot{q}_i^2 - \frac{1}{2}\frac{\partial}{\partial q_i}\mathscr{I}_i(\dot{\theta} + \dot{\phi}_i)^2 + c_{q_i}\dot{q}_i + \frac{\partial V_i}{\partial q_i} + \frac{\partial V_{\text{Coul}}}{\partial q_i} + \frac{\partial V_N}{\partial q_i} = 0$$

$$i = P, T \qquad (2.3.15)$$

Here, c_{q_i} are the additional elements of the friction tensor associated with the nuclear vibrations.

It should be noted that Equations (2.3.8) and (2.3.15) are only approximately consistent with the model assumptions. As will be discussed further below, the intrinsic motion reacts to the changing collective coordinates, with the appearance of dissipative forces being only one of the consequences of this interplay. In addition, neglect of considering intrinsic degrees of freedom explicitly is equivalent (see Section 2.6) to a generation of induced nondissipative forces. Their presence can approximately be accounted for by introducing effective mass parameters (Deu 77) or potentials (Zin 75, Bon 74b). Presumably, however, the effects of such induced forces on the relative motion are comparatively small.

There are, of course, many ways in which the nuclear surface can be parametrized, all corresponding to different sets of equations of motion, inertia, and friction tensors. For not too extreme distortions, the nuclear surface can be expressed by an expansion of the nuclear radius in terms of spherical harmonics:

$$R(\hat{r}) = R_0\left[1 + \sum_{\lambda,\mu}\alpha_{\lambda\mu}Y_\mu^{\lambda *}(\hat{r})\right] \qquad (2.3.16)$$

where $\hat{r}$ is the unit radius vector. The amplitudes of $\alpha_{\lambda\mu}$ of multipolarity λ describe the time-dependent displacement of the nuclear surface from its equilibrium shape. Since nuclei interact with their surfaces, it is, in principle, conceivable that surface oscillations and vibrations can be excited in damped nuclear collisions, as has been advocated by Broglia and collaborators (Bro 74, Bro 76, Bro 76c, Bro 78c, Bro 78e). Figure 8 provides a qualitative impression of the variation of the nuclear shapes that may occur in a nuclear collision.

In the case of a harmonic potential with stiffness C_λ associated with surface modes of multipolarity λ, the time development of such modes is classically determined by an equation

$$B_\lambda \ddot{\alpha}_{\lambda\mu} + c_\lambda \dot{\alpha}_{\lambda\mu} + C_\lambda \alpha_{\lambda\mu} + F_{\lambda\mu}(\mathbf{r}(t)) = 0 \tag{2.3.17}$$

describing a damped harmonic oscillator excited by the time-dependent external multipole force $F_{\lambda\mu}$. Equations such as Equation (2.3.17) for the important multipole modes for each of the interacting nuclei, as well as for the whole system, have to be added to the remaining equations of motion involving the distance r between the two centers of gravity and the orientation angles of the system. As in the case of dynamical deformations discussed earlier, models are necessary to estimate the appropriate inertia, stiffness, friction, and force parameters. Some rough evaluation of these quantities is possible employing experimental values of excitation energies, cross sections, and widths of such vibrational nuclear modes if excited in other, more elementary and well understood reactions. It is then still necessary to extrapolate the parameters to typical conditions prevalent in heavy-ion reactions. Whether such oscillations can, in fact, develop during damped reactions, depends strongly on the external driving forces $F_{\lambda\mu}$ and the strength of the damping parameters C_λ is relation to the frequency of these modes. Strong damping will prevent development of oscillations and lead to slowly varying dynamical deformations instead, as described, for example, by the equations of motion (2.3.15).

In the dynamical models discussed so far, it was always assumed that the surfaces of the two interacting nuclei are well defined and remain the intact enclosures of the nucleons within each nucleus throughout a reaction.

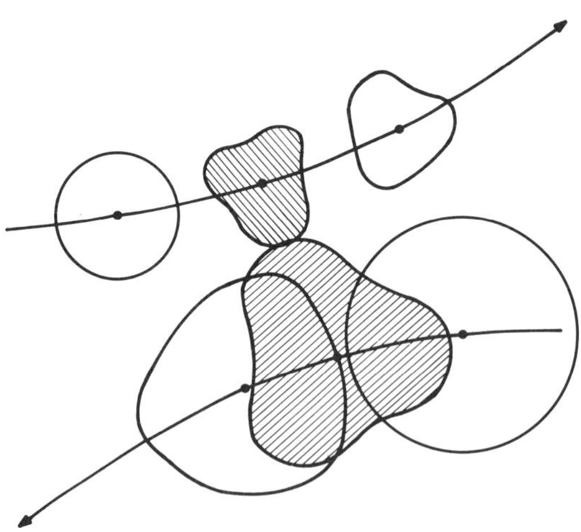

Figure 8. Illustration of the development of octupole nuclear deformations induced by the surface interaction of the reaction partners.

While this may be true for near-grazing collisions, a realistic treatment of processes with smaller impact parameters requires the consideration of a further coordinate: the degree of freedom for neck formation leading to communication between the nuclear interiors. The origin of this effect is illustrated in Figure 9. Here, the sum of the two single-particle potentials is plotted in the form of a two-dimensional contour diagram for ^{209}Bi and ^{136}Xe nuclei facing each other at a separation distance of 15.2 fm. This distance corresponds to the experimental (Sch 78) interaction or strong-absorption radius (cf. Section 2.1), where the two matter distributions only overlap with their tails at the 6% level. The potentials are assumed to be of Woods–Saxon shape:

$$V(r) = V_0\left[1 + \exp\left(\frac{r-R}{a}\right)\right]^{-1} \tag{2.3.18}$$

The parameters chosen are $V_0 = -45.3$ MeV and $a = 0.65$ fm (Boh 69), and the half-potential radius is calculated from $R = 1.25A^{1/3}$ fm. Although at the distance of 15.2 fm the nuclear surfaces are separated by approximately 3 fm, the combined single-particle potential exhibits a saddle in the region between the two nuclei with a height of $\sim$ 22 MeV above the bottom of the potential. Since the nucleon Fermi kinetic energy is equal to about $T_F \approx 37$ MeV (Boh 69), nucleons at the Fermi level are energetically permitted to pass over this saddle. This situation of a "potential breakthrough" (Joh 77) will eventually lead to an accumulation of matter between the nuclei in the region of the potential depression. For separations smaller than that depicted in Figure 9, the saddle disappears, and a dinuclear configuration with contiguous interior and surface is favored. The region of interface between the nuclei is commonly called "neck" and its cross section perpendicular to the plane of Figure 9 is termed "window." It should be noted that the

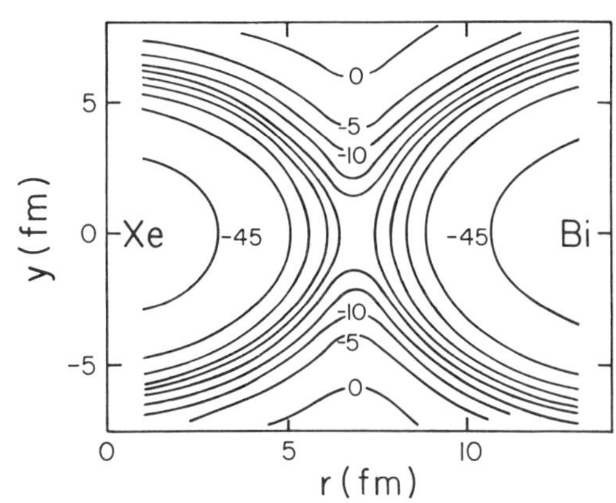

Figure 9. Equipotential contour lines of the sum of the nuclear single-particle potentials of a ^{209}Bi and a ^{136}Xe nucleus at a center separation distance corresponding to the experimental strong-absorption radius $R_{SA} = 15.2$ fm.

situation depicted in Figure 9 is only qualitatively correct. In reality, the interaction of the nuclear surfaces is expected to modify the total potential considerably (cf. Section 2.5).

In order to accommodate a neck-degree of freedom in dynamical calculations, a variety of phenomenological configurations of the interacting system has been proposed (Bon 74b, Swi 79, Swi 80, Blo 80, Spe 79). Figure 10 gives a schematic illustration of a sample of such shapes. A rather simple way (cf. Figure 10a) to incorporate a neck has been described by Swiatecki (Swi 79) approximating it by a cylinder of radius ρ connecting the spherical target and projectile nuclei. The neck radius ρ and the distance r between the centers of the two fragments or the distance s between the undisturbed surfaces, along with the time-dependent radii R_T and R_P of target and projectile, describe the family of shapes completely.

Dynamical calculations employing this family of dinuclear configurations have been performed in several cases (Sch 80, Sch 81b) approximating the neck development in terms of an overdamped motion. In a more recent model (Swi 80), the neck is approximated by the form of a cone with semiopening θ (cf. Fig. 10b). These shapes are described by a set of three variables: a time-dependent asymmetry variable $\Delta = (R_T - R_P)/(R_T + R_P)$, a separation coordinate $\rho = r/(R_T + R_P)$, and a window-opening parameter $\alpha = (\sin\theta/\sin\theta_{\max})^2$, where $\theta_{\max}$ is the maximum opening angle possible at a given distance ρ. Defining new variables, a reduced neck radius $\nu = \alpha^{1/2}$ and surface separation variable $\sigma = (\rho^2 - 1)/(1 - \Delta^2)$, and expanding the potential energy in terms of these coordinates, the equations of motion for central ($l = 0$) collisions take on the simple form (Swi 80)

$$\bar{\mu}\frac{d^2\sigma}{dt^2} + \nu^2\frac{d\sigma}{dt} + \nu - X = 0$$

$$\frac{d\nu}{dt} = \frac{2\nu - 3\nu^2 - \sigma}{\nu(\sigma + \nu^2)} \tag{2.3.19}$$

Here, $\bar{\mu}$ and t are the reduced mass of the system and the time variable, respectively, expressed in dimensionless units, and X is a generalized fissility parameter. In the derivation of Equations (2.3.19), the inertia of the neck has been identified with the reduced mass of the dinuclear system. For the dissipation function $\mathscr{F}$, the "wall-plus-window formula" (cf. Section 2.6) has been adopted describing the damping of the relative motion and that of the growth of the neck. Although applicability of this simple formalism is restricted to central collisions, it is useful in obtaining a qualitative insight into the variation of the reaction dynamics with changing initial mass asymmetry or incident energy.

The configuration exhibited in Figure 10c represents another example of how the neck degree of freedom can be incorporated in a simple

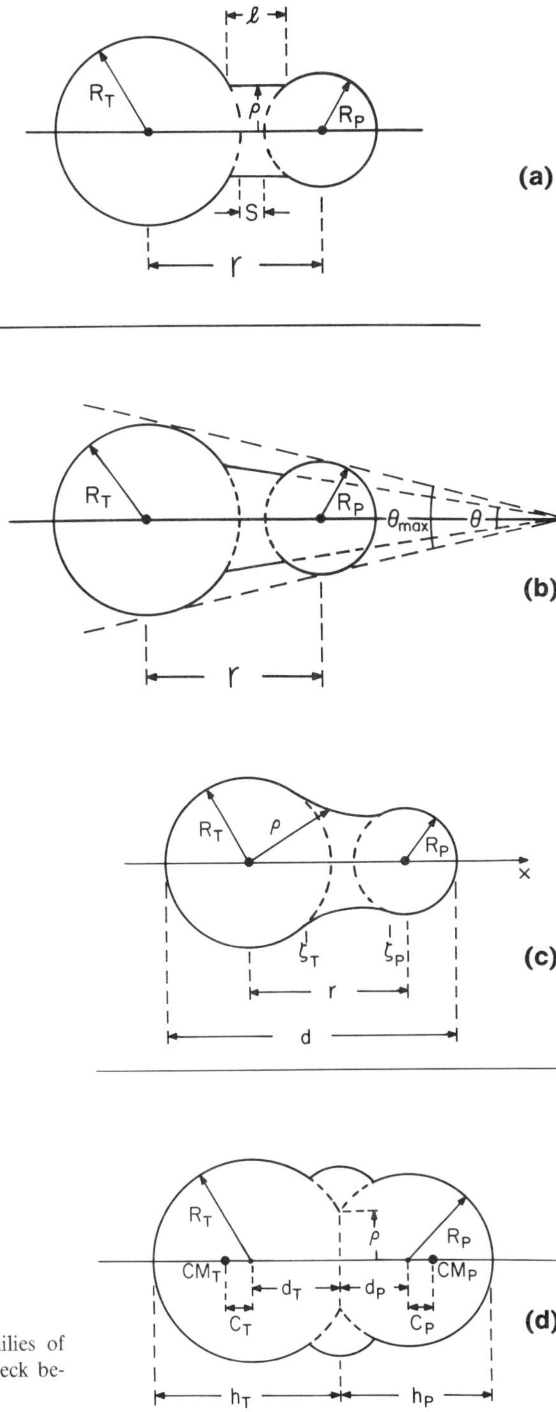

Figure 10. (a–d) Various families of dinuclear shapes featuring a neck between spherical fragments.

phenomenological way (Blo 80, Bec 80). This shape is given by two spheres connected by a parabolical neck attaching smoothly to the spheres at $x = \zeta_T$ and $x = \zeta_p$. As a neck variable, one may adopt the ratio of the neck volume Δv as defined by the neck envelope and the opposing boundaries of the undisturbed nuclei and their volume v_0, $\sigma = (\Delta v / v_0)^{1/2}$. By specifying the radial distance r or the elongation d, as well as the excess volume Δv and a time-dependent mass asymmetry parameter such as $\alpha = (R_T / R_P)^3$ the shapes are completely determined.

While dinuclear configurations displayed in Figures 10a–10c have been used in dynamical calculations implementing a neck degree of freedom as an independent dynamical variable, a class of different shapes has been introduced (Bon 74b, Nix 74, Fab 79b, 80, 81) to obtain theoretical estimates of the inertial and potential parameters of significance to an interpretation of damped reactions. One of these shapes (Bon 74b) is shown in Figure 10d. Here, it is assumed that the excess mass dislocated from the region of overlap of the two initially spherical nuclei forms an extra ring filling in the crevice between their surfaces.

An alternative procedure (Zoh 78) to define a neck coordinate starts from a consideration of the combined single-particle potential for the system of two interacting nuclei such as shown in Figure 9. In an adiabatic picture of a nuclear collision, use is made of the assumption that the nuclear density remains constant and follows the form of the combined potential such that the dinuclear surface is determined by an equipotential contour line. The latter can be derived from a comparison of the total volume with that of the two separated nuclei. An evolution of the neck degree of freedom can then be described in terms of a deviation of the potential saddle between the two reaction partners (cf. Figure 9) from the form corresponding to a simple superposition. In a first approximation, simply the relative height of the saddle point may be regarded as a neck variable. Such a procedure has been adopted (Zoh 78) in an asymmetric two-center shell model (Mar 72) employing two-dimensional harmonic-oscillator potentials allowing for ellipsoidal deformations of the reaction partners. However, up to now, full dynamical calculations have not been performed in such an approach.

The selections of collective coordinates discussed above have been used to give model representations of a variety of nuclear reactions. They range from peripheral collisions of nearly spherical nuclei to deeply interpenetrating reactions in which first a neck is assumed to be a dominant feature of the configuration of an intermediate dinuclear system, until the contact becomes so intimate that the individual identities of projectile-like and target-like constituents are lost. This is the "mononucleus" domain (Swi 80) branching out into the compound-nucleus regime where the collective dynamics may be adequately described by the rotating liquid-drop model (Coh 74). If, however, the repulsive forces such as centrifugal and Coulomb forces are strong enough to prevent a permanent coalescence, the mononuclear system will eventually reseparate into two or more fragments. Such transitions are normally not addressed in dynamical reaction studies relying on the distinguishability of the system constituents during the entire reaction. Some progress has been made (Swi 80), however, to establish an approximate connection between dinuclear and mononuclear domains.

Other simplifying assumptions made usually in dynamical models concern the neglect of the finite diffuseness of the nuclear surface and the possible changes of inertia, potential, as well as of friction parameters with the amount of excitation energy acquired by the system. For example, calculations are often performed under the assumptions that moments of inertia are those of rigid bodies, as specified by the adopted configuration, and that friction forces are associated with form factors that are independent of excitation energy. It appears reasonable to assume that the finiteness of the nuclear surface can be adequately accounted for by some effective form factors of the forces, because the surface layer can be regarded as thin compared to the nuclear dimensions and varies insignificantly with mass number. It is also important to realize (Mye 76) that multipole moments of different nuclear matter distributions that are related to each other by folding with a normalized convolution function are identical. The corresponding moments of inertia differ only by the trivial term $2Mb_c^2$, where M is the total mass and b_c is the width of the convolution function. Appropriateness of neglecting other effects mentioned above, however, is difficult to estimate and requires justification by microscopic theory.

2.4. Collective Modes of Mass Transfer in Damped Reactions

From the previous discussion of dynamical reaction models, it is obvious that the outcome of a nuclear collision depends significantly on the mass asymmetry of the initial projectile–target system. Consequently, one can also expect the collision dynamics to be strongly coupled to mass transfer processes occurring in the reaction. It is not self-evident, however, whether or not the mass asymmetry constitutes a suitable collective coordinate. This depends on the details of the underlying microscopic mass transfer mechanisms. Analogously to the discussion on the general form of the damped-reaction cross section presented in Section 2.2, one expects that the cross section found for the various final mass and charge asymmetries can be interpreted in terms of coherent and fluctuating or stochastic properties of the microscopic reaction amplitudes. The task is then to discover whether coherent or stochastic processes determine the dominant features of the observed fragment mass and charge distributions. In the following, some model views of a collective mass or charge transfer in damped reactions are discussed.

The possibility that mass and charge transfer in damped reactions may occur as a collective process has been considered relatively early by Greiner and collaborators (Fin 74, Fin 74b, Zoh 75, San 76, Yam 76,76b,77, Gup 77,77b,c,78,79,80, Mar 80). In their fragmentation theory, the evolution of the system along the asymmetry coordinates $\eta_A = (A_1 - A_2)/(A_1 + A_2)$ and $\eta_Z = (Z_1 - Z_2)/(Z_1 + Z_2)$ is treated quantum-mechanically, where A_ν and Z_ν are the time-dependent mass and atomic numbers of the two constituents of the interacting system. These are defined via the volumes associated with the two parts of an asymmetric two-center potential as discussed in Section 2.3 and illustrated in Figure 9. Although the formal framework of this quantal theory is rather general and includes shape degrees of freedom, besides radial-separation and asymmetry coordinates, phenomenological approximations are usually made in actual applications. For example, in the work of Yamaji *et al.* (Yam 76), only mass asymmetry ($\eta = \eta_A$) and radial-separation (r) coordinates are considered explicitly. The total energy of the system is approximated by

$$H(r, \eta, \dot{r}, \dot{\eta}) = \tfrac{1}{2}m_r\dot{r}^2 + \tfrac{1}{2}m_\eta\dot{\eta}^2 + V(R, \eta, \tau) + \tilde{E}(\tau) \qquad (2.4.1)$$

where the inertia parameters m_r and m_η depend on r and η. A coupling term $\sim m_{r\eta}\dot{r}\dot{\eta}$ has been neglected in Equation (2.4.1) on grounds of the smallness of $m_{r\eta}$ as compared to the other masses. These inertia parameters are calculated (Zoh 75) microscopically in the two-center shell model (Mar 72) employing the cranking formula (Ing 56) which is applicable for adiabatic, i.e., relatively slow motion. Potential energies V are deduced as functions of mass asymmetry and radial separation from the liquid-drop model with shell corrections (cf. Section 2.5), which are assumed to disappear exponentially with increasing nuclear temperature τ. It is further assumed that the system follows the minimum of the potential energy with respect to neck parameter and fragment deformations incorporated in a two-center shell model as discussed in Section 2.3. The quantity $\tilde{E}(\tau)$ in Equation (2.4.1) denotes the energy stored in those collective modes and in intrinsic degrees of freedom not explicitly treated. The latter are specified by a nuclear temperature τ.

In the above model, classical equations are assumed for the relative motion (cf. Section 2.3). A time-dependent Schrödinger equation for a wave packet with initial spread σ_0^2 in mass asymmetry is subsequently integrated along such classical trajectories, for a number of impact parameters. As emphasized before (cf. Section 2.2), restriction to individual impact parameters or adoption of an adiabatic picture implies the neglect of fluctuating contributions to cross section or probability. Taking an average inertia parameter $m = \overline{m}_\eta$ and approximating the mass-asymmetry dependence of the potential by a parabola

$$V(\eta) = \tfrac{1}{2}C(\eta - \eta_m)^2 \qquad (2.4.2)$$

one obtains (Yam 76, Yam 77) for the time-dependent average value $\bar{\eta}(t)$ and the quantal dispersion (variance) $\sigma_\eta^2(t)$

$$\bar{\eta}(t) = \eta_m(1-\cos\omega t)$$

and (2.4.3)

$$\sigma_\eta^2(t) = \frac{1}{2\sigma_0^2}\left(\frac{\hbar}{m\omega}\right)^2\left[1+\beta^4-(1-\beta^4)\cos 2\omega t\right]$$

Here, $\omega=(C/m)^{1/2}$ is the collective frequency and $\beta=\sigma_0(m\omega/\hbar)^{1/2}$ is a parameter depending on the initial width σ_0 (standard deviation) acquired in noncollective processes such as nucleon transfer preceding the onset of collective motion along the mass-asymmetry coordinate. In such an adiabatic picture, one expects then an oscillation of the average mass asymmetry around the position η_m of the potential minimum and also an oscillating quantal dispersion. For a potential energy V exhibiting structure such as that due to shell closures, the wave packet will be more complex and show enhancements for fragmentations associated with local potential minima.

Along with the potential energy, the inertia parameters m_η and their development with time play a crucial role in determining the features of collective mass transfer processes as has been reemphasized recently (Bro 78, Bro 79–79b, Bro 80b, Her 80, Gro 81). It will be discussed below that one expects the masses m_η to become singular when the intermediate system undergoes scission such that motion along the mass-asymmetry degree of freedom ceases. If the approach towards the scission point is adiabatic, then the collective frequency ω vanishes with growing inertia. Consequently, the dispersion in η regains its initial value $\sigma_\eta^2=\sigma_0^2$. Similarly, if the system was initially in an η eigenstate, it will asymptotically be in the one characterized by η_m. Hence, in the case of a harmonic driving potential, a finite contribution of quantal collective dispersions to the total width of final fragment mass or charge distributions can only be expected to the extent that the motion along the mass asymmetry degree of freedom is nonadiabatic. Such a situation may occur if the rupture of the neck of the intermediate system is sufficiently rapid to prevent the wave function from readjusting (Her 80). It should be noted, however, that this qualitative discussion of the harmonic-oscillator model serves illustrative purposes only. It is not expected that mean value and variance of mass asymmetry given by Equations (2.4.3) can be related to experimental distributions. So far neglected effects of a coupling of different channels due to the variation of the mass parameters requiring a dynamical calculation (Zoh 76, Zoh 81) may lead to significantly different final populations.

In Figure 11 inertia parameters (Zoh 75) divided by the nucleon mass M are plotted for the system $^{238}U+{}^{238}U$ vs. mass asymmetry, for various separation distances. They are obtained from the two-center shell model (Mar 72). The characteristic peak patterns of the inertia observed in Figure 11 arise because only integer fragmentations are possible after separation of the intermediate system. At small radial separations shown at the bottom of the figure, inertia values are small and, hence, large quantal dispersions are expected in the model, making the population of a broad range of asymmetries equally probable. In the exit channel, because of increasing inertia, the probability amplitude will concentrate around mass asymmetries corresponding to local potential minima. Because of the assumed shell structure in the potential energy surface, this coherent process of mass transfer leads to final fragment mass distributions of finite width even in an adiabatic approach to singular inertias. This is quite in contrast to a narrow distribution centered around an asymmetry η_m corresponding to the potential minimum as expected in this limit for a harmonic potential. As can be observed in Figure 11, the inertia predicted from the cranking formula remains finite even for large separation distances leading to spurious intercorrelations of channels with different mass asymmetries after breakup of the intermediate system. This complicates the interpretation of theoretical results, a problem relevant also to other models based on a common adiabatic potential, such as TDHF models (Koo 79b).

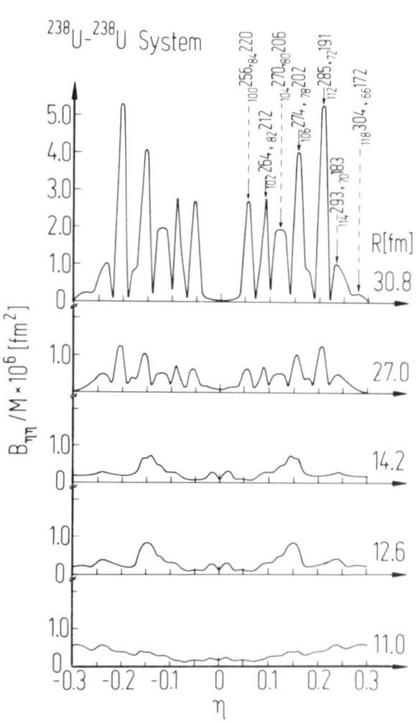

Figure 11. Inertia parameters $B_{\eta\eta}(=m_\eta)$ divided by the nucleon mass M, associated with the motion of the $^{238}U+^{238}U$ system along the mass asymmetry degree of freedom η. R is the distance between the nuclear centers. Various final fragmentations are indicated for the largest distance R. From (Zoh 75).

The concept of a collective mode of mass transfer has been applied in a simplified form by Berlanger *et al.* (Ber 79c) to processes of mass-to-charge equilibration in damped reactions, interpreting experimental isobaric fragment distributions in terms of quantal zero-point fluctuations. In this static approach, the variance $\sigma_Z^2(A)$ of the fragment Z distribution for a fixed final mass asymmetry reflects the dispersion

$$\sigma^2 = \hbar\omega/(2C) \tag{2.4.4}$$

of the wave function of a harmonic isospin oscillator in its ground state. Here, the oscillator energy $\hbar\omega$ and the stiffness C of the driving potential [cf. Equation (2.4.2)] are assumed to be given by the systematics of giant dipole resonance energies and by the liquid-drop model for two spherical nuclei, respectively. The above model has been subsequently refined by Hofmann *et al.* (Hof 79b, Gre 79b, Gre 80–80b) to include interactions between this collective and intrinsic degrees of freedom (cf. Section 2.6). However, important dynamical aspects are neglected such that the above approaches are of limited applicability.

Several more recently developed models (Bro 78, Bro 79–79b, Bro 80b, Mor 79, Mor 80, Her 80, Gro 81) study collective mass transfer incorporating hydrodynamical concepts such as those underlying the Goldhaber–Teller model (Gol 48) of nuclear giant dipole resonances. In particular, they address the phenomenon of mass-to-charge or isospin equilibration in damped reactions. Asymmetries η_I in the isospins $I_\nu = (A_\nu - 2Z_\nu)/2$ of projectile and target usually correspond to an excess of not more than a few nucleons of one kind. Since, furthermore, the mass asymmetry is a relatively slowly varying quantity (cf. Section 3.4), it appears appropriate to treat isospin modes for fixed values of the mass asymmetry. In a rather crude approximation, one may then consider the uniform laminar flow patterns of neutron and proton excess currents $\mathbf{j}_n$ and $\mathbf{j}_p$, respectively, within a cylindrical neck between fragments such as depicted in Figure 12. Since the isospin asymmetry $\eta_I = (I_1 - I_2)/(I_1 + I_2)$ changes with a rate $\dot{\eta}_I = 2(j_n - j_p)(A - 2Z)^{-1}$, and $j_n + j_p = 0$ for constant mass asymmetry, one obtains

$$\mathcal{T}_I = \tfrac{1}{2} m_{\eta_I} \dot{\eta}_I^2$$

$$m_{\eta I} = \frac{M}{16} \frac{l}{\pi\rho^2} \frac{n_0}{n_p n_n} (A - 2Z)^2 \tag{2.4.5}$$

for kinetic energy and inertia parameters associated with the isospin flow within the neck cylinder. The quantities A and Z are the mass and atomic numbers of the total system. As can be inferred from Equation (2.4.5), the inertia becomes singular for vanishing neck radius ρ. Here, n_0, n_p and n_n

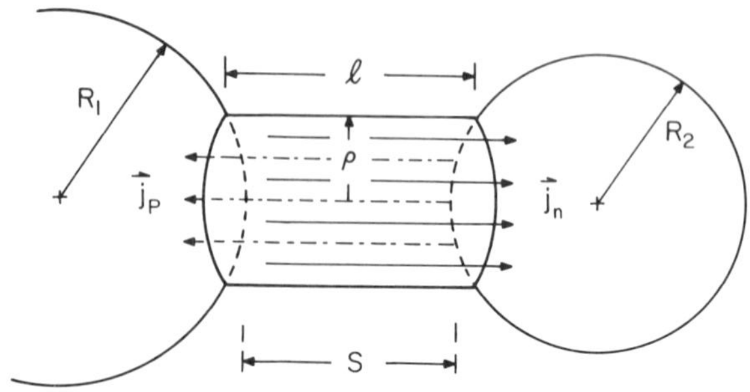

Figure 12. Schematic illustration of proton and neutron exchange currents j_p and j_n, respectively, through the cylindrical neck connecting two spherical nuclei.

are the densities of nucleons, protons and neutrons assumed to be constant throughout the system, for purposes of calculating the inertia, and M denotes the nucleon mass. For small neck radii, the inertia can be approximated by

$$m_{\eta_I} \approx \frac{M(A-2Z)^2}{2\pi n_0 R}\left(1+2\frac{\overline{R}s}{\rho^2}\right), \qquad \rho>0 \tag{2.4.6}$$

in the notation of Figure 11. The radius of curvature is given by $\overline{R} = R_1R_2/(R_1+R_2)$. Together with the parabolic dependence of the potential energy on isospin asymmetry suggested by empirical mass formulas (Mye 74), the motion along this asymmetry coordinate could be described in terms of a harmonic oscillator. Using a characteristic stiffness C_I of the potential of a few hundred MeV per unit in η_I (Her 80), one obtains $T_I \approx 10^{-21}$ s for the period and $\hbar\omega \approx 5$ MeV for the energy of such an undamped isospin oscillation. These values apply to a typical system such as ^{165}Ho$+^{56}$Fe with the two nuclei touching at their matter half-density radii (cf. Section 2.5).

In reality, the situation is expected to be more complex than suggested by Figure 12 and the above discussion. The flow pattern fans out beyond the neck region and exhibits velocity gradients dependent on shear viscosity and friction with the nuclear walls. Brosa *et al.* (Bro 78) and Grossman *et al.* (Gro 81) have studied mass transfer in damped reactions in somewhat more realistic models. These studies stress the importance of inertial effects on mass transfer in damped reactions and their dependence on the relative motion and the configuration of the intermediate system.

Of similar importance is a proper estimate of the driving potential $V(\eta_I)$ associated with the isospin modes. Most model approaches infer this potential energy statically from applying the liquid-drop model to the particular configurations studied. From a schematic model, Moretto *et al.* (Mor 80e), however, conclude that the effective curvature or stiffness of this potential depends strongly on the nature of the oscillatory mode and that dynamical polarizations of the fragments should be considered in its evaluation. The effect is illustrated in Figure 13 (Mor 80e) comparing the stiffness $C_{\text{sph}} = [\partial^2 V(Z_1, Z_2)/\partial Z_1^2]_{A=\text{const}}$ obtained from the liquid drop model for two touching spheres (dashed curve) with that corresponding to the lowest-mode isospin wave in a cylinder that breaks nonadiabatically at a fraction Q of its total length (full curve in Figure 13). If the convergence of the two curves at symmetry ($Q = 1/2$) can be taken as an indication of a sufficiently accurate approximation of the two-spheres configuration by a cylinder, large deviations of the actual potential stiffness from the usual evaluation may be expected due to dynamical fragment polarizations, in particular for very asymmetric final charge splits. However, further studies of more realistic configurations are needed to establish such effects.

Common to the approaches discussed above is a rather rudimentary account of relative motion and evolution of the nuclear configurations during the reaction, although the importance of these aspects is realized by

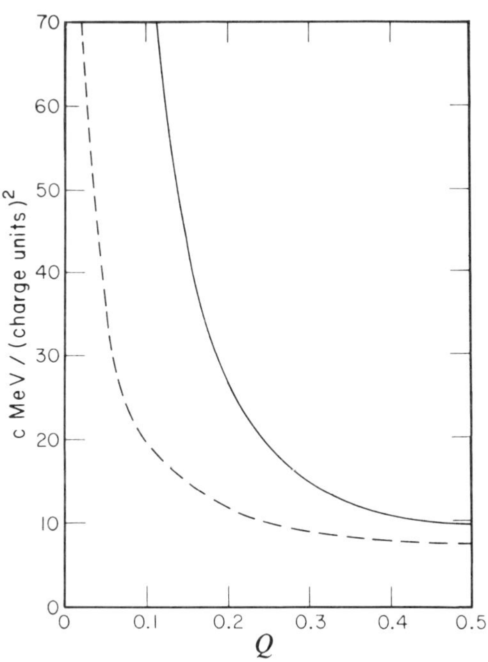

Figure 13. Model stiffness parameter C for a configuration of two touching spheres (dashed curve) and that of cylinder (solid curve) breaking at a fraction Q of its total length. From (Mor 80e).

their proponents. An exception is represented by the recent work of Hernandez *et al.* (Her 80). These authors employ a dynamical model based on dinuclear configurations illustrated in Figure 10a and the proximity approximation (cf. Section 2.6) of conservative and dissipative forces in order to describe classical system trajectories. A "frictional Schrödinger equation" (Kos 75, Kan 74)

$$i\hbar \frac{\partial}{\partial t}\psi = [\mathscr{H}_0(t) + \mathscr{W}(t)]\psi \tag{2.4.7}$$

is adopted for the wave function ψ associated with isospin asymmetry. In Equation (2.4.7), $\mathscr{H}_0$ is a harmonic-oscillator Hamiltonian incorporating a time-dependent potential and an inertia similar to the one of Equation (2.4.5). The friction term $\mathscr{W}$ depends on a friction coefficient determined from the observed widths of giant dipole resonance states [cf. (Aue 75)]. Solution to Equation (2.4.7) is a Gaussian wave function for the isospin asymmetry with a dispersion determined by a set of differential equations (Her 80). For an adiabatic motion, where inertia as well as strength factors of conservative and dissipative forces vary slowly over the period of an isospin oscillation, the wave function represents the ground state of the harmonic oscillator.

Typical results of the above model calculations are shown in Figure 14 for the reaction $^{86}\text{Kr} + ^{92}\text{Mo}$ at $E_{\text{lab}} = 430$ MeV. The full width Γ_{FWHM} of the predicted isobaric distribution of the light reaction fragments is plotted in perspective view vs. reaction time and initial angular momentum. Some of the displayed features are already expected from the previous discussion; new aspects are due to the introduction of a damping $\mathscr{W}$ of the isospin mode and a better account of the dynamics of neck formation determining the associated inertia. Dots on the curves in Figure 14 mark the instant in time when the neck between the two reaction fragments breaks, according to the dynamical model. Because of a rapid, nonadiabatic neck shrinking near this point, the quantal dispersion cannot follow the correspondingly fast increase of the inertia and, hence, is "frozen in." As can be observed from Figure 14, a transition from overdamped motion, associated with a continuous growth of the width, to an overdamped one showing oscillatory behavior is expected in this model, proceeding from high to low angular momenta. An analogous behavior is observed for the centroids of the distributions.

Generally, the models of collective mass and charge transfer discussed above suffer from a lack of internal consistency and/or completeness. The specialization of some of these theories to singular aspects such as mass-to-charge equilibration and their reliance on a proper description of relevant reaction features by auxiliary theories presents a somewhat unsatisfactory situation in which it is difficult to judge the validity of the basic concepts

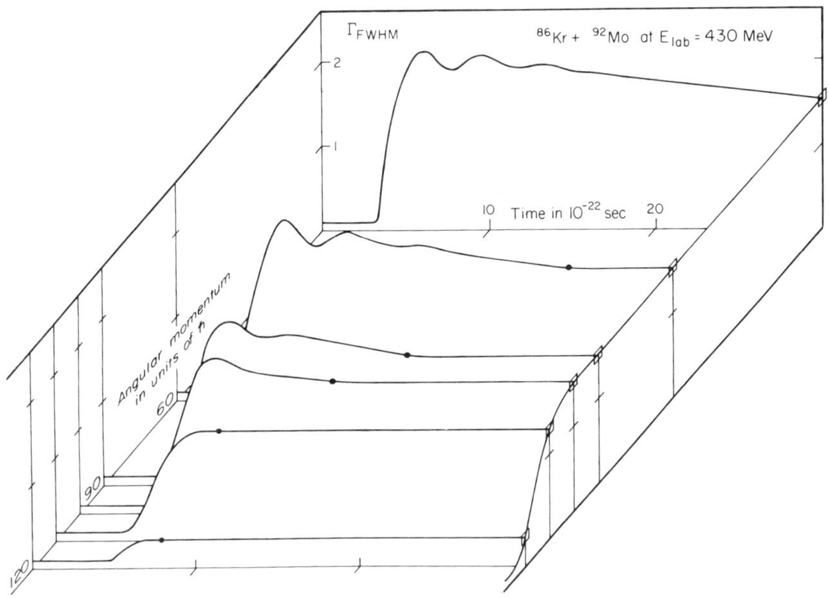

Figure 14. The predicted time evolution of the fragment charge width Γ_{FWHM} for various values of initial angular momentum. The point of neck rupture is indicated by a dot on each curve. From (Her 80).

from a comparison to experiments. Clearly, this area deserves further theoretical efforts.

2.5. Macroscopic and Microscopic Potential Energies

In a fully consistent microscopic description of nucleus–nucleus collisions, in general no explicit reference is made to notions such as nuclear interaction potential, configuration of the interacting system or dissipation. Adoption of a nucleon–nucleon interaction is, in principle, sufficient to generate the macroscopic features of a nuclear reaction which, subsequently, may or may not be describable in terms of macroscopic nuclear forces and shapes. It is one of the virtues of microscopic theories such as self-consistent Hartree–Fock models that they lend some insight into what significant properties these macroscopic entities can be expected to show. Conceptually, it is always possible to define a macroscopic potential energy by averaging over all interactions of the individual nucleons in the system. However, an interaction potential obtained in this way is only useful for a deeper understanding of the reaction mechanism, if it can be related to a few macroscopic parameters connected to the size and the configuration of the interacting system. In addition, there should be sufficiently universal and smooth scaling laws describing its dependence on the other reaction conditions such as imposed by a variation of the incident energy.

In view of the difficulties and ambiguities that are encountered in analyzing the relatively gentle interactions in elastic scattering of heavy nuclei, and the additional complications arising from dynamical effects (Gla 74, Lov 77, Bal 78, Don 79, Fra 80b), it is not self-evident that universal potentials applicable to damped nuclear reactions can be found. But even if the hope to deduce a unique central interaction potential describing damped reactions for a wide range of systems and incident energies is vain, it should still be possible to gain a qualitative understanding of the gross properties of the reaction mechanism from the application of models including sufficiently realistic configurations. Such phenomenological studies are simplified significantly, if they can be based on general principles derived in microscopic or semimicroscopic theories. Folding and proximity models to be discussed below are examples of approaches based on such principles. As emphasized earlier, since there is no unique way by which to select the most proper sets of collective and intrinsic degrees of freedom, it may turn out to be impossible to discriminate between different model views entertaining similar internal consistency.

The least conceptual difficulties are expected for evaluation of a realistic Coulomb interaction potential V_{Coul}. Because of the long range of the Coulomb force, it cannot be very sensitive to small details of the nuclear configuration, justifying the use of somewhat cruder approximations than permissible in the evaluation of the short-range nuclear force. Nevertheless, because of the large magnitude of the Coulomb potential in reactions between heavy nuclei, it has to be examined whether a given approximation matches the intended application, which, in practice, presents a difficult task.

In Figure 15, several simple models for the charge distributions of two interacting spherical nuclei with frozen configurations are indicated. Distributions (a) and (b) refer to the cases where either or both nuclei can be represented by point-like charge distributions. In the distribution (c) the "equivalent sharp radius" [cf. Equations (2.5.4)–(2.5.7) below] has been renormalized in order to simulate the spatial extension of both nuclei. Cases (d) and (e) reflect the situations where both nuclei are representable by either homogeneous or Fermi-type charge distributions, respectively. Obviously, the latter approximation describes the charge distribution of undistorted nuclei best. Unfortunately, in this case, the potential can only be evaluated numerically. Alternative and more convenient descriptions of diffuse nuclear charge distributions have also been introduced, e.g., by Krappe (Kra 76). Coulomb potentials associated with the simple distributions shown in Figure 15 can be expressed as

$$V_{\text{Coul}}(r) = \begin{cases} e^2 Z_1 Z_2 / r, & r \geq R_c \\ \dfrac{e^2 Z_1 Z_2}{2R_c}\left[3 - \left(\dfrac{r}{R_c}\right)^2\right], & r < R_c \end{cases} \tag{2.5.1}$$

where $R_c = 0$, $R_c = R_2$ and $R_c = (R_1^3 + R_2^3)^{1/3}$ for the cases (a), (b), and (c), respectively. A formula for the Coulomb potential associated with two homogeneous charge distributions (cf. Figure 15d) has been derived by DeVries *et al.* (Dev 75).

The different approximations for the Coulomb potentials are compared in Figure 16 where the ratios with respect to the "exact" potential (cf. Figure 15e) are plotted vs. separation distance r for two heavy-ion systems. To put this comparison in perspective it should be noted that the strong-absorption radii are roughly 14 and 15 fm for the systems ^{290}Bi + ^{84}Kr and ^{209}Bi + ^{136}Xe, respectively (cf. Section 3.1). From Figure 16, one observes that the deviations of the approximative formulas from the "exact" solution

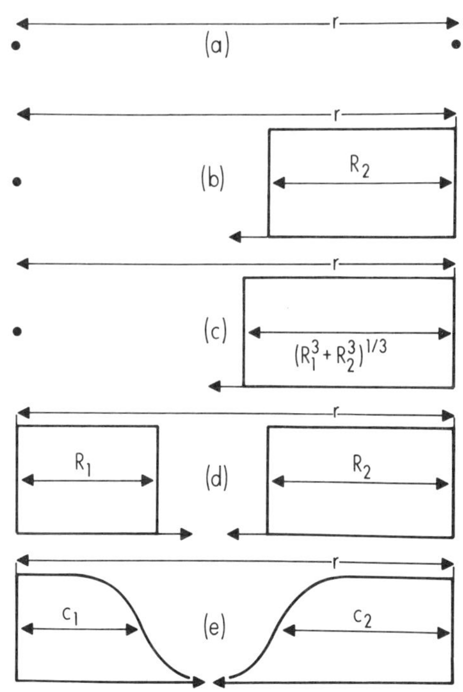

Figure 15. Different approximations of the nuclear charge distributions used to calculate the corresponding Coulomb potentials shown in Fig. 16. From (Hui 76b).

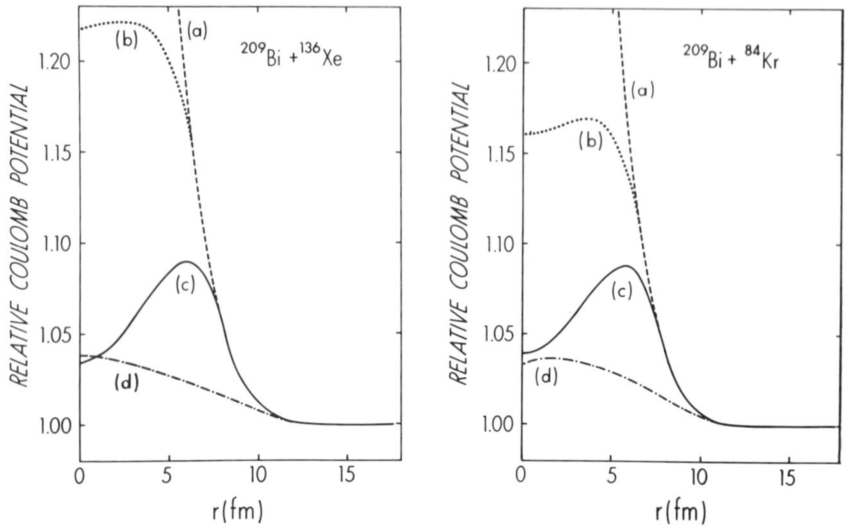

Figure 16. Coulomb potentials for the models a, b, c, and d for the nuclear charge distributions shown in Fig. 15, relative to that of two Fermi distributions. From (Hui 76b).

are small and comparable for distances between the charge centers of more than 10 fm. For much smaller separation distances, these deviations may become significant and of the order of the nuclear potential.

The equivalent sharp radii R referred to in the above discussion are related to the half-density or central radii C of a Fermi distribution

$$\rho(r) = \rho_0\left\{1 + \exp\left(\frac{r - C}{a}\right)\right\}^{-1} \tag{2.5.2}$$

according to (Mye 73)

$$C = R\left[1 - (b/R)^2 + \cdots\right] \tag{2.5.3}$$

The radii R refer to those of uniform sharp distributions having the same asymptotic values in the center and the same volume integrals as the corresponding Fermi distributions. The "surface thickness" b (≈ 1 fm) entering Equation (2.5.3) represents a second moment (Sus 73) of the Fermi distribution and is related to its diffuseness, a, via $b = \pi a/3^{1/2}$. Due to the excess of neutrons over protons leading to a "neutron skin" of thickness t_A (Mye 73, Mye 80), i.e., a relative excess of neutrons in the nuclear surface layer, the equivalent sharp radii R_N and R_Z of neutron and proton distributions are different and connected to the corresponding matter radius

R_A according to

$$R_N = R_A + (Z/A)t_A$$
$$R_Z = R_A - (N/A)t_A \tag{2.5.4}$$

Fits to experimental values for mass and charge distribution parameters yield (Mye 73, Blo 77)

$$R_A = 1.28A^{1/3} - 0.76 + 0.8A^{-1/3} \approx 1.15A^{1/3} \text{ fm} \tag{2.5.5}$$

The neutron skin thickness is given in terms of the droplet-model mass parameters (Mye 77b):

$$t_A = (2/3)r_0 A^{1/3}(I_A - \bar{\delta}_A) \tag{2.5.6}$$

with $I_A = (N - Z)/A$ and

$$\bar{\delta}_A = \left(I_A + \frac{3c_1}{16Q} ZA^{-2/3}\right) \Big/ \left(1 + \frac{J}{4Q} A^{-1/3}\right)$$
$$r_0 = 1.2249 \text{ fm}, \qquad J = 36.8 \text{ MeV} \tag{2.5.7}$$
$$c_1 = 0.73219 \text{ MeV}, \qquad Q = 17 \text{ MeV}$$

The radius parameters R_i referred to in Figure 15 and Equation (2.5.1) should be identified with the corresponding values of R_Z as defined by Equations (2.5.4)–(2.5.7).

The above discussion pertains to the Coulomb interaction potential of two spherical undisturbed nuclear charge distributions. Because of the low compressibility of nuclear matter, this assumption of frozen configurations leads to an overestimation of the Coulomb repulsion for deeper interpenetrations. An attempt to provide a smooth interpolation between the Coulomb potential energy of separated spherical nuclei and that corresponding to the combined mononucleus has been given by Bondorf *et al.* (Bon 74b). These authors propose a potential of the form

$$V_{\text{Coul}}(r) = \begin{cases} e^2 Z_1 Z_2 / r, & r \geq R_c \\ V_0 - kr^n, & r < R_c \end{cases} \tag{2.5.8}$$

with

$$V_0 = \frac{3}{5}e^2\left[\frac{(Z_1 + Z_2)^2}{(R_1^3 + R_2^3)^{1/3}} - \frac{Z_1^2}{R_1} - \frac{Z_2^2}{R_2}\right]$$

The quantities k and n entering Equation (2.5.8) are determined by the requirement of continuity of potential and force at $r = R_c$ yielding

$$k = \left(V_0 - e^2 Z_1 Z_2 / R_c\right) / R_c^n$$

and (2.5.9)

$$n = e^2 Z_1 Z_2 \left[R_c \left(V_0 - e^2 Z_1 Z_2 / R_c\right)\right]^{-1}$$

The charge radius R_c in the above equations is the sum of the equivalent sharp charge radii of the two nuclei as given by Equations (2.5.4)–(2.5.7). Earlier parametrizations (Mye 69) of these radii are also frequently used (Bir 78–79). In Figure 17, the Bondorf Coulomb potential for the system $^{165}\mathrm{Ho} + {}^{84}\mathrm{Kr}$ is compared (Bir 79) to the other approximations discussed previously. As can be observed, this potential gives values considerably

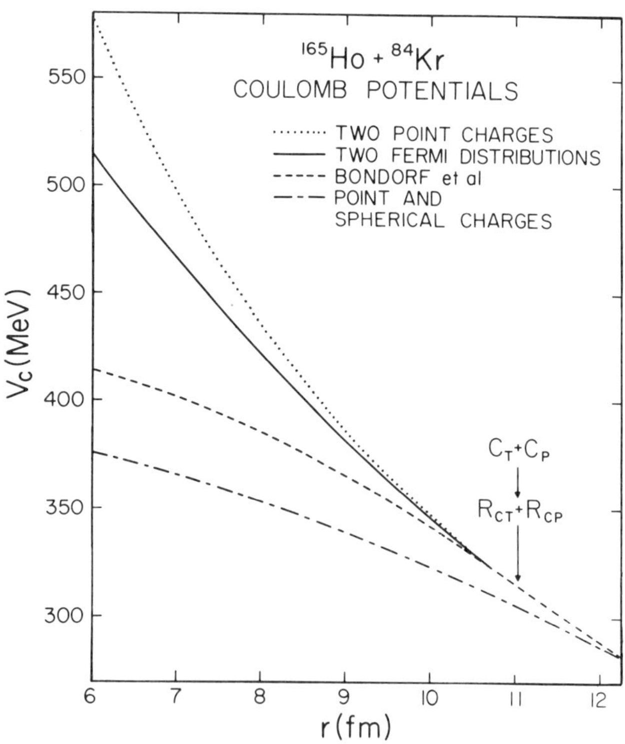

Figure 17. Various Coulomb potentials for the system $^{165}\mathrm{Ho} + {}^{84}\mathrm{Kr}$ plotted vs. center separation distance. From (Bir 79).

lower than deduced for point charges and Fermi charge distributions. The dash-dotted curve in Figure 17 corresponding to a point charge interacting with a homogeneous distribution should be disregarded in this comparison, even though such a potential has been used in optical-model calculations, since an arbitrarily large charge radius has been used in this case.

As pointed out already in Section 2.3, it appears necessary to consider deformations of nuclei undergoing damped collisions under certain conditions. General methods of calculating the multipole Coulomb interaction between deformed nuclei with sharp or diffuse surfaces have been devised by several authors (Nix 64, Arn 76, Kra 76). In the simple case of two juxtaposed coaxial spheroids with sharp surfaces, an approximate expression for the Coulomb potential has been derived by Wong (Won 68). As usual, one describes the nuclear surface by a radius vector of length

$$R(\phi) = R_0\left[1 + \beta Y_0^2(\phi)\right] \tag{2.5.10}$$

where the radius parameter R_0 and the deformation parameter β are related to the semiaxes a and b of the spheroid as $R_0 = (a+2b)/3$ and $\beta = (16\pi/5)^{1/2}(a-b)/(a+2b)$. In Equation (2.5.10), ϕ is the angle with respect to the symmetry axis a. Denoting with r the distance between the centers of the coaxial spheroids, one obtains (Won 68) for the Coulomb potential

$$V_{\text{Coul}}(r) = \frac{e^2 Z_1 Z_2}{r}(1+K) \tag{2.5.11}$$

where

$$K = \frac{3}{\pi r^2}\left[\sum_i \left(\frac{1}{20^{1/2}}\beta_i + \frac{1}{7}\beta_i^2\right)R_{0i}^2 + \frac{3}{r^2}\left(\frac{1}{28}\sum_i \beta_i^2 R_{0i}^4 + \frac{3}{10}\beta_1\beta_2 R_{01}^2 R_{02}^2\right)\right]$$

including terms up to second order in the deformation parameters. Here, β_i and R_{0i} apply to the individual nuclei, and the sums run over both nuclei. An inclusion of a finite surface diffuseness of the nuclear charge distributions is not expected to result in a significantly different Coulomb energy for slightly overlapping distributions. This is suggested by inspection of Figure 17 applying to spherical nuclei and is also borne out in studies (Kra 76) of interactions between deformed charge distributions. As discussed earlier, for deeper interpenetrations other effects present much larger uncertainties than associated with a neglect of the surface diffuseness.

In comparison with model calculations of the Coulomb interaction potential discussed above, the evaluation of the short-range nuclear interaction V_N requires closer scrutiny. A variety of approaches has been pursued in the literature that are either based on a microscopic description of the interacting nuclei or incorporate macroscopic phenomenological energy parameters allowing to circumvent some of the difficulties encountered in fully microscopic models.

A relatively obvious method to deduce a nuclear interaction potential V_N microscopically consists in the summation of the interactions of all pairs of nucleons in different nuclei. This has resulted in the development of several folding procedures for the calculation of V_N (Bro 72, Var 73, Sat 74,75, Bri 74,75, Sin 75, Dov 75, Gro 78, Clo 78, Sch 79). In the single-folding model (Sat 74, Bri 75, Gro 78), the nuclear interaction potential is deduced from folding the matter density distribution $\rho_1(\mathbf{r}_1)$ of one nucleus with an experimentally determined nucleon–nucleus potential $V_2(\mathbf{r}_2)$ corresponding to the other nucleus:

$$V_{12}(\mathbf{r}) = \int d^3\mathbf{r}'\rho_1(\mathbf{r}')V_2(\mathbf{r}-\mathbf{r}') \tag{2.5.12}$$

Since this expression is nonsymmetric in the two nuclei, one may take (Gro 78) $V_N = (V_{12} + V_{21})/2$ as a better estimate of the nuclear interaction. If the density distribution and the nucleon–nucleus potential are both given by spherically symmetric Fermi or Woods–Saxon forms [cf. Equations (2.3.18) and (2.5.2)], the radial form factor in Equation (2.5.12) can be evaluated approximately analytically (Bri 74) for sufficiently large separations yielding

$$\begin{aligned} V_{12}(r) &= (\pi\rho_0 V_0/4)f(r)/r \\ f(r) &\approx 2a(r+a)^2(r+a-R_1-R_2)\exp[-(r-R_1-R_2)/a] \end{aligned} \tag{2.5.13}$$

Here, R_1 and R_2 are the half-value radii of the matter distribution and the nucleon–nucleus potential. Their diffuseness parameters a_1 and a_2 determine the parameter a in Equation (2.5.13), according to $a = 2a_1a_2/(a_1 + a_2)$. Gross and Kalinowski (Gro 78) arrive at the following parametrization of the nuclear interaction based on the single-folding procedure:

$$\begin{aligned} V_N(r) &= -\sum_{n=1}^{3} A_n(r-R)^{n-1}\ln\{1+\exp[(R-r)/a]\} \\ R &= r_0\left(A_1^{1/3} + A_2^{1/3}\right) \end{aligned} \tag{2.5.14}$$

From a fit of this formula to the numerical values of the folding potentials, the parameters $A_1 = 33$, $A_2 = 2$, and $A_3 = 3$ were obtained (Gro 78) for the choice of $r_0 = 1.3$ fm and $a = 0.61$ fm. This potential is shown in Figure 18 (short-dashed curve) for the system ^{209}Bi + ^{84}Kr. For a better orientation, the experimental (Bir 76) strong-absorption radius R_{SA} is indicated by an arrow. As can be observed from the figure, the folding potential is rather steep and has already a significant depth (≈ -10 MeV) at the strong-absorption radius.

The behavior described above is observed (Sat 75b, 76b) not only for single-folding potentials but also for double-folding models, where an effective nucleon–nucleon potential is folded between the densities of both interacting nuclei. The magnitude and slope of these potentials appear

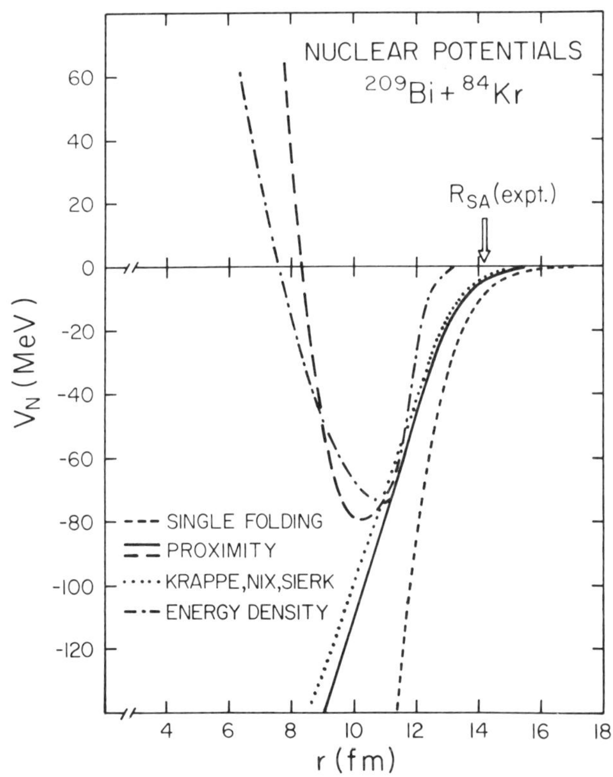

Figure 18. Nuclear interaction potentials for the system ^{209}Bi + ^{84}Kr plotted vs. the center separation distance. The short-dashed curve represents the single-folding model of Gross and Kalinowski; the dotted curve corresponds to the folding potential of Krappe *et al.* The dashed-dotted curve illustrates the energy-density potential suggested by Ngô *et al.* The long-dashed and solid curves describe the normal and the modified nuclear proximity potentials, respectively. The arrow denotes the experimental strong-absorption radius.

unrealistic at least for separation distances where phenomenological potentials obtained from fits to elastic-scattering data are available for comparison. This failure may, in part, be due to a lack of knowledge of strength or geometry of the nucleon potentials used in the folding procedure. There is, however, also a more fundamental criticism, as has been raised, for example, by Mosel and collaborators (Mos 76, Zin 75, 75b, 76). From their microscopic self-consistent Hartree–Fock calculations performed for systems ranging from $^{12}C + ^{12}C$ to $^{40}Ca + ^{40}Ca$, they were able to identify significant contributions to the nucleus–nucleus interaction potential neglected in folding-type models. It is mainly exchange effects due to the Fermi statistics, which are important already for small overlaps of the matter distributions, that lead to alterations of the intrinsic structure of both nuclei and, eventually, to a relaxation of their shapes. Consequently, there are induced forces (cf. Section 2.6) that reduce the strength of the nuclear potential as obtained in a folding approach assuming frozen configurations. This represents a fundamental objection to folding models neglecting exchange effects but does not exclude the possibility that such models in combination with particular Coulomb potentials and choices of radial constants may be able to describe certain reaction or scattering features [see, e.g., (Sat 76b, Clo 78)].

Fully microscopic calculations of the potential energy such as performed (Zin 76) for relatively light systems become prohibitively elaborate for heavier systems and relatively complex nucleon–nucleon interactions. Alternative, though less accurate, views are adopted in semiphenomenological evaluations making explicit reference to empirical nuclear properties as represented, for example, by mass formulas. By adjusting the overall strength of a model potential energy to its empirical value, some of the antisymmetrization or exchange effects are implicitly taken into account (Mos 76). Such calculations (Bas 74, Swi 75, Blo 77, Blo 80, Kra 74, Kra 79) are often inspired by the liquid-drop (Wei 35, Boh 39, Hil 53, Str 64, Hil 66, Swi 50, 51, Mye 66, 67) or droplet (Mye 70, 77b, Has 71b) models.

A phenomenological model of the nuclear potential energy has been described by Krappe *et al.* (Kra 74, Kra 79, 79b) generalizing the liquid-drop model for arbitrary nuclear configurations. In this model, the volume elements of a sharply defined density distribution interact with each other via a Yukawa force generating a potential with a diffuse edge. The corresponding surface energy can then be written as a double-folding integral over a Yukawa force of range a:

$$\mathscr{E}(a, c_s) = -\frac{c_s}{8\pi^2 r_0^2 a^3} \int d^3\mathbf{r}\, d^3\mathbf{r}' \frac{\exp[-|\mathbf{r}-\mathbf{r}'|/a]}{|\mathbf{r}-\mathbf{r}'|} + \frac{2}{3}\frac{r_0}{a} c_s A \tag{2.5.15}$$

Here, $r_0 = 1.18$ fm is the nuclear radius constant, and c_s is the liquid-drop surface energy coefficient. Equation (2.5.15) describes the surface energy of a nucleus with a diffuse surface, since the second term in this equation cancels the volume contribution of the folding integral. In the limit $a \to 0$, the above expression assumes the ordinary liquid-drop value for a sharp surface. However, since the expression (2.5.15) implies (Kra 79) a violation of the requirement of minimum potential energy for saturated nuclear matter, it appeared necessary to calculate the interaction energy from the difference between the folding expressions for two Yukawa functions with slightly different ranges a_i and energy constants c_{si}. This is equivalent to folding the sum of a Yukawa and an exponential function. The resulting nuclear interaction energy for two spherical nuclei with a distance s between their equivalent sharp surfaces can then be represented by

$$V_N(r) = \begin{cases} -DR_{12}(F + s/a)\exp(-s/a)/r, & s > 0 \\ -V_0 + (B/R_{12})r, & s \leq 0 \end{cases} \tag{2.5.16}$$

The parameters entering this potential are functions of the equivalent sharp nuclear radii R_1 and R_2, their sum $R_{12} = R_1 + R_2$, the average range a ($= 0.65$ fm) of the Yukawa forces, and a depth V_0 (≈ 85 MeV) obtained from the continuity requirement of the first term in Equation (2.5.16) with its linear extrapolation at $s = 0$. An overall strength of the above potential was adjusted to optimally reproduce elastic-scattering data. The potential given by Equation (2.5.16) is represented for the system $^{209}\mathrm{Bi} + {}^{84}\mathrm{Kr}$ by the dotted curve in Figure 18. It is observed to be a factor of ~ 2 less deep at the strong-absorption radius than the folding potential (dashed curve in Figure 18) and also in general less steep than the latter. For heavy systems such as exhibited in this figure, the nuclear potential of Equation (2.5.16) compares well with the proximity potential discussed below. As pointed out by the authors (Kra 79), the potential should be best suited for a description of elastic scattering with strong absorption, because the model implies frozen densities and neglects possible polarization of the surface as well as neck formation.

When, as in the discussion above, reference is made to the surface and the interior of a nucleus, it is implicitly assumed that such a differentiation finds its justification in the actual radial dependence of the nuclear mass or charge density distribution. In fact, a surface is only considered to be well defined if the radial range over which the distribution drops to zero is small compared to that over which it maintains its nearly constant bulk value, i.e., it is required that $b \ll R$. This is only approximately true for heavy nuclei; lighter nuclei such as $^{12}\mathrm{C}$ cannot at all be viewed as "leptodermous" in this sense. The idealization of a leptodermous quality of nuclear systems, where it is justified, offers the opportunity to classify the various contributions to the potential energy of the system in terms of physically transparent categories and to derive some of their general features without any further specific assumptions. Efforts to study the properties of leptodermous nuclear systems are mainly due to Swiatecki and collaborators (Swi 75, 79, 80, Blo 77, 80, 81b, Ran 76, 80b).

Following the work cited above, the total potential energy V of a nuclear system can be expressed as a sum of three terms:

$$V = c_A A + \int d^3\mathbf{r}(\varepsilon - c_A \rho) + V_{\text{Prox}} \tag{2.5.17}$$

Here, A denotes the total number of nucleons in the system, c_A is the volume energy coefficient, and ρ and ε are the matter and energy densities, respectively. Obviously, the first term in Equation (2.5.17) cancels the contribution from the second part of the integrand. If the energy density ε were given by its bulk value throughout the system, the integral in Equation (2.5.17) would vanish and, hence, it represents the surface energy. An additional energy, the proximity energy, V_{Prox}, arises when juxtaposed parts of the surface interact with each other across gaps or crevices. The usefulness of a distinction between bulk and surface energies depends on how well an energy-equivalent nuclear surface $S(\mathbf{r})$ defined by the function $\Delta\varepsilon(r) = \varepsilon(\mathbf{r}) - c_A\delta(\mathbf{r})$ is actually localized in space, i.e., how fast an expansion of the integral in Equation (2.5.17) in powers of the ratio of surface diffuseness b and average radial distance R converges (Blo 77, Ran 80b). With reference to such an equivalent sharp surface $S(\mathbf{r})$, the integral in Equation (2.5.17) can be transformed into a surface integral

$$V_s = \int d^3\mathbf{r}(\varepsilon - c_A\rho) = \int dS\,\gamma(S) \tag{2.5.18}$$

where the local surface tension $\gamma(S)$ describes the reaction of a surface element dS to shape distortions. Describing the local surface of a leptodermous system in terms of a Taylor expansion in Cartesian coordinates yields (Blo 77) a representation of the surface energy V_s in terms of the total surface area S and various curvature integrals. Blocki *et al.* (Blo 77) conclude that the term γS associated with the total surface area and the constant surface tension coefficient γ of a plane surface gives the largest contribution to V_s, in particular as it is possible to define S in terms of a "surface of tension" (Mye 66) which is shifted slightly outwards of the equivalent sharp surface and for which the curvature correction vanishes. In the light of recent droplet-model mass formulas (Mye 77b), the curvature corrections to V_s appear negligible. One has then (Blo 77) for the nuclear surface energy of a system with an equivalent sharp surface area S

$$V_s \approx \gamma S$$

$$\gamma = 0.951\left[1 - 1.7826 I_A^2\right] \text{ MeV fm}^{-2} \tag{2.5.19}$$

Here, $I_A = (N - Z)/A$, as defined earlier. The coefficients entering the above expression for the surface tension γ are determined by mass formula fits (Mye 67).

In order to arrive at an expression for the nuclear energy, the proximity term V_{Prox} in Equation (2.5.17) has to be specified. For illustrative purposes, Figure 19 shows a schematic view of possible configurations of a dinuclear system where juxtaposed surfaces of the two constituents interact across a gap or a crevice. For a crevice to occur, it is assumed that the system has a continuous interior characterized by one bulk density. The shapes indicated are understood to represent the equidensity contours associated with the half-density or central surfaces. The nuclear surface is taken to possess a single radial density profile. Although the force lines between the surfaces are curved, the effect of this nonlinearity on the total energy is expected to be small for gently curved surfaces because of the short range of the nuclear force. Under such conditions, the interaction can be obtained from integrating over the surface elements defined by the cylinders indicated in Figure 19. Approximately, the surface elements dS_1 and dS_2 are parallel to each

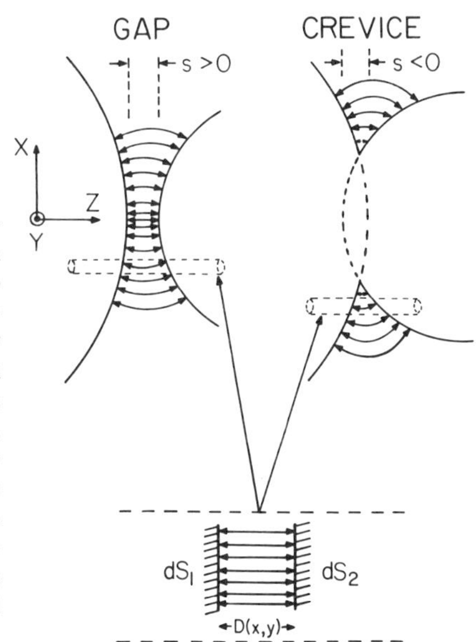

Figure 19. Schematic illustration of the interaction region of a dinuclear system with spherical fragments. A gap corresponds to positive surface separations s and also to negative values of s, if the matter density distributions are simply superimposed on each other. Crevices arise when the overlap region is characterized by the normal matter density. The curved arrows represent the proximity force field lines. Note that the strength of the force decreases with increasing angle between juxtaposed surface portions. The dashed cylinders define the juxtaposed surface elements for integration which are taken to be parallel to each other as illustrated at the bottom.

other and separated by a distance $D(x, y)$. Denoting with $e(D)$ the interaction energy per unit area for two parallel flat surfaces, the proximity energy can be expressed (Ran 77, Blo 77,80b) as an integral

$$V_{\text{Prox}}(s) \approx \int\int dx\, dy\, e(D(x, y)) = \int dD J(D) e(D) \tag{2.5.20}$$

where the lower integration limit, the minimum distance between the surfaces, is equal to s for gaps and zero for crevices (cf. Figure 19). In Equation (2.5.20), J denotes the "gap width distribution function" given by $J(D) = d(S_1 + S_2)/dD$. The upper limit of integration can approximately be extended to infinity for gently curved surfaces, again because of the short range of the nuclear force. This proximity approximation is hard to justify when the nuclei overlap to an extent that the opening angle of juxtaposed interacting surface elements becomes large.

Considering the case of two coaxial paraboloids facing each other across a gap with their principal axes aligned with respect to the x and y directions, $D(x, y)$ is given by a quadratic form involving the principal radii of curvature R_x and R_y of the gap. Then, the proximity energy reduces to the form

$$V_{\text{Prox}}(s) = 4\pi\gamma\bar{R}b \int_{s/b}^{\infty} dD' \phi(D') = 4\pi\gamma\bar{R}b\Phi(s/b) \tag{2.5.21}$$

with $D' = D/b$ and $\phi(D') = e(D')/2\gamma$. The reduced radius is given by $\bar{R} = (R_x R_y)^{1/2}$. For nonaligned paraboloids, Equation (2.5.21) is still valid with a more complicated expression for $\bar{R}$. The resulting force has radial as well as nonradial components leading to torques exerted on the interacting nuclei. In the special case of two juxtaposed spheres, $\bar{R} = C_1C_2/(C_1 + C_2)$, where the quantities C_ν are the central or half-density radii [cf. Equation (2.5.3)]. Correspondingly, the (radial) proximity force is given by

$$F_{\text{Prox}}(s) = 4\pi\gamma\bar{R}b\phi(s/b) \tag{2.5.22}$$

Obviously, proximity potential and force can be expressed in terms of products of a geometrical factor, $4\pi\gamma\bar{R}b$, and the universal, dimensionless form factors $\Phi(s)$ and $\phi(s)$, respectively, that depend only on the nature of the interaction between two parallel surfaces at a distance s. This is the content of the so-called Proximity Theorem. It has been generalized (Blo 80b) to include surfaces with large curvatures, provided the interacting portions form relatively small angles with each other.

It should be noted in passing that the nuclear forces discussed so far arise mainly from the interactions of surface elements with each other and, hence, the effective points of attachment of these forces are on the surfaces and not necessarily localized. Nevertheless, it appears natural, but is not undisputed (Str 78d, Alb 79), to decompose the forces into components directed towards the centers of gravity of the collision partners and others that lead to the exertion of torques on the nuclear circumference and warp the surface. Normally, only the former, radial components are considered, whereas the latter are accounted for in models including dynamical deformations (cf. Section 2.3) or bending and twisting modes (Zie 74, Mor 80).

From Equation (2.5.21) it follows that the proximity energy associated with a crevice is approximately proportional to $\Phi(0)$ and, hence, independent of the degree of overlap, as long as it is not too large. For this configuration to apply, it is assumed that the region of overlap becomes part of the nuclear interior characterized by the normal bulk matter density. Therefore, a nuclear surface area, given by $2\cdot(2\pi\overline{R}s)$ for small overlaps, is destroyed yielding a corresponding drop in the nuclear surface energy. Subtracting from the total potential energy V [cf. Equation (2.5.17)] the asymptotic value of the potential energy of two nuclei at infinity, one obtains the nuclear interaction potential (Blo 80c):

$$V_N(s/b)=\begin{cases}4\pi\gamma\overline{R}b\Phi(s/b), & s\geq 0 \qquad (2.5.23a)\\ 4\pi\gamma\overline{R}b\Phi(0)+4\pi\gamma\overline{R}s, & s\leq 0 \qquad (2.5.23b)\end{cases}$$

Introducing a "critical" surface separation distance $s_{cr}=-b\Phi(0)$, the potential for negative surface separations is transformed into $V_N(s/b<0) = 4\pi\gamma\overline{R}(s-s_{cr})$. This finding allows the physical interpretation that the proximity interaction counterbalances the surface energy of elements facing each other at distances $D<s_{cr}$. This is obvious for a separation $s=0$ of the nuclear tips, where, according to Equation (2.5.23b), the magnitude of the interaction energy is equal to the surface energy of two caps that would be erased at a separation of $s=-s_{cr}$. The conditions in the interface region would then be similar to those inside a neck of radius $\rho=(2\overline{R}s_{cr})^{1/2}$, which would actually form, given sufficient time for the system to readjust its shape. Although illustrated only for the special case of zero tip separation ($s=0$), the appearance of a neck comprising all parts of the dinuclear configuration with juxtaposed surface elements less than $D=s_{cr}$ apart is a general consequence of the proximity force. One expects then that for surface separations $s<s_{cr}$ such a minimum or "geometrical" neck will be formed between two interacting nuclei. Its radius is approximately given by

$$\rho_g\approx\left[2\overline{R}(s_{cr}-s)\right]^{1/2}, \qquad s<s_{cr} \qquad (2.5.24)$$

For large overlaps, this approximate expression for the geometrical neck radius has to be expanded to include higher orders of $s/2\bar{R}$ (Swi 79). The length of the geometrical neck is trivially $l_g = s_{cr}$ for small neck radii. However, since the formation of this minimum neck is entirely due to the action of the proximity force on the surface elements, which must, eventually, disappear for large neck radii, Swiatecki (Swi 79) proposed to apply a modification such that

$$l_g = s_{cr}(1 - \rho/2\bar{R}) \tag{2.5.25}$$

Once the crevice is completely filled in, $l_g = 0$, and the geometrical neck disappears. One can express the nuclear interaction potential [cf. Equations (2.5.23)] in terms of r and the neck variable ρ, as suggested by Swiatecki (Swi 79) and Randrup (Her 80, Ran 81):

$$V_N(r,\rho) \approx -2\pi\gamma\left[\rho^2 - \rho(l - l_g)\right] + 4\pi\gamma\bar{R}b\Phi(l/b)\exp(-\rho^2/2\bar{R}b) \tag{2.5.26}$$

with

$$l \approx s + \rho^2/2\bar{R} = r - C_1 - C_2 + \rho^2/2\bar{R} \tag{2.5.27}$$

The potential described by Equation (2.5.26) represents an approximation to that corresponding to Equations (2.5.23) and allows for additional surface energies associated with necks larger in size than in the geometrical limit. For vanishing neck radius ρ, Equation (2.5.26) is identical to Equation (2.5.23a). Along the geometrical path where $\rho = \rho_g$, the surface energy terms for the two potential expressions correspond to each other. However, whereas the proximity term in Equation (2.5.26) arises from an attempt to correct for a decrease of this contribution for large neck radii, where the angle between the opposing surfaces is large, no such correction is made in the derivation of the potential given by Equation (2.5.23b).

It should be remembered that the above discussion pertaining to the formation of a neck and the erasure of nuclear surfaces implies an adiabatic situation to be prevalent in a nuclear collision, where the system has sufficient time to relax its shape at each stage of the reaction. For higher relative velocities, this adiabatic limit may not provide an accurate enough description of the process. In the proximity picture discussed here, one would then expect a gradual transition with increasing incident energy from a situation characterized by a potential given by Equations (2.5.23) to one where the potential of Equation (2.5.23a) would be applicable also for separations $s < 0$. Nuclear matter densities in excess of the equilibrium value that are possible in the sudden limit would result in a potential with a repulsive core at small separation. Although microscopic calculations (Zin 77, Flo 78,78b, Bon 78b) indicate the presence of dynamical effects in damped reactions leading to repulsive contributions to the interaction potential, as will be discussed in Section 2.6, their energy dependence is as yet not well understood.

Since, in the proximity formalism, the interaction potential and force separate into geometrical factors and universal radial form factors, the latter can be evaluated considering the interaction of two slabs of semi-infinite nuclear matter opposing each other with parallel equivalent sharp surfaces. Randrup (Ran 76) and Blocki *et al.* (Blo 77) have performed calculations of bulk and surface properties of nuclear matter employing a phenomenological momentum-dependent nucleon–nucleon interaction (Sey 61, Sey 63). The many-body problem was solved in a Thomas–Fermi approach. In this method, the local kinetic-energy density t is related to the matter density ρ according to $t(\mathbf{r}) \sim \rho(\mathbf{r})^{5/3}$, while the assumed nucleon–nucleon interaction gives rise to a single-particle potential. The total-energy density may then be varied with respect to the matter density under the constraint of a constant number of particles, in order to derive iteratively the matter and energy density distributions corresponding to the minimum total energy of the system. The surface energy is defined similarly to the second term in Equation (2.5.17). The form factor $\Phi(s)$ representing the specific interaction between two flat surfaces a distance s apart is derived from the difference between the energy corresponding to two density distributions superimposed at a distance s and that for infinite separation. Similar calculations have also been performed by Ngô *et al.* (Ngô 75,75b) in the "energy-density formalism" of Brueckner *et al.* (Bru 68) and by Brink *et al.* (Bri 78), who employed the energy functional of the Skyrme interaction.

The proximity potential and force form factors resulting from their Thomas–Fermi calculation have been tabulated by Blocki *et al.* (Blo 77). For $\zeta = s/b > 2.74$ fm, both form factors are exact exponentials with a constant determined by the range of the Yukawa force and the surface diffuseness. The potential form factor [cf. Equation (2.5.21)] is well approximated (Blo 80c) by the formula

$$\Phi(\zeta) = \begin{cases} -4.41\exp(-\zeta/0.7176), & \zeta > 1.9475 \\ -1.7817 + 0.927\zeta + 0.0169\zeta^2 - 0.0514\zeta^3, & 0 < \zeta < 1.9475 \\ -1.7817 + 0.927\zeta + 0.143\zeta^2 - 0.09\zeta^3, & \zeta < 0 \end{cases} \tag{2.5.28}$$

Slightly different parameterizations have also been proposed (Blo 80). This form factor applies to gap geometries, where the matter density distributions are simply superimposed such that, for $\zeta = s/b < 0$, matter densities of twice the equilibrium value occur. In the adiabatic limit, where crevices are formed instead of gaps, Φ should be replaced by

$$\hat{\Phi}(\zeta) = \begin{cases} \Phi(\zeta), & \zeta \geq 0 \\ -1.7817 + \zeta, & \zeta \leq 0 \end{cases} \tag{2.5.29}$$

The proximity potentials for the system ^{209}Bi + ^{84}Kr employing the above form factors Φ and $\hat{\Phi}$ are displayed in Figure 18 as long-dashed and solid curves, respectively. As can be observed from this figure, the proximity potential is a factor of ~ 2 shallower than the single-folding potential [cf. Equation (2.5.14)] near the strong-absorption radius and compares well with the potential of Krappe *et al.* [cf. Equation (2.5.16)]. In contrast, the "energy-density" potential of Ngô *et al.* (Ngô 75, 75b) has a shorter range and is steeper than the proximity potential. For both these latter potentials, versions with a repulsive core corresponding to gap geometries with overlapping density distributions are included in Figure 18.

Various form factors for the proximity potential are compared in Figure 20. The solid curve represents the "adiabatic" or "modified" form factor of Equation (2.5.29). It is seen to agree well with the effective function $\Phi = V_N/(4\pi\gamma\bar{R}b)$ corresponding to the potential of Equation (2.5.26). This latter potential does not factorize any more into geometry and form factors, and, hence, the effective form factor depends on the radius of curvature, $\bar{R}$. A value of $\bar{R} = 2.9$ fm was used to generate the dashed and dotted curves in Figure 20. Both these curves are calculated for a minimum neck size, with the dotted curve resulting with the assumption $l_g = s_{cr}$. When the proximity correction is made to the geometrical neck length, according to Equation (2.5.25), the potential becomes shallower as observed from a comparison of the dashed curve in Figure 20 to the other form factors exhibited. Obviously, this correction may be quite significant for large overlaps. It is unfortunate, therefore, that it has so far only been roughly estimated (Swi 79, Blo 80).

The potential of Equation (2.5.26) provides a description of the nuclear interaction energy for a variety of shapes of the dinuclear system. Figure 21 displays two-dimensional contour maps of the combined Coulomb and nuclear potentials [cf. Equations (2.5.8) and (2.5.26)] for the system ^{165}Ho + ^{56}Fe plotted vs. radial distance r and neck radius ρ given in units of the projectile radius C_P. The top part of this figure corresponds to a potential where the proximity correction to the length of the geometrical neck has been made in accordance with Equation (2.5.25). The potential at the bottom of Figure 21 was calculated neglecting this correction ($l_g = s_{cr}$). Both potentials are normalized to zero at the strong-absorption radius

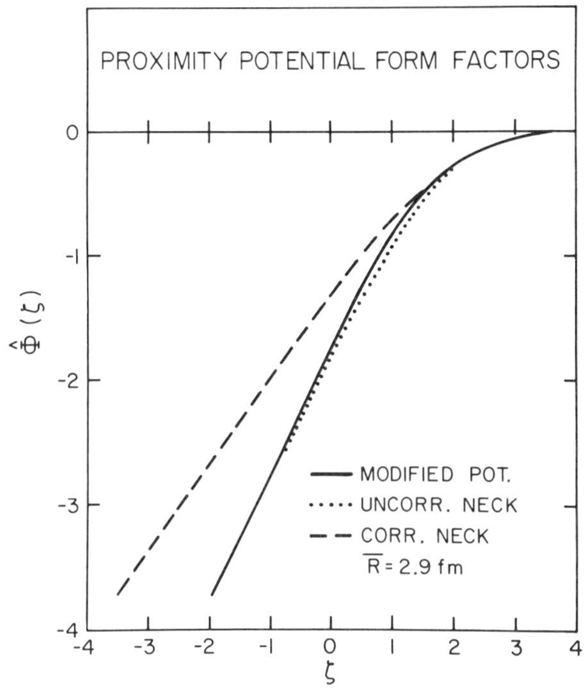

Figure 20. Form factors of various forms of the nuclear proximity interaction. The solid curve represents the modified proximity potential. The dotted curve results from implementation of a neck according to Equation (2.5.26) with $l_g = s_{cr}$. The dashed curve accounts for the vanishing of the proximity correction [cf. equation (2.5.25)] for large angles between juxtaposed surface elements. Both these latter form factors have been calculated for a reduced radius of $\bar{R} = 2.9$ fm and the minimum neck radius.

($R_{SA} = 13.2$ fm). The "forbidden" regions in these plots correspond to neck radii smaller than the geometrical limit ρ_g [cf. Equation (2.5.24)] not allowed in the adiabatic limit.

Obviously, the two potential energy surfaces displayed in Figure 21 are quite different in appearance, due to the different surface energies resulting from the two assumptions for the length l_g of the geometric neck, as discussed previously. The potential at the top of the figure is, in general, resisting larger neck sizes, whereas that shown at the bottom favors neck radii in excess of the geometrical values. The latter potential exhibits a saddle point separating an inner from the outer region. Once a system with a fixed mass asymmetry and zero angular momentum has passed this saddle, the potential will tend to drive it to fusion following the development of neck radii of the order of the nuclear radius of the smaller reaction partner, at deep interpenetrations. For heavier systems or for large angular momenta, where the Coulomb or centrifugal repulsions dominate, the saddle moves

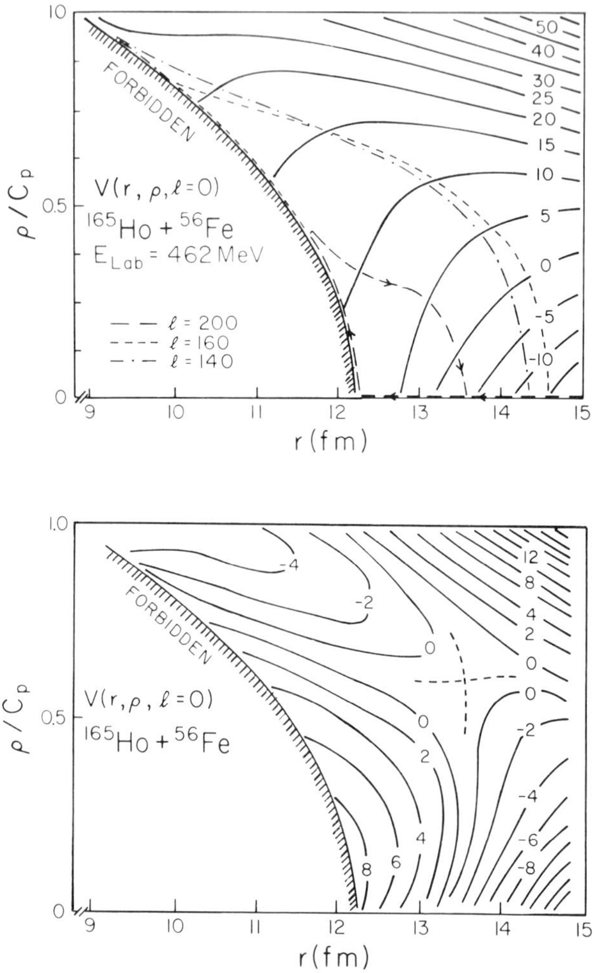

Figure 21. Two-dimensional contour maps of the sum of Coulomb potential and nuclear proximity potential plotted vs. ratio of neck radius ρ to projectile radius C_p and the center separation r. The potential at the top corresponds to a proximity correction given by Equation (2.5.25). The potential at the bottom corresponds to $l_g = s_{cr}$. Both potentials are normalized to zero at $r = R_{SA}$ (13.2 fm). Forbidden regions correspond to neck radii smaller than given by the geometrical overlap. Various dashed curves superimposed on the potential at the top represent trajectories for different l values predicted by dynamical calculations.

towards smaller surface separations, inhibiting fusion. It will eventually disappear, diminishing the differences in the two types of interaction potentials exhibited in Figure 21. In general, however, quite different reaction patterns are expected to be associated with these potentials. For example, the potential exhibiting a saddle should lead to enhanced fusion cross sections in comparison to those expected from repulsive potentials of the form displayed at the top of Figure 21. Swiatecki (Swi 80) has studied the dynamics of coalescence and reseparation in central collisions for various systems employing a potential based on dinuclear shapes illustrated in Figure 10b. For not too heavy systems, these potential energy surfaces also exhibit saddles in (ρ, r) space such as shown at the bottom of Figure 21.

It has been indicated above that the potential energies depend strongly on the system under consideration. Hence, a discussion of potential energy surfaces would remain incomplete, if mass and charge asymmetries, which are known to be variables in a damped nuclear reaction (cf. Section 3.4), were not included as additional coordinates. More adequately, the effective potential energy

$$V(1,2) = V_{\text{Coul}} + V_N + V_l + V_1 + V_2 \tag{2.5.30}$$

should be considered for an interacting dinuclear system. Here, the different terms refer to the Coulomb and nuclear interactions, the centrifugal potential for an orbital angular momentum l, and the remaining potential energies V_1 and V_2 reflecting changes in the total binding energies of the two reaction partners, normalized to zero for the initial fragmentation. All quantities entering Equation (2.5.30) are dependent on the shapes of the system, the radial distance, the division of the total mass and charge between the reaction partners, and, possibly, other parameters such as the nuclear temperature (Sto 73, Kup 74, Sau 76, Gah 77).

Frequently, the liquid-drop model (Mye 66, 67, 77b) is used to calculate the additional energies of the collision partners assuming that they remain well distinguishable throughout the reaction. In this model, the binding energy of a spherical nucleus i is described by

$$V_i = -\left(a_1 - a_2 A_i^{2/3}\right)\left[1 - \kappa\left(\frac{N_i - Z_i}{A_i}\right)^2\right] + c_3 \frac{Z_i^2}{A_i^{1/3}} \tag{2.5.31}$$

The coefficients are given by (Mye 67)

$$a_1 = 15.4941\ \text{MeV}, \qquad c_3 = 0.7053\ \text{MeV}$$

$$a_2 = 17.9439\ \text{MeV}, \qquad \kappa = 1.7826 \tag{2.5.32}$$

The modulation of the nuclear binding energies due to shell effects can be approximately accounted for (Mye 67) by adding to the liquid-drop energies V_i the correction terms

$$\Delta E_i = C\left\{(2/A_i)^{2/3}\left[F(Z_i)+F(N_i)\right]-cA_i^{1/3}\right\} \quad (2.5.33)$$

Here, $C = 5.8$ MeV and $c = 0.325$. The quantity F denotes the shell function

$$F(x) = \frac{3}{5}(x - M_{i-1})\frac{M_i^{5/3} - M_{i-1}^{5/3}}{M_i - M_{i-1}} - \frac{3}{5}\left(x^{5/3} - M_{i-1}^{5/3}\right) \quad (2.5.34)$$

where M_{i-1} is the lower magic number closest to x. Shell closures are assumed to occur for the magic numbers $\{M_i\} = \{0, 2, 8, 14, 28, 50, 82, 126, \dots\}$. Effects of these shell closures are noticeable not only in the nuclear ground-state binding energies but influence, for example, also the moment of inertia [see, e.g., (Boh 75)] and the density of levels [see, e.g., (Hui 72)]. They vary with deformation (Str 67,68; Bra 72) and generally diminish (Mor 72b) with increasing excitation energy. Theories such as the fragmentation model referred to in Section 2.4 take explicit account of these shell effects for the interacting system.

In Figure 22 examples of phenomenological potential-energy surfaces [cf. Equation (2.5.30)] derived from the above liquid-drop binding energies are shown for a range of possible fragmentations of a system with total mass number $A = 221$ and atomic number $Z = 93$. The fragmentations are characterized by proton and neutron numbers Z_1 and N_1, respectively, of one of the (spherical) fragments. The potentials are calculated according to Equation (2.5.30) with the Coulomb potential of Equation (2.5.8) and the modified nuclear proximity potential given by Equation (2.5.29). The total potential is normalized to zero for the fragmentation ^{165}Ho + ^{56}Fe indicated by a cross in the top part of Figure 22, for a radial separation equal to the strong-absorption radius ($R_{SA} \approx 13.2$ fm). This part of the figure exhibits the surface corresponding to an angular momentum $l = 0$ and a constant separation distance of $r = 13.2$ fm. The potential energies displayed at the bottom of Figure 21 are evaluated for $l = 110$ and $r = 9.6$ fm. The centrifugal term is calculated as $V = \hbar^2 l(l+1)/2\mu r^2$ in this latter case, where μr^2 is the moment of inertia of each dinuclear fragmentation with reduced mass μ. As visualized by the equipotential contour lines, the potential energy surface calculated in this approximation represents an elongated valley with relatively steep slopes in the Z_1 direction, whereas its gradient in the direction $A_1 = Z_1 + N_1$ is somewhat shallower. The point of minimum potential energy with respect to this energy surface corresponds to symmetric fragmentations: $Z_1 = Z/2$ and $A_1 = A/2$. The effect of the $N_1 = 28$ shell closure is seen in the top part of Figure 22 only as a slight local depression. This indentation is not noticeable in the potential energy surface valid for small separations and larger angular momenta such as indicated at the bottom of Figure 22.

A somewhat different perspective of the potential energy surface is depicted in Figure 23 (Mol 77) for a combined system similar to the one considered above. Here, the potential is plotted vs. radial distance r in units of the radius R_0 of the combined spherical nucleus and the mass asymmetry

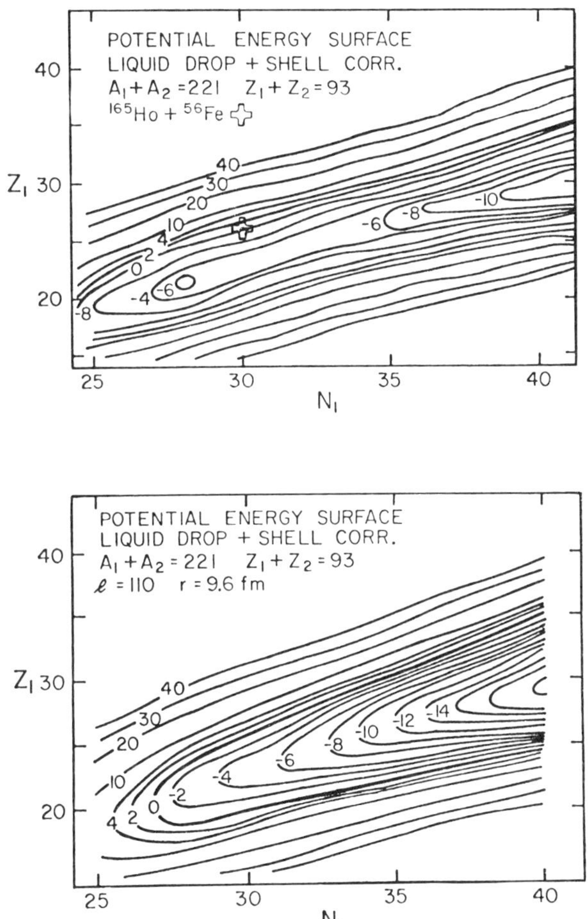

Figure 22. Contour maps of the liquid-drop model potential energy surface for dinuclear fragmentations specified by Z and N, of one of the spherical fragments, for a system with total $A = 221$ and total $Z = 93$. Shell corrections are included. The potential at the top is for $l = 0$ and $r = R_{SA}$ $(= 13.2$ fm); the potential shown in the lower half is calculated for $l = 110$ and $r = 9.6$ fm. The latter potential may be more representative for the situation at the distance of closest approach predicted by model calculations to be reached for an initial value of $l = 150$. The cross marks the fragmentation $^{165}Ho + ^{56}Fe$.

parameter η_A, in a two-dimensional contour diagram. The Coulomb potential is calculated for a uniformly charged configuration with a sharp surface described by two spheres connected by a conical neck. The nuclear interaction energy is determined from folding of a Yukawa interaction with the nuclear volume as referred to earlier in this section. Shell and pairing effects are not accounted for. The potential shown in Figure 23 corresponding to zero angular momentum exhibits a valley centered around symmetric fragmentations, for relatively large separations, and develops a saddle for smaller distances. The saddle is bordered by two "Businaro–Gallone peaks" (Bus 57) at large asymmetries. Their slopes tend to drive a system with not too large asymmetry towards symmetry, whereas large asymmetries will be increased further. These peaks occur (Kra 74) for systems with $Z^2/A > 23$ as a result of the balance between Coulomb and surface energies. For somewhat lighter systems they merge into one peak, whereas the peak positions move to larger asymmetries with increasing Z^2/A of the system.

Although it can be of illustrative value to consider various cuts of the multidimensional potential energy surface, one should realize that some of the particular features appearing in these projections, such as minima, saddles, and Businaro–Gallone peaks, are contingent upon the individual

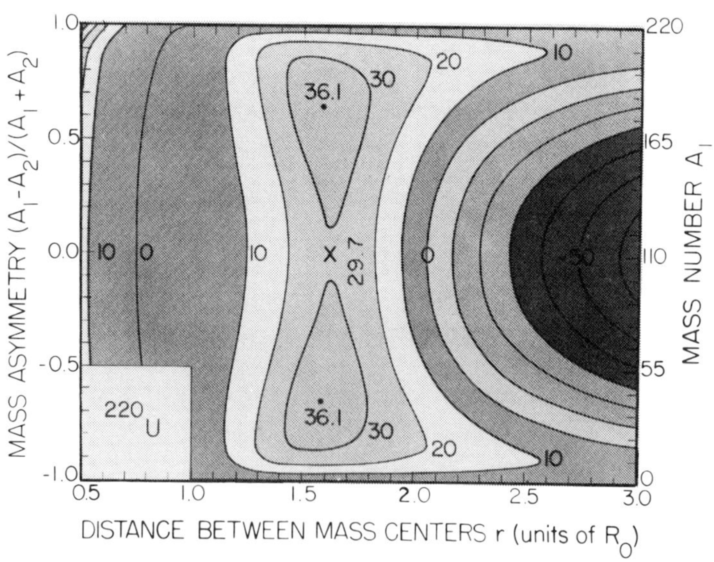

Figure 23. Contour diagram of the potential energy surface for the uniformly charged dinuclear configuration, ^{220}U, where two spherical nuclei are connected by a conical neck, plotted vs. center separation and mass asymmetry. From (Mol 77).

fixations of other degrees of freedom assumed. To the extent that such features are local (conditional), they may not be too important for the outcome of a reaction, except for coordinates along which the system evolves slowly.

A relatively complex structure of the multidimensional potential-energy surface valid for damped reactions is suggested by the observation (cf. Section 3.5) of final kinetic energies of the products corresponding to a Coulomb repulsion for very elongated fragments emerging from the reaction. The associated distances between the two mass or charge centers can be so large that approximately spherical fragments would not experience the short-range nuclear attraction. Hence, the configurations of the interacting system must be different in entrance and exit channels. If the reaction were, in fact, determined in terms of an adiabatic motion governed by a multidimensional potential, one would be led to conclude that this surface contains a valley corresponding to entrance-channel configurations, separated by a ridge from the exit-channel or fission valley. The existence of such misaligned potential valleys was postulated by Strutinsky (Str 63) and has been confirmed in other studies as well (see, e.g., Swi 72, Kra 76b, Mar 80).

From the above discussion, it is evident that the potential-energy surface can change its appearance quite drastically with changing values of the coordinates such as radial distance, orbital angular momentum, deformation, and inertia parameters. More general shapes of the interacting system than considered above are accessible within the framework of the droplet model for arbitrary shapes developed by Myers and Swiatecki (Mye 74) or by using expansion of the energy in terms of moments of the density distributions (Has 71b, Koo 76, Kra 76, Mol 77, Led 77, Tre 80). However, in judging the significance of the potential-energy surface for the outcome of a nuclear reaction, one should realize that the potential is only part of a dynamical problem, where the evaluations of inertia and dissipation present even more difficult tasks. It can hardly be expected that ground-state Q values (Q_{gg}) evaluated for fixed spherical two-fragment configurations provide a sufficient basis for an understanding of the macroscopic reaction energetics, although discussion of these Q_{gg} values can often be found in the literature.

2.6. Mesoscopic Dynamics and Induced Forces

In the preceding sections, emphasis was placed on the discussion of macroscopic and collective features of damped reactions, although it was already mentioned that dissipative interactions may arise from the microscopic nucleonic motion. On a microscopic level, the development of an interacting system is determined by the time-dependent Schrödinger equation

$$i\hbar \frac{d}{dt} \Psi = \mathscr{H} \Psi \tag{2.6.1}$$

involving the Hamiltonian operator $\mathscr{H}$ and the many-body wave function Ψ or, alternatively, by the Liouville equation

$$i\hbar \frac{d}{dt} \rho = [\mathscr{H}, \rho] =: \mathscr{L} \rho \tag{2.6.2}$$

for the corresponding density matrix ρ. Although it is, in principle, conceivable to solve these equations for a system of interacting complex nuclei, it is, in practice, not feasible to do so unless serious approximations are made, except, perhaps, for very light systems. In addition, such a completely general microscopic theory would process much redundant information that is inaccessible to experiments studying only average properties of the final reaction fragments such as reaction angle, fragment mass and charge, total kinetic energy, etc. As discussed earlier, such considerations have motivated model approaches characterized by a choice of "important" collective coordinates and intrinsic degrees of freedom whose development is not followed in any detail. Apart from the advantages in terms of physical transparency that such models offer, the penalty to be paid is the appearance of irreversibility and of induced forces due to the abandonment of information on the microscopic processes.

In order to understand the origin of these effects and to assess how their macroscopic consequences depend on details of nucleonic processes, it is necessary to consider the mesoscopic dynamics of nuclear reactions. On this level of consideration, certain averages are taken over nucleonic variables to define collective or macroscopic coordinates, but enough information about the former is retained to allow a nearly consistent dynamical study. Although, usually, approximations are made in the description of collective and intrinsic motion, this procedure is contrasted with approaches where the general properties of induced forces are calculated from a given microscopic model but then applied in an environment that may only partially or temporarily be consistent with the microscopic postulates.

A considerable number of theoretical studies have been devoted (Bec 73, Gla 74b, 76, Gro 75b, 77, 79, 79b, Hof 76b, 77b, c, 79, Ko 76, 79, Koo 76b, 77, 77b, Sin 76b, 81, Alb 77, 79, Ran 78, Blo 78, Aga 78, 79, Sat 78, Sie 78, Cas 79, Pha 79, Tak 79, Moh 80, Koh 80, Mat 80b, Yam 80, Wer 81, Pal 81, Tak 81b) to investigate by what means the reaction dynamics of the many-body system can lead to irretrievable transmutation of relative kinetic energy. Most of this work is concerned with the derivation of frictional forces; others (Bro 74, 76, 76c, 78c, 78e, Tak 79, Mun 80) emphasize the importance of particular collective doorway states in the dissipative process. Relatively little is known (Bec 73, Gro 75b, Gla 76, Joh 77, Zin 77, Flo 78, 78b, Bon 78b, Nor 79, 80d) about nondissipative forces induced in nuclear reactions through the reversible response of nucleonic degrees of freedom to changing macroscopic variables.

The various theoretical attempts to describe the forces that are dynamically induced in damped reactions differ in the extent to which coherent or stochastic processes (cf. Section 2.2) are assumed to dominate in the reaction and in the microscopic nuclear model believed to be applicable for the interacting nuclei. As will be referred to further below, most approaches are based on a statistical description of the intrinsic system and are, hence, not capable of providing a self-consistent evaluation of the interplay of coherent and stochastic processes in damped nuclear reactions. In general, a selection of a set of collective coordinates implies (Buc 79) also that not all intrinsic degrees of freedom can behave statistically independently. However, more recently, formal studies have been undertaken (Man 80) that address this rather interesting problem in an almost self-consistent way. The approach of Mantzouranis and Pauli (Man 77, 80) goes beyond TDHF models in that it takes account of a schematical two-body residual interaction V to a rather good approximation. Use is made of the property of V to divide into one part, V_L, that has a relatively long range and a smooth state dependence, and a relatively short-range contribution, V_S, whose matrix elements exhibit drastic sign fluctuations but are typically two orders of magnitude smaller than those of V_L. While V_L can be regarded, in many respects, as an effective one-body operator, though state-dependent and nonlocal, certain matrix elements involving V_S vanish on the average. Summation of the most important graphs in the perturbation series yields (Man 80) the following reduced equation for the one-body density matrix ρ:

$$i\hbar\, \partial\rho/\partial t = [T + \tilde{U}, \rho(t)] + i\sum_{m} |m\rangle \left\{ \sum_{m'} W_{mm'}(P_{m'} - P_m) \right\} \langle m| \tag{2.6.3}$$

Here, T is the kinetic-energy operator and $\tilde{U}$ denotes a self-consistent average potential including an average induced shift of the single-particle states $|m\rangle$. While the first (commutator) term in Equation (2.6.3) constitutes a problem similar to what is normally solved in TDHF models, the sum term involving the populations P_m of and the transition probabilities $W_{mm'}$ between single-particle states corresponds to a "collision term." Equation (2.6.3) can be regarded as a generalization of a master equation, by means of which the development of coherent and stochastic modes and their competition in damped reactions can be studied. It is estimated (Man 80) to be applicable for the interesting range of reaction times between $\sim 10^{-22}$ and 10^{-14} s and, hence, encompasses both, fast particle–hole excitation and slower collective modes. Although detailed calculations employing this formalism have not been performed as yet, one expects that for low excitation energies or early in the reaction, one-body effects governed by the commutator term in Equation (2.6.3) dominate, while for longer times and higher excitations the residual interaction may lead to strong configuration mixing and, hence, stochastic reaction properties described by the gain–loss terms in Equation (2.6.3) may acquire importance. However, just how well those two domains are predicted to be separated with respect to interaction time or excitation energy depends strongly on the model Hamiltonian and the degree of self-consistency imposed in the calculations. In general, the propagation of an average density matrix $\bar{\rho} = \langle \rho \rangle$ is not given by an equation analogous to Equation (2.6.2) where $\mathscr{H}$ is replaced by its collective or coherent part $\mathscr{H}_c$, but an additional term $\langle [\mathscr{H} - \mathscr{H}_c, \rho] \rangle$ appears (Zoh 79) on the right-hand side, describing the feedback of the incoherent part $\mathscr{H} - \mathscr{H}_c$ on the evolution of $\bar{\rho}$. Furthermore, as alluded to before, in order to retain the fundamental symmetries of the system, certain constraints have to be met by what can be termed a suitable "residual" interaction in a generalized sense.

In order to gain a more heuristic perception of the interactions induced on the mesoscopic level, it appears useful to discuss as an illustrative example the results of a quantal, time-dependent shell model calculation (Gla 76) applied to a relatively light system.

In this model calculation, Glas and Mosel (Gla 76) solve the time-dependent Schrödinger equation [cf. Equation (2.6.1)] for the reaction $^{16}O + ^{16}O$, describing the intrinsic system by a spherical two-center, harmonic-oscillator Hamiltonian $\mathscr{H}_0$ and a pairing-type residual interaction $\mathscr{H}'$ coupling the ground state with certain states comprising an even number of particle–hole (ph) pairs. The many-body wave function Ψ is written as a superposition

$$\Psi(\mathbf{x}, t) = \sum_n c_n(t) \Phi_n(\mathbf{x}, \mathbf{r}) \tag{2.6.4}$$

of single-particle Slater determinants Φ_n, which are eigenfunctions of $\mathscr{H}_0$ with eigenvalues E_n. These wave functions depend on the separation $\mathbf{r}$ of the two oscillators comprising $\mathscr{H}_0$. As collective variables, the separation distance $r = |\mathbf{r}(t)|$ between the oscillator centers and the orientation angle $\theta(t)$ of their common symmetry axis are chosen in the model. The ground-state energy is renormalized to match (Gla 76) the liquid-drop potential energy V corresponding to the oscillator shapes considered and corrected for the appropriate shell energies. Hence, the total energy of the $^{16}O + ^{16}O$ system is written as (Gla 76)

$$E = \langle \mathscr{H} \rangle \approx V + 2 \sum_{n,m} c_n^* c_m \left(E_n \delta_{nm} + \mathscr{H}'_{nm} \right) \tag{2.6.5}$$

Here, $\mathscr{H}'_{nm}$ are the matrix elements of the residual interaction, and the factor 2 in Equation (2.6.5) is due to the neutron–proton symmetry of the particular system. Hence, the sum runs only over neutron states.

Figure 24 shows the energies of $(0p,0h)$, $(2p,2h)$, and $(4p,4h)$ neutron configurations considered in the model involving only the two neutrons outside each of the inert ^{12}C cores, as obtained from the model Hamiltonian for different center separations. At large separations, one observes essentially the energy-doubled single-particle spectrum of ^{16}O, which develops into that of the ^{32}S-compound nucleus, for small distances. A considerable density of states occurs at intermediate separations, with various ph levels assuming successively the roles of the ground and low-lying states. The situation becomes even more complicated when more general ph combina-

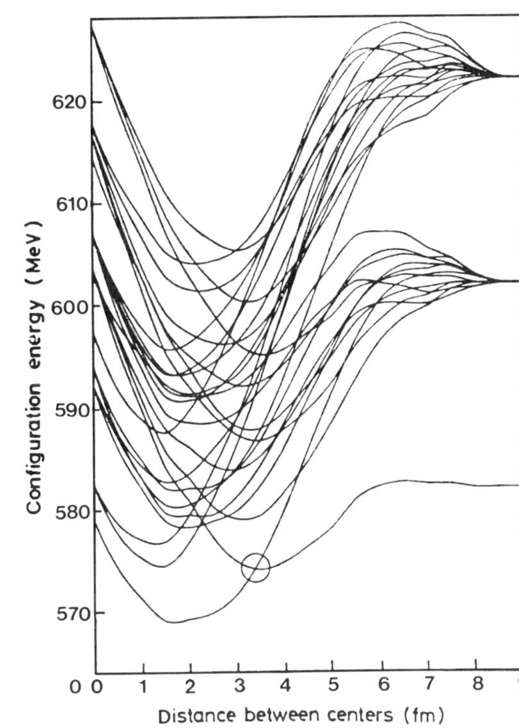

Figure 24. Variation with separation distance of the energies of $(0p,0h)$, $(2p,2h)$, and $(4p,4h)$ configurations of valence neutrons outside the two inert ^{12}C cores of a $^{16}O + ^{16}O$ system. The circle emphasizes the first crossing of the asymptotic ground state level by that corresponding to another configuration.

tions are allowed for. These configurations are coupled by both the relative collective motion and the residual interaction. One derives (Gla 76) from Equations (2.6.1) and (2.6.4) the coupled equations

$$i\hbar\dot{c}_n = E_n c_n + \sum_m \mathscr{H}'_{nm} c_m - i\hbar \sum_m \left(\dot{r} \langle \Phi_n | \partial \Phi_m / \partial r \rangle + \dot{\theta} \langle \Phi_n | \partial \Phi_m / \partial \theta \rangle \right) c_m \tag{2.6.6}$$

for the probability amplitudes c_n. In a perturbative evaluation of Equation (2.6.6), valid in the adiabatic limit of low energies, it is shown (Gla 76) that the terms involving the single-particle operators $\partial/\partial r$ and $\partial/\partial\theta$ give rise to the collective kinetic energy

$$E_{\text{kin}} \approx (m_r/2)\dot{r}^2 + (m_\theta/2)\dot{\theta}^2 = (m_r/2)\dot{r}^2 + L^2/(2\mathscr{I}) \tag{2.6.7}$$

Here, m_r and m_θ are the associated inertias, and the classical relation $L = \mathscr{I}\dot{\theta} = \text{const}$ is assumed for the angular momentum. $\mathscr{I} = \mathscr{I}(r)$ is the moment of inertia. The total energy of the system is then expressed as $E = V + E_{\text{kin}} + E^*$, where the intrinsic excitation energy $E^* = \langle \mathscr{H} \rangle - 2E_{gs}$ is calculated relative to the ground-state energy $2E_{\text{gs}}$. From total energy balance, $dE/dt = \partial E/\partial t + \dot{r}\partial E/\partial r \equiv 0$, Glas and Mosel derive the following equation of radial motion:

$$\begin{aligned} m_r \ddot{r} = &- \frac{\partial V}{\partial r} + \frac{L^2}{2\mathscr{I}^2}\frac{\partial \mathscr{I}}{\partial r} + 2\frac{\partial E_{\text{gs}}}{\partial r} - 2\sum_n \frac{\partial E_n}{\partial r}|c_n|^2 \\ &- 2\sum_{n,m} \frac{\partial \mathscr{H}'_{nm}}{\partial r} c_n^* c_m - \frac{1}{2}\frac{\partial m_r}{\partial r}\dot{r}^2 \end{aligned} \tag{2.6.8}$$

The first two terms in this equation are the familiar conservative and centrifugal forces, while the quantity $\partial E_{\text{gs}}/\partial r$ is due to the renormalization of the ground-state energy. It is the last three terms that are of most interest for the present discussion, since they represent the forces induced by the coupling between collective and intrinsic motion. Of these, the inertial force $F_I = -(\partial m_r/\partial r)\dot{r}^2/2$ is mostly reversible, i.e., nondissipative. It leads mainly to an additional repulsion which may become dominant for the radial motion at higher energies. Obviously, the contributions to the force involving the probability amplitudes c_n affect the conversion of collective radial energy into intrinsic excitation. Although a fraction of the excitation energy may be regained in the exit channel as collective energy, the high density of states (see Figure 24) and a large number of level crossings, at which transitions can be easily induced by the residual interaction $\mathscr{H}'$, lead to a quite efficient depletion of the dinuclear ground state which is essentially irreversible within the short interaction time (cf. Section 3). It is basically the long Poincaré recurrence time (Cha 43) $t_{rec} \approx 2\pi\hbar\rho$, roughly given (Nör 75) by the level density ρ, which it takes a dynamical system to return to its initial state, that introduces an effective macroscopic irreversibility of processes that are reversible on a microscopic scale. Hence, to a certain extent, one may view the effect of the residual interaction as equivalent to that of a randomizing, fluctuating force in purely statistical models.

The above, somewhat schematic example illustrates how the coupling between collective and intrinsic degrees of freedom induces dissipative and nondissipative forces. Some of the basic conclusions drawn from these model calculations are confirmed in more elaborate microscopic TDHF calculations (Flo 78,78b, Bon 78b), where, however, residual interactions are generally not considered [but see, e.g., (Koh 80)]. Induced forces are also calculated in the single-particle model of Gross (Gro 75b) and of Johnson *et al.* (Joh 77), where they are due to the response of the nucleons in one nucleus to the tail of the single-particle potential of the approaching other nucleus. The latter model is based on the linear-response theory (Hof 76b) and leads to the prediction of a largely coherent nuclear response at the earliest times which is associated, however, with only a negligible contribution to the total energy transfer, followed by incoherent processes effecting irreversible energy dissipation.

From microscopic model calculations such as discussed above, one is led to expect the appearance of nondissipative and dissipative induced interactions associated with coherent or incoherent excitations. The extent to which one or the other type of process is predicted to dominate in a damped reaction is, of course, a consequence of the model assumptions. In addition, as can be seen from the example of Equation (2.6.8), it is not obvious whether the dissipative forces are accurately representable by classical-proportional friction forces as commonly assumed in classical dynamical calculations (cf. Section 2.3). For high enough excitation energies, where individual nuclear structure is less important, one may at least expect the induced forces to exhibit a universal character depending on such average quantities as level density parameters. However, in the initial phase of the reaction, the details of nuclear structure and interaction may play a more distinctive role. For example, the nature of the doorway states through which the system passes may well be of influence for the subsequent processes. In the model of Glas *et al.* (Gla 76), these doorway states are represented by $(2p,2h)$ configurations, whereas this role is ascribed to low-lying and giant isoscalar surface modes by the model of Broglia *et al.* (Bro 74,76,76c,78c,78e), implying a high degree of coherence in damped reactions. In general, dissipation of kinetic energy through excitation of collective vibrational modes is not expected (Tak 79) to be equivalent to the effect of classical frictional forces. Such a correspondence would result only from a rather strong coupling between these vibrations and more complicated ph states, which, however, may preclude the full development of collective oscillations, as pointed out in Section 2.3.

In the following, several models will be discussed that have resulted in the derivation of dissipative forces whose properties can be tested in a comparison of dynamical calculations with experimental data. In these model calculations, usually, rather stringent assumptions are made, explicitly or implicitly, about the statistical properties of the intrinsic system and its coupling to the collective degree of freedom. Some of these theories relate also to mass transport phenomena, which are a subject of Section 2.7.

A relatively obvious way to evaluate dissipative forces follows folding procedures analogous to those employed in the derivation of conservative potentials and forces, as outlined in Section 2.5. In this picture, two-body scattering of individual nucleons of target and projectile at each other, within the region of density overlap, is considered to be the origin of dissipative forces in damped reactions. From classical analogy, one is led to expect a velocity-proportional friction force of the form

$$F(\mathbf{r}) = -c\int d^3\mathbf{r}_1\, \rho_P(\mathbf{r}_1)\mathbf{u}(\mathbf{r}_1)\rho_T(\mathbf{r}-\mathbf{r}_1) \tag{2.6.9}$$

to first approximation. Here, c is a constant, ρ_P and ρ_T denote the density distributions of projectile and target, respectively, and $\mathbf{u}$ is the average local velocity mismatch of the two ensembles of nucleons, due to the relative velocity of the two interacting nuclei. By construction, this friction force is isotropic, i.e., the friction tensor defined by Equation (2.3.3) is diagonal, with $c_{11} = c_{22} = c_{33} \sim c$. It is strongly dependent on the radial separation $\mathbf{r}$ between the two nuclei because of the exponential variation of the density distribution near the surface. As discussed already in Section 2.5, the ansatz of Equation (2.6.9) does not account for effects of the Pauli exclusion principle and, hence, may be regarded as a reasonable assumption within the general context of the model only for very dilute nuclear matter distributions, i.e., for peripheral overlaps. The above form of a friction force was first proposed by Tsang and Swiatecki (Tsa 74) and later applied by Wilczyński *et al.* (Wil 76b, Swi 76) in dynamical calculations adjusting the overall friction coefficient c to experimental data.

Albrecht *et al.* (Alb 77, 79) have considered the microscopic nucleon–nucleon scattering process in somewhat more detail, although the Pauli principle is not accounted for in these calculations, either. Assuming quasifree nucleon–nucleon scattering characterized by the momentum-dependent total cross section σ_{NN}, they derive a velocity-dependent friction force of the form

$$\mathbf{F}(\mathbf{r}) \sim \int d^3\mathbf{r}_1\, \rho_P(\mathbf{r}_1)\rho_T(\mathbf{r}-\mathbf{r}_1)\langle \mathbf{k}k\sigma_{NN}\rangle \tag{2.6.10}$$

where $\hbar\mathbf{k}=\hbar(\mathbf{K}+\mathbf{k}_1-\mathbf{k}_2)$ is the relative momentum of two scattering nucleons having intrinsic momenta $\hbar\mathbf{k}_1$ and $\hbar\mathbf{k}_2$, respectively, and $\hbar\mathbf{K}$ is the specific momentum of relative motion. The angle brackets in Equation (2.6.10) denote an average over the intrinsic nucleon momenta $\hbar\mathbf{k}_1$ and $\hbar\mathbf{k}_2$. The dissipative character of this force is implied by the assumption of a random nucleonic motion and the description of the scattering process in terms of densities and averages, in contrast to a self-consistent microscopic derivation. In order to establish the friction coefficient, Fermi-gas distributions were assumed for the intrinsic nucleon momenta, as determined by the density-dependent Fermi momenta in the semiclassical local density approximation (Alb 77). Equation (2.6.10) has the structure $F\sim j\cdot\Delta p$, where $j\sim\rho k\sigma_{NN}$ is the collision rate and Δp the momentum change per collision. Hence, k enters quadratically in this force. Consequently, the dependence of the friction force on the relative velocity u is generally more complicated than suggested by the simple ansatz of Equation (2.6.9). However, the first approximation, the force of Equation (2.6.10), is also seen (Alb 77) to be linear in u. The form factor resulting from such a linearization depends on the bombarding energy and, therefore, nonuniversal friction strength factors c are predicted with typical magnitudes ranging between 1 and 5×10^{-20} MeV fm s.

In anticipation of subjects to be discussed later, it is mentioned already here that the above classical two-body friction model implies a rather sharply defined region where nucleon interactions take place and, therefore, a nucleon mean free path that is short compared to the dimensions of the system. Consequently, although not calculated in detail, the model assumptions would lead to development of a strong temperature gradient over the extension of the system, with the possibility of very high temperatures in the center of the interaction region. Mass and charge transport (cf. Section 2.7) would be the result of an uncorrelated diffusion process in which relatively few nucleons are exchanged back and forth between the collision partners, uninfluenced by global nuclear structure as represented by the potential energy surface. The highly excited interaction region could partially decay (Wei 77c, Got 77, 79) by preequilibrium particle emission, and a final equidivision of the remaining excitation energy could be expected to occur without regard to the sizes or heat capacities of the emerging fragments.

A possible reason for the lack of an extension of the above two-body model into the areas indicated may be seen in the fact that the validity of the underlying basic assumption of a short nucleonic mean free path has not been demonstrated for damped nuclear reactions in the interesting bombarding-energy range up to a few MeV/u above the barrier. Hydrodynamical model calculations (Sie 76, Dav 76) of nuclear reactions and fission are subject to the same criticism. In fact, it has been suggested early by Gross and collaborators (Bec 73, Gro 75b, 77, 77c) that it is the influence of the time-dependent mean nuclear field of one nucleus on the individual nucleons of the reaction partner that induces the important microscopic processes mediating dissipation. More recently, Bartel and Feldmeier (Bar 80d) have performed microscopic calculations for peripheral reactions, solving the many-body Schrödinger equation in second-order perturbation theory, employing a schematic two-body interaction. It is their finding that the excitation probability for $(1p, 1h)$ states due to the mean field, which is determined by the coherent sum over two-body matrix elements, exceeds by far the $(2p, 2h)$ excitation induced by the residual two-body interaction.

Further, more illustrative support for the probable dominance of one-body interactions in damped collisions can be derived from considerations of the nucleonic mean free path λ in nuclear matter at finite temperature. Theoretical estimates of this quantity have been given in various papers (Kin 53, Weg 74b, Ber 78g, Ran 79c, Neg 81). In Figure 25, recent results obtained by Collins *et al.* (Col 80b) are exhibited. Here, the full curves represent contour lines of constant mean free path plotted in a

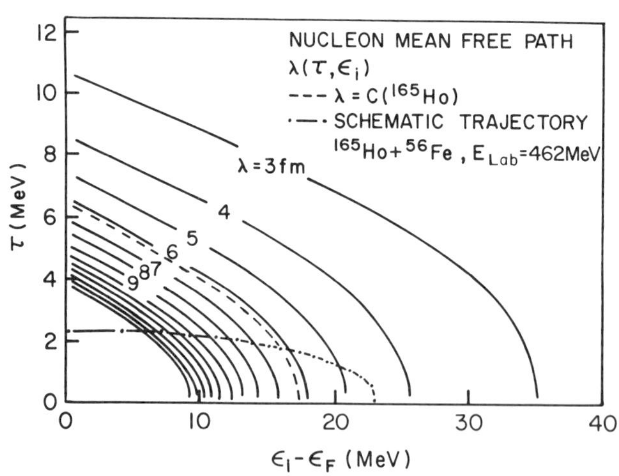

Figure 25. Contour diagram of the nucleonic mean free path λ plotted vs. nuclear temperature τ and nucleon kinetic energy ε_i above the Fermi energy ε_F according to Collins *et al.* (Col 80b). The dashed-dotted curve represents a typical path explored by the reaction $^{165}Ho + ^{56}Fe$ at $E_{Lab} = 462$ MeV for different values of initial angular momentum (see text).

two-dimensional diagram vs. nuclear temperature τ and nucleonic kinetic energy $\varepsilon_i = mv^2/2$ above the Fermi energy ε_F. In the model, the mean free path λ of a nucleon with velocity v is given by

$$\lambda(\varepsilon_i, \tau) = vR^{-1}(\varepsilon_i, \tau) = vt_{\text{rel}}(\varepsilon_i, \tau) \tag{2.6.11}$$

where R is the transition rate due to nucleon–nucleon collisions in a Fermi gas corresponding to the free nucleon–nucleon cross section, modified by the Pauli exclusion principle. t_{rel} denotes a characteristic thermal relaxation time decreasing approximately quadratically with increasing temperature τ. The resulting model λ values should be taken (Col 80b) as lower limits to the actual mean free path.

As can be observed from Figure 25 the mean free path generally decreases with increasing temperature τ or nucleon kinetic energy ε_i. In order to estimate the domain of λ values sampled in a typical damped nuclear reaction, the dashed-dotted curve in Figure 25 represents characteristic conditions experienced in the exchange of a nucleon between a ^{165}Ho and a ^{56}Fe nucleus at a ^{56}Fe bombarding energy of $E_{\text{Lab}} = 462$ MeV. The nucleon kinetic energy above the Fermi energy $\varepsilon_F = P_F^2/(2m)$ is assumed to be given by $\varepsilon_i - \varepsilon_F \approx P_F u$, where P_F is the Fermi momentum, and m is the nucleon mass. The representative relative velocity u is calculated at a radial separation $r = R_{\text{SA}}$ in the exit channel, assuming various values of energy loss E_{Loss} to have occurred, corresponding to an equilibrium Fermi-gas temperature $\tau \approx (8E_{\text{Loss}}/A)^{1/2}$. The "schematic trajectory" in Figure 25 can then be regarded as reflecting a representative relation between the temperature τ that a transferred nucleon encounters in the recipient nucleus and the maximum kinetic energy it possesses in that nucleus. Obviously, the values of λ sampled along this "trajectory" are of the order of or larger than the radius C of a ^{165}Ho nucleus given by the dashed contour line in Figure 25. Although, in the particular reaction considered, the extreme one-body limit $\lambda \gg C$ is not realized on the average, the mean free path appears to be not short enough to suggest a dominance of two-body interactions, in particular, not even at the highest temperatures attained in typical damped reactions. However, on the other hand, the above conclusions should not be taken too literally because they rest on the assumptions of a particular nuclear model and an adiabatic reaction mechanism which have to be justified by experiment.

Although anticipated to a certain extent already by Kaufmann *et al.* (Kau 61), the notion of friction was first developed and applied to damped reactions by Gross and collaborators (Gro 73, Bec 73, Gro 74, 75, 75b, 77, 77c, 77d, 78, 80d, 81). Part of the formal work of these authors does not pertain to a particular structure of the interaction between collective and intrinsic degrees of freedom that generates dissipative forces. However, it is argued in their work (see, e.g., Gro 75b) that $(1p, 1h)$ excitations induced by the time-dependent mean field are the dominant contributions to energy dissipation. In order to better visualize some of the properties of this type of "one-body" friction, it appears useful to discuss the classical analog of a rigid wall moving at velocity u through a gas of fermions, on the average at rest (see Figure 26), also termed "piston model" (Gro 75b).

In the notation of Figure 26 the classical flux $dj_+(\mathbf{p})$ of particles with relative momentum $\mathbf{p} = m\mathbf{v}$ impinging per unit area and time on the wall from the left is given by (Gro 75c)

$$dj_+(\mathbf{p}) = (2\pi\hbar)^{-3}(p_x/m)f(\mathbf{p} + m\mathbf{u})d^3\mathbf{p} \tag{2.6.12}$$

Here $f(\mathbf{p}) = f(-\mathbf{p})$ is the isotropic single-particle occupation probability. Assuming a unit reflection coefficient, a flux

$$dj'_+(\mathbf{p}) = dj_+(\mathbf{p})[1 - f(\mathbf{p}' + m\mathbf{u})] \tag{2.6.13}$$

of particles can actually undergo reflection at the wall, where the bracket modifier in Equation (2.6.13) is the Pauli blocking factor ensuring that the

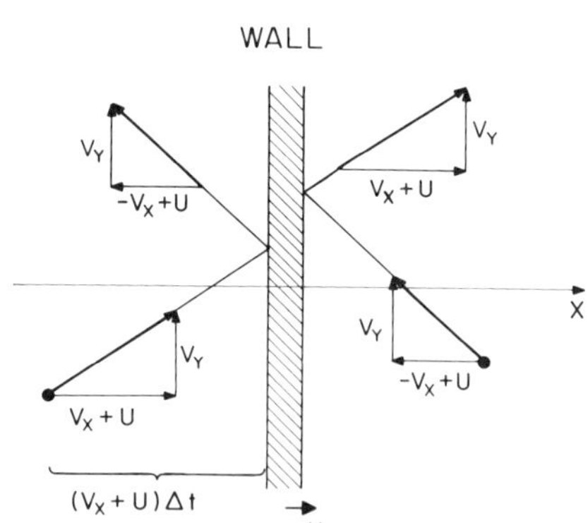

Figure 26. Illustration of the classical "piston model" for friction. Velocity diagrams of two collisions with a wall moving at velocity u are shown for nucleons impinging on it from the right and left, respectively. The net effects on the wall motion by the two types of collisions cancel, as long as these collisions occur with the same frequencies. This simulates the influence of the Pauli blocking in a quantal model.

final state for a relative component $\mathbf{p}'$, differing from $\mathbf{p}$ by a reversal of the x component, is unoccupied. The momentum imparted on the wall in a single collision is given by $\Delta p_x = 2p_x$, and it is assumed that this does not lead to a significant instantaneous change in the wall velocity u. The corresponding flux dj'_- reflected from the wall at the right is obtained from Equation (2.6.13) by replacing $\mathbf{p} \to \mathbf{p}'$. The total force per unit area exerted on the wall is then given by

$$\begin{aligned} dF &= \int \Delta p_x \left[dj'_+(\mathbf{p}) - dj'_-(\mathbf{p}) \right] \\ &= 2(2\pi\hbar)^{-3} m^{-1} \int d^3\mathbf{p}\, p_x^2 \left[f(\mathbf{p}+m\mathbf{u}) - f(\mathbf{p}-m\mathbf{u}) \right] \end{aligned} \tag{2.6.14}$$

It is interesting to note that in Equation (2.6.14) the Pauli blocking factors [cf. Equation (2.6.13)] have canceled and, hence, the same net forces would have resulted from a treatment neglecting the Pauli exclusion principle. This is so because the push the wall receives from particles scattered on the left is offset by that from particles on the right, as long as corresponding scattering events are equally likely for any given relative momentum with respect to the wall. However, both the number of scattered particles and the average momentum transfer per particle would be different in the two cases. But, as can be seen from Equation (2.6.14) only regions in single-particle space where the occupation probability varies with $\mathbf{p}$, i.e., at the Fermi surface, have a net effect on the wall motion. Obviously, single-particle momenta for which $f(\mathbf{p}+m\mathbf{u}) = f(\mathbf{p}-m\mathbf{u})$ do not contribute to the force dF. Because the difference term in the curly brackets can be approximated by the quantity $2mu\,df/dp_x$, the force of Equation (2.6.14) is a velocity-proportional friction force.

The above simple classical estimate of a friction force has been supported (Gro 75b) by a quantal perturbation calculation which is seen as a justification for the application of classical friction forces in the description of damped nuclear reactions. In this model, a classical equation of motion for the separation $\mathbf{r}$ between the centers of the interacting nuclei is obtained in terms of expectation values, making use of Ehrenfest's theorem:

$$\mu \frac{d^2}{dt^2} \langle r_\nu \rangle_t + \langle \nabla_\nu U(\mathbf{r}) \rangle_t + \langle \nabla_\nu V \rangle_t = 0 \tag{2.6.15}$$

where the expectation values $\langle \mathcal{O} \rangle = \mathrm{Tr}\{\rho(t)\mathcal{O}\}$ of operators $\mathcal{O}$ are defined in terms of the density matrix ρ solving the Liouville equation (2.6.2). In Equation (2.6.15), $U(\mathbf{r})$ is the average collective interaction potential, and $V = \mathscr{H} - \mathscr{H}_0$ is the residual interaction. Approximating ρ in first-order perturbation theory for V yields the two contributions to the induced force $F_\nu = \langle \nabla_\nu V \rangle_t$ already discussed earlier: a correction to the conservative interaction and a dissipative force corresponding to a friction tensor [cf. Equation (2.3.3)]

$$c_{\nu\mu}(t) = 2\pi\hbar\langle \mathbf{p}(t),0|F_\nu \frac{\delta(\mathscr{H}_0 - E_0)\mathscr{P}}{\mathscr{H}_i - E_{i0}} F_\mu|\mathbf{p}(t),0\rangle \tag{2.6.16}$$

that is in general anisotropic. Here, the state $|\mathbf{p}(t),0\rangle$ denotes the initial wave packet of relative motion with the intrinsic system in its ground state. $\mathscr{H}_i$ is the intrinsic part of the total unperturbed Hamiltonian $\mathscr{H}_0$ corresponding to ground-state energies E_{i0} and E_0, respectively. The appearance of the delta function in Equation (2.6.16) indicates that friction in this model is due to real excitations of intrinsic states (Gro 75b). Since friction leads to removal of flux from the elastic channel, it was argued by Gross (Gro 75b) that it can be related to the imaginary part of the optical potential. However, it is felt that such a correspondence may only exist on a phenomenological level, because friction represents a classical concept that may be related to an averaging over incoherent microscopic processes, whereas the optical potential describes the coherent scattering of a quantum-mechanical wave.

As in other approaches, irreversibility arises in the above theory because of the restriction to average values of the interesting collective and intrinsic motion. In addition, certain "memory" effects associated with the finite width in time of correlation functions appearing in the derivation of Equation (2.6.16) are neglected, resulting in further incoherence of intrinsic excitations. Some of the loss of coherence and correlation is directly associated with the application of first-order perturbation theory (Die 76, Nor 76b). Justification of such a perturbative approach depends on the choice of the residual interaction V. It has been shown by Gross *et al.* (Gro 77,78) that it is consistent with a one-body model, where excitations are induced by the single-particle potential. Although the influence of this potential corresponds to only a small distortion of each single-particle component of the many-body wave function that is tractable in first-order perturbation theory, the combined effect of many single-particle excitations constitutes a significant change of the total wave function.

From Equation (2.6.16), one expects the friction form factor to be roughly proportional to the square of the gradient of the potential V corresponding to the residual interaction. The further approximations and assumptions made for V determine decisively the character of the predicted friction force. If, as proposed by Gross *et al.* (Gro 77,78), a single-folding-type residual interaction $V \sim V_N$ [cf. Equation (2.5.14)] is assumed, one obtains (Gro 78) a radial friction form factor

$$c_r^{\mathrm{GK}}(r) \approx 2\pi\hbar g^2 |\langle dV/dr\rangle|^2 \approx c_r^0 [dV_{12}/dr]^2 \tag{2.6.17}$$

Here, $g \approx 6A/(8\pi^2)$ is the single-particle level density at the Fermi surface of a nucleus with mass A, and V_{12} is the potential defined as in Equation (2.5.14). Because of the lesser variation of the potential in the tangential direction, the tangential friction form factor c_t^{GK} is expected to be much smaller than c_r. Numerical values of the friction constants have been estimated (Gro 78) at $c_r^0 = 4\times10^{-23}$ s MeV^{-1} and $c_r^0/c_t^0 = 400$. Both radial and tangential form factors are displayed in Figure 27 for the $^{165}\mathrm{Ho}+{}^{56}\mathrm{Fe}$ reaction. Obviously, the model assumptions lead to the prediction of a very fast radial dissipation of kinetic energy during the approach phase. The friction form factors are seen to be of significant magnitude already for separation distances larger than the strong-absorption radius R_{SA}.

In contrast to the above description, the microscopic model of Bartel *et al.* (Bar 80d) carried out also in perturbation theory suggests that one-body dissipation can be rather substantial also in the exit channel. In addition, this calculation predicts significant amounts of energy in virtual excitation that return to the collective motion in the later reaction stages. Also in deviation from models such as that of Gross *et al.*, it indicates that dissipation for increasing incident energies can be weaker than expected from an energy-independent friction form factor.

Quite similar in spirit to the "piston model" of Gross *et al.* discussed above is the one-body "wall-dissipation" model proposed by Swiatecki and collaborators (Swi 75,79, Bon 76, Ran 77c,78, Blo 78). It represents a classical second-order calculation of the kinetic-energy transfer between a classical gas of independent particles and the walls of its deforming container. As observed above, it is not necessary to include the effect of Pauli blocking in the calculation of energy dissipation in such a one-body picture. Particles with velocity v_s normal to a wall surface element dS moving with velocity u_s collide with it at a rate $dj(v_s) = f(v_s)(v_s - u_s)\theta(v_s - u_s)\,dS$ transferring an energy $\Delta\varepsilon = -2m(v_s - u_s)u_s$ in each collision. For a uniform sharp velocity density distribution $f(v) = \rho_0(4\pi v_F^3/3)^{-1}\theta(v_F - v)$, where ρ_0 is the particle density and v_F the Fermi velocity, integration over all velocities and the wall area yields a decrease of the wall kinetic energy according to (Blo 78)

$$dE/dt = (1/3)m\rho_0\langle v^2\rangle\oint dS\,u_s - m\rho_0\langle v\rangle\oint dS\,u_s^2 \tag{2.6.18}$$

where the brackets denote averaging with respect to the distribution function f. Obviously, for volume-preserving deformations of the container, the first term in Equation (2.6.18) vanishes. It represents the reversible energy transfer given by the ideal-gas law. The term quadratic in the surface velocities u_s is dissipative and describes the rate of intrinsic excitations due to the dynamical change of the shape of the container.

As discussed by Blocki *et al.* (Blo 78) and Swiatecki (Swi 79), Equation (2.6.18) applies to a situation, where the container surface is randomly

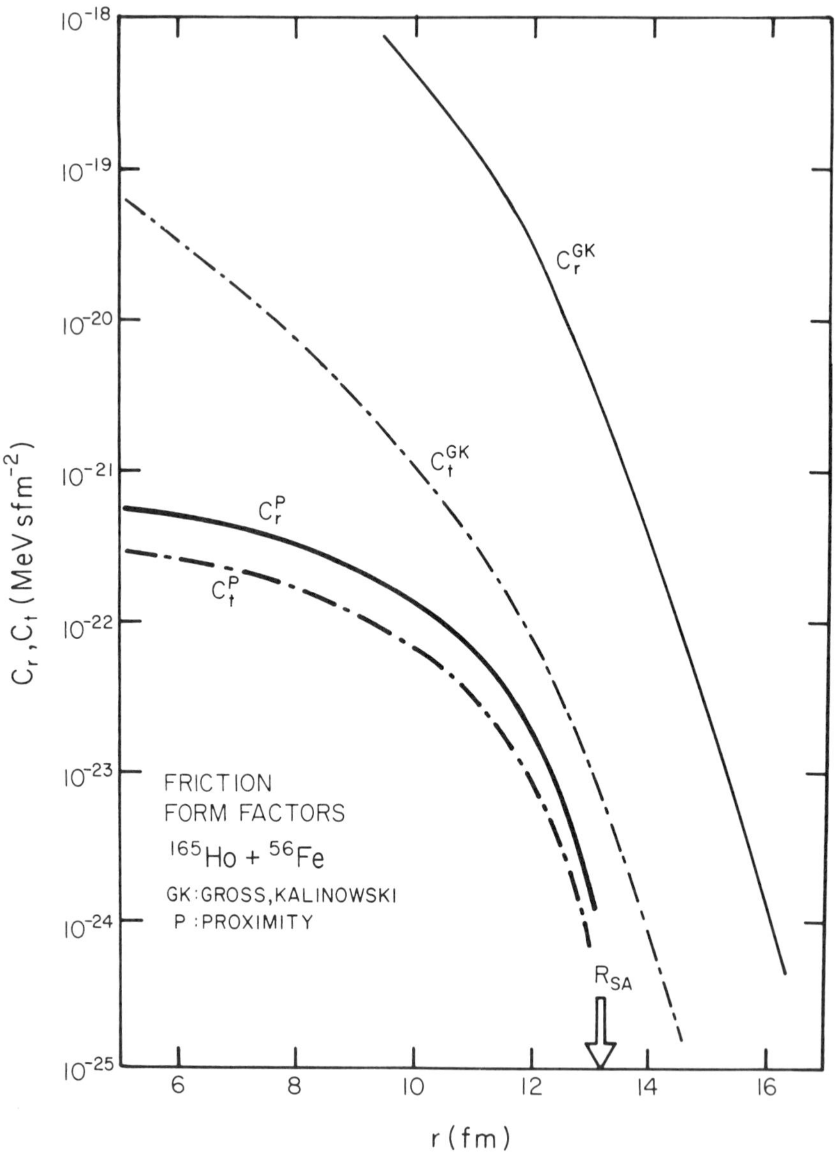

Figure 27. Radial and tangential form factors of the friction forces proposed by Gross *et al.*, (c_r^{GK}, c_t^{GK}), and those of the proximity model, (c_r^P, c_t^P).

bombarded by particles characterized by the momentum distribution of the undistorted gas at rest. These requirements are not fulfilled if there are regularities in the motion of the surface elements. For example, uniform translations or rotations of the whole container would result in spurious excitations of the gas, according to Equation (2.6.18). Such general drifts are taken into account by the "generalized wall formula" for the dissipated collective kinetic energy

$$dE/dt = -m\rho_0\langle v\rangle \oint dS\,(u_s - D_s)^2 \tag{2.6.19}$$

Here, D_s is the normal component of the average velocity $\mathbf{D} = \mathbf{u} + \boldsymbol{\Omega} \times \mathbf{R}$ at the surface of the gas container, where $\mathbf{R}$, $\mathbf{u}$, and $\boldsymbol{\Omega}$ are the radius vector, the translational velocity, and the rotational frequency of a surface element dS with respect to the center of gravity of the gas. $\mathbf{D}$ is defined such that the surface integral over the relative velocities $\oint d\mathbf{S}\cdot(\mathbf{u}-\mathbf{D})$, and that over the rotational velocities, $\oint(d\mathbf{S}\times\mathbf{R})\cdot(\mathbf{u}-\mathbf{D})$, vanish. The appearance of the wall formula as the semiclassical limit of a quantal theory has been discussed by Koonin *et al.* (Koo 77,77b) in the context of the linear-response theory (Hof 76b). There it has been shown that in the quantal analog of "wall dissipation," couplings may exist between different points on the nuclear surface. Modifications of the "wall formula" due to diffuseness and curvature of the nuclear surface and the spatial variation of the velocity field $\mathbf{u}$ are discussed in the above work and, in a classical context, by Randrup (Ran 80b).

In the classical derivation of the "wall formula" [cf. Equation (2.6.19)], a unit reflection coefficient corresponding to an infinitely high wall has been assumed. However, for a nuclear well of a realistic depth, it is possible that the additional energy $\Delta\varepsilon \approx 2mv_s u_s$ a particle receives upon scattering at the moving wall is sufficiently high to correspond to an excitation into the continuum. In this situation, there is a certain probability that the particle is emitted from the system. Such a mechanism has been considered by Bondorf *et al.* (Bon 76b,79) and others (Gro 77b, Rob 79, Bon 79c).

In the linear-response theory (Hof 76b,77b,c,79, Koo 77,77b, Wer 81), excitations of the intrinsic system are considered at each point of a classical trajectory specified by average separation $\mathbf{r}_0$ and relative momentum $\mathbf{p}_0$ which are induced by the residual one-body interaction

$$V'(\{\mathbf{x}_i\},\mathbf{r},\mathbf{r}_0) = \sum_i \{V(\mathbf{x}_i,\mathbf{r}) - V(\mathbf{x}_i,\mathbf{r}_0)\} \tag{2.6.20}$$

Here, $\mathbf{x}_i$ are the coordinates of the individual nucleons. It is further assumed that it is possible to separate V' in terms of products of form factors F_μ and coupling constants A_μ, according to

$$V' = \sum_\mu A_\mu(\mathbf{r},\mathbf{r}_0)\,F_\mu(\{\mathbf{x}_i\},\mathbf{r}_0) \tag{2.6.21}$$

Such an assumption has been shown (Hof 76b) to be justified for very peripheral collisions with a small matter overlap. The energy transfer between collective and intrinsic degrees of freedom is then given by an average of V' with respect to the intrinsic system, calculated in perturbation theory. This average is defined in terms of a density matrix that represents an intrinsic system always close to local statistical equilibrium. Accordingly, the unperturbed density matrix ρ_0 is assumed to be given by $\rho_0 = \exp(-\mathscr{H}_i/\tau_0)Z^{-1}(\tau_0)$, where $\mathscr{H}_i$ is the intrinsic Hamiltonian, τ_0 the local temperature, and $Z(\tau_0) = \mathrm{Tr}\{\exp(-\mathscr{H}_i/\tau_0)\}$ is the partition function. The intrinsic density matrix ρ is assumed to evolve in time as determined by the Liouville equation (2.6.2) for the intrinsic Hamiltonian only. Such an approach neglects certain, potentially important correlations between relative and intrinsic motion. In this approximation, the rate of dissipation of collective energy is calculated as (Koo 77) as

$$\begin{aligned} dE/dt &= -\mathrm{Tr}\{\rho\,\partial V'/\partial t\} \\ &= -\mathrm{Tr}\{\rho_0\,\partial V'/\partial t\} + i\int_0^t dt'\,\mathrm{Tr}\{\rho_0[\partial\hat{V}'(t)/\partial t, \hat{V}'(t')]\} \end{aligned} \tag{2.6.22}$$

Here, the carets denote operator representation in the interaction picture. The first term on the right-hand side of Equation (2.6.22) describes an adiabatic perturbation associated with reversible, induced forces. The integral represents inertial terms and a dissipation rate. Its character and that of the resulting dissipation function is mainly determined by correlation functions of the type $\chi(t-t') \sim \mathrm{Tr}\{\rho_0[\hat{A}(t)\hat{F}(t), \hat{A}(t')\hat{F}(t')]\}$ and how they are further evaluated. For example, it was pointed out by Gross *et al.* (Gro 78) that if, as done in the work of Hofmann *et al.* (Hof 76b), the time dependence of collective variables is neglected for the interval over which the intrinsic correlation function ($\sim \mathrm{Tr}\{\rho_0[\hat{F}(t), \hat{F}(t')]\}$) varies, a dissipation function of the form Equation (2.3.3) can be derived. However, in the limit of zero temperature, the friction force vanishes identically for a system with a discrete energy spectrum, an effect due to the adoption of a canonical intrinsic density matrix ρ_0. This deficiency is eliminated in the extended linear-response theory of Takigawa *et al.* (Tak 81b), where it is shown that the finite decay time of more realistic correlation functions is equivalent to a smearing of the system energy levels.

In some of the models discussed so far, emphasis is placed on the creation of ph excitations in either one of the two interacting nuclei. While such a mechanism may be regarded as probably dominant at very large internuclear separations, inspection of Figure 9 showing the superposition of the single-particle potentials of ^{209}Bi and ^{136}Xe nuclei separated by the strong-absorption radius suggests that already at relatively large separations neck formation may set in. The associated "potential breakthrough" is expected to lead to a more intimate communication allowing the exchange of particles between the reaction partners. Since this may occur already quite early in the reaction, it has been proposed (Swi 72, 75, 79, 80, Bas 76, Bon 76, Sch 77d, Ran 77, 77c, 78, 78c, 79d, Blo 78) that energy transfer associated with the exchange of independent particles may, in fact, constitute a major, if not the dominant, dissipation mechanism. This type of one-body dissipation has been associated with the term "window formula," as it is characterized by an interface between two interacting nuclei that is transparent to nucleon exchange.

A formulation of this "window-dissipation" mechanism has been given by Blocki *et al.* (Blo 78), in a classical context. In this model, two nuclei A and B containing each a gas of independent particles of mass m have a common interface of cross section area ΔS, a "window" (cf. Figure 28). The containers move with velocities $\mathbf{u}^A$ and $\mathbf{u}^B$ with respect to the window. In a transfer of particles with velocity $\mathbf{v}$ from A to B, nucleus A experiences an average momentum loss per particle

$$\Delta \mathbf{p}^A = m\left[\mathbf{u}^A + (\mathbf{v} - \mathbf{u}^A)\cdot\hat{z}\right] = m\left[\mathbf{u}_t^A + v_n\hat{z}\right] \qquad (2.6.23)$$

where $\mathbf{u}_t^A$ is the component of the velocity $\mathbf{u}^A$ in the (x, y) plane, and v_n is the component of $\mathbf{v}$ in the direction of the unit vector $\hat{z}$. Because of the random orientations of the component $\mathbf{v}_t$, its effect averages out in Equation (2.6.23). The force exerted on nucleus A is then calculated as the average $\mathbf{F}_{BA} = \Delta S m \rho_0 \langle(\Delta\mathbf{p}^A - \Delta\mathbf{p}^B)v_n\rangle$, taken over all particle velocities $\mathbf{v}$ with $|v| > u_n$, yielding (Blo 78)

$$\mathbf{F}_{BA} \approx \Delta S m \rho_0\left\{\tfrac{1}{3}\langle v^2\rangle\hat{z} + \tfrac{1}{4}\langle v\rangle\left[\left(\mathbf{u}_t^A - \mathbf{u}_t^B\right) + 2\left(u_n^A - u_n^B\right)\hat{z}\right]\right\} \qquad (2.6.24)$$

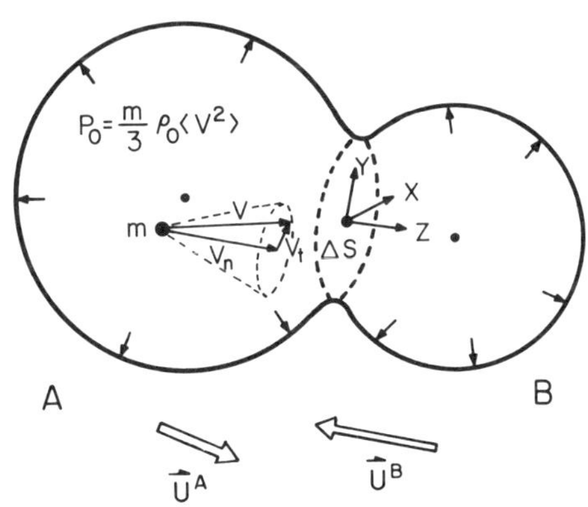

Figure 28. Illustration of the "window-dissipation" mechanism for two nuclei with an interface ΔS transparent for nucleon exchange. P_0 is the static pressure due to individual nucleons colliding at random with the nuclear circumference. Because of the random distribution of the intrinsic tangential velocity components v_t of exchanged nucleons, their effect cancels in the dissipation mechanism.

Here, the first term corresponds to the static pressure P_0 of the gas which would be present also in the absence of a window [cf. Equation (2.6.18)]. The terms proportional to the components of the relative velocity $\mathbf{u} = \mathbf{u}^A - \mathbf{u}^B$ effect an irreversible transfer of kinetic energy of relative motion into the gas of particles. Alternatively, with a small probability, nucleons may leave the system as promptly emitted particles (Bon 79) or "Fermi jets" (Rob 79, Koh 79). In general irreversibility is assumed in the model to be induced by the randomness of the local intrinsic motion, due to either irregularities in the motion of container surface elements or effected by residual two-body interactions. The total friction force acting on the relative motion is then given by the "window formula"

$$\mathbf{F} \approx n_0 \Delta S(\mathbf{u}_t + 2u_n \hat{z}) \tag{2.6.25}$$

Here, $n_0 = m\rho_0\langle v \rangle$ can be regarded as a universal bulk nucleon current for which a value of $n_0 = 2.63 \times 10^{-23}$ MeV s fm^{-4} has been estimated (Ran 78). The effective window opening ΔS is in a nuclear reaction, of course, a form factor depending on the distance and surface shape of the reaction partners, as well as on the nuclear matter density profile.

The above window dissipation mechanism has been treated by Randrup (Ran 78, 78d, 79d) in the first-order perturbation theory. In this model, the two interacting nuclei are described by separate single-particle Hamiltonians, an approach valid at most for peripheral interactions. Nucleon transitions between the reaction partners are assumed to be induced by a time-independent interaction V. In an adiabatic picture, where the single-particle motion is rapid compared to the macroscopic motion, V can induce transitions only between single-particle levels of comparable energy ε in both nuclei. Then (Ran 78d), the quantity

$$N'(\varepsilon) = 2\pi|V(\varepsilon)|^2\rho^A(\varepsilon)\rho^B(\varepsilon) \tag{2.6.26}$$

characterizes an average differential rate of transfer between orbitals of the same energy ε, where ρ^A and ρ^B are the single-particle level densities in nucleus A and B, respectively. N' is to be compared with the energy differential of the bulk current n_0 appearing in the classical equation (2.6.25). The current of actually transferred nucleons is calculated by modifying the current N' by the Pauli blocking factor, e.g., $f^A(1-f^B)$, where f^i is the occupation probability in nucleus i, and by integration over all single-particle energies. It is further assumed that the interacting nuclei are describable in terms of Fermi gases at equilibrium with a common temperature τ. The occupation probabilities f for the two nuclei moving with a relative velocity $\mathbf{u}$ can then be written in terms of the average, $\varepsilon_F = (\varepsilon_F^A + \varepsilon_F^B)/2$, and the difference of the Fermi energies, $F_{AB} = \varepsilon_F^B - \varepsilon_F^A =$

$-F_{BA}$, according to

$$f^{B}_{A}(p)=\left\{1+\exp\left[\left(\varepsilon-\varepsilon_F\pm\left(F_{AB}-\mathbf{u}\cdot\mathbf{p}\right)\right)/\tau\right]^{-1}\right. \tag{2.6.27}$$

Here, $p=(2m\varepsilon)^{1/2}$ is the nucleon momentum. The quantity $\omega=\omega_A=F_{AB}-\mathbf{u}\cdot\mathbf{p}=-\omega_B$ denotes the change in excitation energy associated with a transfer of a nucleon of momentum $\mathbf{p}$ between two orbits with the same energy ε, which is assumed to be balanced by a corresponding change of the energy of relative motion. Hence, one calculates a rate of dissipation

$$-\frac{dE}{dt}=\int d\varepsilon\, N'(\varepsilon)\left\langle\omega\left[f^{B}-f^{A}\right]\right\rangle\approx N'(\varepsilon_F)\langle\omega^2\rangle \tag{2.6.28}$$

Evaluating the flux average over all nucleon momenta indicated by the brackets $\langle\,\rangle$, Randrup (Ran 78d) arrives at the dissipation rate

$$-dE/dt\approx N'(\varepsilon_F)\left[F_{AB}^2+\left(p_F^2/4\right)\left(u_t^2+2u_n^2\right)\right] \tag{2.6.29}$$

Here, the second term, proportional to the square of the Fermi momentum p_F, corresponds to the recoil term already familiar from the classical evaluation of the "window formula" [cf. Equation (2.6.25)] discussed above. The term involving F_{AB} describes the loss of energy due to the change in binding energies. Both contributions are, obviously, irreversible, because of the quadratic dependence on velocities and the "static driving force" F_{AB}. As mentioned earlier in the discussion of the "piston model," the effects of the Pauli blocking factors cancel in the derivation of the energy dissipation rate also in this case.

Concerning the appearance of dissipation in the above model, similar arguments can be given as applied earlier to other model approaches: the coupling between collective and intrinsic motion is not treated self-consistently, but only on the average and in first-order perturbation theory. Approximations based on the assumption of very different time scales of intrinsic and collective motion lead to a neglect of coherence and correlations. Furthermore, randomization of the nucleonic degrees of freedom is postulated without further specification of the responsible mechanisms. These are some of the reasons why the above one-body model cannot be directly compared to other one-body approaches such as TDHF models. However, the model offers a rather broad range of applications, because it makes explicit connections between mass exchange and dissipation in nuclear reactions. Unfortunately, the effect of particle–hole excitations without nucleon exchange which lies in the realm of the general approach has not been considered thus far.

The nucleon current form factor N' entering the dissipation rate [cf. Equation (2.6.29)] has been evaluated by Randrup (Ran 77c, 78) in a Thomas–Fermi model considering two opposing semi-infinite slabs of nuclear matter. In this approach the local particle density is proportional to $P_F^3(r)$, where $P_F(r)$ is the local Fermi momentum. Hence, the local flux of particles $n(r)$ is given by $n(r) \sim P_F^4(r)/m$. Consequently, the minimum value $\psi(r) = n(r)/n_0$ occurring at the barrier between the two slabs, where n_0 is the flux in the bulk of the matter distribution, defines the transparency of their interface. Since ψ depends quadratically on the Fermi kinetic energy $T_F = P_F^2/(2m)$, the relative energy-differential current is connected via $\chi(r) = \partial\psi/\partial T_F = (2/T_F)\psi$ to the integral current. The evaluation of these dimensionless form factors χ and ψ allows a separation of geometry in the proximity formalism as discussed in Section 2.5. With $\mathscr{X}$ and Ψ denoting the integrals of χ and ψ, respectively, over gaps or crevices such as indicated in Figure 18, the dissipation rate [cf. Equation (2.6.29)] can be written, in the notation introduced in Section 2.5, as

$$-dE/dt = 2\pi n_0 \overline{R} b\left[m^{-1}\mathscr{X}(\zeta)F_{AB}^2 + \Psi(\zeta)\left(u_t^2 + 2u_n^2\right)\right] \quad (2.6.30)$$

These form factors have been tabulated by Randrup (Ran 78). Approximate expressions for parabolic gaps or crevices are (Ran 78d)

$$\Psi(\zeta) = \begin{cases} 0, & 3.2 \le \zeta \\ 1.6 - 0.5\zeta - \dfrac{1.8}{\pi}\sin\left(\dfrac{\zeta + 0.4}{3.6}\right), & -0.4 < \zeta < 3.2 \\ 1.4 - \zeta, & \zeta \le -0.4 \end{cases} \quad (2.6.31)$$

and

$$\mathscr{X}(\zeta) = \begin{cases} 0, & 3.35 \le \zeta \\ 2.1 - \zeta + 0.050\zeta^3 - 0.005\zeta^4, & 0 < \zeta < 3.35 \\ 2.1 - \zeta, & \zeta \le 0 \end{cases} \quad (2.6.32)$$

For comparison with the friction form factors c_r^{GK} and c_t^{GK} of the model by Gross *et al.* (Gro 78), the above proximity friction form factors $c_r^P = 4\pi n_0 \overline{R} b \Psi$ and $c_t^P = c_r^P/2$ are included in Figure 27 for the system $^{165}\mathrm{Ho} + {}^{56}\mathrm{Fe}$. As can be observed from the figure, the proximity friction form factors are generally smaller than those of Gross *et al.* and start to become significant only at separation distances inside the strong-absorption radius. The proximity friction force anisotropy of a factor of 2 is also significantly smaller than that of the latter model (~ 400).

The above friction form factors are analogous to the potential form factor of Equation (2.5.28) and are evaluated for gap and crevice geometries corresponding to frozen-density distributions. Allowance for the development of a neck of radius ρ, and length l enclosing nuclear matter at equilibrium density can be made approximately by the replacement

$$\Xi(\zeta) \to \rho^2 \left(2\pi n_0 \bar{R} b\right)^{-1} + \Xi(l/b)\exp\left(-\rho^2/2\bar{R}b\right) \qquad (2.6.33)$$

where Ξ stands for either $\mathscr{X}$ or Ψ, in analogy to the form of the nuclear potential [cf. Equation (2.5.26)]. Of course, such an adaptation of a neck degree of freedom is not quite in accord with the perturbative evaluation described above which is based on the distinguishability of the two interacting nuclei.

While the form factors of Equations (2.6.31) and (2.6.32) have been calculated in a classical Thomas–Fermi model, several authors (Ko 78b, 79, Bro 80f, Gro 81b, Pra 81) have considered the quantal tunneling probability of nucleons through the barrier formed by the superposition of the single-particle potentials of two interacting nuclei. The significance of this penetration effect can be estimated from the calculations of Ko *et al.* (Ko 79), who derived the barrier transparency classically, quantally, and in WKB approximation for Woods–Saxon-type single-particle potentials, assuming a degenerate Fermi-gas model for the nucleons. The resulting transparency functions Ψ are plotted in Figure 29 vs. the distance s between the half-potential points. It is observed that while quantal (solid curve) and WKB (long-dashed curve) calculations give similar results, the classical step-like penetrability yields a transparency (short-dashed curve) that is significantly smaller, in particular, for larger separations. However, although these results suggest an important contribution to the particle flux due to barrier penetration, they are strongly dependent on the model potentials assumed and not directly comparable to self-consistent Thomas–Fermi calculations based on a microscopic nucleon–nucleon interaction.

Certainly, one also has to consider the different contributions of neutrons and protons to the total flux. In barrier-penetration models (Bro 80f, Gro 81b), this difference is due to the additional Coulomb potential suppressing the proton flux. In the Thomas–Fermi approach, the presence of a "neutron skin" (cf. Section 2.5) gives rise to a relative enhancement of neutron flux which can be accounted for in calculations by evaluating the respective form factors corresponding to neutrons and protons at slightly different distances ζ reflecting the thickness t_A of the "neutron skin" [cf. Equation (2.5.6)].

Dissipation has also been considered in the statistical theories of Weidenmüller and collaborators (Ko 76, 79b, Aga 77b, 78, 78b, 79, 79b, Bar

78, Shl 79b) in various approximations of random-matrix models to be discussed in Section 2.7 and in that of Dietrich and Leclercq-Willain (Die 81b, 81c, Lec 81), but most frequently, phenomenological friction forces are used in actual calculations. More formal treatments of dissipation in quantal theory have been undertaken by several authors (Kan 48, 74, Kos 72, 75, Alb 75b, Imm 75, Has 75, 78, 79b, 80, Yas 76, Eck 77, Sto 79, Cas 79, Gho 81, Dor 81). These studies are of interest in a rather fundamental way but have seen, thus far, relatively few detailed applications to damped nuclear reactions.

From the discussion of current models for the description of mesoscopic dynamics given in this section, it is apparent that there is yet no

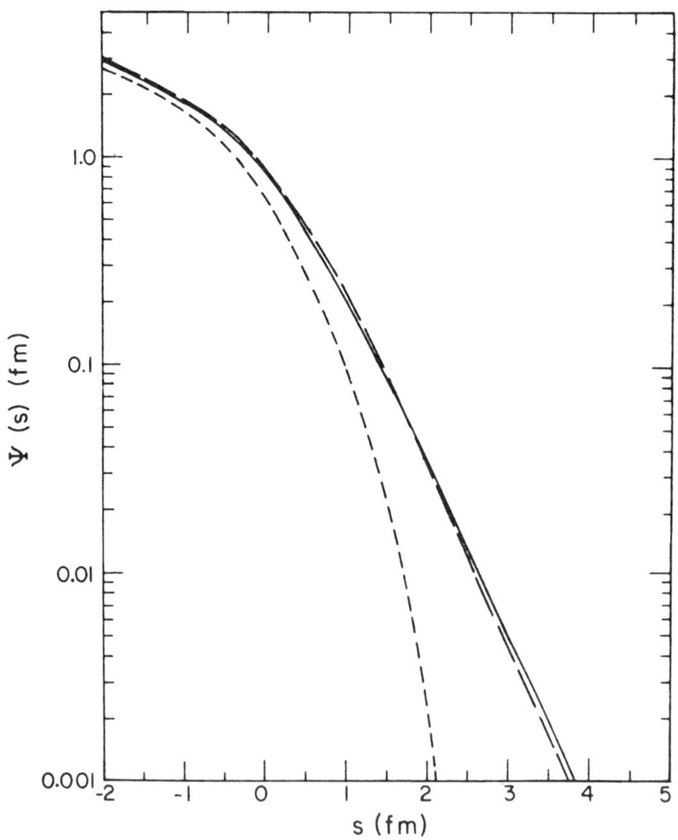

Figure 29. Flux transparency functions Ψ calculated in quantal (solid curve), semiclassical (long-dashed curve), and classical (short-dashed curve) models. Due to penetration effects, the quantal and semiclassical transparencies are much larger than the classical one. From (Ko 79).

unified theory treating coherent and incoherent reaction features in a consistent way. The various approaches differ mainly in the underlying assumptions about the relative importance of these two types of processes, about the dominant collective modes and the associated intrinsic motion, as well as about the relative coupling strength between collective and intrinsic degrees of freedom. Frequently, crucial model assumptions are made ad hoc, without a satisfying physical justification rendering an assessment of the range of applicability of various models difficult. The subject to which reactions between complex nuclei provide a unique access can certainly be regarded as one of the most interesting ones encouraging more intensive research.

2.7. Stochastic Transport Theories

The subject of this section is intimately related to some of the aspects of damped nuclear reactions already considered in the preceding sections. For example, the transformation of relative kinetic energy in damped reactions is, under certain conditions, due to an underlying energy transport mechanism. As has been discussed, such conditions may prevail when the intrinsic motion is sufficiently adiabatic such that the multiple microscopic interactions result in a smooth average evolution of the collective motion, which, in turn, determines the global environment for the intrinsic system. However, stochastic transport theories embrace a wider range of phenomena, for example, the relaxation of the initial mass and charge asymmetries, in addition to effects of kinetic-energy dissipation. With regard to the latter, transport theories predict the occurrence of friction or friction-like forces only in certain approximation schemes. By construction, such models neglect coherent nuclear excitations which is the basis of some of the approaches discussed in Section 2.6.

A typical scenario of a transport model includes a description of the average values of all relevant collective coordinates in terms of coupled equations of motion, whereas their fluctuative spread and its dependence on the average values of the set of collective variables is deduced from transport equations derived microscopically for slow or quasistationary collective motion. Depending on the extent to which microscopic statistical correlations are retained in a model, different types of transport equations such as premaster equations, master equations, Fokker–Planck, Langevin, or diffusion equations are obtained.

The derivation of transport equations usually entails an averaging over the matrix elements of the interaction between collective and intrinsic motion (Ko 76, 79b, Aga 77b, 78, 79, 79b, Bar 78, Shl 79b, Bri 79, Sal 80), an averaging over certain parts of phase space (Nor 75, 76, 79c–e, Ayi 76, 76b, 78, 78b, 79, 80, 80b, Muk 81, Wer 81) termed "coarse-graining," or assumptions about a thermodynamic distribution of the intrinsic degrees of freedom (Hof 76b,77,77b,c,79, Gro 79b), or a combination of the above.

A convenient theoretical scheme in which to visualize the progressing complexity of stochastic processes makes use of the Zwanzig–Mori projection operator techniques (Zwa 61, Mor 65,65b) in the derivation of a hierarchy of reduced equations of motion corresponding to different characteristic time scales. It has been applied in the description of frictional processes in damped interactions by Dietrich (Die 76), Mukamel *et al.* (Muk 81), and Werner *et al.* (Wer 81). In the following, this scheme is discussed in some detail in order to present a transparent example of the approximations made in a similar way also in the derivation of other transport models. According to this method (Muk 79c), first a set of operators A_{nn} is defined, whose expectations values are of interest for the description of collective degrees of freedom in a reaction. The time-dependent density matrix $\rho(t)$ can be expressed in terms of these operators according to

$$\rho(t) = (P+Q)\rho(t) = \sum_n \sigma_n(t) A_{nn} + Q\rho(t) \tag{2.7.1}$$

Here, $\sigma_n(t) = (P\rho(t)P)_n = \mathrm{Tr}[\rho(t)A_{nn}]$ are the components of the density matrix with respect to the macroscopic coordinates of interest, and $P = 1 - Q$ is the Mori projection operator. One may (Muk 79c, d), for example, construct the operators A_{nn} out of eigenstates $|n\nu\rangle$ of the free Hamiltonian $\mathscr{H}_0 = \mathscr{H} - \mathscr{H}'$, where $\mathscr{H}'$ is the interaction between macroscopic (e.g., collective) and intrinsic states described by the quantum numbers n and ν, respectively:

$$A_{nn} = (d_n)^{-1/2} \sum_\nu |n\nu\rangle\langle n\nu| \tag{2.7.2}$$

With d_n being the dimensionality of the intrinsic space compatible with the macroscopic state n, Equation (2.7.2) implies an averaging ("coarse graining") over the intrinsic space. The above set can be designed to include operators corresponding to specific states such as ground and low-lying or Giant-resonance states to which one might wish to attach particular attention.

Applying the Liouville equation (2.6.2) to the partitioned density matrix defined in Equation (2.7.1), one obtains (Muk 79d) the reduced equations of motion for the projection vectors $\boldsymbol{\sigma} = (\sigma_n)$

$$i\hbar\, d\boldsymbol{\sigma}(t)/dt = \langle \mathscr{L} \rangle \boldsymbol{\sigma}(t) - i\int_0^t dt' \langle \mathsf{R}(t-t') \rangle \boldsymbol{\sigma}(t') \tag{2.7.3}$$

The quantity $\mathsf{R}(t) = \mathscr{L}\exp\{-iQ\mathscr{L}t/\hbar\}Q\mathscr{L}$ in Equation (2.7.3) is a tetradic operator describing the propagation in time of successive transitions of the system. $\mathsf{R}(t)$ involves the Liouville operator $\mathscr{L}$ [cf. Equation (2.6.2)] and the Mori projection operator Q, and the brackets in Equation (2.7.3) indicate projection onto the space spanned by the set A_{nn}, i.e., $\langle o \rangle = PoP$. The reduced equation of motion, Equation (2.7.3), is quite general and valid for a large class of projection operators (Die 76, 81b). With a Mori operator projecting on the space of the diagonal operators A_{nn} of Equation (2.7.2) the first term in Equation (2.7.3), $\langle \mathscr{L} \rangle \sigma = [P\mathscr{H}P, P\rho P]$, vanishes. The "correlation matrix" R can be expanded in powers of $Q\mathscr{L}'$, where $\mathscr{L}'$ is the Liouville operator corresponding to the interaction $\mathscr{H}'$. To lowest order, one obtains for the population $P_n = d_n^{1/2}\sigma_n$ of a given macroscopic state $|n\rangle$

$$\frac{dP_n}{dt} = \sum_m \Omega_{nm}^2 \int_0^t dt' I_{nm}(t-t')\left[d_m^{-1}P_m(t') - d_n^{-1}P_n(t')\right] \tag{2.7.4}$$

Here, Ω_{nm}^2 is a matrix element for the transition between macroscopic states $|n\rangle$ and $|m\rangle$, averaged over all compatible intrinsic states. The function I_{nm} is a memory kernel depending on a sum over phase factors $\exp[i(E_{n\nu} - E_{m\mu})(t-t')/\hbar]$ weighted by transition matrix elements, where $(E_{n\nu} - E_{m\mu})$ are the transition energies for the states of the unperturbed system involved in the transition. It takes account of the interference between different transitions coupling two given macroscopic states. Figure 30 illustrates the regular excitation pattern due to a schematic interaction $\mathscr{H}'$ that connects states with $\Delta n = \pm 1$ and $\Delta \nu = \pm 1$ (solid arrows). Here, the energies $E_{n\nu} = \langle n\nu|\mathscr{H}_0|n\nu\rangle$ are plotted for sequences of substates $|n\nu\rangle$ belonging to each macroscopic state $|n\rangle$ which is indicated by an energy band on the left of the figure. The excitation pattern becomes significantly less structured when there is an additional residual interaction (wavy curves) that couples different substates. It gives rise to transitions (broken arrows) that were otherwise forbidden and, hence, leads to a more "random" excitation pattern, and significantly more interfering transitions contributing to the kernel I_{nm}. Thus, the figure can be used to visualize strong- and weak-coupling limits. In the weak-coupling limit, the probability for transitions between macroscopic states is low enough to allow an equilibration of the populations of the substates belonging to each macroscopic state to be accomplished by the residual interaction. In the strong-coupling limit, higher transition probabilities prevent such an equilibration from occurring.

The derivation and further evaluation of reduced equations of motion such as represented in Equation (2.7.3), where the information on the microscopic physical processes is contained in the correlation function (or operator) R, is a fundamental step in transport theories to be discussed below. By considering only projections, i.e., averages of the full density

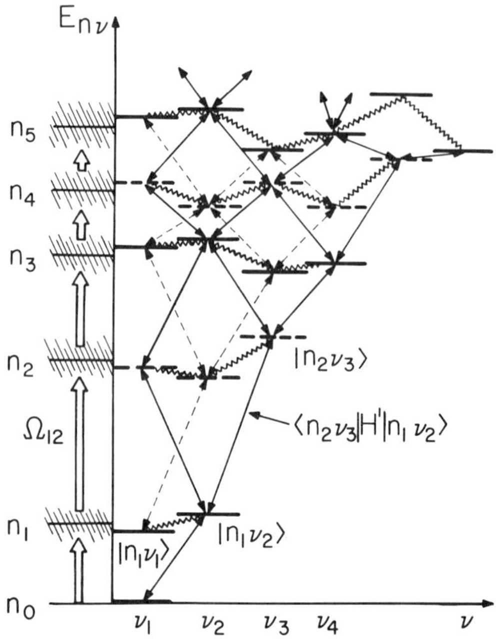

Figure 30. Illustration of various paths leading to the excitation of macroscopic states $|n_i\rangle$ and their microscopic substates $|n_i\nu_k\rangle$ by a schematical interaction H' connecting states with $\Delta n = \pm 1$ and $\Delta\nu = \pm 1$. In the strong-coupling limit, the solid arrows define the excitation pattern. In the weak-coupling limit, a residual interaction (wavy curves) effects a dissipation of the excitation probability among the various substates, before subsequent interactions (solid and broken arrows) occur.

matrix, a large class of coherence and correlation effects is already discarded. Further drastic reductions are usually implied in the derivation of equations for populations or probabilities involving more or less complex memory kernels, for the evaluation of which additional simplifying approximations are made. It is on this level that the various transport theories deviate from each other most fundamentally.

The transport equation (2.7.4) is a premaster equation for the populations P_n, containing gain and loss terms. The memory kernel $I_{nm}(t)$ has a characteristic decay or "memory" time $t_{\rm dec}$, i.e., a finite first moment, because of a destructive interference of the various phase factors at large times. If the average transition probabilities are much smaller than the inverse of this characteristic time, i.e., $\Omega^2_{nm} \ll t^{-1}_{\rm dec}$, the memory kernel effectively degenerates to a delta function in time. Consequently, the memory of the past history of excitations and deexcitations is lost, and one has a Markov process, where the time development of the populations is governed by a master equation (Pau 28)

$$dP_n(t)/dt = \sum_m \{W_{nm}P_m(t) - W_{mn}P_n(t)\} \tag{2.7.5}$$

Here, the average transition probabilities, or "transport coefficients," $W_{ij} \sim$

$\Omega_{ij}^{2} t_{\rm dec} d_j^{-1}$, contain the characteristic time $t_{\rm dec}$, which may also depend on the states involved in the transition.

The above example corresponds only to the lowest-order expansion of the correlation matrix R, i.e., direct transitions between two states. Higher orders lead to equations analogous to Equation (2.7.4), but with nested, time-ordered integrals over products of memory kernels (Muk 79d) reflecting the different paths by which a macroscopic level can be reached (see Figure 30). The description can also be extended to the nondiagonal counterparts of the operators defined in Equation (2.7.2). If successively more operators and their time derivatives are included, the method becomes more and more equivalent to a full solution of the Liouville equation (2.6.2). As pointed out by Dietrich (Die 76), it is essential to describe all slowly varying coordinates in terms of Mori operators, otherwise the technique does not in general lead to a simplification of the problem. Apart from such reservations, the theoretical scheme proves to be a powerful technique in dealing with multiple-interaction processes, avoiding the use of perturbation theory and making transparent what physical properties states and interactions must have in order for the process under consideration to show statistical features. Its success in the application to damped reactions will, however, depend on how fast the series converges in actual cases.

In contrast to the above scheme, the transport theory of Agassi, Ko, and Weidenmüller (Wei 77, 77b, 78, Aga 77b, 78, 78b, 79, 79b, Bar 78, Ko 79b, Shl 79b, Nem 80, Sal 80) is based on specific statistical assumptions on the matrix elements coupling macroscopic and intrinsic degrees of freedom. Collective excitation processes are, hence, excluded from the outset. The model is applicable for peripheral collisions where the division of the Hamiltonian $\mathscr{H} = \mathscr{H}_c + \mathscr{H}_i$ with an intrinsic part $\mathscr{H}_i = \mathscr{H}_1 + \mathscr{H}_2$ describing separated fragments is justified. The collective Hamiltonian contains the kinetic energy, an average potential U, and an interaction $V(\mathbf{r}, \xi_1, \xi_2)$ coupling relative and intrinsic motion specified by the separation $\mathbf{r}$ and the intrinsic coordinates ξ_i. The reduced multipole-matrix elements of the interaction, $\langle sI \| V_L(\mathbf{r}) Y^L \| s'I' \rangle$ for intrinsic states $|sI\rangle$ characterized by quantum number s, energy ε_s, and total spin I are assumed to form a Gaussian distribution with zero average and variance

$$\overline{\langle sI \| V_L(\mathbf{r}) Y^L \| s'I' \rangle \langle s'I' \| V_{L'}(\mathbf{r}') Y^{L'} \| s''I'' \rangle} \sim \delta_{ss''} \delta_{II''} \delta_{LL'} \alpha_L(s, s', I, I', r, r') \qquad (2.7.6)$$

which is diagonal in the representation of eigenstates of $\mathscr{H}_i$. A spin factor has been omitted in Equation (2.7.6). For the correlation factor, the ansatz is made

$$\alpha_L = \frac{W_L}{[\rho(s,I)\rho(s',I')]^{1/2}} f\left(\frac{r+r'}{2}\right)\exp\left[-\frac{(\varepsilon_s-\varepsilon_{s'})^2}{2\Delta^2}-\frac{(r-r')^2}{2\sigma^2}\right] \tag{2.7.7}$$

for which some justification is derived from shell-model calculations (Bar 78, Shl 79b). In Equation (2.7.7), $W_L \approx 10$ MeV is an overall strength factor, the radial form factor f describes the density overlap of the two fragments, and ρ denotes the joint level density. Typical values for correlation length and energy are (Bar 78, Shl 79b) $\sigma \approx 3.5$ fm and $\Delta \approx 7$ MeV, respectively, for heavy nuclei. The interaction strength W_L translates (Wei 78) into a mean free path for the reaction of typically $\lambda_R \approx 0.5$ fm.

In this model, cross sections are calculated by summing up infinite Born series, in which pairwise averaging (contraction) over all products of two reduced matrix elements leads, at each order, to only a few important terms. The symmetry achieved allows a compact formulation in terms of optical-model wave functions ψ and Green's functions $G(r,r')$. For example, the average density matrix is represented as (Aga 77b)

$$\begin{aligned}\bar{\rho}_s(\mathbf{r},\mathbf{r}') &= \delta_{si}|\psi_i^{(+)}(\mathbf{r})\rangle\langle\psi_i^{(+)}(\mathbf{r}')| \\ &\quad + \sum_t \int d^3\mathbf{r}_1\, d^3\mathbf{r}_2\, G_s(\mathbf{r},\mathbf{r}_1)\, \overline{V_{st}(\mathbf{r}_1)\bar{\rho}_t(\mathbf{r}_1,\mathbf{r}_2)V_{ts}(\mathbf{r}_2)}\, G_s^*(\mathbf{r}_2,\mathbf{r}')\end{aligned} \tag{2.7.8}$$

Here, i denotes the initial intrinsic state, and the bar indicates contraction of two matrix elements V_{st}, where the spin summation has already been carried out. In terms of this density matrix, the differential cross section can be written as

$$\frac{d^2\sigma_{i\to s}}{dE_s\, d\Omega_s} = \frac{v_s}{v_i}\lim_{r\to\infty} r^2\bar{\rho}_s(\mathbf{r},\mathbf{r})\rho_0(E_i - E_s) \tag{2.7.9}$$

where E_i and E_s are initial and final fragment energies, v_i and v_s the corresponding velocities, and ρ_0 is the intrinsic level density. Via its Wigner transform (Mes 65), $\bar{\rho}_s$ is interpreted in terms of the probability $P_s(\mathbf{r},\mathbf{k})$ of a classical statistical mixture for a given separation $\mathbf{r}$ and relative momentum $\hbar\mathbf{k}$. It obeys, according to the model of Agassi *et al.*, the transport equation

$$\begin{aligned}\left[\frac{\hbar^2}{\mu}\mathbf{k}\cdot\nabla_r - (\nabla_r U)\cdot\nabla_k\right]P_s(\mathbf{r},\mathbf{k}) &= \sum_t \int d^3\mathbf{r}'\, d^3\mathbf{k}'\big[\tilde{G}_{st}(\mathbf{r},\mathbf{k},\mathbf{r}',\mathbf{k}')P_t(\mathbf{r}',\mathbf{k}') \\ &\quad - \tilde{L}_{st}(\mathbf{r},\mathbf{k},\mathbf{r}',\mathbf{k}')P_s(\mathbf{r}',\mathbf{k}')\big]\end{aligned} \tag{2.7.10}$$

Here, U is the mean nuclear interaction potential governing collective motion, μ is the reduced mass assumed to be constant and $\tilde{G}_{st}$ and $\tilde{L}_{st}$ are gain and loss functions, respectively, involving the optical-model Green function. Equation (2.7.10) is a stationary transport equation valid in the semiclassical limit. The left-hand side corresponds to the total derivative taken along the classical trajectory, and the "collision terms" on the right describe the statistical spread of the probability.

The transport model can be extended (Aga 77b, 79b) to include mass exchange in an approximate manner. The problem of implementing the collective motion in a consistent way has been discussed in the work of Agassi (Aga 79). However, even without these extensions, a full model solution proves impractical (Ko 79b, Aga 79b). Simplified calculations have, nevertheless, provided insight into certain reaction features and suggest that quantum fluctuations, associated with the off-energy shell behavior of the microscopic transitions, may be important. It is then, in general, not possible to relate the intrinsic excitation energy to changes of the collective energies, or to calculate a nuclear temperature from the kinetic-energy loss. Furthermore, Einstein relations (see below) between transport coefficients are not expected to hold (Aga 77b, Wei 78, 80). It is concluded from these studies that the "strong-coupling limit," where the correlation length σ is long compared to the mean free path λ_R, is more likely realized in damped reactions than the opposite "weak-coupling limit." The latter can be simulated, however, in certain respects, if model parameters are reinterpreted. This allows the use of a Fokker–Planck equation for the probabilities in actual calculations, subject to certain reservations.

The strict applicability of the "strong-coupling limit" would render classical concepts such as trajectories and friction inappropriate for damped nuclear reactions. However, in this context, it appears worthwhile to remember the arguments given by Gross *et al.* (Gro 77, 78) referred to in Section 2.6, according to which perturbation theory (Van Hove limit) can be applicable for the individual single-particle components of the many-body wave function, although it is invalid for the total wave function. Consequently, it may also be permissible to employ the weak-coupling limit on this level, as proposed recently by Ayik *et al.* (Ayi 80).

Nörenberg (Nör 74) was the first to analyze damped reaction features in terms of transport equations. Master equations were applied by Moretto *et al.* (Mor 75) in the description of experimental fragment charge distributions. Microscopic models have been developed by Nörenberg and collaborators (Nör 75, 76, 79c, 80, Ayi 76, 76b, 78, 78b, 79, 80, Sch 78c, Wol 77, 78, 78b, 78c, Rie 79, 79b, 80) for the derivation of transport equations and transport coefficients, subsequently applied in the interpretation of various experimental observations.

In his quantum-statistical approach, Nörenberg (Nör 75, 79c) starts from the Liouville equation for a semiclassical Hamiltonian $\mathscr{H}(t) = \mathscr{H}_i(\xi)$

$+\mathscr{V}(\xi,\mathbf{r}(t))$ that contains an intrinsic part $\mathscr{H}_i$ corresponding to separated fragments and the interaction $\mathscr{V}$ depending on the average classical trajectory $\mathbf{r}(t)$. The many-body density matrix ρ is expanded in terms of channel wave functions $\phi_i(\xi)$, the eigenfunctions of $\mathscr{H}_i$, restricting applicability of the model to peripheral collisions. In a similar fashion as discussed previously, "coarse-graining" projection operators are defined for macroscopic states m characterized by multidimensional bins in intrinsic excitation energy, mass and charge asymmetry, etc. The resulting memory kernel R [see Equation (2.7.3) for comparison] is evaluated by taking averages over substates μ and ν within macroscopic bins m and n, respectively, connected by the nondiagonal parts $V_{m\mu,n\nu}$ of the interaction. The kernel elements R_{mn} are then written as products

$$R_{mn}(t,t') \approx \hbar^{-2}\langle V_{m\mu,n\nu}(t)V_{n\nu,m\mu}(t')\rangle G_{nm}(t,t')+\text{c.c.} \tag{2.7.11}$$

where G_{nm} describe the average propagation of the macroscopic states. This entails a rather drastic reduction of the problem and implies neglect of correlations and coherence effects that are retained in the model of Weidenmüller and collaborators, where averaging is performed for each term of an infinite Born series, giving rise to an optical-model Green's function in the corresponding kernel. The above approximation leads to a propagator that decays as a Gaussian in $\Delta t = t - t'$, with a characteristic decay or memory time $t_{\text{dec}} = t_{\text{mem}} \sim (2\text{–}3)\times 10^{-23}$ s (Nör 79c).

The form of Equation (2.7.11) for the kernel suggests consideration of two characteristic times, the autocorrelation time t_{cor}^{mn} of the matrix elements, and the decay time of the propagator, t_{dec}^{mn}, visualized as typical times for intrinsic equilibration among substates, and the time between two successive interactions, respectively. Hence, $t_{\text{dec}}^{mn} \ll t_{\text{cor}}^{mn}$ is expected in the "strong-coupling limit," and $t_{\text{cor}}^{mn} \ll t_{\text{dec}}^{mn}$ in the "weak-coupling limit." Such a separation of time scales allows time integration of R_{mn} in either limit, because of the simplified form of Equation (2.7.11), resulting (Nör 79c) in a master equation [cf. Equation (2.7.5)] in both cases. In the "strong-coupling limit," transition probabilities acquire the form (Nör 79c)

$$d_n^{-1}W_{mn}(t) = d_m^{-1}W_{nm}(t) \approx (2\pi)^{1/2}\hbar^{-2}\langle |V_{m\mu,n\nu}(t)|^2\rangle t_{\text{dec}}^{mn}(t) \tag{2.7.12}$$

where d_n, as before, is the microscopic dimensionality, or the level density in the case of a continuous variable n. It should be mentioned that the simple form of the master equation (2.7.5) with the "transport coefficients" of Equation (2.7.12) arises in the model only after further assumptions have been made about the nature of matrix elements and average propagators, by which Nörenberg's approach appears to describe a situation intermediate between the two coupling limits.

Further simplifications of the master equation (2.7.5) at the expense of additional loss of information on statistical correlations are effected by a moment expansion (Nör 74, 76, 79c) of the transition probabilities $W(x', x'', t)$ for continuous state variables x. Denoting $x = (x' + x'')/2$ and $y = x' - x''$, the second moments are defined as

$$\mu_{ij}(\mathbf{x},\mathbf{t}) = (1/2)\int d\mathbf{y}\, y_i y_j W(\mathbf{x},\mathbf{y},t) \tag{2.7.13}$$

The master equation then reduces to a Fokker–Planck equation

$$\frac{\partial}{\partial t}P(\mathbf{x},\mathbf{t}) = -\sum_i \frac{\partial}{\partial x_i}\{v_i(\mathbf{x},t)P(\mathbf{x},t)\} + \sum_{ij}\frac{\partial^2}{\partial x_i \partial x_j}\{D_{ij}(\mathbf{x},t)P(\mathbf{x},t)\} \tag{2.7.14}$$

The transport coefficients v_i and D_{ij} are known as drift and diffusion coefficients, respectively. They are defined as

$$v_i(\mathbf{x},t) = \rho^{-1}(\mathbf{x})\sum_j \frac{\partial}{\partial x_j}\{\rho(\mathbf{x})D_{ij}(\mathbf{x},t)\}$$

$$D_{ij}(\mathbf{x},t) = \rho(\mathbf{x})\mu_{ij}(\mathbf{x},t) = D_{ji}(\mathbf{x},t) \tag{2.7.15}$$

where $\rho(\mathbf{x})$ is the density of states corresponding to a particular set of macroscopic variables $\mathbf{x}$. For a diffusion tensor $\mathsf{D} = (D_{ij})$ that is independent of $\mathbf{x}$, one has the "fluctuation-dissipation theorem"

$$\mathbf{v}(\mathbf{x},t) = \mathsf{D}(t)\nabla_x[\ln\rho(\mathbf{x},t)] \tag{2.7.16}$$

If further, the state density corresponds to that of a classical canonical ensemble, i.e., $\rho(\mathbf{x},t) \sim \exp[H_i(\mathbf{x},t)/\tau]$, where $H_i(\mathbf{x},t)$ is the intrinsic energy and τ a slowly varying temperature, the above relation is reduced to

$$v(\mathbf{x},t) \approx \tau^{-1}\mathsf{D}(t)F(\mathbf{x},t) \tag{2.7.17}$$

Here, $\mathbf{F} = \nabla_x H_i$ is the "dynamical driving force" which, under the assumption of $H \approx H_c + H_i \approx \text{const}$, can also be written in terms of the collective Hamiltonian H_c as $\mathbf{F} = -\nabla_x H_c$. Equation (2.7.17) is known as the "Einstein relation."

Ayik *et al.* (Ayi 78b) have incorporated the relative motion into the transport model discussed above by expressing the density matrix terms of asymptotic channel wave functions $|\mathbf{p}, j\rangle = (2\pi\hbar)^{-3/2}\exp[i\mathbf{p}\cdot\mathbf{r}/\hbar]\phi_j(\xi)$ containing plane waves for the relative motion, specified by relative position vector $\mathbf{r}$ and momentum $\mathbf{p}$. The relation between density matrix and the joint probability distribution $P_m(\mathbf{r}, \mathbf{p}, t)$ is established by a Wigner transformation. It leads (Ayi 78b, Nör 79c) to a transport equation

$$\left[\frac{\partial}{\partial t} + \frac{\mathbf{p}}{\mu}\nabla_r - (\nabla_r U)\cdot\nabla_p\right]P(\mathbf{r},\mathbf{p},t) = -\sum_i \frac{\partial}{\partial p_i}\left[v_i(\mathbf{r},\mathbf{p},t)P(\mathbf{r},\mathbf{p},t)\right] + \sum_{ij}\frac{\partial^2}{\partial p_i\,\partial p_j}\left\{D_{ij}(\mathbf{r},\mathbf{p},t)P(\mathbf{r},\mathbf{p},t)\right\} \tag{2.7.18}$$

with drift and diffusion coefficient v_i and D_{ij}, respectively, corresponding to the momentum components p_i. The mean phase space distribution function $P(\mathbf{r}, \mathbf{p}, t)$ is obtained from summing $P_m(\mathbf{r}, \mathbf{p}, t)$ over all states m, and the probabilities $P_m(t)$ result from integration over phase space. Equation (2.7.18), together with either the master equation (2.7.5) or the Fokker–Planck equation (2.7.14) for $P_m(t)$, forms a set of coupled equations for the time evolution of the joint probability distribution for the relative motion and macroscopic observables m signifying, for example, mass or charge asymmetry. Cross sections are finally calculated from products of the probability function with classical cross sections, using methods involving classical deflection functions outlined in Section 2.1. Dietrich and Leclercq-Willain (Die 77, 81b, Lec 81) have discussed methods allowing one to retain certain quantum-coherent effects of the relative motion in the derivation of "coarse-grained" macroscopic cross sections. The transport equation (2.7.18) can be generalized to other suitable macroscopic variables by relacing the left-hand side of Equation (2.7.18) by the proper total derivative of the probability with respect to these coordinates and by inserting the associated transport coefficients on the right-hand side. The procedure has been applied by Riedel *et al.* (Rie 79), and Wolschin (Wol 79b, 79d) in the description of shape relaxation phenomena.

The statistical transport model of Hofmann and collaborators (Hof 75, 76b, 76c, 77, 77b, 77c, 79, Joh 77, Ngo 77, 79, 80b, Ber 78d, 78e, 78f, Wer 81) employs methods rather similar to the projection operator techniques outlined previously. Again, a reduced density operator (or matrix) d is defined by averaging over all microscopic (intrinsic) states associated with a given macroscopic state characterized by a set of macroscopic observables $\mathbf{x}$, i.e., by taking the trace $d(\mathbf{x}, t) = \mathrm{Tr}_i\{\rho(\mathbf{x}, t)\}$ over the intrinsic part of the full density matrix $\rho(\mathbf{x}, t)$. Although it has been emphasized (Hof 77b) that it is not essential for the applicability of the general theoretical framework, in most of the work of this group the intrinsic system is assumed to be always close to local equilibrium, i.e., describable by the density matrix ρ_i of a canonical ensemble at a slowly varying temperature τ. Crucial prerequisites of the model are, however, adiabatic motion and smallness of statistical fluctuations of collective coordinates around their mean values defining a classical trajectory. This allows expansion of the Hamiltonian around these average coordinates and a factorization of the interaction into coupling constant and form factor in analogy to Equation (2.6.21). Time correlation functions (memory kernels) appearing in the further theoretical evaluation are expressed in terms of functions $\chi_{\mp}(t, t') = (1/2)\mathrm{Tr}\{\rho_i[\hat{F}(t), \hat{F}(t')]_{\mp}\}$, where minus and plus signs stand for the commutator and anticommutator, respectively, and $\hat{F}$ is the force form factor in the interaction representation. The response function χ_- represents the average or net change, and χ_+ gives rise to a fluctuative width of collective variables due to interactions with the intrinsic system. The Fourier transforms of response and correlation functions with respect to the collective energy $\hbar\omega$ obey the relation

$$\chi_-(\omega) = (1/2)\hbar\omega\chi_+(\omega)/\tau^*(\omega) \tag{2.7.19}$$

with

$$\tau^*(\omega) = (1/2)\hbar\omega \coth[\hbar\omega/(2\tau)] \tag{2.7.20}$$

Here, τ is the intrinsic temperature. Equation (2.7.19) represents a generalized "fluctuation-dissipation theorem." The quantity τ^* is well known (Pat 78) in statistical mechanics of a linear harmonic oscillator coupled to a heat bath of temperature τ, where it describes the mean energy, i.e., $\tau^* = \langle \mathscr{H}_c \rangle$, of the oscillator corresponding to a Hamiltonian $\mathscr{H}_c = -(\hbar^2/2m)\partial^2/\partial x^2 + (m/2)\omega^2 x^2$. The probability density for an oscillator amplitude x is determined by the Gaussian

$$\langle x|\rho_i|x\rangle = \left(\frac{c}{2\pi\hbar\tau^*}\right)^{1/2} \exp\left(-\frac{x^2}{2\tau^*/c}\right) \tag{2.7.21}$$

with the variance

$$\sigma_x^2 = \tau^*/c \to \begin{cases} \hbar\omega/(2c), & \hbar\omega \gg \tau \qquad (2.7.22a) \\ \tau/c, & \hbar\omega \ll \tau \qquad (2.7.22b) \end{cases}$$

Here, c is the restoring-force constant, analogous to the form factors F in the linear-response theory. In the quantal, high-frequency or low-temperature limit of Equation (2.7.22a), the dispersion of the collective coordinate x is entirely due to quantum fluctuations already referred to in the discussion of collective modes of mass exchange (cf. Section 2.4). In the classical, high-temperature limit [Equation (2.7.22b)], this dispersion is caused by thermal fluctuations. In this latter case, Equation (2.7.19) represents a classical dissipation–fluctuation theorem or Einstein relation.

From the above simple example, it is expected that the model of Hofmann and Siemens (Hof 76b, 77c) based on the linear-response theory describes statistical features in the high-temperature limit. In the model, it is actually required that $\hbar\omega \ll 3.5\tau$, under which conditions a Fokker–Planck equation is derived for the reduced probability density $d = d(\mathbf{x}, \mathbf{p}, t)$ for macroscopic variables $\mathbf{x}$ and associated canonical momenta $\mathbf{p}$:

$$\frac{\partial d}{\partial t} = -\sum_i \left\{ \frac{p_i}{\mu}\frac{\partial}{\partial x_i} + \left[K_i^{\text{eff}} - \sum_j c_{ij}(x_j - \bar{x}_j) \right] \frac{\partial}{\partial p_i} \right\} d$$
$$+ \sum_{ij} \gamma_{ij} \frac{\partial}{\partial p_i}\left(\frac{p_i}{\mu} + \tau \frac{\partial}{\partial p_j} \right) d + \sum_{ij} \frac{m_{ij}}{\mu} \frac{\partial}{\partial p_i} \left(K_j - \tau \frac{\partial}{\partial x_j} \right) d \tag{2.7.23}$$

In this equation, μ is the collective inertia, m_{ij} describe changes in the inertia tensor, $\mathbf{K}$ is the conservative force associated with the free collective Hamiltonian, and $\mathbf{K}^{\text{eff}}$ includes the conservative forces induced by the interaction (cf. Section 2.6). The effective restoring-force constants c_{ij} contain the response function χ_-. The quantities γ_{ij} are friction coefficients describing the removal of probability from the classical trajectory at each time. All coefficients depend on the average values $\bar{x}_i$ defining this trajectory. By integrating the Fokker–Planck equation (2.7.23) over phase space, it can be shown (Hof 77c) that the time evolution of these average values $\bar{x}_i$ follows from classical equations of motion [cf. Equation (2.3.1)]. The second moments of the probability distribution are seen to obey a set of linear differential equations. Together they allow construction of the reduced density d. To perform this explicitly, however, requires microscopic calculation of the transport coefficients entering Equation (2.7.23) which has, so far, not been achieved. Rather, in actual calculations, phenomenological coefficients are used (see, e.g., Ngô 77) obtained from analysis of experimental data. Such a procedure does, of course, not provide a consistent test of the model.

The coupling of single-particle modes to an intrinsic heat bath has been considered by Brosa (Bro 79b) in a schematic model. This work illustrates the importance of randomizing, i.e., dissipative effects due to the heat bath for an equilibration of the initial mass-to-charge asymmetry of an interacting dinuclear system in the course of successive exchanges of independent nucleons. This study suggests that equilibration can, in general, not be achieved by coherent single-particle motion alone. The presence of additional, dissipative processes is essential and is implicitly assumed in transport theories postulating randomness of matrix elements or fast intrinsic equilibration. In microscopic approaches (see, e.g., Man 80), they are seen to be due to the short-range parts of residual two-body interactions (cf. Section 2.6). As pointed out by Brosa (Bro 79b) in the particular application to charge equilibration in damped reactions, collective and single-particle reaction modes can lead to indistinguishable results in the asymptotic classical limit of high temperatures.

A transparent formulation of transport phenomena in damped reactions has been given by Gross (Gro 79, 79b) in terms of the theory of Brownian motion. Here, the interacting dinuclear system is viewed as a Brownian particle with a (constant) reduced mass μ much larger than the mass m of nucleons constituting the heat bath, emphasizing the inertia and, hence, the collectivity of the motion. In this model, the relative motion specified by separation r and relative momentum p is described classically, whereas the intrinsic system is represented by the density matrix of a canonical ensemble at slowly varying temperature τ. The square-root of the mass ratio, $\xi=(m/\mu)^{1/2}$, serves as a suitable small parameter for an expansion of the propagator $\exp(iQ\mathscr{L}t/\hbar)$ such as discussed in the context of Equation (2.7.3). As before, correlations are assumed to be important only during time intervals shorter than a characteristic short memory time t_{mem}. For times $t \gg t_{\mathrm{mem}}$, the evolution of the relative motion is then observed (Gro 79b) to follow a Langevin equation

$$\partial p/\partial t \approx \exp(i\mathscr{L}t/\hbar)\langle F\rangle + F^{+}(t) + \kappa p(t)/\mu \qquad (2.7.24)$$

where $\mathscr{L}$ is the Liouville operator, $\langle F\rangle$ denotes the average, i.e., the driving force, and $F^{+}(t)$ the fluctuating force, with $\langle F^{+}\rangle = 0$. The friction term proportional to κ arises again because of the postulate of rapid randomization of the intrinsic system after distortions induced by the relative motion. Under the additional requirement that collective variables do not change appreciably during the memory time, a Fokker–Planck equation is derived (Gro 79b) for the classical phase space density $d(r, p, t)$, which is similar in structure to Equation (2.7.23) but, by construction, lacks a term due to changes in the inertia. The classical analog of the Langevin equation (2.7.24) has been applied by Rehm *et al.* (Reh 79, 79b) and Grossmann *et al.* (Gro

81) in a description of the relaxation of the mass asymmetry in damped reactions.

As in other approaches referred to earlier, in the above model the system is described in terms of well-separated nuclei, and a constant inertia is assumed for the relative motion, the only collective degree of freedom considered. It is due to these simplifications and assumptions about characteristic memory and equilibration times that the model provides a compact description of transport phenomena in damped reactions. However, it is not clear whether all approximations made in the model are valid simultaneously. For example, for the entrance channel, it is not obvious that the relative motion changes so slowly and the intrinsic system equilibrates so rapidly that correlations such as memory effects can be neglected. In order to estimate the range of applicability of such a Brownian-motion reaction model, it appears necessary to prove the fundamental assumptions in individual cases.

There have also been a number of attempts (De 78, 78b, 78c, She 78, Ran 79d, Gri 81) to describe stochastic features of damped reactions in random-walk models. The method is similar to a solution of a dynamical master equation and more general than the Fokker–Planck approximation, which assumes a relatively smooth and definite behavior of the multidimensional system trajectory such that an average trajectory is well defined in terms of average microscopic matrix elements. In fact, it is one of the virtues of random-walk models that they allow a test of such underlying assumptions. In these models, the time evolution of collective variables is usually followed in terms of classical equations of motion between microscopic transitions such as the exchange of nucleons made at random, but with the relative probabilities given by the microscopic transition rates. Coherent excitations can, of course, not be treated in a model propagating probabilities, rather than amplitudes. Although it is, in principle, possible in a random-walk approach to consider equilibration processes in all variables, usually thermal equilibrium is postulated in actual calculations.

An interesting theoretical scheme in which transport phenomena of a certain class can be treated, combines Feynman's path integral method (Fey 81) with statistical arguments (Avi 78, Bri 79, Moh 80). It has the advantage of being applicable to both weak- and strong-coupling limits and does not necessarily refer to concepts such as nuclear temperature or fluctuating forces. So far, however, detailed calculations that can be compared to experiments have not been performed.

Because of their statistical nature, transport theories predict the reaction flow to proceed in the general direction of increasing entropy, i.e., increasing intrinsic level density. While several approaches attempt to describe microscopically the way in which an interacting system that is initially far from equilibrium eventually approaches a local thermal equilibrium, others assume such an equilibrium situation to be realized at all times. Ultimately, the dinuclear complex may be expected to achieve the global equilibrium determined by the total energy content and the number of particles within the system. Such a view is held by Friedman (Fri 81), who proposes that the asymptotic probability distribution P for macroscopic variables $\mathbf{x}$ resulting from a damped reaction is always determined by the maximum of the combined intrinsic level densities ρ_i of two interacting nuclei at equilibrium, according to

$$dP(\mathbf{x})/d\mathbf{x}\, d\xi_1\, d\xi_2 = \rho_1(\xi_1)\rho_2(\xi_2)\delta(\mathbf{x} - \xi_1 - \xi_2) \qquad (2.7.25)$$

Here, ξ_i are the sets of intrinsic coordinates of the two separated nuclei that correspond to the macroscopic variables $\mathbf{x}$. The dynamics of collective motion are not followed in the model. It appears reasonable, however, to expect that such dynamics impose some constraints on the reaction, e.g., due to shape deformations, inertias, etc., that preclude the establishment of a global thermal equilibrium. A convenient way to discover from experimental data what the most important constraints are employs an information-theoretical analysis scheme (Lev 74, 78b), briefly termed "surprisal analysis." By this method, applied recently (Lev 78, 80, Alh 78, 79, 79b, 80d) to damped nuclear reactions, deviations of data from the predictions of a particular statistical model can be made more apparent.

In the discussion presented in this section, it has been emphasized that the derivation of relatively simple transport equations within the various theoretical models relies crucially on the relation between typical time scales characterizing the different aspects of a damped reaction. Although estimates of intrinsic equilibration or memory times are sometimes offered in theoretical approaches, it appears that the detail in which intrinsic equilibration or "thermalization" processes are considered does not, by far, match the fundamental significance these mechanisms have in damped nuclear reactions. Their significance results from the requirement that all degrees of freedom not participating in a thermal equilibrium be treated explicitly by realistic theories. The identification of such degrees of freedom presents a difficult task, which is not attempted in various theoretical models.

The derivation of characteristic equilibration times requires consideration of the microscopic interactions of individual nucleons within the nuclear system, as determined by a nucleon–nucleon interaction potential. In order to specify a thermal-equilibrium situation microscopically, it is necessary to adopt concepts of entropy and temperature defined in terms of microscopic quantities such as particle occupation numbers. This approach has been pursued by Wong and collaborators (Won 78c, 79, 79d, 79f) in the framework of an extended TDHF model taking account of nucleon–nucleon collisions induced by the residual interaction. In this theory, the approach to thermal equilibrium is studied through the time dependence of the entropy

$$S = -k_B(2\pi\hbar)^{-3}\int d\mathbf{p}[f(\mathbf{p})\ln f(\mathbf{p}) + \bar{f}(\mathbf{p})\ln \bar{f}(\mathbf{p})] \tag{2.7.26}$$

defined in terms of single-particle occupation numbers $f(\mathbf{p})$ and $\bar{f}(\mathbf{p}) = 1 - f(\mathbf{p})$, where $\mathbf{p}$ is a nucleon momentum and k_B the Boltzmann constant. It is shown (Won 79) that through microscopic collisions this entropy always increases (H-theorem). Under the assumption that the mean field varies slowly in time, the time dependence of the occupation numbers corresponding to the Hartree–Fock single-particle states can be approximately described by

$$\begin{aligned}\frac{\partial f(\mathbf{x}_1)}{\partial t} = {} & t_0^{-1}\int d\mathbf{x}_2\, d\mathbf{x}_3\, d\mathbf{x}_4\, \delta(\mathbf{x}_i)\delta(\mathbf{x}_i^2) \\ & \times[f(\mathbf{x}_3)f(\mathbf{x}_4)\bar{f}(\mathbf{x}_1)\bar{f}(\mathbf{x}_2) - f(\mathbf{x}_1)f(\mathbf{x}_2)\bar{f}(\mathbf{x}_3)\bar{f}(\mathbf{x}_4)]\end{aligned} \tag{2.7.27}$$

Here, $\mathbf{x} = \mathbf{p}/p_F$ is the nucleon momentum in units of the Fermi momentum p_F, and the two δ functions in Equation (2.7.27) describe the momentum and energy balance in the scattering of the initial particles 3 and 4, resulting in particles 1 and 2. The characteristic time t_0 is proportional to the inverse of the collision matrix element which is taken to be approximately constant. It is estimated to be of the order $t_0 \approx 2.6 \times 10^{-22}$ s. If the rate of energy change in such nucleon–nucleon collisions is small, the occupation probabilities are observed (Won 78c, 79, 79d, f) to redistribute, such that an irregular initial distribution eventually approaches that of a Fermi gas at a finite temperature. This behavior is illustrated in Figure 31 for a schematic model of infinite nuclear matter (Won 79f), which is initially ($t/t_0 = 0$) disturbed by a spherical shell around the Fermi momentum sphere. In this model, it is observed that the entropy [cf. Equation (2.7.26)] subsides to increase appreciably, once $t \gtrsim t_{\rm rel} = 0.5t_0$, corresponding to a thermal relaxation time of $t_{\rm rel} \approx 1.3 \times 10^{-22}$ s. The more general finding is that thermalization occurs when the collective velocities are sufficiently small and when the single-particle level density at the Fermi surface is sufficiently high. This implies also that the rate of thermalization increases with increasing temperature of the system.

Formally, Equation (2.7.27) can also be written as a nonlinear master equation

$$\frac{\partial f(\mathbf{x}_1)}{\partial t} = \int d\mathbf{x}_3[\bar{f}(\mathbf{x}_1)W(\mathbf{x}_3,\mathbf{x}_1)f(\mathbf{x}_3) - f(\mathbf{x}_1)W(\mathbf{x}_1,\mathbf{x}_3)\bar{f}(\mathbf{x}_3)] \tag{2.7.28}$$

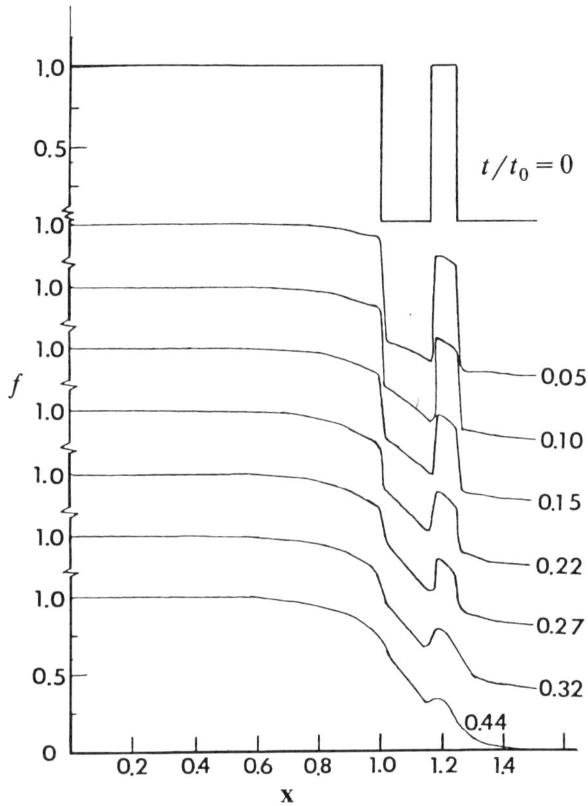

Figure 31. The time evolution of the nucleon momentum distribution $f(x = p/p_F)$, following an initial distortion in form of an extra shell around the Fermi momentum distribution. The times indicated at the curves are given in terms of the characteristic time $t_0 = 2.6 \times 10^{-22}$ s. From (Won 79).

with the transition probabilities

$$\begin{Bmatrix} W(\mathbf{x}_1, \mathbf{x}_3) \\ W(\mathbf{x}_3, \mathbf{x}_1) \end{Bmatrix} = t_0^{-1} \int d\mathbf{x}_2 \, d\mathbf{x}_4 \, \delta(\mathbf{x}_i) \delta(\mathbf{x}_i^2) \begin{Bmatrix} f(\mathbf{x}_2) \bar{f}(\mathbf{x}_4) \\ \bar{f}(\mathbf{x}_2) f(\mathbf{x}_4) \end{Bmatrix} \tag{2.7.29}$$

Here, the nonlinearities arise from the Pauli exclusion principle leading to an explicit dependence of the transition probabilities W on the time-dependent occupation numbers themselves and to the appearance of products of these numbers in the master equation (2.7.28). An analogous formulation has been given by Wolschin (Wol 81b) for the description of thermal equilibration processes in finite fermion systems. A moment expansion as described earlier yields then nonlinear equations of the Fokker–Planck type for the time evolution of initially disturbed occupation number distributions towards the equilibrium Fermi distribution. An estimate (Wol 81b) of the corresponding transport coefficients which are assumed to be constants in this work suggests values for thermal relaxation times that are generally somewhat smaller than, but still consistent with, the estimates of Wong *et al.* quoted above. Considerably larger values of these characteristic equilibration times t_{rel} are found by Bertsch (Ber 78g) in the evaluation of the collision integral [compare Equation (2.7.27)] for a momentum distribution given by a Fermi bisphere. For heavy-ion collision with an incident energy E_p in units of MeV/nucleon, he estimates

$$t_{\text{rel}} \approx 3.3\times 10^{-20} E_p^{-1}\ \text{s} \tag{2.7.30}$$

In this model, depending on the bombarding energy, deformations of the nucleon momentum distribution with respect to a single Fermi sphere, equivalent to anisotropies in the associated pressure tensor (Ber 78g), may survive until late in a collision and significantly affect the dynamics of separation of the intermediate system. Such memory effects have also important consequences for the frictional processes of energy dissipation in damped reactions as discussed by Nörenberg (Nör 81) in a study of the combined effect of diabatic ph excitations and their subsequent equilibration by two-body nucleon–nucleon collisions.

The general conclusion drawn from the model calculations discussed above is that the characteristic thermal relaxation time of a fermion system decreases with increasing nuclear temperature reflecting the decreasing efficiency of the Pauli blocking effect on the two-body collision rate at higher temperature. However, the estimated magnitude of this relaxation time is rather different in the various approaches considering different initial distortions of the Fermi momentum distribution. These distortions are assumed to exhibit rather high symmetries in momentum space, a condition that may not be too realistic for damped reactions in which a number of nucleons can successively be exchanged between the reaction partners with momentum distributions that reflect the direction of relative motion. A large number of variables may be necessary in order to describe such a deviation from thermal equilibrium which includes the possibility (Bet 38) of a temperature gradient over the nuclear system. One may then expect that the decay in time of such accumulating thermal distortions of the momentum distribution is more complicated and involves more steps than indicated by the calculations concerned with much simpler disturbances. Consequently, the approach to thermal equilibrium in damped reactions may be governed by different laws from those suggested by the schematical models discussed above.

Classical studies of thermalization processes have been performed by Kind *et al.* (Kin 53) and Weiner *et al.* (Wei 77c) considering the propagation of local thermal distortions of the temperature field $\tau(\mathbf{r}, t)$ mediated by nucleon-nucleon collisions. For example, Weiner *et al.* (Wei 77c) solve a classical heat transport equation

$$\frac{\partial \tau(\mathbf{r}, t)}{\partial t} = \frac{1}{\rho c_p} \nabla \cdot [\kappa \nabla \tau(\mathbf{r}, t)] \tag{2.7.31}$$

for an initial spatially localized "hot spot." In Equation (2.7.31), ρ is the nuclear matter density, c_p denotes the specific heat, and κ the thermal conductivity which are defined in terms of a Fermi gas. Such extreme initial conditions lead to the emission of fast particles into the continuum of the early stages of the heat diffusion process, as has been discussed by Gottschalk *et al.* (Got 77, 79) in the case of the reaction $^{58}Ni + ^{16}O$. Kind *et al.* (Kin 53) have calculated the thermal nuclear equilibration time t_R in a single-particle model, as given by the product of the nucleon transit time R/v_R, where R is the nuclear radius and v_F the Fermi velocity, and the number $R/\lambda(t)$ of nucleon–nucleon collisions:

$$t_R \approx R^2 [v_F \lambda(\tau)]^{-1} \tag{2.7.32}$$

Here, the temperature dependence of the nucleon mean free path λ is calculated from exciton transition rates. Results for this equilibration time are included in Figure 32 (Hil 79c), showing also characteristic reaction times t_{Int} and equilibrium particle evaporation times t_p for the system $^{165}Ho + ^{56}Fe$, as a function of nuclear temperature. According to these calculations, the thermal equilibration time t_R increases with increasing temperature and, hence, exhibits a dependence opposite to the approximately τ^{-2} dependence of the relaxation time t_{rel} predicted by models referred to earlier. This behavior can be understood from Equation (2.6.11) relating the nucleon mean free path to the thermal relaxation time t_{rel}.

These studies of equilibration processes have been made in the context of light-particle-induced preequilibrium reactions (see, e.g., Bla 75). Such reactions result in the well-known consequences of preequilibrium particle emission which can be studied experimentally. Hence, it appears worthwhile to establish theoretically similarities and differences between equilibration processes occurring in nucleon-induced and damped heavy-ion reactions.

The higher complexity of equilibration processes in heavy-ion reactions is immediately obvious, as their progress in time has to be evaluated, in principle, on at least two levels. The above estimates of relaxation times may approximately be applied separately to each of the two nuclei participating in a damped reaction or to the total intermediate system if it forms a mononucleus. However, the transport models discussed above postulate that the constituents of a transient dinuclear complex formed in a damped collision largely retain their separate identities. Hence, even though thermal equilibration may be achieved within each constituent on a fast time scale, a thermal disparity may conceivably exist between the two interacting nuclei. A temperature gradient over the dinuclear complex produced dynamically by the peculiarities of the energy dissipation mechanism is expected to decay within a time t_τ that is characteristic of the processes that effect a heat transport between nuclei in contact with each other. In a simple model, Randrup (Ran 81b) has considered the effect of convection due to the currents of nucleons exchanged between the reaction partners. Because of the restrictions imposed on the exchange process by Pauli blocking, transfer probabilities are relatively enhanced for nucleons from orbits corresponding to the high-energy tails of the single-particle occupation probability distribution. Since the Fermi surface of a nucleus at the higher temperature τ_1 is more diffuse than that of a nucleus at a lower temperature τ_2, the resulting asymmetry in the nucleon exchange currents tends to decrease the tempera-

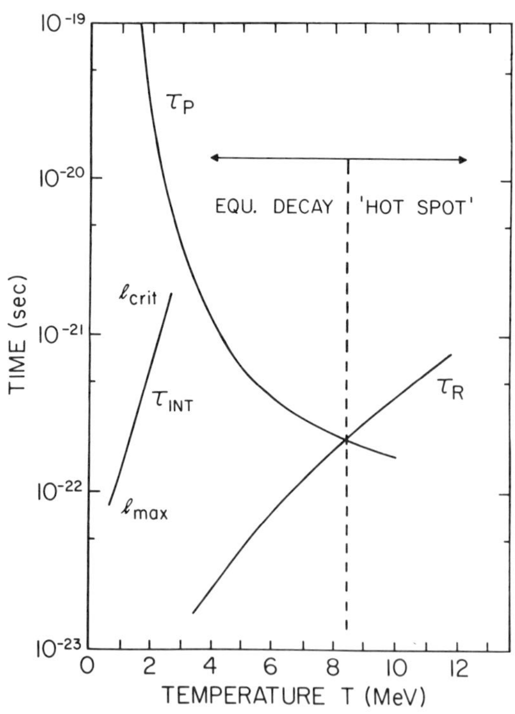

Figure 32. Dependence of characteristic times on the nuclear temperature. The quantity t_R is the thermal relaxation time given by Equation (2.7.32), whereas t_p is the compound-nucleus lifetime as determined from systematics of particle evaporation widths. Also shown are the phenomenological interaction time $t_{\rm Int}$ (cf. Section 3.3) for the ^{165}Ho+^{56}Fe reaction at $E_{\rm Lab} = 465$ MeV which are related to the temperature because of the dependence of the total excitation energy on angular momentum.

ture difference according to (Ran 81b)

$$d\left(\tau_1^2-\tau_2^2\right)/dt=-t_\tau^{-1}\left(\tau_1^2-\tau_2^2\right) \tag{2.7.33}$$

Here,

$$t_\tau=A_1A_2(A_1+A_2)^{-1}\left[\varepsilon_F N'(\varepsilon_F)\right]^{-1} \tag{2.7.34}$$

is the characteristic thermal decay time given by the mass numbers A_1 and A_2 of the interacting nuclei assumed to be at thermal equilibrium individually, and $N'(\varepsilon_F)$ is the form factor of the nucleon exchange current referred to in Section 2.6. This heat convection mechanism is expected to lose efficiency with increasing relative kinetic energy and increasing average nuclear temperature of the interacting nuclei.

Summarizing the present discussion of transport theories, it is apparent that the various microscopic models yield invariably descriptions that are too complex to allow a numerical execution and a comparison to experimental data on damped reactions. Consequently, successive approximation schemes are adopted that eventually lead to transport equations of a simpler structure, such as master, Fokker–Planck, or Langevin equations. The more detailed microscopic derivations of such equations have the advantage of providing estimates of applicability and a transparent interpretation of the corresponding transport coefficients. However, the similarity or even equivalence of the final structure of different theories, the result of successive approximations, reduces the chance that remaining differences can be readily detected from comparison to experiment. This is, in particular, true when model parameters are determined phenomenologically or when certain significant phenomena have to be excluded from consideration, because of obvious oversimplifications of certain aspects of the postulated reaction mechanism. However, the theoretical models discussed offer, in principle, interesting physical alternatives that require more detailed studies.

2.8. Applications of Transport Equations

As has been discussed in the preceding section, various theoretical models based on quite different views of the underlying microscopic reaction mechanism result in transport equations for the probability distribution of macroscopic observables, that are of similar structure. The interesting physical implications of each model are, hence, mainly contained in the transport coefficients entering these equations and the relations that connect them. Properties of transport equations and microscopically derived coefficients deserve, therefore, particular attention.

The physical significance of transport coefficients can be assessed by considering the simple case of a one-dimensional Fokker–Planck equation [cf. Equation (2.7.14)] for the probability $P(x,t)$ for a macroscopic observable x, with constant drift and diffusion coefficients $x(x,t) \equiv v$ and $D(x,t) \equiv D$, respectively. Then, Gaussian solutions

$$P(x,t) = \left(2\pi\sigma_x^2\right)^{-1/2} \exp\left[-(x-\bar{x})^2/\left(2\sigma_x^2\right)\right] \tag{2.8.1a}$$

$$\bar{x}(t) = \bar{x}(t=0) + vt, \qquad \sigma_x^2(t) = \sigma_x^2(t=0) + 2Dt \tag{2.8.1b}$$

result with average value $\bar{x}$ and variance σ_x^2 that increase linearly with time t. The time evolution of the probability distribution given by Equations (2.8.1a) and (2.8.1b) is pictured at the bottom of Figure 22, where $\Gamma = 2.35\sigma_x$ is the FWHM of P. If, as turns out to be the case (cf. Section 3.4) when mass exchange between reaction partners is considered, both temperature τ and collective kinetic energy depend relatively weakly on x, then the drift coefficient is proportional to the gradient of the collective "driving" potential U, i.e., $v \sim -\partial U/\partial x$, according to Equation (2.7.17). Constant drift and diffusion coefficients, hence, imply that U be approximately a linear function of the coordinate x, such as indicated at the top of Figure 33. In turn, in this approximation, a constant potential gradient gives rise only to a displacement of the probability distribution as a whole, but does not distort its Gaussian shape. For a planar multidimensional driving potential $U(\mathbf{x})$, Equation (2.7.14) is equivalent to a one-dimensional Fokker–Planck equation for each separate macroscopic coordinate, describing an uncorrelated transport process (Bec 78d, where the total probability is a product of the separate probabilities, i.e., $P(\mathbf{x},t) = \Pi_i P(x_i,t)$.

The solutions of the one-dimensional Fokker–Planck equation are still Gaussians (Sch 78i) in the case of a constant diffusion coefficient and a harmonic potential, i.e., for a linear drift coefficient $v(x) = -\kappa x$, with

$$\bar{x}(t) = \bar{x}(t-0)\exp(-\kappa t) \tag{2.8.2a}$$

$$\sigma_x^2(t) = \sigma_x^2(t=0)\exp(-2\kappa t) + (D/\kappa)[1-\exp(-2\kappa t)] \tag{2.8.2b}$$

For times t much smaller than the characteristic relaxation time $t_x = \kappa^{-1}$ for the macroscopic observable x, average value and variance change linearly with time. However, now correlations result (Bec 78d) for the transport along different degrees of freedom. Whereas for a constant driving force F, the probability distribution broadens indefinitely with time, a linear force $\mathbf{F} = -c\mathbf{x}$ enforces asymptotically a limiting distribution, corresponding to the statistical equilibrium defined by $\partial P/\partial t = 0$. In one dimension, this "stationary solution" of the Fokker–Planck equation reads (Pat 78)

$$P_\infty(x) = \lim_{t \to \infty} P(x,t) = \left(\frac{c}{2\pi t}\right)^{1/2} \exp\left[-\frac{(x-\bar{x})^2}{2(\tau/c)}\right] \tag{2.8.3}$$

i.e., it describes a Gaussian with an equilibrium variance $\sigma_\infty^2 = \tau/c$ determined by the temperature τ and the driving-force constant c. It is interesting to note that the equilibrium distribution of Equation (2.8.3) can also be written in terms of a generalized effective Q value, $Q_{\text{eff}}(x) = U(x) - U(\bar{x}) = (c/2)(x-\bar{x})^2$, as a simple exponential:

$$P_\infty(x) \sim \exp(-Q_{\text{eff}}/\tau) \tag{2.8.4}$$

For a damped reaction, of course, Q_{eff} is expected to be quite different from the ordinary ground-state Q value for two separated nuclei. It is determined by the effective total driving potential applicable to a possibly highly deformed, hot dinuclear complex as it may exist close to the end of a damped interaction. However, it should be remembered that Equation (2.8.3) results from a Fokker–Planck equation only in the limit of a constant

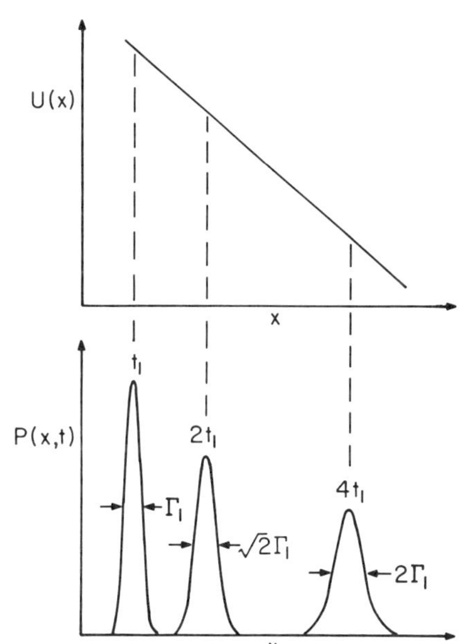

Figure 33. Schematic illustration of the driving potential $U(x)$ governing the transport along a coordinate x and the time evolution of the corresponding probability function $P(x, t)$.

diffusion coefficient and a constant temperature and under the assumption that the dynamical driving force is reasonably well represented by the static force given by the gradient of the effective potential. These may, in fact, be rather crude approximations except possibly for deeply interpenetrating nuclear collisions associated with high excitation energies and long interaction times.

Whereas in the stationary limit of long interaction times, experimental probability distributions for macroscopic observables such as the final mass and charge asymmetries of the reaction fragments are predicted to reflect only the energetics of the underlying potential energy surface, they depend, via the transport coefficients, on dynamic transition probabilities for shorter reaction times. Hence, these transport coefficients contain the relevant physical information about the reaction. They determine the properties of the final probability distributions for macroscopic variables, such as the mass or charge asymmetry of the reaction fragments. Knowing the transport coefficients and assuming an initial boundary condition for the probability distribution, for example, a delta function, one can integrate (Bro 80c, Lej 81) the Fokker–Planck equation (2.7.14) exactly. The resulting distributions are in general not single Gaussians. They may reflect the structure of the underlying driving potential imposed, for example, by shell closures to a different extent than the results obtained by approximate methods of solution. Figure 34, for example, represents the total fragment mass distributions $P(A, t)$ as a function of interaction time $t = \tau_{int}/10^{-21}$ s obtained

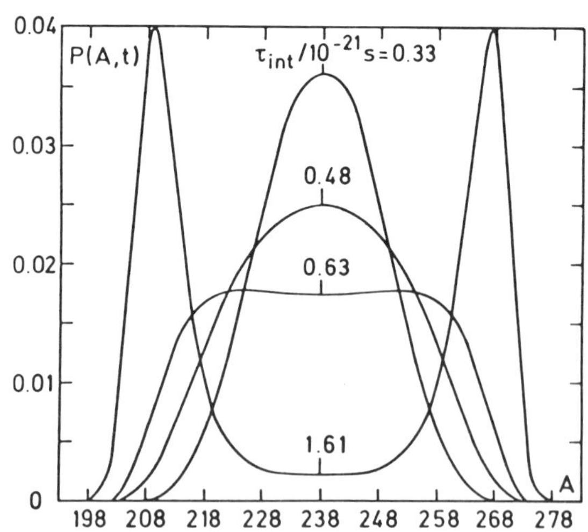

Figure 34. Time evolution of the probability density $P(A, t)$ for the reaction $^{238}U + ^{238}U$ at $E_{Lab} = 7.4$ MeV/nucleon. From (Bro 80c).

by Brosa (Bro 80c) for the symmetric system $^{238}U+{}^{238}U$. The calculation demonstrates the evolution of a bimodal A distribution that is strongly influenced by the $A=208$ shell closure, from the initial delta function centered at $A=238$.

In a more approximate way, one can expand (Sch 81b, Gro 81, Mer 81) the probability function P in moments and derive differential equations for their time evolution. This latter procedure is exemplified below for a two-dimensional Fokker–Planck equation describing the exchange of mass and charge between the constituents of the dinuclear complex in damped reactions.

Denoting with N and Z the number of neutrons and protons, respectively, found in one of the reaction fragments at a given time, one defines the time-dependent first and second moments of the probability distribution

$$\begin{Bmatrix} \bar{N}(t) \\ \bar{Z}(t) \end{Bmatrix} = \int dN'dZ' \begin{Bmatrix} N' \\ Z' \end{Bmatrix} P(N',Z',t)$$

$$\begin{Bmatrix} \sigma_N^2(t) \\ \sigma_N^2(t) \end{Bmatrix} = \int dN'dZ' \begin{Bmatrix} [N'-\bar{N}(t)]^2 \\ [Z'-\bar{Z}(t)]^2 \end{Bmatrix} P(N',Z',t)$$

$$\sigma_{NZ}(t) = \int dN'dZ' [N'-\bar{N}(t)][Z'-\bar{Z}(t)] P(N',Z',t) \quad (2.8.5)$$

For the mass variable $A=N+Z$, one calculates from Equation (2.8.5) a variance

$$\sigma_A^2(t)=\sigma_N^2(t)+\sigma_Z^2(t)+2\sigma_{NZ}(t)=\sigma_N^2(t)+\sigma_Z^2(t)+2\kappa(t)\sigma_N(t)\sigma_Z(t) \quad (2.8.6)$$

where the coefficient $\kappa=\sigma_{NZ}(\sigma_N\sigma_Z)^{-1}$ with $-1\le\kappa\le+1$ describes correlations between transfer processes of neutrons and protons. For uncorrelated neutron and proton transfer, the probability function separates (Bec 78d), i.e., $P(N,Z,t)=P(N,t)P(Z,t)$, and $\sigma_{NZ}(t)\equiv 0$. This situation ($\kappa=0$) is depicted schematically in Figure 35a, where equiprobability contour lines are shown in the (N,Z) plane assuming that the moments of Equation (2.8.5) define approximately a symmetric (e.g., Gaussian) probability distribution. Obviously, in this case, the principal axes of the probability ellipsoid are aligned with respect to the N and Z axes. In contrast, if there are correlations between mass and charge exchange indicated by a covariance $\sigma_{NZ}(t)\neq 0$, the joint probability distribution will be tilted with respect to

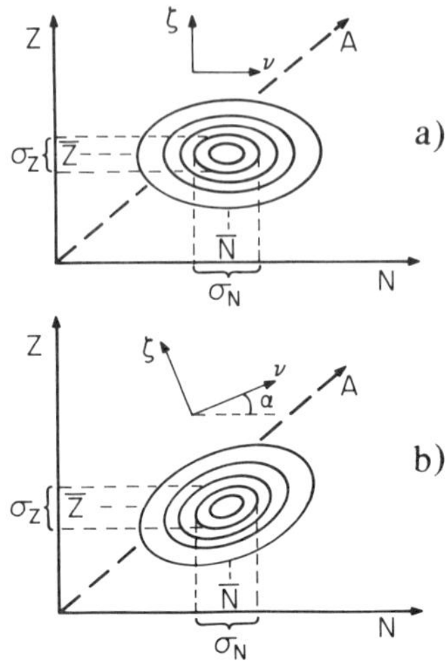

Figure 35. Equiprobability contour lines in the (N, Z) plane for one of the reaction fragments. ζ and ν denote the principal axes of the probability ellipsoid $P(N, Z, t)$. The distribution at the top shows no average correlations, i.e., $\sigma_{NZ} = 0$. The inclination of the probability distribution at the bottom with respect to the N and Z axes reflects the macroscopic correlations imposed on the transport process by variations in the driving forces.

these axes (cf. Figure 35b) by an angle α determined by

$$\cot \alpha = \left\{ \left(\sigma_N^2 - \sigma_Z^2 \right)/2 + \left[\left(\sigma_N^2 - \sigma_Z^2 \right)^2/4 + \sigma_{NZ}^2 \right]^{1/2} \right\} / \sigma_{NZ} \qquad (2.8.7)$$

as one can deduce from a transformation of the covariance tensor to its principal axes ν and ξ (cf. Figure 35b). For a two-dimensional Gaussian probability distribution, cross sections taken along straight lines in the (N, Z) plane are again Gaussians with variances depending only on the orientation of these lines with respect to the axes of the probability ellipsoid. Hence, one predicts for the experimentally interesting variances of fragment isobaric and isotopic distributions

$$\sigma_Z^2(A \equiv \text{const}) = \left(\sigma_N^2 \sigma_Z^2 - \sigma_{NZ}^2 \right) \sigma_A^{-2}$$

$$\sigma_Z^2(Z \equiv \text{const}) = \left(\sigma_N^2 \sigma_Z^2 - \sigma_{NZ}^2 \right) \sigma_Z^{-2} \qquad (2.8.8)$$

The corresponding first moments, i.e., the centroids of isobaric and isotopic fragment distributions, are in this approximation, obviously, linear functions of A and Z, respectively, with slopes that depend on the angle α [cf. Equation (2.8.7)].

Equations of motion for the variances can be derived from Equations (2.8.5) making use of the Fokker–Planck equation (2.7.14). They become particularly simple, if drift coefficients are well approximated by their first-order expansions around the average values $\overline{N}$ and $\overline{Z}$, in which case the relations

$$\partial\sigma_N^2/\partial t \approx 2\left[D_{NN} + \sigma_N^2 \partial v_N/\partial N + \sigma_{NZ}\partial v_N/\partial Z\right]$$

$$\partial\sigma_N^2/\partial t \approx 2\left[D_{ZZ} + \sigma_Z^2 \partial v_Z/\partial Z + \sigma_{NZ}\partial v_Z/\partial N\right]$$

$$\partial\sigma_{NZ}/\partial t \approx 2D_{NZ} + \sigma_N^2 \partial v_Z/\partial N + \sigma_Z^2 \partial v_N/\partial Z + \sigma_{NZ}\left(\partial v_N/\partial N + \partial v_Z/\partial Z\right) \tag{2.8.9}$$

govern the time evolution of the variances. For constant drift coefficients and in the absence of microscopic correlations ($D_{NZ} = 0$), $\sigma_{NZ}(t) \equiv 0$, and the variances depend only on the corresponding diffusion coefficients, according to separate one-dimensional Fokker–Planck equations [cf. Equations (2.7.14) and (2.8.2)]. In general, however, correlations are expected to occur if the driving forces [cf. Equation (2.7.16)] governing the exchange of mass are not constant. This would be expected, for example, if the driving potential had a large curvature leading to an alignment of the probability distribution with respect to the potential valley.

The above scheme can be readily extended to a description of multivariate probability distributions in many macroscopic variables that can, in principle, include fragment spin and various shape degrees of freedom. Of course, for each degree of freedom, it has to be proven that the nondiagonal interaction matrix elements coupling different macroscopic states fulfill the statistical conditions reflected in short memory times of the corresponding correlation functions, in order that a Fokker–Planck equation be applicable. Knowledge of these matrix elements and of the dependence of the intrinsic level density on the considered macroscopic variables allows the calculation of diffusion and drift coefficients, according to Equation (2.7.15). That such a program represents, in practice, a formidable task has been demonstrated, e.g., by Agassi (Aga 79) in the case of shape degrees of freedom. In the following, transport coefficients obtained in relatively simple models will be discussed.

Ayik *et al.* (Ayi 76, 76c, 78) have derived transport coefficients for the exchange of mass and charge between two interacting nuclei, as well as for the transfer of relative kinetic energy and orbital angular momentum into intrinsic degrees of freedom. In the model, both ph excitation and nucleon exchange are induced by the off-diagonal matrix elements of a single-particle interaction potential whose diagonal elements constitute the collective interaction potential. The transition matrix elements are parametrized as products of Gaussians in single-particle energy, spin, and spin projection. Characteristic width parameters and an overall strength are assumed to be constant for a given system and estimated from elastic transfer. As seen from Equations (2.7.12)–(2.7.15), evaluation of the transport coefficients involves calculation of memory or decay times, second moments of transition matrix elements averaged over allowed single-particle configurations, and level densities. The averages are calculated (Ayi 76) with methods of spectral distributions applicable to large shell-model spaces.

The resulting diffusion coefficients depend only weakly on the excitation energy, via $D_{ij}(\mathbf{x}) \sim \tau(\mathbf{x})^{1/2}$, where the temperature τ is calculated from the local intrinsic excitation energy

$$E^*(\mathbf{x}) = E_0^* - [U(\mathbf{x}) - U_0] - \hbar^2 l^2/(2\mathscr{I}) \tag{2.8.10}$$

Here U is the energy of the ground state of the dinuclear system approximated by a sum of liquid-drop-model binding energy, Coulomb and centrifugal energies with a minimum value U_0, and E_0^* is the excitation energy relative to this minimum. l is the instantaneous orbital angular momentum, and $\mathscr{I}$ is the moment of inertia. The characteristic model memory time corresponding to t_{dec} in Equation (2.7.12), which is proportional to the inverse of a root-mean square matrix element, is plotted in Figure 36a (Nor 79c) vs. excitation energy for various systems represented by the single-particle level densities $g = A/12$. It is seen to be of the order of 10^{-23} s and decreasing with increasing excitation energy and increasing level density as expected. For the mass diffusion coefficient, one obtains (Ayi 76c, 78, Nor 79c)

$$D_{AA}(\mathbf{x}) = 0.21\tilde{\gamma}\tilde{\Delta}^{1/2}[E^*(x)/A]^{1/4}(A_1A_2)^{2/3}(A_1^{1/3} + A_2^{1/3})^{-2} \times 10^{22}\ \mathrm{s}^{-1} \tag{2.8.11}$$

Here, $\tilde{\Delta}^2$, denoting the variance of the transition matrix elements in single-particle energy, is slightly excitation-energy dependent, and $\tilde{\gamma}$ is a reduced transition strength. The mass numbers A_i of the reaction partners and the total mass A enter the expression for the diffusion coefficient through the single-particle level densities. Similarly, the energy diffusion coefficient due to ph excitation and nucleon exchange is given by

$$D_{EE}(\mathbf{x}) = 0.053\tilde{\gamma}\tilde{\Delta}^{5/2}[E^*(\mathbf{x})/A]^{1/4}(A_1A_2)^{1/3} \times 10^{22}\ \mathrm{MeV}^2\,\mathrm{s}^{-1} \tag{2.8.12}$$

Mass and energy diffusion coefficients are also plotted in Figure 36. Both ph excitation and nucleon exchange are affected by the variation of the level density with intrinsic angular momentum. In addition, the transfer of a nucleon of mass m between the reaction partners introduces a change Δl of

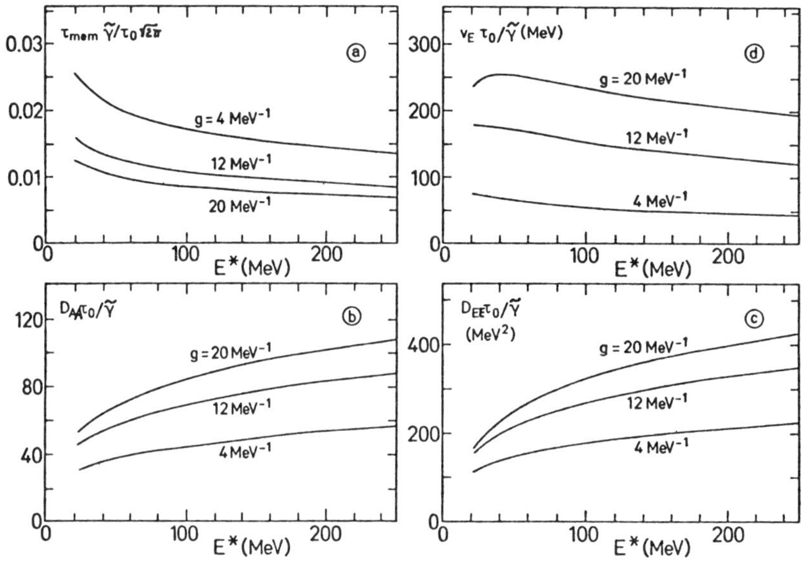

Figure 36. Characteristic memory times t_{mem} and transport coefficients are plotted vs. local excitation energy E^* for various systems represented by their single-particle level density parameters $g = A/12$. From (Nor 79c).

orbital angular momentum due to recoil which is given for peripheral transfer by (Bri 72)

$$\Delta l \approx lm/(2\mu) \tag{2.8.13}$$

Hence, there is a dispersion of the total intrinsic spin projection M which is counterbalanced by a similar induced spread in orbital angular momentum that can be described (Ayi 78) in terms of a diffusion coefficient

$$D_{MM}(\mathbf{x}) = \frac{(\Delta j)^2}{(\Delta j)^2 + 0.033A}\left[0.002\frac{A_1 A_2}{(\Delta j)^2}D_{AA} + 0.033\frac{A}{\tilde{\Delta}^2}D_{EE}\right] + \frac{D_{AM}^2}{D_{AA}} \tag{2.8.14}$$

where D_{AM} is a mixed diffusion coefficient proportional to Δl that vanishes for symmetric fragmentations. The quantity $(\Delta j)^2$, the variance of the transition matrix elements with respect to the spin projections of the accessible states, is, in the model of Ayik *et al.* (Ayi 78), roughly estimated from the extension of the interaction region. The corresponding drift coefficients can be derived from the above diffusion coefficients employing the generalized Einstein relations of Equations (2.7.15) or (2.7.16). In particular, the model predicts for the energy drift coefficient v_E related to the kinetic-energy dissipation rate $-\dot{E}$

$$-\dot{E} = v_E = D_{EE}\tau^{-1} \sim E^*(\mathbf{x})^{-1/4} \tag{2.8.15}$$

Such an energy dissipation rate is quite different in character from that due to a classical, velocity-proportional friction force which yields qualitatively $-\dot{E} \sim E$ [see, e.g., Equation (2.6.29)], where E is the kinetic energy available at each time during the reaction. As observed from Figure 36, the dissipation rate calculated by Ayik *et al.* (Ayi 78) varies very little with excitation energy (or available kinetic energy) but is strongly increasing with total mass of the system.

Since transport coefficients in the above microscopic model depend so strongly on the single-particle level densities, it can be expected that the individual single-particle level schemes of the interacting fragments, i.e., their shell structure, has a certain effect on these coefficients, as long as the nuclear states excited in the reaction contain relatively few ph configurations. Schürmann *et al.* (Sch 78c) have illustrated the dependence of mass transport coefficients on shell effects present in one of the interacting nuclei, calculated with Strutinsky's (Str 67,68) level-bunching prescription. The result is that the mass diffusion coefficient D_{AA} [cf. Equation (2.8.8)] is reduced by typically 20%, due to the appearance of a gap in the single-particle spectrum. Shell effects on drift coefficients favoring the population of closed-shell fragmentations can be considerably stronger. The corrections are seen to disappear (Sch 78c) for total excitation energies E_0^* [cf. Equation (2.8.10)] above a critical energy

$$E_0^*(A_1) \approx 0.25A \text{ MeV} \tag{2.8.16}$$

where A_1 is the mass number of the closed-shell fragment and A is the total mass of the system. Hence, shell effects may be discernible in damped reactions for relatively low kinetic energy losses, most notably in a drift of the probability distribution into the direction of shell minima of the potential energy surface. However, the effect could be masked by a different property of the total driving force [cf. Equation (2.7.16)] if all degrees of freedom are considered.

Whereas the above model incorporates both ph excitation and nucleon exchange induced by the mean single-particle field, the model of Randrup (Ran 78,78d,79d) already referred to in Section 2.6 considers transport phenomena related only to the exchange of individual nucleons between colliding nuclei. Because of the assumption that the reaction partners retain approximately their separate identities underlying Randrup's single-particle model, applicability is restricted to peripheral collisions, similar to other models discussed previously. In this theory, drift and diffusion coefficients are calculated (Ran 79d) according to

$$v_{\mathcal{O}} = \int d\varepsilon\, N'(\varepsilon)\langle (f^B - f^A)\mathcal{O}\rangle \tag{2.8.17a}$$

$$D_{\mathcal{O}_1\mathcal{O}_2} = \int d\varepsilon\, N'(\varepsilon)\langle [f^B(1-f^A)+f^A(1-f^B)]\,\mathcal{O}_1\mathcal{O}_2\rangle \tag{2.8.17b}$$

Here, $\mathcal{O}_i$ stands for the change of any macroscopic observable such as mass or charge asymmetry, relative kinetic energy, etc., induced by the exchange of a single nucleon. The current N' contains the form factor $\mathscr{X}$ defined in Equation (2.6.33) or Equation (2.6.32), depending on whether the dinuclear shape has a neck or not. $N'(\varepsilon)$ is an average differential nucleon current defined in Equation (2.6.26), and f^A and f^B are the single-particle occupation probabilities of nucleus A or B, respectively, [cf. Equation (2.6.27)]. The brackets denote again the flux average taken over all velocity directions of the exchanged nucleons. As discussed already in Section 2.6, in the calculation of flux averages of one-body operators, the effects of the Pauli blocking factors $(1-f)$ cancel. They remain present in the expressions [cf. Equation (2.8.17b)] for the diffusion coefficients. An example for a drift coefficient has already been given in the form of the energy dissipation rate of Equations (2.6.28) and (2.6.29) resulting from Equation (2.8.17a) with $\mathcal{O} = \omega = F_{AB} - \mathbf{u}\cdot\mathbf{p}$, where F_{AB} is the change of the ground-state energies induced by the exchange of a nucleon with momentum $\mathbf{p}$, and $\mathbf{u}$ is the relative velocity of the reaction partners.

In the limit of a common nuclear temperature τ of the system that is small compared to the Fermi kinetic energy ($T_F \approx 37$ MeV), the expressions of Equations (2.8.17) can be simplified to

$$v_{\mathcal{O}} \approx N'(\varepsilon_F)\langle \omega\mathcal{O}\rangle \tag{2.8.18a}$$

$$D_{\mathcal{O}_1\mathcal{O}_2} \approx N'(\varepsilon_F)\langle \tilde{\tau}\mathcal{O}_1\mathcal{O}_2\rangle \tag{2.8.18b}$$

where $\tilde{\tau}$ is an energy measuring the unoccupied single-particle space participating in the exchange of a nucleon of momentum $\mathbf{p}$. It has the same form as the energy of a harmonic oscillator coupled to a heat bath [cf. Equation (2.7.20)] and is given by

$$\tilde{\tau}(F_{AB}, \mathbf{u}\cdot\mathbf{p}, \tau) = \frac{\omega}{2}\coth\frac{\omega}{2\tau} \to \begin{cases} \omega/2, & \omega \gg \tau \qquad (2.8.19a) \\ \tau, & \omega \ll \tau \qquad (2.8.19b) \end{cases}$$

but now with the strongly velocity-dependent quantity $\omega = F_{AB} - \mathbf{u}\cdot\mathbf{p}$, instead of the oscillator energy $\hbar\omega$. For mass (or charge) transport, $\mathcal{O} = 1$ and one has

$$v_A = N'(\varepsilon_F)\langle\omega\rangle = N'(\varepsilon_F) F_{AB} \tag{2.8.20a}$$

$$D_{AA} = N'(\varepsilon_F)\langle\tilde{\tau}\rangle = N'(\varepsilon_F)\tau^*(F_{AB}, \mathbf{u}, \tau) \tag{2.8.20b}$$

where τ^* is the flux average of $\tilde{\tau}$. Analogously, the model predicts transport coefficients for momentum with $\mathcal{O} = \mathbf{p}$ and other observables of interest, where, however, the quantity τ^* cannot be separated out, in general. According to Equations (2.8.20), the mass and, analogously, the charge transport coefficients obey generalized Einstein relations of the form

$$\mathbf{v} = \tau^{*-1}\mathsf{D}\mathbf{F}_{AB} \tag{2.8.21}$$

which differ from the classical relations [cf. Equation (2.7.17)] by a replacement of the normal nuclear temperature τ by the "effective temperature" τ^*. In the high-temperature limit of Equation (2.8.19b) both relations are identical. The relation between τ and τ^* is depicted in Figure 37 (Sch 79f), where both quantities evaluated at the classical turning point are plotted vs. the total loss of kinetic energy, for the system $^{209}Bi + {}^{136}Xe$ at two incident energies. The results are obtained from dynamical calculations discussed in Section 2.4. According to the large differences between τ and τ^* observed in Figure 37, quite different relations between centroid and variance of fragment mass distributions are expected in the classical and quantal limits. Similarly, the stationary solution of a one-dimensional Fokker–Planck equation with the above transport coefficients corresponds to a Gaussian with asymptotic variance $\sigma_\infty^2 \approx \tau^*/c$, where c is the restoring-force constant associated with a harmonic driving potential. In general, these asymptotic variances will be influenced by the coupling to other degrees of freedom such as exemplified in Equation (2.8.9). As before, the driving force can approximately be expressed in terms of the collective Hamiltonian H_c as $\mathbf{F}_{AB} \approx -\nabla H_c$. It is, hence, a dynamical force containing inertial terms, in addition to the usually dominating static driving force. It is also of dissipative character and contributes a term $\frac{1}{2}F_{AB}v \sim F_{AB}^2$ to the proximity dissipation function of Equation (2.6.30).

The transport models of Ayik *et al.* and Randrup have a similar range of validity. They apply only to peripheral reactions assuming that the intrinsic system is describable in terms of a sum of two independent-particle Hamiltonians. The approach of Ayik *et al.* is nonperturbative, includes ph excitations and nucleon exchange, and is capable of accommodating certain aspects of the individual nuclear structure. In Randrup's model, the inter-

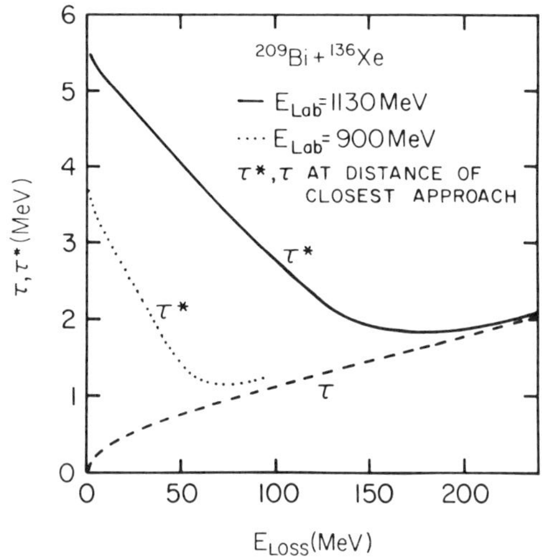

Figure 37. Relation between total kinetic energy loss E_{Loss} and the "effective temperature" τ^* and the normal nuclear temperature τ at the distance of closest approach, as predicted by model calculations for the reaction $^{209}\text{Bi} + {}^{136}\text{Xe}$ at two bombarding energies. From (Sch 79f).

acting nuclei are approximated by Fermi–Dirac gases allowing one to account for the nuclear structure in a more average way, via the ground-state energy and level density parameters. Particle–hole excitations are neglected in this approach considering only nucleon exchange. On the other hand, this one-body proximity transport model includes dynamical effects that are neglected by Ayik *et al.*, who adopt universal values for parameters such as correlation energies that are, in principle, dynamical quantities. Furthermore, since conservative forces and transport phenomena are treated by Randrup on the same basis, his model allows a rather consistent test of a variety of reaction features.

Instead of solving a Fokker–Planck equation (2.7.14), one can also parametrize the master equation (2.7.5) by employing a model for the average transition probabilities $W_{nm}(t)$. Such an approach has been taken by Moretto *et al.* (Mor 75, 75b, 76, 76b, Sve 76) and others (Bra 78, Bec 78d, Zoh 78b). In the cited work of Moretto *et al.* describing the transport of charge in damped reactions, these probabilities are parametrized as

$$W_{nm} = w_0\left[\rho\left(E_0^* - U_n\right)/\rho\left(E_0^* - U_m\right)\right]^{1/2} \tag{2.8.22}$$

Here, $w_0 = 2\pi\overline{R}b\kappa$ is an overall transition strength depending on an average size of the window open between two interacting nuclei, κ is a characteristic nucleon current density comparable to the bulk flux n_0 discussed earlier, ρ is the level density of the dinuclear complex evaluated at the local excitation energy [cf. Equation (2.8.10)] $E_n^* = E_0^* - U_n$, measured for each charge fragmentation n with respect to the minimum of the driving potential U_n. The latter is expressed in terms of the liquid-drop and centrifugal potentials as $U_n = V_{LD}(n) + \hbar^2 l^2/[2\mathscr{I}(n)] - U_0$, where U_0 provides normalization to zero at the minimum potential energy. The level densities are expanded as $\rho(E_0^* - U_n) = \rho(E_0^*)\exp(-U_n/\tau)$, where τ is the nuclear temperature. Hence, $W_{nm} = w_0 \exp[(U_m - U_n)/(2\tau)]$ is mainly dependent on the potential gradient along the charge-asymmetry coordinate, reflecting the underlying assumption of a fast and complete dissipation of the available radial kinetic energy. Taking as initial boundary condition for the probabilities a delta function located at the projectile–target fragmentation, the master equation (2.7.5) can be integrated numerically, for a given choice of the free parameter κ. Figure 38 (Mor 76b) shows the evolution of the fragment charge distribution for the 620-MeV ^{197}Au + ^{86}Kr reaction for four values of the orbital angular momentum l assumed to remain constant. The figure shows equiprobability contour lines in a two-dimensional plot vs. atomic number Z of the fragments and time t elapsed in a collision as calculated for a parameter $\kappa = 5 \times 10^{20}$ fm^{-2} s^{-1}. With increasing time, the two components of the bimodal Z distribution are observed to broaden and to merge at the potential energy minimum corresponding to a symmetric fragmentation. The distributions are wider for smaller values of l, because the potential energy surface U_n is shallower and the radial kinetic energy assumed to be transformed into excitation energy is larger for smaller l values. Of course, for long enough times, the probability distribution will always assume its stationary form. Hence, a separate phenomenological model has to be adopted for the relative motion determining the interaction time, i.e., the lifetime of the dinuclear system, for which Moretto *et al.* (Mor 76b) assume an exponential decay law. Mass and charge transport have also been considered by Beck *et al.* (Bec 78d) in the framework of a master equation describing the exchange of independent particles constituting a classical gas. The dynamics of relative motion is treated (Bec 78b) in the one-body proximity model.

The general transport equations, as given by Equation (2.7.18) for the example of relative motion, describe the full dynamical evolution of macroscopic coordinates with conservative, inertial, and dissipative terms, based on the assumptions made for collective and intrinsic Hamiltonians and the interaction. These equations for the phase space distribution function $P(\mathbf{r},\mathbf{p},t)$ are equivalent [see, e.g., (Hak 75,78)] to Langevin equations [see, e.g., Equation (2.7.24)] for the motion along the collective variables $\{\mathbf{r},\mathbf{p}\}$ under the influence of stochastic forces. In contrast, the applications of transport equations discussed so far in this section pertain to the much simpler Fokker–Planck equation (2.7.14) which is based on an overdamped motion and, hence, does not account for the retardation effects associated with a collective inertia. For studies of dynamical collective nuclear defor-

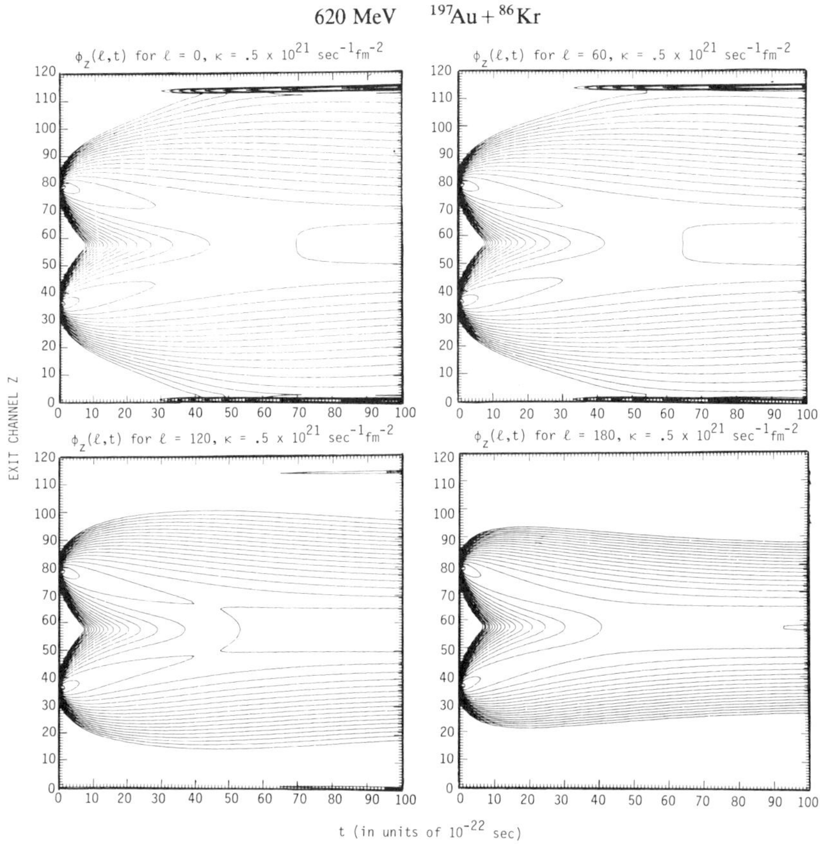

Figure 38. Time evolution of the Z distributions of fragments from the reaction ^{197}Au + ^{86}Kr at $E_{\text{Lab}} = 620$ MeV for different l values predicted by the diffusion model of Moretto *et al.* (Mor 76b).

mations and vibrations in damped reactions, one would suspect that considerations neglecting inertia effects can be highly inadequate. In the case of mass exchange between the reaction partners, such a conclusion may be less justified on fundamental grounds, although there are approaches (cf. Section 2.4) that emphasize collective aspects of mass exchange in damped reactions. However, in processes leading to a significant redistribution of the nucleons within the system among the reaction partners, mass exchange must eventually be accompanied by changes in the collective nuclear shapes which may effect an inertial retardation of mass transport. Already the finite transparency of the interface between two interacting nuclei gives rise to an inertia associated with mass exchange, as discussed in the context of

Equation (2.4.5). Furthermore, induced driving forces that are velocity dependent are equivalent to an effective inertia for the mass exchange. Attempts to estimate such effects have been discussed in Section 2.4.

Descriptions of mass transport in damped reactions in terms of a Langevin equation

$$m_A\, d^2A/dt^2 + \gamma_A\, dA/dt - F(A) = \gamma_A F_L(t) \tag{2.8.23}$$

have been given by Rehm *et al.* (Reh 79, 79b, 81) and Grossmann *et al.* (Gro 81). In Equation (2.8.23), A is the mass number of one of the interaction partners, m_A and γ_A are mass and friction parameters, respectively. The quantity $F(A)$ is the collective driving force, and F_L is the fluctuating Langevin force with zero average and a correlation function $\langle F_L(t) F_L(t')\rangle \approx 2D_A\,\delta(t-t')$. The solutions of Equation (2.8.23) correspond to Gaussian probability distributions $P(A,t)$ with average values $\bar{A}(t)$ and variances $\sigma_A^2(t)$ whose time evolution depends on the relative magnitudes of inertia, friction, and force coefficients. Grossmann *et al.* (Gro 81) have discussed in detail the case of a linear driving force $F(A) = -CA$ and have estimated m_A from a hydrodynamical model. The characteristic time scales of interest are the diffusion time $t_D = D_A^{-1}$, the relaxation time $t_{\text{rel}} = \gamma_A/C$, and the damping time $t_d = m_A/\gamma_A$. For an overdamped motion, i.e., $t_{\text{rel}} > 4t_d$, average values and variances behave analogously to Equations (2.8.26) or (2.8.1b). However, for the general case, it is inferred (Gro 81) that the variance increases as $\sigma_A^2 \propto t^3$ for $t \ll t_d$. Hence, a finite inertia m_A inhibits mass exchange during the initial stages of the reaction. Both average value and variance can show an oscillatory behavior during times of the order of a few times t_d. However, in order to arrive at predictions that are comparable to experiments that do not measure the interaction time directly, it appears necessary to consider the coupling of the mass asymmetry degree of freedom described by Equation (2.8.23) with the other reaction variables in a dynamical reaction model. Such a program has, thus far, not been carried out in sufficient detail.

The examples provided in this section for the application of transport equations to damped-reaction phenomena are characteristic of the level on which theoretical concepts are currently tested in comparison to experiment. Section 3 will be devoted to a discussion of the experimental results and their interpretation in terms of the various models. In view of the apparent similarities of the final reduced equations of motion for the reaction parameters deduced in rather different approaches, it is premature to hope for an unambiguous identification of the microscopic reaction mechanism on the basis of such comparisons or analyses. This obviously establishes the need for more complete and detailed model calculations than currently available.

3. Experimental Features of Damped Reactions and Their Implications

In the preceding Section 2, an attempt is made to summarize the current status of theoretical concepts of the damped-reaction mechanism and their execution in terms of reaction models. This provides a tentative basis from which to appreciate experimental observations of reactions between complex nuclei, which form the major concern of this section. However, as remarked earlier, this basis is currently still too narrow to yield a consistent prediction for the majority of the reaction properties derived from a hypothetical underlying mechanism. Hence, comparisons between data and model predictions can be referred to only occasionally and in particular cases. The main emphasis will be placed on the establishment of the systematics of damped-reaction properties, their intercorrelations, and a more heuristic, phenomenological understanding of the dominant trends. As shown in this section, it is already possible to draw tentative general conclusions about the origin of several features of damped nuclear interactions. However, these inferences are contingent mainly on the validity of classical and statistical concepts. They may, hence, require a later revision and, possibly, even a reinterpretation.

The subsequent discussion starts out with reference to the systematics of the total reaction cross section obtained mainly from studies of elastic scattering. Other important aspects of elastic scattering such as the inference of optical potentials and wave-mechanical considerations exceed the scope of this paper. Consecutively, the basic damped-reaction features are discussed, including the multiplicity of fragments, as well as their angular, mass, and energy distributions. In addition, the rather limited number of experimental results on the dissipation of orbital angular momentum and on reactions that are associated with light-particle emission and other sequential deexcitation processes are exhibited. Although in each subsection, similarities and differences between damped-reaction features and those of other nuclear reactions such as fission are pointed out, a separate subsection is devoted to discuss the transition from damped reactions to fusion–fission-like processes that are of particular current interest. As a general limitation, only heavy-ion reactions with bombarding energies of up to a few MeV/nucleon above the interaction barrier $V_{\mathrm{Coul}}(R_{\mathrm{SA}})$ are considered, although occasional references to reactions at higher energies are made.

3.1. Systematics of the Total Reaction Cross Section and Interaction Radius

The total cross section σ_R associated with a nuclear reaction is a fundamental observable characterizing the process. Classically, it measures the sum over the geometrical cross sections $\sigma(l) = (2l+1)\pi\lambda\!\!\!^{-2}$ corresponding to each of the contributing partial waves l, weighted by the transmission coefficients T_l [cf. Equation (2.1.5)] describing the reaction dynamics. Strictly, σ_R comprises the probabilities for all inelastic processes induced by the nuclear interaction between projectile and target nuclei. Elastic scattering (EL) and inelastic events due to Coulomb excitation (CE) have to be excluded. In practice, however, one often resorts to a somewhat arbitrary operational definition of the reaction cross section, due to the considerable experimental difficulties in distinguishing events corresponding to the various processes. For most applications, the resulting uncertainty of the deduced cross section is probably of minor consequence for the further interpretation of experimental observations.

Experimental total reaction cross sections are usually analyzed in terms of an effective maximum initial angular momentum l_{max} leading to a nuclear reaction, e.g., by employing the sharp cutoff model represented by Equation (2.1.22). In extension of this or similar model views (see, e.g., Section 2.2), characteristic limiting values of angular momentum are associated with different parts of the total reaction cross section. As is illustrated in Figure 39, various reaction types such as compound-nucleus (CN), fusionlike (FL), damped (D), and quasielastic (QE) processes are assumed

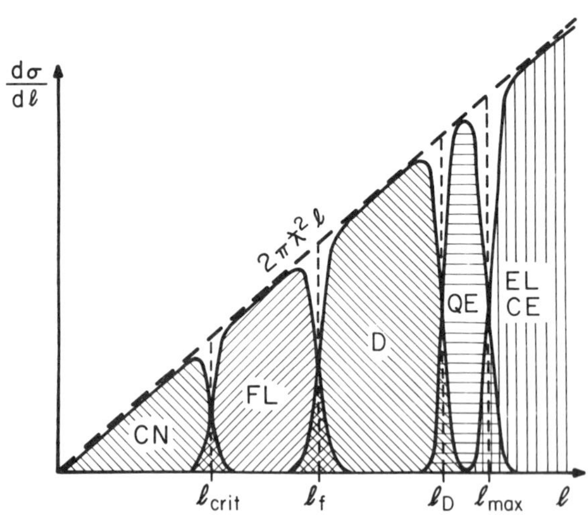

Figure 39. Schematic illustration of the l dependence of the partial cross section for compound-nucleus (CN), fusion-like (FL), damped (D), quasielastic (QE), Coulomb-excitation (CE), and elastic (EL) processes. The long-dashed line represents the geometrical partial cross section $d\sigma/dl = 2\pi\lambda\!\!\!^{-2} l$. Vertical dashed lines indicate the extensions of the various l windows in a sharp cutoff model with the characteristic l values noted at the abscissa. Hatched areas represent the diffuse l windows assumed in a smooth cutoff model.

to correspond to successive angular momentum windows exhausting the total reaction cross section. In a smooth cutoff model, different reaction types may compete for l values in the transition region between adjacent l windows.

An unambiguous determination of the experimental total reaction cross section σ_R entails an exclusive measurement of all possible reaction products, which constitutes a difficult experimental task. Therefore, the method most commonly used for determination of σ_R consists of a measurement and analysis of elastic scattering data. In practice, the cross section ratio $\sigma_{el}/\sigma_{Ruth}$ is measured as a function of scattering angle, and the results are analyzed with optical and Fresnel models [see, e.g., (Bir 76)]. As discussed in Section 2.1, the total reaction cross section calculated from an optical-model analysis of the elastic scattering angular distribution is given by Equation (2.1.5), whereas in the generalized Fresnel model (Fra 78) σ_R is given by Equation (2.1.30).

Elastic scattering experiments between heavy nuclei are scarce where the elastic channel is measured with sufficient energy resolution to isolate it from the inelastic channels (Ric 79, Che 81). High-resolution spectra of inelastically scattered ^{6,7}Li ions from ^{238}U and ^{232}Th showed considerable cross section for exciting the low-lying 2^+ and 4^+ states of these deformed targets (Fre 75). However, comparison of the elastic and inelastic cross sections with predictions of a Coulomb excitation code indicate that the inelastic scattering to these states is predominantly due to Coulomb excitation. Similar results have been reported (Tho 77) for the scattering of 90-MeV ^{18}O on ^{184}W. The experimental cross section ratio, $\sigma_{el}/\sigma_{Ruth}$, for the elastic scattering as a function of center-of-mass angle is shown in Figure 40. The scattered ions were detected and identified by a proportional counter in the focal plane of a QDDD spectrometer with an energy resolution (FWHM) of 80 keV. This resolution was sufficient to separate the yields of the first excited (2^+, 111 keV) and ground states of ^{184}W (Tho 77). The ground-state (elastic) cross section, divided by the appropriate Rutherford cross section, is represented by open circles in Figure 40. One observes that these data exhibit a dramatic deviation from the typical Fresnel shape measured for many heavy-ion reactions. The theoretical curve, which agrees well with the data, is calculated with a generalized Fresnel model (Fra 78c) where the depletion of the elastic channel due to Coulomb excitation of the low-lying states is included. Hence, it appears that the reduction in the elastic cross section over a considerable angular range is due to Coulomb excitation. This explanation is convincing for the ^{184}W + ^{18}O reaction at $E_{lab} = 90$ MeV, where the projectiles are expected to follow Coulomb trajectories because of the large Sommerfeld parameter $\eta = 42$. For example, removal of flux from the elastic channel at a c.m. angle of 45° requires an imaginary potential whose range extends to greater than

19 fm. Since this distance is well beyond the range of the nuclear potential, one concludes that the primary mechanism for absorption is due to the long range of the Coulomb force. Addition of the elastic and first excited state (2^+) cross sections for the $^{184}W + ^{18}O$ reaction gives back the familiar Fresnel shape (Tho 77), illustrated by the top curve of Figure 40 which is calculated using the generalized Fresnel model (Fra 78c).

In contrast to the above observations, a recent high-resolution study (Che 81) of the $^{208}Pb + ^{86}Kr$ reaction at $E_{lab} = 695$ MeV suggests that a separation of inelastic scattering to excited projectile states can yield significantly augmented total reaction cross sections as compared to analyses (Bir 76, Van 76d) of earlier experiments incapable of resolving such inelastic excitations. Consequently, also the phenomenological interpretation of the different sets of data results in somewhat different optical potentials and interaction radii. In order to establish the details of experimental trends of these quantities, one should, therefore, consider only data including com-

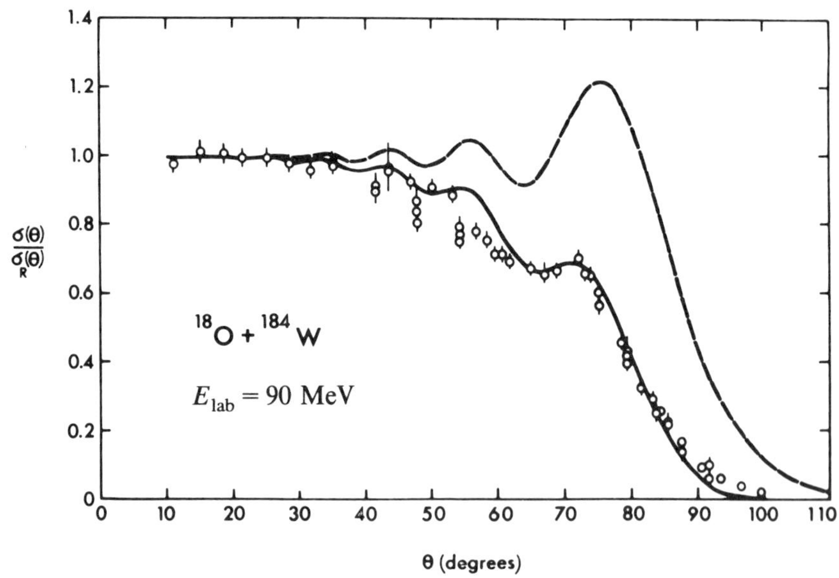

Figure 40. Angular distribution for elastic scattering of ^{18}O by ^{184}W at $E_{Lab} = 90$ MeV. The experimental data (circles) were taken with sufficient energy resolution to separate the first excited state at 111 keV (Tho 77). The theoretical curve through the data is calculated with a generalized Fresnel model where the depletion of the elastic channel and excitation of low-lying states due to Coulomb excitation are included (Fra 78c). The other curve is the generalized Fresnel model prediction without Coulomb excitation. Similar fit to the experimental data has been obtained by others (Lov 77). From (Fra 78c).

parable classes of events and use a corresponding operational definition of the reaction cross section.

Based on limited experimental and theoretical work of the type discussed above, it appears safe to conclude that heavy-ion elastic scattering data, where the energy resolution is insufficient to resolve the low-lying states, is useful at least for an approximate determination of the total reaction cross section. On the other hand, as illustrated for the $^{184}W + ^{18}O$ reaction in Figure 40, analyses of high-resolution measurements without proper correction for Coulomb excitation give an overestimate of the total reaction cross section. It is the sum of the cross section for the Coulomb excitation of low-lying states and the experimentally measured elastic scattering cross section that has to be known, in order to derive the "true" quarter-point angle from which the reaction cross section may be calculated.

The competition of Coulomb excitation with elastic scattering is expected to be relatively more important for the heaviest systems [see, e.g., (Lov 77)]. Insofar as the experimental energy resolution of most beams of

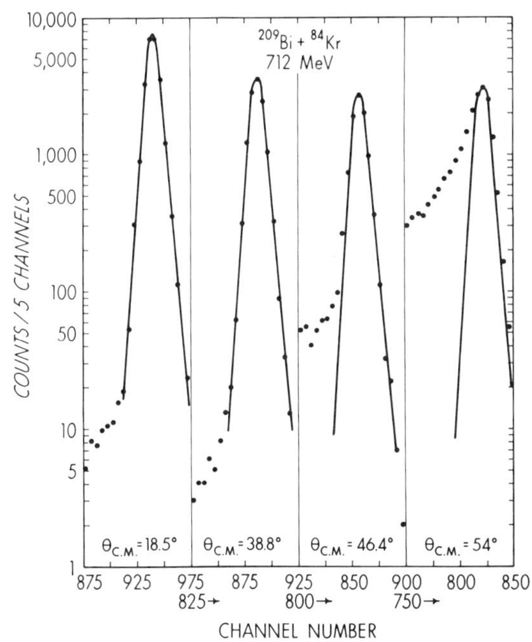

Figure 41. Experimental laboratory energy spectra (points). The data at 18.5° are used as a standard elastic spectrum (see text). This spectrum is shown as a solid curve on the other three spectra. The angles 38.8°, 46.4°, and 54° represent a forward angle, an angle near the grazing angle, and a very backward angle, respectively. The counts in the 18.5° spectrum have been divided by 10. From (Bir 76).

very heavy ions used to date for elastic scattering measurements is insufficient to resolve inelastic scattering to low-energy excited states, it is assumed in all analyses discussed below that the measured "elastic" cross section is a sum of the elastic scattering and Coulomb-excitation cross sections. Analysis of data for 712-MeV ^{84}Kr on a ^{209}Bi target, where the measured energy width (FWHM) was 1.2%, is illustrated in Figure 41. In this case, the measured laboratory energy spectrum at $\theta_{\text{c.m.}} = 18.5°$ is used as a standard spectrum to define an elastic-peak shape. This spectrum is measured at an angle sufficiently far forward to be unaffected by inelastic events, but not so far forward as to be affected by resolution degradation brought about by high count rates. Use of this standard spectrum to separate the "elastic"

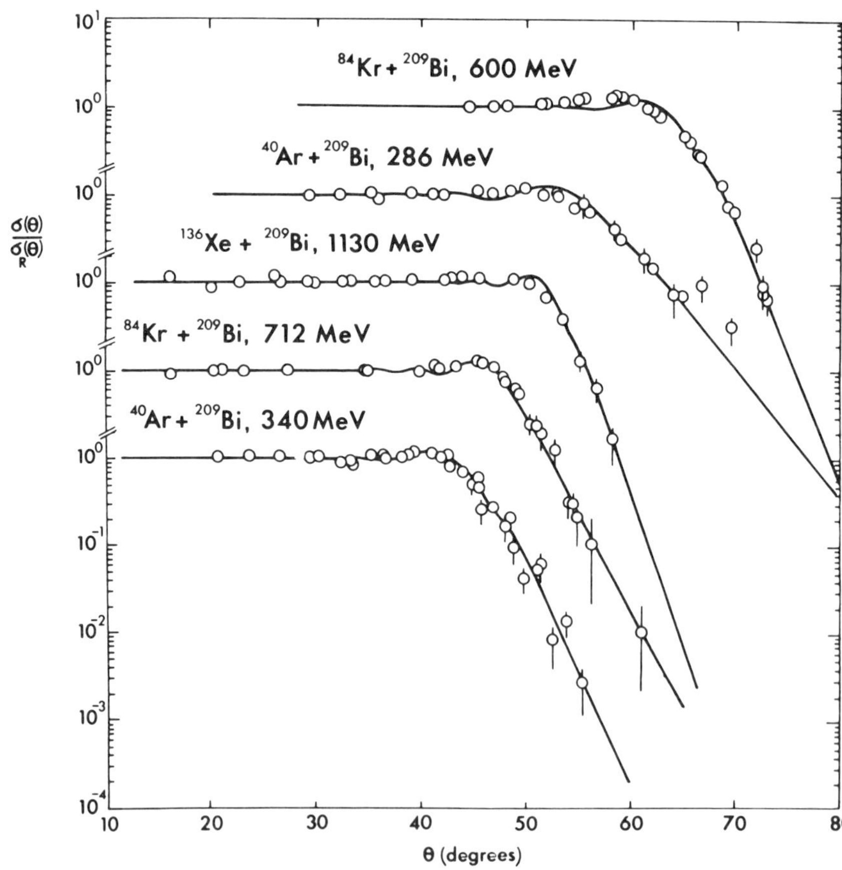

Figure 42. Angular distributions for elastic (see text) scattering of ^{40}Ar, ^{84}Kr, and ^{136}Xe by ^{209}Bi (Hui 76b, Bir 76, Sch 78g). The solid lines are fits to the experimental data with the generalized Fresnel model (Fra 78–78c).

events from nuclear reaction events at larger angles is illustrated in Figure 41. With the limited resolution inherent in this technique it is obvious that the "elastic" cross section contains the contributions due to excitation of low-lying states.

The angular distributions for elastic scattering of ^{40}Ar, ^{84}Kr, and ^{136}Xe by ^{209}Bi (Hui 76b, Bir 76, Sch 78g), obtained by the method outlined above, are shown in Figure 42. The solid lines demonstrate the excellent fits to the experimental data obtained with the generalized Fresnel model (Fra 78–78c) discussed in Section 2.1. Information about the quarter point angle ($\theta_{1/4}$), the grazing angular momentum ($l_{\max}$), the interaction or strong-absorption radius (R_{SA}), and the total reaction cross section (σ_R) is listed in Table 1 for some typical heavy-ion reactions. On the basis of the good agreement for the Fresnel and optical models, one concludes that the Fresnel model is extremely useful for estimating the total reaction cross section for heavy-ion collisions. The generalized Fresnel model, despite its much greater simplicity, gives information about the interaction between heavy ions equivalent to that given by the optical model. The strong damping of the elastic

Table 1. Some Parameters Determined from a Generalized Fresnel Model[a] Fit to Elastic-Scattering Angular Distributions of Selected Heavy-Ion Reactions. Also Included for Comparison Are Optical-Model Reaction Cross Sections

Projectile	Target	E_{lab} (MeV)	$\theta_{1/4}^{c.m.}$ (°)	l_{max} ($\hbar$)	R_{SA} (fm)	σ_R^{GFM} (mb)	σ_R^{OM} (mb)	Ref.
^{20}Ne	^{235}U	175	56.2	92.1	12.83	2078	2060	(Mey 76)
	^{235}U	252	33.8	134.8	12.66	3007	2920	(Mey 76)
^{40}Ar	^{40}Ca	191	42	67	10.26	1739	1736	(Was 79)
	^{40}Ca	236	32	82	10.26	2068	2038	(Was 79)
	^{40}Ca	272	26	93	10.26	2250	2202	(Was 79)
	^{209}Bi	286	60.0	153.0	13.44	1977	1937	(Bir 76)
	^{209}Bi	340	47.2	185.4	13.18	2503	2491	(Bir 76)
	^{238}U	286	68.5	143.8	13.52	1654	1803	(Bir 76)
	^{238}U	340	51.7	185.3	13.49	2345	2416	(Bir 76)
^{84}Kr	^{208}Pb	494	98.0	167.3	14.04	887	885	(Van 76d)
	^{208}Pb	718	49.0	350.2	14.18	2683	2509	(Van 76d)
	^{209}Bi	600	66.7	268.5	14.17	1938	1924	(Bir 76)
	^{209}Bi	712	50.4	344.7	14.18	2621	2610	(Bir 76)
^{136}Xe	^{208}Pb	1120	55.5	463.5	14.8	2610	2630	(Van 76c)
	^{209}Bi	940	70.1	383.0	15.5	2060		(Wil 80)
	^{209}Bi	1130	54.3	479.2	15.1	2750	2780	(Sch 78g)
	^{209}Bi	1422	38.0	633	15.2	3810		(Wol 81)

[a] (Fra 78–78c).

scattering cross section for very heavy-ion reactions due to the long-range Coulomb interaction complicates probing the radial dependence of the magnitude of the nuclear potential in such experiments.

The energy dependences of the total reaction cross section for two reactions, $^{209}Bi + ^{136}Xe$ and $^{40}Ca + ^{40}Ar$, as determined from a generalized Fresnel model analysis of elastic scattering data, are shown in Figure 43. The solid lines are calculated with the formula (Fra 78b)

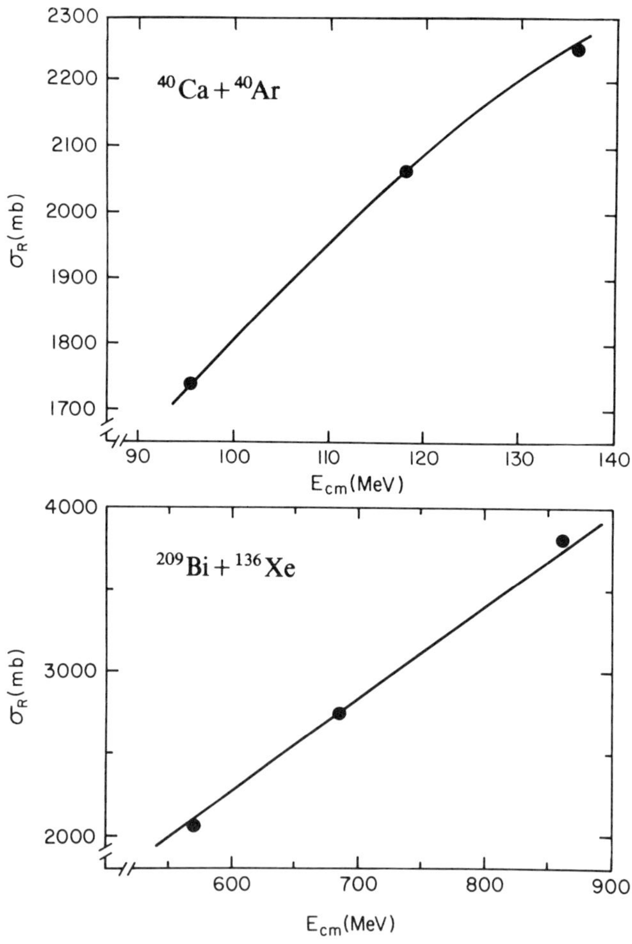

Figure 43. Energy dependence of the total reaction cross section as determined from a generalized Fresnel model analysis of elastic scattering data. The solid curves are from Equation (3.1.1) (see text).

$$\sigma_R = \pi R_{SA}^2\left(1-\frac{V_c}{E}\right)\left[1+c_1\frac{a}{R_{SA}}\left(\frac{1-0.5V_c/E}{1-V_c/E}\right)+c_2\left(\frac{a}{R_{SA}}\right)^2\left(\frac{1-0.5V_c/E}{1-V_c/E}\right)\right] \tag{3.1.1}$$

where $V_c = Z_1Z_2e^2/R_{SA}$, $c_1 = 2.54$, $c_2 = 3.26$, and $a = 0.065$ fm for the $^{209}Bi + ^{136}Xe$ reaction, and $a = 0.26$ fm for the $^{40}Ca + ^{40}Ar$ reaction. The values of the quantity a were adjusted here in order to fit the intermediate energy in each case. The theoretical slopes are in good agreement with the data, and only slightly altered from the familiar semiclassical result given by the first term of Equation (3.1.1).

As stated before, the total reaction cross section for heavy-ion reactions is composed of several reaction types including fusion (CN and FL), damped or deep inelastic collisions and few-nucleon transfer or quasielastic events. The competition between these different processes depends on several factors such as Z_TZ_P and the degree of mass asymmetry of the reaction partners. Fusion is the dominant reaction channel for reactions induced by lighter heavy ions and gives a good estimate of their reaction cross sections. However, the fusion process has essentially disappeared for krypton-induced reactions on heavy targets (Bir 79). Hence, for very heavy systems, like the $^{209}Bi + ^{136}Xe$ reaction, it is possible to measure the total reaction cross section by integrating the angular distribution of the damped reaction events. Such an angular distribution for the lighter Xe-like fragments is shown in Figure 44 where binary events of all degrees of energy damping are included, in a manner consistent with the definition of elastic scattering events discussed above. Integration of the curve drawn through the data points yields a total reaction cross section of $\sigma_R = (2.84 \pm 0.15)$ b. This value is in excellent agreement with that obtained from analysis of the elastic-scattering angular distribution (see Table 1), and gives support to the method described earlier for extracting the total reaction cross section from a generalized Fresnel model analysis of low-resolution heavy-ion elastic scattering angular distributions.

Heavy-ion elastic scattering angular distributions have been interpreted either in terms of Fresnel diffractive scattering (Fra 78) or Coulomb rainbow scattering (Chr 76). The physical mechanism responsible for the elastic scattering angular distribution is quite different in these two models. Whereas in the Fresnel model the diffraction is associated with strong nuclear absorption, the second model assumes weak absorption and a rainbow refraction associated with the real part of the nuclear potential. An intermediate position states that the characteristics of elastic scattering are not determined in general by either of the imaginary or real potentials alone, but that both potentials are required for good fits to the data (Sat

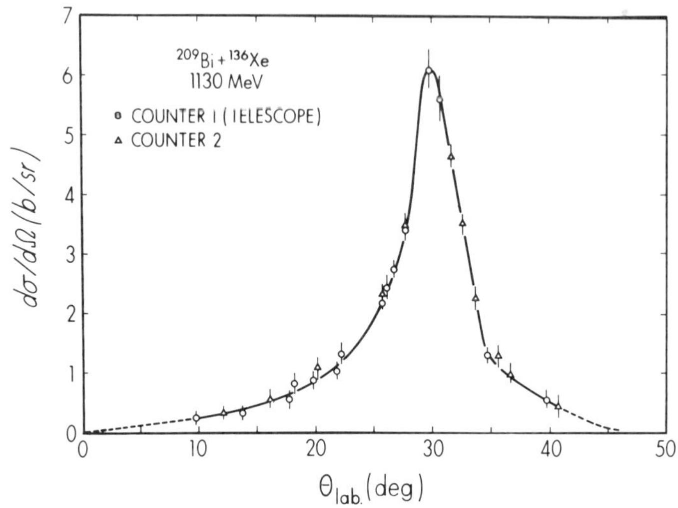

Figure 44. Laboratory angular distribution of the total reaction cross section. All of the reaction channels are included. From (Sch 78g).

79c). Authors have claimed a transition from rainbow scattering to strong absorption as a function of $Z_T Z_P$ (Vaz 78b) and bombarding energy (Mad 80b). Angular momentum values l_{CR} associated with Coulomb rainbow scattering (see Figure 3) have been calculated with the nuclear proximity potential and compared with values of l_{max} (Hui 76b) for systems with a wide range of $Z_T Z_P$ values and bombarding energies. Although there is a tendency for l_{max} to be slightly larger than l_{CR} for heavier systems, the difference decreases with bombarding energy. For the lightest systems $l_{max} \approx l_{CR}$.

Interaction or strong-absorption radii R_{SA} have been compiled (Bir 76, Chr 76, Bir 77, Vaz 78b, Sat 79c) for a number of heavy-ion reactions. A plot of $\zeta = R_{SA} - C_T - C_P$ vs. $(C_T + C_P)$ is shown in Figure 45. Values of R_{SA} on this plot are from three types of analyses; the generalized Fresnel model, the optical model and the Coulomb rainbow scattering model. The optical-model strong-absorption radius $D_{1/2}$ is defined as the distance of closest approach for the classical Rutherford orbit of angular momentum l for which the transmission coefficient $T_l = 0.5$. It corresponds to an elastic partial wave amplitude $\eta_l = (1/2)^{1/2}(\sigma_{el}/\sigma_{Ruth} = 0.086)$ and is expected to be slightly larger than the Fresnel radius R_{SA} ($\eta_l = 0.5$ and $\sigma_{el}/\sigma_{Ruth} = 1/4$). Even so for purposes of Figure 45, the three slightly different distances are not distinguished. The distance between the half-density radii at the point where strong absorption sets in, $\zeta(R_{SA}) = R_{SA} - C_T - C_P$, decreases with

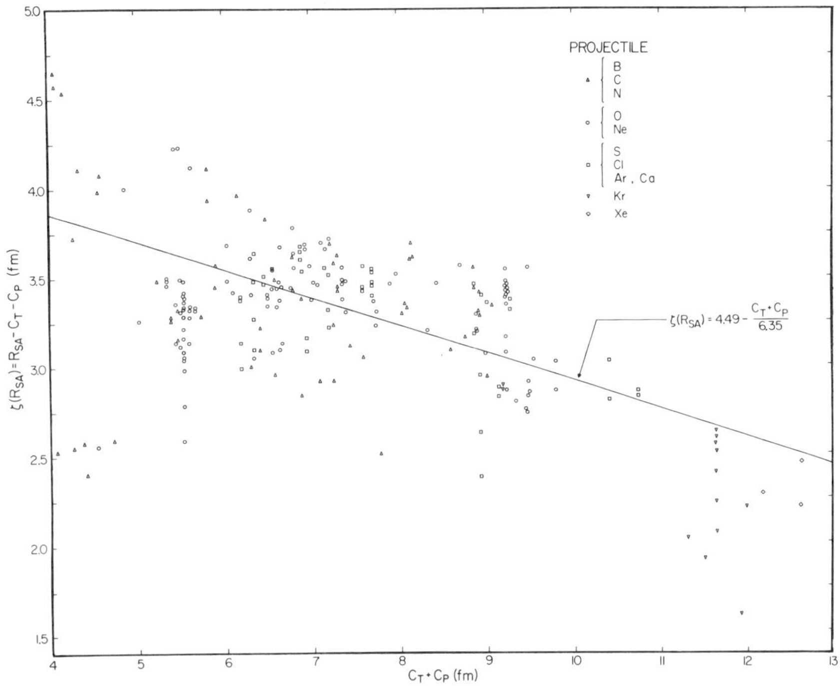

Figure 45. Correlation of strong-absorption radii for heavy-ion reactions (Bir 77). From J. R. Huizenga, US ERDA Progress Report C00-3496-65 (1977), p. 131.

$C_T + C_P$ (heavier systems) and is given by

$$\zeta(R_{SA}) = 4.49 - (C_T + C_P)/6.35 \text{ fm} \tag{3.1.2}$$

(for the surface diffuseness of $b = 1$ fm, see Section 2.5). However, one observes a wide spread in the data about the line with a root-mean-square deviation of 0.33 fm. Part of the spread in the data is no doubt associated with the bombarding energy dependence of R_{SA} which is not explicitly accounted for in Figure 45. The actual values of R_{SA} can vary by several tenths of 1 fm as a function of bombarding energy. This is illustrated in Figure 46 for the $^{40}Ca + ^{16}O$ reaction where R_{SA} (actually $D_{1/2}$) values have been determined from analyses of elastic scattering angular distributions for laboratory bombarding energies ranging from 40 to 214 MeV (Sat 79c). The values of R_{SA} decrease with energy linearly with $E^{-1/3}$ (Mad 80b). For the $^{40}Ca + ^{16}O$ reaction they are well reproduced by the expression $R_{SA}(\text{fm}) = 7.61 + 6.38E_{\text{c.m.}}^{-1/3}$. Different reactions require slightly different values of the two constants.

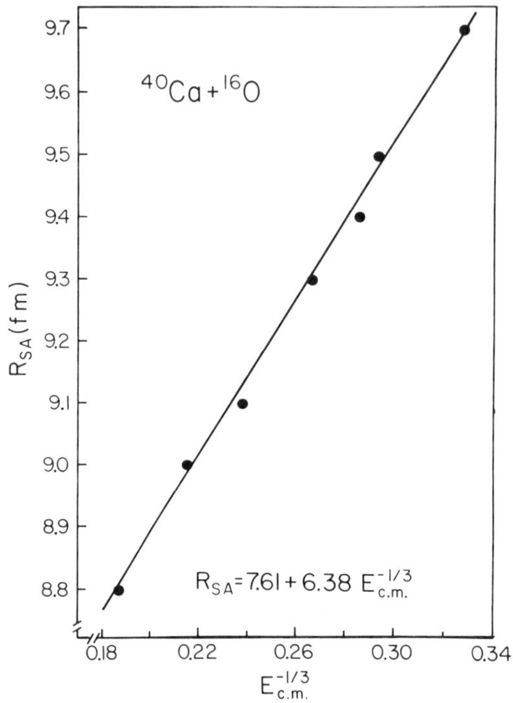

Figure 46. Correlation of the strong absorption radius R_{SA}, as determined from elastic scattering (Sat 79c), with bombarding energy $E_{c.m.}^{-1/3}$.

From the discussion presented above, it appears that some of the main experimental trends of the total reaction cross section σ_R and of the related quantities are fairly well established for heavy-ion reactions in the limited bombarding energy range of interest here. For example, realistic estimates for σ_R can be obtained from a Fresnel model analysis of experimental elastic-scattering angular distributions. On the other hand, it is also clear that the question of the relative degrees to which Coulomb excitation and nuclear inelastic scattering contribute to a "damping" of the elastic channel is far from being settled definitively. Further elucidation of this matter requires systematic high-resolution studies of heavy-ion elastic and inelastic scattering for a range of projectile–target systems and bombarding energies. The fundamental importance of such investigations is also seen for a better understanding of the damped reaction mechanism, in particular of the initial stages of the dissipative processes occurring in such reactions.

3.2. Kinematics and Fragment Multiplicity

In this section, experimental evidence is discussed for the characteristic two-body nature of damped nuclear reactions, a property that is assumed in the theoretical work outlined in Section 2. For reactions between complex nuclei at bombarding energies of the order of GeV, a dominant production of only two massive fragments in the exit channels is not necessarily expected. In fact, high-efficiency measurements covering a large fraction of the total solid angle suggest a rather complex reaction pattern, in certain cases. This is illustrated in Figure 47 for the reaction $^{nat}U + {}^{238}U$ at $E_{Lab} = 1785$ MeV, where ionization tracks of events with final fragment multiplicities of 4 and 5 are shown at the top and the bottom, respectively, as measured (Got 79b) with glass and mica track detectors. The circles in the figure represent tracks of beam particles penetrating the detector perpendicular to the plane of view shown. For this reaction, events with final fragment multiplicities of 3 (≈ 58%) and 4 (≈ 42%) essentially exhaust the total reaction cross section. However, despite this apparent complexity of the reaction, it can be inferred (Got 79b) from the measured lengths and directions of correlated fragment tracks that the U + U reaction pattern is consistent with a binary first reaction step, followed by the sequential fission of one or both of the two primary products. It is such a restricted definition of a binary reaction mode that is adopted here, with the understanding that the two primary products escape the range of their mutual nuclear interaction prior to an eventual decay via the sequential processes of fission and/or emission of light particles and γ rays. In practice, however, it may not always be feasible to distinguish an instantaneous high-multiplicity

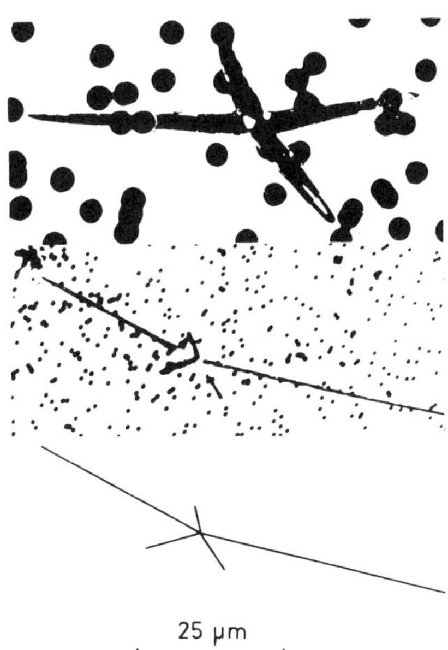

Figure 47. Microphotographs of a four-pronged and an extremely rare five-pronged event in a glass and mica track detector shown at the top and in the middle, respectively, as obtained for the reaction $^{nat}U +$ (1785 MeV)^{238}U. The dots are due to beam projectiles traversing the detector perpendicular to the plane of view. The drawing at the bottom represents a reconstruction of the five-prong event. From (Got 79b).

event from a fast two-step process where the decay of the primary reaction products occurs under the influence of a strong Coulomb repulsion.

The dominant primary fragment multiplicity can be established more directly than possible with the technique described above by utilizing experimental in-beam procedures where several reaction products are measured in coincidence. Such exclusive measurements test the degree to which total initial energy, momentum, mass, and charge of the system is balanced by quantities associated with the observed fragments and allow a reconstruction of the primary event history. A variety of options exists for the reconstruction of a truly binary event from the measurement of kinematical correlations between two coincident reaction fragments (Nic 59, Min 72), since it is determined already by three independent kinematical quantities. The relevant quantities of a three-body reaction can also be fully deduced from a kinematical coincidence measurement of two of the correlated fragments (Ohl 65, Arm 76). An approximate determination of the reaction kinematics for multiparticle events is still experimentally feasible, if the majority of particles have comparatively small masses and velocities and are emitted sequentially from primary reaction fragments in random directions. This is the case in damped nuclear reactions at relatively low bombarding energies, as will be discussed in Section 3.7.

A modern and rather sophisticated apparatus (San 75, Har 79) for kinematical coincidence experiments is illustrated in Figure 48. It consists of a multielement, position-sensitive gas ionization chamber combined with a fast parallel-plate avalanche counter and a large position-sensitive avalanche coincidence counter. With this arrangement, the emission angles of two heavy fragments and their arrival times in the counters with respect to a beam pulse, as well as the specific ionization and the total energy of the correlated fragments, can be obtained. Other arrangements relying on solid-state detectors lack the large solid angle of the setup shown in Figure 48.

Kinematic-coincidence experiments have been performed for damped reactions induced by a range of projectiles with $A \geq 40$ and for various energies (Lef 73, Han 74b, Wol 74,76, Pet 75,75c, San 77, Har 79,82, She 79). Here, the measured reaction angles and energies of two correlated fragments allowed inference of primary fragment masses and center-of-mass energies. Typical examples of in-plane angular correlations of fragments from the reaction $^{209}\text{Bi} + {}^{84}\text{Kr}$ at $E_{\text{Lab}} = 600$ MeV (Wol 76) are represented in Figure 49, for various values of the mass number of the light (projectile-like) reaction fragment. The correlation angles expected from two-body kinematics, as indicated by the arrows in this figure, are well matched by the centroids of the measured correlation functions. Their finite width is understood to be entirely due to the recoil broadening induced by neutrons

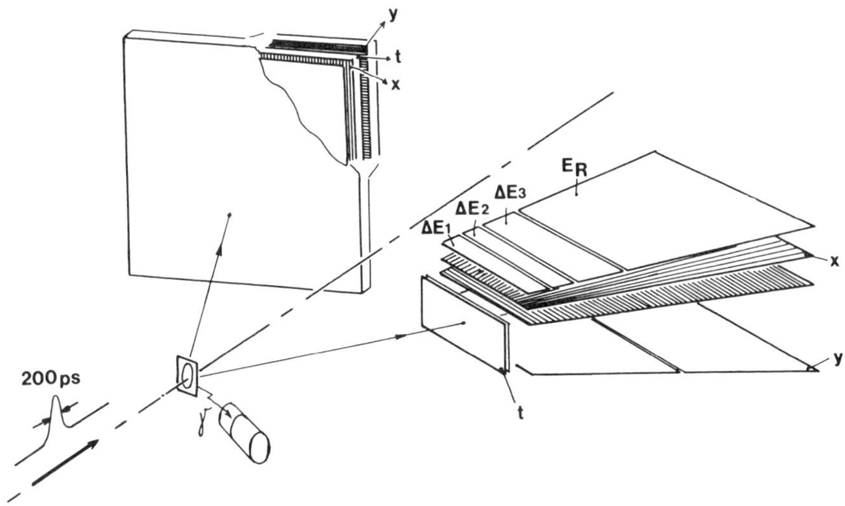

Figure 48. Schematics of a coincidence apparatus for measuring correlations between heavy fragments produced in damped reactions. The large counter on the left is a position-sensitive parallel-plate avalanche detector with fast timing characteristics for a measurement of fragment reaction angle and velocity which is derived from the time of arrival with respect to a beam pulse. The front counter of the array on the right of the figure is also an avalanche detector for the determination of the velocity of the correlated fragment, followed by a multielement, position-sensitive ionization chamber measuring its reaction angle, specific energy loss, and total energy. Also shown is a γ-ray detector close to the target. From GSI report 80-4.

evaporated from the two primary fragments in flight. The good agreement between the data of Figure 49 with evaporation calculations (curves) precludes a significant contribution of three-body primary events involving a third heavy participant such as an α particle, for the reaction $^{209}Bi + ^{84}Kr$.

Experiments where, in addition to angle–energy coincidence information, charges and velocities of one or both heavy fragments were determined, have been performed for reactions induced by ^{12}C (Nat 76, Nov 78), ^{16}O (Bon 76h), ^{20}Ne (Nat 77, Sch 78h), ^{32}S (Gel 76, Bra 76), ^{40}Ar (Bab 78), (Cau 78b), ^{63}Cu (Pet 77b), and ^{132}Xe (Heu 78). Of the two methods of deducing the primary fragment masses mentioned above, the time-of-flight technique measuring the final velocities v_i is the more direct one. Here, the primary fragment masses M_3 and M_4 are inferred employing the relation

$$M_i = \frac{M_1 v_1}{v_i} \frac{\sin(\theta_{34} - \theta_i)}{\sin \theta_{34}}, \qquad i = 3,4 \tag{3.2.1}$$

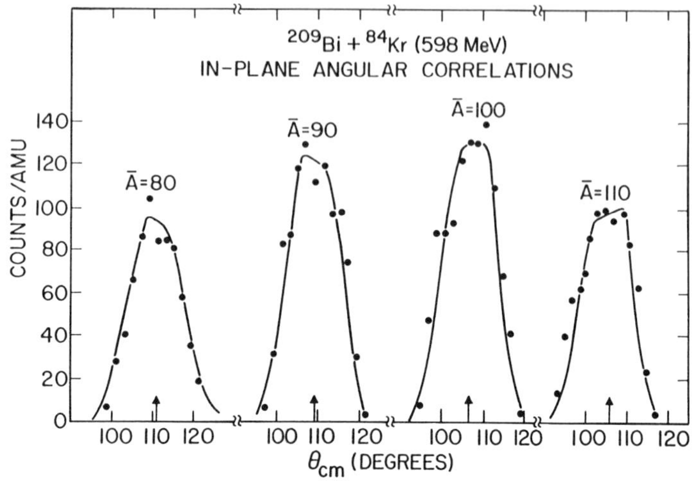

Figure 49. In-plane angular correlation of fragments from the reaction $^{209}Bi + ^{84}Kr$ at 600 MeV, where the light fragment of mass A was measured at $\theta_L = 48.5°$. The average emission angles expected from two-body kinematics are indicated by arrows. The curves represent an evaporation calculation. From (Wol 76).

derived from linear-momentum balance for a two-body reaction. It involves the projectile mass M_1, its laboratory velocity v_1, and the measurable laboratory reaction angles θ_3 and θ_4 of the outgoing fragments, with $\theta_{34} := \theta_3 + \theta_4$. In deexcitation processes, where a primary fragment i decays mainly via evaporation of light particles with an angular distribution that is isotropic in the fragment's rest frame, the primary velocity v_i and reaction angle θ_i remain on the average unchanged. Hence, by inserting into Equation (3.2.1) the first moments $v_i = \langle v_i' \rangle$ and $\theta_i = \langle \theta_i' \rangle$ of the measured distributions over secondary velocities and angles v_i' and θ_i', respectively, one is able to deduce the correct primary fragment masses. In contrast, if the secondary fragment energies E_i' are measured instead of the velocities v_i', explicit corrections for the evaporation process are required, in order to make such a deduction. To first order, the secondary kinetic energies are reduced with respect to the primary ones by the ratio of primary to secondary masses, i.e., $E_i'/E_i = (M_i - \Delta M_i)/M_i$. An estimate has to be made for the evaporated masses ΔM_i, in order to derive the primary masses M_i from the relation

$$M_i = \frac{M_1 E_1}{2E_i'} \frac{\sin^2(\theta_{34} - \theta_i)}{\sin^2\theta_{34}} \left\{ 1 \pm \left[1 - \frac{4\Delta M_i E_i'}{M_1 E_1} \frac{\sin^2\theta_{34}}{\sin^2(\theta_{34} - \theta_i)} \right]^{1/2} \right\} \tag{3.2.2}$$

The angles appearing in Equation (3.2.2) are again to be identified with the first moments of the corresponding secondary-fragment distributions. In the most interesting cases of $\Delta M_i \ll M_i$, the plus sign in Equation (3.2.2) applies.

The methods for deriving primary fragment masses and multiplicities discussed above are based on the assumption that particles are emitted from the hot primary fragments in a statistical evaporation process that does not show any directional preference in the rest frame of the emitter or leads to an angular emission pattern that is at least symmetric around 90° with respect to the fragment flight direction. These requirements are no longer fulfilled if preequilibrium particle emission (Bla 75) from the primary fragments becomes important, which is associated with a preferential emission direction. In such cases, particle emission patterns must be established by an extended correlation experiment including the detection of the light associated particles, before definite conclusions on the primary masses of the reaction fragments and their multiplicity can be drawn. Anticipating the discussion of Section 3.7, it may already be mentioned here, that preequilibrium emission from fragments produced in damped reactions at energies of at most a few MeV/nucleon above the interaction barrier is not a likely process.

With the reservation made above with respect to light-particle emission, all of the experiments reported for bombarding energies lower than about 10 MeV/nucleon support strongly the binary nature of damped heavy-ion collisions. There is, however, considerable evidence that, when the target or projectile is heavy ($A \gtrsim 190$), sequential fission of heavy reaction fragments is an important decay mode (Dye 77, 79b, Woz 78b, Pui 79, Got 79b, Har 79, Gla 79, Spe 79b) in strongly damped collisions.

In the case of reactions where multibody final channels are seen to be important (see, e.g., Figure 47), the sequential, two-step character of the reaction mechanism is established by exclusive experiments which are complete both with respect to the measurement of kinematical determinants and the acceptance of the total phase-space distribution of the events (Har 79, Gla 79, Spe 79b). Results have been reported for reactions of ^{208}Pb projectiles on targets of ^{58}Ni and ^{90}Zr, and for ^{238}U-induced reactions on ^{58}Ni, ^{90}Zr, ^{238}U, and ^{248}Cm targets at incident energies of 7.5 MeV/nucleon. The evidence for sequentiality derived from these experiments is based on a sharply peaked probability distribution for the vector difference $\mathbf{v}_R = |\mathbf{v}_1 - \mathbf{v}_2|$ of the fission fragment laboratory velocities $\mathbf{v}_1$ and $\mathbf{v}_2$ measured in coincidence with target-like fragments in a setup similar to that of Figure 48. The quantity $E_K = (\mu/2)v_R^2$ determines the total kinetic energy of the fission fragments in the rest frame of the fissioning system, where μ is the reduced mass of the two fission fragments. Examples for the sharply peaked probability distribution for v_R are shown in Figure 50 for two reactions (Har 79).

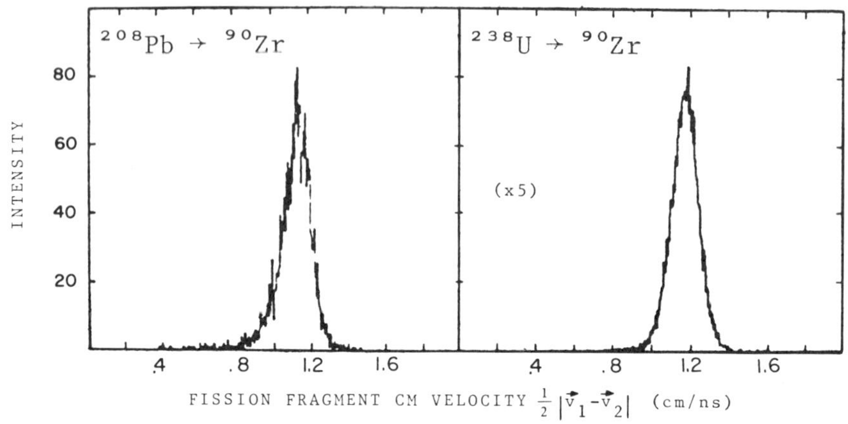

Figure 50. Distribution of the vector difference $|\mathbf{v}_1 - \mathbf{v}_2|$ of the fission fragment laboratory velocities, integrated over all fission directions. This figure demonstrates the existence of an intermediate fissioning system by resonance-like velocity distributions. (Har 79.)

The first and second moments of these distributions, i.e., average fission fragment velocities and their dispersions, are consistent with other conventional fission data (Van 73, Han 74c). Within experimental errors, the velocity distributions are found to be essentially independent of the fission direction and the Q value of the first reaction step. Since fast three-body breakup or instantaneous fission processes (Deu 76, 78) are expected to lead to a significant influence of the nuclear and/or Coulomb interaction forces on the fission fragment trajectories, the above experimental results may be taken as an indication that, instead, a sequential fission process is observed in these cases, occurring after full acceleration of two primary fragments by their mutual Coulomb interaction. Coulomb trajectory calculations establish a lower limit on the primary-fragment separation distance of 70–100 fm at the instance of scission of one of them, corresponding to a minimum time of 3×10^{-20} s (Gla 79) after the damped reaction has subsided.

While the above observations with respect to a two-step character of damped reactions leading to fragment multiplicities larger than two apply to bombarding energies of less than 10 MeV/nucleon, the onset of a transition towards one-step multifragment processes may have been seen in ^{84,86}Kr- and ^{129}Xe-induced reactions at slightly higher energies. Experiments reported by Olmi *et al.* (Olm 80) in which ^{166}Er and ^{89}Y targets were bombarded with 12.1 MeV/nucleon ^{86}Kr projectiles yielded evidence for a fission probability of projectile-like reaction fragments much higher than expected for a slow, equilibrium-statistical fission process. These results were corroborated by exclusive measurements of ternary processes occurring in ^{84}Kr-induced reactions on ^{90}Zr, ^{166}Er, ^{197}Au, and ^{238}U (Gla 81), and in the reaction $^{122}\text{Sn}+{}^{129}\text{Xe}$ (Spe 81) at bombarding energies of 12.5 MeV/nucleon. Distinct correlations observed in these experiments for the relative velocity v_R of correlated fragment pairs with the emission angle of a hypothetical fissioning primary reaction fragment were interpreted (Spe 81) in terms of a Coulomb distortion of the sequential fission process, due to the close proximity of the third observed reaction fragment. Although no direct evidence for an instantaneous ternary reaction mode has been established yet, the above studies clearly emphasize the necessity of multiparameter exclusive measurements for a correct interpretation of reaction properties in the bombarding-energy domain beyond ~ 10 MeV/nucleon and indicate that new reaction modes can be expected to occur at these higher energies.

3.3. Angular Distributions of the Reaction Fragments

The angular distribution of the differential cross section, $d\sigma(\theta)/d\Omega$, associated with a reaction constitutes an important piece of information about the nature of the reaction mechanism. The pronounced oscillatory pattern of the angular distributions of few-nucleon transfer processes induced by relatively light heavy ions, as discussed, e.g., by Glendenning (Gle 75, 77) and reviewed by Bass (Bas 80), presents an example of a reaction feature that is indicative of the directness of the considered transition, the amount of orbital angular momentum converted in the process, and the structure of the nuclear states it connects. However, as noted (Wil 75b, Oer 75, Vig 76, Gol 79) already in the case of such comparatively simple nuclear reactions, angular distributions lose their distinctive structure, once relatively heavy projectiles and targets and incident energies well above the Coulomb barrier are involved. The large number of open channels, important kinematical influences such as recoil effects present under such conditions, as well as the impossibility of resolving individual nuclear states in the region of high level densities, restrict thc access to these reactions to studies of their average, gross properties. Although methods developed for a description of individual transitions in light-ion-induced reactions are sometimes employed [see, e.g. (Bal 77)] in the interpretation of angular distributions of damped-reaction fragments, their direct application to damped reactions exhausting a major fraction of the geometrical reaction cross section [cf. Equations (2.1.21) and (2.1.23)] appears to be highly questionable. On the other hand, classical and semiclassical concepts developed for the former reaction type can provide guidelines also in the interpretation of damped nuclear reactions. As discussed in Section 2, the questions one hopes to address with studies of damped-reaction angular distributions pertain to the relative importance of coherent and incoherent aspects of the reaction mechanism (cf. Sections 2.2, 2.6, and 2.7), the average deflection function (cf. Section 2.1), and its interpretation in terms of the conservative and dissipative (cf. Sections 2.5–2.7) forces associated with the motion along macroscopic degrees of freedom modeled in phenomenological reaction theories.

In order to characterize the damped-reaction events of main interest for the following discussion, Figure 51 illustrates laboratory kinetic-energy spectra of fragments from the reaction $^{139}La + {}^{86}Kr$, obtained by Webb *et al.* (Web 76b) for $E_{\rm Lab} = 710$ MeV in a solid-state counter experiment. Experimental intensities are plotted in arbitrary logarithmic units vs. pulse height measured in channel numbers. As seen from Figure 51, at forward laboratory reaction angles θ_L, one observes a dominant high-energy elastic peak merging with a broad distribution of damped events extending several hundred MeV down in energy. As the angle is increased from $\theta_L = 11°$, the relatively structureless damped distribution evolves into a double-humped spectrum, where the higher-energy bump approaches the corresponding elastic energy, while the centroid of the low-energy peak decreases in energy. For the spectrum taken at $\theta_L = 23°$, which is near to the grazing or quarter-point angle [$\theta_{1/4\ \rm c.m.} = 38.8°$ (Web 77)], the high-energy component has degenerated into a tail of quasielastic and partially damped events on the left of the elastic peak. At even more backward angles, it disappears with only the fully damped component surviving. These energy spectra illustrate a particular reaction type that can be interpreted (Web 76b, 77, Van 78) in terms of the nuclear orbiting phenomenon discussed below. Most other reactions reported so far lead to fragment energy spectra with a single damped component that exhibits an angular dependence corresponding to that of the high-energy bump in the $^{139}La + {}^{86}Kr$ spectra displayed in Fig. 51. It should be noted that in this, as in most other experiments, only the lighter, projectile-like reaction fragments are measured. Due to the two-body kinematics (cf. Section 3.1), there is an angular range where the corresponding heavy reaction partners are well separated in laboratory energy from the projectile-like fragments.

Disturbing contaminations of the laboratory energy spectra can arise from sequential fission (cf. Section 3.7) and from secondary scattering of projectile particles at detector apertures and supporting structures. Laboratory events measured in inclusive experiments are usually transformed into the center-of-mass system assuming the kinematic conditions of the original projectile–target combination to be representative of all events, or by making use of supplementary information on atomic or mass number obtained for the projectile-like fragments. Corrections to the measured

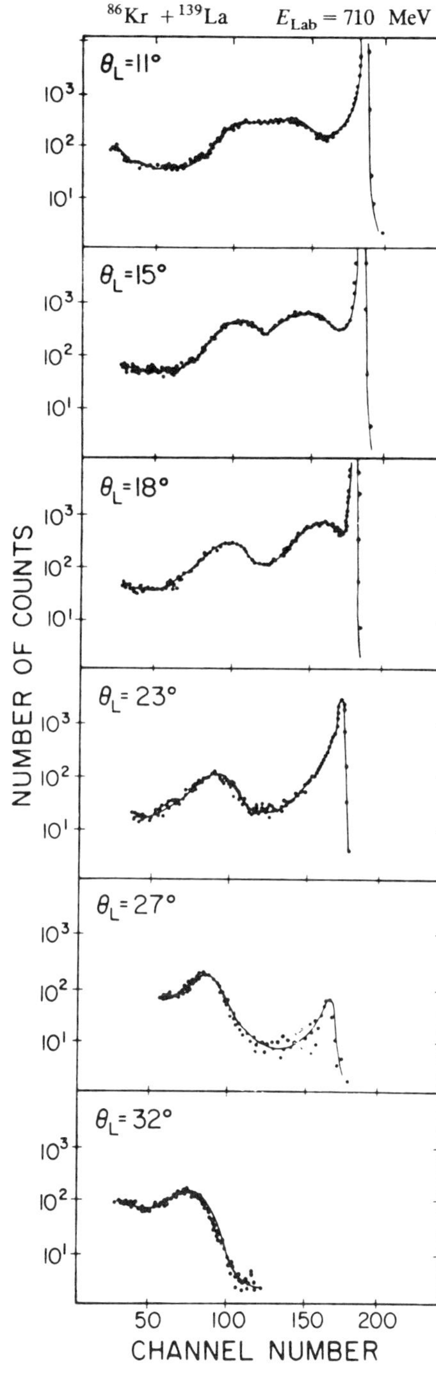

Figure 51. Singles energy spectra for ^{86}Kr + ^{139}La at $E_{\text{Lab}} = 710$ MeV. Only the shapes and not the absolute yields are indicated. The solid lines serve to guide the eye. From (Web 76b).

fragment energy accounting for the evaporation of particles off the fragments in flight (cf. Sections 3.2 and 3.7) and for pulse-height defects [see, e.g., (Mou 78)] have to be made, in order to derive the kinetic energies of the primary fragments and to perform a reasonably accurate transformation of data. The experimental procedures applied by various groups in the past have been rather different and evolved with time, such that it seems worthwhile to point out that care should be exercised in the comparison of different sets of data and in the judgement of their details.

Particle evaporation from the excited primary reaction fragments leads to a broadening of the measurable angular distributions. Fortunately, the effect is small in most interesting cases. Its size can be estimated from the angular distributions of evaporation residues produced in compound-nucleus reactions. For example, excited nuclei of mass number A_{CN} and initial linear momentum $\mathbf{P}_{\text{CN}}$ that decay via isotropic evaporation of neutrons with mean energy $\langle E_n \rangle$ can be shown (Sik 63) to result in evaporation residues with emission angles normally distributed around the original direction of flight. The width of this distribution is characterized by the angular variance

$$\sigma_\theta^2 = nm \frac{\langle E_n \rangle}{|\mathbf{P}_{\text{CN}}|^2} \left(1 + \frac{n+1}{A_{\text{CN}}}\right) \tag{3.3.1}$$

where m is the mass and n the number of evaporated neutrons. The mean multiplicity n is roughly proportional to the nuclear excitation energy, whereas $\langle E_n \rangle$ varies with the nuclear temperature. One estimates from Equation (3.3.1) that evaporative broadening of fragment angular distributions is negligibly small for weakly or partially damped reactions. It may result in an averaging over several degrees for strongly damped reactions at bombarding energies of a few MeV/nucleon above the barrier, where a medium-weight primary fragment may acquire excitation energies in excess of 100 MeV. For lighter systems, however, and dominant charged-particle evaporation, when the particle momenta are not small compared to that of the emitter, considerable distortions of the fragment angular distributions and, in particular, of correlation functions may result [see, e.g., (You 80)].

As already suggested at the beginning of this section, one expects relatively smooth and featureless angular distributions for fragments produced in damped and fusion-like heavy-ion reactions. Nevertheless, their overall shape should reflect the dynamics of production and decay of the transient intermediate system (Gat 75b, Vol 76) formed in a collision. The existence of an intermediate system, which under the conditions of complete equilibrium of all intrinsic and collective degrees of freedom is equivalent to the familiar notion of a compound system, is postulated because of the multistep microscopic character of collisions between complex nuclei. It manifests itself in a host of reaction features to be discussed below.

In the limit of very long interaction times, an intermediate compound system may actually be produced that eventually decays via fission or the emission of light particles and γ rays. It is currently not known how long it takes a dinuclear system to achieve such a complete statistical equilibrium, starting from initial conditions very far from this limit. Crude estimates for the lifetime of a nearly spherical compound nucleus, once it has been formed, are available from studies (Bla 76, Sto 77c, Bas 80) of compound-nucleus decay products. For a nucleus at equilibrium with excitation energy E^* and spin J, this lifetime is roughly given (Bla 52) by $t_{\mathrm{CN}} \approx \hbar\rho(E^*, J)/N(E^*, J)$, where ρ is the compound-nucleus level density and N the number of open decay channels. Hence, depending on mass and excitation energy of the system, these characteristic times may range from 10^{-16} to 10^{-21} s. Possibly, due to a preformation of the final reaction fragments in the intermediate system, a highly excited nonspherical dinucleus in local thermal equilibrium may have a reduced stability and, hence, may decay on a time scale much faster than given by the above estimates. It may then become difficult, if not meaningless, to distinguish a presumably more direct multistep damped reaction from the compound-nuclear fission process. Current studies emphasizing the relation between these reaction types are the subject of Section 3.8.

Experimentally, one is then interested in establishing to what extent damped-reaction fragment angular distributions are different from those of fusion–fission reactions, on the one hand, and from the type associated with quasielastic few-nucleon transfer reactions, on the other hand. For extremely long interaction times, angular distributions of damped-reaction fragments are expected to approach those of a binary fission reaction (Hil 53, Swi 72, 72b). Early experiments on heavy-ion-induced fission have been performed by Britt and Quinton (Bri 60, Qui 60) and by Sikkeland and collaborators (Gor 60, 60b, Sik 62, 64, 70b, 71, Vio 62, 63, 63b). Theoretical interpretations have been given by Bohr (Boh 56), Strutinsky *et al.* (Str 56, Hal 58), Griffin (Gri 59), and more recently by Back *et al.* (Bac 78) for sequential fission.

According to a simple classical model (Gri 59), the angular distribution $d\sigma_{\mathrm{ff}}/d\Omega$ of fission fragments is determined by the spatial probability distribution of the preferred fission axis, as given by the rotator function $D^I_{MK}(\theta)$:

$$\frac{d\sigma(\theta, I, M, K)}{d\Omega} \propto \left(I + \frac{1}{2}\right)|D^I_{MK}(\theta)|^2$$

$$\sim \left[\sin^2\theta - \frac{K^2 + M^2}{I(I+1)} + \frac{2KM}{I(I+1)}\cos\theta\right]^{-1/2} \quad (3.3.2)$$

Here, θ is the angle with respect to the beam direction, I is the spin of the fissioning system with projections M and K on beam direction and preferred fission axis, respectively. For large spins, $I \gg 1$, and perfect alignment ($M \approx 0$), Equation (3.3.2) simplifies to

$$\frac{d\sigma_{\text{ff}}(\theta, I, K)}{d\Omega} \propto \left(\sin^2\theta - K^2/I^2\right)^{-1/2} \qquad \text{for } \sin\theta > K/I \quad (3.3.3)$$

If, furthermore, the preferred fission axis is nearly perpendicular to the nuclear spin I (i.e., $K \ll I$), the angular distribution becomes independent of spin,

$$\frac{d\sigma_{\text{ff}}(\theta)}{d\Omega} \propto (\sin\theta)^{-1} \qquad (3.3.4)$$

In general, however, proper averages of the angular distribution in Equation (3.3.2) over the contributing values of I, M, and K have to be taken, in order to obtain a meaningful fit to experimental distributions. An extensive discussion of fission fragment angular distributions has been given by Vandenbosch and Huizenga (Van 73).

As a typical example, Figure 52 shows the c.m. angular distribution of fission fragments from the reaction $^{197}\text{Au} + ^{12}\text{C}$ at $E_{\text{Lab}} = 93.3$ MeV (Gor 60b). As expected, the distribution is symmetric around $\theta_{\text{c.m.}} = 90°$ but shows large anisotropies. A $1/\sin\theta$-type angular distribution (dashed curve in Figure 52) fits the data well in the vicinity of the symmetry angle but exhibits large deviations at forward and backward angles. The solid curve in Figure 52 represents a fit to the data with a model angular distribution (Hal 58), derived by accounting for the proper statistical weights of the contributions with different spins and spin projections. For a compound-nuclear reaction, these weights are determined by the spin-dependent nuclear level densities [see, e.g., (Hui 72, Van 73)] which are functions of the excitation energy and spin cutoff parameters related to various moments of inertia of the compound system. One can, hence, relate [see, e.g., (Van 73)] the features of the fragment angular distributions, such as their anisotropies, to the conditions prevalent in the exit channel.

Recent experiments on fission induced by relatively light ($A \leq 40$) heavy ions abound. They include angular-distribution measurements obtained with the projectiles $^{6,7}\text{Li}$ (Fre 75, Vig 80, Kar 80b), ^{11}B (Zeb 74), $^{12,13}\text{C}$ (Gal 70, Kow 71, Zeb 74, Nat 76, 77b, Mil 78b, Nov 78), ^{14}N (Nat 77b), $^{16,18}\text{O}$ (Kow 71, Zeb 74, Nat 77b, Vid 77, Aga 80b, Bac 80b), ^{19}F (Mil 78b), $^{20,22}\text{Ne}$ (Zeb 74, Bab 76, Nat 77), ^{32}S (Bar 75, 77, 77b, Gel 76, Bac 81),

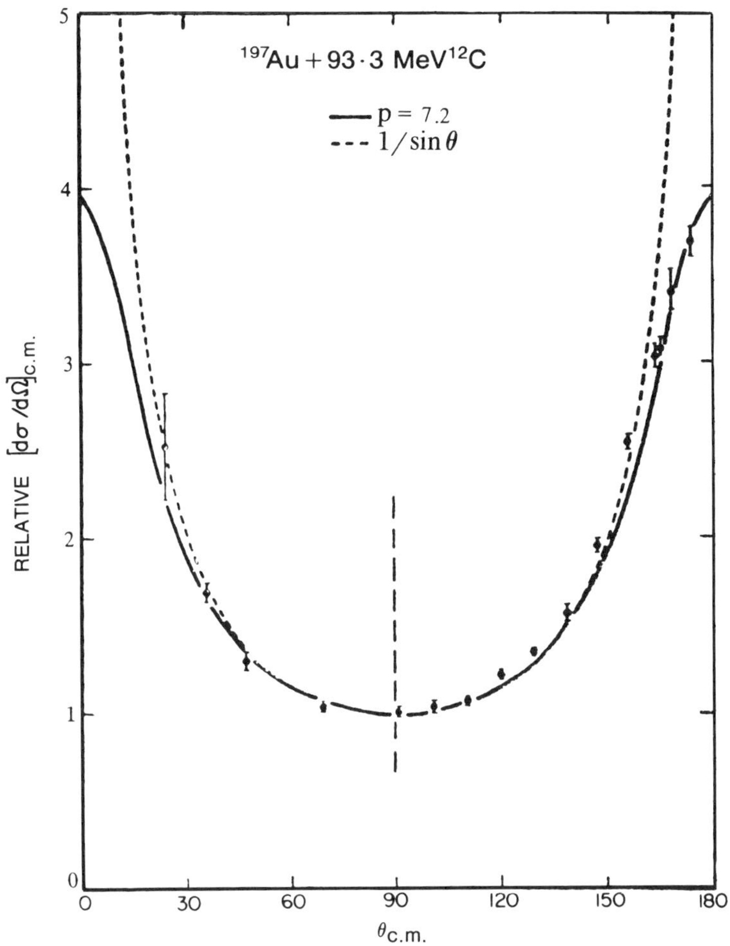

Figure 52. Center-of-mass angular distribution of fission fragments from the reaction $^{197}Au+$ ^{12}C at $E_{Lab}=93.3$ MeV. The solid curve represents a model fit (Hal 58), the dashed curve depicts a $1/\sin\theta$ dependence. From (Gor 60b).

^{35}Cl (Sco 76, Bis 78), and ^{40}Ar (Han 74c, Tam 75, Gal 75, Gat 75, Bri 76, Luc 76,77, Ngô 77b, Kal 77d, Bim 79, Log 80b) at bombarding energies ranging from a few MeV/nucleon up to more than 10 MeV/nucleon and a variety of targets.

In order to illustrate the general observations on the angular distributions made in the above studies, characteristic results obtained by Babinet *et al.* (Bab 76) for the reaction $^{107,109}Ag+^{20}Ne$ at $E_{Lab}=175$ and 252 MeV are exhibited in Figure 53. As can be seen from this figure, damped-reaction fragments with atomic numbers $Z > 15$ are associated with angular distributions that are well described by a $1/\sin\theta_{c.m.}$ dependence, as can be expected for a compound-nuclear fusion–fission reaction where the intermediate nucleus is produced on the average with a large spin. Deviations from such a dependence due to the finite spin ratio K/I [cf. Equation (3.3.3)] could occur [see, e.g., (Kow 71)] at far-forward or far-backward angles not measured in the above experiments. In contrast to this behavior, a conspicuous forward peaking of the angular distribution in excess of the $1/\sin\theta_{c.m.}$ dependence is observed in Figure 53 for the lighter fragments from the reaction ^{107,109}Ag + ^{20}Ne, in particular for those with $Z \leq 10$. This forward rise is not due to quasielastic events which have been excluded (Bab 76) from the data shown. One must then conclude that there is a distinctively different behavior of the angular distributions for fragments with charges (and masses) in the vicinity of the projectile, as compared to those products

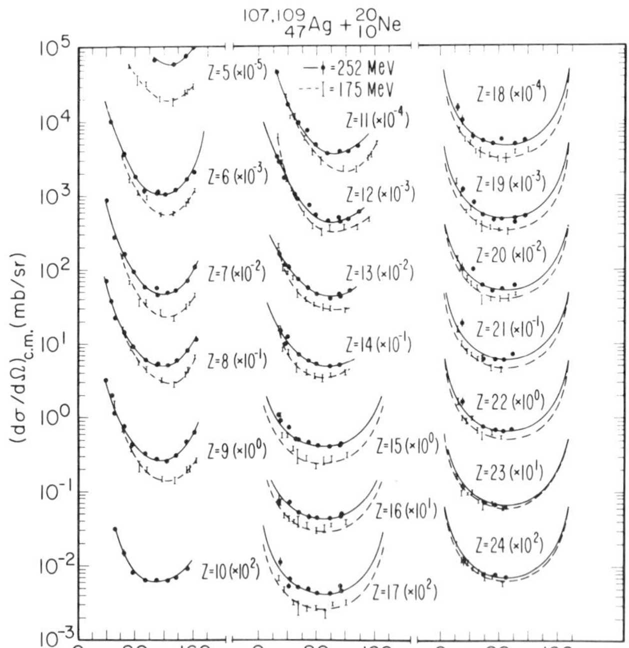

Figure 53. The c.m. angular distributions of fragments from the reaction ^{107,109}Ag + ^{20}Ne at $E_{Lab} = 175$ and 252 MeV. For fragment atomic numbers of $Z > 15$ curves drawn through the data correspond to $d\sigma/d\Omega \sim (\sin\theta_{c.m.})^{-1}$. From (Bab 76).

that are closer to symmetry, $Z \approx (Z_P + Z_T)/2$, which may be more representative of a high-energy fission process. The distinction between the two groups of events is more pronounced at the lower bombarding energy of $E_{\text{Lab}} = 175$ MeV.

To summarize the finding of the experiments mentioned above, fusion–fission-like angular distributions are observed in reactions induced by relatively light heavy ions. They are rather well established, in particular for lighter projectiles and lower energies. However, additional damped components with nonfission characteristics are also present that become dominant for projectiles as heavy as or heavier than ^{40}Ar, with a yield increasing with bombarding energy. The properties of these components differ also in other respects from what is expected from a compound-nuclear fission reaction, as will be discussed in subsequent sections. In the context of this discussion, however, it suffices to note that the forward-peaked events must correspond to faster, more direct processes which are also associated with comparatively little mass exchange between projectile and target.

Qualitatively, one can correlate the appearance of different reaction types and the dependence of their relative weight on projectile–target system and bombarding energy already with the characteristics of the elastic-scattering deflection function. This is demonstrated in Figure 54 (Hui 76b) for a relatively light system, ^{28}Si + ^{16}O, and a heavier one, ^{209}Bi + ^{40}Ar. In each case, the dependence of the distance of closest approach $D(l)$ or interpenetration depth, corresponding to the classical turning point r_0 in Equation (2.1.7), is plotted vs. angular momentum l, for two incident energies. The effective nuclear interaction potential V_{eff} is here assumed to be given by the sum of Coulomb and nuclear proximity potentials [cf. Equation (2.5.23a)]. The arrows in Figure 54 indicate the grazing angular momenta $l_{\max}$ determined by $r_0(l_{\max}) = R_{\text{SA}}$ (see Section 2.1). As demonstrated in this figure, the interpenetration depth for the reaction ^{28}Si + ^{16}O at $E_{\text{Lab}} = 47$ MeV exhibits a pronounced discontinuity for l values slightly below $l_{\max}$, quite in contrast to the relatively smooth dependence $D(l)$ predicted for the same system at the higher energy of $E_{\text{Lab}} = 100$ MeV, as well as for ^{209}Bi + ^{40}Ar at two incident energies. This effect is understood to be due to the presence of a barrier in the effective potential V_{eff} (cf. Figure 3) which is more pronounced for lighter systems than for heavier ones where the Coulomb repulsion overcompensates the nuclear attraction. If one trusts the elastic-scattering deflection function to be qualitatively representative also for weakly damped processes, one would expect for relatively light systems and low bombarding energies only a small band of l waves below $l_{\max}$ to contribute to fast, partially damped reactions, with the major fraction of the total reaction cross section corresponding to relatively deep interpenetrations and long interaction times. With increasing bombarding

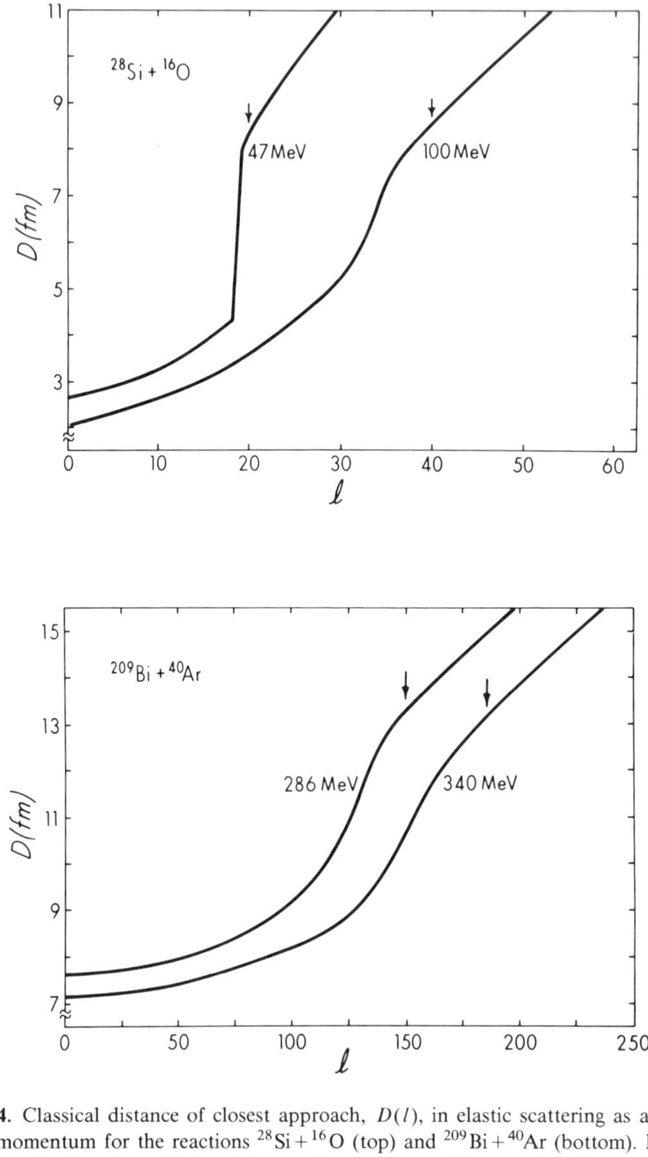

Figure 54. Classical distance of closest approach, $D(l)$, in elastic scattering as a function of angular momentum for the reactions $^{28}Si + {}^{16}O$ (top) and $^{209}Bi + {}^{40}Ar$ (bottom). Results were obtained with Equation (2.1.7) with an effective potential given by the sum of nuclear proximity potential [cf. Equation (2.5.23a)] and Coulomb potential. From (Hui 76b).

energy and total charge of the system, the transition between fast and slow reaction types is predicted to occur more gradually. In effect, the more repulsive interaction potential retards the fusion process under these conditions. This expectation is qualitatively corroborated by the measurements quoted above, although the discussion in Section 3.8 will show that the aspects of this transition are in general more complex than suggested by these simple considerations. The above arguments have validity for reactions of light projectile–target systems where the interaction barrier (cf. Figure 3) is well developed and located at relatively large internuclear distances, i.e., only slightly inside the strong-absorption radius R_{SA}.

In order to be able to investigate the damped-reaction mechanism over a broad range of l waves, it then appears advantageous to utilize the effect of static fusion inhibition due to the Coulomb repulsion by emphasizing reactions induced by heavy ($A_P > 40$) projectiles and heavy ($A_T > 100$) targets. Although there are models (cf. section 2.5) that predict pronounced minima in the effective radial potential even for heavy systems, the corresponding barrier is reached only in deeper interpenetrating collisions associated with angular momenta significantly below l_{max}. It is then possible to study the dynamical effects occurring in damped collisions such as frictional energy loss and other relaxation phenomena. As an example, Figure 55 illustrates the influence of the strength of the frictional forces on the ability of a system to become trapped within a minimum of the effective potential $V_{eff}(r)$. As shown for a trajectory with initial kinetic energy E_i, fusion is inhibited both for very weak and very strong friction associated with small and large energy losses $E_{Loss} = E_i - E_f$, respectively. In contrast to the static fusion inhibition referred to previously, this latter effect represents an example of dynamic fusion inhibition. In general, both the multidimensional potential energy surface and the dissipative collision dynamics influence the outcome of a reaction. Different combinations of conservative and dissipative forces may lead to similar results.

For reactions with interaction times comparable to those of a low-energy fission process, there is no well-defined reaction angle θ. The concept

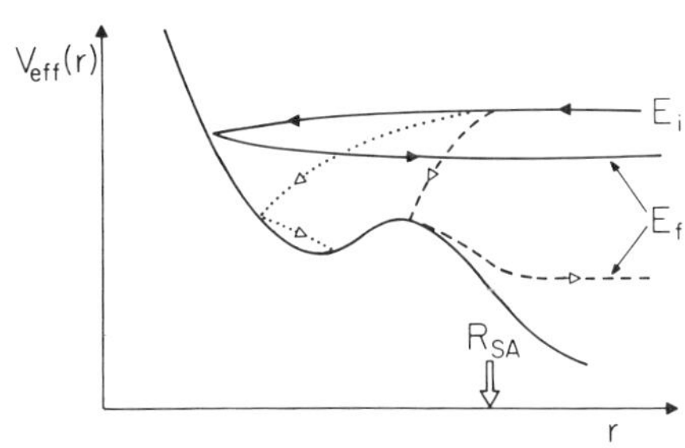

Figure 55. Schematic diagram of the effective interaction potential V_{eff} for a medium-heavy projectile–target system with a barrier located inside the strong-absorption radius R_{SA}. The trajectories drawn for an initial relative kinetic energy E_i illustrate that weak (solid curve) as well as strong (dashed curve) friction prevent trapping of the system inside the potential minimum.

of a deflection function $\theta(l)$ introduced in Section 2.1 becomes useful for the interpretation of fragment angular distributions only when the lifetime of the intermediate system is of the order of or less than a typical rotation period. Then, the correspondence between average classical angular distribution $d\sigma/d\Omega$ and average deflection function $\theta(l)$ represented by Equation (2.1.7) is well defined. Classical dynamical models (cf. Section 2.3) attempt to relate this deflection function to phenomenological conservative and dissipative forces influencing the motion of the system along a selected set of collective coordinates, whereas a description of the dispersion of system trajectories about the average path is provided by transport models (cf. Section 2.7).

An early classical picture of damped nuclear reactions was introduced by Wilczyński (Wil 73c), who interpreted experimental results (Art 73b) from the reaction $^{232}Th + ^{40}Ar$ in terms of an orbiting process. As schematically illustrated in Figure 56, it is assumed in this model that the attractive nuclear interaction force traps the intermediate system temporarily while it rotates and loses energy, E, via frictional processes. The reaction angle θ is then determined by the lifetime of the system and its l-dependent rotational

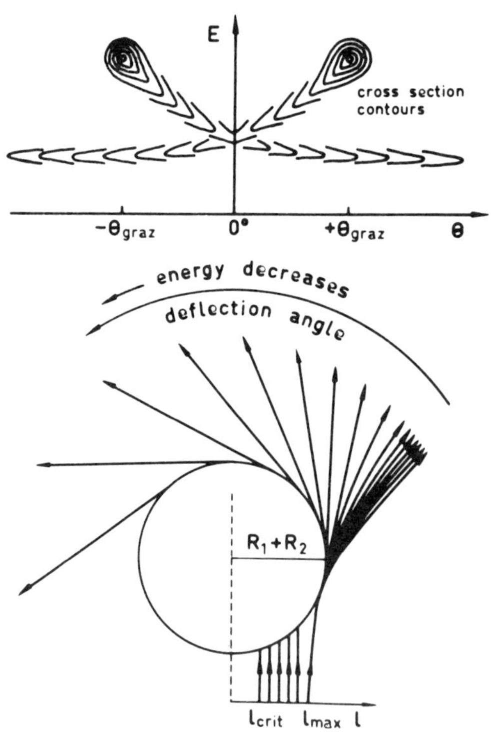

Figure 56. Illustration of the orbiting phenomenon in damped nuclear reactions. Trajectories for a band of l waves between l_{max} and the critical angular momentum l_{crit} for fusion are depicted at the bottom. The associated cross section pattern is displayed at the top as a contour diagram plotted vs. final fragment energy E and deflection angle θ. Negative reaction angles correspond to rotation of the intermediate system through the beam direction ($\theta = 0°$). From (Wil 73c).

velocity. More quantitatively, the orbiting phenomenon has been studied by Gross *et al.* (Gro 73,74,75,78,81, De 76,77) and Bondorf *et al.* (Bon 74b) in classical dynamical models. In the latter reaction model, the dinuclear configuration depicted in Figure 10d and the Coulomb potential of Equation (2.5.8) were employed. For the nuclear potential a Saxon–Woods form was assumed with a depth adjusted to the liquid-drop surface energy. The friction tensor was taken to be diagonal describing radial, tangential, and rolling friction with a friction form factor depending linearly on the radial separation. It is shown in this work that for strong friction and l values close to the critical l value for fusion, l_{crit}, the angular distribution of damped reaction fragments can be approximated by the relation (Bon 74b)

$$d\sigma/d\theta \propto 2\pi\lambda\!\!\!^{-2} l_{\text{crit}}^2 \exp[(\theta - \theta_c)/Y] \tag{3.3.5}$$

Here, θ_c corresponds to the sum of Coulomb deflection angles in entrance and exit channels, as can be calculated from Equation (2.1.7) for $l \approx l_{\text{crit}}$, and Y is a quantity that depends on the relative strengths of the friction and nuclear forces. The predicted exponential decrease of the angular distribution for negative reaction angles becomes more gradual with increasing relative strength of the friction force.

Angular distributions consistent with the exponential behavior suggested by the above orbiting model are often observed for the lighter systems already referred to. Figure 57 demonstrates this for the damped component of fragments from the reaction $^{58}\text{Ni} + {}^{40}\text{Ar}$ at $E_{\text{Lab}} = 280$ MeV measured by Gatty *et al.* (Gat 75) for individual fragment atomic numbers Z. As can be inferred from this figure, the data follow the predicted logarithmic straight lines, with slopes that decrease slightly with decreasing fragment Z. In the model of Bondorf *et al.* (Bon 74b), this dependence is interpreted as due to a relative strength of the friction force that increases with increasing distance of the final fragment charge from the projectile Z. In support of this view is the observation (Gat 75) that the total kinetic energy loss in the reaction increases accordingly.

An orbiting behavior for reactions such as $^{232}\text{Th} + {}^{40}\text{Ar}$ at $E_{\text{Lab}} = 379$ MeV illustrated in Figure 56 is predicted in various classical reaction models based on rather different combinations of conservative and dissipative forces. The condition for this phenomenon to occur in such models is that the radial acceleration provided by the effective interaction potential is damped by a sufficiently strong friction force such that the dinuclear system spends a relatively long time in the vicinity of the classical turning point while it rotates around its center of gravity. For a given conservative force, the tangential component of the friction force [cf. Equation (2.3.8)] mediating the dissipation of orbital angular momentum enhances the tendency for orbiting, since it reduces the repulsive centrifugal contribution to the

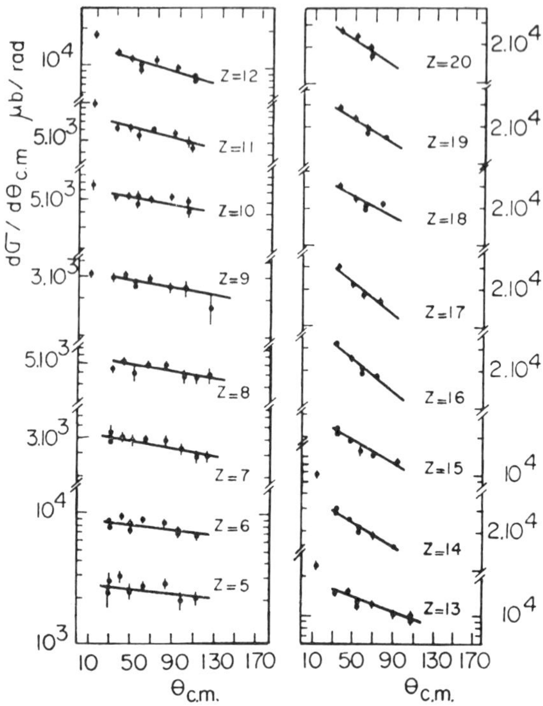

Figure 57. Angular distributions $d\sigma/d\theta_{\text{c.m.}}$ of strongly damped fragments from the reaction $^{58}\text{Ni}+^{40}\text{Ar}$ at $E_{\text{Lab}}=280$ MeV. The fragment atomic number Z is indicated at each data set. From (Gat 75).

effective interaction potential (compare Figure 3). However, whether orbiting occurs or not depends on the relation between conservative and dissipative forces to a much higher degree than on the shape of the individual form factors. For example, both the combination of a steep folding-type nuclear potential [cf. Equation (2.5.14) and Figure 18] with a strong friction force [see Equation (2.6.17) and Figure 27] employed by Gross *et al.* (Gro 78) and that of the shallower proximity potential [cf. Equation (2.5.23) and Figure 18] with the weaker proximity dissipation [see equation (2.6.29) and Figure 27] lead to the prediction of orbiting in the case of the $^{232}\text{Th}+^{40}\text{Ar}$ reaction. The deflection function obtained by Gross *et al.* (Gro 78) for this system is depicted on the left of Figure 58. In the vicinity of the grazing angular momentum indicated by an arrow, the theoretical deflection function is rather flat, i.e., $d\theta/dl \approx 0$. Consequently, the theoretical angular distribution shown on the right of Figure 58 becomes singular at the corresponding deflection angle. This distribution is to be compared to the experimental differential cross section $d\sigma/d\theta_{\text{c.m.}}$ for the particular reaction channel that

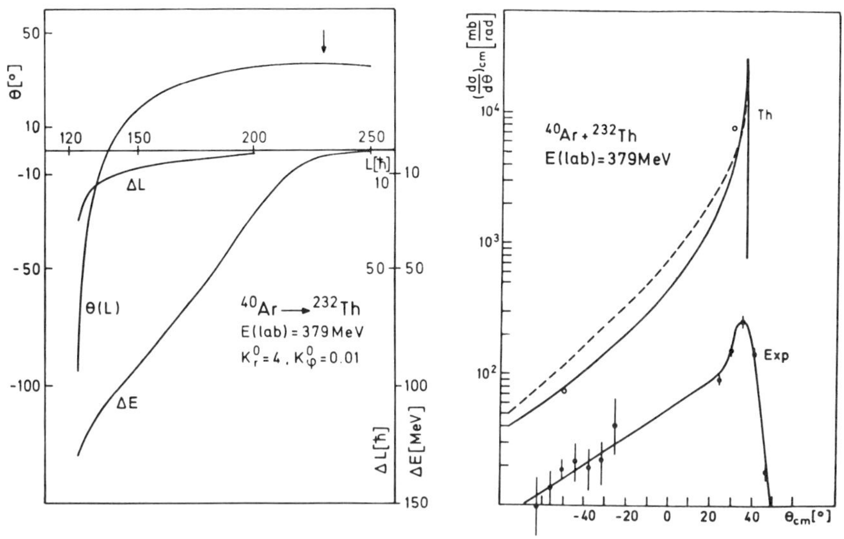

Figure 58. Predictions of a classical dynamical reaction model (Gro 78) for deflection function $\theta(L)$, energy loss $\Delta E(L)$, and angular-momentum loss $\Delta L(L)$ are plotted on the left vs. angular momentum L, for the reaction $^{232}\mathrm{Th}+{}^{40}\mathrm{Ar}$. The model friction constants correspond to the quantities c_r^0 and c_t^0 discussed in the context of Equation (2.6.17). An experimental angular distribution (points) for K fragments is shown on the bottom right, where weakly damped events are plotted at positive and strongly damped at negative reaction angles. The solid curve at the top represents the theoretical angular distribution. The broken curve illustrates the predicted effect of fragment deformations. From (Gro 78).

leads to projectile-like fragments with $Z=19(K)$. The experimental angular distribution (points in Figure 58) is the result of a decomposition of experimental data, where strongly damped events have been interpreted to be produced in collisions associated with a rotation of the intermediate system through the beam direction, corresponding to negative reaction angles. For weakly damped events, the system is assumed to proceed along a grazing-like trajectory, and, hence, the reaction angle is counted positive in these cases (cf. Figure 56). The validity of this experimental procedure is supported by results on fragment spin polarizations, as will be discussed in Section 3.6. As is apparent from Figure 58, the general character of the "experimental" angular distribution is well reproduced by the model calculation of Gross *et al.* (Gro 78), although, of course, the data do not show a singularity near the grazing angle. According to the discussion in Section 2.1, a closer correspondence between data and a classical trajectory model is not expected because of quantum penetration effects that may become important at rainbow angles. In addition, quantum and statistical fluctuations can lead to a dispersion of trajectories and, hence, avoid a cross-section singularity, as is further discussed below.

The ambiguity in the interpretation of fragment angular distributions with respect to the strength of conservative and dissipative forces would be somewhat restrained if supplemental information on the nuclear interaction potential could be deduced independently. This is, in general, not possible, but one may as a first approximation use those potentials that have proven successful in the description of elastic scattering or are derived on a microscopic basis (cf. Section 2.5). For example, Deubler and Dietrich (Deu 75, 77, 77b) have investigated how the classical deflection function depends on the strength of the friction force and its anisotropy by performing classical dynamical calculations (cf. Section 3.3) for the scattering of spherical or deformable nuclei using an interaction potential similar to that obtained in the energy-density formalism (compare Figure 18). Figure 59 displays a sample (Deu 77) of their results on deflection functions for the $^{232}Th + ^{40}Ar$ reaction, where curves 1 to 5 correspond to different strengths of the radial and tangential friction constants c_r and c_t, respectively, and curve 6 is calculated for deformable nuclei and weak friction. In these particular model studies, it is deduced that for weak friction ($c_r \leq 1.10^{-21}$ MeV fm s) the deflection function exhibits an inner, nuclear rainbow, in

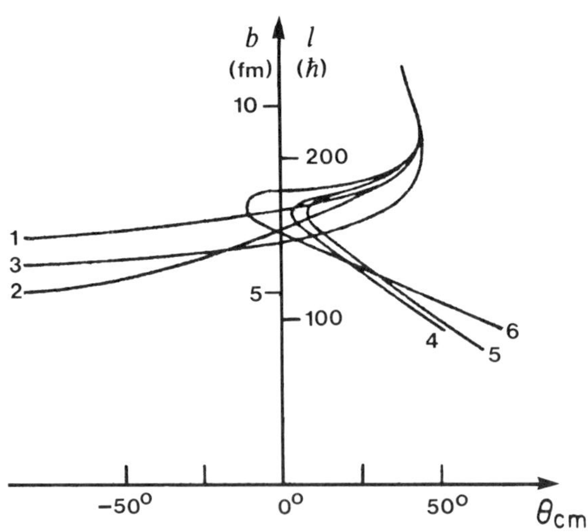

Figure 59. Deflection functions for $^{40}Ar + ^{232}Th$ at $E_{Lab} = 388$ MeV for different strengths of the friction constants c_r and c_t. The parameters are ($c_r[10^{-21}$ MeV fm s], $c_t[10^{-21}$ MeV fm^2 s]) = (3,3), (3,0), (20,20), (1,1), and (1,0) for curves 1–5, respectively, calculated in the model for two rigid spherical nuclei. Curve 6 is calculated in the deformed model with (c_r, c_t) = (0.5,0.5). From (Deu 77).

addition to the Coulomb rainbow located at larger values of angular momentum l or impact parameter b, instead of the orbiting behavior resulting in the cases (1, 2, 3) with stronger friction. For friction strengths in excess of $c_r = 30 \times 10^{-21}$ MeV fm s scattering to only positive angles results again, since classical turning points do not reach the top of the effective potential barrier (compare Figure 55). The deformability of projectile and target nuclei does not alter (Deu 77) these conclusions significantly, although the reduced effective Coulomb potential in the case of spheroidal nuclei touching at their tips leads in general to somewhat larger deflection angles than experienced by spherical nuclei. The type of a double-rainbow deflection function expected from the model for particular combinations of conservative and dissipative forces gives rise to a cross section peak at forward angles associated with strongly damped events, in addition to the quasielastic peak close to the grazing angle. From experimental angular distributions alone, such a rainbow peak cannot be distinguished from the superposition of positive-angle and negative-angle cross section branches at very forward angles, as illustrated at the top of Figure 56. It took the additional measurement of the fragment spin polarizations (cf. Section 3.6) in order to decide against a double-rainbow-type deflection function, in the particular case of the $^{232}Th + ^{40}Ar$ reaction.

As is obvious from the comparison of theoretical predictions and experimental results shown in Figure 58, it is necessary to invoke fluctuations in order to understand the shape of experimental angular distributions. This raises general questions as to the validity of an interpretation of the reaction in terms of average classical system trajectories. It was elaborated in some detail in Section 2.2 and illustrated in Figure 5 that the shape of the angular distribution does not necessarily reflect the mean scattering angle $\bar{\theta}(l)$ associated with an l window contribution to the reaction. According to Strutinsky (Str 78d), even with a mean deflection angle that does not vary at all with initial angular momentum, i.e., $\bar{\theta}(l) \equiv \theta_0$ [cf. Equation (2.2.6)], the character of the angular distribution of the associated reaction events can change from a Gaussian centered at $\bar{\theta} = \theta_0$ to a forward-rising distribution that gives the impression of a varying mean reaction angle. The factor controlling the form of the angular distribution in this model is a combination of quantal, statistical, and dynamical dispersions which cannot be inferred separately from the shape of the broad, relatively featureless experimental angular distributions that are characteristic for relatively light systems. Purely classical dynamical models relate this shape solely to the dynamical dispersion, neglecting potentially important quantal and statistical fluctuations. Only in those cases, where narrow angular distributions are observed, i.e., where average reaction angles are well defined and fluctuations are small, do both types of analysis lead to corresponding results. For other reactions, one has to realize that an interpretation of angular distributions in terms of classical potential scattering under the influence of friction represents only one possibility that has to be justified independently. On the other hand, analyses of data employing Strutinsky's model as reported (Har 78, 80, Rud 79b, Hoo 82) for a few reactions, do not lead to further insights into the reaction mechanism unless they are followed by an interpretation of the extracted model parameters in terms of microscopic theory. Such reservations should be kept in mind in the continuing discussion of the systematics of damped reaction features which refers mainly to classical and statistical reaction models for an explanation of experimental results.

For the two types of angular distributions, fusion–fission-like and orbiting-like distributions, discussed so far, the associated classical interaction times are presumably very long as compared to the fast (10^{-22} s) grazing reactions, because the intermediate system has to rotate through a large angle. As already alluded to in the context of Figure 54, the abrupt transition between quasielastic and strongly damped processes should become more gradual for systems with a more repulsive effective interaction potential, e.g., for the Coulomb-dominated heavy systems. In fact, one observes experimentally the transition from $1/\sin\theta_{\text{c.m.}}$ dependences to orbiting-type and grazing-like angular distributions by increasing the masses of projectile and target. Examples for the former transition are represented by the lighter systems referred to earlier, to which reactions induced by ^{40}Ca (see, e.g., Roy 77) and ^{86}Kr (Bla 79b) can be added. The transition from orbiting to grazing-like collisions can be followed in reactions involving ^{40}Ca (Aga 77, 80), ^{52}Cr (Aga 80, 80b, Mor 76c), ^{56}Fe (She 79, Heu 78), and ^{84,86}Kr (Rus 77b, Sch 77f, Woz 78, Cau 78) ions. Figure 60 demonstrates this for the reaction $^{86}\text{Kr}+{}^{181}\text{Ta}$ at $E_{\text{Lab}} = 620$ MeV, where the gradual evolution from sharply sideways peaked angular distributions, for fragments with Z close to the projectile ($Z = 36$), to forward-rising ones is noticeable with increasing distance from the projectile.

Following the arguments given in the discussion of Figure 54, raising the bombarding energy should have an effect on the fragment angular distributions that is to a certain extent analogous to an increase of the Coulomb repulsion. This is exemplified in Figure 61 for the reaction (Kau 61) $^{103}\text{Rh}(^{16}\text{O}, {}^{15}\text{O})^{104}\text{Rh}$ at various bombarding energies. Apparently, with increasing bombarding energy, a grazing-like character of the reaction emerges, although a forward-rising cross section component reminiscent of the orbiting picture also grows with energy. Qualitatively, this latter component is understood to arise from low or intermediate l waves for which the centrifugal barrier is not high enough to prevent significant interpenetration depths from being reached. Consequently, these l values lead to orbiting-like reactions associated with long interaction times. Of course, unlike the centrifugal repulsion, a dominant Coulomb interaction restricts collisions with all impact parameters to surface interactions. However, as already

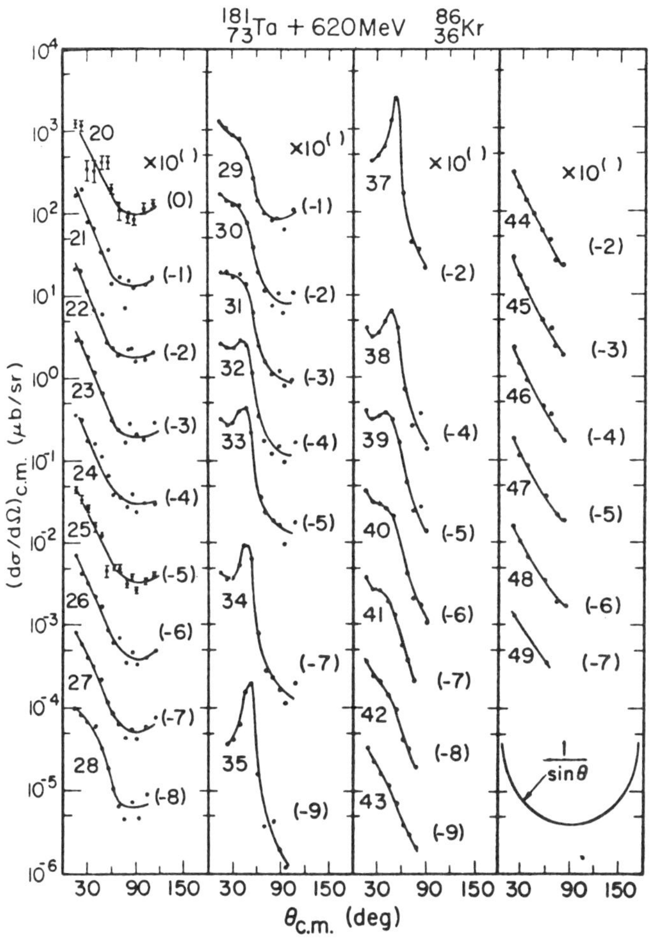

Figure 60. Center-of-mass angular distributions of fragments from the reaction ^{181}Ta + ^{86}Kr at E_{Lab} = 620 MeV as a function of fragment atomic number. Curves are drawn through the data points to guide the eye. A $1/\sin\theta_{c.m.}$ curve is shown for comparison. From (Cau 78).

emphasized, a more detailed understanding of the particular shapes of fragment angular distributions requires the inclusion of dissipative processes in the considerations.

The wealth of physical phenomena that have emerged from studies of damped and fusion-like reactions induced by relatively light heavy ions has challenged and deepened the understanding of the mechanisms governing the interactions between complex nuclei. The foregoing discussion of these systems has given a rather brief and incomplete account of this but is to be

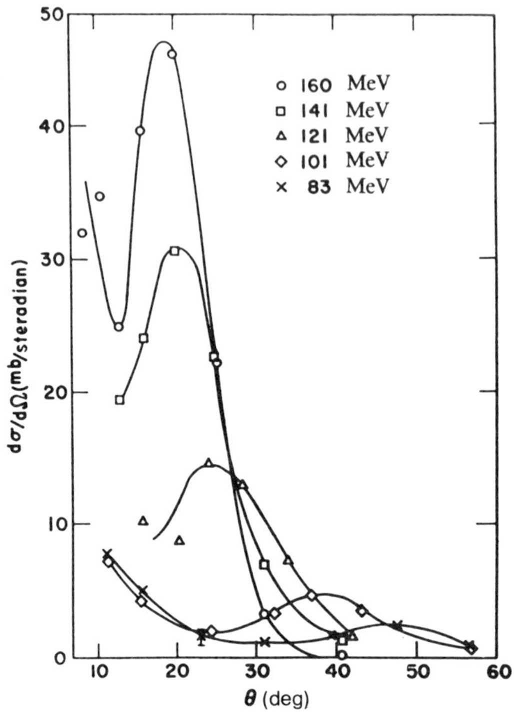

Figure 61. Differential cross section for the reaction ^{103}Rh(^{16}O, ^{15}O)^{104}Rh at various bombarding energies. From (Kau 61).

amended by related considerations presented in subsequent sections. There is no doubt, however, that significant advancements in the interpretation of damped-reaction properties have emerged from a confrontation of the characteristics of reactions induced by very heavy projectiles and targets with those known of the lighter systems. For heavy projectile–target systems, one is less hesitant to apply the classical concepts introduced in Section 2, on which most current reaction theories rely to some extent. Classicality, as represented, for example, by a large asymptotic Sommerfeld parameter, is realized for such systems already at relatively small relative velocities. In addition, the significant Coulomb repulsion (see, e.g., Figure 17) exerted on the interacting heavy nuclei provides a sufficiently strong static fusion inhibition and, consequently, a broad band of l waves contributing to damped peripheral interactions. Furthermore, as an important difference to the very light heavy-ion systems, both constituents of heavy systems can easily sustain the excitation energies required for the safe application of statistical arguments, without undergoing a complete disintegration. This, of course, implies also a different sensitivity to quantal and coherent aspects of the reaction, at least in practice. In the following, examples for the characteristic properties of angular distributions for fragments from reactions induced by very heavy ions will be discussed. They represent the general trends observed for systems involving ^{52}Cr (Aga 80, 80b, Riv 80), ^{56}Fe (Riv 77, 80, Gar 80c, Hoo 82), ^{63}Cu (Lef 76, Pet 76, 77b, Tam 76, Ber 77, Riv 77), ^{84,86}Kr (Lef 72, 76, Col 74, Han 74b, 74c, Cab 75c, Wol 74, 76, 76b, Mor 76c, Van 76, 76d, Web 76b, 77, Bri 76, Rus 77b, San 77, Sch 77f, Woz 78, Cau 78, Mil 78, Olm 78b, 80, Bla 79b, Rud 79, 81, Lyn 79, Luc 79b, Pui 79, 80, Dye 80, Eya 80, Chr 80, Sob 81), ^{132,136}Xe (Sch 76, 78g, Wol 76, 76b, 81, Van 76c, Rus 77, San 77, Heu 78, Ess 79c, Wil 80), ^{165}Ho (Laz 81), ^{208}Pb (San 77, Got 79b, Tan 80c), and ^{238}U (Hil 77, San 77, 80c, Fre 79b, Gla 79, Got 79b) projectiles.

Striking examples for the shape of angular distributions obtained for very heavy systems are displayed in Figure 62 where the experimental differential cross section is plotted vs. laboratory reaction angle for projectile-like fragments from the reaction ^{209}Bi + ^{136}Xe measured at incident energies of 940 (Wil 80), 1130 (Sch 76), and 1422 (Wol 81) MeV. These energies are between a factor of 1.3 to 2.0 larger than the Coulomb energy calculated for spherical nuclei touching at the strong-absorption radius (cf. Section 2.1). The associated large asymptotic Sommerfeld parameter ranging from $\eta = 268$ to 218 suggests that these systems are close to the classical limit. As is the case for most reactions induced by heavy systems, an insufficient experimental energy resolution of only 1%–2% (Bir 76, Van 76d) prevents a clear separation of quasielastic from damped events. Hence, quasielastic reactions are included in the angular distributions of Figure 62. As is obvious from the figure, the cross section is strongly focused into a narrow peak located a few degrees forward of the corresponding grazing or quarter-point angle [cf. Equation (2.1.19)] indicated by an arrow. The concentration of the cross section and its rigid correlation with the quarter-point angle is characteristic of a very fast peripheral reaction. It is important to note that the angular distributions shown in Figure 62 exhaust the total reaction cross section in each case and, hence, comprise several hundred l waves (see Table 1). A similar trend in the bombarding-energy dependence of the angular distributions of projectile-like fragments has been observed for the ^{208}Pb + ^{84}Kr (Van 76) and the ^{209}Bi + ^{84}Kr (Han 74b, Wol 74, 76) reactions.

In the case of the reaction ^{209}Bi + ^{136}Xe at $E_{\text{Lab}} = 940$ MeV, Wilcke *et al.* (Wil 80) were able to measure both the projectile-like and the target-like fragments simultaneously. Figure 63 shows the laboratory angular distributions for both fragment types, where target-like events are attributed to negative reaction angles. The asymmetry of the distribution of the heavy reaction partners is understood to be due to a kinematical effect. The integrated cross section for these events is somewhat reduced from that of the light fragments because of sequential fission (cf. Section 3.7) of highly excited Bi-like primary fragments.

Although it is rather difficult to separate experimentally fusion–fission events, that might be present in the reaction $^{209}Bi + {}^{136}Xe$, from very strongly damped processes, the fast falloff of the cross section on both sides of the peak (cf. Figure 62) signalizes a negligible contribution of interaction times that are long enough to allow for a complete rotation of the interacting system. At least for the two lower energies, there is no indication of negative-angle scattering or orbiting in the reaction $^{209}Bi + {}^{136}Xe$. However, one observes a shift of the slight back angle asymmetry of the cross section,

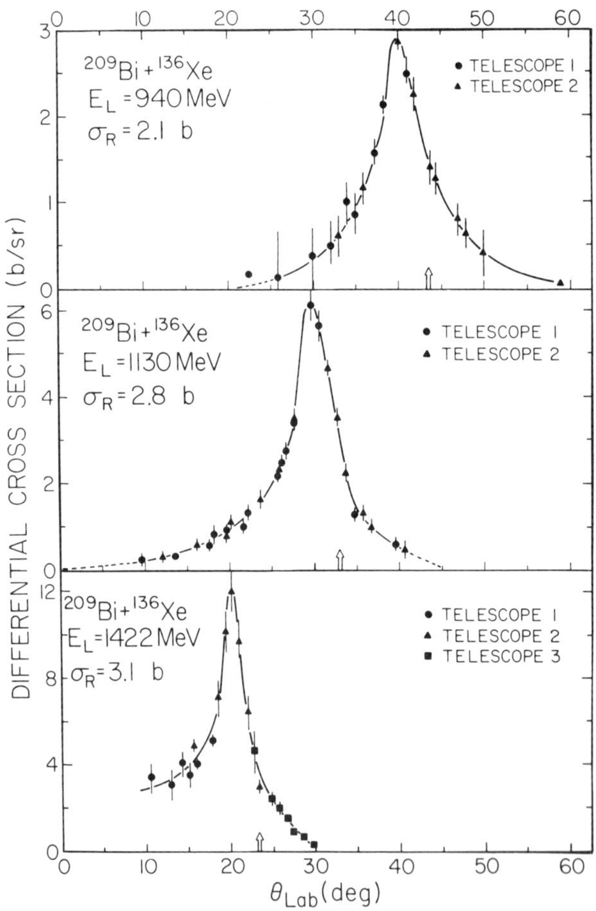

Figure 62. Laboratory angular distributions of the $^{209}Bi + {}^{136}Xe$ reaction at three energies (Wil 80, Sch 76,78g, Wol 81). The arrow indicates the angle of the quarter point in the laboratory. The center-of-mass energies above the Coulomb barrier for the three cases are 1.75, 3.14, and 5.29 MeV/nucleon.

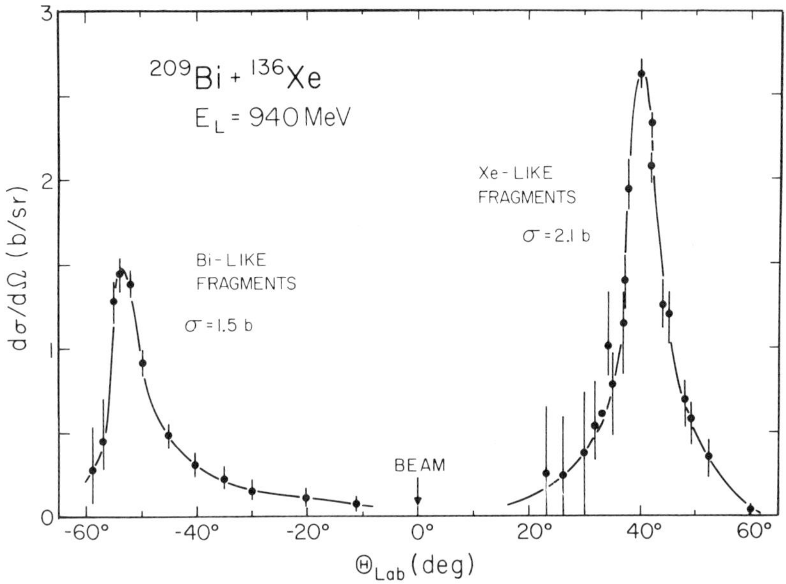

Figure 63. Laboratory angular distributions of light (Xe-like) and heavy (Bi-like) reaction products observed for the $^{209}\text{Bi}+{}^{136}\text{Xe}$ reaction at $E_{\text{Lab}} = 940$ MeV (Wil 80).

that exists for this system at $E_{\text{Lab}} = 940$ MeV, to forward angles, as the bombarding energy is raised. At the highest energy of $E_{\text{Lab}} = 1422$ MeV, a small orbiting contribution to the reaction cannot be ruled out.

The phenomenon of a strong angular focusing (Wol 74, Bec 76) of the cross section so clearly demonstrated for the reaction $^{209}\text{Bi}+{}^{136}\text{Xe}$ has an origin that is rather different from that of the grazing peak observed with lighter systems for the quasielastic events. For the latter systems, the cross section peak is due to l waves in the vicinity of the grazing angular momentum (or l_{max}) that experience very little deflection by the nuclear interaction, whereas smaller angular momenta lead to trajectories that are deflected forward by increasing amounts, as pictured in Figure 56. In contrast, for the heavy system $^{209}\text{Bi}+{}^{136}\text{Xe}$, the total reaction cross section is seen to be focused into a narrow angular range. Hence, in this case, the balance between conservative and dissipative forces leads to a classical deflection function that is approximately independent of l, i.e., $\bar{\theta}(l) \approx \bar{\theta}_0$, for several hundred l waves. As an immediate consequence, one concludes that the classical interaction times

$$t_{\text{Int}}(l_i) = \hbar^{-1}\int_{\phi(r_i)}^{\phi(r_f)} d\theta \mathscr{I}(l_i,\theta)/l(\theta) \geq (\hbar l_i)^{-1}\bar{\mathscr{I}}(l_i)\Delta\theta(l_i) \quad (3.3.6)$$

vary drastically with initial angular momentum l_i. Here, $\phi(r_i)$ and $\phi(r_f)$ are the angles in the coordinate system of Figure 2 of first and last nuclear contact, respectively, of the reaction partners, and $l(\theta)$ is the instantaneous angular momentum which may be reduced from l_i due to tangential friction. The average moment of inertia $\bar{\mathscr{I}}$ for orbital motion is not expected to decrease dramatically with decreasing l_i because of the peripheral character of the collision enforced by the strong Coulomb repulsion. In fact, it is more likely to show the opposite dependence, due to the possibility of neck formation and the resulting elongated shapes of the intermediate system in the exit channel. Furthermore, the angle $\Delta\theta(l_i) = \phi(r_i) - \phi(r_f)$ through which the system rotates under the influence of the nuclear interaction increases with decreasing l_i, in the case of angular focusing. The combination of these effects suggests a variation of the interaction time which is much stronger than given by $\sim l_i^{-1}$. A more detailed analysis of the data in terms of interaction times requires their partial-wave decomposition. Methods to achieve this will be explained further below.

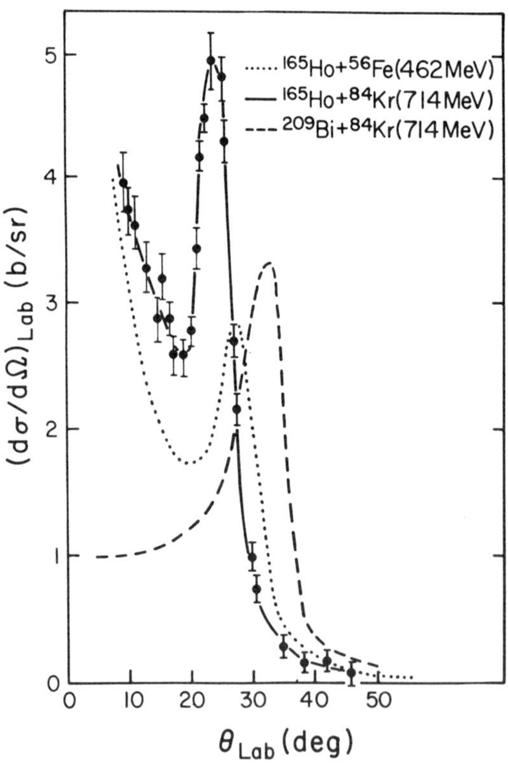

Figure 64. Laboratory differential cross section of light fragments from the reactions $^{165}Ho + ^{56}Fe$, $^{165}Ho + ^{84}Kr$, and $^{209}Bi + ^{84}Kr$ (Hoo 82, Wol 76).

Table 2. Values of the Sommerfeld and Modified Sommerfeld Parameters for Several Heavy-Ion Systems[a]

Reaction	$E_{\text{c.m.}}$	V_c	$(E_{\text{c.m.}}-V_c)/\mu$	η	η'	Ref.
$^{238}U+^{238}U$	892	735	1.32	486	1159	(San 77, Fre 79b)
$^{208}Pb+^{208}Pb$	780	605	1.68	386	816	(Tan 80c)
$^{209}Bi+^{136}Xe$	569	427	1.72	268	537	(Wil 80)
	685		3.13	245	399	(Sch 76, 78g)
	861		5.27	218	307	(Wol 81)
$^{209}Bi+^{84}Kr$	374	302	1.20	188	429	(Han 74b)
	428		2.10	176	324	(Wol 74)
	508		3.44	161	254	(Wol 76)
$^{166}Er+^{86}Kr$	339	255	1.48	157	316	(San 77)
	463		3.67	135	201	(Rud 81)
	685		7.59	111	140	(Olm 80)
$^{165}Ho+^{56}Fe$	345	187	3.78	95	141	(Hoo 82)
$^{232}Th+^{40}Ar$	246	174	2.11	95	175	(Art 73b)
	323		4.37	83	122	(Art 73b)
$^{40}Ca+^{40}Ca$	139	55	4.20	24	31	(Col 75)
	200		7.25	20	23	(Fra 80)

[a] See text for definitions.

The angular focusing of at least a significant fraction of the total reaction cross section is a common feature observed for all reactions between heavy nuclei, most notably for Xe-induced reactions on heavy targets (Van 76c, San 77, Rus 77, Ess 79c). However, the magnitude of this fraction varies as expected with the mass (or charge) of projectile and target, as is exemplified in Figure 64 for three reactions (Hoo 82, Wol 76) at comparable bombarding energies of 1.7 to 1.9 times the Coulomb barrier. With decreasing mass of projectile or target, the degree of angular focusing diminishes, and the angular distributions approach the orbiting type familiar from the light systems discussed earlier in this section. Attempts have been made (Gal 76, Mor 76b, Sch 77e, Mat 77) to classify the type of fragment angular distributions in terms of a single parameter that depends on bombarding energy and the charges and masses of projectile and target. As an example, the modified Sommerfeld parameters

$$\eta' = e^2 Z_P Z_T \left\{\mu/\left[2\hbar^2(E_{\text{c.m.}} - V_{\text{Coul}})\right]\right\}^{1/2} \tag{3.3.7}$$

are collected in Table 2 for several heavy-ion reactions. Qualitatively, three types of angular distributions are represented by the reactions in Table 2. For the very largest η' values, the angular distribution has a component at angles larger than the grazing angle. The effect begins to be noticeable for values $\eta' > 500$ as indicated in Figure 65 for the ^{209}Bi + ^{136}Xe reaction at the lowest energy. This figure shows the center-of-mass angular distributions corresponding to those of Figure 62, where, however, some of the quasielastic events have been excluded. Both the ^{208}Pb + ^{208}Pb and the ^{238}U + ^{238}U reactions listed in Table 2 have angular distributions with sizable backward

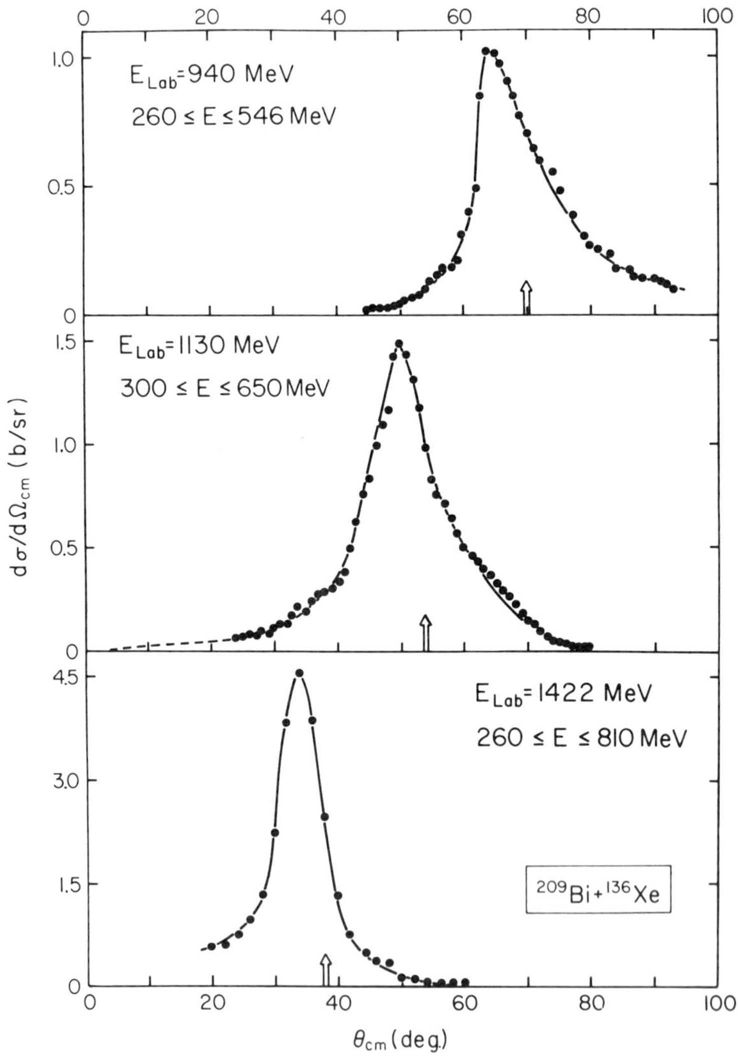

Figure 65. Center-of-mass angular distributions for the ^{209}Bi + ^{136}Xe reaction at three energies. Some of the quasielastic events have been excluded.

components. This is to be expected for systems with dominant Coulomb repulsion and correspondingly small interpenetration depths, resulting in classical deflection functions that are more Coulomb-like (see dashed curves in Figure 3). For intermediate values, $250 \lesssim \eta' \lesssim 400$, angular focusing is expected, according to the systematics, whereas the orbiting component should become dominant for η' values much smaller than 150. For comparison, the angular distributions shown in Figure 64 correspond to modified Sommerfeld parameters ranging from (Wil 81) $\eta' = 141$ for the $^{165}Ho + ^{56}Fe$ reaction at $E_{Lab} = 462$ MeV to $\eta' = 254$ for the $^{209}Bi + ^{84}Kr$ reaction. This comparison also illustrates that the above simple classification in terms of a single parameter has only a rather qualitative predictive power.

A more quantitative analysis of damped-reaction angular distributions has been performed by Wolschin *et al.* (Wol 78b) and Riedel *et al.* (Rie 79, 80). In this work, experimental angular distributions are fitted with a classical average deflection function parameterized as

$$\bar{\theta}(l_i) = 2\arctan\left(\frac{e^2 Z_P Z_T k}{2E_i l_i}\right) - \beta\theta_c \frac{l_i}{l_{gr}}\left(\frac{\delta}{\beta}\right)^{l_i/l_{gr}} \tag{3.3.8}$$

In Equation (3.3.8), the first term corresponds to the Coulomb deflection function [compare Equation (2.1.12)] involving the atomic numbers of projectile and target, Z_P and Z_T, respectively, the asymptotic wave number k, the initial kinetic energy E_i, and the orbital angular momentum l_i. The second term in Equation (3.3.8) parameterizes the nuclear part of the deflection function. Here, l_{gr} is the grazing angular momentum, and $\theta_c = \theta_{Coul}(l_{gr})$. The quantities β and δ are free parameters adjusted to fit the data. Figure 66 (Rie 79) displays a sample of deflection functions obtained from a fit of Equation (3.3.8) to experimental angular distributions for the reactions $^{209}Bi + ^{136}Xe$ and $^{166}Er + ^{86}Kr$ at two bombarding energies each. The deduced parameter values β range from 9 to 70 with a fairly constant δ of 0.11–0.17, for the four reactions illustrated in Figure 66. Since $\delta \ll \beta$, Equation (3.3.8) approaches the expression for the Coulomb deflection function for $l > l_{gr}$.

The quality of a fit of experimental angular distributions by the parameterized deflection function of Equation (3.3.8) is exemplified in Figure 67 for the reaction $^{166}Er + ^{86}Kr$ at $E_{Lab} = 8.18$ MeV/nucleon. Here, the dashed curve corresponds to the deflection function shown in Figure 66. Because this deflection function is of the double-rainbow type, the classical angular distribution derived from it through Equation (2.1.4) exhibits a nuclear rainbow peak plotted at $\theta_{c.m.} = +60°$ in Figure 67. As noted before, such singularities associated with classically forbidden scattering regions are expected to be washed out due to the effect of quantum penetration, as well

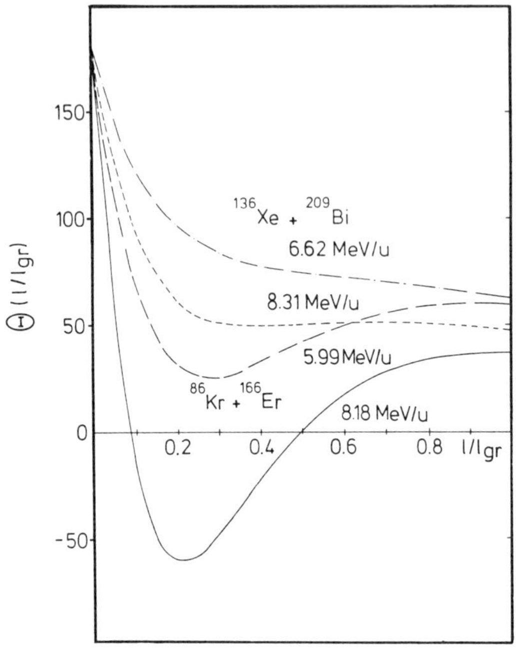

Figure 66. Average deflection functions for the reactions $^{166}Er + ^{86}Kr$ ($E_{Lab} = 5.99$ and 8.18 MeV/nucleon) and $^{209}Bi + ^{136}Xe$ ($E_{Lab} = 6.62$ and 8.31 MeV/nucleon) determined from a fit of Equation (3.3.8) to experimental angular distributions. From (Rie 79).

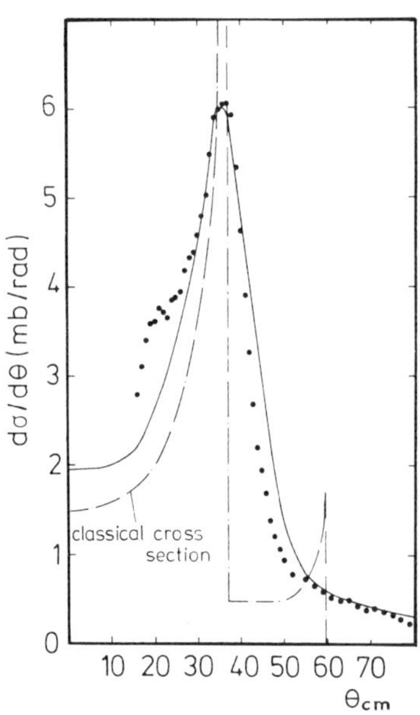

Figure 67. Angular distribution for the reaction $^{166}Er + ^{86}Kr$ at $E_{Lab} = 8.18$ MeV/nucleon. Experimental data are represented by dots. The dashed curve is calculated from the corresponding deflection function of Fig. 66 using Equation (2.1.4). The solid curve results from this deflection function when dispersed with the probability function P [cf. Equation (3.3.9)]. From (Rie 79).

as owing to quantum and statistical fluctuations. Considering only statistical fluctuations, one may assume approximately a normal dispersion of reaction angles θ associated with a given initial angular momentum l_i, around a mean angle $\bar{\theta}(l_i)$:

$$P(\theta, l_i) = \left(2\pi\sigma_\theta^2\right)^{-1/2} \exp\left\{-\left[\theta - \bar{\theta}(l_i)\right]^2 / 2\sigma_\theta^2(l_i)\right\} \qquad (3.3.9)$$

One can then express the classical cross section as

$$\frac{d\sigma(\theta)}{d\Omega} = \frac{\lambda\!\!\!^{-2}}{\left(2\pi\sigma_\theta^2\right)^{1/2} \sin\theta} \int dl_i\, l_i \exp\left\{-\frac{\left[\theta - \bar{\theta}(l_i)\right]^2}{2\sigma_\theta^2(l_i)}\right\} \qquad (3.3.10)$$

where the integration has to be performed over the l windows contributing to the reaction. For vanishing variance σ_θ^2 of the angular dispersion, the Gaussian in Equation (3.3.9) degenerates to a delta function $\delta(\theta - \bar{\theta}(l_i)) = \delta(l - l_i)\, dl_i / d\bar{\theta}$, and one regains the classical expression of Equation (2.1.4). The solid curve in Figure 67 has been obtained (Rie 79) by dispersing the classical deflection function [cf. Equation (3.3.8) and Figure 66] with a Gaussian probability function [cf. Equation (3.3.9)], where the angular variance σ_θ^2 was assumed to increase monotonically with decreasing value of l_i. It gives a reasonable account of the data, except at forward angles, where the observed cross section is underestimated. One notices, of course, that the above fitting procedure of experimental angular distributions does not yield unambiguous results. The data can be decomposed into a sum of Gaussians in various ways associated with rather different deflection functions. In support of this view, Hofmann and Ngô (Hof 76c) have shown in a study of the influence of statistical fluctuations on the damped-reaction angular distribution that the angular variance σ_θ^2 depends strongly on the assumed form factors of conservative and dissipative forces and emphasized the necessity to derive the associated transport coefficients from a microscopic model.

Whereas then the information on the damped-reaction mechanism contained in the total fragment angular distributions appears to be somewhat ambiguous, one may hope to elucidate the nature of the microscopic processes further by studying correlations of the angular distributions with various other reaction observables such as fragment mass, charge, spin, etc. As already illustrated in Figure 60 for the reaction $^{181}Ta + {}^{86}Kr$, correlations of angular distributions with the final fragmentation can be rather conspicuous. Similarly strong dependences have been reported relatively early by Kaufmann and Wolfgang (Kau 61,61b) and Artukh *et al.* (Art 73b) for lighter systems.

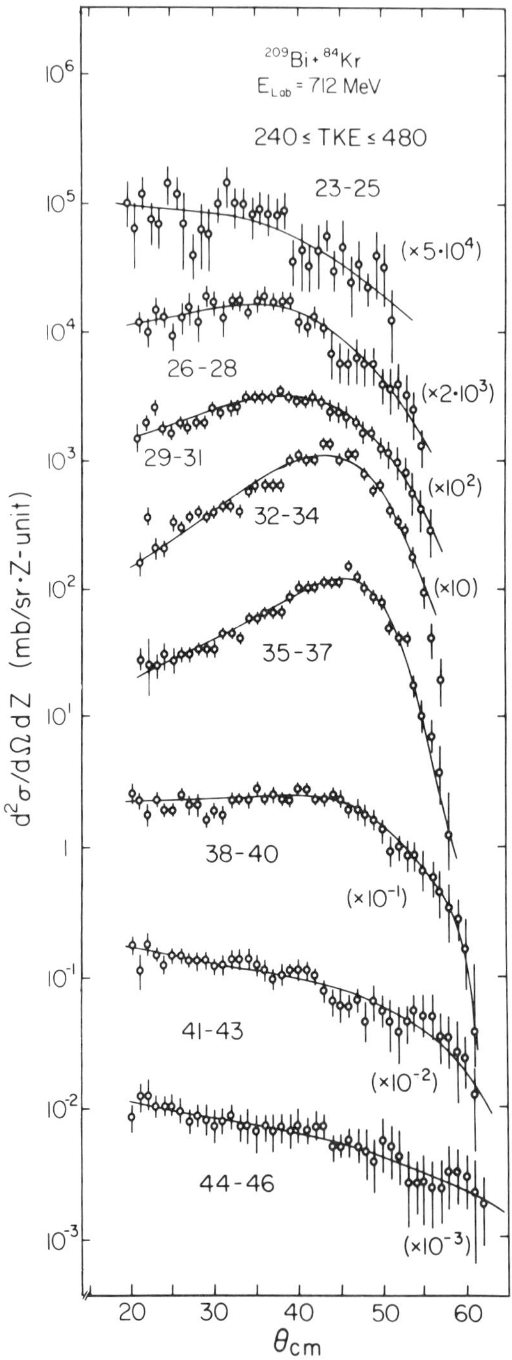

Figure 68. Center-of-mass angular distributions of the light fragments from the damped reaction ^{209}Bi + ^{84}Kr at $E_{\mathrm{Lab}} = 712$ MeV as a function of Z (Bir 82). The experimental double-differential cross sections are multiplied by the factors listed on the right before plotting.

Figures 68 and 69 represent examples of such correlations for the heavier systems $^{209}Bi + ^{84}Kr$ (Bir 82) and $^{209}Bi + ^{136}Xe$ (Wil 80), respectively. As usual, the angular distributions are plotted for the light (projectile-like) products only and given in terms of the double-differential cross section $d^2\sigma/d\Omega\, dZ$ for various Z bins. The data are integrated over all final fragment kinetic energies, but quasielastic events are suppressed. As seen from the figures, angular distributions are peaked just inside of the grazing angle, for fragments with atomic numbers Z close to that of the projectile. As the distance $|\Delta Z|$ between final fragment and projectile increases, this peak disappears and the angular distributions become increasingly broader. However, whereas the angular distributions become more forward-tending in the case of the $^{209}Bi + ^{84}Kr$ reaction, the distributions of fragments from the reaction $^{209}Bi + ^{136}Xe$ develop a backward slope for larger distances

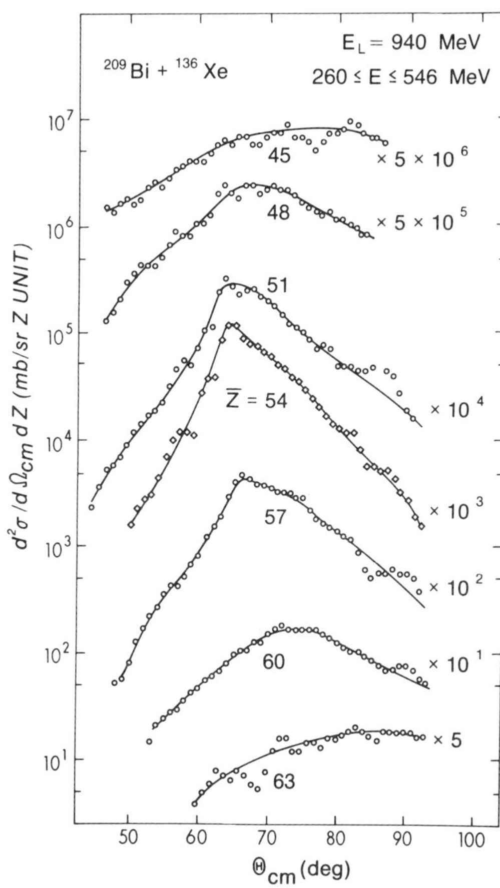

Figure 69. Center-of-mass angular distributions of the light fragments from the damped reaction $^{209}Bi + ^{136}Xe$ at $E_{Lab} = 940$ MeV as a function of Z (Wil 80). The experimental double-differential cross sections are multiplied by the factors listed on the right before plotting.

from the projectile Z. One, hence, observes a reflection of the features of the overall angular distributions (cf. Figures 62 and 64) weighted differently for different final fragmentations. Events with Z values close to the projectile charge contribute the focusing character of the total angular distribution, although also these fragments exhibit angular distributions with long tails. The forward or backward asymmetries of the total angular distributions are seen to be largely due to a corresponding behavior of final fragmentations that are very different from the initial projectile–target combination.

The angular distributions associated with multiproton pickup reactions are rather similar to those of multiproton stripping reactions for the two systems discussed above. However, closer inspection of Figure 68 reveals that for the $^{209}\mathrm{Bi}+{}^{84}\mathrm{Kr}$ reaction, the focusing tendency disappears slightly faster in pickup reactions as compared to collisions leading to an increased fragment charge asymmetry. A corresponding behavior is observed for the angular distributions of fragments from the reaction $^{181}\mathrm{Ta}+{}^{86}\mathrm{Kr}$ shown in Figure 60. Here, for example, a remnant of the side-peak is still discernible as a small shoulder on the forward slope of the distribution for $Z=29$ ($\Delta Z=-7$), whereas it is absent for $Z=43$ ($\Delta Z=+7$). This effect is indicative of slightly longer reaction times for more symmetric fragmentations. An analogous result is obtained for much lighter systems such as demonstrated in Figure 53, for the reaction $^{107,109}\mathrm{Ag}+{}^{20}\mathrm{Ne}$, where very asymmetric fragmentations are found to be associated with angular distributions that contain a forward-peaked component, whereas more symmetric fragmentations yield $1/\sin\theta_{\mathrm{c.m.}}$-type distributions characteristic of very long reaction times. Obviously, for the heavier systems such as displayed in Figures 68 and 69, differences between pickup and stripping angular distributions are rather minute. In any case, the large net charge transfers of more than 10 units in either direction observed for the $^{209}\mathrm{Bi}+{}^{84}\mathrm{Kr}$ reaction correspond to differences in the Coulomb barrier of two spherical nuclei touching at the strong-absorption radius of the order of 100 MeV. Hence, in such a simple model, one would expect that symmetric fragmentations are connected with shorter interaction times and more Coulomb-like deflection functions than asymmetric fragmentations for which the Coulomb repulsion is much weaker. The data clearly contradict such a picture. Obvious resolutions of this paradox could consist in a development of the total effective interaction potential in which the change in Coulomb repulsion is canceled by a simultaneous change of collective shape degrees of freedom, or in a reaction mechanism where significant charge and mass exchange takes place in the exit channels, when the separation process is already well underway. Later discussion will return to both these alternatives for the production of fragmentations far removed from the projectile–target combination. For the present purpose, however, it suffices to note that the systematics of angular distributions indicates an average increase in interaction time with the net amount of charge (or mass) transferred in a reaction.

The interrelation between angular distributions and mass or charge exchange in damped collisions has been studied in detail by Moretto and collaborators (Mor 75, 75b, 76–76d, 77, 78d, Sve 76, 78, Bab 76, Rus 77, 77b, Sch 77f, Cau 78, Woz 78) for a number of damped reactions induced by ^{14}N, ^{20}Ne, ^{40}Ar, $^{84,86}Kr$, and ^{136}Xe projectiles. Describing mass and charge exchange in terms of a stationary diffusion process briefly discussed in Section 2.8, they were able to understand semiquantitatively the features of the fragmentation-dependent angular distributions. In this phenomenological model, the double-differential cross section with respect to solid angle Ω and charge Z of the light reaction partner is approximated by

$$\frac{d^2\sigma(\theta)}{d\Omega\, dZ} = \frac{\lambda\!\!\!^{-2}}{\sin\theta} \int_0^\infty dt \sum_{l=0}^{l_D} l|d\theta/dl|^{-1} P_Z(l,t) P_t(l) \tag{3.3.11}$$

Here, the sum extends up to the limiting l value, l_D, for damped reactions (cf. Figure 39), as inferred from experimental values of $l_{\max}$ and of the magnitude of the quasielastic cross section. The probability function $P_Z(l,t)$ for the time-dependent fragment Z distribution is a solution of the master equation (2.7.5) with transition probabilities [cf. Equation (2.8.22)] that depend on angular momentum l through the local excitation energy [cf. Equation (2.8.10)]. The function $P_t(l)$ describes the distribution of lifetimes of the rigidly rotating intermediate system, for a given l value. Different distribution functions P_t have been employed in the various model calculations. For a rigidly rotating system, the deflection function is determined by the time at which the system breaks up [cf. Equation (3.3.6)] and, hence the folding with a lifetime distribution function is equivalent to dispersing the classical deflection function $\theta(l) = \bar{\theta}(l)$. Assuming a linear increase of the interaction time with decreasing angular momentum, i.e.,

$$P_t(l) = \delta(t - t_0(1 - l/l_{\max})) \tag{3.3.12}$$

the angular distributions represented by the curves in Figure 70 result, (Rus 77b) for various fragment charges produced in the reaction $^{197}Au + {}^{84}Kr$ at $E_{Lab} = 620$ MeV. Here $t_0 = 4.10^{-21}$ s was used as the longest interaction time (for $l = 0$). As can be observed from this figure, a reasonable description of the data (points) is obtained with the model. The underestimation of the magnitude of the sideways peaked component is attributed to a contamination of the data by quasielastic events, as well as to a neglect of partial damping (cf. Section 3.5) in this particular calculation.

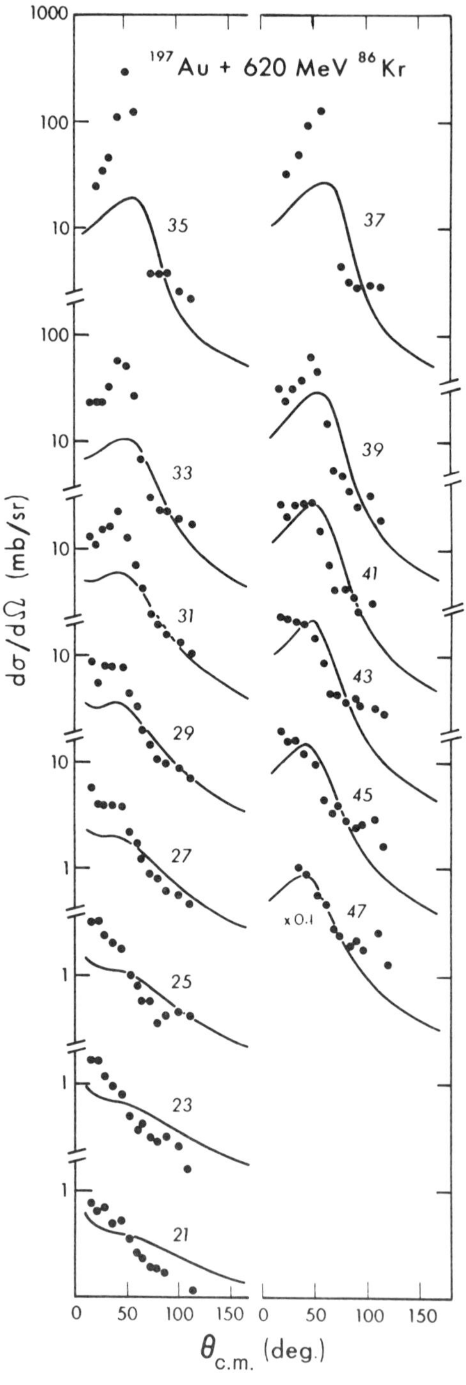

Figure 70. Experimental (dots) angular distributions for fragments with the Z values indicated at the data sets. The curves are fits obtained with a classical diffusion model. From (Rus 77b).

In order to explore further possibilities of correlations between fragment angular distributions and other reaction observables, Figures 71 and 72 display their dependence on the final total kinetic energy, for the reactions ^{165}Ho + ^{56}Fe (Hoo 82) at $E_{\text{Lab}} = 462$ MeV and ^{209}Bi + ^{136}Xe (Wil 80) at $E_{\text{Lab}} = 940$ MeV, respectively. For the bins of highest energy E, the angular distributions are peaked at angles slightly forward of the quarter-point angles of 43° and 70°, for the ^{165}Ho + ^{56}Fe and ^{209}Bi + ^{136}Xe reactions, respectively. As the final energy decreases or as the total kinetic energy loss E_{Loss} increases, the angular distributions for the two reactions show different trends. In the case of ^{165}Ho + ^{56}Fe, the double-differential cross section $d^2\sigma/d\theta_{\text{c.m.}}\, dE$ is forward peaked for large energy losses. As before, this trend is interpreted classically as an orbiting tendency of the

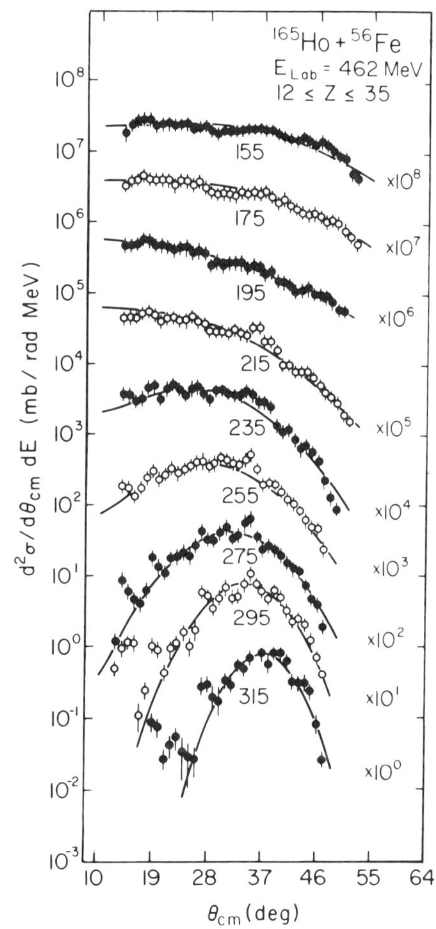

Figure 71. Angular distribution as a function of total kinetic energy for the ^{165}Ho + ^{56}Fe reaction at $E_{\text{Lab}} = 462$ MeV. Each energy bin is 20 MeV wide and integrated over the Z range indicated. The centroid energy of each bin is given at each curve. The solid lines are theoretical fits to the data. (Hoo 82.)

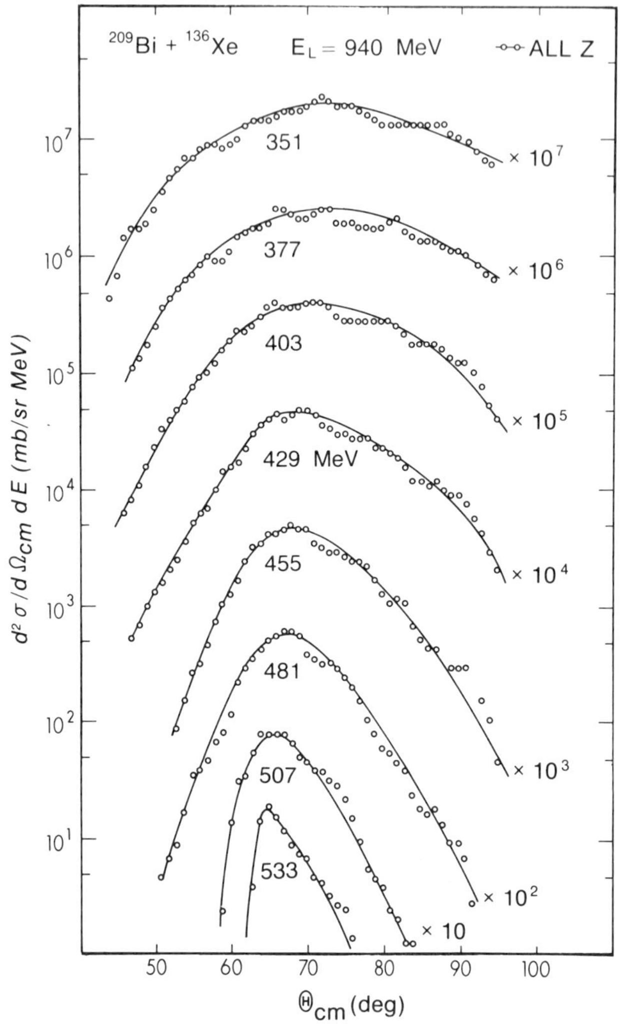

Figure 72. Angular distribution as a function of total kinetic energy for the $^{209}Bi + ^{136}Xe$ reaction at $E_{Lab} = 940$ MeV. Each energy bin is 26 MeV wide and is integrated over Z of the light fragments. The centroid energy of each bin is given at each curve. The solid lines are drawn through the data points. (Wil 80.)

intermediate system that increases with increasing energy loss. Correspondingly, classical interaction times must increase with energy loss.

In contrast to this behavior, the double-differential cross section $d^2\sigma/d\Omega_{c.m.}\,dE$ for the reaction $^{209}Bi + ^{136}Xe$ at $E_{Lab} = 940$ MeV (cf. Figure 72) remains sideways peaked for all values of E (or E_{Loss}), with the cross section peak moving slightly to backward angles with increasing E_{Loss}. Although the sideways peaking persists for this system up to high energy losses, the significant broadening of the angular distributions of Figure 72 with increasing E_{Loss} may be taken as an indication for a corresponding increase of the interaction time.

The angular distributions for the $^{165}\text{Ho} + {}^{56}\text{Fe}$ reaction displayed in Figure 71 have also been analyzed in terms of Strutinsky's phenomenological model discussed in Section 2.2. The solid curves drawn through the data points in this figure represent fits applying Equation (2.2.8) to each of the indicated final kinetic energy windows, implying that there is a monotonic relation between angular momentum and energy loss. These fits allow one to deduce the average deflection angle $\theta_E = \bar{\theta}(l_E)$, the slope $\kappa_0(l_E)$ of the average deflection function [cf. Equation (2.2.6)] and the total angular dispersion

$$\sigma_\theta^2(E) = \frac{1}{2}\xi^2 = \frac{1}{(\Delta l_E)^2} + \frac{1}{4}\kappa_0^2(l_E)(\Delta l_E)^2 \tag{3.3.13}$$

associated with a Gaussian angular momentum window with a centroid l_E and a variance $\sigma_E^2 = (\Delta l_E)^2/2$, that is assumed to contribute to events in the energy bin E. The results of the fits to the $^{165}\text{Ho} + {}^{56}\text{Fe}$ reaction shown in Figure 71 yield an average value $\overline{\Delta l_E} \approx 30$ consistent with similar results obtained (Har 78) for the $^{232}\text{Th} + {}^{40}\text{Ar}$ reaction. They indicate that the increase of σ_θ^2 with energy loss is caused by dynamical dispersion corresponding to a large slope $\kappa_0 = d\theta_E/dl_E$ of the average deflection function. Qualitatively similar conclusions have been reached (Rud 81) for the $^{166}\text{Er} + {}^{86}\text{Kr}$ reaction at $E_{\text{Lab}} = 8.18$ MeV/nucleon, where a linear relationship between dispersion $\sigma_\theta(E)$ and θ_E was deduced.

Applying the above procedure to angular distributions such as shown in Figure 72 for the $^{209}\text{Bi} + {}^{136}\text{Xe}$ reaction, where the average reaction angle changes relatively little with energy loss, while the width increases drastically with E_{Loss}, yields results at variance with the conclusion, drawn above. Since in such cases $\kappa_0 \approx 0$, the observed angular dispersion must be of nondynamical origin, i.e., reflecting quantal or statistical fluctuations. As already stated earlier in this section, a microscopic model is needed in order to disentangle the different effects. One may, however, conclude tentatively from the above analyses that angular distributions for heavy systems are not dominantly due to quantum fluctuations. Then, one may interpret the characteristic broadening of fragment angular distributions with increasing energy loss as owing to an increase of the classical interaction time.

The experimental data reviewed thus far indicate that the shape or width of fragment angular distributions is strongly correlated with both the closeness of the final fragmentation to the initial projectile–target combination and the degree to which the initial kinetic energy has been dissipated in a reaction. In order to discover a possible connection between these two types of correlation, a finer differentiation of data is necessary. Inspecting triple-differential cross sections $d^3\sigma/d\Omega_{\text{c.m.}}\,dE\,dZ$ for the reaction ^{209}Bi + ^{136}Xe at $E_{\text{Lab}} = 1130$ MeV (Sch 77e, 78g), it appears inescapable to assign a more fundamental role to the total kinetic energy loss as an indicator of the complexity of a reaction and, hence, of the interaction time. These distributions are displayed in Figure 73 for a 3-Z units wide bin centered around

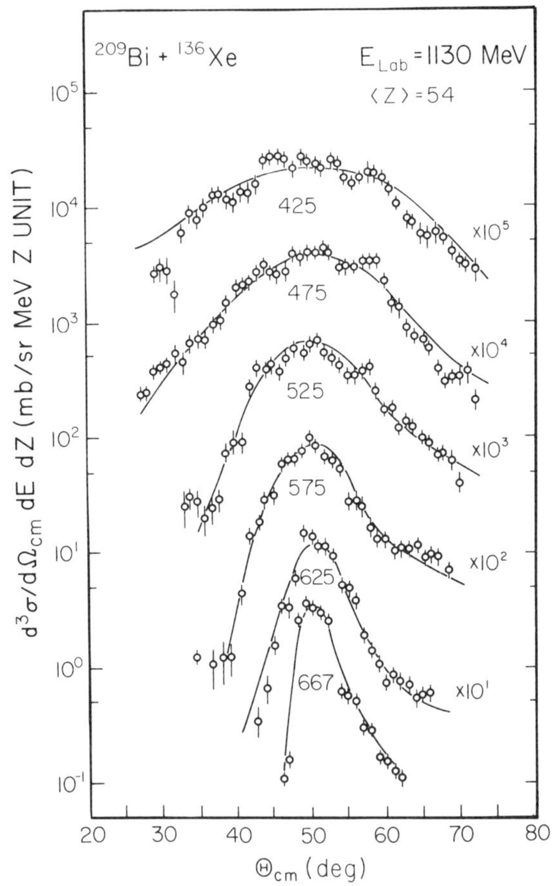

Figure 73. Triple-differential cross section $d^3\sigma/d\Omega_{\text{c.m.}}\,dE\,dZ$ plotted as a function of the total kinetic energy E for fragments near the projectile Xe ($53 \leq Z \leq 55$). These data are from (Sch 78g). The solid lines represent the double-differential cross section $d^2\sigma/d\Omega_{\text{c.m.}}\,dE$ as a function of total kinetic energy E, integrated over all fragments. Each solid curve is normalized and displayed here to show that the shape of the angular distribution for each energy bin is independent of Z.

the projectile Z ($Z_P = 54$). They are grouped from bottom to top according to increasing energy loss. Although also for this reaction, the angular distributions for fragments with $Z \approx Z_P$ are sharply sideways peaked when integrated over all final kinetic energies, the decomposition shown in Figure 73 demonstrates that there is still the familiar broadening of the angular distributions with increasing energy loss. In fact, the data points for $\langle Z \rangle = 54$ follow very closely the common angular dependence of all reaction fragments represented by the curves in Figure 73, for each final kinetic energy bin. Only the relative normalization of the curves had to be readjusted to fit the triple-differential cross section. It is, hence, evident from Figure 73 that the shape of the angular distribution depends on the energy loss but is independent of Z. The full range of angular distribution types is realized for fragments with $Z \approx Z_P$. This illustrates the primary role of the energy loss as an indicator of the progress of the reaction, i.e., of the interaction time. However, elements near the projectile are dominant in the angular distributions with strong side-peaking, where interaction times are short, whereas fragments further removed from the projectile dominate the broader or forward-peaked angular distributions associated with longer reaction times. This variation with energy loss of the relative contributions of the different fragmentations has to be regarded as the origin of the correlations observed between shape of the angular distribution and final fragmentation. Similarly, it accounts for the angular dependence of fragment mass or charge distributions as illustrated in Section 3.4. The progressive dissipation of kinetic energy of relative motion during a collision forms, of course, the basis for dynamical reaction models (cf. Section 2.3) employing frictional forces, for which the above data provide independent support. Similar analyses of other reactions [e.g., (Rus 77b, Wil 80)] corroborate these conclusions which can be summarized by stating that, on the average, the energy loss increases with increasing interaction times which, in turn, increases with decreasing initial orbital angular momentum. However, how strong and well defined these correlations are, is currently not precisely known. This problem remains an active topic of research. Experimental observations discussed below and in subsequent sections will further demonstrate the utility of the energy loss (or reaction Q-value) parameter for a systematic classification of damped reaction features.

In Figure 74, the double-differential cross section $d^2\sigma/d\Omega_{\mathrm{c.m.}}\,dE$ integrated over all fragments is plotted vs. final total kinetic energy and reaction angle in the form of a contour diagram (Wilczyński plot) for the reaction $^{209}\mathrm{Bi} + {}^{136}\mathrm{Xe}$ at three bombarding energies. Such Wilczyński diagrams are instructive in that they provide an overview over the evolution of the reaction dynamics with increasing energy loss and, correspondingly, increasing interaction time and decreasing initial angular momentum. The contour plots in Figure 74 demonstrate a strong focusing behavior for this

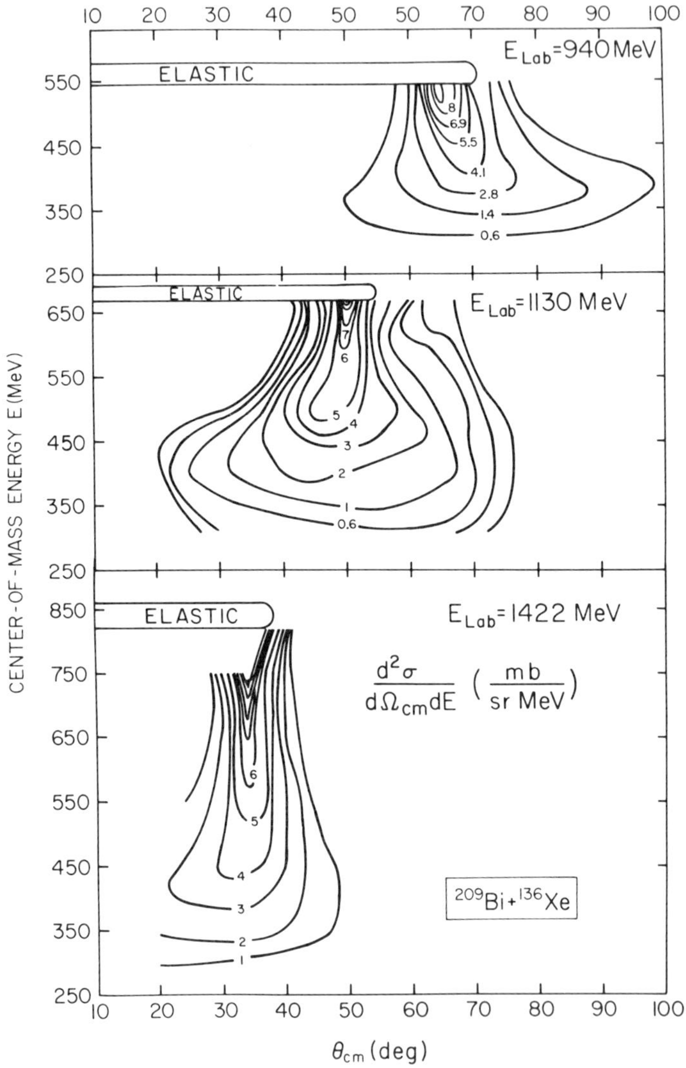

Figure 74. Double-differential cross section $d^2\sigma/d\Omega_{\text{c.m.}}\,dE$ for the $^{209}\text{Bi} + {}^{136}\text{Xe}$ reaction plotted versus the total kinetic energy E and reaction angle $\theta_{\text{c.m.}}$ in a contour (Wilczyński) diagram (Sch 78g, Wil 80, Wol 81). This figure illustrates the energy dependence of the Wilczyński plot for a heavy system where strong focusing dominates.

reaction at all three bombarding energies. For better orientation, the elastic cross section ridge is schematically drawn from forward angles up to the respective quarter-point angles $\theta_{1/4}$. For the intermediate energy of $E_{\text{Lab}} =$ 1130 MeV, this angle corresponds to 54°. Slightly forward of this angle, near $\theta_{\text{c.m.}} = 50°$, a cross section ridge develops extending down in energy almost parallel to the energy axis. With decreasing energy, the cross section is more and more dispersed, resulting in a growing width of the angular distribution. This behavior is more explicitly seen in Figure 73, where the solid curves correspond to a sum over all fragment Z values and represent projections of 50-MeV-wide cuts of the Wilczyński diagram, parallel to the angle axis. As the bombarding energy is reduced to $E_{\text{Lab}} = 940$ MeV, the quarter-point angle moves back to $\theta_{1/4} = 70°$. A cross-section ridge of damped events, starting out now at $\theta_{\text{c.m.}} = 64°$, moves slightly backward in angle, with decreasing energy E. Qualitatively, this result is expected in a classical model, due to the angular momenta and corresponding angular velocities which are lower at the lower bombarding energies, for any given impact parameter. The Coulomb deflection towards backward angles is more pronounced at the lower energy, and the rotation towards forward angles during the nuclear interaction is less such that trajectories with lower angular momenta do not, on the average, rotate past the quarter-point angle. For the $^{209}\text{Bi} + ^{136}\text{Xe}$ reaction at the higher bombarding energy of $E_{\text{Lab}} = 1422$ MeV illustrated at the bottom of Figure 74, a ridge of damped events develops around $\theta_{\text{c.m.}} = 35°$ and evolves almost parallel to the energy axis for energy losses up to 300–400 MeV. With increasing energy loss, the distribution broadens somewhat. It shows a tendency to move towards forward angles for the highest energy losses, which correspond to final energies that are smaller than the Coulomb energy ($V_{\text{Coul}} \approx 420$ MeV) for spherical nuclei touching at the strong-absorption radius. Hence, in Figure 74, one observes for a very heavy system an evolution of the cross section pattern from a slightly backward-going ridge towards one that tends slightly forward, as the bombarding energy is raised. It is a remarkable fact, however, that the cross section remains well focused in all three cases, for events with final energies well above the Coulomb barrier for spherical nuclei. A strong angular focusing has also been observed [cf. (Gob 81)] for the reaction $^{154}\text{Sm} + ^{154}\text{Sm}$ at 6.3 MeV/nucleon.

The backward-going tendency seen to be only slight in the reaction $^{209}\text{Bi} + ^{136}\text{Xe}$ at $E_{\text{Lab}} = 940$ MeV (cf. Figure 74) is dramatically increased for even heavier systems at relatively low bombarding energies. This is demonstrated in Figure 75 for the reaction $^{208}\text{Pb} + ^{208}\text{Pb}$ measured (Tan 80c) at a c.m. bombarding energy of approximately 200 MeV above the entrance-channel Coulomb barrier of $V_{\text{Coul}} \approx 600$ MeV. Here, the Wilczyński diagram is shown in terms of $d^2\sigma/d\theta\, d(\text{TKE}^*)$ plotted vs. $\theta_{\text{c.m.}}$ and a final kinetic energy parameter TKE* that is modified for the difference between initial

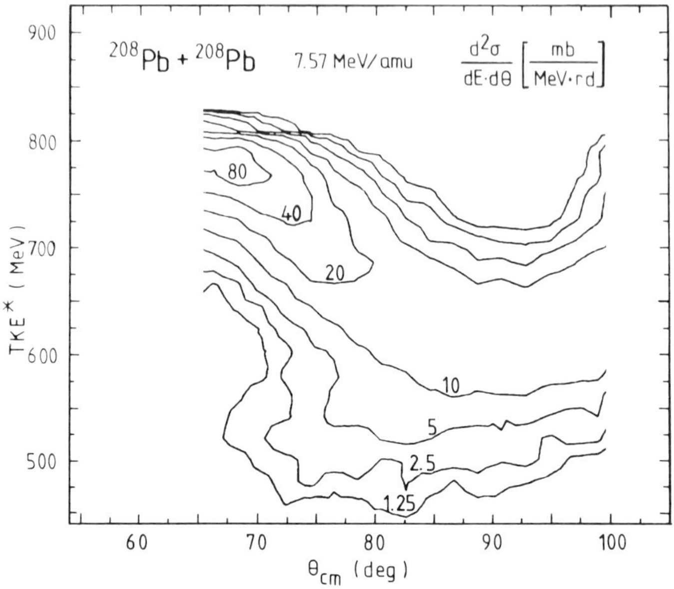

Figure 75. Contour diagram of the double-differential cross section $d^2\sigma/d\theta_{\text{c.m.}}\, d(\text{TKE}^*)$ for the reaction $^{208}\text{Pb} + ^{208}\text{Pb}$ at $E_{\text{Lab}} = 7.57$ MeV/nucleon, integrated over all fragments with $Z > 62$. TKE* is the final c.m. total kinetic energy adjusted by the difference between entrance and exit channel Coulomb energies. From (Tan 80c).

and final Coulomb energies for each fragmentation. Such a cross section pattern is very characteristic of a Coulomb-dominated deflection function.

The opposite behavior, an enhanced orbiting tendency, is deduced for the reaction $^{139}\text{La} + ^{86}\text{Kr}$ measured (Van 78) at laboratory energies of $E_{\text{Lab}} = 505$, 610, and 710 MeV. Corresponding Wilczyński diagrams $d^2\sigma/d\theta_{\text{c.m.}}\, dE$ are exhibited in Figure 76. As seen from this figure, at all three energies the cross section pattern shows a forward-going ridge and one that extends towards backward angles, consistent with the orbiting picture illustrated in Figure 56. The latter ridge corresponds to final kinetic energies well below the entrance-channel Coulomb energy V_{Coul}, indicated by arrows in Figure 76, and may contain a contribution from the fusion–fission process. Whether or not this ridge is actually due to orbiting is currently not known. Polarization measurements to be discussed in Section 3.6 for similarly heavy systems indicate that the system $^{139}\text{La} + ^{86}\text{Kr}$ may be close to a transition region where the character of the classical deflection function changes from an orbiting behavior to a double-rainbow type (cf. Figure 59), as the mass of the system is increased.

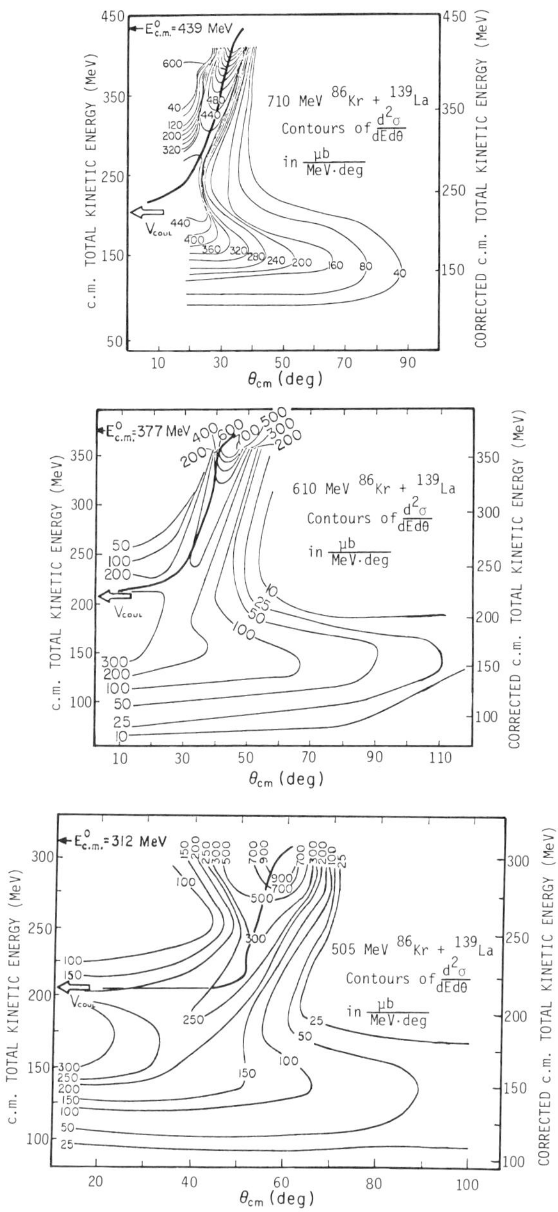

Figure 76. Double-differential cross section $d^2\sigma/d\theta\, dE$ plotted versus the total kinetic energy E and reaction angle $\theta_{\text{c.m.}}$ in a contour (Wilczyński) diagram (from Van 78). This figure illustrates the energy dependence of the Wilczyński plot for a lighter system where some orbiting occurs. The heavy solid lines represent the theoretical correlation between average deflection angle and final energy (see text).

A less ambiguous determination of the type of the deflection function than possible from a measurement of angular distributions appears to be highly desirable for systems as heavy as or heavier than the system ^{139}La + ^{86}Kr, because of the consequences for a better establishment of the dissipative and conservative forces acting on the relative motion. Their influence is modeled in classical dynamical trajectory calculations or in the more extensive dynamical transport models that predict average classical deflection functions and their correlations with other observables such as the energy loss and mass transfer. Results of a dynamical transport calculation are represented by the solid curves in Figure 76. They describe the average evolution of the cross section ridge rather well, down to final energies in the vicinity of the entrance-channel Coulomb barrier where the predicted orbiting tendency is much stronger and average energy losses are smaller than consistent with the data.

In the model calculations represented in Figure 76, the equations of motion (2.3.8) were solved numerically employing a family of shapes of the intermediate system featuring a neck as shown in Figure 10a. Correspondingly, the liquid-drop nuclear potential accounting for the proximity correction [cf. Equations (2.5.25) and (2.5.26)] was used, whereas the Coulomb potential was taken as suggested by Bondorf *et al.* [cf. Equations (2.5.8) and (2.5.9)]. Energy as well as mass and charge transport were accounted for by incorporating the transport model of Randrup (cf. Section 2.8). Accordingly, the window formula in the form of Equations (2.6.30)–(2.6.33) was adopted for that part of the dissipation function due to the exchange of individual nucleons, whereas the neck motion was assumed to be overdamped by virtue of the wall formula [cf. Equation (2.6.19)]. Transport coefficients as given by Equations (2.8.17) were integrated numerically along the classical trajectories, and the average changes of the collective variables induced by mass and charge exchange were properly accounted for in each step of the calculation. This represents a consistent modeling of a variety of reaction phenomena which, in the model, are all assumed to be due to the interaction between single-particle exchange modes and the collective degrees of freedom, as explained in Section 2. Once a choice of dinuclear shapes has been made, there are no adjustable parameters.

In Figure 77, the development of several interesting parameters with angular momentum l is depicted as resulting from the model calculations compared in Figure 76 to data for the ^{139}La + ^{86}Kr reaction at $E_{\text{Lab}} = 505$, 610, and 710 MeV. Here, the predicted classical distance of closest approach, r_0, and the interaction time elapsed between the initial contact at $r = R_{\text{SA}}$ and the final snapping of the neck are plotted vs. the ratio l/l_{max} or relative impact parameter. The rather weak l dependence of r_0 observed in the lower part of Figure 77 demonstrates the peripherality of the reaction, for a major fraction of the cross section, that is caused by the strongly repulsive Coulomb and centrifugal forces. These forces are also understood to be partially responsible for the small variation of r_0 with bombarding energy for a given ratio l/l_{max}, especially for the higher incident energies. The action of friction forces further reduces this variation. The upper half of Figure 77 depicts the model interaction times t_{Int} which are seen to vary almost exponentially with l. Again, there is very little dependence on the bombarding energy predicted for interaction times associated with a given value of l/l_{max}.

The comparison between model predictions and data on the reaction $^{139}La + ^{86}Kr$ provided in Figure 76 may be taken as representative for the results obtained with this model for other systems such as the $^{232}Th + ^{40}Ar$ reaction. Usually, the cross-section ridge of partially damped events is well described. The failure to reproduce the patterns for completely damped events may be due to an inadequacy of the adopted dumbbell-type dinuclear shapes. It may, in addition, reflect the possibility that a mono-

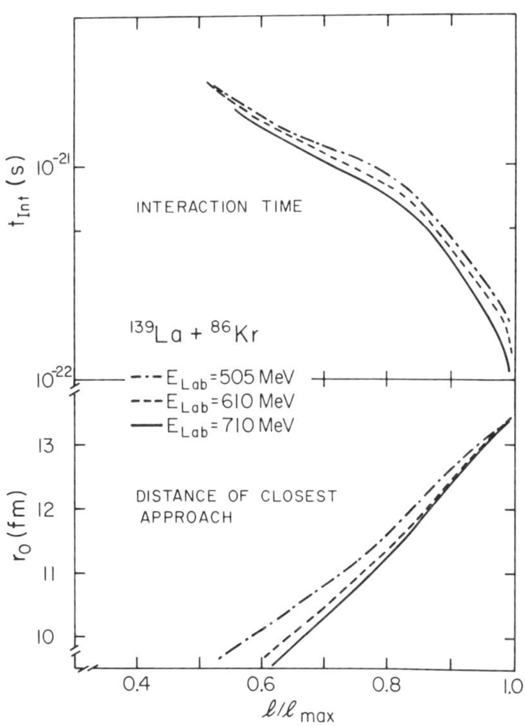

Figure 77. Interaction time t_{Int} and distance of closest approach r_0 plotted vs. l/l_{max} for the reaction $^{139}La + ^{86}Kr$ at three bombarding energies, as predicted by the dynamical transport calculations described in the text.

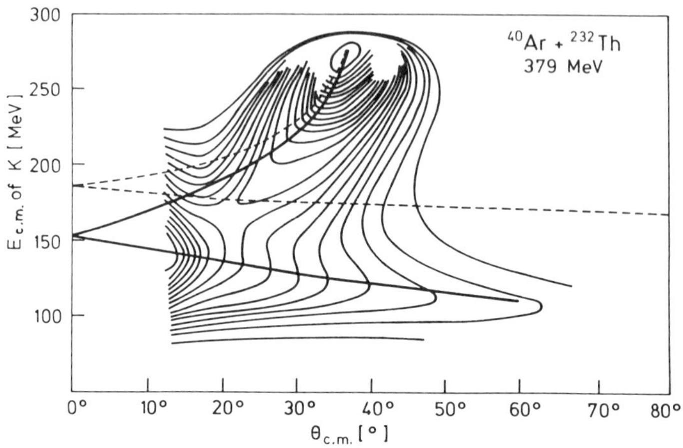

Figure 78. Final energy vs. scattering angle (Wilczyński plot) for $^{40}Ar + ^{232}Th$ at $E_{Lab} = 388$ MeV. This energy corresponds to 379 MeV, if the target thickness is taken into account. Contours: experimental data of (Wil 73c), solid line: model prediction with deformation, dashed line: spherical model. From (Gro 81).

nucleus with a continuous interior has been formed in such deeply penetrating collisions. Further comparisons of the predictions of this model with other data will be made in subsequent sections.

The effect on the predicted cross section pattern that the inclusion of spheroidal deformations of the interacting nuclei produces can be estimated from Figure 78, where results of model calculations by Gross *et al.* (Gro 81) are compared to data (Wil 73c) on the reaction $^{232}Th + ^{40}Ar$ at $E_{Lab} = 379$ MeV. In the model, the reaction partners are represented either by two overlapping spheres or by two aligned spheroids with equal deformation parameters. The deformed nuclear potential is derived from a single-folding model analogous to Equation (2.5.12), with a Woods–Saxon nucleon–nucleus potential V_2. Friction force form factors are taken as described in the context of Equation (2.6.17). Their strength coefficients have been deduced (Gro 75) from an analysis of experimental heavy-ion fusion cross sections. No account is given of the influence of mass exchange on the reaction dynamics. With all other parameters unchanged, one observes a superior description of the $^{232}Th + ^{40}Ar$ data by the deformed model (solid curve) as compared to the results obtained in the calculation neglecting nuclear deformations (dashed curve in Fig. 78). On the other hand, the strong focusing effect characterizing the $^{209}Bi + ^{136}Xe$ reaction at $E_{Lab} = 1130$ MeV is not too well reproduced by either model.

Apparently, the various dynamical reaction models referred to above have certain deficiencies and a limited range of applicability. It is currently not known whether these deficiencies are due to some oversimplifications in the models that can, in principle, be improved upon or if they reflect a more significant misrepresentation of the basic reaction mechanism. Further comparisons between model predictions and experimental data will be discussed below.

The generation of average system trajectories is part of the problem addressed by transport theories discussed in Sections 2.7 and 2.8. At least in the weak-coupling limit, average values of collective coordinates evolve in time as described by classical equations of motion with conservative and dissipative forces. For example, Agassi *et al.* (Aga 79b) have applied a simplified version of their transport model to the reaction $^{232}\mathrm{Th}+{}^{40}\mathrm{Ar}$ at 388 MeV, where the joint probability function for the coordinates $(\mathbf{r},\mathbf{p})$ of relative motion is described by a stationary Fokker–Planck equation. The differential cross section is calculated in a manner similar to what is illustrated by Equation (3.3.10), however, now with a more complex multivariate Gaussian distribution function. Friction and diffusion terms are taken to be proportional to the density overlap of the interacting spherical nuclei. A typical value for the strength coefficient of the isotropic friction force is estimated as $c_0=9\times10^{-22}$ MeV s fm^{-2}, which is comparable in magnitude to the proximity friction coefficient (see Figure 27). In the model of Agassi *et al.*, however, it is not a constant but decreases with decreasing kinetic energy of relative motion. For the collective interaction potential, the proximity potential with a repulsive core [cf. Equation (2.5.23a)] was assumed in the calculations.

Results of this model calculation (Aga 79b) are displayed in Figure 79 for the $^{232}\mathrm{Th}+{}^{40}\mathrm{Ar}$ reaction in form of a Wilczyński plot that should be compared to the data of Figure 78. One observes a general reproduction of the experimental cross-section pattern by the calculation. The forward-going ridge of partially damped events is reasonably well described also with respect to the magnitude of the cross section; however, the cross section for the peak at forward angles is underestimated in the calculation by a factor of 10. The same is true for the lower-energy orbiting ridge. It results from the calculation only if the distribution function is assumed to remain Gaussian even for complete energy damping of the average trajectory, where the Coulomb energy presents a barrier that cannot be penetrated by classical trajectories. Taking account of the deformability of the reaction partners, as is done in the models of Gross *et al.* (Gro 81) and Ko (Ko 79c), alleviates this problem in the context of the simplified transport model [see, e.g., (Ko 79c)]. Calculations such as described above have also been performed (Aga 78b) for the heavier systems $^{208}\mathrm{Pb}+{}^{84}\mathrm{Kr}$ and $^{209}\mathrm{Bi}+{}^{136}\mathrm{Xe}$, where a rather qualitative explanation of the corresponding data was achieved. However, in view of the significant simplifications made in order to cast the more detailed transport theory into a tractable form and because

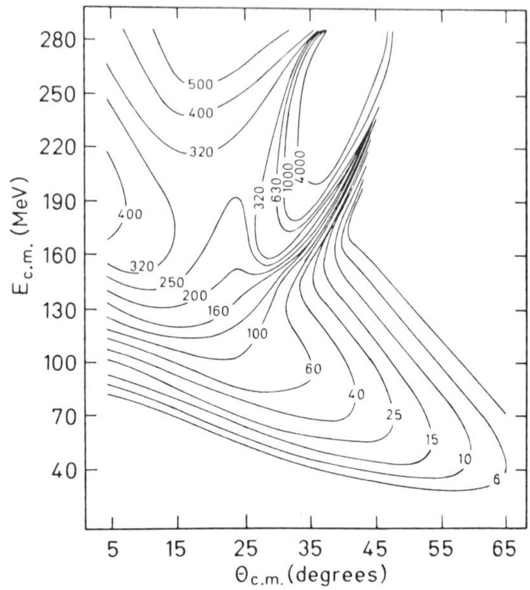

Figure 79. Wilczyński plot for the reaction ^{232}Th(^{40}Ar, K) at $E_{\text{Lab}} = 388$ MeV predicted by the transport model of Agassi *et al.* (Aga 79b.)

of the very limited number of case studies performed, it seems currently not possible to give a definitive account of the validity of the basic reaction mechanism assumed in this theory.

Applications of the transport model of Hofmann *et al.* (cf. Section 2.7) in a description of angular-energy cross-section patterns for various damped reactions have been performed (Hof 76c, Ngô 77, 77b, Ber 78d–f) on a more phenomenological level. In these calculations, various modifications of the classical dynamical models of Gross *et al.* (Gro 75, 81) or Siwek–Wilczyńska *et al.* (Siw 76) were used in the description of average system trajectories. The formalism of the transport theory was then applied to provide their probability distribution, e.g., via generalized fluctuation–dissipation relations. This procedure allows the calculation of cross sections in analogy to Equation (3.3.10). The close correspondence between model predictions and experimental data suggests that the spread in the experimental cross section can be interpreted as being due to the statistical fluctuations of system trajectories around a mean path. However, such comparisons do not allow one to draw more direct conclusions on the validity of the basic transport theory.

The current scarcity of full-fledged, yet tractable transport theories and the reliance of available models on phenomenological input emphasizes the significance of the less ambitious classical trajectory calculations which aim at an understanding of average angle–energy correlations of the reaction cross section in terms of classical conservative and dissipative forces. The lack of an explicit consideration of a sufficient number of collective and intrinsic degrees of freedom in these models is sometimes compensated by an emulation of expected effects achieved by a renormalization of model parameters. For example, the development of fragment deformations in the exit channel can be simulated (Wil 76, Siw 76, De 77) by augmenting the model interaction potential by the corresponding deformation energy. The finite response time of intrinsic degrees of freedom to the relative motion can be approximately accounted for (Nör 79, 79b, 79d, Gre 81, Ngô 81, Gnu 81) by switching in a smooth manner from a diabatic, sudden nuclear potential valid at early reaction stages to an adiabatic form reflecting the relaxation of shape degrees of freedom. Finally, the feedback effect of mass transfer on the scattering dynamics can, to a certain extent, be incorporated in terms of an effective reduced mass for the relative motion or in terms of an effective interaction potential (Bon 74b). The value of such model interpretations of data in which phenomenological parameters are adjusted to fit experimental results consists in a separation and ordering of experimental trends and a visualization of the underlying mechanisms. Systematical studies by which this could be achieved are currently in progress.

Even more phenomenological are approaches where experimental data are used to generate average deflection functions from which, subsequently, information on other interesting quantities such as classical interaction times are deduced. One of these methods of analysis (Wol 78b, Rie 79, 80), where a parametrized theoretical deflection function [cf. Equation (3.3.8)] is fitted to experimental angular distributions, has already been discussed. The second method (Hui 77b, Sch 77c), which yields very similar results (Reh 81), employs the relation between initial angular momentum and energy loss suggested by the data discussed earlier, according to which the average energy loss increases with decreasing angular momentum l. Employing then a classical model (e.g., Hui 75, Bon 75) for the motion of the intermediate dinuclear system, one can derive the dependence of the interaction time on l from the experimental average deflection angle.

Assuming that the damped reaction cross section is related to the initial angular momentum by a relation analogous to Equation (2.1.21), one can deduce from the cross section $\Delta\sigma_{ij}$ measured for a given total kinetic energy window $[E_i, E_j]$, a corresponding l window $[l_i, l_j]$, where

$$l_i = \left[(l_j+1)^2 - \Delta\sigma_{ij}/(\pi\lambda\!\!\!^{-2})\right] - 1 \qquad (3.3.14)$$

Starting with $l_1 = l_{\max}(E_{\text{c.m.}})$, one can successively generate the relation

$l_i(E_i)$ or $l_i(E_{\text{Loss}})$. The only assumption to be made is that there is a monotonic relationship between initial angular momentum and energy loss E_{Loss} (or final energy E). Since the mean reaction angle $\bar{\theta}$ for a given energy loss is also an experimental observable, one obtains immediately the average deflection function $\bar{\theta}(l_i)$. An example for such a deflection function is shown in Figure 80 for the reaction ^{165}Ho$+^{56}$Fe (Hoo 82). The hatched band illustrates the range of derived values associated with different assumptions made in the conversion of the cross section to an angular momentum scale. The character of the deflection function implying orbiting is typical of lighter systems.

The procedure used to deduce interaction times (Hui 75, Bon 75) is illustrated in Figure 81. Here, the two (spherical) nuclei are assumed to follow Coulomb trajectories prior to nuclear contact and after breakup of the intermediate system. The corresponding Coulomb deflection angles $(\theta_c^i - \phi^i) = \phi_{\text{Coul}}(l_i, r_i)$, $(\theta_c^f - \phi^f) = \phi_{\text{Coul}}(l_f, r_f)$ in entrance and exit channels, respectively, can be calculated from Equation (2.1.9), if the corresponding contact and breakup radii r_i and r_f are known. The angle $\Delta\theta$ of rotation

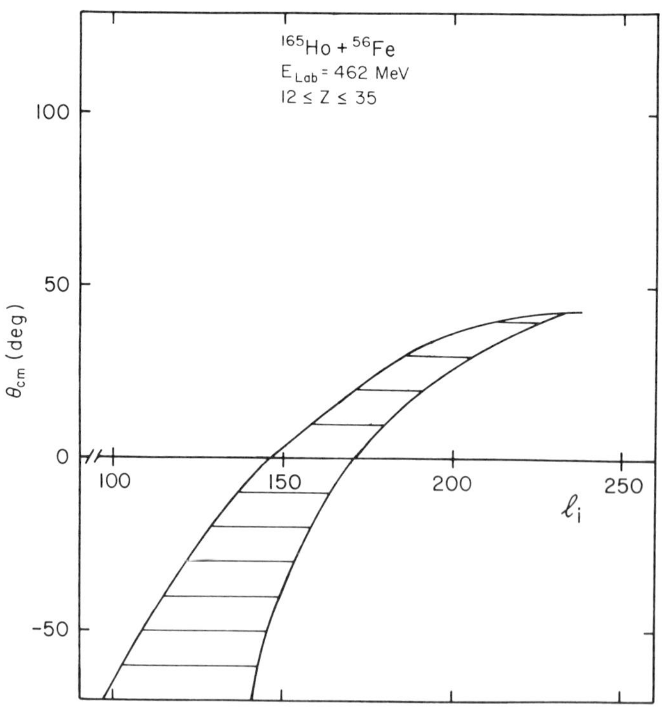

Figure 80. Deflection function for the ^{165}Ho$+^{56}$Fe reaction. In this system, because of the substantial amount of orbiting, it was possible to determine accurately the cross section maxima for the higher and lower energy losses only. (Hoo 82).

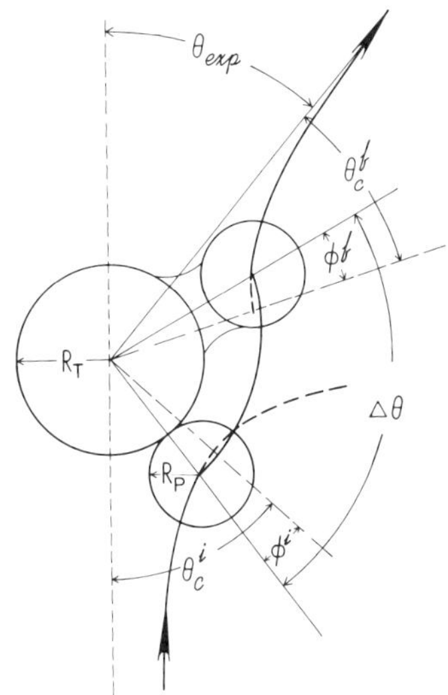

Figure 81. Schematic illustration of the classical model used for the calculation of interaction times (Bon 75, Hui 77b). For definitions of angles, see text.

of the intermediate system is then determined by the measured reaction angle as

$$\Delta\theta = \pi - \theta_{\text{exp}} - \phi_{\text{Coul}}(l_i, r_i) - \phi_{\text{Coul}}(l_f, r_f) \tag{3.3.15}$$

Using Equation (3.3.6), one can then deduce a relation between interaction time and initial angular momentum l. For a rough estimate, one may use

$$t_{\text{Int}}(l_i) \approx \Delta\theta \bar{\mathscr{I}}/(\hbar l_f) \tag{3.3.16}$$

as an approximation to Equation (3.3.6), where $\bar{\mathscr{I}} \approx \mu R_{\text{SA}}^2$ may be taken as a typical value of the moment of inertia associated with the orbital motion of the intermediate system. The quantity μ denotes its reduced mass. Whereas one would naturally identify r_i with R_{SA}, a determination of r_f, l_f, and $\bar{\mathscr{I}}$ requires the adoption of a more detailed reaction model. Using the above choice for $\bar{\mathscr{I}}$ and assuming that there is no radial motion at the point of breakup of the dinuclear system, the above procedure yields (Wol 81) the l-dependent interaction times plotted in Figure 82 vs. the ratio l/l_g $(=l_i/l_{\text{max}})$ for the reaction $^{209}\text{Bi} + {}^{136}\text{Xe}$ at $E_{\text{Lab}} = 940$, 1130, and 1422 MeV. Here, the nonsticking (NS) and sticking (S) cases refer to conservation of orbital angular momentum $(l_i = l_f)$ or to instant reduction of l_i to the sticking value [cf. Equation (2.3.11)]. In either model, the interaction times t_{Int} are seen to increase approximately exponentially with decreasing angular momentum l, except in the vicinity of the grazing angular momentum $l \approx l_g$. This behavior has also been inferred for several other heavy-ion reactions (Hui 77b, Sch 77e, Wol 77, 78, 78b, Rie 79). As can be seen from Figure 82, for a given ratio l/l_g, the interaction time does not depend much on the bombarding energy. Variations are less than ~ 30%. This interesting

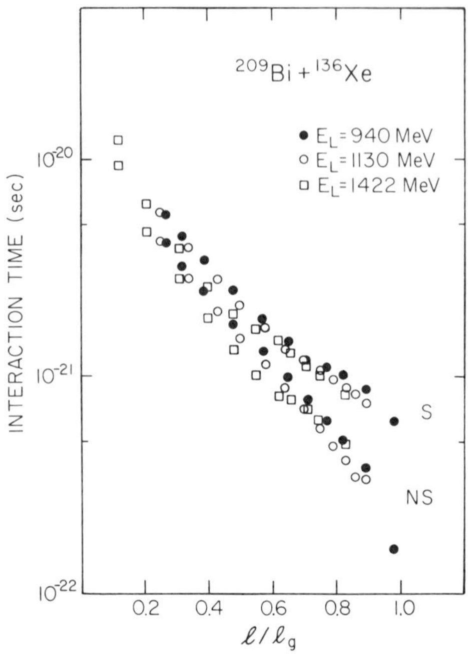

Figure 82. The mean interaction time $t_{\rm Int}$ is plotted as a function of the initial relative angular momentum l/l_g for ^{209}Bi + ^{136}Xe reaction at $E_{\rm Lab}$ = 940, 1130 and 1422 MeV. The final angular momentum is calculated in the limits of the "nonsticking" (NS) and "sticking" (S) models. (Wol 81.)

result, which has also been obtained (Rie 79) for the reaction ^{166}Er + ^{86}Kr at $E_{\rm Lab}$ = 515 and 703 MeV, is consistent with the predictions of the transport model discussed in the context of Figures 76 and 77. It is understood to be mainly due to the dynamical balance of Coulomb, centrifugal, and frictional forces for heavy systems and bombarding energies well above the interaction barrier.

In summary, in this section the characteristic features of experimental fragment angular distributions have been discussed. It has been shown how the shape of the angular distribution varies with total mass and mass asymmetry of the projectile–target combination, with bombarding energy and final fragmentation and energy loss. Although this question is not as yet definitively settled, model analyses suggest that the shape of experimental angular distributions can be ascribed to dynamical and statistical dispersions of system trajectories around an average path. The recourse to phenomenological input and significant reductions made in the application of various transport theories has so far, however, prevented a deeper appreciation of the validity of the basic reaction mechanisms proposed. Similarly, it has as yet not been possible to identify the detailed properties of conservative and dissipative forces. As demonstrated by the results of dynamical model calculations, several combinations of rather different forces provide descriptions of data of comparable accuracy. It is certainly indicated to submit the various models to more crucial tests of intercorrelations between measurable reaction parameters. This task will be pursued in subsequent sections.

3.4. Mass and Charge Transfer

There are few features of damped nuclear reactions that have yielded a deeper insight into the microscopic reaction mechanism than those associated with the observed division of the total mass and charge of a system between the fragments emerging from a binary heavy-ion collision. Experimental studies of various moments of A, Z, isobaric, and isotopic distributions of damped reaction fragments and their variation with remaining reaction variables such as scattering angle and total kinetic energy loss have initiated the development of theoretical models involving methods that are new to nuclear reaction theory. It is the aim of this section to present a survey over available data on fragment (A, Z) distributions and to discuss the extent to which the observations are presently understood in terms of the models outlined in Section 2. First, however, a few remarks addressing problems often associated with the experimental determination of the A and Z values of heavy reaction fragments will be added to the considerations of experimental procedures given in the preceding subsections.

Although, as will be demonstrated below, damped reaction fragment A and Z distributions are relatively featureless, with continuous spectra, it is nevertheless important to have command over an experimental accuracy which is sufficiently high to resolve the sometimes narrow structure of interest. The lack of a sufficient Z or A resolution in early experiments or those studying very heavy systems can severely impede or limit the validity of detailed information on the fragment distributions derived from experimental data.

The identification of heavy ions with respect to their atomic number Z is usually accomplished by sampling their characteristic ionization pattern, employing a detector telescope consisting of one or several transmission detectors measuring specific energy losses, ΔE, and a stop detector for the determination of the residual kinetic energy E. Figure 83 shows an experimental distribution (Hoo 81) of projectile-like fragments plotted vs. energy loss (ΔE) and total energy $(E + \Delta E)$ obtained for projectile-like fragments from the reaction ^{165}Ho$+^{56}$Fe, using a silicon surface barrier detector telescope. The blown-up area in the insert of this figure demonstrates that the resolution of the telescope is sufficient to distinguish the characteristic $\Delta E - (E + \Delta E)$ ridges of different fragment atomic numbers Z. The information necessary to convert the distribution of Figure 83 to Z scale is contained in Figure 84 depicting the dependence of the specific energy loss $\Delta E/\Delta x$ in silicon, measured (Hoo 81) for a number of heavy ions as a

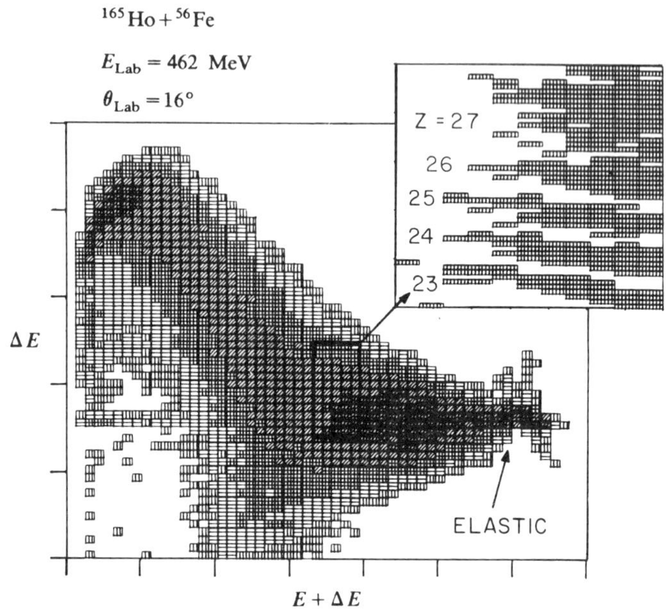

Figure 83. Diagram of energy loss, ΔE, vs. total energy, $E + \Delta E$, of projectile-like fragments from the reaction ^{165}Ho + ^{56}Fe at $E_{\text{Lab}} = 462$ MeV and $\theta_{\text{Lab}} = 16°$. A solid-state detector telescope with a 15.9-μm-thick transmission detector was used to obtain the data. The blow-up shown in the insert illustrates the quality of the Z identification. From (Hoo 81).

function of the ion kinetic energy per nucleon, E/A. Deviations of predictions (Nor 70, Bra 78c, Hub 80b) for the energy loss of heavy ions in solids, represented by various curves in Figure 84, from the measured values emphasize the necessity of a calibration of the particular detection system used in an experiment with a variety of heavy-ion beams and a large range of kinetic energies.

The Z resolution $\Gamma_Z(\text{FWHM}) \sim 0.5$ Z units typically achieved for fragments as heavy as Fe is not sufficient to resolve the different isotopes of each element which is possible for lighter ions (see, e.g., Gou 79). Hence, additional experimental information is required for a precise determination of the fragment mass number A. Detection systems incorporating a magnetic analysis (see Vol 79), i.e., determining the magnetic rigidity of incident particles, can provide that information. More conveniently, time-of-flight (TOF) methods are used to deduce the fragment mass from the measured velocity and total energy. Although fast plastic-scintillator detectors have been used (Gel 71, Cor 74) to measure the flight time of heavy ions over distances of the order of 1 m, the superior time resolution of better than $\Gamma_t(\text{FWHM}) \sim 0.1$ ns achievable with solid-state detectors and their more linear energy response make these latter devices more suitable for a simulta-

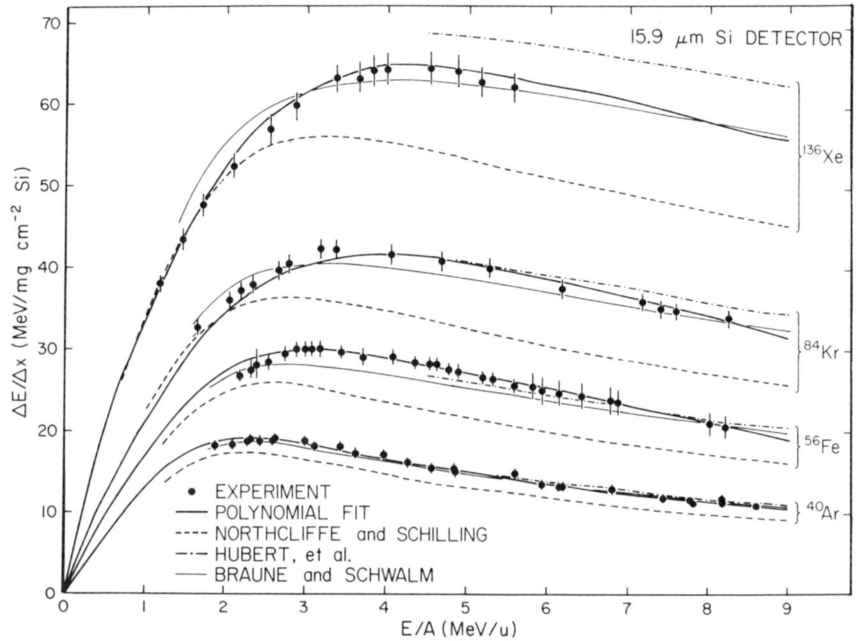

Figure 84. Specific energy loss $\Delta E/\Delta x$ of ^{40}Ar, ^{56}Fe, ^{84}Kr, and ^{136}Xe ions of kinetic energy per nucleon E/A in silicon of thickness Δx expressed in mg cm^{-2}. The dots represent measurements performed with a detector telescope using a 15.9-μm silicon transmission detector. The heavy solid curves are the results of polynomial fits to the data. Predictions of theoretical models are indicated by the thin solid curves (Bra 78c), the dashed curves (Nor 70), and the dashed-dotted (Hub 80b) curves. From (Hoo 81).

neous A determination using the TOF method. Figure 85 displays, as an example, the isotopic distribution of Cr fragments observed (Bre 80) in a bombardment of a ^{238}U target with a 8.3-MeV per nucleon ^{56}Fe projectiles. A TOF distance of 1 m between a 17-μm surface barrier transmission detector and a 100-μm stop detector was used in the measurements. Usually, the energy resolution of typically $\sim 1\%$ limits the mass resolution obtainable with TOF systems also to $\sim 1\%$ [see, e.g., (Bet 79c)].

Due to the facility to construct detection systems with a large solid angle and their insensitivity to radiation damage, gas-filled counters such as proportional detectors, ionization chambers, and avalanche counters are being used to an increasing extent in heavy-ion reaction studies [see (Ste 81b) for a recent survey]. Due to the incompatibility of simultaneous good time and energy resolution inherent in such detectors, usually different types are employed in conjunction, in order to supply simultaneous $Z-E$ and velocity information for the reaction fragments. As an example, a $\Delta E-E$–TOF telescope (Sch 81g) consisting of a three-element gas ionization chamber providing good Z and E resolution and an array of several

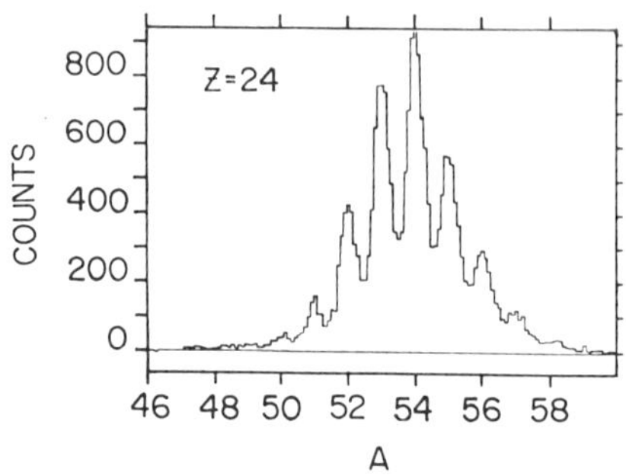

Figure 85. Mass spectrum of Cr fragments produced in the reaction $^{238}U + ^{56}Fe$ at $E_{Lab} = 8.3$ MeV per nucleon. The measurement was performed with a solid-state $\Delta E - E$ TOF detector telescope (Bre 80).

channel plate detectors for the TOF measurement is schematically depicted in Figure 86. Mass resolutions of ~ 2% have been achieved (Sch 81g) with this instrument, employing a TOF distance of 2.7 m. A large-solid-angle detection system for Z and A identification of reaction fragments is displayed in Figure 48. It allows the determination of average preevaporation masses, e.g., via Equation (3.2.1) in a kinematical-coincidence setup already discussed in Section 3.2.

Apart from the above more direct methods, reaction fragments can also be identified by detecting the characteristic radiation from the decay of the initially excited nuclei, such as γ rays, conversion electrons, or α particles. So far, the method of in-beam γ coincidence measurement has been applied only in experiments studying fusion (Sar 76, 78, Wes 78, 78d) and incomplete fusion (Zol 78, Wil 80b) reactions where γ rays emitted from slowly moving heavy residual nuclei have been detected. While the effects of Doppler shift and broadening render the application of in-beam spectroscopy of radiation from the fast projectile-like damped reaction products relatively less suitable for A and Z identification, such methods belong to

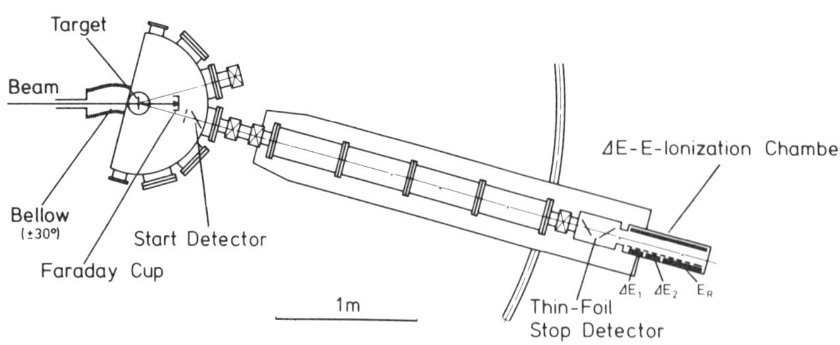

Figure 86. Schematic layout of scattering chamber (left) and $\Delta E - E$ TOF telescope. $\Delta E - E$ information is obtained with a three-element ionization chamber. The TOF is measured with start and stop detectors using channel plate electron multipliers measuring secondary electrons produced in the passage of charged particles through thin carbon foils. From (Sch 81g).

the standard repertory of off-line radiochemical studies [see, e.g., (Poi 81, Kra 81)]. In particular for very heavy systems, such radiochemical techniques constitute the only means currently available to achieve individual A and Z resolution, although at the expense of the accurate kinematic information that is vital for the reconstruction of the primary event.

Several of the difficulties to infer the primary fragment distributions from the measured secondary ones, resulting from the decay of the excited fragments prior to their detection, have already been mentioned in previous sections. These problems may become quite severe for the interpretation of experimental A and Z distributions (Bon 73, Bla 79b, Loc 82). Figure 87 (Bon 73) demonstrates the effect of light-particle evaporation on the mass and charge distribution of light fragments measured (Art 73) in the ^{232}Th$+^{22}$Ne reaction at $E_{\text{Lab}}=174$ MeV. In this figure, the relative yield of each fragmentation is plotted vs. neutron and proton numbers of the projectile-like fragment, in a two-dimensional logarithmic contour diagram. Starting from the assumed primary distribution shown in Figure 87b, Bondorf *et al.* (Bon 73) arrive at the post-evaporation distributions depicted in Figure 87d for quasielastic and in Figure 87c for damped events. A simplified evaporation code considering up to six sequential n, p, and α decay channels was employed in the calculation. It is seen that the secondary distribution can shift and become rather deformed, as compared to the initial one. Particularly apparent is a tendency of the secondary distributions to become aligned with respect to the valley of stability which increases with increasing amount of kinetic-energy damping.

A broadening of fragment (A, Z) distribution does not always result from the evaporation process, as has been pointed out recently by Lock *et al.* (Loc 82). For several systems, these authors deduce post-evaporation probability distributions $P_f(Z_f, N_f, E^*)$ from hypothetical initial three-dimensional Gaussians $P_i(Z_i, N_i, J_i)$ according to

$$P_f\left(Z_f, N_f, E^*\right)=\sum_{Z_i, N_i, J_i}\Gamma\left(Z_f, N_f, Z_i, N_i, E^*, J_i\right)P_i\left(Z_i, N_i, J_i\right) \tag{3.4.1}$$

Here, the transformation matrix Γ, a sum over possible evaporation chains, is calculated using a Monte Carlo evaporation code (Gav 80). In the calculation, the (fixed) excitation energy E^* is used as a parameter, and estimates are made for the initial spin (J_i) distribution. The solid curves in Figure 88 represent the predicted relations between post- and pre-evaporation variances σ_Z^2 of Z distributions, assuming an uncorrelated (cf. Figure 35a) initial distribution centered around ^{59}Fe. As can be seen from Figure 88, for small pre-evaporation variances, evaporation processes lead to a broadening, whereas the opposite is the case for large initial variances. It is

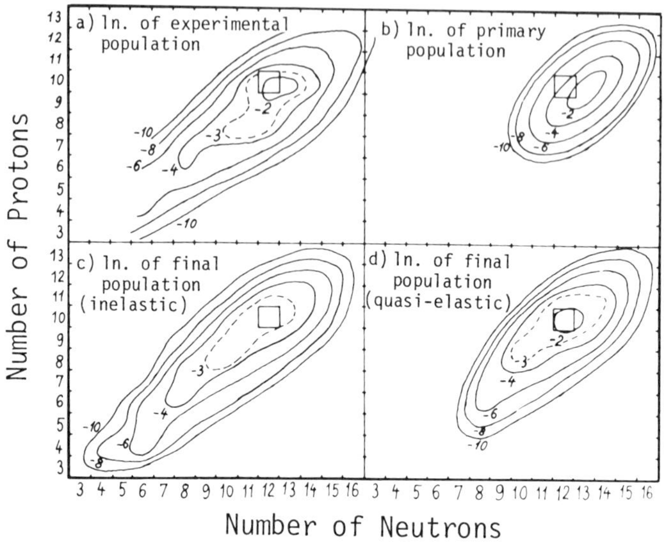

Figure 87. Logarithmic two-dimensional contour diagrams of the relative yield of fragmentations observed (Art 73) in a study of the ^{232}Th$+^{22}$Ne reaction at $E_{\mathrm{Lab}} = 174$ MeV. Part (a) reproduces the data, (b) a hypothetical primary distribution transformed by particle evaporation processes into the distributions (c) and (d) for damped and quasi-elastic events, respectively. The square indicates the position of the projectile ^{22}Ne. From (Bon 73).

pointed out in this work that charged-particle emission from the excited primary fragments, in particular, leads to sizable distortions of the fragment probability distribution in N and Z. The effect is most noticeable for isobaric and isotopic distributions of relatively light fragments that have acquired a significant amount of excitation energy during the reaction.

The probabilities and branching ratios for particle evaporation depend drastically (see, e.g., Gro 67d) on the poorly known properties of the initial nucleus. Hence, it seems a difficult and, in some cases, an impossible task to reconstruct the details of the primary distributions of damped reaction fragments from the measurements without experimental knowledge about the multiplicities of the associated light particles (see Section 3.7). However, an inspection of the final product yields measured for fusion–evaporation reactions leading to relatively light (Boh 80, Cof 78, Col 80, Con 78, Cor 77, Dou 79, Fer 78,80, Fro 76, Gar 75, Gom 79,79b, Gon 80, Her 78b, Heu 81, Jor 79, Kov 79, Lee 81, Nam 76, Nat 73, Obe 80, Puh 75,77b, Van 80e, Via 79) or medium-weight and heavy (Aga 76, Bus 79, Dou 78, Fuk 80, Gal 77, Gom 80, Hor 78, Nat 76, Pla 78,78d, Sai 78, Sch 81g, Sko 79, Spe 80b) intermediate nuclei can be of help for the interpretation of data. The problem will certainly be less severe for relatively heavy and neutron-rich fragments, where a high Coulomb barrier strongly hinders the evaporation of charged particles and, hence, preserves at least the primary fragment Z

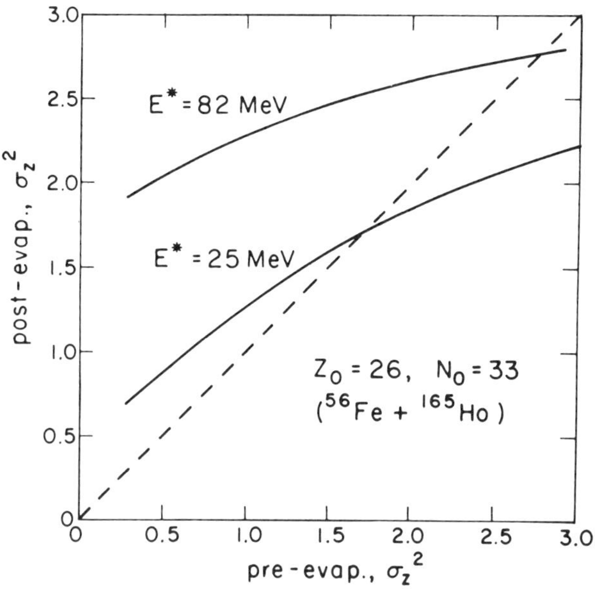

Figure 88. Relation between post- and pre-evaporative variances σ_Z^2 of fragment Z distributions. The primary distribution is assumed to be uncorrelated in N and Z and centered around $(Z_0 = 26,\ N_0 = 33)$. Solid curves represent results for initial excitation energies of $E^* = 25$ and 82 MeV associated with estimated average spins of $J_i = 17\hbar$ and $32\hbar$, respectively. From (Loc 82).

distribution. The above arguments suggest that proper caution has to be taken in judging details of the data discussed in this section.

The complexity of the mass spectrum of all fragments resulting from a heavy-ion reaction is exemplified in Figure 89 for the system $^{208}Pb + ^{48}Ca$. The reaction has been studied (Mor 78b) for mean projectile energies of $E_{Lab} \approx 255$ and 300 MeV corresponding to ~ 33 and 70 MeV above the interaction barrier, respectively. In this experiment, the induced radioactivities in the 42–45-mg cm^{-2}-thick targets were identified by measuring the γ rays from the decaying products. The mass distribution shown in Figure 89 is decomposed into various components $(A-F)$, as illustrated by the curves, a procedure relying on plausible assumptions about the shapes of the individual components. The mass distributions of this figure are interpreted (Mor 78b) to contain contributions from projectile-like (E) and target-like (F) fragments from quasielastic reactions, the corresponding components, C and D, respectively, from damped interactions, a contribution (A) centered around symmetry, from fission of an intermediate composite nucleus, and a distribution (B) arising from sequential fission of strongly damped target-like fragments. The shaded bands represent the sums of the components A–D and the uncertainty assigned to the overall fits. Additional systematic

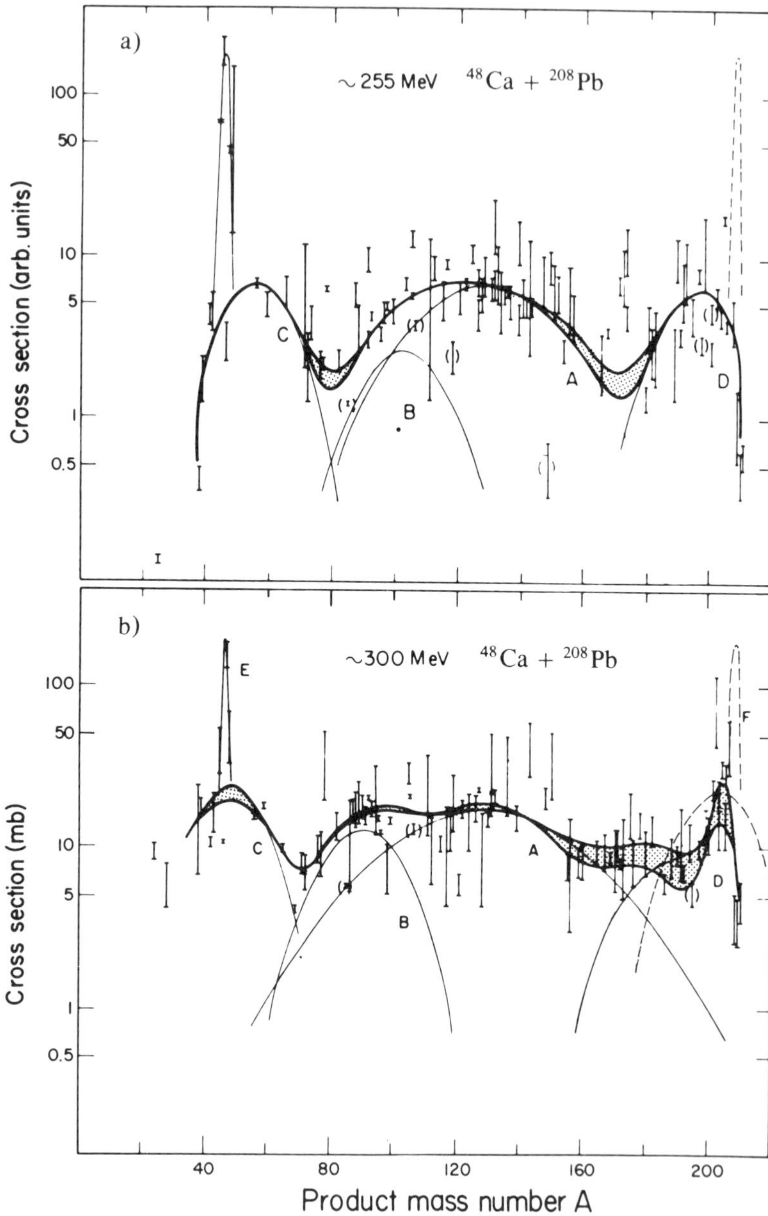

Figure 89. Mass distributions of fragments from the reaction $^{208}Pb + ^{48}Ca$ for two incident energies obtained from a thick-target experiment using radiochemical identification of the fragments. The total distribution is decomposed into: fusion–fission events (A), sequential fission of damped target-like fragments (B), damped (C) and quasielastic (E) projectile-like fragments, and their reflections about symmetry (dashed curves D and F, respectively). The shaded band indicates the total fit and its estimated uncertainty. From (Mor 78b).

uncertainties arise from the dependence of the product yield distribution on the projectile energy which varies appreciably through the target depth. For instance, the fission-like mass distribution can change from symmetric to asymmetric for the energy range explored in this experiment (Kal 77c). These examples clearly illustrate the ambiguities inherent in the interpretation of data obtained with off-line radiochemical methods lacking the benefits of kinematic information and of various correlations between reaction fragments. More detail of the reaction properties is preserved in radiochemical experiments (e.g., Poi 81, Kra 81) employing a thin target and a stacked-catcher foil technique for the collection of reaction products in incremental angle intervals.

Following the scheme developed in Section 3.3, in the following, the mass and charge distributions of fragments produced in damped reactions will be contrasted to those resulting from a fission process. Fission fragment A and Z spectra have been discussed by Vandenbosch and Huizenga (Van 73) referring to fission induced by photons, neutrons, deuterons, and α particles. In the present context, fission induced by heavy ions is emphasized, because here, reaction conditions with respect to mass of the fissioning system, angular momentum, and excitation energy are more similar to those under which damped reactions occur. Fragment mass or charge distributions have been measured (Bla 61, Nat 73,76, Ber 75,77, Rei 76, Gel 76, Bra 76, Cor 76c, Kal 77c, Oes 79b, Bac 80b) for a number of heavy-ion-induced fission reactions.

Figure 90 compares the mass yield curves (Bla 61) of fission fragments from the reaction ^{197}Au$+^{12}$C at $E_{\text{Lab}}=112$ MeV with those corresponding to deuterium-induced fission of a very similar composite nucleus produced at two rather different excitation energies. The angular dependence of fragment yields from the same ^{197}Au$+^{12}$C reaction at a slightly lower bombarding energy has the symmetric appearance characteristic of CN processes, as has already been discussed in the context of Figure 52. As can be seen from Figure 90, the fission fragment mass distributions are single-humped. In contrast to the distribution corresponding to low-energy d-induced fission which is rather narrow and centered close to symmetry ($A_{\text{sym}}=105.5$), those resulting from both the high-energy d-induced and the ^{12}C-induced processes are much broader and peaked at $A\approx 99$ and 98, respectively. This shift is attributed (Bla 61) to the emission of neutrons and charged particles during the fission process. A more quantitative discussion of centroid and width of the distributions displayed in Figure 90 is presently not possible because the multiplicity of particles emitted prior to scission and, hence, the identities of the fissioning systems and their excitation energies are not known. These instantaneous and sequential particle emission mechanisms are the subject of Section 3.7.

One concludes from data such as shown in Figure 90 that A and Z distributions of fragments from fission induced by heavy ions correspond to

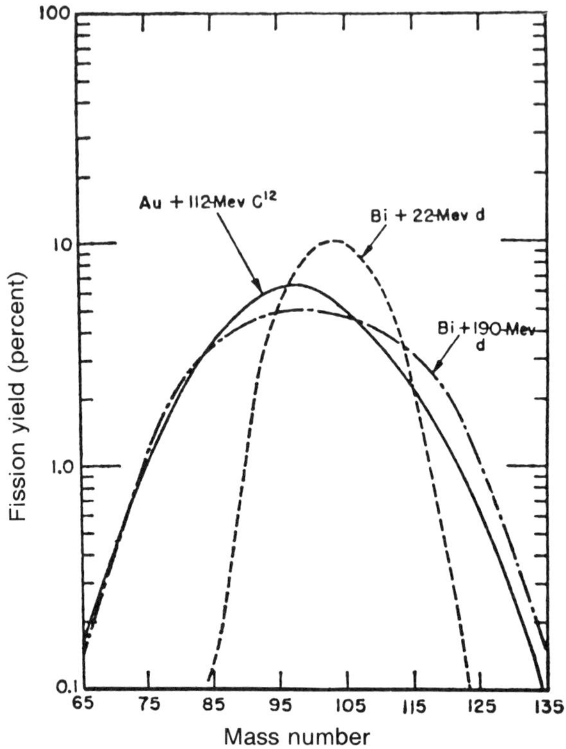

Figure 90. Mass yield curves of fission fragments produced in the reactions $^{197}Au+^{12}C$ at $E_{Lab}=112$ MeV (solid curve) and $^{209}Bi+^{2}H$ at 22 MeV (dashed curve) and 190 MeV (dashed-dotted curve). The measured masses are not corrected for evaporation. From (Bla 61).

preferentially symmetric splits but exhibit sizable widths. This is what would be expected (Van 73) from the systematics established for the normal fission process at moderate excitation energies. In contrast, asymmetric mass distributions of fragments from fission-like processes have been deduced from two-body coincidence experiments on the reactions $^{208}Pb+^{48}Ca$ (Kal 77c) and $^{243}Am+^{40}Ar$ (Kal 77d) at excitation energies $E^* \lesssim 50$ MeV of the composite nucleus. A similar behavior has been claimed (Oes 79b) for the relaxed reaction products measured in reactions induced by 160-MeV ^{32}S projectiles on targets of ^{65}Cu and ^{59}Co. For heavy systems, Sandulescu *et al.* (San 78) predict such asymmetric fission mass distributions on the basis of their fragmentation model (cf. Section 2.4). Referring to experimental studies of the systems $^{238}U+^{40}Ar$ and $^{238}U+^{48}Ca$, however, Otto *et al.* (Ott 78b) and others (Sai 78, Mat 79) argue that Kalpakchieva *et al.* (Kal 77c, d) observe damped reactions rather than the fission process of a fully equilibrated compound nucleus. Certainly, as exemplified in Figure 38, a mass diffusion process operating in damped reactions (cf. Sections 2.7 and 2.8) can in principle also lead to symmetric or nearly symmetric mass fragmentations, but without a complete relaxation of the shape of the intermediate dinuclear system that would be expected for a true compound nucleus.

Although it appears plausible to require $(1/\sin\theta)$-type angular distributions [cf. Equations (3.3.2)–(3.3.4)], fission-like kinetic energies (Vio 66) and masses, as well as the conservation of linear momentum for the reaction products measured in a fusion–fission-like reaction, these requirements are regarded only as necessary but not sufficient conditions to identify a compound-nucleus (CN) fission process. They will be adopted in the following to characterize fusion–fission-like (FL, compare Figure 39) and CN reactions. To discuss the finer distinctions between these two reactions types is the objective of Section 3.8.

Experimental information is available on the Z distributions of fragments from damped reactions induced by the relatively light projectiles ^{11}B (Art 79), ^{12}C (Nat 77b, Day 80, Wu 80c), ^{14,15}N (Tho 74, Vol 74,74b,76, Mor 75d,76b, Nat 77b, Bil 80b), ^{16,18}O (Art 71, Vol 74,76, Gel 76b,78, You 76, Nat 77b, Web 77, Aga 80b), ^{20,22}Ne (Art 70c,73,77, Vol 74,74b,76, Mor 76b, Bab 76, Nat 77, Obe 78, Sch 78h, Hom 79,79b, Van 80e, Day 80), ^{32}S (Bar 77, Oes 79b, Zie 79, Nov 80b, Pel 81, Win 81), and ^{35,37}Cl (Mig 82). Similarly, fragment A distributions have been measured for ^{12}C (Vol 78), ^{14,15}N (Vol 74,74b,76,78, Gre 79), ^{16,18}O (Art 70,71, Vol 74,76,78, Gel 76b,78, You 76, Cor 76), ^{20,22}Ne (Art 70c,73,77, Vol 74,74b,76,78), ^{28}Si (Bet 81), ^{32}S (Bra 76, Gel 76, Oes 79b, Dou 79, Chi 79, Zie 79), and ^{35,37}Cl (Mig 82) induced reactions. Damped reactions induced by ^{40}Ar ions have been studied particularly well, both with respect to Z distributions (Art 71b,73b,78,78b,79, Oga 74, Tho 74, Vol 74b,76,78,80, Gat 74,75,75b, Gal 75,76,76b,77, Mor 76b, Bri 76, Kra 76c, Oui 76, Sai 78, Bab 78, Bar 78d, Cau 78b, Aug 79,79b, Bim 79, Mat 79,80d, Was 79b, Bil 80b) and A distributions (Art 71b, Vol 74b,78, Gat 74,75,75b, Gal 76,76b,77, Oui 76, Kra 76c, Pet 76, Kal 77d,79b, Luc 77, Ngô 77b, Bab 78, Bar 78d, Sai 78, Aug 79,79b, Chi 79, Gue 80, Riv 80) of the damped reaction fragments. It is beyond the scope of interest here to discuss in any detail the mass distributions of fragments produced in quasielastic transfer reactions which are perceived to constitute the initial stages of damped reactions. The subject has been reviewed recently by Goldfarb and Von Oertzen (Gol 79). It is, nevertheless, instructive to observe the characteristic structure in the mass spectrum of fragments from the ^{28}Si + ^{28}Si reaction (Bet 81), displayed in Figure 91, disappear with increasing bombarding energy. The data have been acquired (Bet 81) for two-body events, where the masses M_3 and M_4 of the measured reaction partners add up to the total mass of the system, i.e., sequential or breakup processes are excluded from the data. Whereas an enhancement of multiple four-nucleon and two-nucleon cluster transfer is striking at low and intermediate bombarding energies, this feature gives way to an increasingly smoother spectrum at the higher energies. This develop-

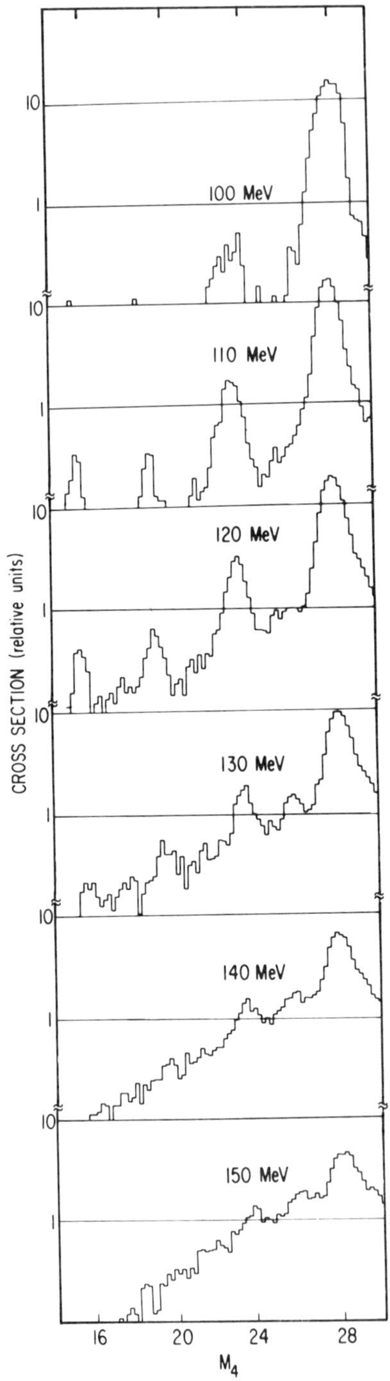

Figure 91. Mass distributions of fragments from the reaction $^{28}Si + {}^{28}Si$ at the various bombarding energies, plotted vs. mass of one of the final fragments. In these experiments, two final fragments were measured in coincidence. The distributions shown are selected such that the total mass M_{tot} of the detected fragments lies in the interval $55.5 \leq M_{tot} \leq 56.5$. From (Bet 81).

ment is accompanied by an evolution of the energy loss or Q-value spectrum from one showing the selective population of individual final nuclear states to a broad, continuous distribution indicating a degree of damping that increases with bombarding energy. Of course, the distributions in Figure 91 represent only fractions of the total $^{28}Si + ^{28}Si$ reaction cross section, due to the selectivity of the experimental method.

Inclusive fragment mass distributions for the slightly heavier system $^{50}Ti + ^{32}S$ (Bar 77) are depicted in Figure 92 for various incident energies and a detection angle of $\theta_{Lab} = 35°$. Only strongly damped reaction events are shown, selected with reference to the final energy spectrum (compare Figure 51). Calibration mass lines are included in the bottom portion of this figure. For low and intermediate bombarding energies, the mass distributions show two well-defined components, centered in the vicinity of projectile and target masses, respectively. Raising the bombarding energy results in a distinctive broadening of the individual components and a shift of their maxima towards more asymmetric mass splits, indicating an increasing preference for nucleon stripping off the lighter ^{32}S nucleus. Simultaneously, however, the cross section for symmetric fragmentations increases more than proportionally, filling in the valley between the

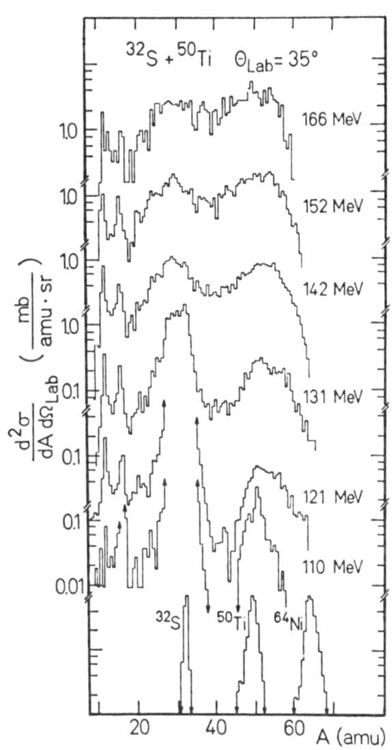

Figure 92. Mass distributions of fragments from the reaction $^{50}Ti + ^{32}S$ at various incident energies, measured at a laboratory reaction angle $\theta_L = 35°$. The events are selected by integrating over a 50-MeV-wide window centered at the maximum of the damped fragment energy spectrum. The mass lines pictured at the bottom serve for calibration purposes. From (Bar 77).

projectile-like and target-like components. Although Figure 92 shows data only at one reaction angle, this latter feature is also observed (Bar 77) for the angle-integrated mass distributions. The angular distributions for the fragments from this reaction show only slight deviations from a $(1/\sin\theta)$-type variation of the differential cross section, as can be expected (compare Figure 53) for a relatively light system at moderate bombarding energies.

It is interesting to explore whether the mass distributions discussed above show correlations with the structure of the underlying driving potential energy surface V. As noted in Section 2.4 [cf., e.g., Equation (2.8.4)], if mass exchange in damped reactions is due to a diffusionlike, equilibrium-statistical nucleon exchange process, the distribution $P(A_1)$ of final fragmentations should amplify the shape of this surface exponentially, according to the Boltzmann factor, i.e.,

$$P(A_1) \propto \exp[-V(A_1)/\tau] \tag{3.4.2}$$

Here, τ is the nuclear temperature. The potential energy surface V for the system $^{50}Ti + ^{32}S$ at $l = 0$ is shown in Figure 93 as a contour diagram plotted vs. neutron and proton numbers N_1 and Z_1, respectively, of one of the fragments. V is calculated from Equations (2.5.30)–(2.5.34). It incorporates the liquid-drop model binding energies corrected for shell effects, the Coulomb interaction potential of Bondorf *et al.* [see Equation (2.5.8)], and the nuclear proximity potential in the form of Equation (2.5.29). The dinuclear configuration assumed is that of two spherical nuclei touching at a

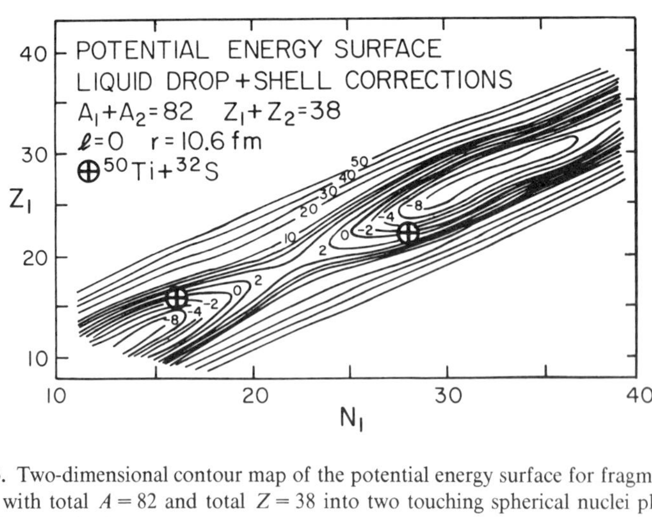

Figure 93. Two-dimensional contour map of the potential energy surface for fragmentations of a system with total $A = 82$ and total $Z = 38$ into two touching spherical nuclei plotted vs. N and Z of one of the fragments.

center separation distance equal to the strong-absorption radius $r = R_{SA}$ ($^{50}Ti + ^{32}S$) = 10.6 fm.

The most striking feature of the surface displayed in Figure 93 is the appearance of a Businaro–Gallone saddle (see Section 2.5) located at symmetry, signifying an instability for symmetric fragmentations of the $^{50}Ti + ^{32}S$ system. This general shape of the surface is preserved for the whole range of angular momenta explored in the experiments, i.e., addition of a centrifugal energy V_l with $l \leq 75\hbar$ does not destroy its structure. The injection points corresponding to the two initial fragments $^{50}Ti + ^{32}S$ are marked by crosses in Figure 93. They are located some 4 MeV below the Businaro–Gallone saddle at the slopes of the valleys on either side. Hence, one expects a tendency for the system of two interacting nuclei to be driven towards asymmetry, with a very small probability for symmetric fragmentations. This is basically the feature observed in the data (cf. Figure 92) for the lower bombarding energies. In such a picture, one would also qualitatively understand the relative gain of symmetric fragmentations at the expense of asymmetric ones, as the bombarding energy and, hence, the excitation energy introduced into the system is increased. According to Equation (3.4.2), the peak-to-valley ratio of the mass distribution should vary exponentially with the ratio of the potential energy difference ΔV of the corresponding fragmentations and the nuclear temperature τ, i.e., $P(A_{sym})/P(A_{asym}) \approx \exp(-\Delta V/\tau)$. Consequently, with increasing bombarding energy and temperature, the peak-to-valley ratio is expected to decrease drastically. In addition, the diminishing importance of shell effects at higher excitation energies should result in a gradual disappearance of the Businaro–Gallone saddle in the surface depicted in Figure 93, an effect that should also enhance the probability for symmetric fragmentations at higher bombarding energies.

A quantitative analysis of the experimental mass (and charge) distributions for the $^{50}Ti + ^{32}S$ system discussed above has been performed by Barrette *et al.* (Bar 77b) and Braun-Munzinger *et al.* (Bra 78) employing a phenomenological diffusion model similar to the one of Moretto *et al.* (Mor 75, Sve 76) outlined in Section 2.8. Assuming a liquid-drop driving potential, this model was found unable to account simultaneously for the relatively small drift of the asymmetric mass components and the enhancement of the probability for symmetric mass splits observed (cf. Figure 92) with increasing energy. A good representation of the data by the model was, however, achieved when the Einstein relation [cf. Equation (2.7.17)] between drift and diffusion coefficients, required by the model, was disregarded and a phenomenological bombarding-energy dependence of the diffusion coefficient introduced. This example demonstrates for a relatively light system that an interpretation of the features of fragment mass distributions in terms of a classical statistical diffusion process is probably inadequate, although good model fits to data can be obtained by adjusting parameters.

A further characteristic dependence of the fragmentation probability of a colliding heavy-ion system on entrance channel conditions is illustrated in Figure 94. Here, experimental Z spectra $d^2\sigma/d\theta\, dZ$ are shown for two systems of identical total mass and atomic numbers but rather different initial asymmetries, as reported by Agarwal *et al.* (Aga 80b). The bombarding energies are chosen to produce similar values of l_{max} for the two reactions. Not too surprisingly, the Z distribution of fragments from the symmetric system $^{56}Fe+^{52}Cr$ at $E_{Lab}=261$ MeV are observed to be symmetric around $Z_{sym}=25$. As can be seen from Figure 94, at forward angles ($\theta_{Lab}\leq 15°$), the Z distributions for this reaction are sharply peaked around the charges of projectile and target, with tails extending towards larger and smaller Z values. At further backward angles, these distributions attain a flat and broad appearance. In contrast, the charge distribution of fragments from the $^{92}Mo+^{16}O$ reaction shown on the right of Figure 94 exhibits clearly two distinctive groups of fragments. The dominant component for damped projectile-like fragments peaked around $Z=6$ increases in width and decreases in cross section, as the observation angle is changed from 15° to 80°. In contrast, the broad fission-like distribution of fragments centered at $Z_{sym}=25$ shows no discernible angle dependence. It is then very suggestive to interpret the Z distributions of fragments from the $^{56}Fe+^{52}Cr$ reaction on the left of Figure 94 as similar superpositions of a damped and a fission-like component which are both centered around $Z=25$, because of the symmetry of the system.

Another interesting observation made (Aga 80b) from the data shown in Figure 94 pertains to the rather different cross sections associated with the fission-like components in the two complementary reactions. The cross sections for evaporation residues believed to result from a true compound-nucleus reaction have been found (Aga 80b) to constitute approximately 40% of the total reaction cross section σ_R, for both the $^{56}Fe+^{52}Cr$ and the $^{92}Mo+^{16}O$ reactions. Because of the very similar ranges of l values explored in these reactions, the low cross section of only 0.5% of σ_R deduced for the fission-like component in the $^{92}Mo+^{16}O$ reaction suggests that compound nucleus formation in the reaction $^{56}Fe+^{53}Cr$ is also not likely to be followed by fission. It is then tempting to conclude that the broad component of fragments from the $^{56}Fe+^{52}Cr$ reaction is not associated with fission of a completely equilibrated system but, rather, that one observes the signature of a process associated with rather long interaction times and other characteristics reminiscent of fission. Although it is impossible to distinguish, for this symmetric system, damped reaction products from those due to a fusion–fission-like reaction (FL, see Figure 39), the persistence (Aga 80b) of the liquid-drop fission barrier up to the l values of interest

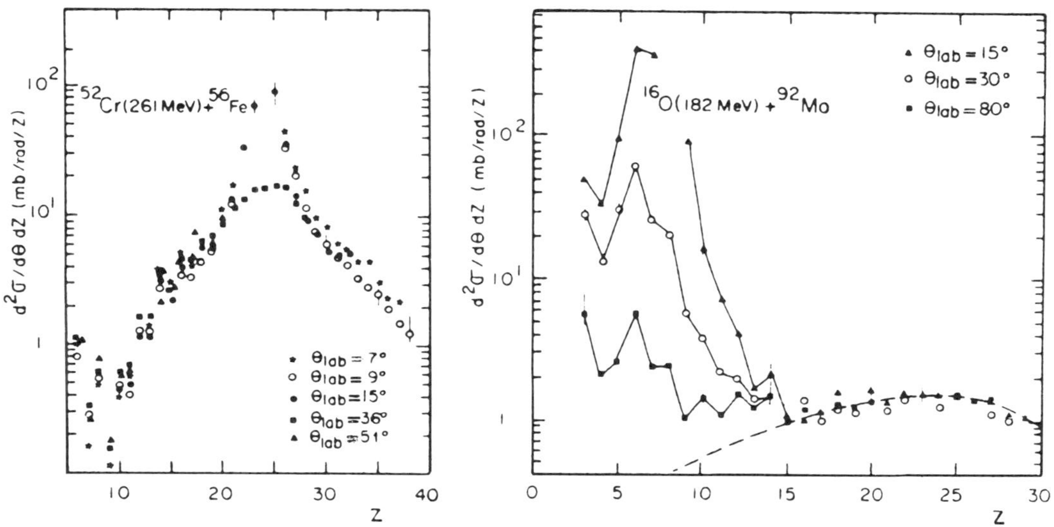

Figure 94. Charge distributions $d^2\sigma/d\theta\,dZ$ of fragments from the complementary reactions $^{56}Fe + ^{52}Cr$ and $^{92}Mo + ^{16}O$ leading to the same composite systems. The selected bombarding energies correspond to very similar values of l_{max}. The dashed curve on the right represents an attempt to extrapolate the distribution of fission-like fragments centered around $Z_{sym} = 25$ into the region of damped reaction events. From (Aga 80b).

here suggests a negligible contribution of FL events to the cross section for the $^{56}Fe+^{52}Cr$ reaction.

While the transition from damped to FL processes will be discussed in Section 3.8, conclusions on the peculiarities of the mass exchange mechanism operating in damped reactions can be drawn from the comparison of the two distributions shown in Figure 94. Since the same driving potential energy surface applies for the two systems $^{56}Fe+^{52}Cr$ and $^{92}Mo+^{16}O$, the significant difference in the fragment Z distributions discovered for these systems provides further arguments against the presence of a classical stationary mass diffusion process in damped reactions that would lead to fragment distributions P that are approximately described by Equation (3.4.2). Apparently, there must be dynamical qualities of the mass exchange mechanism or significant variations of the driving potential during a reaction that simultaneously prevent the $^{92}Mo+^{16}O$ system from approaching a symmetric critical point and the $^{56}Fe+^{52}Cr$ system from being driven away from it.

As one proceeds from lighter to heavier projectile–target systems, the overall character of the fragment A and Z distribution changes appreciably. An appearance of a bimodal distribution corresponding to two distinct classes of projectile-like and target-like events, respectively, becomes more and more accentuated. This development is illustrated by the Z distributions shown in Figure 95. The distribution for the $^{56}Fe+^{132}Xe$ reaction on the right of this figure has been measured by Heusch *et al.* (Heu 78). The associated composite nucleus is $^{188}_{80}Hg$, only slightly lighter than $^{193}_{83}Bi$ formed in the reaction $^{107,109}Ag+^{84,86}Kr$ for which the experimental fragment Z distribution (Sch 77f) is depicted in the middle of Figure 95. Here, the open circles are obtained from a reflection of the measured data (solid points) about symmetry. On the left of Figure 95, only the measured (Woz 78) portion of the charge distribution of fragments from the reaction $^{159}Tb+^{86}Kr$ leading to the $^{245}_{101}Md$ composite nucleus is reproduced by the dots. Gaussian curves centered at symmetry with similar widths of $\Gamma_Z(\mathrm{FWHM}) \approx 20$–$25$ Z units are drawn for each of the Z distributions, suggesting a hypothetical decomposition into a symmetric component and projectile- and target-like contributions. Obviously, the relative importance of the symmetric fragment component diminishes significantly, as the projectile–target system becomes heavier.

The tendency illustrated by the succession of the fragment Z distributions of Figure 95 is accompanied by an analogous development of the angular distributions of the differential reaction cross section (see Section 3.3), in which a $(1/\sin\theta)$- or orbiting-type, long-lifetime component loses importance to the extent to which symmetric fragmentations become improbable in a reaction. It should be emphasized that the validity of the decomposition of the fragment distribution illustrated in Figure 95 is

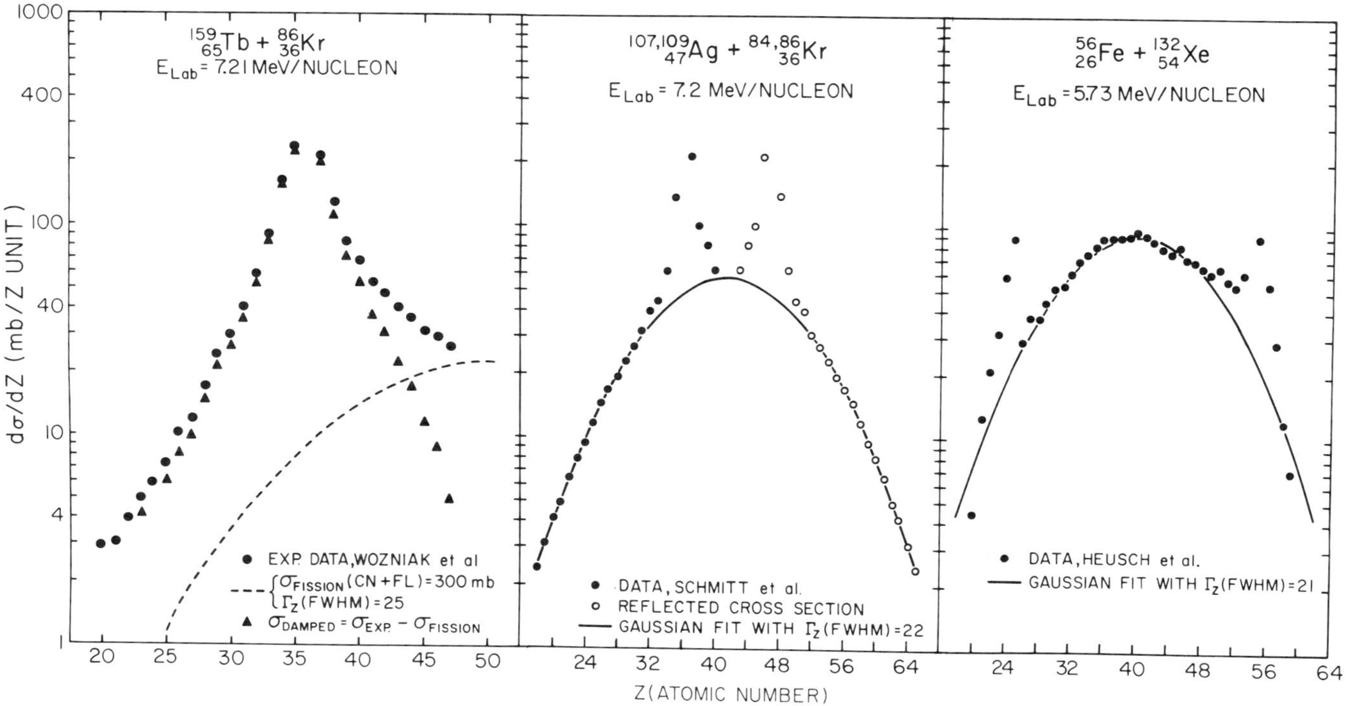

Figure 95. Fragment charge distributions for the reactions $^{56}Fe + ^{132}Xe$ (Heu 78), $^{107,109}Ag + ^{84,86}Kr$ (Sch 77f), and $^{159}Tb + ^{86}Kr$ (Woz 78). Solid dots represent measured data; open circles result from a reflection of the data about symmetry. The solid curves are Gaussians centered at symmetry fitted to the data. The dashed curve in the case of Tb + Kr indicates an estimated symmetric charge component; the solid triangles represent the difference between the measured charge distribution and the dashed curve.

currently merely suggested but not firmly established in the sense that it represents different classes of reaction mechanisms. The experimental results are also basically consistent with a single statistical mass transport mechanism operating in damped collisions (see, e.g., Bar 77b, Sch 77f, Aga 80b), if phenomenological model parameters are introduced. By such a procedure, however, the problems to be solved are only deferred to a different reference frame.

The features of the fragment mass and charge distributions appear to be more clearly defined in damped reactions induced by very heavy ions. This is exemplified in Figure 96 by the Z distributions of fragments from damped interactions induced by ^{136}Xe projectiles on a ^{209}Bi target at three

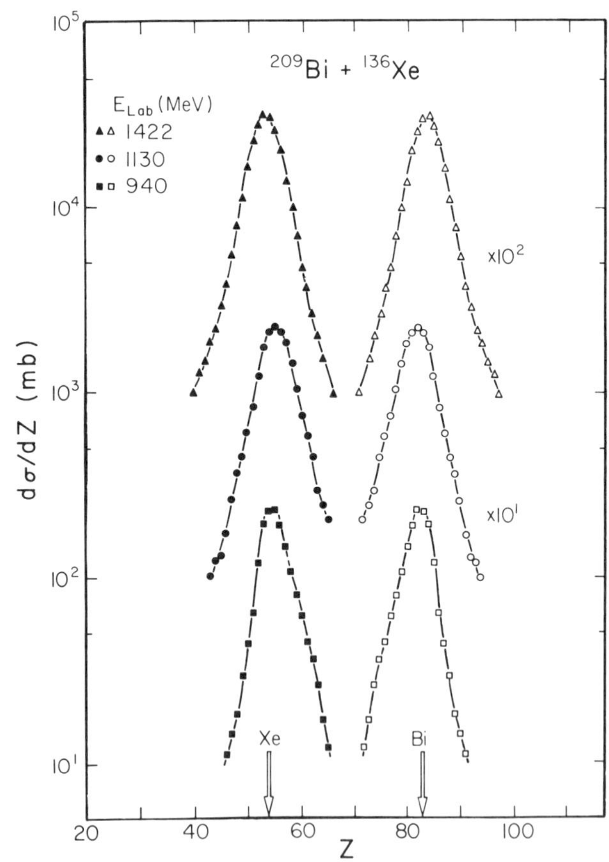

Figure 96. Z distribution of fragments from the reaction $^{209}Bi + ^{136}Xe$ measured at laboratory bombarding energies of 940 MeV (Wil 80), 1130 MeV (Sch 78g), and 1422 MeV (Wol 81). The solid symbols mark experimental cross sections; the open symbols represent their reflection about symmetry, assuming a binary reaction.

incident energies. The data (Wil 80, Sch 78g, Wol 81) are given by solid symbols; their open counterparts represent inferred cross sections, assuming binary collisions. As can be seen from this figure, the charge distributions resemble those of a low-energy fission process, in contradistinction to the symmetric distributions (cf. Figure 90) resulting from heavy-ion-induced fission at these energies. The cross section is almost symmetrically distributed about the initial projectile–target asymmetry. There is very little, if any, drift of the centroids of the bimodal distribution with changing bombarding energy. In addition, there is no yield of significance found for symmetric fragmentations, very much in contrast to the mass distributions for the lighter systems discussed earlier. Apparently, the very heavy interactions partners in such a collision retain the memory of the initial conditions to a large extent. Although considerable transfer of charge (and mass) can occur, the probability for such fragmentations drops nearly exponentially with the distance from the initial asymmetry.

In an attempt to trace the salient features of the $^{209}\text{Bi}+^{136}\text{Xe}$ Z distributions of Figure 96 to the corresponding driving potential, it is useful to note that the potential energy surface, calculated in a manner described previously, exhibits a shell structure-related minimum for fragmentations in the vicinity of the projectile–target asymmetry. For $l=0$, this minimum is about 14 MeV deep and persists up to l values in excess of $400\hbar$. For comparison, $l_{\text{max}} \approx 480\hbar$ for $^{209}\text{Bi}+^{136}\text{Xe}$ at $E_{\text{Lab}}=1130$ MeV. Hence, the structure is of similar magnitude to that appearing in the surface predicted for lighter systems such as $^{50}\text{Ti}+^{32}\text{S}$ (cf. Figure 93). Assuming the same reaction mechanism to be in operation, the effect on the fragment mass and charge distributions should be similar for relatively light and heavy systems. This appears to be qualitatively true, if one restricts the consideration to the projectile–target-like fragmentations which are the only ones observed for heavy systems such as $^{209}\text{Bi}+^{136}\text{Xe}$. However, the expected disappearance of shell effects [cf. Equation (2.8.16)] at the enormous excitation energies introduced (cf. Section 3.5) into the system renders such qualitative arguments incomplete and unconvincing. The driving potential derived from the liquid-drop model exhibits always a minimum for symmetric fragmentations of a heavy system, which is not reflected in the Z or A distributions.

Because the damped reaction component has a more distinctive appearance for interactions induced by heavier ions, as is illustrated in Figures 95 and 96, the following discussion of fragment A and Z distributions will concentrate on processes induced by projectiles heavier than ^{40}Ar. Fragment charge distributions have been measured with beams of $^{40,48}\text{Ca}$ (Col 75, Roy 76, 77, Aga 77, Ame 80, Mig 82), ^{52}Cr (Aga 80b), ^{56}Fe (Bre 79, 80, Sch 79f, Mig 80, Vio 80, 80b, 81, Gob 81, Bri 82), $^{63,65}\text{Cu}$ (Tas 76), ^{74}Ge (Oga 74b), $^{84,86}\text{Kr}$ (Wol 74, 76, 76b, Kra 74c, Mor 76b, 76c, Hui 76c, 77, 77b, 79b, Bri 76, Sch 76b, 77f, 78, Rus 77b, San 77, Web 77, Woz 78, Cau 78, Rud 79, 81, Ber 79c, Bla 79b, Luc 79b, Lyn 79, Eya 79, 80b, Dye 80, Chr 80, Olm 80, Gob

81, Sob 81b), ^{129,132,136}Xe (Sch 76, 76b, 77e, 78, 78g, 79f, 81h, Wol 76, 76b, 81, Hui 76c, 77, 77b, Van 76c, Mor 76b, Ott 76, Kra 77, 79d, Rus 77, San 77, Poi 79, Wil 80, Oes 80, Gob 81), ^{144,154}Sm (Wu 81), ^{208}Pb (San 77, Ess 79c, Reh 79, 81, Tan 80c), and ^{238}U (San 77, Hil 77, Sch 78f, Fre 79b, Gla 79). Mass distributions are available for fragments resulting from bombardments of a variety of targets with ^{40,48}Ca (Roy 77, Mig 82), ^{52}Cr (Riv 80), ^{56}Fe (Reu 77, She 79, Bre 79, 80, Mig 80, Vio 80, 80b, 81, Riv 80, Gob 81, Bri 82), ^{63,65}Cu (Pet 75c, 76, 77b, Lef 76, 79, Ngô 76, Tas 76, Ber 77, Chi 79), ^{74}Ge (Oga 74b), ^{84,86}Kr (Wol 74, Oga 74, Kra 74c, Lef 76, 79, Luc 79b, Bla 79b, Eya 79, Ber 79c, Spe 81, Har 82, 82b), ^{129,132,136}Xe (Oga 74, Ott 76, Kra 79d, 81, Poi 79, Est 80, Oes 80, Sch 81h, Har 82b), ^{144,154}Sm (Wu 81), ^{208}Pb (Ess 79c, Reh 79, 81, Gob 81), and ^{238}U (Sch 78f) projectiles.

As has been qualitatively discussed previously for a number of cases, it appears difficult to understand the features of mass and charge exchange in damped reactions in terms of a stationary nucleon diffusion process, where the final fragment A and Z distributions simply reflect the structure of the underlying driving potential as it may exist at the end of the interaction. Rather, it seems that the mass exchange mechanism proceeds on a time scale that is significantly longer than the dinuclear interaction time. Such a conjecture is also supported by the observation of a strong influence of the initial projectile–target asymmetry on the final distribution of fragmentations. It is then obvious to explore the possibility of explaining the data in an approach using the master or Fokker–Planck Equations (2.7.5) and (2.7.14), respectively, which describe a broader class of statistical processes. Because of their time dependence, the application of these equations requires either a rather detailed modeling of the collision history, or, in a more approximate and phenomenological fashion, the establishment of relations between the total time spent on an average trajectory and experimental observables other than the mass asymmetry. The latter possibility is realized, e.g., by relatively crude classical phenomenological reaction modes (Bon 75, Hui 77b, Sch 77c, 77e, Wol 77, 78, 78b, Rie 79) connecting interaction time with average scattering angle. A roughly exponential dependence of the interaction time on angular momentum results from such a phenomenological analysis, as is illustrated in Figure 82.

Russo *et al.* (Rus 77b) have applied the master equation approach of Moretto *et al.* (Mor 75, 75b, 76, 76b, Sve 76) to the reaction ^{179}Au + ^{86}Kr, assuming a linear increase of the interaction time with l. The predicted distribution $d^2\sigma/d\Omega\, dZ$ of the double-differential cross section depends then on the parameter describing this linear relationship and a diffusion coefficient D. These model parameters can be deduced from experimental data by fitting simultaneously fragment angular and charge distributions. Comparisons between data and fits for the ^{197}Au + ^{86}Kr reaction have already been made in Figure 70. Experimental and theoretical best-fit results for the charge distribution of Kr-like fragments from this reaction are displayed in Figure 97. A qualitative description of the data, which look

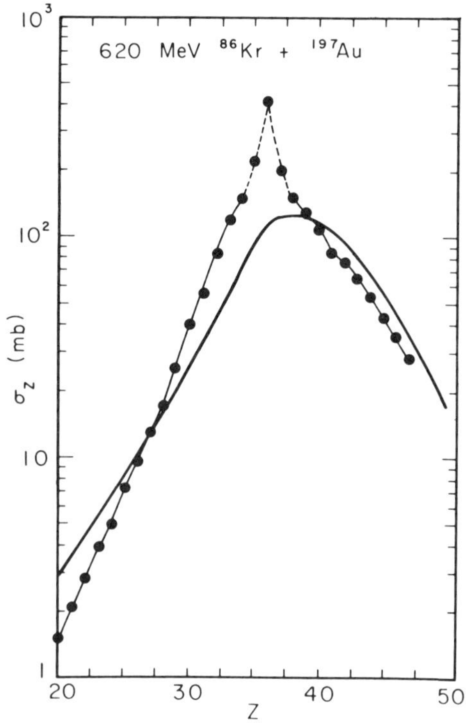

Figure 97. Comparison between experimental (solid dots) and theoretical (heavy curve) charge distribution for fragments from the reaction $^{197}Au + ^{86}Kr$ at $E_{Lab} = 620$ MeV. From (Rus 77b).

similar to those on the left of Figure 95, is achieved. However, large discrepancies remain for the interesting range of fragments with Z values close to that of the projectile or below. A value of the diffusion coefficient of $D_Z = 3.3 \times 10^{21}$ s^{-1} is derived (Rus 77b) from this phenomenological analysis. Similar studies have been performed for the systems $^{159}Tb + ^{86}Kr$ (Woz 78) and $^{181}Ta + ^{86}Kr$ using this diffusion model. The quality of the model fits obtained in these cases corresponds to the one shown in Figure 97.

Mass transfer in damped heavy-ion collisions has also been investigated applying the Fokker–Planck approach of Nörenberg and collaborators (Nör 74, 75, 76, 79c). As explained in Section 2.8, the transport coefficients entering this transport model have been derived microscopically (Ayi 76, 76c, 78). However, since this model does not incorporate an explicit description of the dynamics of the relative motion, its application also requires a fit to data in terms of an l-dependent interaction time $t_{\text{Int}}(l)$. The total fragment A (or Z) distribution is then determined as a superposition of Gaussian solutions $P(A, t_{\text{Int}})$ of the one-dimensional Fokker–Planck equation for individual values of l. For a given l value, the variance σ_Z^2 of the fragment Z distribution is related to the interaction time according to

$$\sigma_Z^2 = 2\int_0^{t_{\text{Int}}(l)} dt\, D_{ZZ}(\mathbf{r}(t)) = 2\, D_{ZZ}^{(0)} f(l, l_g) t_{\text{Int}}(l) \tag{3.4.3}$$

Here, the curve integral over the diffusion coefficient is written as a product of a strength factor $D_{ZZ}^{(0)}$ and a form factor f which is characteristic of the trajectory. The quantity l_g is the grazing-angular momentum. This form factor is determined (Wol 78b) from the variation of σ_Z^2 with reaction angle. The dependence of t_{Int} on l is extracted from a fit of the fragment angular distribution, as has been discussed Section 3.3. The cross section for a given charge fragmentation Z_1 can then be written as an integral

$$\frac{d\sigma}{dZ_1} = 2\pi\lambda\!\!\!^{-2} \int dl\, l P(Z_1, t_{\text{Int}}(l)) \tag{3.4.4}$$

over all l values contributing to the reaction.

Examples where the above model has been used to describe fragment Z distributions are given on the right of Figure 98 (Wol 77) for the reactions $^{166}\text{Er} + {}^{86}\text{Kr}$, $^{120}\text{Sn} + {}^{132}\text{Xe}$, and $^{238}\text{U} + {}^{238}\text{U}$. The solid symbols mark the experimental cross sections. In the case of the $^{238}\text{U} + {}^{238}\text{U}$ reaction (Hil 77), the open squares describe the primary Z distribution obtained by correcting the measured data for sequential fission of the primary reaction fragments. The dashed curves represent model calculations based on the theoretical mass transport coefficients v_A and D_{AA} discussed in Section 2.8. The transport coefficients for the charge are assumed to scale with the global Z/A ratio as $D_{ZZ} = (Z/a)^2 D_{AA}$ and $v_Z = (Z/A) v_A$, respectively. Obviously, for $^{166}\text{Er} + {}^{86}\text{Kr}$ and $^{120}\text{Sn} + {}^{132}\text{Xe}$, the experimental data are reasonably well described by these calculations, although a larger drift towards symmetry is predicted than observed for these systems. For $^{238}\text{U} + {}^{238}\text{U}$, the theoretical Z distribution is clearly too narrow. The solid curves drawn through the experimental points in Figure 98 represent best-fit results, where drift and diffusion coefficients have been adjusted to the data. On the left-hand side of Figure 98, for each system a number of Gaussian probability distributions $P(Z_i, t_{\text{Int}}(l))$ are displayed illustrating the evolution of the fragment Z distribution with decreasing l or increasing interaction time.

Similar to what is experienced with simpler diffusion models discussed previously, the model calculations of Wolschin *et al.* (Wol 77,78b) prove unable to explain the relatively small drifts towards symmetry of the experimental charge distributions. Theoretical mass drift coefficients are (Nör 79c) approximately a factor of 2 larger than predicted by microscopic calculations. Not too much can be learned from the large discrepancy between model and experiment in the case of $^{238}\text{U} + {}^{238}\text{U}$. Here, the significant distortions of the data by the sequential fission process prevent a reasonably accurate determination of the primary Z distribution. Even a drift of this symmetric system away from symmetry, such as illustrated by the calculations of Brosa (Bro 80c, cf. Figure 34), cannot be excluded.

For purposes of a numerical comparison, mass diffusion coefficients, D_{AA}^{EXP}, derived from a phenomenological analysis of experimental fragment

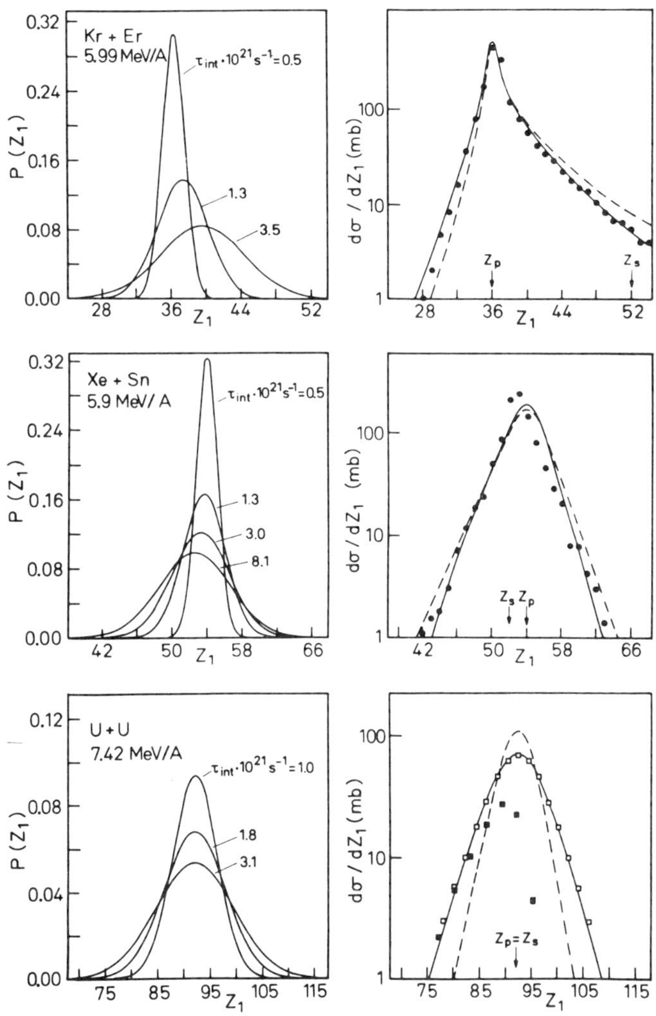

Figure 98. Fragment Z distributions are shown for the reactions $^{166}Er + ^{86}Kr$, $^{120}Sn + ^{132}Xe$, and $^{238}U + ^{238}U$. The data are indicated by solid symbols. The open squares for the $^{238}U + ^{238}U$ reaction are obtained through corrections of measured cross sections for sequential fission. Dashed curves represent model predictions based on theoretical transport coefficients; in the calculations resulting in the solid curves, these coefficients have been adjusted to fit the data. On the left, for each system the evolution of the probability distribution P with interaction time is illustrated. From (Wol 77).

Table 3. Diffusion Coefficients for Mass, Neutron and Proton Exchange D_{AA}, D_{NN}, and D_{ZZ}, Respectively[a]

Target	Projectile	E_{Lab} (MeV)	D_{AA}^{EXP}	D_{AA}^{TH}	D_{AA}^{EFF}	D_{NN}^{TH}	D_{ZZ}^{TH}
^{166}Er	^{86}Kr	515	2.0	2.4	2.5	2.6	0.7
		619	2.7	3.1	3.4	3.4	0.9
		703	3.7	3.7	4.0	4.0	1.0
^{209}Bi	^{84}Kr	712	2.9	3.5	3.6	3.8	0.9
^{120}Sn	^{132}Xe	779	1.6	2.5	2.8	2.8	0.7
^{209}Bi	^{136}Xe	940	3.5	2.9	2.6	3.1	0.6
		1130	4.2	4.2	3.8	4.5	0.9
^{208}Pb	^{208}Pb	1456	2.5	3.1	2.4	3.2	0.5
		1560	2.9	3.6	2.8	3.8	0.6
^{238}U	^{208}Pb	1560	3.9	3.5	2.6	3.7	0.5
^{238}U	^{238}U	1766	7.5	3.4	2.4	3.5	0.5

[a] Values D_{AA}^{EXP} are derived from a phenomenological fit to data. The superscript TH denotes values predicted by microscopic calculations. D_{AA}^{EFF} denotes an effective theoretical mass diffusion coefficient taking correlations between neutron and proton diffusion into account. The coefficients are given for a touching dinuclear configuration and in units of 10^{22} s^{-1}. From (Mer 81).

distributions are collected in Table 3 (Mer 81), together with values D_{AA}^{TH} predicted from microscopic calculations (cf. Section 2.8). These values are derived for two spherical nuclei in the touching configuration, where $r = C_P + C_T$. Also included in this table are theoretical neutron, proton, and effective mass diffusion coefficients D_{NN}^{TH}, D_{ZZ}^{TH}, and D_{AA}^{EFF}, respectively, derived by Merchant and Nörenberg (Mer 81) in a slightly different model approach that accounts for correlations [cf. Equations (2.8.6)–(2.8.9)] imposed on the neutron and proton diffusion processes by the curvatures of the two-dimensional driving potential $V(N_1, Z_1)$. One observes a satisfactory correspondence between experimental and theoretical diffusion coefficients, although there are some significant differences.

From the discussion presented above, one concludes that experimental features of overall fragment A or Z distributions are somewhat indicative of a stochastic mechanism of nucleon exchange in damped reactions. This may be concluded from the qualitative reproduction of these features by diffusion-type model calculations taking the effects of the collective motion to some extent into account. However, these models rely on the prior adjustment of important parameters to experimental data, such that agreement between predictions and measurements may not be too conclusive. For example, unexplained relations between deduced transport coefficients, which are in violation of those characteristic of these statistical theories, as well as surprising dependences on bombarding energy could potentially be of serious consequence for the validity of such approaches. On the other hand, there is currently little evidence for a collective mode of mass transfer in damped reactions (cf. Section 2.4). Available models (see, e.g., Mar 80, 81b) are not yet fully developed and independent of phenomenological parameters, such that definite inferences from theoretical fits to experimental fragment A or Z distributions [see, e.g., (Mar 80)] seem premature. A consideration of correlations between Z, A distributions and other experimental observables provides additional means to distinguish between possible reaction mechanisms. Such correlations are discussed next.

Correlations between fragment mass asymmetry and reaction angle have already been noted in Section 3.3 (cf. Figures 68–70). They have also been discussed in the context of Figure 94, for relatively light systems. Figure 99 summarizes the findings for the medium-weight system $^{50}Ti+^{208}Pb$ (Gob 81). Here, the double-differential cross section $d^2\sigma/dA\,d\theta$ is plotted vs. fragment mass, A, and center-of-mass scattering angle, θ, in a two-dimensional contour diagram. The measurements were done in a coincidence arrangement where both fragments from a two-body collision are detected simultaneously. A strong focusing of projectile-like and target-like reaction products close to the corresponding quarter-point angles is indicated in this figure by the appearance of two well-defined cross section peaks. These peaks fan out towards symmetry, where the angular distribution for a given mass becomes rather broad. In this plot, a $(1/\sin\theta)$-type angular distribution $d\sigma/d\Omega$ manifests itself in iso-cross-section contour lines that are parallel to the angle axis. The figure, hence, provides an

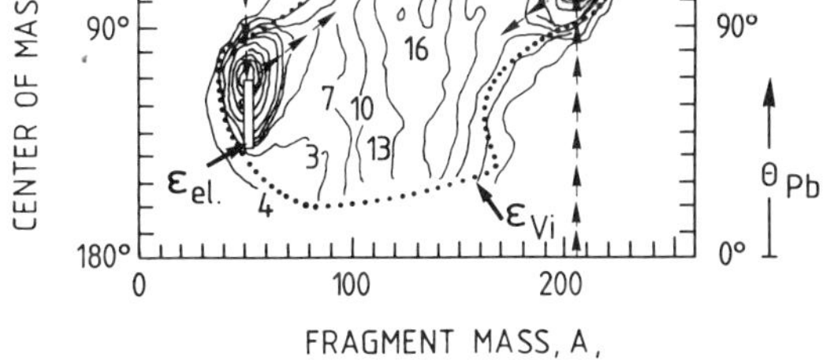

Figure 99. Two-contour lines of the double-differential cross section $d^2\sigma/dA\,d\theta$ vs cm reaction angle θ and mass number A for the reaction $^{50}Ti+^{208}Pb$ at $E_{Lab}=5.5$ MeV per nucleon. In the experiment, two-body reaction products were measured in coincidence. The dashed curves outline the region where the cross section pattern may appear distorted because of a low coincidence efficiency. From (Gob 81).

impressive example of the existence of two distinct classes of events. Those with mass asymmetries close to that of the initial projectile–target combination have an angular distribution that is peaked in the vicinity of the grazing angle, a feature characteristic of fast, peripheral interactions. The other class of more symmetric events has angular distributions which show little, if any, dependence on the angular distance from the grazing peak and correspond to long lifetimes of the system. Similar plots with respect to charge distributions have been given, e.g., by Galin *et al.* (Gal 77) for the reaction ^{197}Au + 220-MeV ^{40}Ar.

For the heavier systems, the memory of the initial mass asymmetry is preserved for the entire reaction cross section and shows up as a strong correlation between fragment mass and reaction angle. Figure 100 exhibits these correlations for the reaction ^{197}Au + ^{63}Cu at $E_{\text{Lab}} = 365$ MeV (Ngô 77), again in form of a contour plot of $d^2\sigma/dM\,d\theta$. The top part of this figure represents experimental results. Here, the peak cross section for the projectile-like fragments is located at an angle very close to the quarter-point angle of $\theta_{1/4} = 105°$ and at a mass a few units above the projectile mass. At a given angle, the mass distribution is roughly of Gaussian shape with an asymmetric tail extending towards symmetry (Ngô 76); with decreasing angle, the centroid of the mass distribution moves to more symmetric fragmentations. Simultaneously, the width increases. In a picture of statistical diffusion-like mass transfer mechanism (cf. Figure 33), such a dependence implies interaction times that increase in this direction.

Ngô *et al.* (Ngô 76) have made an analysis of the data shown in the top part of Figure 100 and of similar results for $E_{\text{Lab}} = 443$ MeV, that is inspired by such a mass diffusion picture. In Figure 101, experimental values for centroid, α, and width, Γ(FWHM), of the dominantly Gaussian fragment mass distributions deduced for various c.m. scattering angles are plotted as a $\alpha - \Gamma^2$ correlation. As can be seen, these correlations are linear, with a slope decreasing with increasing bombarding energy. This observation has been taken by Ngô *et al.* to be indicative of a nucleon diffusion process governed by a one-dimensional Fokker–Planck equation. According to Equation (2.8.1b), applying to such a simple statistical process, both average displacement α and variance σ^2 [or Γ^2(FWHM)] are linear functions of the diffusion time t, such that the ratio

$$[\alpha(t) - \alpha(0)]/[\Gamma^2(t) - \Gamma^2(0)] = (16\ln 2)^{-1} v_A/D_A \propto F_A/\tau \quad (3.4.5)$$

becomes independent of time. The proportionality on the right-hand side of this equation involving the driving force F_A and the nuclear temperature τ is invoked, because the validity of classical Einstein relations [cf. Equation (2.7.17)] is expected for such a mass diffusion process. Assuming that the

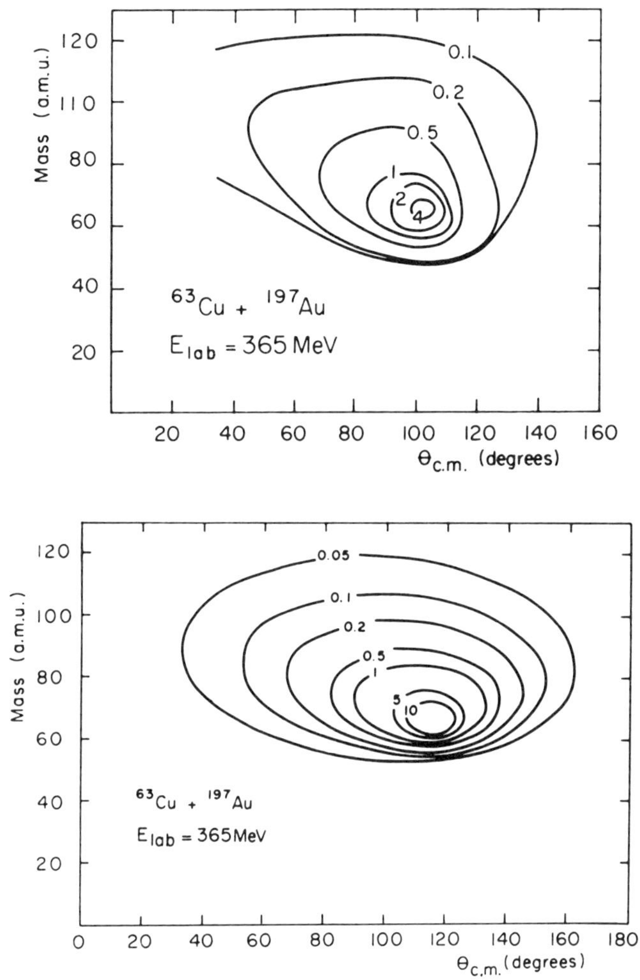

Figure 100. Contour plot of $d^2\sigma/dM\,d\theta_{\text{c.m.}}$ (mb rad^{-1} amu^{-1}) of projectile-like fragments from the reaction ^{197}Au + ^{63}Cu. Experimental results are presented at the top. The bottom part displays the corresponding theoretical distribution obtained by employing the linear-response theory. From (Ngô 77).

reaction angle is a monotonic function of the interaction time t, the linear correlations of the ^{197}Au + ^{63}Cu mass data depicted in Figure 101 may be interpreted (Ngô 77) as lending support to the existence of a classical diffusion process where the ratio F_A/τ is constant, although it is somewhat disturbing that the straight lines do not pass through the point ($\alpha = 63$, $\Gamma^2 = 0$).

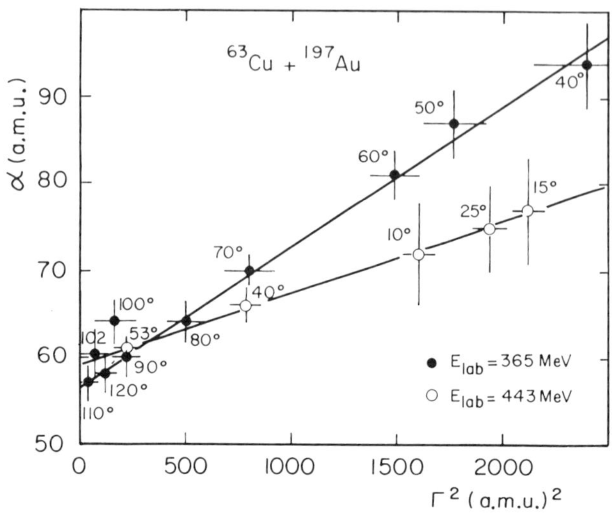

Figure 101. The position, α, of the maximum of the projectile-like mass distribution is plotted vs. the square of its width, Γ(FWHM), for fragments from the $^{197}Au + ^{63}Cu$ reaction at two bombarding energies. Solid points and circles represent data corresponding to the c.m. angles indicated. The straight lines are fits to the data. From (Ngô 76).

In order to assess the compatibility of the slopes of the data displayed in Figure 101 with Equation (3.4.5), one notices (Ngô 76) that excitation energies are on the average approximately $E^* = 20.5$ and 79.5 MeV for $E_{\text{Lab}} = 365$ and 443 MeV, respectively. These excitation energies correspond to temperatures of $\tau(365) = 0.8$ MeV and $\tau(443) = 1.6$ MeV, assuming a relation $\tau \approx (8E^*/A)^{1/2}$ between temperature and excitation energy of a system of mass A. Assuming for simplicity that the driving force F_A is similar for both bombarding energies, one expects a decrease by a factor of 2 of the slopes $d\alpha/d\Gamma^2$, when the bombarding energy is increased from 365 to 443 MeV. This is consistent with the experimental ratio of (2 ± 0.5) found in the experiments. In addition, the absolute magnitude of the correlation $\alpha(\Gamma^2)$ comes out as expected from such a model.

The above result supporting the presence of a classical mass diffusion process in the reaction $^{197}Au + ^{63}Cu$ is at variance with the more detailed analyses of other systems discussed previously. It is presently difficult to reconcile these various findings. However, it should be noted that the conclusions drawn from the analysis of the $^{197}Au + ^{63}Cu$ data described above rest not only on the assumption of a monotonic relation between scattering angle and reaction time but also on rather drastic simplifications that neglect the dependences of driving force and temperature on initial angular momentum and reaction history, whose effects may, in this particular case, cancel fortuitously. Later discussion will return to this point.

For the system $^{197}Au + ^{63}Cu$, Ngô and Hofmann (Ngô 77) have derived theoretical correlations between mass distribution and reaction angle, based on the linear-response theory outlined in Section 2.7. The quantity from which the corresponding double-differential cross section is deduced in this model is the distribution function d whose evolution is governed by a Fokker–Planck equation in the form of Equation (2.7.23). For simplicity, in this calculation the assumption is made that relative motion and mass exchange are statistically independent, i.e., that the solutions of the multidimensional Fokker–Planck Equation (2.7.23) are products of Gaussians. Only average correlations for the first moments of d are taken into account by a set of coupled differential equations describing their evolution. In addition, overdamped motion is assumed for the mass-asymmetry degree of freedom η. Then, the double-differential cross section is given by (Ngô 77)

$$\frac{d^2\sigma}{d\theta\, d\eta} = \frac{1}{2}\lambda\!\!\!^{-2} \int dl\, l \lim_{t\to\infty} \left\{ \Delta(t)^{-1/2} \exp\left[-\frac{[\theta-\theta_c(t)]^2 \chi_{\eta\eta} + [\eta-\eta_c(t)]^2 \chi_{\theta\theta}}{4\Delta(t)} \right] \right\} \tag{3.4.6}$$

Here, $\chi_{\eta\eta} = \sigma_\eta^2/2$ and $\chi_{\theta\theta} = \sigma_\theta^2/2$ are second moments of the distribution function with respect to mass asymmetry and scattering angle, respectively, which are coupled to the corresponding first moments η_c and θ_c via the fluctuation–dissipation relations [cf. Equation (2.7.19)] of response and correlation functions, as discussed in Section 2.7. In Equation (3.4.6), the quantity 4Δ is the product of the two variances.

Adopting the dynamical model developed by Gross (Gro 74, 78) for the relative motion and a step-function-like dependence of the adjustable friction coefficient for the mass motion, the model predicts the distribution displayed in the bottom part of Figure 100, which is to be compared with the experimental contour plot $d^2\sigma/d\theta\, dM$ shown at the top. As can be seen, the qualitative features of the data are roughly reproduced. Qualitatively, however, the model underestimates considerably the width of the mass distribution at a given angle and its further details. In addition, the predicted location of the maximum cross section differs from the observation by some 10°. Fits to the experimental angular distribution of the damped reaction cross section, obtained by integrating the double-differential cross section of Equation (3.4.6) over all mass asymmetries, are of corresponding quality. Linear relationships between centroid and variance of the mass distribution as a function of scattering angle (cf. Figure 101) are not predicted by the model.

Nevertheless, the approach is illustrative in that it models the spread of system trajectories away from the average classical path into several degrees of freedom, induced by the combined effect of quantal and thermal fluctuations. Remaining inadequacies in the description may be due to the dynamical model adopted for the relative motion, the neglect of correlations between mass transfer and reaction angle, as well as to the crude assumptions made about the friction and inertia coefficients for the motion along the mass asymmetry degree of freedom. In particular, the restriction to overdamped η motion makes it difficult to estimate the relative importance of quantal vs. statistical fluctuations in mass exchange processes, from comparisons such as shown in Figure 100.

As concluded in Section 3.3, correlations between fragment mass or charge division and reaction angle are, in general, not very pronounced and reflect only indirectly the strong dependence of the cross section features on the amount of kinetic energy dissipated in a reaction. Hence, a further discussion of the former relations will be omitted in favor of an evaluation to what extent fragment Z and A distributions are determined by the total kinetic energy loss.

An illustration of the general evolution of the reaction with energy loss, E_{Loss}, or final total kinetic energy, E, for a medium-weight system, is given in Figure 102 depicting the iso-contour lines of the double-differential cross section $d^2\sigma/dZ\,dE$ measured by Hoover *et al.* (Hoo 82) for the reaction $^{165}\text{Ho} + {}^{56}\text{Fe}$ at $E_{\text{Lab}} = 462$ MeV, corresponding to 3.8 MeV/nucleon above the interaction barrier. The data plotted in this figure vs. final total kinetic energy and charge of one of the fragments have been integrated over the measured angular range. It is observed that, starting from the projectile Z ($=26$), the cross section develops a ridge with a centroid in Z moving slightly towards smaller Z values as E decreases or E_{Loss} increases. In the same direction, its width in Z increases. These events comprise the damped reaction cross section. In addition, there is a cross section peak located around symmetric charge divisions ($Z_{\text{sym}} = 46.5$) which are, as before, attributed to the FL process. In the transitional region between these two features, it is not possible to disentangle the components. The dashed curve in Figure 102 represents the Coulomb energy V_c of two spherical nuclei touching at their strong-absorption radius. This energy varies by ~ 75 MeV through the domain of damped events and by more than 100 MeV over the total region displayed. Obviously, the Coulomb energy for touching spheres does not constitute a natural barrier for the development of the reaction features, since the cross section drops only gradually for final energies below V_c.

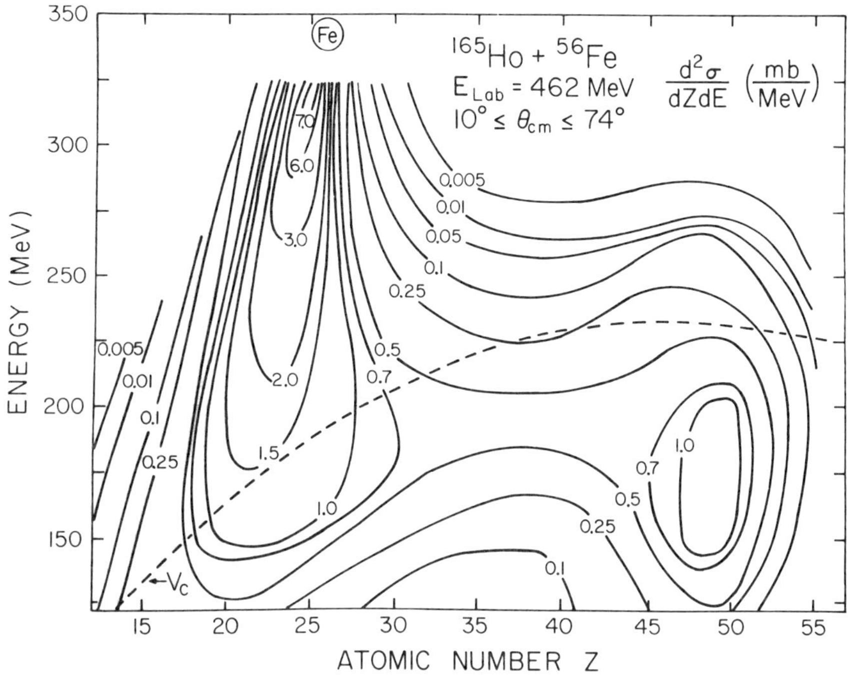

Figure 102. Iso-cross-section contour lines $d^2\sigma/dZdE$ for fragments from the reaction ^{165}Ho + ^{56}Fe at $E_{\text{Lab}} = 462$ MeV, plotted vs. final total kinetic energy in the center of mass frame and the fragment charge. The data have been integrated over the angular range indicated. V_c denotes the Coulomb energy for spherical nuclei touching at their strong-absorption radius. From (Hoo 82).

Similar data for the somewhat heavier system ^{166}Er + ^{86}Kr measured at $E_{\text{Lab}} = 8.18$ MeV/nucleon by Rudolf *et al.* (Rud 79) appear in Figure 103. Here, TKE stands for the final total kinetic energy, the quantity Q_{gg} (open circles) denotes the maximum ground-state Q value deduced for each measured fragment atomic number, with an associated mass asymmetry parameter given by the top scale. The thin solid curves in Figure 103 are drawn at equal intervals parallel to V_c. Below V_c, these curves correspond to the Coulomb repulsion of touching spheroids with equal axis ratios b/a as indicated. Only projectile-like fragments were detected in the experiment. Hence, the right half of the plot in Figure 103 is deduced assuming a binary process. As in the ^{165}Ho + ^{56}Fe reaction, also in the case of ^{166}Er + ^{86}Kr, the cross section ridge for projectile- or target-like fragments broadens with energy loss and favors more asymmetric fragmentations. The component of FL events, which is well defined in the ^{165}Ho + ^{56}Fe reaction, is noticeable in Figure 103 only as a distortion of the wings of the distributions towards

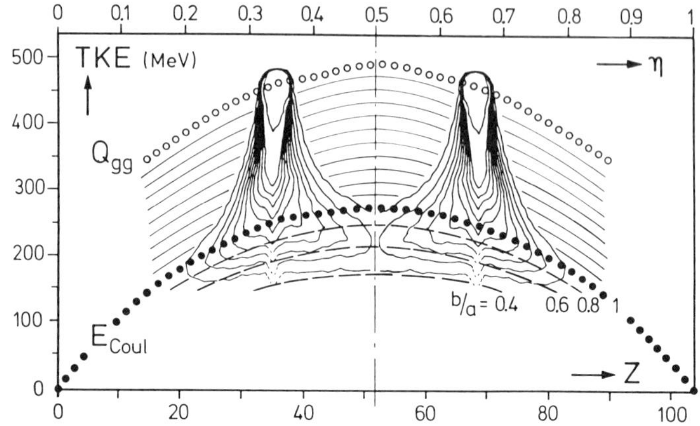

Figure 103. Contour diagram of the cross section $d^2\sigma/d(\text{TKE})\,dZ$ for projectile- and target-like fragments from the reaction $^{166}\text{Er}+{}^{86}\text{Kr}$ at $E_{\text{Lab}} = 8.18$ MeV per nucleon, plotted vs. final total kinetic energy (TKE) and fragment Z. The right half of the distribution is reconstructed from the measured data. The top scale denotes the mass asymmetry $\eta(Z)$ maximizing the ground-state Q value (Q_{gg}) for each Z value; V_c indicates the Coulomb repulsion of two spheres touching at $r = 1.2(A_1^{1/3} + A_2^{1/3}) + 2$ fm. Below V_c, dashed curves represent Coulomb energies of touching spheroids with various axis ratios b/a. The open circles marked Q_{gg} defined the reference from which to measure the ground-state Q value. From (Rud 79).

symmetry. Such a behavior is expected if a decomposition of the total Z distribution as suggested in Figure 95 is correct.

The evolution of the cross section ridge $d^2\sigma/dE\,dZ$ with increasing amount of dissipated kinetic energy for purely damped events is more clearly discernible for very heavy systems, where FL processes do not contribute to the reaction cross section. Such an example is given in Figure 104 exhibiting data reported by Wollersheim *et al.* (Wol 81) for the 1422-MeV $^{209}\text{Bi}+{}^{136}\text{Xe}$ reaction. It has already been noted from the integrated fragment Z distributions for this system (cf. Figure 96), that there is no symmetric FL component of significance observed at any of the three studied bombarding energies. As depicted in Figure 104, the cross section ridge $d^2\sigma/dE\,dZ$ for the $^{209}\text{Bi}+{}^{136}\text{Xe}$ reaction extends down in energy, almost parallel to the energy axis, broadening significantly only for final kinetic energies approaching or dropping below the Coulomb energy V_c.

It was first noticed (Sch 76, 76b, Hui 76c, Wol 76, 76b) for ^{136}Xe-induced reactions that cross sections of distributions such as depicted in Figure 104 cut at a fixed final energy or energy loss, have nearly a Gaussian shape. This is exemplified in Figure 105, where Z distributions of projectile-like fragments from the $^{209}\text{Bi}+{}^{136}\text{Xe}$ reaction at $E_{\text{Lab}} = 940$ MeV (Wil 80)

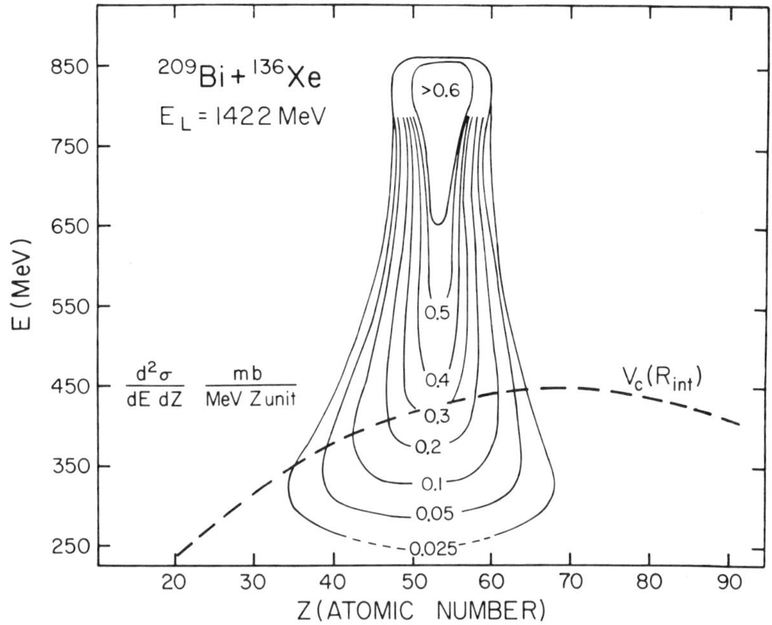

Figure 104. Double-differential cross section $d^2\sigma/dE\,dZ$ of projectile-like fragments from the reaction ^{209}Bi + ^{136}Xe at $E_{\text{Lab}} = 1422$ MeV plotted vs. final total kinetic energy and fragment Z. The dashed curve indicates the Coulomb repulsion of two spheres touching at their interaction radius. From (Wol 81).

are plotted for a succession of 26-MeV-wide bins in total kinetic energy, with centroids as indicated at each distribution. These are angle-integrated data. It is rather striking how well the experimental spectra are described by the Gaussian fit curves drawn through the data points. Some contamination of the data by fragments (FF) from sequential fission of the heavy, Bi-like reaction partner is present on the low-Z side of the distributions.

In agreement with similar data taken for this system at higher bombarding energies, it is found that the centroids of the distributions in Figure 105 stay close to the charge of Xe and drift only very slowly, by not more than 3 Z units, towards symmetry, as the energy loss increases. At the same time, one observes a rather dramatic broadening of the distributions. This latter development is, indeed, very suggestive of a classical diffusion process in which nucleons are randomly exchanged between projectile and target nuclei, leading to increasingly wider distributions as the interaction time proceeds. That such a view is deceptive also in the case of a very heavy system such as ^{209}Bi + ^{136}Xe, is revealed upon closer scrutiny of the data with respect to the relation between average displacement ΔZ and variance

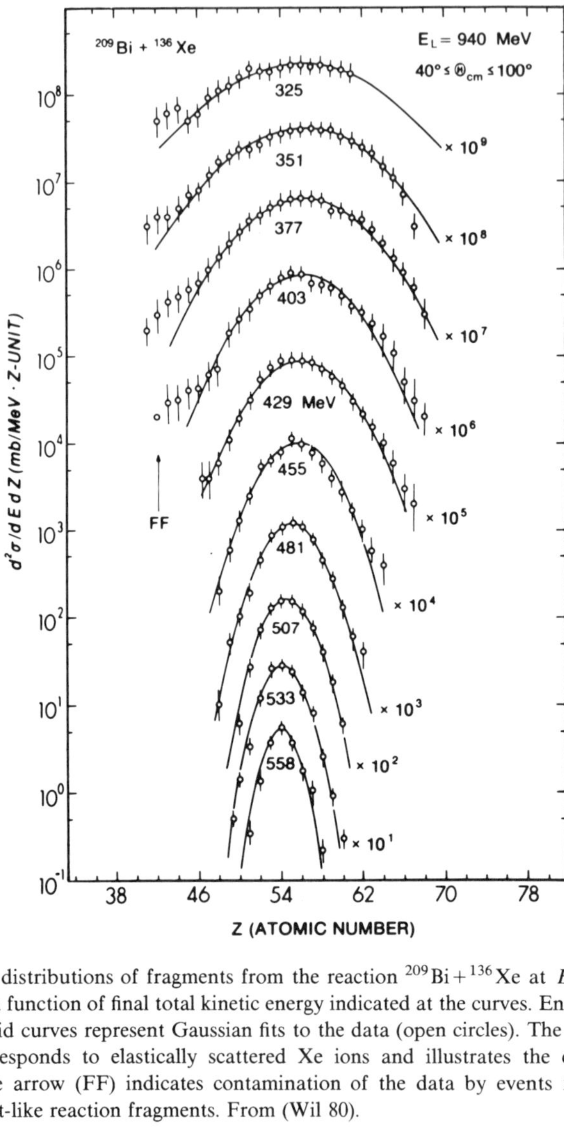

Figure 105. Z distributions of fragments from the reaction $^{209}\mathrm{Bi} + {}^{136}\mathrm{Xe}$ at $E_{\mathrm{Lab}} = 940$ MeV are plotted as a function of final total kinetic energy indicated at the curves. Energy bins are 26 MeV wide. Solid curves represent Gaussian fits to the data (open circles). The distribution for 558 MeV corresponds to elastically scattered Xe ions and illustrates the experimental Z resolution. The arrow (FF) indicates contamination of the data by events from sequential fission of target-like reaction fragments. From (Wil 80).

σ_Z^2 of the charge distributions. Although the ratio $\Delta Z/\sigma_Z^2$ decreases with increasing energy loss and temperature for $E_{\mathrm{Loss}} \gtrsim 80$ MeV, as required by the Einstein relations [cf. Equations (2.7.17) and (3.4.5)], this variation is found to be typically an order of magnitude weaker than expected. This behavior appears to be a consistent pattern associated with mass and charge transfer processes in damped reactions.

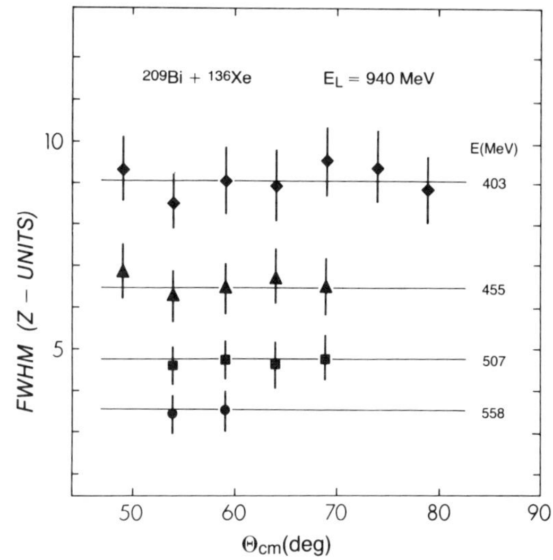

Figure 106. The FWHM of Z distributions $d^3\sigma/d\Omega\, dE dZ$ for the indicated final kinetic energies is plotted vs. c.m. reaction angle, for projectile-like fragments from the reaction $^{209}\mathrm{Bi} + {}^{136}\mathrm{Xe}$ at $E_{\mathrm{Lab}} =$ 940 MeV. The horizontal lines represent the fits to the angle-integrated Z distributions $d^2\sigma/dZ dE$. From (Wil 80).

It has already been concluded in Section 3.3 that the degree to which kinetic energy is damped is an important parameter determining the shape of the angular distribution of the differential cross section. No significant additional influence of the asymmetry of the fragmentation on the angular distribution has been discovered. Further differentiation of the Z distributions shown in Figure 105 with respect to reaction angle displays this lack of correlation in a complementary way. In Figure 106, the independence of the width Γ_Z(FWHM) of the fragment Z distribution on the reaction angle, once a final kinetic energy or energy loss has been chosen, is evident. It is, hence, permitted to write the damped reaction cross section in a separated form, i.e.,

$$\frac{d^4\sigma(E_0, E, \theta, A, Z)}{d\Omega\, dE_{\mathrm{Loss}}\, dA\, dZ} \approx \frac{d^2\sigma(E, E_{\mathrm{Loss}}, \theta)}{d\Omega\, dE_{\mathrm{Loss}}} \cdot \frac{d^2\sigma(E_0, E_{\mathrm{Loss}}, A, Z)}{dA\, dZ} \tag{3.4.7}$$

with

$$E_{\mathrm{Loss}} = -Q = E_0 - E \tag{3.4.8}$$

Here, E_0 and E are the initial and final asymptotic total kinetic energies, respectively, A and Z characterize the mass and charge fragmentation. It is then possible to study the average time evolution of Z and A distributions, using the relations between time, angular momentum, and energy loss derived, for example, from an analysis of the angular distribution of the differential reaction cross section, as discussed in Section 3.3. Whether or not the reaction angle θ can also be used as an indicator of the interaction time depends on the degree to which θ and E_{Loss} are correlated in a reaction.

For damped reactions involving somewhat lighter systems such as $^{165}\text{Ho}+{}^{56}\text{Fe}$ or $^{166}\text{Er}+{}^{86}\text{Kr}$, energy-differential fragment mass or charge distributions are expected to show the influence of the group of FL events centers around symmetry. This is demonstrated on the left of Figure 107 illustrating the evolution of Z distributions of fragments from the $^{165}\text{Ho}+{}^{56}\text{Fe}$ reaction at $E_{\text{Lab}}=462$ MeV (Hoo 82) with decreasing final kinetic energy, i.e., increasing interaction time. In contrast to the distributions shown in Figure 105, in the case of $^{165}\text{Ho}+{}^{56}\text{Fe}$, fragment Z spectra have non-Gaussian shapes, as can be seen from Figure 107. The deviation is emphasized by the solid curves in this figure, representing Gaussian fits to the data for Z values in the vicinity of that for the cross section maxima and below. Generally, the relative strength of the asymmetry grows with energy loss, as is expected from a comparison with Figure 102 depicting the same data in a different way.

So far, correlations have been discussed with reference to the final total kinetic energy, energy loss, or reaction Q value defined in Equation (3.4.8). Several authors (Mor 76b, Aga 77, Sve 78, Rud 79), however, propose to consider similar correlations with respect to the kinetic energy above the Coulomb barrier or that in excess of the sum of Coulomb and centrifugal energies. The objective of these studies is to identify a parameter which is more directly and unambiguously related to reaction time or l value than the quantity E_{Loss} describing the total amount of initially available kinetic energy transformed into intrinsic excitation, spin, or deformation energy of the fragments. For example, Rudolf *et al.* (Rud 79) employ the modified energy loss parameter

$$E^{*}_{\text{Loss}}=(E_0-V_0)-(E-V) \tag{3.4.9}$$

in order to classify Z distributions of fragments from the $^{166}\text{Er}+{}^{86}\text{Kr}$ reaction. Here, the potential energies V_0 and V are approximated by the Coulomb energies in the entrance channel and at the instant of breakup of the intermediate complex, respectively. A similar analysis performed (Hoo 82) for the $^{165}\text{Ho}+{}^{56}\text{Fe}$ reaction yields the fragment Z distributions depicted on the right of Figure 106. Here, each Z distribution comprises events with equal final kinetic energies above the Coulomb energy for spherical nuclei

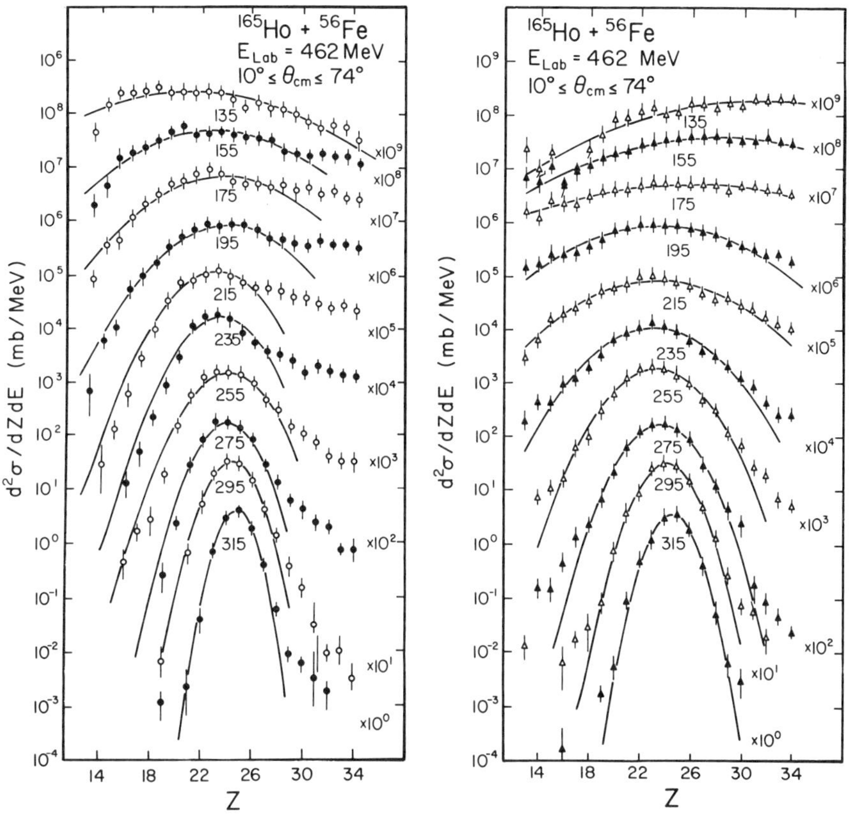

Figure 107. Z distributions of fragments from the reaction $^{165}Ho + ^{56}Fe$ plotted as a function of total kinetic energy. The distributions on the left are projections of data corresponding to 20-MeV-wide bins in total kinetic energy, whose centers are included on the curves. The distributions on the right are obtained for similar bins of equal kinetic energy above the respective Coulomb energies of spherical fragments touching at the strong-absorption radius. The energy values given on the curves correspond in this case only to the initial Z fragmentation. Solid curves drawn through the data points are the results of Gaussian fits. From (Hoo 82).

touching at the strong absorption radius. That is, events in bins defined by curves parallel to the one marked "V_c" in Figure 102 are grouped together.

It is quite remarkable how well Gaussian fit curves included on the right of Figure 107 describe the fragment charge spectra binned with respect to the modified energy loss parameter E^*_{Loss}. Asymmetric high-Z tails featured by the Z distributions on the left of this figure have almost completely disappeared. An analogous observation has been reported by Rudolf *et al.* (Rud 79). It is a consequence of these features that deduced experimental parameters, such as mean displacement ΔZ and variance σ_Z^2, characterizing a Z distribution can depend drastically on the assumptions made in the analysis, especially for intermediate and high energy losses, where systematical uncertainties in deduced σ_Z^2 values may amount to 100%.

Despite the more Gaussian-like appearance of fragment Z distributions resulting from binning with respect to the modified energy loss parameter, a convincing justification for such an analytic procedure remains to be given. Parameters such as E_{Loss}^* defined in Equation (3.4.9), where V may include the Coulomb repulsion of spherical or deformed fragments, the sum of Coulomb and centrifugal potentials, and the binding energy of the system, are believed to provide, although in a model-dependent way, a measure of the most probable intrinsic excitation energy of the dinuclear system at the instance of breakup. A second step in the chain of arguments is to assume that the reaction will proceed in the direction of maximum attainable intrinsic heat energy, as determined by the minimum potential energy. Hence, on these grounds, the magnitude of E_{Loss}^* could be expected to be indicative of the reaction time, which may then be related to an initial angular momentum, as explained in Section 3.3. However, such a picture is equivalent to reducing the damped reaction mechanism to an equilibrium-statistical process governed by thermodynamical entropy. Various examples given in this section appear to contradict such a concept. It is clear from the discussion in Section 2.6 that kinetic energy of relative motion can be converted, at least transiently, by various modes of initial excitation and into different forms of excitation energy. Energy transformation processes due to induced forces or doorway excitation of collective modes (cf. Section 2.6) are not expected to proceed on a path to maximum entropy, i.e., comply with the laws of equilibrium-statistical theory. Experimental data, for example those shown in Figures 102–104, provide no hint to a dominant influence of a potential barrier such as the Coulomb barrier V_c which may vary by up to ~ 100 MeV over the region of interest.

In contrast to the quantity E_{Loss}^*, the parameter E_{Loss} defined in Equation (3.4.8) comprises the effects of all energy transformation processes. It is a parameter that can be deduced from experiment in a model-independent fashion. To the extent, however, to which the relative contributions of these mechanisms and their time scales differ or change either with l or in the course of a reaction, E_{Loss} may not be a unique indicator of this time either, and a procedure such as described in Section 3.3 may be quite inaccurate. Due to the current lack of a more complete understanding of dissipation and equilibration processes, it is presently not possible to settle this controversial problem. With these reservations in mind, the following discussion returns to the correlations between energy dissipation and mass exchange in damped reactions.

Results of Gaussian fits to fragment Z distributions for the reaction $^{209}Bi + ^{136}Xe$ at 940, 1130, and 1422 MeV (Sch 78g, Wil 80, Wol 81, Hui 81b) are plotted in Figure 108 in the form of a correlation between variance σ_Z^2 and E_{Loss} [cf. Equation (3.4.8)]. The variation of σ_Z^2 with E_{Loss} observed in this figure is typical for all systems studied so far and is analogous to the behavior of fragment A distributions. For small variances, the rate of change in E_{Loss} for a unit increase in σ_Z^2 is rather large but decreases significantly for energy losses close to or in excess of the kinetic energy initially available above the interaction barrier. Although the correlations $E_{Loss}(\sigma_Z^2)$ are rather similar for small variances, the dynamic influence of the bombarding energy is clearly visible for larger variances or energy losses where the curves in Figure 108 separate from each other. Generally, a larger amount of kinetic energy has to be dissipated at a higher bombarding energy than at a lower one, in order to produce the same variance in Z or A.

The approximate universality of the functional relation between energy loss and σ_Z^2 is demonstrated by Figure 109 (Wol 82) displaying a superposi-

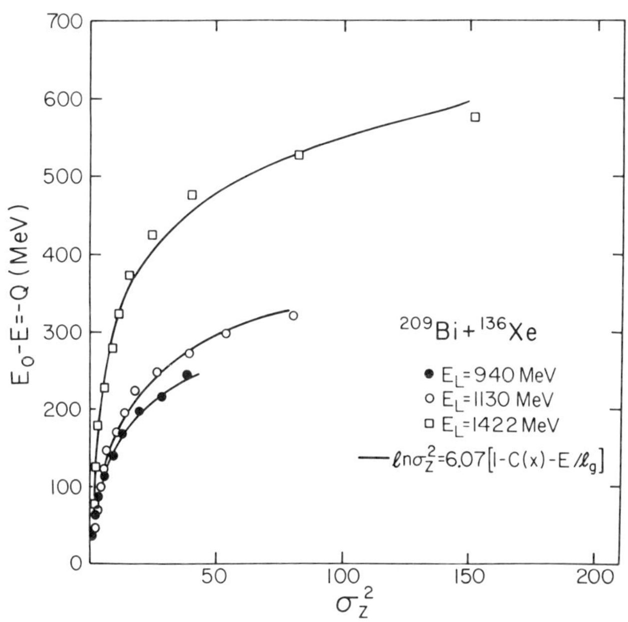

Figure 108. Correlation between variance σ_Z^2 of the Z distribution and total kinetic energy loss $E_{Loss} = -Q$ for projectile-like fragments from the reaction $^{209}Bi + ^{136}Xe$ at three laboratory bombarding energies E_L. The curves drawn through the data points represent fits discussed in the text. From (Hui 81b).

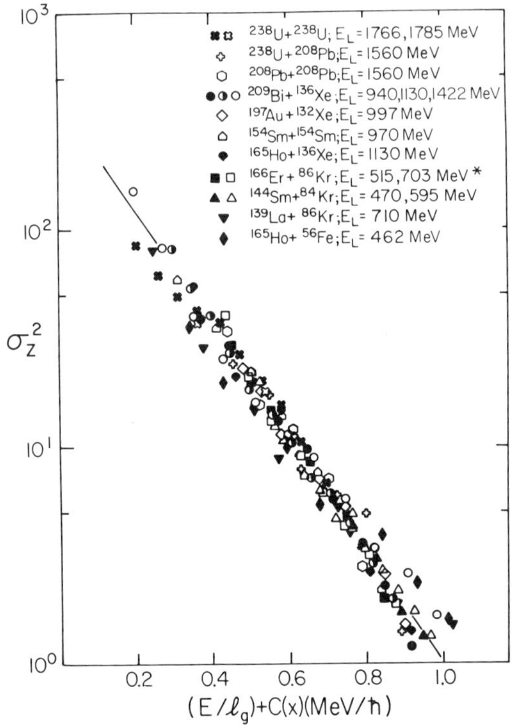

Figure 109. Dependence of the variance σ_Z^2 of the fragment Z distribution on the ratio E/l_g of final total kinetic energy and grazing-angular momentum, for a number of heavy-ion reactions. A system- and bombarding-energy-dependent constant $C(x)$ is added to E/l_g, in order to demonstrate the universal slope of the correlation. From (Wol 82).

tion of data for various systems, some of which have been measured at several bombarding energies. The logarithmic straight line drawn through the data points represents the average parameterization

$$\ln \sigma_Z^2 = -6.07\left[E/l_g + C(x) - 1\right] \tag{3.4.10a}$$

$$= -6.07\left[E/l_g - 1.393 E_0/l_g + 0.651\right] \tag{3.4.10b}$$

Here, E_0 and E are the initial and final total kinetic energies in units of MeV, respectively, l_g is the grazing-angular momentum in units of $\hbar$ to be identified with $l_{\max}$ in Equation (2.1.22). The quantity C in Equation (3.4.10a) depends on the projectile–target combination x and describes the value of E for which $\sigma_Z^2 = 1$.

The rigidity of the correlation depicted in Figure 109 is remarkable also with respect to the fact that data are included for reactions ranging in character from a strongly orbiting to a well-focused behavior of the angular distribution. Typical scattering of the data points around the average trend corresponds to roughly 10%, although larger deviations do occur. Wollersheim *et al.* (Wol 82) have traced the functional form of the correlation given by Equations (3.4.10) back to the nearly exponential relation between model interaction time and angular momentum illustrated in Figures 82 for the reaction $^{209}\mathrm{Bi}+{}^{136}\mathrm{Xe}$. An approximate independence of the average charge diffusion coefficient of the initial angular momentum or impact parameter and the absence of correlations between mass and charge transport is also required, in order to understand the universal relation between σ_Z^2 and E/l_g in terms of the phenomenological model analysis described in Section 3.3. Different ways to analyze the fragment Z and angular distributions and their dependence on energy loss are certainly conceivable, which may lead to different conclusions about the properties of the diffusion coefficient. Dynamical transport model calculations discussed below also give a qualitative account of the data, however, with transport coefficients that have a definite dependence on angular momentum, energy loss, and further details of the collision history. Nevertheless, the strong influence of both the incident kinetic energy and the energy loss on the fragment Z distributions demonstrates clearly and in a model-independent way a dynamical quality of the mass transfer mechanism operating in damped reactions, which is uncharacteristic of a classical diffusion process.

Various simple models have been devised in attempts to elucidate the microscopic origin of the mass transfer mechanism in damped nuclear interactions by interpreting the $E_{\mathrm{Loss}}(\sigma_Z^2)$ or $E_{\mathrm{Loss}}(\sigma_A^2)$ correlations. Despite the arguments given above against the presence of a simple classical mass diffusion process, the rather featureless appearance of fragment Z distributions exemplified in Figures 105 and 107 and their broadening with increasing energy loss or interaction time are very suggestive of a stochastic, though nonequilibrium, process.

Huizenga and collaborators (Hui 76,76c,77,77b,82, Sch 77d,77e,78g, Wil 80, Wol 81) have studied the interrelation between mass transfer and kinetic energy dissipation in damped reactions, in terms of classical phenomenological models for the stochastic exchange of independent nucleons between two interacting heavy ions.

Assuming a Fokker–Planck equation [cf. Equation (2.7.14)] to govern the exchange process, in the simple case of uncorrelated exchange of protons and nucleons, both fragment Z and A distributions are described in terms of Gaussians according to Equations (2.8.1) with first and second moments that increase linearly with reaction time t. It is further assumed that dissipation of kinetic energy T of relative motion occurs on the same time scale and is mediated by a classical friction force (cf. Section 2.3) characterized by a force constant k such that

$$-dT/dt = 2(k/\mu)T \tag{3.4.11}$$

where μ is the reduced mass. Using Equation (2.8.1b) in the form $d\sigma_Z^2/dt = 2D_Z$, one arrives (Hui 76) at the relation

$$\ln(T_0/T) = [k/(\mu D_{ZZ})]\sigma_Z^2 \tag{3.4.12}$$

under the condition that the ratio $k/\mu D_{ZZ}$ does not vary along a trajectory. Here, T_0 and T are the initial and final kinetic energies above the Coulomb energy $V_{\text{Coul}}(R_{\text{SA}})$, taken as a constant representative value of the potential energy that, in principle, varies along a collision trajectory.

Plotting experimental values of $\ln(T_0/T)$ vs. measured variances σ_Z^2 for energy losses not exceeding T_0, as done in Figure 110 for the $^{209}\text{Bi} + {}^{136}\text{Xe}$ reaction at three bombarding energies, results in correlations that are linear, with a slope $[k/(\mu D_{zz})]$ depending only on the bombarding energy. For a given incident energy, this quantity does not change with E_{Loss}, i.e., angular momentum.

Exploring the extent to which the tight correlations of mass and charge distributions with the amount of dissipated energy are intrinsic manifestations of a stochastic nucleon exchange process, one notices from the

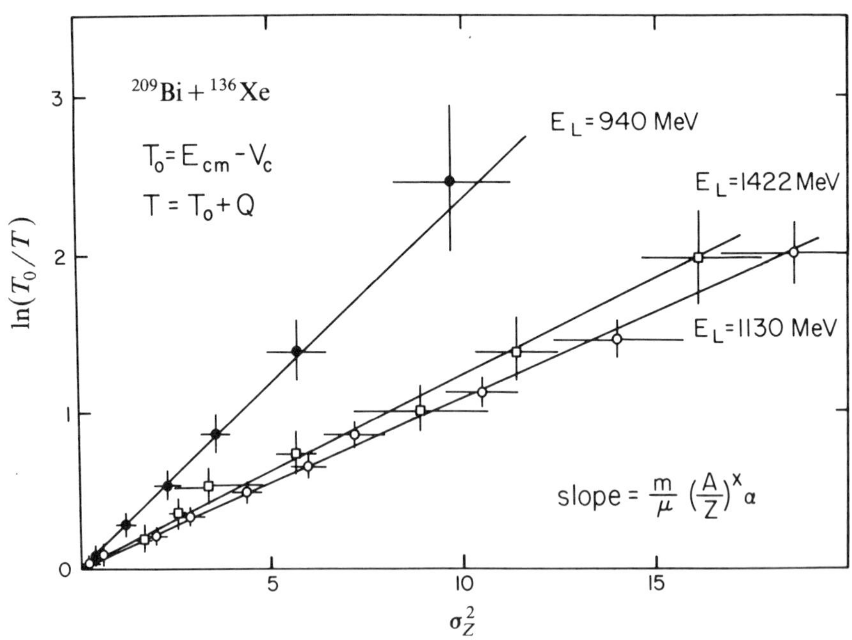

Figure 110. The relation between $\ln(T_0/T)$ and the fragment charge variance σ_Z^2 for the reaction $^{209}\text{Bi} + {}^{136}\text{Xe}$ at three laboratory bombarding energies E_L. T and T_0 are the initial and final kinetic energies above the Coulomb barrier, respectively. The straight lines are fits to the data. From (Wol 81).

discussion in Section 2.6 that such an exchange gives rise to a classical friction force. In a simple classical model for the transfer of nucleons of mass m between moving nuclei, it is observed (Sch 77d) that, in fact, the mass diffusion coefficient $D_{AA} = j/2$ is proportional to the friction force coefficient $k_{ex} = mj/2$. Here, $j = dN_{ex}/dt$ is the nucleon exchange rate and $N_{ex} = \sigma_A^2$. Such a classical model, neglecting the nucleonic Fermi motion, obviously, predicts a linear relationship similar to that given by Equation (3.4.12), but now between $\ln(T_0/T)$ and σ_A^2.

From numerical comparisons with data such as shown in Figure 108, however, it turns out (Sch 78,80, Dak 79,80, Dye 80) that the number of exchanged nucleons deduced from either the mass or charge variance is, in general, too small to account for the observed energy loss simply in terms of the recoil imposed on the fragments by a transferred nucleon. Hence, one arrives at a parametrization

$$E_{\text{Loss}}(\sigma_Z^2) = (E_{\text{c.m.}} - V_{\text{Coul}})\left\{1 - \exp\left[-\alpha \frac{m}{\mu}\left(\frac{A}{Z}\right)_{\text{cs}}^x \sigma_Z^2\right]\right\} \tag{3.4.13}$$

which is equivalent to inserting friction and diffusion coefficients describing the stochastic nucleon exchange process into Equation (3.4.12), with two important differences. First, the total number N_{ex} of nucleons exchanged has to be known to derive the total energy loss from the above model. In Equation (3.4.13), this number is parametrized in terms of the total mass and atomic numbers of the composite systems, A_{cs} and Z_{cs}, respectively, and an exponent x describing the strengths of correlations [cf. Equation (2.8.6)] present in the fragment (A, Z) probability distribution. Secondly, a scaling parameter α has been introduced denoting the ratio of the coefficient k of the total friction force to that due to the exchange process, k_{ex}.

Experimental correlations $E_{\text{Loss}}(\sigma_Z^2)$ for a given bombarding energy are well described (Sch 78,80, Dye 80) by the functional form of Equation (3.4.13) with a single value of $\alpha(A/Z)^x$, up to energy losses close to the initially available kinetic energy T_0 above the barrier. However, in the more numerous cases where only σ_Z^2 is known experimentally, it is impossible to deduce both α and x from a fit to experimental data. Assuming a fully correlated fragment (A, Z) distribution and a uniform charge density for the system, i.e., $x = 2$, one obtains (Sch 80) fit values for α ranging from 6 to 7, for bombarding energies of the order of 1–2 MeV per nucleon above the barrier, with α approaching unity for energies in excess of ~ 4 MeV per nucleon. Furthermore, α is found to be somewhat system dependent.

Such a bombarding energy dependence of α is expected (Sch 80) for a stochastic exchange of nucleons with Fermi motion, when account is taken of the Pauli blocking effect in each transfer. This basic mechanism underlies the dynamical transport model of Randrup (Ran 78,78d,79d) discussed in

Section 2.8. It has in a simpler form also been discussed by Wilcke *et al.* (Wil 80). Combining Equations (2.8.20b) and (2.6.28) under the assumption of negligible correlations between fragment A and Z distributions, one predicts (Sch 80) the approximate relation

$$\alpha = (\mu/mT)\, dE_{\text{Loss}}/d\sigma_A^2 \approx T_F/(2\tau^*) \tag{3.4.14}$$

Here, T_F is the Fermi kinetic energy and τ^* is the effective temperature parameter discussed in the context of Equations (2.8.19). The strong dependence of τ^* on both the bombarding energy and the energy loss is illustrated in Figure 37. In particular, this latter dependence renders the conclusion drawn earlier from the successful description of fragment Z and A distributions with Equation (3.4.13) and an E_{Loss}-independent value of α unjustified, due to a fortuitous cancellation of the effect of several unrealistic assumptions in the formulation of the simple nucleon exchange model.

Following the general framework of Randrup's one-body transport model, a somewhat more satisfactory approximation of the nucleon exchange process results (Hui 82) in a relation

$$E^{1/2} = E_0^{1/2} - \frac{3\pi}{16}\left(\frac{m}{\mu}T_F\right)^{1/2}\left(\frac{\sigma_N^2+\sigma_Z^2}{\sigma^2}\right)\left\langle\frac{\Psi(\zeta)}{\chi(\zeta)}\right\rangle\sigma^2 \tag{3.4.15}$$

involving the initial and final kinetic energies above the barrier, E_0 and E, respectively. Equation (3.4.15) is valid for peripheral collisions and for either $\sigma^2 = \sigma_A^2$ or $\sigma^2 = \sigma_Z^2$. In this equation, the quantity in brackets denotes an average over a collision trajectory of the ratio of form factors of dissipation function and mass diffusion coefficient Ψ and χ, respectively (cf. Section 2.6). The number of nucleons exchanged is approximated by the sum of neutron and proton variances, σ_N^2 and σ_Z^2, respectively.

On grounds of the above formulation, one expects a linear dependence of σ_Z^2 or σ_A^2 on the square-root of the final total kinetic energy E. In Figure 111, data for the ^{209}Bi + ^{136}Xe reaction are plotted in a way to make such correlations obvious. Indeed, the data follow roughly the predictions of Equation (3.4.15), i.e., form almost parallel straight lines of constant slope. The solid lines in Figure 111 are obtained (Hui 82) by assuming constant values $(\sigma_N^2+\sigma_Z^2)/\sigma_Z^2 = 3$ and $\langle\Psi/\chi\rangle = 0.8$. It is pointed out by De and Kapoor (De 83), taking a similar approach, that relations such as $E_{\text{Loss}}(\sigma_Z^2)$ or $E^{1/2}(\sigma_Z^2)$ are independent of the degree of average correlations associated with the exchange of protons and nucleons. Under such conditions, the functional dependence of the measured relationship between final energy and variance depicted in Figure 111 lends strong support to the presence of a stochastic mass transport mechanism based on the exchange of independent nucleons in damped reactions, whose most important properties are successfully modeled in Randrup's theory.

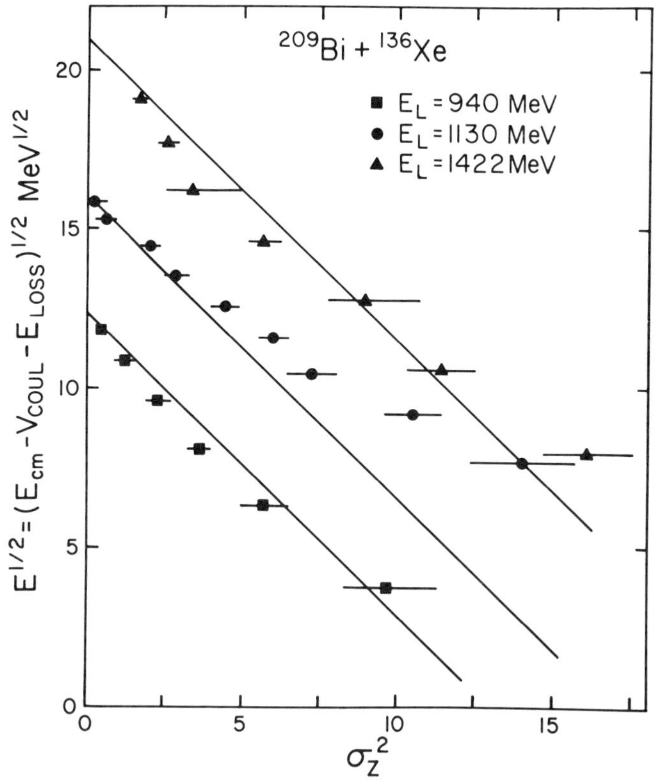

Figure 111. Square-root of the final total kinetic energy E plotted vs. the variance σ_Z^2 of Z distributions of fragments from the reaction $^{209}Bi + ^{136}Xe$ at the indicated bombarding energies. Solid symbols are data points. The parallel straight lines are predicted by Equation (3.4.15) with $(\sigma_N^2 + \sigma_Z^2)/\sigma_Z^2 = 3$ and $\langle \Psi/\chi \rangle = 0.8$. From (Hui 82).

Comparison with more detailed dynamical transport model calculations, where transport coefficients given by Equations (2.8.18)–(2.8.20) are integrated along classical trajectories that are determined by the interaction potentials of Equations (2.5.8) and (2.5.26) and the one-body dissipation function [cf. Equations (2.6.30)–(2.6.33)], suggest, however, that the above conclusions derived from a simplified adaptation of the model are not entirely correct. In the extensive transport calculations, macroscopic correlations (cf. Figure 35) of the joint fragment probability distribution $P(A, Z)$ are induced by the curvature of the underlying driving potential energy surface through the equations of motion (2.8.9) for the corresponding covariance tensor. These correlations are eliminated by a local linearization of the driving potential (uncorrelated case). Model predictions for the dependence $E_{\mathrm{Loss}}(\sigma_Z^2)$ are illustrated in Figure 112 for the system ^{144}Sm +

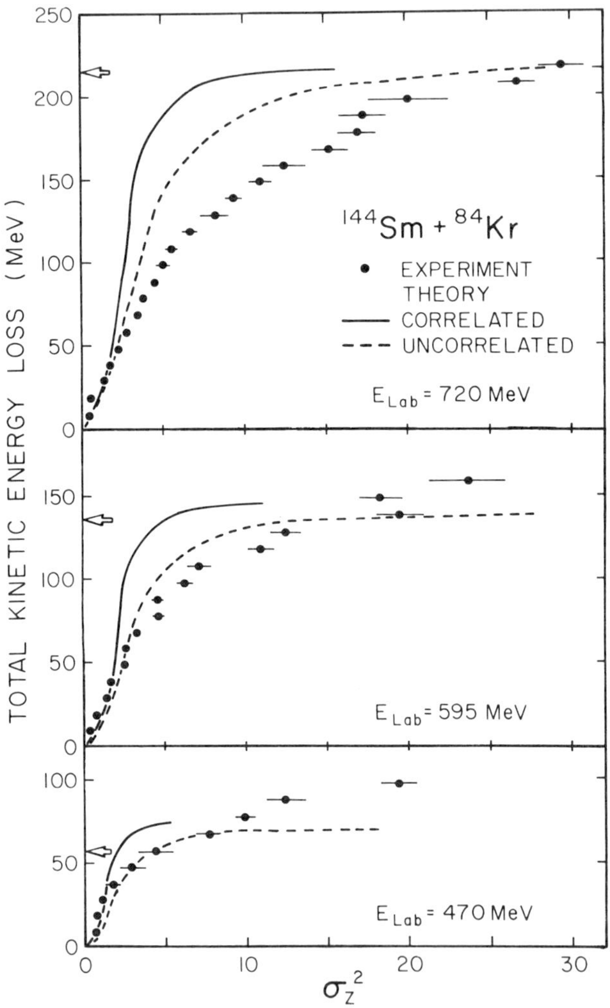

Figure 112. Total kinetic energy loss E_{Loss} plotted vs. variance σ_Z^2 of the charge distribution of projectile-like fragments from the reaction $^{144}\text{Sm} + {}^{84}\text{Kr}$ measured at three bombarding energies. Solid points are experimental data. The dashed and solid curves, respectively, represent model calculations neglecting or accounting for macroscopic correlations in the nucleon exchange mechanism induced by the driving potential. Arrows indicate the initially available total kinetic energy above the barrier, in each case. From (Mig 84).

^{84}Kr studied (Mig 84) at bombarding energies of 470, 595, and 720 MeV. Obviously, there are significant differences in the results of the calculations accounting for or neglecting the presence of such correlations, as can be seen from comparing the solid (correlations) to the dashed (no correlations) curves.

Overall, one observes a reasonably good reproduction of the data in Figure 112 by the calculations employing Randrup's transport model. In appreciating the agreement achieved between predictions and data, which is typical for heavy systems, one should remember that there are not adjustable model parameters. Nevertheless, a better description of the data is provided by the dashed curves in this figure, obtained by neglecting details of the calculated liquid-drop model driving potential [cf. Equation (2.5.30)] such as its curvature. However, as will be seen below, such details are important for a description of fragment A and Z distributions for somewhat lighter systems. In addition, the influence of the gradient of the potential energy surface on the mean displacements in A and Z of the fragments is somewhat overestimated by the model calculations. Partially, these discrepancies could be due to particle evaporation from the primary fragments, i.e., proton emission from the Kr-like reaction products in the case illustrated in Figure 112. However, it is a repeating pattern of the reaction model to overemphasize the bombarding-energy dependence of the relation $E_{\text{Loss}}(\sigma_Z^2)$. The dynamical properties of this transport theory, which distinguish it from a classical diffusion model, are mainly contained in the quantity τ^* [cf. Equations (2.8.19)–(2.8.21)]. They are found to be essential for even a qualitative reproduction of the experimental results. By contrast, a classical treatment neglecting, for example, the velocity-dependent Pauli blocking effect, results (Sch 80) in dramatically increased variances for a given energy loss, in gross violation of experimental data.

Although the presence of dynamical features in the mass exchange mechanism is suggested by the insensitivity of fragment A and Z distributions to the details of the driving potential, the intrinsic correlations imposed on the exchange of individual nucleons discussed above may be partially responsible for it. It has been briefly noted at the beginning of Section 2.8 that the familiar stochastic appearance of the probability distribution $P(x, t)$ is expected only for the limit of reaction times that are long compared to the relaxation or damping time characteristic of a given macroscopic degree of freedom x. That mass exchange between the partners in a damped reaction is associated with changes in shape and volume of the nuclei seems to be a reasonable assumption in view of the incompressibility of nuclear matter. On the other hand, phenomenological studies (Rie 79) discussed in Section 3.5 indicate that deformations belong to the most slowly relaxing degrees of freedom in damped reactions. Hence, a retarding effect of the comparatively slow evolution of deformations on mass exchange is conceivable.

Employing a classical model for dissipation and fluctuations induced by the exchange of independent nucleons, Beck *et al.* (Bec 78d) and Feldmeier (Fel 80) have studied the influence of slow volume relaxation on the correlations $E_{\text{Loss}}(\sigma_Z^2)$. In their random-walk model, no account is given of the effect of the driving potential or of the Pauli blocking on the exchange process. Under these assumptions, Beck *et al.* (Bec 78d) derive

$$\sigma_A^2(t) = \frac{A_1 A_2}{A_1 + A_2}\left\{1 - \exp\left[-\frac{2(A_1 + A_2)}{A_1 A_2}\int_0^t dt'\, I(t')\right]\right\} \tag{3.4.16}$$

for the equation of motion for the mass variance σ_A^2, involving the mass numbers A_1 and A_2 of the interacting (spherical) nuclei and the mean nucleon current I through the window between them. Employing the window-dissipation formula [cf. Equation (2.6.25)] and a combination of liquid-drop surface and proximity energies, the intrinsic exchange mode is coupled to the classical equations for the relative motion.

Results of these classical dynamical calculations (Fel 80) are compared to data for the $^{166}Er + ^{86}Kr$ reaction in Figure 113. A reasonable reproduction of experimental $E_{\text{Loss}}(\sigma_Z^2)$ correlations is achieved for not too large energy losses, which is somewhat worse at the lowest bombarding energy of $E_{\text{Lab}} = 516$ MeV. Although the model calculations suggest that slow volume

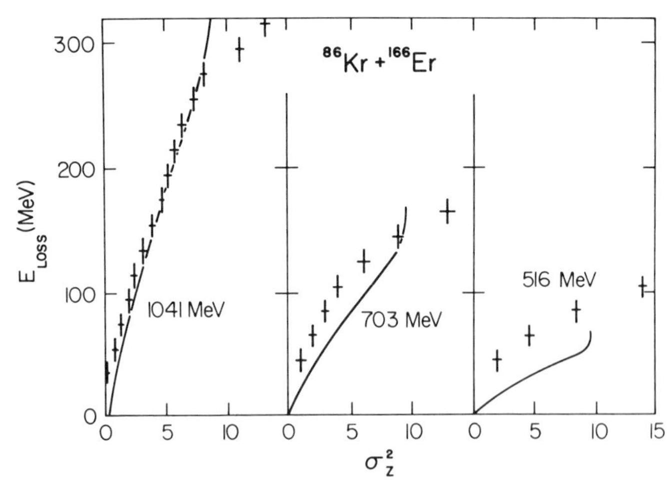

Figure 113. Energy loss E_{Loss} vs. variance σ_Z^2 for the reaction $^{166}Er + ^{86}Kr$ at $E_{\text{Lab}} = 516$, 703, and 1041 MeV. Experimental results are indicated by crosses. The curves represent dynamical calculations employing a random-walk-like mass exchange mechanism with volume conservation. From (Fel 80).

relaxation can produce a bombarding-energy dependence of correlations between energy loss and fragment A and Z distributions that is in agreement with experimental observations, the effect is probably unrealistically large in this model. Strict volume conservation of the individual nuclei implies an interaction potential with a repulsive core (see Section 2.5) for which there is no experimental support, as discussed in Section 3.3. However, it is conceivable that transient diabatic effects at the early stages of a reaction as proposed by Nörenberg (Nor 79, 79b, 79c) may induce a retardation of the mass flow that is successfully simulated in the above model.

Further details of the mass transfer mechanism operating in damped nuclear collisions can be explored by considering the relations between neutron and proton exchange. Fragment mass and charge distributions for very heavy systems at energies close to the interaction barrier have been studied by Rehm *et al.* (Reh 81), who bombarded targets of ^{94}Zr, ^{110}Pd, ^{148}Sm, and ^{170}Er with 1180- and 1280-MeV ^{208}Pb beams. Results for the interrelation between energy loss (TKEL) and mass or charge variances obtained by this group are depicted in Figure 114 at the top and bottom, respectively. These variances are deduced from Gaussian fits to energy-differentiated fragment A or Z distributions. They are not corrected for secondary particle evaporation from the primary fragments.

The bombarding energies at which the data in Figure 114 were taken correspond to only 20–80 MeV above the interaction barrier. It is, hence, remarkable to note the occurrence of events where several times this energy is dissipated in the reaction. In that respect, these systems are very different from those discussed earlier that were studied at much higher incident energies. Consequently, effects of slow volume or shape relaxation referred to above are expected to be more significant for the reactions illustrated in Figure 114.

From the data shown in Figure 114, one concludes that the variances for a given energy loss depend strongly on the system. This is especially true for the charge variances σ_Z^2. Contrary to what could have been expected, the smallest variances result from the heaviest systems $^{170}Er + ^{208}Pb$ or $^{148}Sm + ^{208}Pb$. While the mass variances for a given energy loss exhibit no significant variation with bombarding energy, the charge variances depend strongly on it. This latter dependence is opposite in character to that seen for higher-energy data, such as shown in Figure 108 for the $^{209}Bi + ^{136}Xe$ system. It appears (Reh 81) that with decreasing bombarding energy the total mass flux remains approximately unaltered but is redistributed among protons and neutrons in a way enhancing the relative importance of neutron exchange. It is a consequence of these experimental results that the energy loss per exchanged nucleon is independent of the bombarding energy for the systems and energies studied, provided σ_A^2 is a measure of the number of transferred nucleons. The neutron-rich flow has also been observed for other

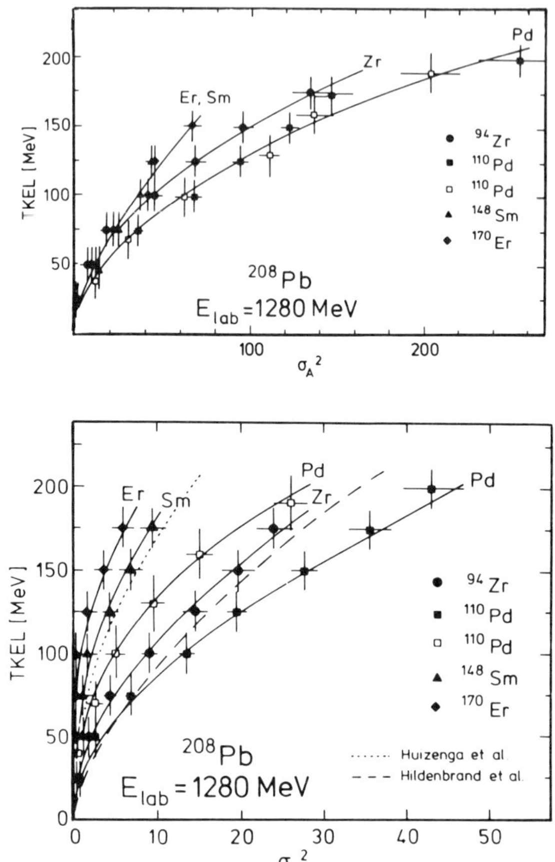

Figure 114. Total kinetic energy loss (TKEL) vs. variances σ_A^2 (top) and σ_Z^2 (bottom) of fragment A and Z distributions, respectively, from 1280-MeV ^{208}Pb-induced reactions on the targets indicated by solid symbols in the figures. Open squares represent additional measurements for the ^{110}Pd + ^{208}Pb reaction at 1180 MeV. Solid curves are drawn through the data to guide the eye. From (Reh 81).

heavy systems, e.g., by Essel *et al.* (Ess 79,79c) and Kratz *et al.* (Kra 81). Apparently, its effect is less pronounced at higher bombarding energies. Hence, the influence of the vicinity to the interaction barrier is suspected to be responsible for this phenomenon, at least to some extent.

Rehm *et al.* (Reh 81) have analyzed their low-energy data in terms of a Langevin equation (2.8.23), inferring interaction times from fitting a parametrized deflection function [cf. Equation (3.3.8) and Figure 66] to angular distributions as proposed by Riedel *et al.* (Rie 79). The effect of the driving force is neglected in this analysis, i.e., $F(A)=0$ in Equation (2.8.23). The solutions of this simplified equation are then characterized by variances that develop in time as (Reh 81)

$$\sigma_A^2(t)=\dot{A}_0^2 t_d^2(1-e^{-t/t_d})^2+2Dt-Dt_d(1-e^{-t/t_d})(3-e^{-t/t_d}) \tag{3.4.17}$$

where t_d is the damping time and D the diffusion coefficient, respectively. The quantity A_0 stands for the initial mass fragmentation. Fits to the data for the $^{170}\mathrm{Er}+^{208}\mathrm{Pb}$ reaction using this equation are illustrated in Figure 115. Plotted vs. interaction time t, the data show a steep rise at small interaction times, followed by a slower linear increase for larger times. Damping times of the order of $\sim 5\times10^{-22}$ s were deduced for the mass exchange process from these fits, assuming constant inertia and friction coefficients.

Although the functional dependence of the variances in Figure 115 is in accordance with what is expected from a simplified Langevin equation, it appears premature to conclude that inertia effects are definitely seen in these data. The dynamical transport calculations described earlier show also (Hui

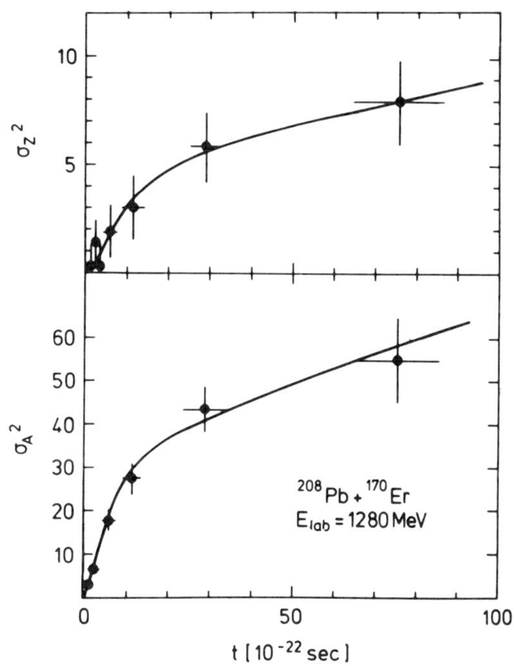

Figure 115. Variances for charge (σ_Z^2) and mass (σ_A^2) distributions of fragments from the reaction $^{170}\mathrm{Er}+^{208}\mathrm{Pb}$ at $E_{\mathrm{Lab}}=1280$ MeV plotted vs. model interaction time. Points are experimental data. The curves represent fits to the data using Equation (3.4.17). From (Reh 81).

82) an initially steep rise of the variances in time, followed by a much slower increase at later stages of the reaction. However, since such calculations are currently incapable of accommodating the rather large deformations produced in reactions close to the interaction barrier, a decisive comparison of model predictions has to await further study.

The interesting phenomenon of the preponderance of the exchange of neutrons over that of protons for incident energies close to the interaction barrier has been investigated by Brosa and Gross (Bro 80f) in terms of a static barrier penetration model. Figure 116 illustrates for the system $^{197}Au + ^{132}Xe$ the dependence on separation distance d of the barriers V_p and V_n resulting for protons and neutrons, respectively, from a superposition of the corresponding single-particle potentials. According to the simple parabolic-barrier formula, the transmission probabilities

$$P = \left\{1 + \exp\left[\frac{2\pi}{\hbar\omega}(V+S)\right]\right\}^{-1}, \qquad \omega = \left[\frac{1}{m}\frac{\partial^2(V_1+V_2)}{\partial r^2}\right]^{1/2} \tag{3.4.18}$$

are vastly different for neutrons and protons, at relatively large center separations d, e.g., at low incident energies. In Equation (3.4.18), S is the average particle separation energy of projectile and target. As can be seen from the figure, with decreasing separation, at first, the barrier for neutrons

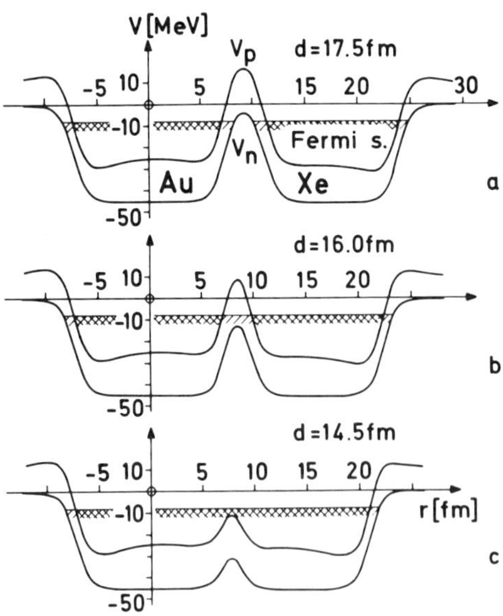

Figure 116. Superposition of proton and neutron single-particle potentials V_p and V_n, respectively, for a collision between ^{197}Au and ^{132}Xe nuclei, pictured at several center separation distances d. The radial distance r is measured from the center of the Au nucleus along the symmetry axis. From (Bro 80f).

drops below the Fermi surface indicated by the hatched band, followed by that for protons at smaller distances. For comparison, the interaction radius for this system amounts to about 15 fm.

Based on the same principle, Gross and Hartmann (Gro 81b) have calculated the exchange of independent nucleons between two Fermi gases in contact, governed by a two-dimensional Fokker–Planck equation. Transport coefficients were calculated taking into account the barrier penetration probability, the Pauli blocking effect, as well as the correlations between neutron and proton exchange induced, on the average by the driving potential. The latter is calculated from the liquid-drop model including shell effects that vanish with increasing temperature. The transport equations were integrated along trajectories defined by the friction model of Gross and Kalinowski (Gro 78). Although dynamical effects of the relative motion of projectile and target nuclei on the exchange of nucleons are included, no feedback of mass exchange on the classical trajectory is taken into account. As a particular feature of this trajectory model, it should be remembered that the friction form factor is of sizable magnitude already at relatively large separations (compare Figure 27) and, hence, very little overlap is allowed between projectile and target. Consequently, the influence of barrier penetrability should be significant. In contrast, it is expected to be of less importance for models employing a softer friction force such as the one-body proximity friction.

Results of the model calculations by Gross *et al.* (Gro 81b) are compared in Figure 117 with data for Xe-induced reactions on Fe and Sn targets at the relatively low bombarding energies of about $E_{\text{Lab}} = 800$ MeV, corresponding to c.m. energies of less than 100 MeV above the barrier. In the experiments (Sch 81h), the heavy Xe-like fragments are detected, and the variances σ_N^2 of the neutron number distribution displayed in Figure 117 relative to the charge variances σ_Z^2 are corrected for neutron evaporation from the primary fragments.

As can be seen from Figure 117, experimental ratios σ_N^2/σ_Z^2 can be as large as 10 and clearly reflect the neutron-richness of the system. For the Sn targets, the ratio σ_N^2/σ_Z^2 drops strongly with increasing excitation energy or energy loss, a property that is well reproduced by the model predictions. For the neutron-poor ^{56}Fe targets, the σ_N^2/σ_Z^2 values are generally smaller and do not show a drastic dependence on the excitation energy, which is also in accord with the model expectations. This system dependence of the correlations of σ_N^2/σ_Z^2 with excitation energy is attributed (Gro 81b) to the effect of the Coulomb potential which leads to smaller charge diffusion coefficients and, hence, smaller charge variances for heavier and more symmetric systems, while the neutron diffusion coefficients are unaffected by it. Although the model reproduces satisfactorily, for these low-energy reactions, the dependence of the ratio σ_N^2/σ_Z^2 on excitation energy, it systematically

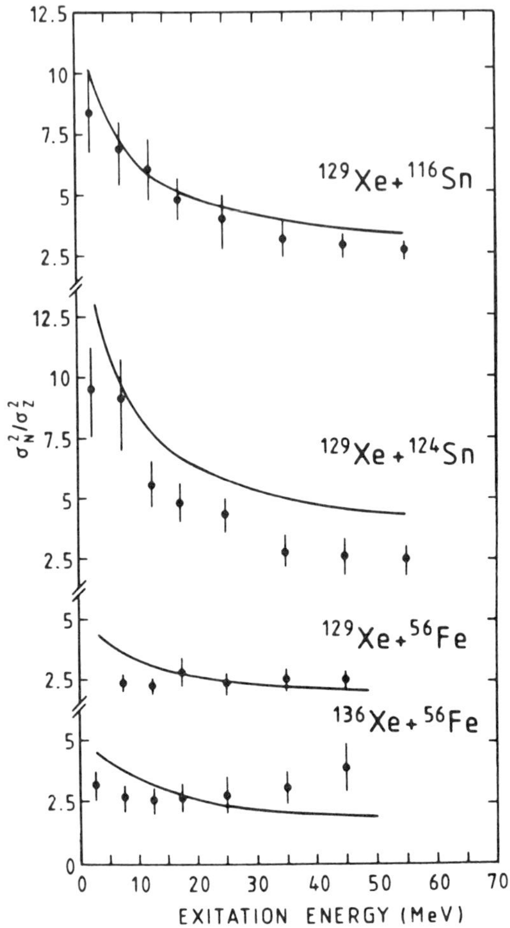

Figure 117. Ratio of neutron to proton variances, σ_N^2 and σ_Z^2, respectively, plotted vs. excitation energy for the reactions ^{56}Fe + ^{136}Xe at $E_{\text{Lab}} = 800$ MeV and ^{56}Fe, ^{116}Sn, ^{124}Sn + ^{129}Xe at $E_{\text{Lab}} = 780$ MeV. Solid points indicate data taken near the grazing angles. Curves represent model calculations by Gross *et al.* From (Gro 81b).

underestimates the magnitude of both σ_N^2 and σ_Z^2 for a given energy loss. Similarly, theoretical values of isotopic and isobaric variances, to be discussed below, turn out to be too small. This could also be interpreted as the result of an overestimated range and strength of the friction force which, in the model, is not directly related to the mass exchange process. From such an interpretation, one may expect discrepancies between calculations and experiments to be larger at increased bombarding energies. This is indeed found (Gro 81b) to be the case for reactions induced by 8.3-MeV per nucleon ^{56}Fe projectiles on targets ranging from ^{56}Fe to ^{238}U discussed further below. Consequently, one may also suspect that the influence of the barrier on the relation between neutron and proton exchange in damped reactions is exaggerated by the model of Gross and Hartmann. It is conceivable that the relative richness in neutrons of the nuclear surface, i.e., the "neutron skin" [cf. Equations (2.5.6) and (2.5.7)], suggested by the droplet model contributes significantly to a relative enhancement of neutron exchange processes in low-energy or grazing-like collisions. This contribution is considered in calculations (Sch 81b) employing Randrup's one-body transport model as exemplified below.

The neutron enhancement in the mass exchange processes, due to the effects referred to previously, should be important only for collisions with radial energies close to the effective barrier. For reactions at incident energies well above the barrier, they should, hence, contribute dominantly in the domain of relatively small energy losses, corresponding to the comparatively small interpenetrations reached in grazing-like collisions. For a further investigation of the statics and dynamics of mass exchange in damped reactions it appears then interesting to consider the correlations within the joint A and Z distributions of fragments produced in damped reactions, spanning a large range of energy losses and corresponding reaction times. The following discussion focuses particularly on the static and dynamic features of the charge density equilibration mechanism in damped reactions, which is a topic of high current interest. More generally, correlations between various moments of the probability distributions depicted schematically in Figure 35 and their variations with reaction conditions will receive attention.

The anticipation of further revelations about the mass exchange mechanism through studies of the correlations between A and Z distributions is based on the different influence of the underlying driving potential expected for the different components of the total mass flows between interacting nuclei. As an illustrative example, Figure 118 shows the values of the liquid-drop driving potential V represented in the top portion of Figure 22 along the lines of constant proton number Z_1 (solid curve), constant neutron number N_1 (dashed-dotted curve), and constant mass number A_1 (dashed curve) of one of the reaction partners, passing through the point ($N_1 = 30$, $Z_1 = 26$). The curves have been shifted in order to position their minima at the origin $\Delta Z_1 = 0$ or $\Delta A_1 = 0$ of the abscissa. Obviously, the steepnesses of the various potential cuts are rather different. The potential $V(Z_1,\ A_1 = \text{const})$ for constant mass asymmetry is the steepest, with a curvature parameter c [compare Equation (2.4.2)] almost an order of magnitude larger than that of the potential $V(Z_1 = \text{const},\ A_1)$ for constant charge asymmetry. This is understood to be due to the strong symmetry force of the liquid-drop model. According to the discussion in Section 2.7 [cf. Equations (2.7.20)–(2.7.22)], the relation between the collective frequency ω associated with the driving potential and the thermodynamic temperature τ of the dinuclear system is expected to determine to a large extent the synergetics of mass exchange in damped reactions. Although neither mass

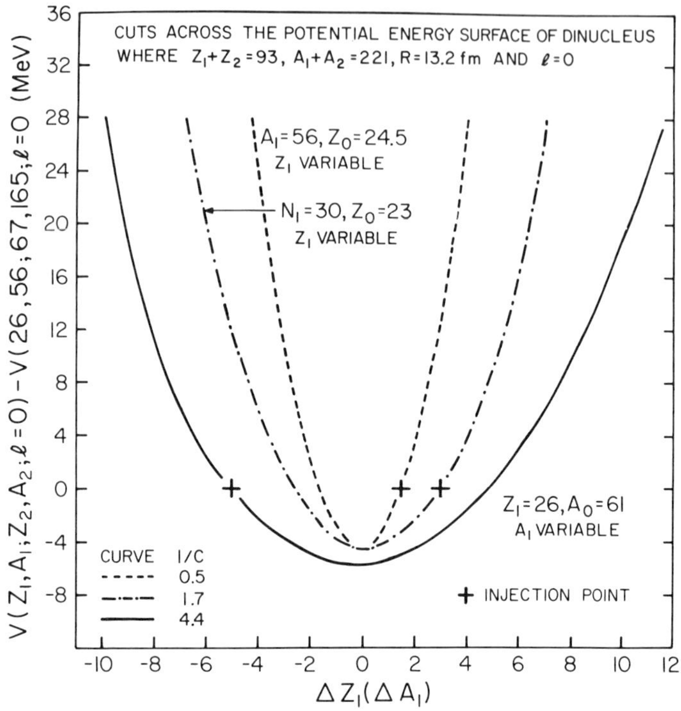

Figure 118. Cuts through the potential energy surface shown in Fig. 22 for $l = 0$ for fixed charge asymmetric (solid curve), fixed neutron number asymmetry (dashed-dotted curve), and fixed mass asymmetry (dashed curve) indicated by Z_1, N_1, and A_1 corresponding to one of the fragments. The abscissa measures the distance in Z_1 or A_1 from the minimum of each of the potential energy curves. The crosses mark the injection points corresponding to the fragmentation $^{165}Ho + ^{56}Fe$. From (Hui 81b).

nor charge asymmetry can be frozen by the choice of experimental parameters, nor can decoupled modes of mass and charge exchange be reasonably assumed to exist, one could hope to find a coexistence of collective (cf. Section 2.4) and stochastic (cf. Section 2.7) exchange modes with different weights for different degrees of freedom.

Required detailed information on the correlations between mass and charge exchange is available in the form of isobaric, isotopic, and isotonic fragment distributions. In a more average way, they may also be inferred from a comparison of moments of the total (A, Z) distributions. For reactions induced by light projectiles, isobaric and isotopic fragment distributions have extensively been studied by the Dubna group (e.g., Art 71, Vol 74, 74b, 75b, 75c, 76, 76b, 78). As an example, Figure 119 shows results ob-

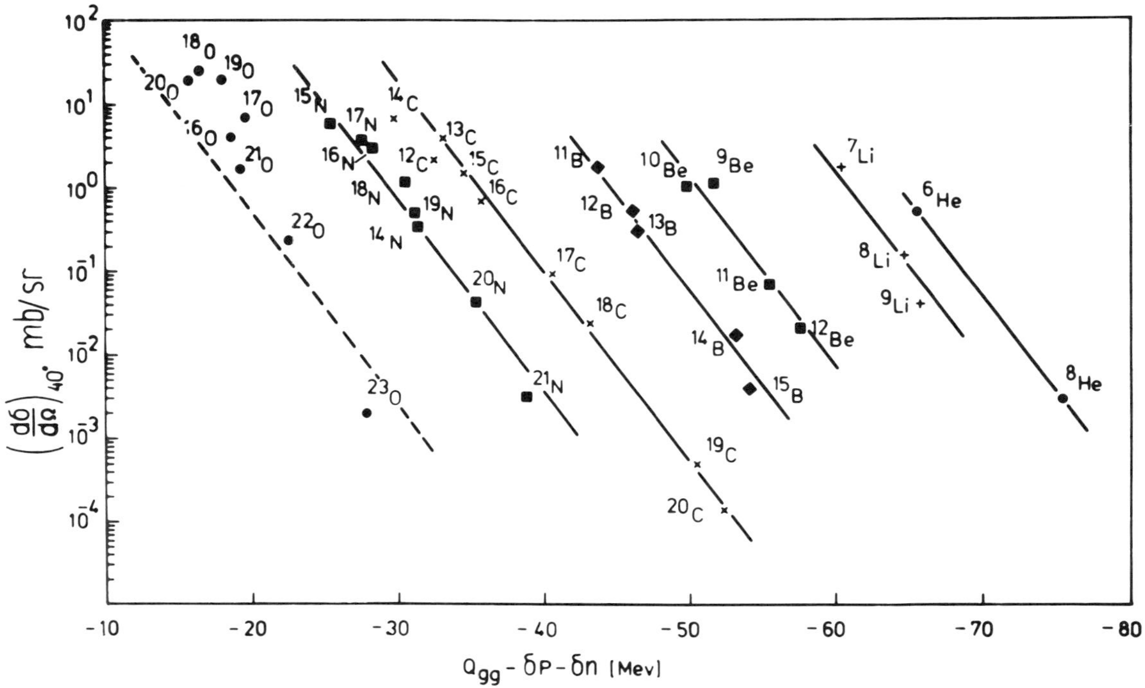

Figure 119. Isotopic distributions for different fragment elements from the reaction $^{232}Th + ^{22}Ne$ at $E_{Lab} = 172$ MeV and a detection angle of $\theta_{Lab} = 40°$. The differential cross section $d\sigma/d\Omega$ is plotted vs. the combination of ground-state Q value and pairing energies required to produce a given fragmentation. Logarithmic straight lines represent fits to the data and demonstrate a common slope in the data which is given by the inverse nuclear temperature. From (Vol 78).

tained (Vol 78) by this group for the reaction ^{232}Th+^{22}Ne at a laboratory energy of $E_{\text{Lab}} = 172$ MeV and a fixed detection angle of $\theta_{\text{Lab}} = 40°$. Here, the differential cross section is plotted for damped events vs. the combination of ground-state Q value Q_{gg} and the pairing energies required to produce the final fragmentations indicated by the light fragment. The overall appearance of the isotopic distributions displayed in Figure 119 is that of a uniform exponential decay of the cross section with increasing negative Q value.

Experiments such as illustrated in this figure have been interpreted (Vol 78) to be indicative of the temporary existence of a thermally equilibrated intermediate system, whose decay produces final fragmentations with probabilities P according to the Boltzmann factor [cf. Equation (2.8.4)], i.e., $P \sim \exp(-Q/\tau)$. The temperature parameters τ deduced from the inverse of the common slope are consistent with total dissipation of the initial kinetic energy of relative motion. The complete damping and the shape of the fragment angular distributions indicate relatively long interaction times, of the order of a few times 10^{-21} s (Vol 74b), characteristic for lighter systems. Such values may be regarded as upper limits for the time it takes to equilibrate the initial mass-to-charge asymmetry for a system where the projectile has an A/Z value of 2.2, while that for the target is 2.6. Here, the global equilibrium is defined in terms of an equilibrium-statistical population of fragmentations in accordance with the underlying potential energy surface.

A similar degree of global mass-to-charge equilibration is observed for the systems ^{64}Ni+^{40}Ca and ^{58}Ni+^{40}Ar studied by Gatty *et al.* (Gat 75), Galin *et al.* (Gal 76,76b), and Chiang *et al.* (Chi 79) at $E_{\text{Lab}} = 280$ MeV. These systems form complementary intermediate systems of similar global A/Z ratios, 2.17 and 2.13, respectively, while the initial A/Z asymmetries are rather different. Projectile A/Z ratios are 2.00 and 2.22 for Ca and Ar, respectively. In Figure 120, the isotopic yields of K fragments from the two reactions are compared. The yield is plotted vs. fragment mass and final laboratory energy in a two-dimensional contour diagram. The data have been taken at angles close to the grazing angle, where the high-energy quasielastic events are well separated from the broad distribution of strongly damped products at reduced final kinetic energies. Whereas the behavior of the quasielastic components indicates a preferential stripping or pick up of one proton in the reaction ^{64}Ni+^{40}Ca and ^{58}Ni+^{40}Ar, respectively, the damped component has rather similar mass distributions in both cases. In addition, it is found (Gal 76b) that the damped process yields such a mass distribution already before the initially available kinetic energy is completely damped. This feature has been taken (Gal 76,76b, Chi 79) to indicate a fast relaxation mechanism for the neutron excess mode. Relaxation times of the order of 10^{-22} s have been estimated by Galin *et al.* (Gal

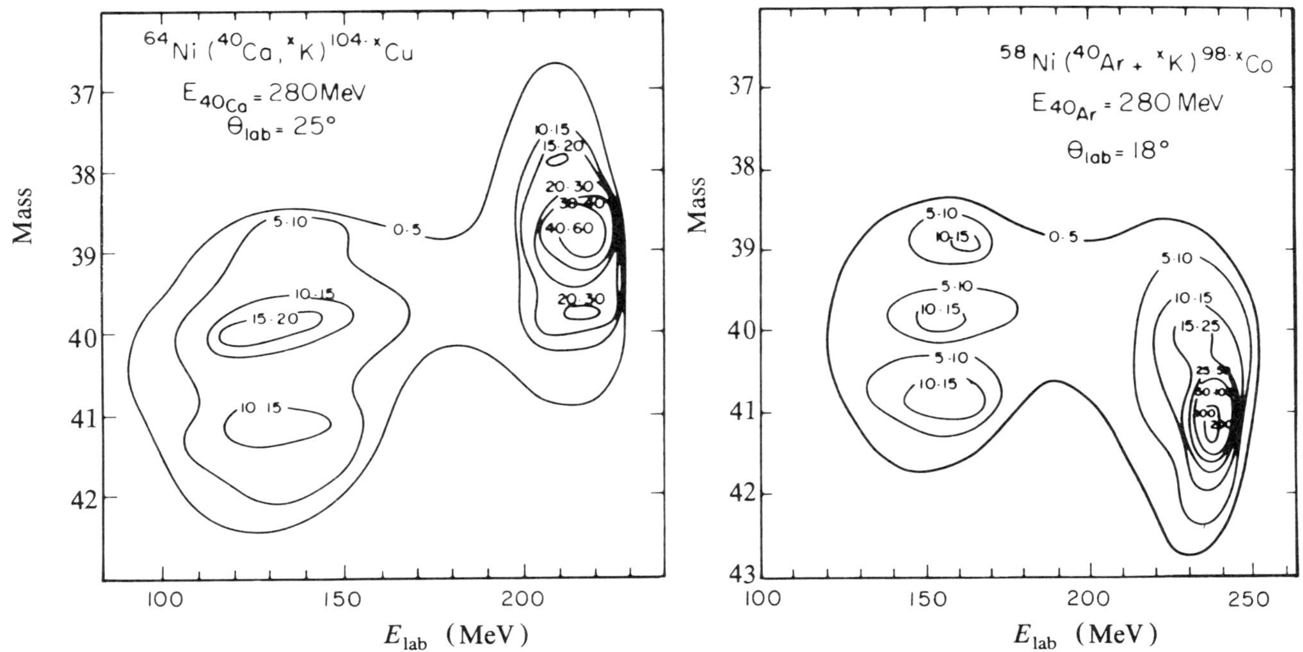

Figure 120. Contour diagrams of the yield of K isotopes produced in the reactions $^{58}Ni + ^{40}Ar$ (right) and $^{64}Ni + ^{40}Ca$ (left) at $E_{Lab} = 280$ MeV measured at angles close to the respective quarter-point angles. The quantity E_{Lab} denotes the final laboratory energy of the fragments. From (Gal 76).

76b), which are of the same order as the periods of giant dipole resonances. However, as one notices from the two quasielastic distributions in Figure 120, only the transfer of a single neutron is required to achieve A/Z equilibration in either case. Hence, one could argue that these systems are initially not far enough from equilibrium to allow definite conclusions to be drawn about the characteristic relaxation time.

Nevertheless, an equilibration of the A/Z asymmetry degree is observed in reactions induced by beams of ^{32}S (Chi 79), ^{40}Ar (Art 71b, Bim 74,75, Gat 75,75b, Gal 76,77, Bar 78d, Aug 79b, Chi 79, Riv 80, Gre 81), and ^{40}Ca (Gat 75b, Gal 76). In these reactions, the mass-to-charge asymmetries of damped events correlate in an obvious manner with the global $(A/Z)_{cs}$ ratio of the composite system. This is exemplified in Figure 121 for ^{40}Ar- and ^{40}Ca-induced reactions on ^{58}Ni and ^{64}Ni targets, corresponding to $(A/Z)_{cs}$ values between 1.04 and 1.26. In this figure, the ratio of the most probable neutron number $\overline{N}$, found for a fragment of given atomic number Z, to this Z value is plotted. Except for the small fragment Z values of $Z \leq 10$, the average neutron access is seen to be rather independent of the choice of the fragment, for each of the given systems. However, fragments from the neutron-rich system $^{64}Ni + ^{40}Ar$ exhibit systematically higher $\overline{N}/Z$ values than those from the neutron-poor system $^{58}Ni + ^{40}Ca$. This property of the damped reaction fragment is found (Gat 75b) to be angle independent. Of course, as pointed out at the beginning of this section, particle

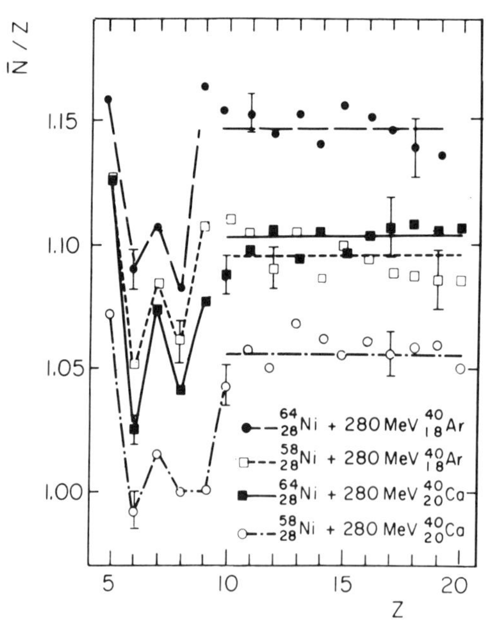

Figure 121. Variation of the most probable neutron excess ratio $\overline{N}/Z$ vs. atomic number Z for fragments from ^{40}Ar and ^{40}Ca induced reactions. Lines are drawn through the data to guide the eye. From (Gat 75b).

evaporation from the excited primary fragments can substantially alter the fragment distribution in A and Z, particularly for systems as light as those illustrated in Figure 121. Evaporation calculations for the four systems of Figure 121 have been performed by Gatty *et al.* (Gat 75,75b), assuming that, for any initial mass asymmetry, the charge fragmentation minimizes the liquid-drop model potential energy of two touching spheres. The evaporation process modeled in those calculations is found to reduce the initial differences in the $\overline{N}/Z$ ratios of fragments from the various reactions. Apart from the details of the odd–even structure present in the data, the calculations reproduce the experiments well for the ^{64}Ni$+^{40}$Ca and ^{58}Ni $+^{40}$Ar systems which have rather similar $(A/Z)_{cs}$ values. Larger discrepancies between experiments and predictions, assuming A/Z equilibration, occur for the other two reactions ^{64}Ni$+^{40}$Ar and ^{58}Ni$+^{40}$Ca in which theoretical equilibrium A/Z values of the fragments are somewhat further removed from the β-stable valley than is the case for the former pair of systems. However, at least qualitatively, the data are consistent with the assumption that charge density equilibrium is reached early in all these reactions.

Due to the drastic alterations of the fragment distributions introduced by the sequential evaporation process, it appears advantageous to study the progress of A/Z equilibration for heavier and neutron-rich systems, as well as for the heavier of the two reaction partners. Unfortunately, due to experimental difficulties in achieving better than unit A and Z resolution for heavy reaction fragments, only a few reactions in this category have been investigated in detail. Among these are reactions induced by projectiles of ^{56}Fe (Bre 79,80, Mig 80, Bri 82), ^{63,65}Cu (Tas 76, Chi 79), ^{86}Kr (Ber 79c, Gre 81), ^{98}Mo (Pet 82), ^{129,132,136}Xe (Kra 77,81, Poi 79, Sch 81h), and ^{144,154}Sm (Mer 81).

A number of reactions have been explored by Chiang *et al.* (Chi 79) with emphasis on the A/Z equilibration mode. Experimental results for the damped fragment isotopic distributions for the reaction ^{238}U$+^{32}$S at E_{Lab} $= 241$ MeV are displayed as solid points in Figure 122, for elements with $10 \leq Z \leq 14$. The relative yield $P(A,\ Z = \text{const})$ is plotted vs. detected fragment mass number A in this figure. It is observed that the most probable mass for each fragment Z decreases with Z. Widths (FWHM) of the isotopic distributions are of the order of 1–3 units. A simple diffusion model is employed to interpret the data, expressing the probability for producing a fragmentation (Z_1, A_1) as

$$P(Z_1, A) \propto P(A_1)\exp[-V(Z_1, A_1)/\tau] \tag{3.4.19}$$

Here, the probability $P(A_1)$ is assumed to be given by a solution of the one-dimensional Fokker–Planck equation [Equation (2.7.15)] for the time

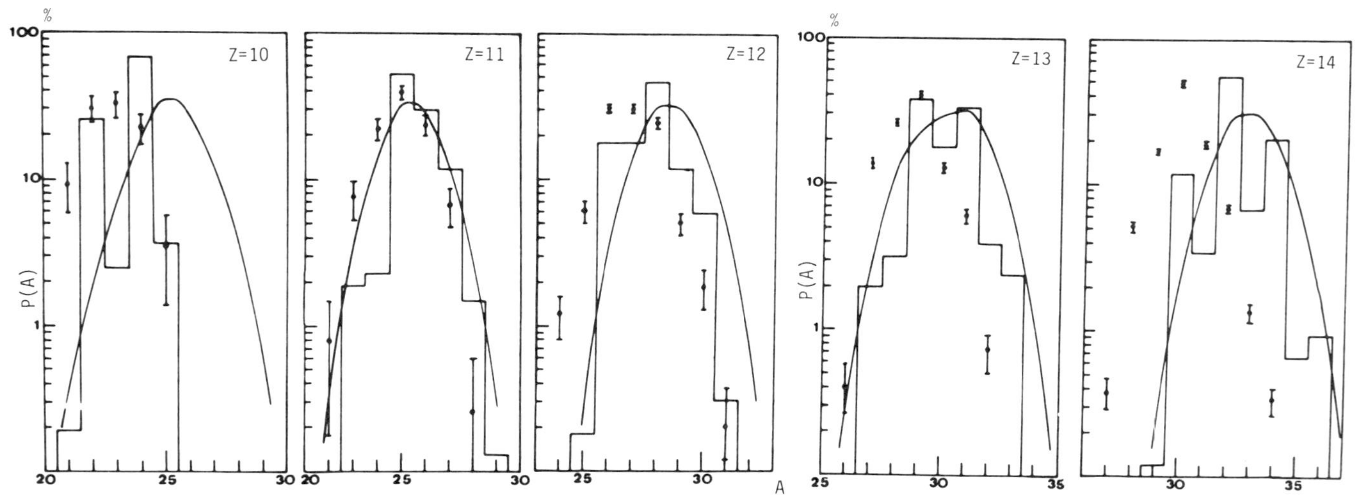

Figure 122. Isotopic distributions of fragments from the reaction $^{238}U + ^{32}S$ at $E_{Lab} = 241$ MeV. Experimental data are given by the points. The smooth curves are predictions for primary distributions derived from a diffusion model. The histograms are the calculated secondary distributions taking neutron evaporation from the primary fragments into account. From (Chi 79).

evolution along the mass asymmetry degree of freedom. The population of charge fragmentations for any given mass asymmetry is approximated by the Boltzmann factor in Equation (3.4.19) involving the driving potential $V(Z_1, A_1)$ as calculated from the liquid-drop energy of two touching spheres, their Coulomb interaction and an average centrifugal potential corresponding to the sticking limit (cf. Section 2.3). The quantity τ denotes the nuclear temperature. Mass drift and diffusion coefficients proposed by Ayik and Nörenberg (see Section 2.8) are adopted to describe the evolution of $P(A_1)$. The reaction time is deduced in the model from the experimental scattering angle.

Results of this simple statistical diffusion model are represented in Figure 122 by the smooth curves. They are taken as the distributions of primary projectile-like fragments whose sequential decay by neutron evaporation leads to the secondary ones given by the histograms in this figure. Although the calculated histograms do not reproduce the details of the measured data, a qualitative agreement is claimed (Chi 79) between experiment and model predictions, except for fragments with Z values close to the projectile where the obviously large discrepancies are attributed to contaminations due to quasielastic processes. Isotopic distributions for fragments from the $^{197}Au + ^{40}Ar$ and $^{197}Au + ^{63}Cu$ reactions studied by the same authors are somewhat better reproduced by the model calculations. It is generally found, however, that theoretical transport coefficients can deviate by as much as a factor of 2 from those consistent with the data.

In accordance with observations made earlier in this section, Chiang *et al.* (Chi 79) conclude that the experimental mass distributions are not representative of a fully equilibrated system. However, the qualitative agreement of experimental isotopic distributions with a description in terms of Equation (3.4.19) is taken as a strong indication of a rapid relaxation of the charge-density degree of freedom and of the statistical nature of this equilibration mechanism.

From the conclusions drawn so far, a picture of the charge density equilibration process operating in damped reactions evolves as of one that is slaved by a slower mass exchange mechanism, always rapidly adjusting to variations in the mass asymmetry. Of course, in order to be able to classify it as a stochastic equilibration mechanism, the rigidity of the correlations in the joint fragment (A, Z) distribution has to be studied under varying reaction conditions, and its consistency with the expectations based on a statistical model should be established. Results for integral fragment distributions from the $^{238}U + ^{32}S$ reaction discussed above (see Figure 122) appear to lend support for such an interpretation.

Experimental data on the $^{92}Mo + ^{86}Kr$ reaction at $E_{Lab} = 430$ MeV obtained by Berlanger *et al.* (Ber 79c), on the other hand, seem to contradict such a simple interpretation. Their data for the dependence of the width

$\sigma_Z(A)$ of the isobaric fragment distribution on the total kinetic energy loss are depicted in Figure 123, for final mass fragmentations indicated by the mass A of one of the fragments. It is seen from the figure that with decreasing TKE, i.e., increasing temperature of the system, $\sigma_Z(A)$ first increases. This is expected for a fast stochastic mode of charge density equilibration that produces an asymptotic equilibrium distribution [cf. Equation (2.8.3)] with a mass-to-charge dispersion increasing with increasing temperature, before the interaction subsides. However, for energy losses in excess of $E_{\text{Loss}} \approx 30$ MeV, the widths displayed in Figure 123 appear to saturate. This effect cannot be easily understood as the result of particle evaporation from the final fragments. Assuming an exponential relation between energy loss and interaction time, a typical relaxation time of $\sim 10^{-22}$ s is estimated from this saturation behavior, a time that is of the order of the period of giant dipole resonances.

From this argument and from the fact that the saturation values for $\sigma_Z(A)$ are much larger than expected from a statistical picture, there appears to be the intriguing possibility that one observes giant dipole oscillations in a dinucleus which are associated with rather large mass-to-charge dispersions. From the systematics of giant dipole resonances in ordinary nuclei, it is known that the phonon energy $\hbar\omega$ is inversely proportional to the linear dimension of the vibrating system. One, hence, estimates (Ber 79c)

$$\hbar\omega \approx 78 \text{ MeV}/\left(A_1^{1/3} + A_2^{1/3}\right) \approx 7\text{–}8 \text{ MeV} \tag{3.4.20}$$

for a dinucleus, assuming that there is a widely open neck between the reaction partners (cf. Section 2.4). This estimated phonon energy is much larger than typical nuclear temperatures attained in a damped reaction and, therefore, one expects the giant oscillator to be closer to the quantal limit than to the classical one. As discussed in Section 2.7, the charge dispersion is then determined by the width of the collective wave function.

Such a collective mode of mass-to-charge equilibration is modeled in the theory of Hofmann (cf. Section 2.7) and collaborators which leads to the quantal Fokker–Planck equation (2.7.23) for the multidimensional probability density. The solid curve in Figure 123 represents the results of a simplified model calculation (Hof 79b), where constant phonon energies of $\hbar\omega_A = 1$ MeV and $\hbar\omega_{N-Z} = 8.5$ MeV are assumed for the motion along the mass and mass-to-charge asymmetry degrees of freedom, respectively. In addition, the friction tensor is taken to be constant and diagonal, with a strength adjusted to fit the estimated relaxation times. Although the experimental values $\sigma_Z(A)$ displayed in Figure 123 are probably some 10% too large, due to insufficient experimental resolution (Ngô 80c), it is concluded that the model can explain the observed experimental correlations. With a similar choice of parameters, it is also possible (Hof 79b) to reproduce data on the ^{197}Au + ^{132}Xe reaction. Further developments of the model (Gre 80b, 81), parametrizing the time-dependent inertias (cf. Section 2.4) associated with the motion along mass and A/Z asymmetry degrees of freedom, yield fits to the data of similar quality to that demonstrated in Figure 123. Basically the same model has been successfully applied by Petrovici *et al.* (Pet 82) to describe the isobaric width of fragment distributions for the reaction ^{154}Sm + ^{98}Mo and $E_{\text{Lab}} = 1176$ MeV.

The magnitude and dependence on excitation energy or energy loss of the dispersions in the isobaric distributions of fragments from both the ^{92}Mo + ^{86}Kr and the ^{197}Au + ^{132}Xe reactions are also well described by the frictional Schrödinger model proposed by Hernandez *et al.* (Her 81). As explained in Section 2.4, this approach employs a classical trajectory model based on the nuclear proximity potential and one-body friction, while inertia and friction coefficients for the mass-to-charge asymmetry motion

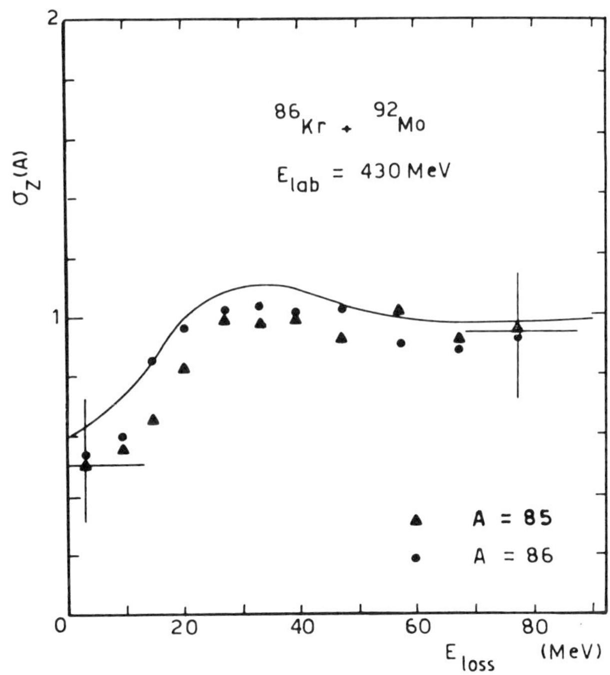

Figure 123. Widths $\sigma_Z(A)$ of the Z distributions for fixed final mass numbers of fragments from the reaction ^{92}Mo + ^{86}Kr at $E_{\text{Lab}} = 430$ MeV plotted vs. total kinetic energy loss E_{Loss}. Symbols represent data (Ber 79c). The curve is the result of a model calculation based on the linear-response theory. From (Hof 79b).

are deduced applying arguments of hydrodynamics. The mass asymmetry degree of freedom is assumed to be fixed, in this approach.

Although the above quantal models are found to be able to account for the data, it appears premature to regard the presence of a collective mode of charge equilibration as established. Owing to the freedom to adjust important model parameters to the data, the calculations may just simulate the experimental trends. A more decisive observation of oscillations in the values of the moments of fragment isobaric or isotopic distributions plotted vs. energy loss or bombarding energy would be needed to support this basic mechanism in a model-independent way.

The nearly saturating behavior at intermediate and large energy losses discovered for the widths of isobaric and, to a lesser extent, of isotopic fragment distributions can, in principle, be due to a number of reasons [see, e.g., (Kra 80)]. However, it seems difficult to reconcile these and previously mentioned experimental results with an equilibrium–statistical nucleon diffusion process, given the current knowledge about the driving potential energy surface and of the temperature attained by the dinuclear system in the course of the reaction. In contrast, as pointed out by Schröder *et al.* (Sch 81b) and Gross *et al.* (Gro 81b), the interplay between static driving forces and dynamic properties of nucleon exchange processes may well result in the observed peculiar correlations in the fragment (A, Z) distributions.

As an example, experimental data (Bre 83) characterizing such correlations are shown in Figure 124 for the reaction $^{165}\mathrm{Ho}+{}^{56}\mathrm{Fe}$ at $E_{\mathrm{Lab}}=465$ MeV. At the top of this figure, experimental ratios σ_A^2/σ_Z^2 of the variances of the A and Z distributions, respectively, are plotted vs. energy loss. For small energy losses, there may be an indication of a slight drop of this ratio with increasing E_{Loss}. After that, there is a definite trend visible in the data, which increase in magnitude with increasing energy loss. At the bottom of Figure 124, the correlation coefficient $\rho=\sigma_{NZ}(\sigma_N\sigma_Z)^{-1}$, to be identified with the parameter κ in Equation (2.8.6), is plotted vs. E_{Loss}. It seems, at first, counterintuitive to observe the correlation coefficient increase drastically with energy loss. For the highest energy losses, ρ approaches unity, indicating completely correlated mass and charge exchange. This is equivalent (see Section 2.8) to reaching the limit $\sigma_A^2=(\sigma_N+\sigma_Z)^2$ by the mass variance.

The solid curves shown in Figure 124 represent results of calculations using the one-body transport mechanism of Randrup discussed in the context of Figure 112. The dashed curves are estimates of the corresponding values for the postevaporative distributions given by Lock *et al.* (Loc 82) in a similar study. Obviously, the theoretical predictions describe the data exceedingly well. The initial decrease in the theoretical σ_A^2/σ_Z^2 values are, in the model, due to the neutron skin of the nuclear surface, enhancing neutron transfer in grazing-like collisions. In the model, average correlations

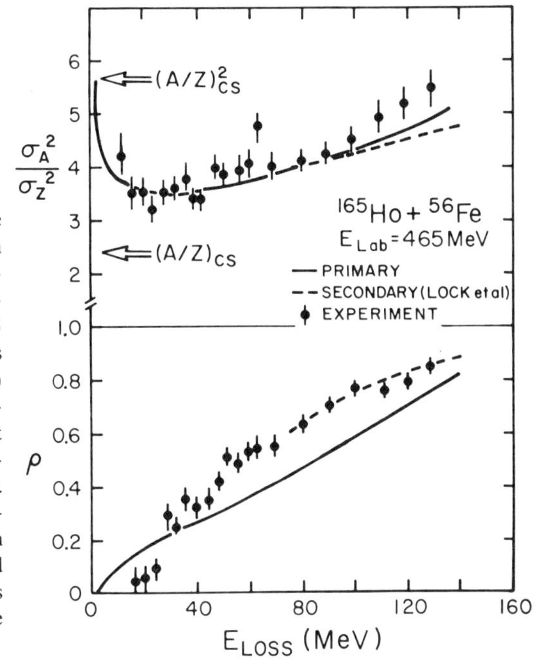

Figure 124. Ratio of mass to charge variances (top) and correlation coefficient for distributions of projectile-like fragments from the reaction $^{165}Ho + ^{56}Fe$ at $E_{Lab} = 465$ MeV, plotted vs. energy loss. Points indicate experimental data (Bre 83) which are for the secondary distributions. Solid curves represent one-body transport model calculations for the primary distributions. Dashed curves indicate corresponding values estimated from evaporation calculations performed by Lock *et al.* (Loc 82). Arrows refer to the A/Z ratio of the composite system.

are imposed on the nucleon exchange process mainly by the curvature of the driving potential resulting in a dependence of drift coefficients on neutron and proton asymmetry [cf. Equations (2.8.9)]. In the absence of such correlations, σ_A^2/σ_Z^2 would drop continuously with increasing E_{Loss} towards the A/Z value of the composite system, $(A/Z)_{\text{cs}}$. Similarly, ρ would become identically zero.

The situation pictured by the data in Figure 124 corresponds to a probability ellipsoid $P(A, Z)$ that rotates and broadens disproportionally in the A and Z directions (cf. Figure 35). From the model calculations [cf. Equations (2.8.9)] it is understood that such correlations grow in importance once mass and charge dispersions become large enough to experience the curvature of the driving potential. Consequently, the widths of isobaric distributions, representing projections of cuts through the integral distribution $P(A, Z)$ along lines of constant $N - Z$, stabilize. The effect is hardly noticeable for isotopic fragment mass distributions because of the much smaller curvature of the potential in the $N + Z$ direction illustrated in Figure 118.

Theoretical results obtained from the above model for isobaric variances $\sigma_Z^2(A = \text{const})$ are compared in Figure 125 to experimental values measured by Breuer *et al.* (Bre 83) for reactions induced by 465-MeV ^{56}Fe

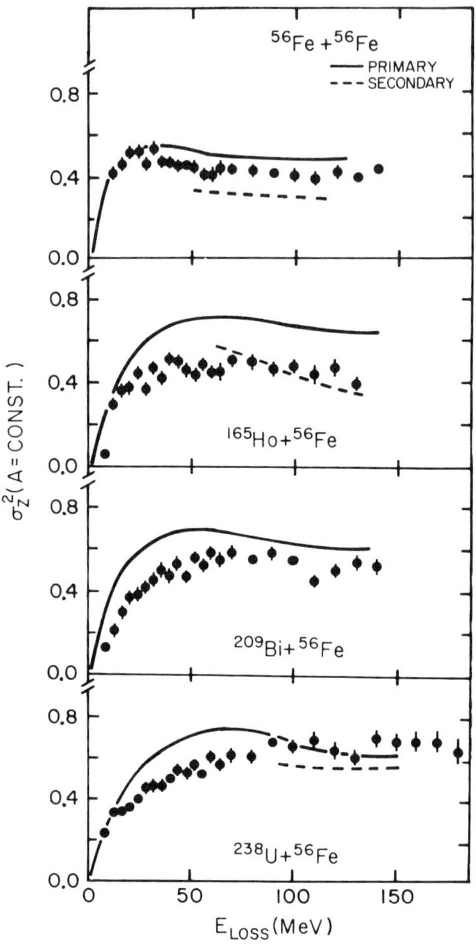

Figure 125. Isobaric variances $\sigma_Z^2(A = \text{const})$ for fragment distributions from 465-MeV ^{56}Fe-induced reactions plotted vs. energy loss. Experimental data (Bre 83) are given by the points. Results from one-body transport calculations for primary fragment distributions are shown by the solid curves. Dashed curves illustrate the effect of particle evaporation.

ions. Again, the predicted primary variances are reproduced by solid curves. Evaporation corrections for ^{56}Fe + ^{56}Fe and ^{165}Ho + ^{56}Fe adapted from the work of Lock *et al.* (Loc 82) yield the dashed curves. This comparison is also very favorable to the transport model. Similar experiences are reported by Britt *et al.* (Bri 82) with the model in a study of the correlations in charge and mass distributions from reactions induced by ^{56}Fe projectiles on targets of ^{58}Ni, ^{64}Ni, and ^{122}Sn.

A shortcoming of the above transport model, which is common also to other theories, is reflected in the fact that absolute values of the first moments of the fragment distributions are frequently only qualitatively reproduced. Experimental mean values in A and Z are usually found to be slowly varying, almost linear functions of the energy loss, with maximum average displacements not exceeding a few mass or charge units. It is, nevertheless, encouraging that predicted relations between average displacement and width of the probability distribution are more like those found in experiments, but rather different from the predictions of classical nucleon diffusion processes discussed earlier in this section. In the transport model, this is due to the particular form of the generalized Einstein relations [cf. Equation (2.8.21)] and the dynamical constituents of the total driving force that act similarly on neutron and proton exchange processes.

It is consistent with such a model view of the dynamics of mass exchange in damped reactions to expect a system to respond more fully to details of the driving potential at very low incident energies above the barrier. Studies of fragment charge density asymmetries performed by Kratz and collaborators (Kra 77, 79d, 81) for Xe-induced reactions at bombarding energies close to the barrier and above are essentially in agreement with this expectation. Figure 126 shows the results (Kra 79d) for fragment N/Z ratios obtained in a radio-chemical experiment, in which a thick ^{197}Au target was bombarded with 761-MeV ^{129}Xe ions. Compared to the experimental data, represented by circles in this figure, are various predictions (solid curves) of primary N/Z ratios, based on the global charge density of the system (a) or on those values that either minimize the driving potential calculated from the liquid-drop model without (b) and with (c) pairing corrections, or from the experimental binding energies (d). As can be seen from Figure 126, the data are consistent with primary fragment distributions reflecting the liquid-drop potential energy including pairing corrections (case c). Accounting for (experimental) shell effects in the calculation of primary N/Z values as illustrated in Figure 126d improves the agreement for fragments in the vicinity of ^{129}Xe and ^{197}Au, which are dominated (Kra 79d) by partially damped events, corresponding to excitation energies of less than 30–40 MeV in the lighter fragment. More symmetric fragmentations, associated on the average with higher excitation energies, have charge densities that do not exhibit the effects of shell closures.

Although a good reproduction of the above experimental N/Z ratios is achieved with a particular choice of the driving potential, conclusions about the characteristic charge density relaxation time can be inferred only rather indirectly from integrated data such as shown in Figure 126. The time evolution of this relaxation process is more clearly suggested by correlations of the fragment A and Z distributions with energy loss, such as illustrated in Figure 127 depicting the ^{56}Fe + ^{136}Xe system trajectory superimposed on the corresponding $l = 0$ driving potential. The crosses in this figure mark experimental values of average neutron and proton numbers of Xe-like fragments for different values of kinetic-energy loss, as reported by Schüll *et al.* (Sch 81h). The solid points connected by a dashed curve represent the

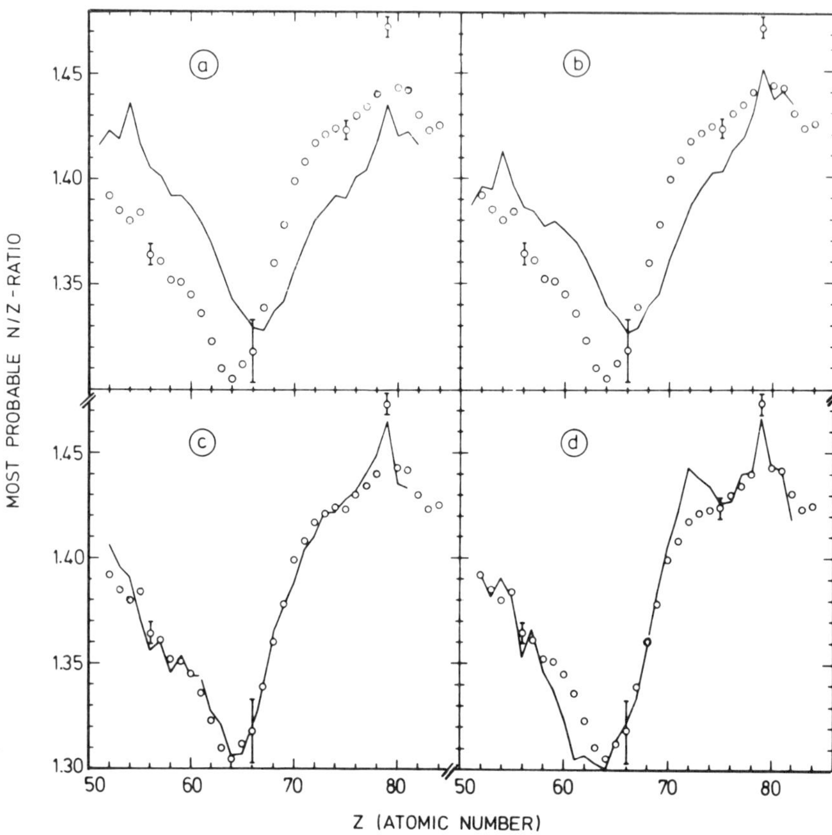

Figure 126. Most probable N/Z ratios (circles) of fragments from the $^{197}Au + ^{129}Xe$ reaction obtained in a thick-target experiment with a bombarding energy of $E_{\mathrm{Lab}} = 761$ MeV. The solid curves represent model predictions for the secondary (postevaporation) N/Z values assuming that the primary most probable N/Z ratios are determined by either the N/Z ratio of the composite system (a), the minimum of the liquid-drop driving potential neglecting (b) or including (c) pairing corrections, or by that minimizing experimental binding energies (d). From (Kra 79d).

same data, corrected for neutron emission from the primary fragments. It is observed that the system, injected with an initial kinetic energy of only ~ 70 MeV above the barrier, follows the steepest gradient of the potential gradually, as energy is dissipated. During the time it takes to dissipate the first 40 MeV of relative kinetic energy, the mass asymmetry varies only little, while the charge asymmetry approaches a value consistent with the local minimum of the driving potential. Obviously, as defined with respect to the $l = 0$ potential energy surface, a charge density equilibrium is not

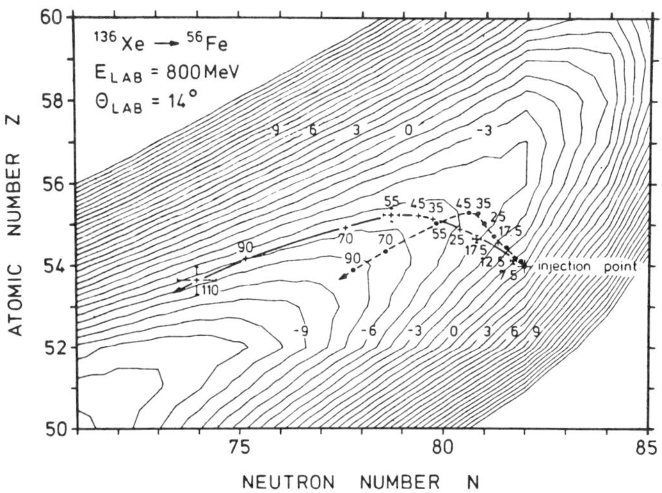

Figure 127. Contour plot of the shell-corrected liquid-drop driving potential, in units of MeV, for fragmentations with the same total mass and charge as the system $^{56}Fe + ^{136}Xe$ (injection point). Crosses connected by the solid curve represent measured average N and Z values for fragments from the reaction $^{56}Fe + ^{136}Xe$ at $E_{Lab} = 800$ MeV for the excitation energies indicated. Solid dots are obtained by applying neutron evaporation corrections to the data. From (Sch 81h).

reached before the reaction subsides, for a dominant fraction of the reaction cross section. This finding is corroborated by examination of data for the $^{165}Ho + ^{56}Fe$ reaction measured (Bre 83, Hoo 82) at the much higher c.m. bombarding energy of ≈ 160 MeV above the barrier. Here, global A/Z equilibration is not reached, except, possibly, for the highest energy losses encountered. These are estimated to correspond to interaction times of the order of $\sim 10^{-21}$ s. Hence, one is led to conclude tentatively that there is nothing intrinsically particular that distinguishes the exchange in the $N-Z$ direction from that along the mass asymmetry degree of freedom.

One is then developing at least a tentative understanding of why the mass and charge exchange occurring between two colliding heavy ions exhibits, in general, relatively little sensitivity to the energetics of the underlying driving potential. Other, as yet unaccounted for properties of the exchange mechanism such as the induced inertial forces should be studied further, however, in order to complete this picture. In any case, one is forced to modify the concept of a generally fast charge density equilibration mechanism suggested by studies of light systems discussed earlier. Both the evolution of a system in mass asymmetry and its variation in mass-to-charge asymmetry appear to be components of the same stochastic nucleon exchange mechanism that is governed by the balance of conservative, inertial, and dissipative forces. Although the experimental observations support reaction mechanisms underlying the dynamical nonequilibrium transport models discussed in this section, a superposition of collective modes cannot be entirely excluded. There is clearly a demand for more systematic and accurate experiments by which the various components of the mass exchange mechanism can be unveiled.

3.5. Dissipation of Kinetic Energy

Certainly the most striking phenomenon associated with the complex interactions between heavy nuclei at several MeV per nucleon above the barrier pertains to the enormous range in kinetic energies that can disappear from the relative motion. It is this feature that distinguishes the damped process most prominently from direct reactions induced by light ions, in which only a minor fraction of the initially available kinetic energy per nucleon is dissipated, and from fusion reactions, in which all of it is converted into other forms of energy. At the high excitation energies that the intermediate system formed in a damped collision may attain, a large number of reaction channels is energetically allowed and, hence, one expects the energy conversion process to exhibit certain statistical features. As discussed in previous sections, the distinct evolution of various reaction observables with progressing loss of kinetic energy of relative motion, discovered experimentally, is consistent with a dissipative–statistical intrinsic nature of the energy transformation mechanism. From such a picture, the concept of nuclear friction has emerged [see, e.g., (Gro 75b)]. Theories outlined in Sections 2.6–2.8 predict a gradual and irreversible transformation of collective energy into intrinsic heat energy, mediated by classical friction forces. Such a dominantly dissipative character of the reaction mechanism would, indeed, provide the opportunity to study the "controlled fusion" of two complex nuclei and image the course that the collision takes with progressing interaction time, indicated by the observable degree of kinetic-energy damping. Although experimental correlations discussed so far with regard to reaction angle, fragment mass (or charge) asymmetry, and energy loss are suggestive of it, they, by no means, require that a thermodynamical weak-coupling limit (cf. Section 2.7) be realized in damped reactions, which would then naturally lead to the appearance of a friction-like mechanism. In fact, as outlined in Sections 2.3, 2.6, and 2.7, there are also theoretical arguments that favor a dominantly coherent or even collective mode of energy transformation that is only in the long-time limit destroyed by growing thermal fluctuations.

It is the purpose of this section to discuss the implications that can be derived from the experimental systematics of fragment energy spectra with respect to the various possible modes of energy transfer. Correlations of average final energies with the projectile–target mass asymmetry, incident energy and angular momentum, reaction angle, and final fragmentation will be discussed insofar as they complement the experimental phenomenology of damped reactions, already presented in previous sections, or add detail to the understanding of the reaction mechanism. Of particular interest in this respect is also the shape of fragment energy distributions which may indicate the presence or absence of collective modes of energy transformation. Finally, experimental data are scrutinized with the aim of inferring information about the destination of the disappeared kinetic energy, i.e., to assess in what collective or intrinsic degrees of freedom this energy is eventually stored and to what extent a thermodynamic equilibrium is achieved at the end of the collision. The subject of this section is closely connected to matters to be discussed in Sections 3.6 and 3.7. In particular, the dissipation of angular momentum is of relevance for the energy transformation process, because it pertains to the orbital component of the kinetic energy of relative motion. It is for reasons of a more convenient presentation that such an arbitrary division of the subject is made here.

Experimental problems associated with a precise determination of fragment kinetic energies are, of course, fundamental also for the identification of atomic and mass numbers and the interpretation of angular distributions. They have, therefore, mostly been delineated already in previous sections. Although high-resolution (~ 0.1%) energy measurements of heavy reaction fragments are possible, for example, with magnetic spectrometers [see, e.g., (Che 81)], in experiments employing gas or solid-state detector systems, intrinsic energy resolutions of typically 1%–2% are achieved. Systematical uncertainties associated with detection nonlinearities, however, can be much larger than that. As an example, the so-called pulse-height defect (Sch 65, Wil 71, Ste 72, Kau 74, Mou 78), an effect due to nonionizing collisions or incomplete charge collection, caused by recombination and trapping of charges produced along the tracks of highly ionizing particles in solid-state and gas detectors, can be as high as ~ 50 MeV for a 940-MeV ^{136}Xe ion (Wil 80). It varies with mass, charge, and energy of the incident particle. Because independent and accurate methods of determining the primary beam energy have been applied only in recent years, earlier reported experimental results should be taken with the proper precaution.

Of greater concern than the effects mentioned above are distortions introduced into fragment energy distributions by the sequential decay of highly excited primary fragments. A similarly drastic influence of particle evaporation processes on fragment A and Z distributions has already been noted previously (see Figures 87 and 88). Again, the disturbing influence of evaporation processes on fragment distributions is expected to be more drastic for lighter heavy ions, as illustrated below.

The result of the emission of a particle of mass m and kinetic energy ε in the rest frame of an emitting fragment of mass A is a reduction of the original fragment laboratory kinetic energy E to a value E', given by [cf., e.g., (Hil 79b)]

$$E' = E\left(1 - \frac{m}{A}\right)\left[1 + \frac{m\varepsilon}{AE} + 2\left(\frac{m\varepsilon}{AE}\right)^{1/2}\cos\theta\right] \tag{3.5.1}$$

Here, θ is the particle emission angle measured with respect to the original direction of flight of the emitting fragment, which is also altered due to the particle recoil. Furthermore, depending on the emission pattern $d^2\sigma/d\varepsilon\, d\theta$ of the particles, fragments with an initially well-defined kinetic energy attain a distribution of final kinetic energies (recoil broadening). Since with increasing total excitation energy E^* or decreasing final laboratory kinetic energy E_{Lab}, decay channels open up successively, structure may appear in the fragment kinetic energy spectrum, despite the complex convolution of the primary energy, mass, and angular distributions with the emission probability for secondary particles. Obviously, such distortions vanish with $m/A \to 0$.

This effect has been studied theoretically for the $^{40}\text{Ca} + {}^{40}\text{Ca}$ reaction at $E_{\text{Lab}} = 400$ MeV by Hilscher *et al.* (Hil 79b). Figure 128 illustrates the final kinetic energy distribution predicted for ^{41}Ca fragments at an observation angle of $\theta_{\text{lab}} = 10°$, under the assumption of a primary yield distribution of the form

$$\frac{d^4P(Z_1, N_1, E^*, E_1^*)}{dZ_1\, dN_1\, dE^*\, dE_1^*} \propto \exp\left\{-\frac{(Z_1 - \langle Z\rangle)^2}{2\sigma_Z^2} - \frac{(N_1 - \langle N\rangle)^2}{2\sigma_N^2}\right\}$$
$$\times E^*\exp\left(-\frac{E^*}{B}\right)\cdot\exp\left[-\frac{\left(E_1^* - E^*A_1/A\right)^2}{2E^*(A_1/A)^2}\right] \tag{3.5.2}$$

with the variances

$$\sigma_Z^2 \approx \sigma_N^2 \approx \frac{Z}{A}\frac{\mu}{\alpha m}\ln\left(\frac{E_{\text{c.m.}} - V_{\text{Coul}} - E^*}{E_{\text{c.m.}} - V_{\text{Coul}}}\right) \tag{3.5.3}$$

In these relations, E_1^*, A_1, Z_1, and N_1 correspond to one of the primary fragments, A is the total mass, and μ is the reduced mass of the system. The quantity B is an adjustable parameter determining the slope of the Maxwellian excitation energy spectrum. The dependence of the variances of the primary Gaussian A and N distributions given by Equation (3.5.3) is suggested by the parametrization in Equation (3.4.13) with $x=1$, involving the phenomenological parameter α. The primary production yield given by Equations (3.5.2) and (3.5.3) serves (Hil 79b) as an input for a simplified evaporation calculation considering only one- or two-particle decay chains.

The decomposition shown in Figure 128 for ^{41}Ca fragments from damped reactions of the relatively light system $^{40}Ca+{}^{40}Ca$ illustrates the possibility that dominant decay channels introduce marked structure in the originally smooth laboratory energy distribution of the final fragments, structure that, in this case, coincides with experimentally observed (Fra 80) features. It is noteworthy that, in the hypothetical laboratory energy spectrum of ^{41}Ca depicted in Figure 128, only the component at the highest kinetic energy, i.e., lowest excitation energy remains from the primary ^{41}Ca production in this reaction.

Although the secondary energy spectrum reflects the shape of the primary fragment distribution in A, Z and excitation energy, the example shown in Figure 128 illustrates another aspect of the difficulty of recon-

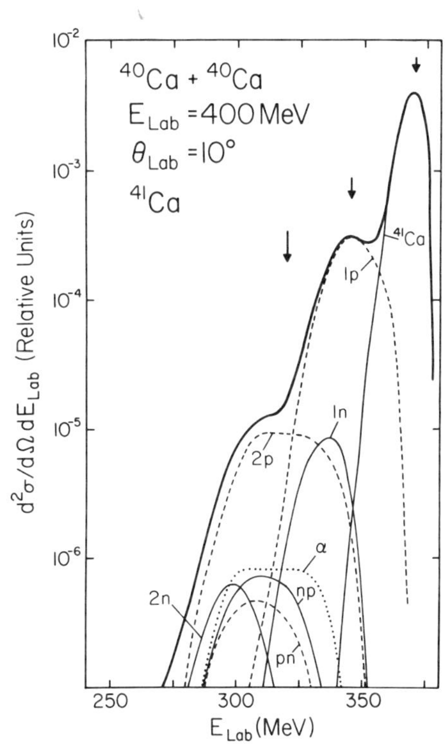

Figure 128. Hypothetical laboratory kinetic energy spectrum $d^2\sigma/d\Omega\, dE_{\text{Lab}}$ at $\theta_{\text{Lab}}=10°$ expected for ^{41}Ca fragments produced in the reaction $^{40}Ca+{}^{40}Ca$ at an incident energy of 400 MeV. The primary, preevaporative probability distribution described by Equations (3.5.2) and (3.5.3) results in the secondary spectrum given by the heavy solid curve after evaporation of up to two light particles. This secondary spectrum is a superposition of various components due to the decay of excited parent fragments in the vicinity of ^{41}Ca, whose identities can be inferred from the evaporation chain indicated at each component. Recoil shift and broadening caused by isotropic emission is accounted for in the calculation of the component spectra. Arrows mark energies at which structure is observed experimentally. From (Hil 79b).

structing unambiguously the preevaporative probability distribution from measured quantities. A more straightforward interpretation of the data is possible through experiments determining the velocities of the fragments, which are on the average unaffected by the evaporation process. For very heavy damped-reaction fragments decaying mainly via neutron evaporation chains, the primary kinetic energies can be inferred by using the relation between energies and missing mass m such as given by Equation (3.5.1) for $(m\varepsilon)/(AE) \ll 1$, in conjunction with kinematical information and assumptions about a statistical partition of the total kinetic energy loss between the reaction partners. Procedures to account for sequential neutron emission through application of simplified evaporation formulas [see, e.g., (Lan 54, 64, Lec 59)] are standard in damped-reaction event analyses.

In the discussion of damped-reaction final energy distributions it is useful to recall the corresponding properties of the fission process. A large number of studies have been performed [see (Van 73)] in which the dependence of the energy release in fission on atomic and mass number of the fissioning nucleus, its excitation energy, and the final fragmentation is explored. For example, a linear relationship between the most probable value $\langle E_f \rangle$ of the total kinetic energy release with the $Z^2/A^{1/3}$ value of the fissioning nucleus has been noticed by Terrell (Ter 59). A more recent fit result obtained by Viola (Vio 66) yields

$$\langle E_f \rangle = (22.3 + 0.1071 Z^2 A^{-1/3})\ \text{MeV} \tag{3.5.4}$$

which describes fission kinetic energies well into the transuranic region (Har 79b). The total kinetic energy varies noticeably with the mass asymmetry of the fission fragments. For example, in the case of spontaneous fission of ^{252}Cf, the variation amounts to about 35 MeV (Van 73). Such a correlation would be expected if the asymptotic fission fragment energies were simply due to acceleration by the Coulomb repulsion. This interpretation, however, neglects possible contributions due to kinetic energy along the fission degree of freedom that may already be present at the saddle, i.e., the barrier top of the effective potential. Up to now, it has not been possible to deduce model-independent information on the relative importance of the two contributions [see, e.g., (Sch 78e, 80b, Ram 80)]. The problem is further aggravated by the effect of dissipative processes acting on the fission mode as illustrated, for example, by extensive calculations of Nix and collaborators (Dav 76, 77b, Sie 78, 80) and others (Pra 79, Pal 79).

Typical values of the r.m.s. width of the energy distributions are of the order $\sigma_{E_f} \sim 8$–12 MeV, increasing only slightly with excitation energy and fissility (Z^2/A) of the fissioning nucleus. These widths are somewhat larger than expected from statistical theory for thermal fluctuations (Van 73, Sam

82). The variations of the most probable kinetic energy are also of that order, although there are definite trends with fissility and fragmentation. Total kinetic energies in fission peak characteristically at fragmentations close to the most probable (asymmetric or symmetric) ones.

While these features of the low-energy fission process may be of relevance in a comparison to damped reactions at energies close to the barrier, fission processes induced by heavy ions at higher energies are certainly of more general interest in the present context. A number of experimental studies of fission induced by projectiles ranging from ^{12}C to ^{40}Ar at energies up to 16 MeV per nucleon have been reported (Gor 60, Bri 60, Vio 63b, Nam 75b, Bor 74, Cab 75b, Nat 76, Vio 76) that emphasize the total kinetic energy release. Since, especially at the higher incident energies, a variety of reactions may occur, including incomplete fusion (Sik 62, 63, 64, Vio 63, 63b, 76, Nat 70, Siw 79, 79b, 79c, Dye 79, Awe 79, 80, 80b, 82) accompanied by the prompt emission of a light projectile remnant, it is important to restrict the consideration to fusion–fission events where transfer of the total linear momentum to the fission fragments and conservation of the total mass is ascertained. This can be inferred from angular correlations and the coplanarity of the velocity vectors for the two fission fragments, determined in a kinematical-coincidence setup such as described earlier.

Typical laboratory kinetic energy spectra of fission fragments from the reaction $^{235}U + ^{20}Ne$ at $E_{\mathrm{Lab}} = 175$ MeV and 252 MeV are depicted in Figure 129 at the bottom and top, respectively. These are coincidence data measured by Viola *et al.* (Vio 76) with one fission fragment detector at a laboratory angle of 55° and a position-sensitive coincidence detector determining the correlation angle corresponding to full momentum transfer. The rather broad, bell-shaped energy distributions shown in Figure 129 correspond (Vio 76) to average total kinetic energies of both fission fragments of $\langle E_f \rangle = (198 \pm 8)$ MeV and (201 ± 8) MeV for incident energies of 175 and 252 MeV, respectively. Both these values are in agreement with the kinetic energy release of 198 MeV predicted (Vio 66) by the systematics [cf. Equation (3.5.4)] for a fissioning ^{255}No nucleus and demonstrate a rather weak dependence on bombarding and excitation energies. These results corroborate similar observations made (Bri 60, Vio 63b) for 125-MeV and 166-MeV ^{12}C-induced reactions on targets ranging from ^{141}Pr to ^{238}U and ^{40}Ar-induced fission (Bor 74) of composite nuclei from ^{136}Nd to $^{278}110$, at several bombarding energies. In these latter experiments, the composite nuclei are produced with excitation energies between approximately 60 and 170 MeV, where shell effects have presumably disappeared. It is concluded from such studies (Bor 74) that the total kinetic energy in fission is not determined simply by the Coulomb energy of the fissioning nucleus but correlated strongly with the fissility parameter. It is also independent of the initial projectile–target mass asymmetry as demonstrated by Britt *et al.* (Bri

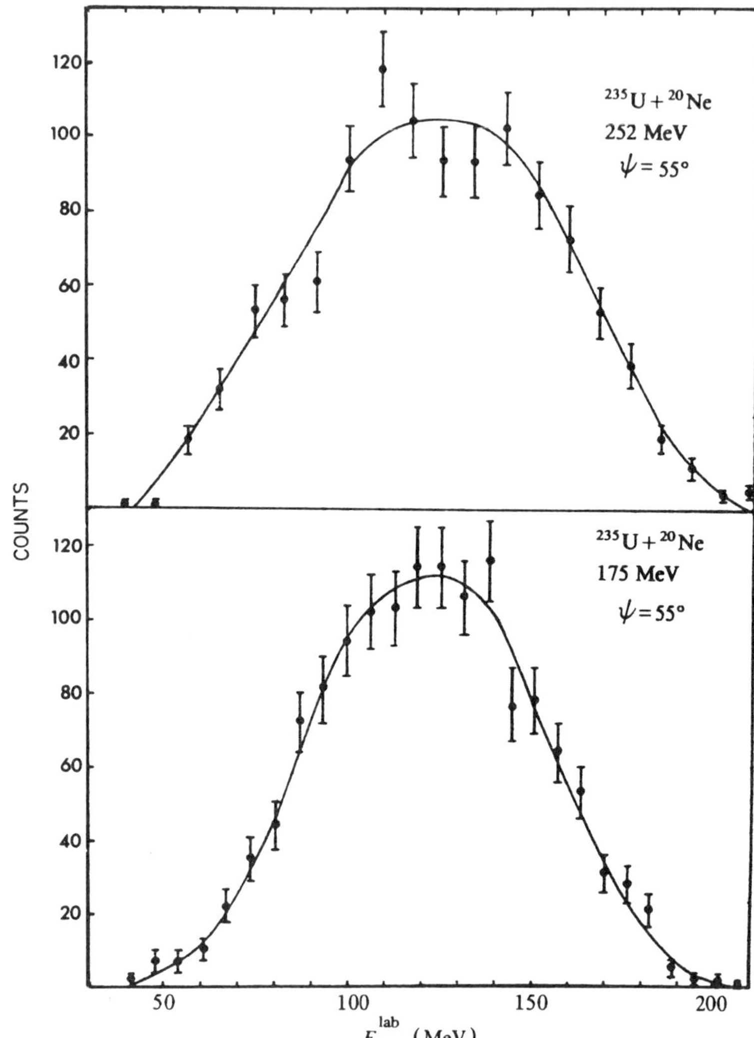

Figure 129. Laboratory kinetic energy spectrum of correlated fission fragments from the reaction $^{235}U + ^{20}Ne$ at 175 MeV (bottom) and 252 MeV bombarding energies. The laboratory detection angle is 55° for one counter, with the coincidence detector on the opposite side of the beam. Events corresponding to the maximum in the coincidence cross section and complete momentum transfer are selected. From (Vio 76).

76) studying fission of the composite nucleus ^{149}Tb produced in the reactions ^{109}Ag + ^{40}Ar and ^{65}Cu + ^{84}Kr. From experiments on ^{12}C-induced fission on targets of ^{89}Y, ^{98}Mo, Ag, and ^{116}Sn, Namboodiri *et al.* (Nam 75b) infer that a possible increase in kinetic energy with increasing angular momentum is exceedingly small, i.e., $d\langle E_f \rangle / d(l^2) \lesssim 0.7\ \text{keV}/\hbar^2$.

As observed in Figure 129, the spread in the kinetic energy release in heavy-ion-induced fission can be considerable. Experimental widths Γ_{E_f} (FWHM) range (Cab 75b) between 15 and 25 MeV for a ^{122}Xe or ^{211}Rn composite nucleus, respectively, produced at excitation energies of the order of 100 MeV. The excitation energy dependence is found to be weak, but a strong increase of Γ_{E_f} with the fissility parameter x is reported (Vio 63b) for values $x \geq 0.7$. This behavior is qualitatively discussed (Vio 63b) in terms of the similarity of saddle and scission shapes for less fissile systems such that, for nuclei with $x < 0.7$, the descent from saddle to scission is too rapid to allow a significant deviation of the fragment properties from their values at the saddle point.

The correlation between the total kinetic energy distribution and the mass asymmetry in fission induced by 126-MeV ^{14}N ions is illustrated in Figure 130 for a Ag (top) and a Ho (bottom) target (Cab 75b). The two-dimensional yield contour plots in this figure represent measured data that are not corrected for sequential emission of light particles, probably mostly neutrons, from the excited primary fragments. Hence, as discussed in the previous subsection, the mass distribution may be somewhat distorted by the evaporation process. To first order, the energy distribution should only be shifted by a constant amount [cf. Equation (3.5.1)]. This is at least expected from an equipartition of the total excitation energy E^* (≈ 100 MeV) among all nucleons of the system and slowly varying particle binding energies B_i. This is because then the mass m_i evaporated from fragment i with initial mass A_i would be approximately proportional to A_i, with $m_i/A_i \approx (E^*/B_i)/(A_1 + A_2)$. Hence, the distributions shown in Figure 130 should give a realistic impression of the shape of the primary distributions. Qualitatively, one concludes that there is no dramatic dependence of the total kinetic energy distribution on the asymmetry of the fragmentation, although energy spectra are somewhat asymmetric for equal fragment masses. There is an obvious lack of correlations (compare Figure 35) in the distributions and the nearly constant mean total energy, in the cases illustrated in Figure 130, coincides with the estimate (Wil 80d) from the systematics for symmetric fission, almost regardless of actual asymmetry. Variations of average total kinetic energies with fission fragment mass asymmetry of about 20 MeV have been found (Bri 76, Aga 77), however, in other reactions.

The weak dependence of the most probable kinetic energy on the asymmetry of the mass split noted above for heavy-ion-induced fission of

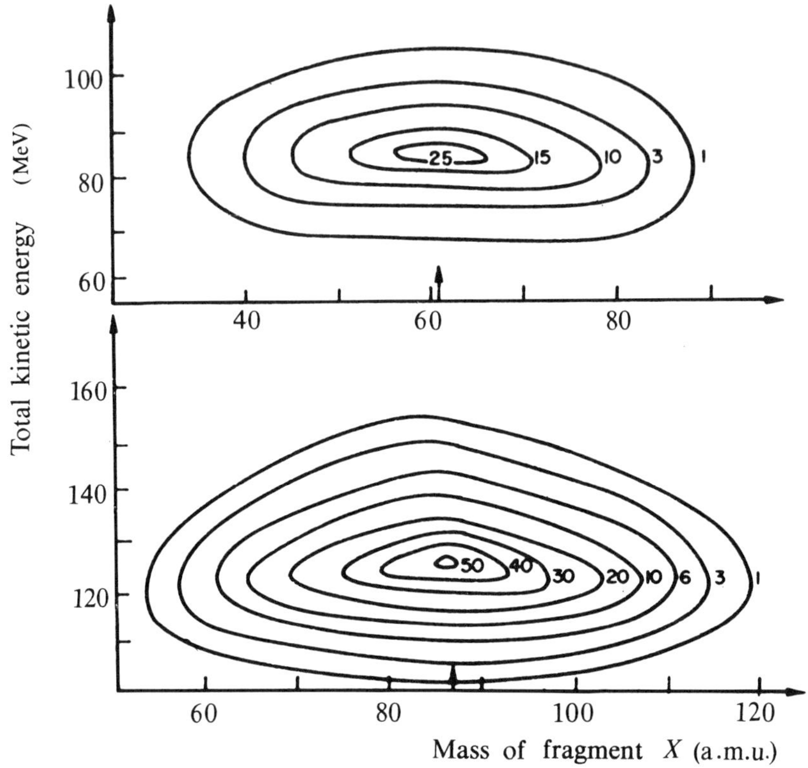

Figure 130. Two-dimensional contour diagram of the yield of correlated fission fragments produced in the reaction of 126-MeV ^{14}N ions with targets of Ag (top) and ^{165}Ho (bottom), plotted vs. total kinetic energy and mass of one of the fragments. The arrows indicate symmetric fragmentations. The distributions are not corrected for particle evaporation from the excited primary fragments. From (Cab 75b).

medium-weight composite nuclei remains a characteristic feature also for the fusion–fission-like (FL) reactions occurring for heavier systems. This is illustrated in Figure 131 (Boc 82) depicting the double-differential cross section $d^2\sigma/dMd(\mathrm{TKE})$ for a number of ^{208}Pb-induced reactions at incident energies $E_{\mathrm{c.m.}}/V_c$ relative to the Coulomb barrier V_c as indicated at the distributions. Curves of constant cross section in μb per MeV and mass unit are plotted in this figure vs. mass M of one of the fragments and final total kinetic energy (TKE). A mass asymmetry-dependent experimental detection efficiency results in a significant suppression of weakly damped events, especially for the higher bombarding energies. In agreement with the data shown in Figure 131, it is observed from Figure 131 that the most probable kinetic energies of FL events do not vary drastically with mass asymmetry,

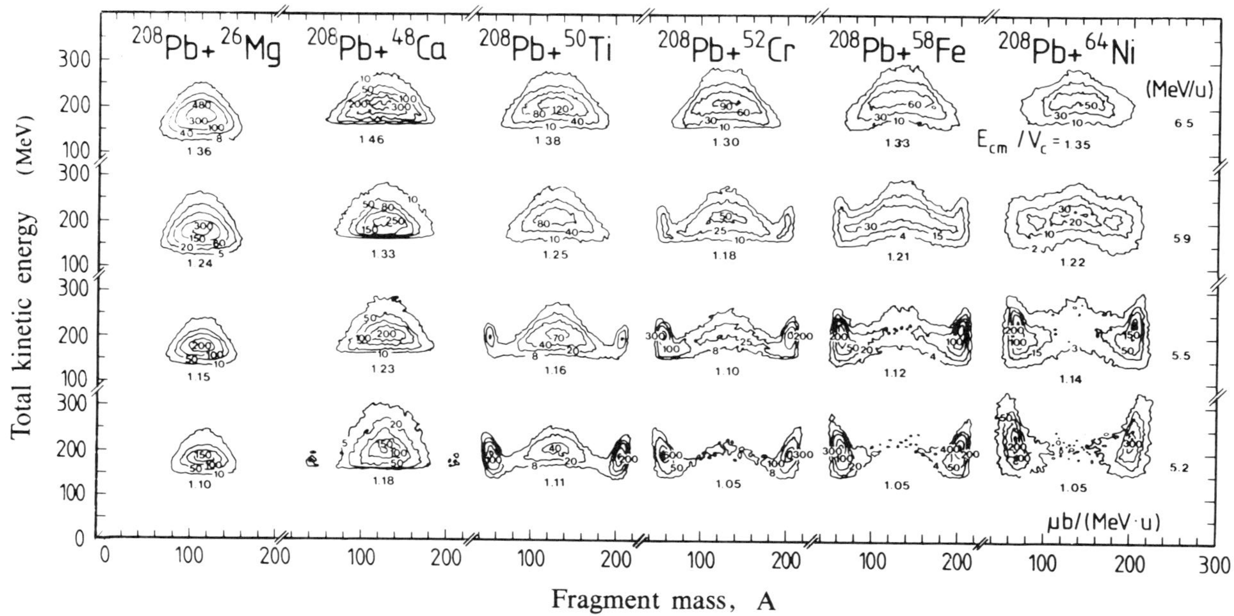

Figure 131. Contour diagrams of constant cross section $d^2\sigma/dM\,d(\text{TKE})$ for several ^{208}Pb-induced reactions plotted vs. mass M of one of the fragments and final total kinetic energy TKE. The incident energy $E_{\text{c.m.}}$ is indicated at each distribution as a fraction, $E_{\text{c.m.}}/V_c$, with respect to the Coulomb barrier. The experimental efficiency is constant only for approximately symmetric mass splits but is much reduced for large asymmetries. Hence, weakly damped events are strongly suppressed, in particular for higher energies. From (Boc 82).

although the energy distributions exhibit high-energy tails for symmetric fragmentations. This finding is corroborated by similar observations made by Heusch *et al.* (Heu 78) for the ^{56}Fe + ^{132}Xe reaction. Only for the heavier systems and lower bombarding energies depicted in the lower right part of Figure 131 are the FL mass–energy distributions reminiscent of the expected variation of the Coulomb barrier for deformed fragments. This is an interesting observation that bears on the understanding of the dynamical evolution of the shape of the fissioning nucleus, which is a topic of current interest, and also on the question of how to calculate the intrinsic excitation energy or the effective energy loss parameter [cf. Section 3.4, Equation (3.4.9)] in a meaningful way. As discussed below, the shapes of energy distributions associated with damped reactions are distinctly different from those of fission and FL reactions.

Following the development of the total kinetic energy or Q-value distributions produced in damped reactions from the opposite extreme, that of quasielastic reactions, one observes an increasing complexity of the spectrum with increasing bombarding energy. This evolution is impressively illustrated by the series of Q-value distributions shown in Figure 132 for fragments with mass $A = 28$ resulting from the ^{28}Si + ^{28}Si reaction (Bet 81) at several bombarding energies. Fragment mass distributions obtained in these measurements have already been discussed in the context of Figure 91. It is instructive to observe in Figure 132 that at low bombarding energies only a few low-lying excited states are selectively populated in either or both of the reaction partners. These states, identified in the 120-MeV spectrum, are mainly yrast states, i.e., those of highest spin for a given Q value. As the bombarding energy is increased, the relative significance of the isolated peaks in the energy spectra displayed in Figure 132 diminishes, and a continuous spectrum component grows in importance, indicating the onset of a more complex reaction mechanism. This trend away from a selective population of states towards more continuous spectra is typical for reactions induced by lighter projectiles (Mik 76, Nov 78,80). It is qualitatively understood to reflect a transition from a discontinuous to a continuous penetration function $D(l)$ with increasing bombarding energy, as is illustrated in Figure 54.

Final kinetic energy spectra of fragments from damped reactions induced by relatively light projectiles have been reported for beams of ^{12,13}C (Gal 70, Lef 70, Ost 76, Nov 78, Day 80), ^{14,15}N (Lef 70, Vol 74b,76, Tho 74, Mor 75d, Bil 78), ^{16}O (Alb 75, Vol 76,78, Cor 76,76b), ^{20,22}Ne (Art 73, Vol 74b,76, Bab 76, Egg 76, Nat 77, Obe 78, Sch 78h, Bil 78, Van 80e, Day 80), ^{28}Si (Oes 77, Nov 80), ^{32}S (Bar 77, Nov 80b, Pel 81), ^{40}Ar (Art 73b,78,78b, Vol 74b,76,78, Tho 74, Gat 74,75, Jac 75, Gal 76b, Mor 76d, Bri 76, Ngô 77b, Art 79, Nam 79, Fra 80f), and ^{40}Ca (Col 74c,75, Fra 77,80, Roy 77,81, Aga 77, Gel 77,78).

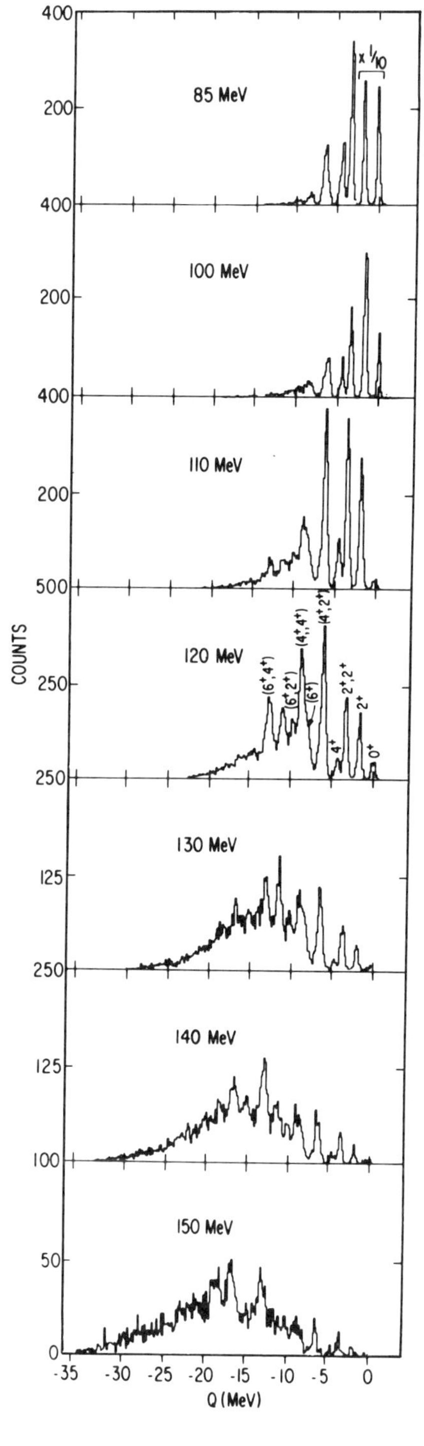

Figure 132. Q-value spectrum of fragments with mass $A = 28$ produced in the reaction $^{28}Si + ^{28}Si$ at the various bombarding energies indicated in the figure. Assignments of levels excited in either of the two reaction fragments are indicated at the 120-MeV spectrum. From (Bet 81).

As an example, the laboratory energy spectra $d^2\sigma/d\Omega\, dE$ of fragments from the reaction $^{232}\text{Th} + {}^{40}\text{Ar}$ measured by Artukh *et al.* (Art 73b) at $E_{\text{Lab}} = 388$ MeV are shown in Figure 133. The distributions are grouped according to fragment atomic number and observation angle. Quite in contrast to the energy distributions shown in Figures 130 and 131 for fusion–fission and FL events, respectively, the damped reaction energy spectra of Figure 133 extend over a range of several hundred MeV. For fragments with the atomic number of the projectile (Ar), the energy spectra are dominated by the high-energy elastic peak, which is most visible at forward angles and diminishes in relative intensity as the angle is increased. For fragments close to the projectile, a quasielastic peak is the most distinctive feature of energy spectra taken in the vicinity of the grazing angle $\theta_{\text{gr}} \approx 33°$. The energy spectra of the damped-reaction fragments from the $^{232}\text{Th} + {}^{40}\text{Ar}$ reaction depicted in Figure 133 exhibit a single, bell-shaped component at forward angles, whereas two components appear at backward angles. The relative intensities and mean kinetic energies for these latter components behave differently, as the observation angle is changed. As already discussed in Section 3.3 (cf. Figures 51, 56, 78), such a behavior is expected from a classical orbiting picture (Wil 73c) where kinetic energy is gradually dissipated as the intermediate dinuclear system rotates about its center of gravity. Hence, one attributes (Wil 73c) the two energy components to the cross section ridge developing in the Wilczyński diagram (Figure 56) from the quasielastic peak in the direction of decreasing final energy and reaction angle, and its reflection about the beam direction. It is also consistent with such a picture that fragments far removed from the projectile show only the strongly damped spectral component, since their production in a reaction requires, on the average, more time and a larger angle of rotation. An alternative view involving a double-rainbow deflection function arising, for example, from a dominantly repulsive interaction force has also been discussed in Section 3.3. As has been noted previously, this latter scattering mechanism can be ruled out experimentally for relatively light and medium-weight systems at bombarding energies of a few MeV per nucleon above the barrier.

The variation with bombarding energy of the characteristic features of the laboratory final energy spectra for damped reaction fragments is illustrated in Figure 134 for the $^{232}\text{Th} + {}^{40}\text{Ar}$ reaction, measured (Art 73b) at $E_{\text{Lab}} = 297$ and 388 MeV. The open circles represent the high-energy data, whereas the 297-MeV data are given by solid dots. Final kinetic energies corresponding to the Coulomb repulsion are indicated in this figure by dashed and solid arrows, respectively. Several interesting observations can be made inspecting Figure 134. First, raising the incident energy from 297 to 388 MeV results in a general increase of the width of the energy spectra by roughly a factor of 2. This is apparently quite in contrast to fission

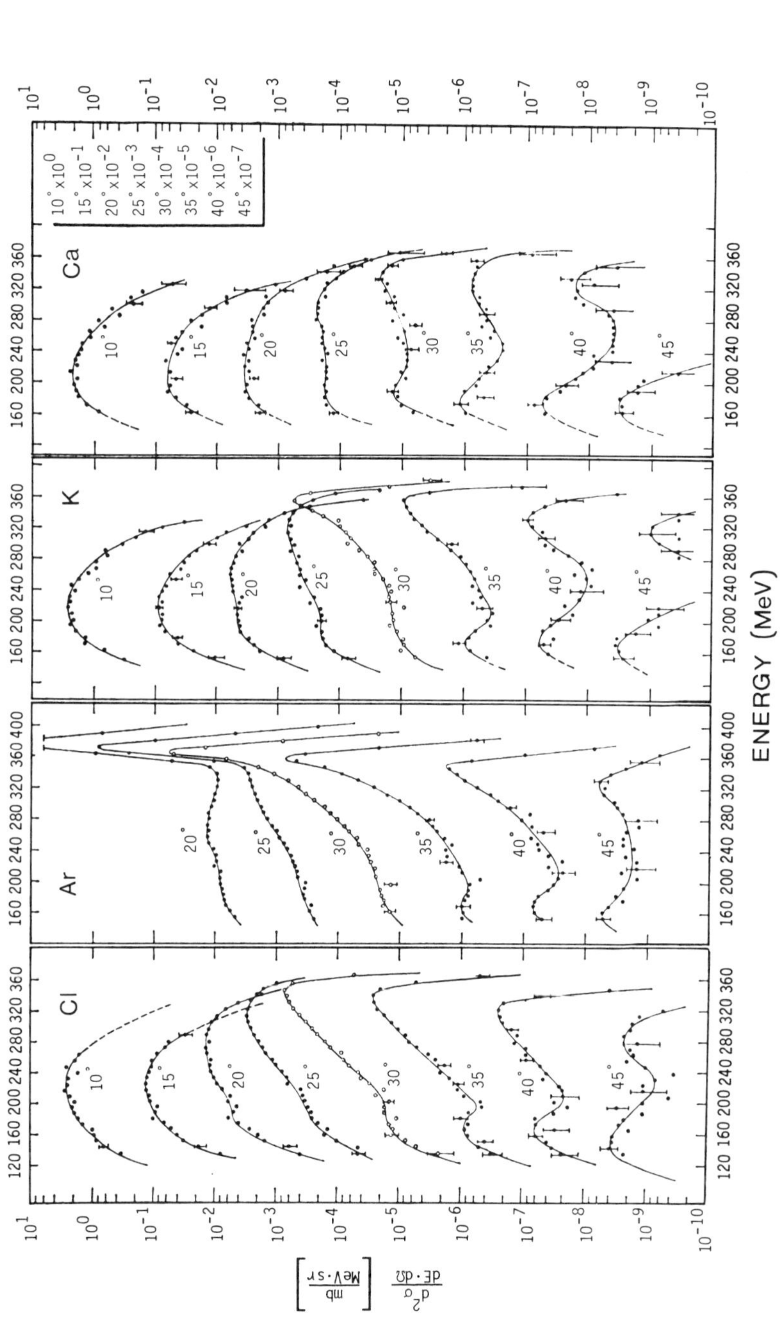

Figure 133. Laboratory energy spectra $d^2\sigma/d\Omega\,dE$ of Cl, Ar, K, and Ca fragments produced in the reaction $^{232}\mathrm{Th} + {}^{40}\mathrm{Ar}$ at $E_{\mathrm{Lab}} = 388$ MeV for various laboratory angles. The cross sections are multiplied by the factors indicated in the insert on the upper right. From (Art 73b).

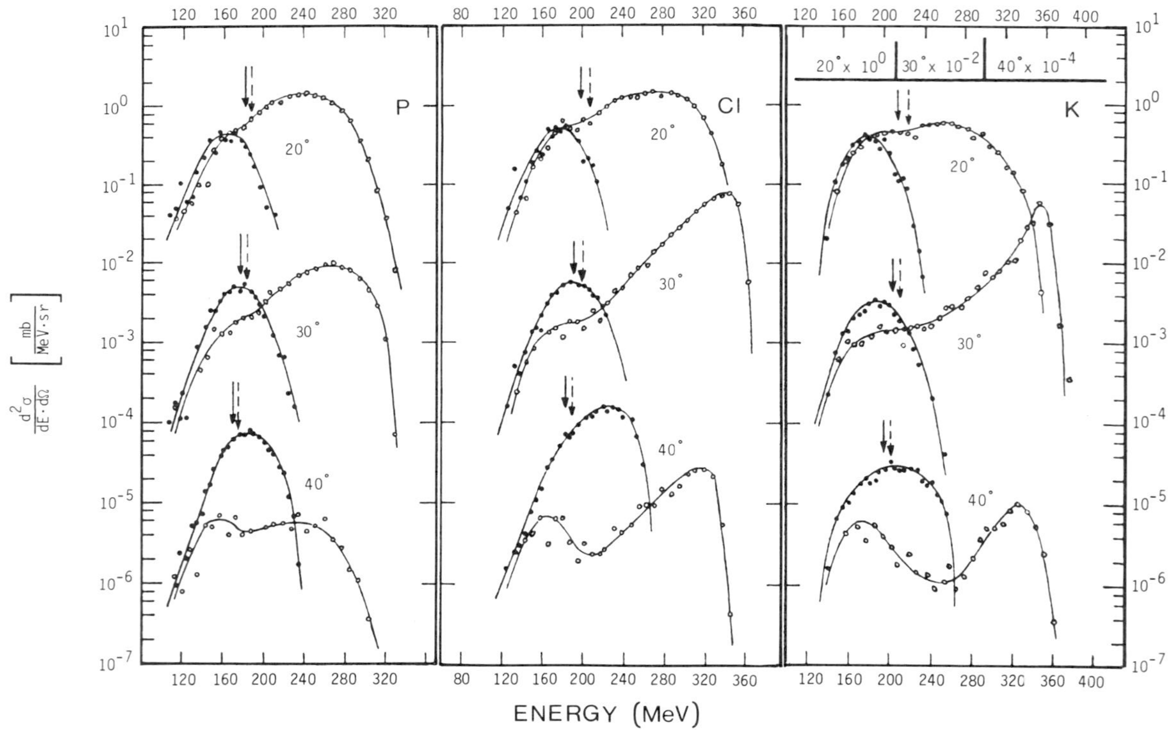

Figure 134. Superposition of kinetic-energy spectra $d^2\sigma/d\Omega\, dE$ of fragments from the reaction $^{232}\mathrm{Th} + {}^{40}\mathrm{Ar}$ measured by Artukh *et al.* for different bombarding energies and laboratory angles. Spectra for P, Cl, and K fragments are shown from the left to the right, grouped according to observation angle. The solid dots represent the measurements at $E_{\mathrm{Lab}} = 297$ MeV, the open circles are data obtained for $E_{\mathrm{Lab}} = 388$ MeV. The solid and dashed arrows mark the laboratory energies corresponding to the Coulomb repulsion of two touching spherical fragments for $E_{\mathrm{Lab}} = 297$ and 388 MeV, respectively. From (Art 73b).

fragment energy spectra discussed earlier. For example, Mg fragments, not included in Figure 134, show energy spectra with a width (FWHM) changing from typically 40–50 MeV at $E_{\text{Lab}} = 297$ MeV to 80–90 MeV at $E_{\text{Lab}} = 388$ MeV (Art 73b).

In addition to the above broadening effect, the spectral shape is drastically altered by raising the bombarding energy. Most fragment energy spectra measured at the lower energy consist of a single bell-shaped component at all angles, whereas two distinct spectral components referred to earlier are seen at the higher energy. However, the appearance of two peaks in the spectra of K and Ca fragments at large angles suggests that the energy distribution could in general be a superposition of two components that can only be resolved at sufficiently high incident energies. In agreement with observations made earlier in this section, it is seen also for the reaction $^{232}\text{Th} + {}^{40}\text{Ar}$ that the most probable fragment kinetic energy does not correlate strongly with the magnitude of the Coulomb repulsion as calculated for spherical nuclei, which is indicated by the arrow at each spectrum in Figure 134. This is true for angles forward and backward of the grazing angle.

This effect is more quantitatively visible in Figure 135, displaying the most probable total kinetic energies for both fragments from the same $^{232}\text{Th} + {}^{40}\text{Ar}$ reaction studied by Jacmart *et al.* (Jac 75) at $E_{\text{Lab}} = 295$ MeV. In this figure, the most probable energies $\langle E \rangle$ are plotted vs. mass number A of the light fragment at laboratory angles of 18° (left) and 40° (right). The data taken at the more forward angle show a staircase-like behavior where $\langle E \rangle$ increases with increasing charge symmetry of the fragmentation, irrespective of the particular isotope measured. In contrast, the 40° data displayed on the right of Figure 135 have a much smoother behavior with A and outline a peak centered around the mass of the projectile. The horizontal dashed lines included in the left portion of the figure correspond to Coulomb repulsion of spherical nuclei touching at a large radial separation calculated from $R = 1.75(A_1^{1/3} + A_2^{1/3})$ fm, thus simulating significant fragment deformations. While the most probable total kinetic energies measured at $\theta_{\text{Lab}} = 18°$ are well described by such a parametrization, energy values observed at $\theta_{\text{Lab}} = 40°$ are much higher than that. Obviously, at angles close to the grazing angle ($\theta_g = 47°$), memory of the entrance channel is largely preserved, in agreement with the relatively narrow mass distributions observed (Jac 75) at such angles. This would again be consistent with an orbiting behavior of the reaction, where the two unresolved cross section branches of different kinetic energy in the Wilczyński diagram (see Figure 78) merge into a single, narrower branch at forward angles.

The above considerations illustrate the difficulty of distinguishing unambiguously between partially and fully relaxed energy components for reactions induced by relatively light projectiles and low bombarding energies. Nevertheless, it is obvious from the data shown in Figures 133–135, that fragments can be produced in damped reactions that have large deformations at the instant of breakup of the intermediate dinuclear system. From inspection of Figure 134, it is rather conspicuous, how well the low-energy slopes of the spectra and their end points remain unaffected by changes in the bombarding energy. The small parallel shifts of the laboratory energy distributions seen in Figure 134 with varying bombarding energy can be attributed entirely to the different velocities of the c.m. system. This feature strongly suggests that in damped reactions the extreme dinuclear shapes that can be formed are limited only by the tensile strength of nuclear matter (Ber 78b).

A less ambiguous evaluation of the limiting final kinetic energies and, hence, of the corresponding dinuclear shapes that can be attained in damped reactions is possible for systems studied at relatively high bombard-

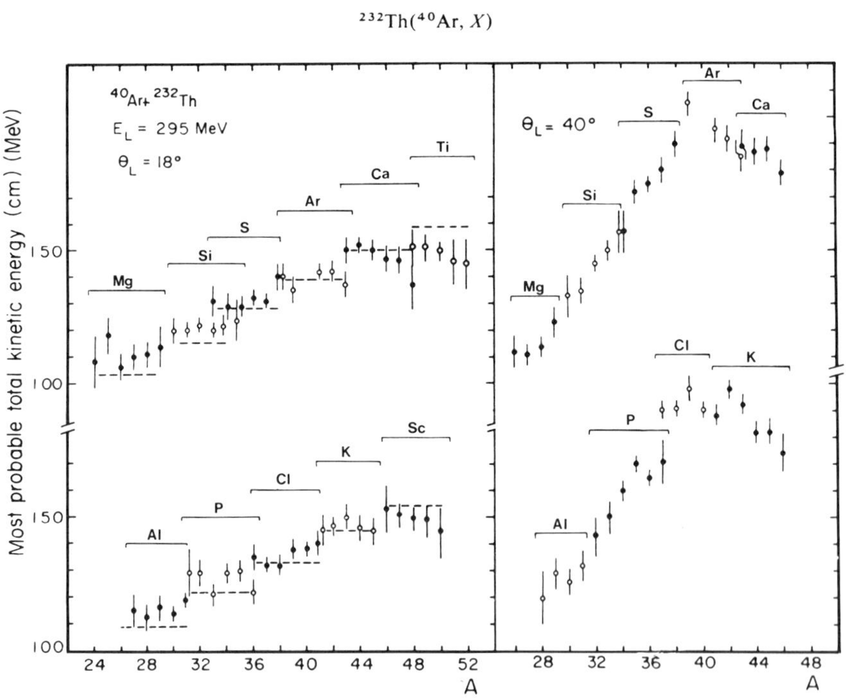

Figure 135. Most probable total kinetic energies of fragments from the reaction $^{232}Th + ^{40}Ar$ at $E_{Lab} = 295$ MeV plotted vs. A of one of the fragments detected at $\theta_{Lab} = 18°$ (left) and $\theta_{Lab} = 40°$ (right). Adjacent elements are distinguished by open circles and solid dots. The horizontal dashed lines represent for each element 80% of the Coulomb repulsion for touching spherical nuclei calculated with a radius parameter $r_0 = 1.4$ fm. From (Jac 75).

ing energies and simultaneously long interaction times. The reaction Ag + ^{20}Ne studied by Babinet *et al.* (Bab 76) at incident energies of $E_{\text{Lab}} = 175$ and 252 MeV presents such an opportunity. Here, partially and fully damped components in the fragment energy spectra are clearly separated for a number of detection angles. Furthermore, as illustrated in Figure 53, the angular distributions of the differential cross section exhibit a transition from forward-peaked to $(1/\sin\theta)$-type shapes, characteristic of very long interaction times, as the projectile-like fragment becomes heavier.

In Figure 136 (Bab 76), the most probable fully damped c.m. kinetic energy $E_{\text{c.m.}}$ of each fragment is plotted vs. Z, for a number of laboratory observation angles. Disregarding the values obtained at relatively forward angles for fragments in the vicinity of the projectile, one observes an approximately 20-MeV variation of $E_{\text{c.m.}}$ with Z, which is approximately independent of detection angle. Included in this figure are the energies obtained by the projectile-like fragment through acceleration by the Coulomb repulsion, calculated for spheres (upper curve) and equilibrium spheroids (lower curve) that are initially at rest with a surface separation of 2 fm. As is inferred (Bab 76) from this model comparison, the interacting fragments

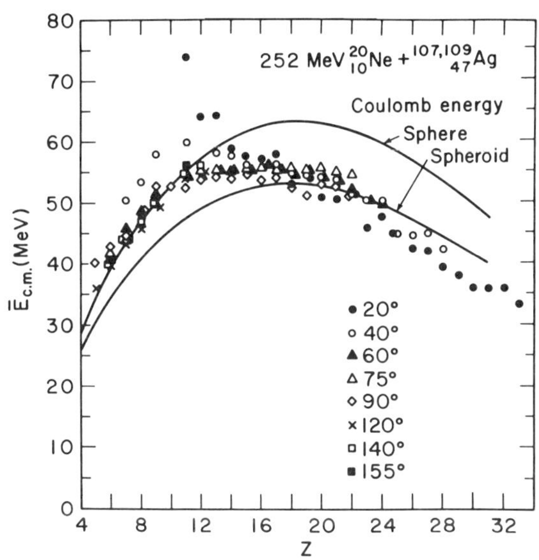

Figure 136. Most probable c.m. kinetic energy $\bar{E}_{\text{c.m.}}$ of the fully relaxed component in each fragment energy spectrum plotted vs. fragment Z, as measured for the reaction Ag + ^{20}Ne at $E_{\text{Lab}} = 252$ MeV and various laboratory angles. These energies are to be distinguished from the most probable total kinetic energies. The two curves represent Coulomb energies for two spheres or equilibrium spheroids with a surface separation of 2 fm. From (Bab 76).

remain rather spherical for large asymmetries but acquire gradually significant deformations with increasing mass transfer to the projectile. A similar observation can be made for the end-point total kinetic energies, as can be inferred from Figure 103 illustrating the Wilczyński diagram for the heavy system $^{166}Er + ^{86}Kr$.

A comparison of data obtained at different bombarding energies shows that the most probable fully damped kinetic energies increase somewhat with increasing incident energy. Similarly, the width of the fully damped component increases. This feature has been observed, for example, by Babinet *et al.* (Bab 76) for the reaction $Ag + ^{20}Ne$ discussed above. A slight bombarding-energy dependence of the average final kinetic energies can be expected if the kinetic energy in the entrance channel is not always completely damped. As has been outlined in Section 2.3, this is expected to hold for the orbital part of the kinetic energy, since classically an initial angular momentum l can at most be reduced to its "sticking limit" l_s [cf. Equation (2.3.11)]. For maximum dissipation of kinetic energy, one expects classically a fully relaxed final total kinetic energy

$$E = V_{\text{Coul}}(R) + V_{\text{Nucl}}(R) + f^2 \frac{\hbar^2 l(l+1)}{2\mu R^2} \tag{3.5.5}$$

given by the sum of Coulomb, nuclear, and centrifugal potential at the breakup distance R. In Equation (3.5.5), μ is the reduced mass of the final fragmentation and $f = l_s/l$ denotes the undamped fraction of the initial angular momentum. In the "sticking limit,"

$$f = \mu R^2/(\mu R^2 + \mathscr{I}_3 + \mathscr{I}_4) \tag{3.5.6}$$

is given by the ratio of the moment of inertia for orbital motion to the rigid-body moment of inertia. In Equation (3.5.6), $\mathscr{I}_3$ and $\mathscr{I}_4$ are the intrinsic moments of inertia of the two outgoing fragments about an axis perpendicular to the reaction plane. Hence, if the range of l values contributing to the reaction increases with incident energy, the minimum energy will increase accordingly. In addition, the angular-momentum term in Equation (3.5.5) predicts a fragmentation dependence of the fully relaxed total kinetic energy.

The above simple model has been used by various groups (Egg 76, Nat 77, Aga 77, Cor 77c, Van 80e) in order to determine the fraction $1/f$ of dissipated angular momentum and the scission radius for strongly damped reactions. Mainly lighter systems were studied, ranging from $^{27}Al + ^{20}Ne$ to $^{64}Ni + ^{40}Ca$, where the band of angular-momentum values contributing is relatively well defined. In Figure 137, results are shown of a fit of Equations

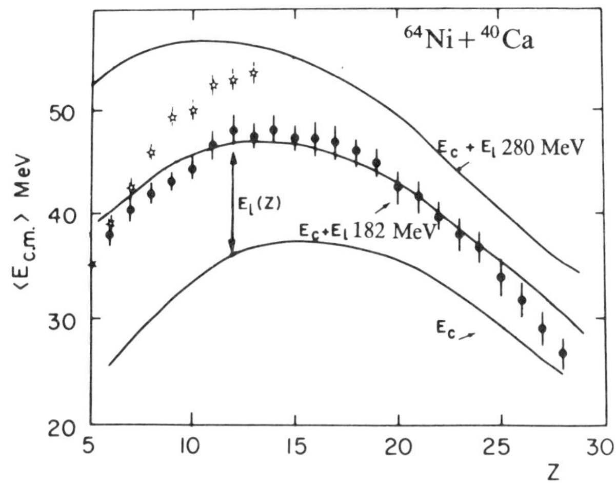

Figure 137. Experimental mean c.m. kinetic energies of fragments produced in the reaction $^{64}Ni + ^{40}Ca$ at $E_{Lab} = 182$ MeV (solid dots) and 280 MeV (open symbols) plotted vs. fragment Z. The curve in the middle represents a fit with Equation (3.5.5) employing the sticking approximation. The curve labeled "E_c" gives the contribution due to Coulomb repulsion. The upper curve represents the predictions for $E_{Lab} = 280$ MeV using the parameters obtained from a fit to the low-energy data. From (Aga 77).

(3.5.5) and (3.5.6), neglecting the small influence of the nuclear potential, to data obtained by Agarwal *et al.* (Aga 77) for the reaction $^{64}Ni + ^{40}Ca$ at $E_{Lab} = 182$ MeV, and by Gatty *et al.* (Gat 75, 75b) at $E_{Lab} = 280$ MeV. In this figure, experimental mean c.m. kinetic energies of projectile-like fragments are plotted vs. fragment charge as solid and open symbols for $E_{Lab} = 182$ and 280 MeV, respectively. Maximum angular-momentum dissipation is assumed, fixing the parameter f for a given R. From the measured cross sections, the range of contributing initial angular momentum is determined to be 58–$69\hbar$ and 75–$100\hbar$ for the low and high incident energy, respectively. The intermediate curve in Figure 137 represents a fit with Equation (3.5.5), where the distance d between the surfaces of spherical fragments with individual radii $R_i = 1.225A_i^{1/3}$ has been adjusted. The curve marked "E_c" represents the contribution of the Coulomb repulsion alone. Although a good fit with a large constant surface separation of $d = 4.5$ fm is obtained (Aga 77) for the 182-MeV data, the high-energy data do not follow the predictions of the model represented by the uppermost curve in Figure 137. Additional model assumptions have to be invoked (Aga 77) such as restrictions in angular momentum (Rus 77b) contributing to the production of the various fragmentations, in order to make the observed discrepancies plausible within the context of this simple model.

For the reaction $^{27}Al + ^{35}Cl$, in the limited bombarding-energy range between $E_{Lab} = 140$–170 MeV, Cormier *et al.* (Cor 77c) have successfully described the fragment kinetic energies in terms of Equations (3.5.5) and (3.5.6), deducing values for the model surface separation of $d \approx 4$ fm. The magnitudes of the surface separations at breakup of the intermediate system derived in these analyses, accounting for angular momentum, are considerably larger than those obtained (e.g., Bab 76, Obe 78) in analyses of final kinetic energies in terms of Coulomb repulsion alone. Even though, as suggested by Braun–Munzinger *et al.* (Bra 76b), it is possible to derive properties of the friction force and the limiting scission shapes in strongly damped collisions from the final fragment energies measured at various incident energies, such a conclusion must be regarded as highly model dependent. Analyses as described above neglect, for example, the possibility of an angular-momentum dependence of the scission configuration as suggested (Coh 74) by the liquid-drop model in the case of fission.

While the above phenomenological attempts to understand the fragment kinetic energies in damped reactions explore the consistency of hypothetical macroscopic conditions of an interacting system with the data, more specific conclusions can be drawn from comparisons with predictions based on a microscopic model. Very simple, semiclassical models derived from more elaborate theories of few-nucleon transfer reactions (e.g., But 71, Bro 72, 72b, 72c) relate the probability for a nucleon exchange process with the degree to which matching conditions with respect to binding and interaction energies, relative velocities, angular momenta, and separation distance are fulfilled in the reaction. While such models do not explicitly describe the intrinsic mechanism by which the corresponding amount of collective kinetic energy is transformed into excitation energy, they identify the transfer of nucleons as a source for dissipation. Several suggestions have been made (But 71, Sie 71, Bro 72c, Bri 72, Kle 73, Sch 73, Wil 73b) to calculate the "optimum Q value" at which the exchange probability has a maximum. A number of experimental studies have been performed (Art 73, 73b, Sch 73, Oer 73, Reh 73, Wil 75b, Pou 75, Mik 76, Bil 78, Gel 78) comparing measured most probable Q values or energy losses with systematics predicted for the optimum Q value in multinucleon transfer reactions induced by relatively light heavy ions.

In order to illustrate the significance of the concept of an optimum Q value in heavy-ion reactions, one may consider the subbarrier transfer of a particle with charge Z from the projectile (Z_P) to the target (Z_T), assumed to be at rest at the classical turning point r_i [cf. Equation (2.1.13)]. Then, the energy, $-Q_{opt}$, necessary to effect the transfer is the sum of the ground-state Q values, Q_{gg}, calculable, for example, from the nuclear binding energies, and the change in interaction energy converted into excitation energy E^*, i.e., (But 71, Sch 73),

$$Q_{\text{opt}} = Q_{gg} - E^* = Q_{gg} - e^2\left[Z_P Z_T/r_i - (Z_P - Z)(Z_T + Z)/r_f\right]$$

$$\approx Q_{gg} - \frac{2E_{\text{c.m.}}}{1 + 1/\sin(\theta_{\text{Coul}}/2)}\left[1 - \frac{(Z_P - Z)(Z_T + Z)r_i}{Z_P Z_T r_f}\right] \quad (3.5.7)$$

In Equation (3.5.7), $E_{\text{c.m.}}$ is the bombarding energy and θ_{Coul} denotes the Rutherford scattering angle associated with r_f, the distance of closest approach for the final charge division. The effect on the Q value of the change in the distances of closest approach is often termed the "recoil effect." Since Equation (3.5.7) applies only to a near-central sub-Coulomb collision, for more general applications, the Coulomb potential in Equation (3.5.7) should be replaced by the total effective interaction potentials $V_{\text{eff}}(i)$ and $V_{\text{eff}}(f)$ in entrance and exit channels, respectively. An additional, important recoil effect arises (Sie 71, Bri 72, Swi 72, Ran 77c, Sch 77d) when the transfer takes place while the nuclei move with relative velocity v. With m being the mass of the transferred particle assumed to be on the average at rest in the donor nucleus, the additional energy required (compare Section 3.4) amounts to $\delta Q = -(m/2)v^2$. Combining the different contributions, one predicts

$$Q_{\text{opt}} = Q_{gg} - E^* = Q_{gg} - \left\{V_{\text{eff}}(i) - V_{\text{eff}}(f) + \frac{m}{\mu}\left[E_{\text{c.m.}} + V_{\text{eff}}(i)\right]\right\} \quad (3.5.8)$$

to characterize the position of maximum cross section in the Q-value spectrum. It is interesting to note that for nucleon exchange between the constituents of a heavy dinuclear complex the effect of Q_{gg} on the optimum Q value is partially canceled by the difference in effective potentials. This can be seen, for example from Figure 103, where the curves for the ground-state Q value, Q_{gg}, and the Coulomb interaction potential for different charge fragmentations are parallel to each other. One expects then approximately a linear variation of the optimum Q value with bombarding energy, $E_{\text{c.m.}}$, and the transferred mass m or the number n of the transferred nucleons, i.e.,

$$Q_{\text{opt}} \approx \alpha n + \beta \quad (3.5.9)$$

with an energy-dependent coefficient α and a system-dependent constant β.

Mikumo *et al.* (Mik 76) have performed an analysis of kinetic energy spectra of fragments produced in the reactions $^{52,53}Cr + {}^{14}N$ and $^{232}Th + {}^{15}N$ (Art 71c, 73) in terms of optimum Q values. Figure 138 displays the variation of the most probable Q value, Q^m ($\approx Q_{opt}$), with the net number n of transferred nucleons for the former reaction at laboratory energies between $E_{Lab} = 64$ and 95 MeV. Indeed, for each bombarding energy, a linear relationship is obtained for $n \leq 4$–5. Fit values of α and β are illustrated in the insert of Figure 138. It is found that, in accord with the above simple considerations, the quantity β (≈ -3 MeV) is a constant, whereas the coefficient α decreases linearly with incident energy above the barrier. The deduced value for the slope of α (Mik 76) with respect to the incident energy above the Coulomb barrier, $d\alpha/d\,[E_{c.m.} - V_{Coul}(i)] \approx -0.1$, is as expected from Equation (3.5.8) with $1/\mu = 0.09$ amu^{-1}. As is also

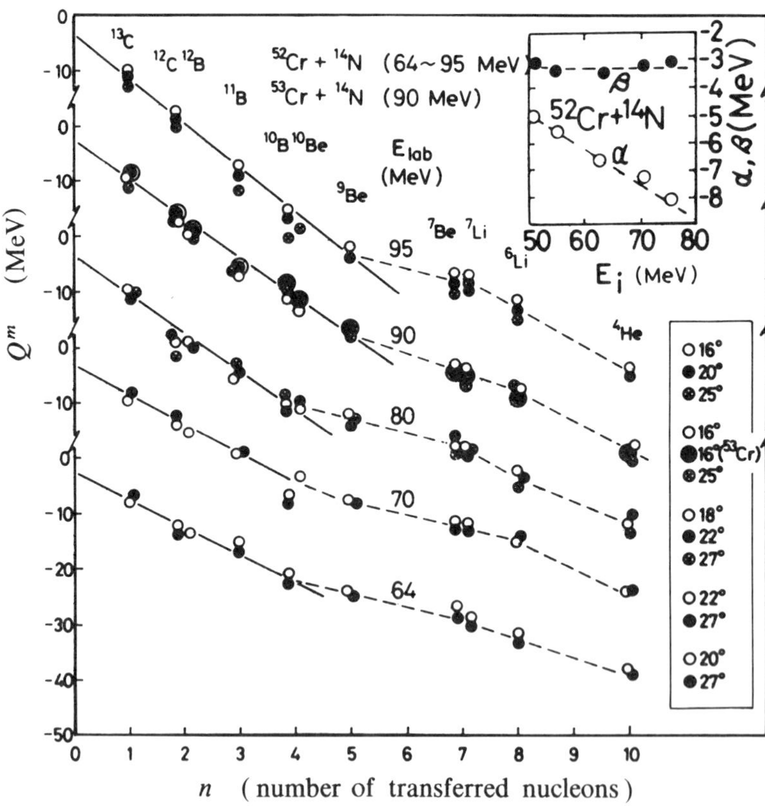

Figure 138. The most probable Q values Q^m of the reactions $^{52,53}Cr + {}^{14}N$ as function of the number of transferred nucleons n. In the upper right part the values of α of Equation (3.5.7) are plotted vs. incident c.m. energy E_i. From (Mik 76).

inferred from Figure 138, the relation of Equation (3.5.9) applied to neutron and proton transfers with the same set of parameters α and β, although the Q_{gg} values are rather different. It is also independent of the target isotope considered and the observation angle. This behavior is interpreted (Mik 76) to indicate that in few-nucleon transfer reactions kinetic energy is mainly dissipated at a rate per exchanged nucleon that is constant and proportional to the available kinetic energy above the barrier, while the contribution of potential energy differences is of less importance. That the proportionality between Q_{opt} and the net number n of transferred nucleons observed in Figure 138 for small n breaks down for larger n is expected from Equation (3.5.8), because of the decrease of the average kinetic energy available above the barrier with an increasing number of transferred nucleons. Mikumo *et al.* (Mik 76) have performed a similar analysis of their data in terms of an effective optimum Q value, where the contribution of the changing Coulomb interaction potential has been subtracted from Q_{opt} [cf. Equations (3.5.8) and (3.5.9)]. This more model-dependent analysis leads to the same functional dependence of the effective Q value on n and bombarding energy and, hence, does not add to the conclusions drawn from the simpler analysis.

The evidence presented above for the dependence of the optimum Q value on the bombarding energy and the net number of nucleons exchanged is very suggestive of a friction-like energy dissipation mechanism, in which the kinetic energy deposited during the exchange is a sum of the change in static potential energies and a dynamical contribution [cf. Equation (3.5.8)]. Its foundation, hence, is very similar to the one-body dissipation mechanism proposed by Randrup (cf. Section 2.6) and simpler realizations of the same principle, as already discussed briefly in Section 3.4. One may, therefore, observe the onset of nuclear friction in few-nucleon transfer processes induced by relatively light heavy ions, which would hint as to the nature of the dominant dissipation mechanism operating in damped reactions.

In fact, interpretations of kinetic energy spectra of fragments produced in interactions of relatively light systems in terms of friction- or diffusion-like processes have been reported by several groups (Nör 74, Wil 75b, Oes 77,78, Bil 78, Gel 78, Mat 80b). Assuming that the radial component of the relative velocity is completely lost before a nucleon transfer takes place, Wilczyński *et al.* (Wil 75b) deduce an average coefficient for tangential friction from the measured angular dependence of the optimum Q value in the reaction $^{58}\text{Ni}+^{16}\text{O}$ at $E_{\text{Lab}} = 60$, 72, and 81 MeV. An approximate explanation of the observed variation of Q_{opt} is obtained (Wil 75b) with a friction coefficient of $c_t \approx 10^{-22}$ MeV s fm^{-2}, which is of the order of that estimated by Gross and Kalinowski (Gro 74, cf. Figure 27) for very peripheral collisions. A similarly qualitative agreement between this friction model and data on ^{12}C-, ^{14}Ni-, ^{16}O-, and ^{22}Ne-induced transfer reactions at incident energies between 1.7 and 3.1 times the interaction barrier is reported by Billerey *et al.* (Bil 78).

Of considerable interest for an understanding of the energy dissipation mechanism is the width of the fragment kinetic energy spectrum, Γ_E, in particular its relation to the mean final energy or Q_{opt}. For example, if the dissipation of kinetic energy were due to a classical equilibrium diffusion process, one would expect a linear relation

$$\Gamma_E^2 = (8\ln 2)\sigma_E^2 = (16\ln 2)(D_{EE}/v_E)E_{\text{Loss}} = (16\ln 2)E_{\text{Loss}}\cdot\tau \quad (3.5.10)$$

between the square of the width (Γ_E^2) or variance (σ_E^2) and the average energy loss (E_{Loss}), due to the Einstein relation [Equation (2.7.17)] between energy drift and diffusion coefficient, respectively, involving the nuclear temperature τ.

Nörenberg (Nör 74) has analyzed the ^{233}Th + ^{40}Ar reaction studied by Artukh *et al.* (Art 73b) at 388 MeV in terms of an energy diffusion model. Relating the interaction time to the angle of rotation of the dinuclear system (see Section 3.3), it is possible to derive effective drift and diffusion coefficients from the variation of centroid and width of the energy spectrum with reaction angle. Values of effective drift and diffusion coefficients, v_E and D_{EE}, respectively, deduced (Nör 74) for this reaction at small and intermediate deflection angles, are

$$v_E \approx 4\cdot 10^{23}\ \text{MeV}\,\text{s}^{-1} \quad (3.5.11\text{a})$$

$$D_{EE} \approx 4\cdot 10^{24}\ \text{MeV}^2\,\text{s}^{-1} \quad (3.5.11\text{b})$$

Although the absolute values of these transport coefficients depend strongly on the phenomenological procedure by which the interaction time is estimated from the data, their ratio should be largely unaffected by it and reflect simply the Einstein relation, i.e., $D_{EE}/v_E = \tau$. It is, hence, significant to notice that the temperature parameter $\tau \approx 10$ MeV, obtained from the ratio of the above energy transport coefficients, exceeds even the Fermi gas temperature calculated from the highest observed (Art 73b) energy losses by more than a factor of 3, suggesting strongly that a description of energy dissipation in terms of a classical equilibrium diffusion model is inadequate. This finding complements similar observations made in Section 3.4 with respect to mass transport.

Very little experimental information on the widths of final kinetic energy spectra is currently available for lighter systems, and what there is appears difficult to understand within any given model. Figure 139, for example, shows the most probable (Q^m) and the half-maximum (Q_1 and Q_2) Q values obtained by Mikumo *et al.* (Mik 76) for ^{14}N-induced

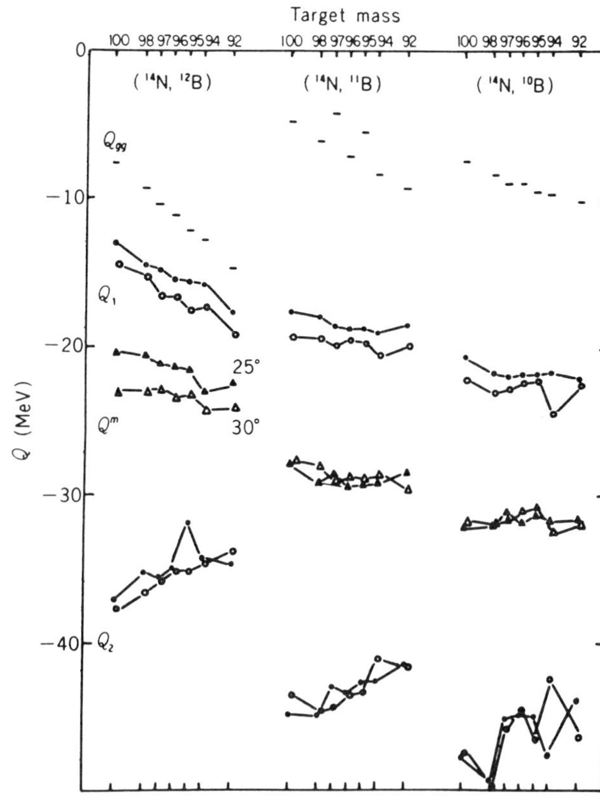

Figure 139. Most probable (Q^m) and half-maximum Q values (Q_1 and Q_2) for fragments produced in the reaction ${}^{A}\text{Mo}+{}^{14}\text{N}$ at $E_{\text{Lab}}=97$ MeV. Open symbols correspond to a laboratory scattering angle of 30°, closed symbols represent $\theta_{\text{Lab}}=25°$ data. The ground-state Q values are indicated by the horizontal bars. From (Mik 76).

reactions on Mo isotopes at $E_{\text{Lab}}=97$ MeV, corresponding to twice the Coulomb barrier. Values for Q_1, Q^m, and Q_2 are given for laboratory angles of 25° (solid symbols) and 30° (open symbols) for 2-, 3-, and 4-nucleon stripping reactions on the seven stable Mo isotopes. One observes from this figure that the energy width $\Gamma_E = Q_1 - Q_2$ generally increases with decreasing value of Q^m or increasing mean energy loss. Furthermore, for a given number of transferred nucleons, Γ_E decreases with decreasing (more negative) ground-state Q value, Q_{gg}, or decreasing excitation energy of the system. This dependence is rather obvious for the (^{14}N, ^{12}B) reaction depicted on the left of Figure 139. Qualitatively, these observations are consistent with the expectations from a diffusion model as represented by Equation (3.5.10). However, as follows from the previous discussion, the effective energy diffusion coefficient is in general not constant but depends on the bombarding energy.

The bombarding-energy dependence of the width Γ_E of the fragment total kinetic energy distribution is exemplified in Figure 140 displaying Γ_E as a function of c.m. scattering angle and final fragmentation for the reaction $^{50}Cr + ^{14}N$ measured by Billerey *et al.* (Bil 78) at $E_{Lab} = 71$ and 100 MeV. It can be observed from this figure that the width Γ_E scales roughly in proportion to the bombarding energy. For example, the energy spectrum of F isotopes has a width of $\Gamma_E = (11 \pm 0.5)$ MeV at $E_{Lab} = 71$ MeV and

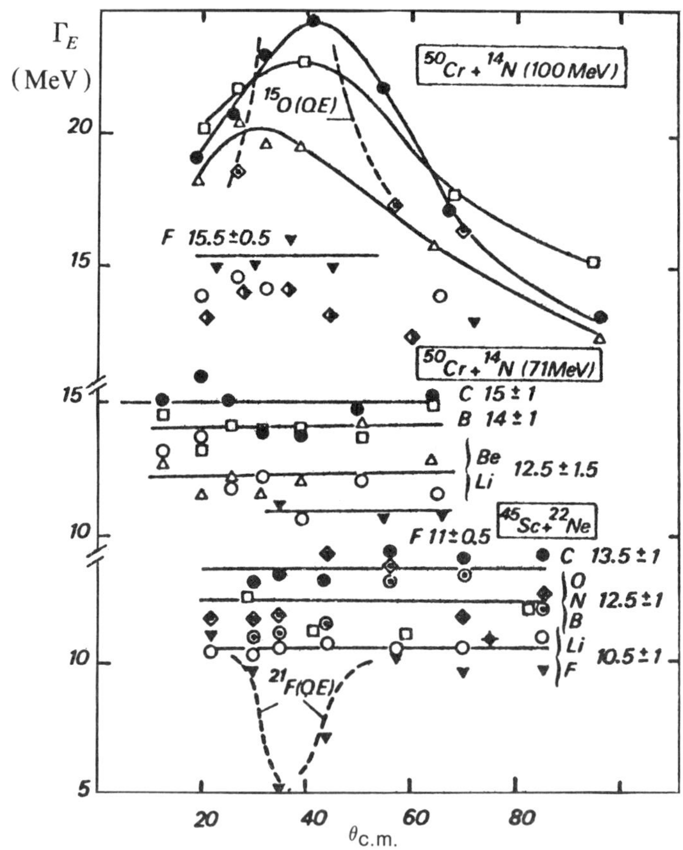

Figure 140. Width Γ_E (FWHM) of fragment kinetic energy spectra vs. scattering angle $\theta_{c.m.}$ for various final fragments indicated at the curves. Data are shown for the $^{50}Cr + ^{14}Ni$ reaction in the top and middle parts for $E_{Lab} = 100$ and 71 MeV, and for $^{45}Sc + ^{22}Ne$ at $E_{Lab} = 84$ MeV at the bottom of the figure. The numbers represent average values of Γ_E and their uncertainties. From (Bil 78).

$\Gamma_E = (15.5 \pm 0.5)$ MeV at $E_{\rm Lab} = 100$ MeV. For fragments close to the projectile, the widths can be rather large and angle dependent, a possible consequence of the existence of two unresolved cross section ridges, as discussed previously. For the reactions illustrated in Figure 140, the dependence of Γ_E on the net number of transferred nucleons is not very obvious. This could lead to the conclusion that the net number of transferred nucleons is only weakly representative of the total number of nucleon exchange processes for such light systems, although the $Q_{\rm opt}$ systematics discussed earlier suggests otherwise. It may also reflect the difficulty of distinguishing quasielastic and damped energy distributions in an unambiguous fashion, except for fragments far removed from the projectile, where the former component is absent. Most likely, sequential evaporation processes do not pose paramount problems for the interpretation of the experimental data, because the light fragments acquire only a small fraction of the total energy loss as intrinsic excitation energy (see below).

In a related study, Oeschler *et al.* (Oes 77,78) have analyzed the dependence on reaction angle and mass asymmetry of the width of the energy distribution for fragments from the reaction ^{130}Te + ^{28}Si at $E_{\rm Lab} = 140$ MeV. In Figure 141 (a–c), their experimental results (Oes 78) for the variance σ_E^2 are plotted vs. angle and number of transferred nucleons. It is seen from these data that, for a given fragmentation, the variance does not vary with angle (cf. Figure 141a, b) in the vicinity of the grazing angle but becomes increasingly larger at forward angles, where an orbiting-like component appears in the angular distribution of the differential reaction cross section. As is demonstrated in Figure 141c, the variance measured close to the grazing angle depends distinctively on the number of exchanged nucleons. At forward angles, characterized by strongly decreasing average Q values, the situation becomes more complicated, as can be expected from the possible simultaneous presence of events produced in reactions of very different reaction times. This transition has been interpreted (Oes 77,78) in terms of the onset of stochastic dissipation and nucleon exchange mechanisms. It is found (Oes 77) that, at forward angles, the energy variance is proportional to the average energy loss, suggesting an analysis in terms of Equation (3.5.10). However, again the ratio of the derived energy diffusion and drift coefficients turns out to be significantly larger than expected from the temperature of the system.

The results of analyses of fragment energy distributions discussed above, although not supporting the concept of a classical equilibrium energy diffusion process in damped reactions, do not contradict a basically stochastic nature of the dissipation mechanism. A more general class of stochastic energy dissipation processes including those giving rise to classical friction forces (cf. Section 2.6) can be described in terms of a Langevin equation for the time dependence of the collective energy E of relative motion, ap-

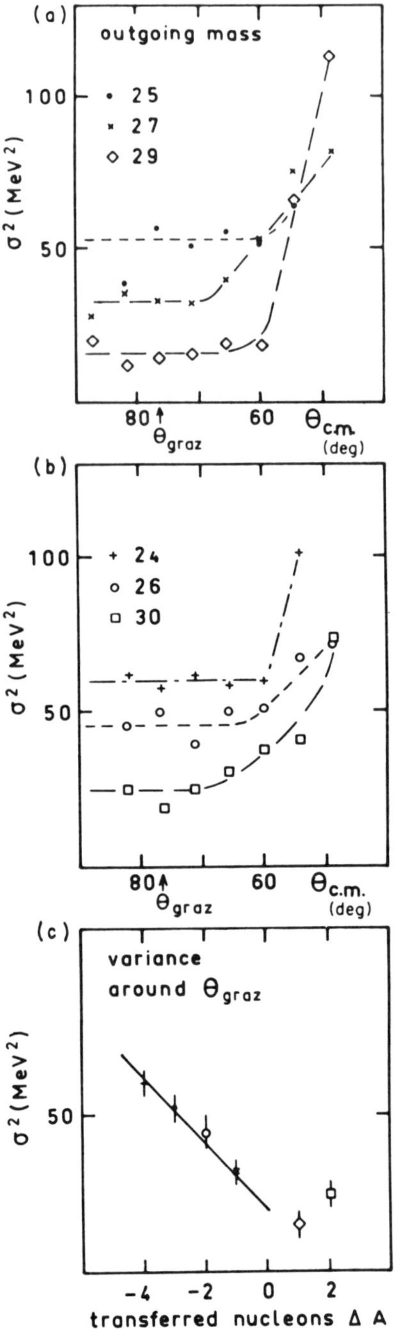

Figure 141. Dependence of the variance σ^2 of the energy distribution of fragments from the reaction $^{130}Te + ^{28}Si$ at $E_{Lab} = 140$ MeV on reaction angle, $\theta_{c.m.}$, and net number of nucleons, ΔA, transferred to the projectile. Different final fragmentations represented by the mass number of the light fragment are indicated by the different symbols. From (Oes 78).

proximated by the simple form

$$dE/dt = -kE + Q^{+}(t) \tag{3.5.12}$$

Here, k is a friction coefficient describing the decrease in the average energy and Q^{+} denotes the fluctuative dispersion of the energy loss caused by a fluctuating force F^{+} [compare Equations (2.7.24) and (2.8.23)]. For normally distributed autocorrelations of this force, Equation (3.5.12) is stochastically equivalent to a Fokker–Planck equation with Gaussian solutions for the probability distribution $P(E,t)$ that can be characterized by first and second moments of the energy loss, $\langle E_{\text{Loss}}\rangle$ and σ_E^2, respectively, given by Equations (2.8.2). Based on such a Langevin mechanism with constant transport coefficients, one would expect a time dependence of the form

$$\langle E_{\text{Loss}}\rangle \approx E_0[1-\exp(-kt)] \tag{3.5.13a}$$

$$\sigma_E^2 \approx D_{EE}k^{-1}[1-\exp(-2kt)] \tag{3.5.13b}$$

where E_0 is the initial kinetic energy available above the barrier. The diffusion coefficient is proportional to the time-integrated autocorrelation function of the fluctuating force F^{+} and related to the friction coefficient k and the nuclear temperature τ via $D_{EE} = k\tau$. For time t smaller than the characteristic relaxation time $t_{\text{rel}} = k^{-1}$, both first and second moments of the energy distribution should be linear functions of time, whereas in the long-time limit, they are expected to level off.

Obviously, Equation (3.5.13a) for the average energy loss has a similar structure as Equation (3.4.13) derived (Sch 77d, Wil 80) under the assumptions that the energy lost in a damped heavy-ion collision is dominantly due to the dynamical part of the Q value [last term in Equation (3.5.8)] associated with the exchange of nucleons, and that the progressing interaction time and the number of transfers can be expressed in terms of the growing width of fragment A or Z distributions. From the slope of the correlation [cf. Equation (3.4.12)] between final kinetic energy and charge variance, an initial dissipation rate of the order of

$$\left.\frac{d\langle E_{\text{Loss}}\rangle}{dt}\right|_{t=0} = kE_0 \propto 10^{24}\ \text{MeV}\,\text{s}^{-1} \tag{3.5.14}$$

considerably larger than that obtained by Nörenberg [cf. Equation (3.5.11a)], has been deduced (Hui 76, Sch 78g, Wil 80) from analyses of heavier systems.

Matsuzaki *et al.* (Mat 80b) have applied the Langevin formulation of energy dissipation in a phenomenological analysis of the ^{232}Th $+$ ^{40}Ar reaction at $E_{\text{Lab}} = 388$ MeV. In their work, the energy loss or excitation energy

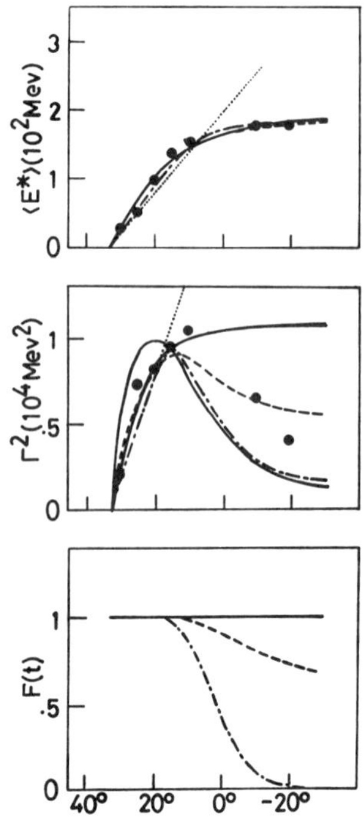

Figure 142. Mean value, $\langle E^* \rangle$, and square width (FWHM), Γ^2, of the energy distribution of Cl fragments produced in the reaction $^{232}\mathrm{Th} + {}^{40}\mathrm{Ar}$ at $E_{\mathrm{Lab}} = 388$ MeV, plotted vs. c.m. reaction angle. The data points are obtained from a deconvolution of the double-differential cross section $d^2\sigma/d\Omega\, dE$ into a higher-energy component associated with positive angles and an orbiting contribution associated with negative deflection angles. The dotted straight lines represent analysis in terms of Nörenberg's diffusion model (Nör 74). The bold solid curves are fit results employing Equations (3.5.13). The thin solid and dashed curves represent analysis in terms of a discrete random-walk-like process with varying fractions F of the cross section corresponding to capture. From (Mat 80b).

(E^*) distribution is obtained either by employing Equations (3.5.13) with phenomenological parameters or by assuming a stochastic energy transport in discrete energy units. To deduce an interaction time scale, a linear relation between time and reaction angle is assumed. In Figure 142, the solid dots represent data (Art 73b) on average values, $\langle E^* \rangle$, and quadratical widths, Γ^2, of the energy loss or excitation energy distribution associated with Cl isotopes produced in this reaction, plotted vs. positive and negative deflection angles. For this representation, the lower-energy, orbiting-like cross-section ridge in the Wilczyński diagram (see Figure 78) has been associated with negative-angle scattering. One observes from Figure 142 an initial linear variation of both $\langle E^* \rangle$ and Γ^2 with reaction angle. The dotted straight lines correspond to an analysis performed by Nörenberg (Nör 74) in terms of the equilibrium energy diffusion model discussed earlier in this section. Obviously, the saturation of the average excitation energy and the decrease in Γ^2 for negative angles, corresponding to long interaction times, are not reproduced by such a model, adding support to the general conclusions drawn in the previous discussion of Equations (3.5.10) and (3.5.11). Evidently, in the approach of the domain of limiting energy loss, the rate of dissipation and the fluctuative dispersion in the energy loss both decrease. It is intuitively clear that, since the total energy must be conserved, a limit in deformation that a dinuclear systems attains in highly damped collisions prevents a growth of the dispersion in the energy loss distribution beyond what is allowed by variations in the Coulomb energy due to thermal shape fluctuations. This effect presents a macroscopic correlation in the dissipation process, in a fashion similar to that imposed on the mass transport mechanism by the walls of the driving potential. It is this feature that appears to be successfully modeled in the stochastic-model analysis of Matsuzaki *et al.* (Mat 80b), accounting for energy balance. The heavy solid curves in Figure 142 corresponding to a fit using Equations (3.5.13), can only reproduce the average energy loss but overestimate the width for large energy losses associated with negative scattering angles. The thin solid curves, obtained from discretizing the stochastic, random-walk-like energy transfer in the allowed kinetic-energy region $E_{\text{Loss}} \leq E_0$, reproduce also the qualitative behavior of the width Γ at negative angles. The dashed curves in Figure 142 correspond to different limits in the number of dissipation steps, simulating the onset of fusion for events with large energy losses. The quantity F shown at the bottom represents the fraction of such trapped events.

The model analyses of average values and widths of fragment energy distributions discussed above serve, in the present context, only illustrative purposes. Not too much significance can be assigned to the values of the fit parameters derived for a particular system and a single bombarding energy. Important aspects of the dissipation mechanism are obviously contained in the bombarding-energy dependence of the energy spectra of damped-reaction fragments. Additional experimental information on the properties of the dissipative reaction mechanism are discussed further below with emphasis on heavier systems.

As is illustrated in Figures 133 and 134, the kinetic-energy distributions of fragments produced in damped interactions of very asymmetric systems have relatively simple shapes, suggesting the presence of a partially damped and a more completely relaxed energy component. There is evidently no trace of a more distinct structure, expected to be introduced by sequential particle evaporation from highly excited primary reaction fragments, as is exemplified in Figure 128 for the $^{40}\text{Ca} + {}^{40}\text{Ca}$ reaction. Such a result may be the consequence of an excitation energy partition process disfavoring the light reaction partner in a very asymmetric colliding system. In addition, since the final kinetic-energy pattern produced by the evaporation process depends drastically on the particular decay channel and, to a lesser extent, on the shape of the primary energy distribution, its effect is expected to be averaged out in data summed over several isotopes of the same element or in those for strongly damped events, where numerous decay channels are open. There are, however, several reactions involving mostly symmetric

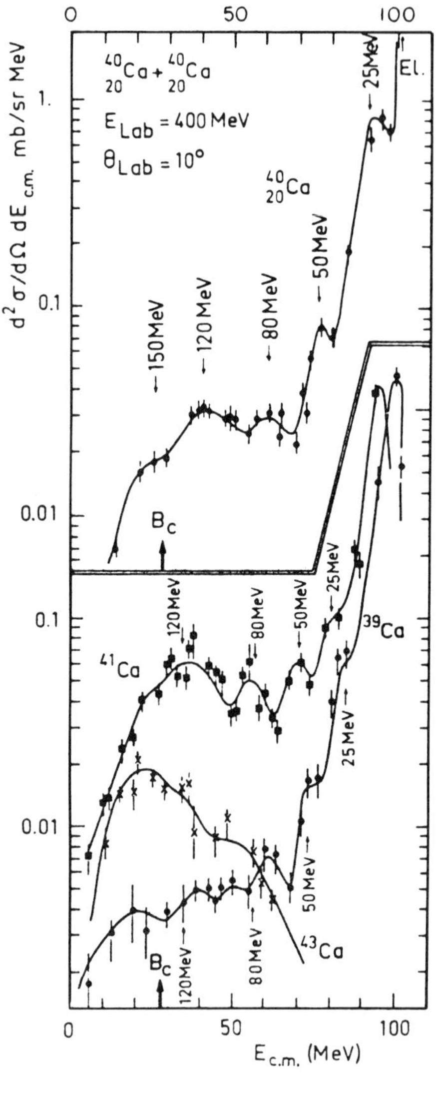

Figure 143. c.m. kinetic energy spectra of ^{40}Ca (top) and $^{39,41,43}Ca$ (bottom) fragments produced in the reaction $^{40}Ca+{}^{40}Ca$ at $E_{Lab} = 400$ MeV. "El." marks the elastic peak; arrows and energy values indicate total excitation energies calculated for a two-body reaction. The Coulomb barrier for spherical fragments is indicated by "B_c." Curves drawn through the data are to guide the eye. From (Fra 80).

medium-weight systems for which distinct structure in the damped fragment energy spectra is reported (Fra 77, 80, 80f, Mig 80b, Roy 81).

An example for this structure is given in Figure 143 displaying c.m. kinetic-energy spectra $d^2\sigma/d\Omega\, dE$ for Ca isotopes produced in the reaction of 400-MeV ^{40}Ca projectiles with a ^{40}Ca target, as observed by Frascaria *et al.* (Fra 80) at a detection angle of $\theta_{Lab} = 10°$. Measurements were performed using magnetic analysis and time-of-flight techniques. The overall shape of these spectra is as expected, emphasizing a strong quasielastic peak for ^{39}Ca, ^{40}Ca, and ^{41}Ca fragments and a much weaker completely relaxed bump, centered at an energy of the magnitude of the Coulomb repulsion for spherical nuclei indicated by the bold arrows (B_c) in the figure. For fragments farther removed from the projectile such as ^{43}Ca, only the completely relaxed component has survived at this angle. Superimposed on the general spectral shape appear additional bumps and shoulders emphasized by the arrows. For the system ^{40}Ca + ^{40}Ca, but not for ^{56}Fe + ^{56}Fe (Mig 80b), the excitation energies at which this structure appears are approximately independent of the particular final fragment considered. At a lower bombarding energy of $E_{\text{Lab}} = 284$ MeV, only the component at ~ 50 MeV of total excitation energy is found (Fra 77) in the energy spectra of fragments from the ^{40}Ca + ^{40}Ca reaction, independent of observation angle in the range $7° \leq \theta_{\text{Lab}} \leq 14°$. For the higher-energy reaction illustrated in Figure 143, additional peaks at total excitation energies of 25, 80, and 120 MeV are discernible. These peaks disappear for angles $\theta_{\text{Lab}} > 15°$, yielding to the familiar smooth appearance of the fully relaxed energy component.

As is illustrated in Figure 128 by the hypothetical energy spectrum of secondary ^{41}Ca ions from the reaction ^{40}Ca + ^{40}Ca at $E_{\text{Lab}} = 400$ MeV, the evaporation process produces structure coinciding with experimentally observed energy peaks indicated by the arrows in this figure. Frascaria *et al.* (Fra 80) have disputed the validity of this interpretation, showing, however, merely that the decay of quasielastic events produced at a sharp excitation energy corresponding to roughly half the total kinetic-energy loss cannot induce the structure in question. This is also borne out by the calculations of Hilscher *et al.* (Hil 79b) allowing for a dispersion in the energy loss partition between the reaction partners. Figure 128 illustrates that point for ^{41}Ca fragments, where a survival of the quasielastic peak is predicted, up to total excitation energies in the vicinity of the sum of proton separation and Coulomb energies, from where on the primary ^{41}Ca yield is depleted exponentially. Hence, the shape of the secondary energy peaks is relatively insensitive to the shape of the primary energy distribution but depends on the relative position of emission thresholds and the excitation energy distribution of the emitting fragment.

Experiments performed by Roynette *et al.* (Roy 81) on light charged-particle emission in coincidence with fragments from the ^{40}Ca + ^{40}Ca reaction at $E_{\text{Lab}} = 400$ MeV show that the structure observed in the inclusive fragment energy spectra (cf. Figure 143) is, indeed, associated with charged-particle emission. This is illustrated in Figure 144 depicting the laboratory energy spectra of Ca and Ar fragments measured (Roy 81) in coincidence with α-particles (top) and protons (bottom), where the fragment detector is located at 10° in the laboratory, with a light charged-particle (LCP) detector placed at a similar angle (11°) on the opposite side of the

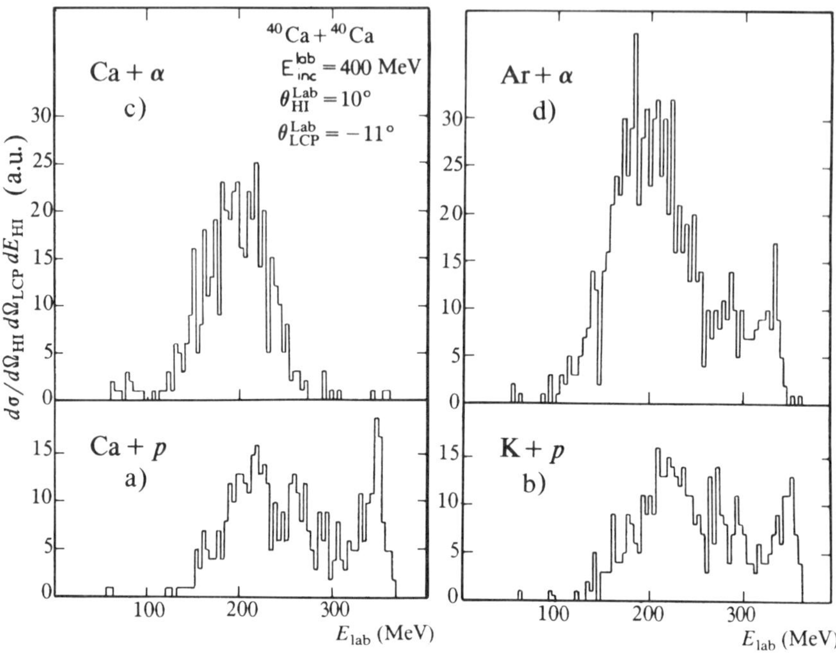

Figure 144. Laboratory energy spectra $d\sigma/d\Omega_{HI}\,d\Omega_{LCP}\,dE_{HI}$ of Ca and K fragments from the reaction $^{40}Ca + ^{40}Ca$ at $E_{Lab} = 400$ MeV, measured in coincidence with α particles and protons. The heavy-ion telescope was placed at an angle of $\theta^{Lab}_{HI} = 10°$, with a light-charged-particle detector (LCP) at $\theta^{Lab}_{LCP} = -11°$ on the opposite side of the beam. From (Roy 81).

beam. The high-energy structure seen in the Ca energy spectrum in coincidence with protons corresponds closely to the structure observed in the inclusive spectra shown in Figure 143. It is absent in the spectrum of Ca fragments in coincidence with α particles (Figure 144c), which emphasizes the completely relaxed energy component. Both the $Ar + \alpha$ and $K + p$ coincidence spectra exhibit high-energy structure. Hence, strong correlations are observed between fragment energy peaks and particle emission probabilities.

However, as discussed in more detail in Section 3.7, while the observed emission patterns of the α particles with respect to the directions of flight of the reaction partners appear to be approximately consistent with sequential evaporation from the primary fragments, the proton angular correlations exhibit larger deviations from such a picture. In contrast, based on a comparison with simplified model calculations, Roynette *et al.* (Roy 81) claim a complete absence of secondary evaporation of protons and α particles for events associated with the high-energy structure in the fragment energy spectra. However, it is felt that the complexity of the particle emission patterns requires a very detailed analysis of a more complete set of coincidence data than presently available for this system, before definitive conclusions as to the origin of the structure can be drawn. In view of the possibility that these discrete energy bumps may indicate a preferential excitation of particular doorway states in the early stages of a damped nuclear reaction, it appears worthwhile to devote further experimental efforts to their investigation. Later discussion will return to this point.

Studies of the properties of the energy dissipation mechanism in damped reactions induced by relatively light and medium-weight projectile nuclei take advantage of the fact that such systems exhibit a strong orbiting tendency. Then, the angle of rotation of the interacting system provides an independent measure of the reaction time, allowing an investigation of the time dependence of various moments of the energy loss distribution. On the other hand, the rapid transition from weakly damped to fusion–fission-like reactions for such systems restricts a relatively unambiguous analysis of the reaction mechanism to a narrow band of angular momenta close to the grazing l value, associated with the exchange of a comparatively small number of nucleons during the rather short interaction times. The size of the angular-momentum window contributing to damped reactions can, in principle, be enlarged by increasing the bombarding energy, as can qualitatively be inferred from the energy dependence of the penetration function $D(l)$ illustrated in Figure 54. Unfortunately, the onset of interfering reaction mechanisms such as breakup or incomplete-fusion reactions at augmented energies makes such a procedure rather impractical. This feature and the lack of more detailed theoretical analyses have, thus far, limited the information obtained on the energy dissipation process occurring in the early stages of a damped reaction between relatively light heavy ions, believed to emphasize the transition from coherent to incoherent reaction mechanisms. Further evaluation has, therefore, to concentrate on reactions induced with heavier systems, which gradually lose the ability to orbit, as the charge product $Z_P \cdot Z_T$ increases. In addition, experimental mass, charge, and energy resolutions are worse for heavy systems and, consequently, details of experimental distributions remain largely unresolved. On the other hand, interfering reaction mechanisms lose importance in collisions between very heavy nuclei at a few MeV per nucleon above the interaction barrier, as demonstrated in previous sections.

Properties of final kinetic energy distributions have been studied with damped reactions induced by the heavy projectiles ^{56}Fe (She 79, Hoo 82), ^{63}Cu (Tam 76, Pet 77b), ^{84,86}Kr (Wol 74,76, Han 74b, Hui 76c, Web 76b, Van 76b,78, Rus 77b, Sch 77f, San 77, Pet 82, Pla 78c, Woz 78, Bla 79b, Dye 80, Rud 81), ^{92,98}Mo (Pet 82), ^{132,136}Xe (Hui 76c, Van 76c, San 77, Sch 78g, Hen 78, Wil 80, Wol 81), ^{208}Pb (San 77, Tam 80c, Reh 81), and ^{238}U (San 77). For these systems, the angular dependences of total kinetic energy spectra are found less informative and emphasis is placed on the dependence of the spectrum on projectile–target combination, bombarding energy, and final fragmentation. In particular, the correlations between energy loss and fragment mass or charge distribution provide a powerful tool for investigating the microscopic origin of energy dissipation in damped nuclear reactions. In the following, however, first some qualitative considerations of kinetic energy distributions for heavy systems and their analysis in macroscopic phenomenological models are given.

As an example, Figure 145 displays the c.m. total kinetic energy loss (TKE_{Loss}) spectra of ^{84}Kr- and ^{136}Xe-induced reactions on targets of ^{165}Ho and ^{209}Bi, reported (Hui 76c) for laboratory bombarding energies between 8.3 and 8.5 MeV per nucleon. These spectra are integrated over all projectile-like fragments in the angular range indicated at each distribution. Similar to what is observed for lighter systems, in each spectrum of Figure 145 there appears a strong quasielastic peak at low energy losses, merging with a very broad, continuous distribution of damped events. The ^{165}Ho +

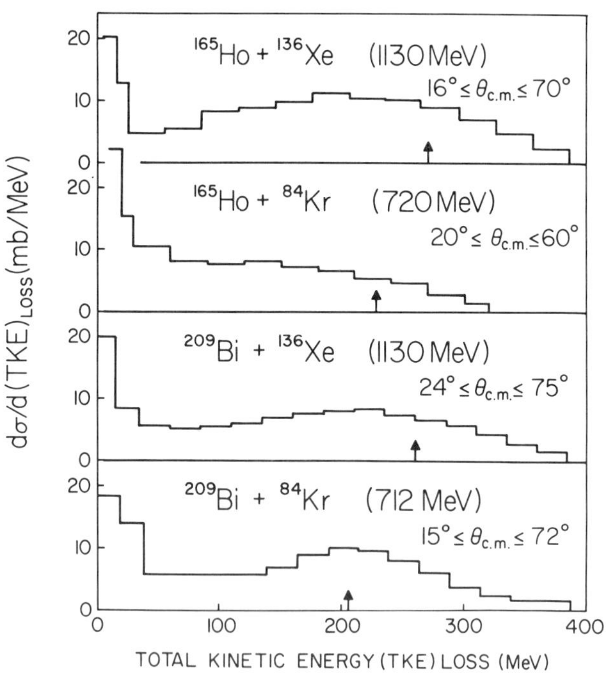

Figure 145. Differential cross section $d\sigma/d\,TKE_{Loss}$ vs. total kinetic energy loss derived from projectile-like fragment energies, measured in ^{84}Ke- and ^{136}Xe-induced reactions at 8.3–8.5 MeV per nucleon. The arrows indicate the kinetic energies available above the Coulomb barrier V_{Coul} (R_{SA}) in the entrance channel. Some cross section is missing at high energy losses, in the case of the ^{165}Ho + ^{84}Kr reaction. From (Hui 76c).

^{84}Kr spectrum is somewhat distorted at high energy losses because an estimated cross section of ~700 mb associated with strongly damped fragments is in the range $\theta_{c.m.} < 20°$ not covered by the experiment. For the other three reactions, the damped energy loss distribution represents a broad bump, whose centroid does not appear to be correlated in any obvious way with the kinetic energy available above the Coulomb barrier in the entrance channel, as calculated for spherical nuclei touching at the interaction radius R_{SA} and indicated by the arrows in Figure 145.

The particular shape of the energy loss distribution illustrated in Figure 145 is typical for heavy systems measured at incident energies of a few MeV per nucleon above the barrier. Until now, it has not been understood in detail, although some phenomenological model approaches are able to reproduce the spectral distributions, as demonstrated further below. In a more qualitative fashion, it is instructive to study the development of the energy loss distributions for incident energies close to the barrier. This is possible with data reported by Rehm *et al.* (Reh 81), who measured ^{208}Pb-induced reactions on targets ranging from ^{94}Zr to ^{170}Er at incident energies from 1.04 to 1.21 times the entrance-channel Coulomb barrier. It is noteworthy that the lighter systems show strong angular focusing, whereas the Wilczyński diagrams for the heavier targets are characterized by Coulomb dispersion, as illustrated in Figure 75.

Figure 146 displays the differential cross section for these reactions, plotted vs. total kinetic energy loss (TKEL). The initially available kinetic energies above the barrier are indicated by the arrows in the figure. Obviously, there is a distinct difference in character of the distributions for the lighter systems shown on the left and those for the heavier ones depicted on the right. The heavy systems ^{148}Sm + ^{208}Pb and ^{170}Er + ^{208}Pb exhibit quasielastic peaks with tails of continuously dropping cross section, reaching energy losses indicative of large deformations of the system at breakup. In contrast, for the lighter systems ^{94}Zr + ^{208}Pb and ^{110}Pd + ^{208}Pb, the quasielastic peak is clearly separated by a valley from the damped distribution, which peaks at final kinetic energies well below the Coulomb repulsion for spherical nuclei. Qualitatively, one can associate the different behavior of the energy distributions with different strengths of the Coulomb interaction that largely overcompensates the attractive nuclear force in the cases of the ^{148}Sm + ^{208}Pb and ^{170}Er + ^{208}Pb reactions. Hence, here the effective force is repulsive for all l values, leading to a monotonic Coulomb-like deflection function, as exemplified by the dashed curves in Figure 3, and very little barrier penetration by the two nuclei. For the lighter systems in Figure 146, particularly for ^{94}Zr + ^{208}Pb, a major portion of the cross section is focused into a narrow angular range, suggesting that the deflection function has a flat top or a nuclear rainbow (cf. Figure 3). Consequently, for these lighter systems the dynamical interaction force must be attractive or, at least, strongly retarding, for a band of low and intermediate l values. For

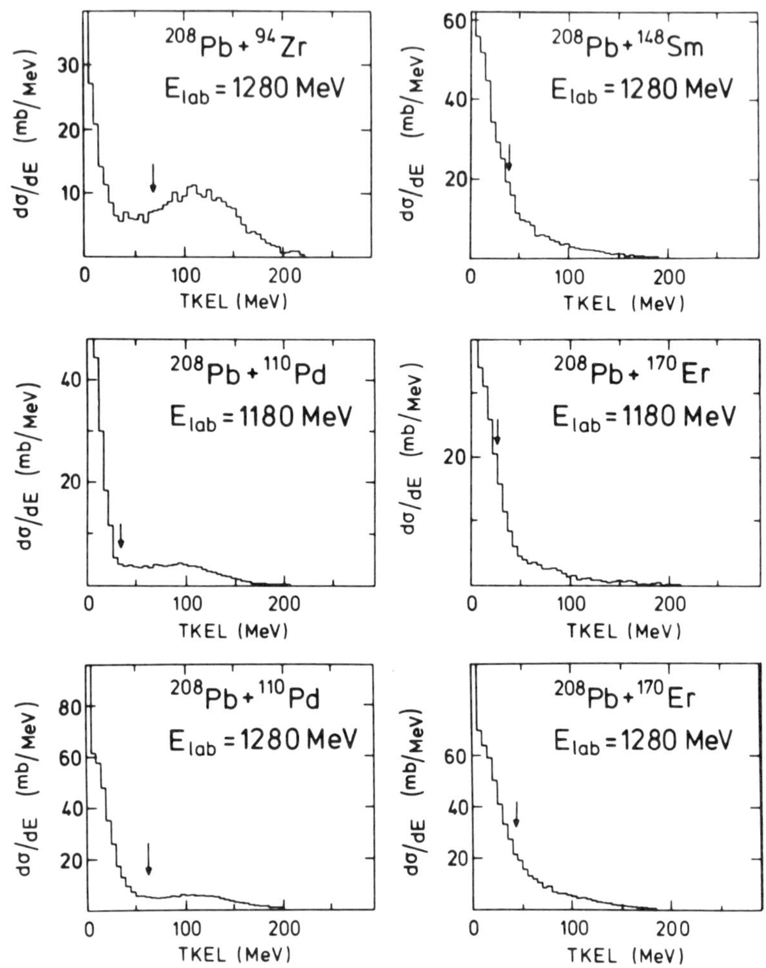

Figure 146. Total kinetic energy loss (TKEL) spectra for ^{208}Pb-induced reactions on various targets for the incident laboratory energies specified. The arrows indicate the kinetic energy initially available above the Coulomb barrier of two touching spheres. From (Reh 81).

relatively light systems exhibiting a strong orbiting tendency, this effect is understood to be due to a barrier in the effective radial potential as shown on the left of Figure 3, which effects a dramatic increase in interaction time for l waves captured behind it. Hence, in these cases, one expects with decreasing angular momentum an exponentially growing fraction of the cross section to disappear into the damped reaction channel, corresponding to an exponential depletion of the quasielastic cross section. Since there appears to be only a limited range of possible kinetic energy losses, as demonstrated earlier for lighter systems, the damped reaction cross section has to pile up and form a bump in the energy distribution. For very heavy systems, such as $^{170}Er + ^{208}Pb$, and at sufficiently high incident energies where there is most probably no barrier in the effective interaction potential, a similar effect can be produced by a friction force, whose radial form factor increases exponentially with decreasing separation distance, as is the case for both the Gross and Randrup friction models at large distances, as can be seen in Figure 27. However, the very fact that the energy loss spectra for the heavy systems illustrated in Figure 146 exhibit a strong quasielastic peak containing most of the reaction cross section, with no distinct damped energy component, may be taken as indication of a relatively short range of the friction force, not extending to the strong-absorption radius. Such an interpretation would, hence, disfavor the long-range friction force proposed by Gross (see Section 2.6).

The bombarding-energy dependence of the total kinetic energy spectrum is illustrated in Figure 147 for the heavy system $^{209}Bi + ^{136}Xe$,

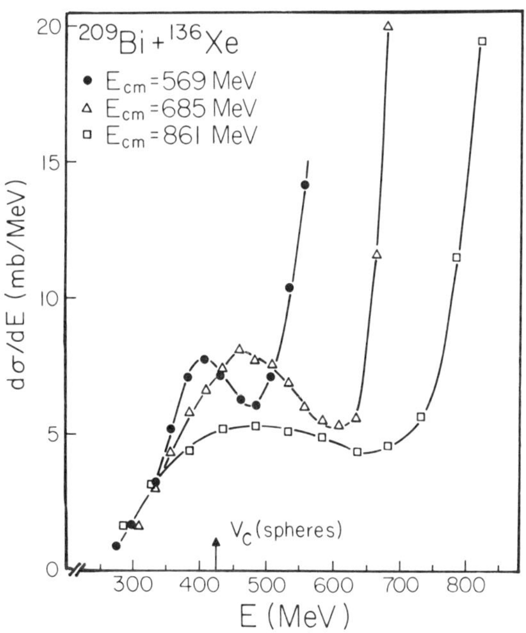

Figure 147. Total kinetic energy distribution $d\sigma/dE$ final fragments produced in the reaction $^{209}Bi + ^{136}Xe$ at three bombarding energies. The energies were calculated from the measured projectile-like fragments assuming two-body kinematics. The energy spectra are integrated over all fragments and reaction angles. The arrow (V_c) indicates the entrance channel Coulomb interaction energy at the strong-absorption radius. From (Hui 82).

measured (Sch 78g, Wil 80, Wol 81) at three incident energies. These distributions are integrated over all fragments and reaction angles. Again, at each energy, the spectrum is dominated by a strong quasielastic peak at high energies and a broad damped energy bump extending down in energy some 150 MeV below the entrance-channel Coulomb barrier, indicated by the arrow (V_c). For the lowest bombarding energy, most of the damped cross section corresponds to final energies below the Coulomb barrier, while this fraction decreases with increasing incident energy. Figure 147 demonstrates the enormous amount of kinetic energy that can be dissipated in a damped reaction, amounting to $\lesssim 600$ MeV in the case of ^{209}Bi + ^{136}Xe at $E_{\text{c.m.}} = 861$ MeV. For comparison, the fission-like kinetic energy, calculated with Equation (3.5.4) with $Z^2 = 4Z_PZ_T$, is estimated to be $\langle E_f(^{209}\text{Bi} + {}^{136}\text{Xe})\rangle = 295$ MeV. It is evident from Figure 148 that, independent of the incident energy, the low-energy end points of the spectra are close to this energy. This is quite in contrast to the energy distributions of fragments from heavy-ion-

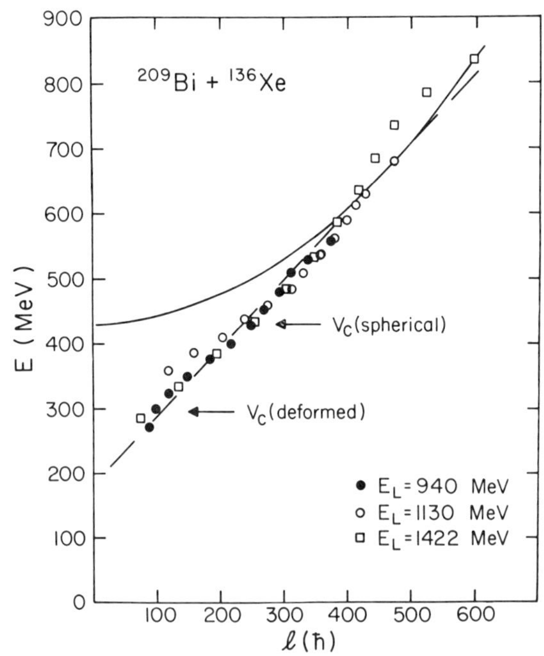

Figure 148. Average final total kinetic energy E for the reaction ^{209}Bi + ^{136}Xe at three incident energies plotted vs. initial orbital angular momentum l. The solid curve represents the Coulomb scattering relation $E(l_{\text{max}})$. The dashed line represents a fit to the data. The Coulomb energy for spherical nuclei touching at the strong-absorption radius and the fission-like energies are indicated by the arrows. From (Wol 82).

induced fission, illustrated in Figures 129 and 130, which are centered around this characteristic energy. Furthermore, it is observed in Figure 147 that the valley between quasielastic and damped spectral components becomes less pronounced at the higher incident energy. This feature is consistent with a broadening of the quasielastic peak seen in Figure 146 for the $^{110}\mathrm{Pd}+{}^{208}\mathrm{Pb}$ and $^{170}\mathrm{Er}+{}^{208}\mathrm{Pb}$ reactions, when the bombarding energy is increased from 1180 to 1280 MeV.

As has been discussed in Section 3.3, from the behavior of angle–energy pattern in damped reactions it appears plausible to assume an approximately monotonic dependence of the average total kinetic energy loss on the initial orbital angular momentum l. This then allows [cf. Equation (3.3.14)] for an l-wave decomposition of the cross section. The results of such a decomposition, reported by Wollersheim *et al.* (Wol 82) for the $^{209}\mathrm{Bi}+{}^{136}\mathrm{Xe}$ reaction at three bombarding energies, are displayed in Figure 148. Here, the average final total kinetic energies E are plotted vs. l. The solid curve corresponds to

$$E(l)=\frac{e^2 Z_P Z_T}{R_{\mathrm{SA}}}+\frac{\hbar^2 l^2}{2\mu R_{\mathrm{SA}}^2} \tag{3.5.15}$$

i.e., Coulomb scattering with a distance of closest approach that coincides with the strong-absorption radius R_{SA}, where the radial component of the kinetic energy is zero. The dashed line presents a fit to the data. The Coulomb energy in Equation (3.5.15), V_c (spherical), and the fission-like kinetic energy, $\langle E_f\rangle = V_c$ (deformed), are indicated by arrows in the figure.

Although the transformation to angular-momentum scale is subject to considerable uncertainty, depending on experimental resolution and the model for $d\sigma/dl$ employed in the analysis, it appears remarkable, how well in Figure 148 the data for different bombarding energies coincide with each other and the dashed straight line. The systematic deviation of the deduced (E, l) values at high energies from Equation (3.5.15) may indicate that, at $E_{\mathrm{Lab}}=1422$ MeV and l values slightly below the grazing angular momentum, the radial component of the kinetic energy is not completely damped when the system passes through the strong-absorption distance in the exit channel. The apparent absence of this behavior for the data taken at lower bombarding energies may possibly be due to the smaller magnitude of the radial energy for a given l value. It is seen from Figure 148 that, for l values below $l\approx 350\hbar$, the data depart significantly from the curve representing Equation (3.5.15), indicating that increasing amounts of radial and orbital energies are dissipated. For $l \lesssim 300\hbar$, final energies are smaller than given by Equation (3.5.15) evaluated with an angular momentum reduced from the initial l value to the rolling limit l_r [cf. Equation (2.3.9)], which, for this system, is similar to the sticking angular momentum. Hence, for $l < 300\hbar$, one concludes that the exit-channel distance at which the radial velocity is zero is larger than the strong-absorption distance R_{SA}, consistent with increasingly larger deformations of the intermediate system and complete damping of the kinetic energy of relative motion at the instance of breakup. This finding puts in perspective the results of analyses of lighter heavy-ion systems discussed in the context of Equation (3.5.5), from which unique deformation parameters are usually derived. Although, for the reactions illustrated in Figure 148 and $l > 300\hbar$, there is a continuous range of possible relative contributions of radial and orbital energies to the measured final total kinetic energy, the data are not consistent with complete damping of both components of the energy of relative motion. Clearly, it would be difficult to understand such a behavior, if the form factors of radial and tangential friction forces would acquire a significant magnitude already at separation distances of the order of or larger than R_{SA}.

The almost linear experimental relation between E and l illustrated in Figure 148 defines a limiting value of the final total kinetic energy in a model-dependent fashion. From the intercept of the dashed fit line with the energy axis, one obtains (Wol 82) a hypothetical end-point energy of $E_{ep} \approx 184$ MeV, some 100 MeV below the fission-like energy referred to earlier. This relation suggests also that the differential reaction cross section can approximately be written as

$$\frac{d\sigma}{dE} = \frac{d\sigma}{dl}\frac{dl}{dE} \approx \frac{\pi\hbar^2}{\mu\xi^2}\frac{E - E_{ep}}{E_{c.m.}} \tag{3.5.16}$$

in terms of the end-point energy and the slope $\xi = dE/dl$. Asymptotically, for high bombarding energies where V_{Coul}, $E_{ep} \ll E_{c.m.}$, the slope is expected to vary slowly with $E_{c.m.}$, as $\xi \to \hbar c(E_{c.m.}/2\mu R_{SA}^2)^{1/2}$. This explains the approximate agreement observed in Figure 148 between the various data sets. Significant deviations from the average can, hence, be expected only for bombarding energies close to the Coulomb barrier, V_{Coul}. Since the area between the straight line in Figure 148 and the ordinate, up to $E = E_{c.m.}$, is normalized to the total reaction cross section, the fluctuations of the data points about this line are understood to reflect the exhaustion of the reaction cross section by partially and strongly damped processes, for a continuous range of angular momenta.

The dependence of the total kinetic energy distribution on the final fragmentation can, in principle, be inferred from the two-dimensional contour diagrams $d^2\sigma/dZ\,dE$ illustrated for several reactions in Figures 103–105. The particular shape of the energy spectra is, however, more obvious from projections of these distributions along cuts for various fixed

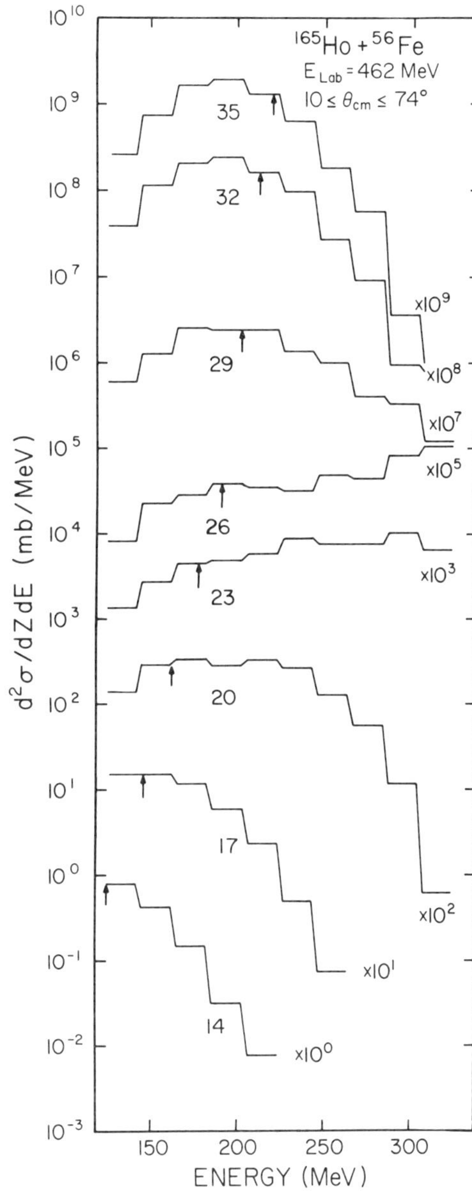

Figure 149. Double differential cross section $d^2\sigma/dZdE$ plotted vs. c.m. final total kinetic energy for the reaction $^{165}\text{Ho} + ^{56}\text{Fe}$ at $E_{\text{Lab}} = 462$ MeV. Histograms correspond to three-Z-units-wide bins in atomic number Z of the detected light fragment, integrated over $10° \leq \theta_{\text{c.m.}} \leq 74°$. The centroid of the bin is indicated at each distribution. Arrows mark the Coulomb energy for spherical nuclei touching at the strong-absorption radius. From (Hoo 82).

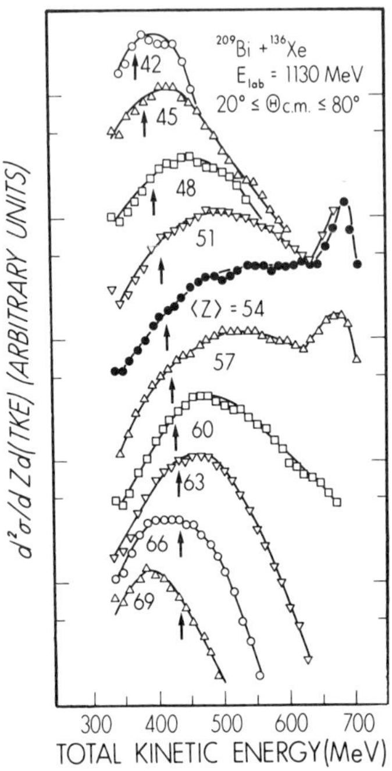

Figure 150. Same as Fig. 150, but for the reaction $^{209}Bi + ^{136}Xe$ at $E_{Lab} = 1130$ MeV, integrated over $20° \leq \theta_{c.m.} \leq 80°$. From (Sch 78g).

atomic numbers Z of the projectile-like fragment, which is usually the one detected by experiment. Such energy spectra are shown in Figures 149 and 150 for the reactions $^{165}Ho + ^{56}Fe$ (Hoo 82) and $^{209}Bi + ^{136}Xe$ (Sch 78g), respectively, grouped in bins that are 3 Z units wide. These spectra are integrated over the angular range studied in the measurements. Hence, for the $^{165}Ho + ^{56}Fe$ reaction, the two energy components characteristic of orbiting-type reactions (cf. Figures 78, 134, 135) are averaged out in the presentation of data given in Figure 149.

One observes from the data in Figure 149 that energy spectra for fragments with charges close ($Z \approx 26$) to the projectile exhibit a cross section that decreases monotonically with decreasing final energy, with a pronounced quasielastic peak, while those with Z values further removed have the characteristic bell-shaped appearance also seen in reactions induced by much lighter projectiles. The distributions for the $^{209}Bi + ^{136}Xe$ reaction depicted in Figure 150 demonstrate this feature more clearly. In neither case are strong correlations of the positions of the damped energy bumps with the (spherical fragment) Coulomb energy, indicated by the arrows in these figures, discernible. However, the maxima and the widths of the damped energy spectra decrease with increasing distance of the charge fragmentation from the initial one. For the $^{209}Bi + ^{136}Xe$ reaction (cf. Figure

150), this occurs in a symmetrical fashion, regardless whether the projectile or the target nucleus has picked up protons. It is also quite remarkable that the end-point energies are, in this case, rather independent of the final fragmentation (Z_3, Z_4), although the fission-like kinetic energies calculated from Equation (3.5.4) with $Z^2 = 4Z_3 \cdot Z_4$ vary by approximately 40 MeV for the displayed range of fragmentations. This is also true for the $^{209}Bi + ^{136}Xe$ reaction measured at lower (Wil 80) and higher (Wol 81) bombarding energies, as well as for the $^{166}Er + ^{86}Kr$ reaction (Rud 79), for which data are displayed in Figure 103. It is a common feature also in the cases of heavy-ion-induced fission and FL reactions discussed earlier in this section. The fact that in the $^{165}Ho + ^{56}Fe$ reaction (cf. Figure 149) the end-point energy does depend on the fragmentation, can be understood in terms of the drift of the system towards asymmetry (Hoo 82), such that more symmetric fragmentations sample relatively short interaction times, that are associated with incomplete energy relaxation.

Although the above conclusions must be considered as somewhat uncertain, since they refer to spectral features at vanishing cross sections, it appears worthwhile to emphasize this marked difference between heavy-ion reactions and low-energy fission, where there is a definite correlation between fragmentation and total kinetic energy. The surprising stability of the end-point energies for all fragmentations of a given projectile–target system could be interpreted as indicative of a strong dependence of the limiting dinuclear shapes, for example, of the size of the neck between the nascent reaction fragments, on the mass or charge division. Alternatively, it could signify a dominantly dynamical origin of the form of the kinetic-energy distributions. In either case, it remains to be understood why the end-point energies track so closely with the fission-like energy of the projectile–target charge asymmetry but not with that of the final fragmentation. Unfortunately, the subject has been studied far too little to allow definitive conclusions to be drawn. Further measurements of fragment intrinsic excitation energies could prove helpful in elucidating this problem.

As exemplified earlier for lighter systems, the widths of the damped component in the energy spectrum provides important clues on the nature of the energy dissipation process. Qualitatively, one observes a broadening of the kinetic-energy distribution with increasing incident energy, demonstrated for the $^{209}Bi + ^{136}Xe$ reaction in Figure 147. This behavior, hence, concurs with the bombarding-energy dependence of the damped spectral component of lighter systems, as well as that of the quasielastic peak. However, for heavy systems, almost no quantitative information on the correlation of the energy width with other reaction variables is reported in the literature.

The dependence of the width of the energy distribution on the charge Z of one of the fragments produced in the FL reaction $^{56}Fe + ^{132}Xe$ at a

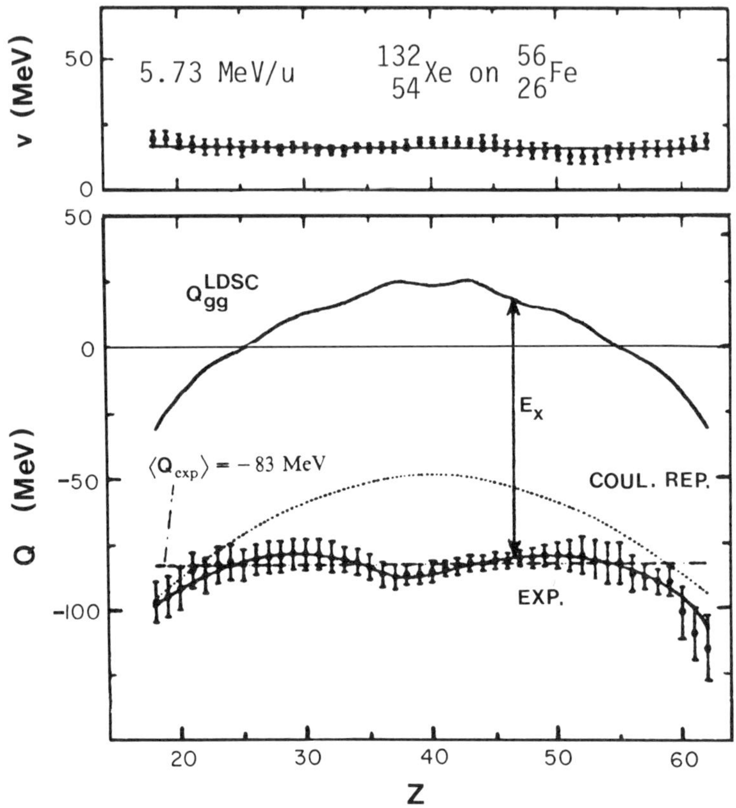

Figure 151. Mean Q values and standard deviations v $(=\sigma_E)$ for the fully relaxed energy component vs. Z of fragments produced in the FL reaction $^{56}Fe+{}^{132}Xe$ at $E_{Lab}=5.73$ MeV per nucleon. The data are averaged over several angles between 12° and 18°. The solid curve is to guide the eye. The thin solid curve labeled Q_{gg}^{LDSC} represents the ground-state Q values calculated from the LD model with shell corrections. The dashed curve indicates the Coulomb energy for spherical fragments relative to the incident energy. The dashed-dotted line represents the average Q values of $\langle Q_{exp}\rangle = -83$ MeV. From (Heu 78).

bombarding of 5.73 MeV per nucleon has been studied by Heusch *et al.* (Heu 78). Figure 151 depicts their results for the root-mean-square width $v(=\sigma_E)$, together with the measured mean Q values already referred to earlier in this section. It is evident from the figure that the width is independent of the fragmentation and has the value $v \approx 15$ MeV, corresponding to Γ_E (FWHM) ≈ 35 MeV. This width is approximately a factor of 2 smaller than the smallest widths observed for any fragmentation in the damped reactions illustrated in Figures 149 and 150, but roughly in agreement with the values obtained (Gal 75) for the damped component in the

Ag + 340 MeV ^{40}Ar reaction. Since FL reactions are believed to correspond to very long interaction times, as evidenced by their broad, $(1/\sin\theta)$-like angular distributions (cf. Section 3.3), the widths of fragment energy spectra from such reactions provide an estimate for the upper limit of the contribution of purely equilibrium-statistical fluctuations. In the ^{56}Fe + ^{132}Xe reaction, the average excitation energy is of the order of 2 MeV per nucleon. This corresponds to an equilibrium temperature comparable to that for the maximum of the damped component for the ^{209}Bi + ^{136}Xe reaction at $E_{\text{c.m.}} = 596$ MeV (cf. Figure 147). In the latter case, however, the width, estimated from the low-energy side of the distribution shown in Figure 147, amounts to the much larger value of $\Gamma_E \approx 135$ MeV. Furthermore, for this system the width Γ_E increases proportionally with the most probable excitation energy $\langle E^* \rangle$ and, hence, does not exhibit the temperature dependence, $\Gamma_E \sim \langle E^* \rangle^{1/2}$, expected to hold for thermal fluctuations.

A rather detailed study of the width of fragment kinetic energy distributions is due to Blann *et al.* (Bla 79b), investigating the ^{124}Sn + ^{86}Kr reaction at bombarding energies of $E_{\text{Lab}} = 440$ and 720 MeV. Data on fragment A and Z values are reported for two laboratory angles, for each energy. Measured results are shown in Figure 152 for average kinetic energy (E_Z), the standard deviation (σ) in energy, and that in mass (σ_A), as functions of atomic number of the detected fragment. The general finding is that both average kinetic energy and dispersion increase with bombarding energy, in agreement with results presented earlier in this section. After significant corrections of the data for sequential neutron and α-particle evaporation, Blann *et al.* (Bla 79b) deduce that the ratio of the standard deviation to average Q value represents a nearly constant fraction of 0.13, independent of the bombarding energy.

These observation are not surprising, in view of the findings for the lighter systems discussed earlier. They add evidence in support of a generally nonequilibrium statistical nature of the dissipation mechanism that depends on the projectile–target systems and its dynamical properties. They do not preclude, however, that its thermal limit can be reached in collisions with very long interaction times, such as occurring for lighter systems and reactions angles backward of the grazing angle, discussed, e.g., by Lefort (Lef 79c). It is very obvious from the data presented so far that the energy dissipation mechanism operating in damped collisions has characteristics depending on the collective dynamics of the relative motion. Clearly, this motion is not overdamped for most l values. The question then poses itself, whether other nuclear collective modes are sufficiently weakly damped to be excited in damped nuclear reactions and effect a transformation of kinetic energy of relative motion.

The potential importance of nuclear surface oscillations as doorway excitation modes in damped nuclear reactions has been emphasized by

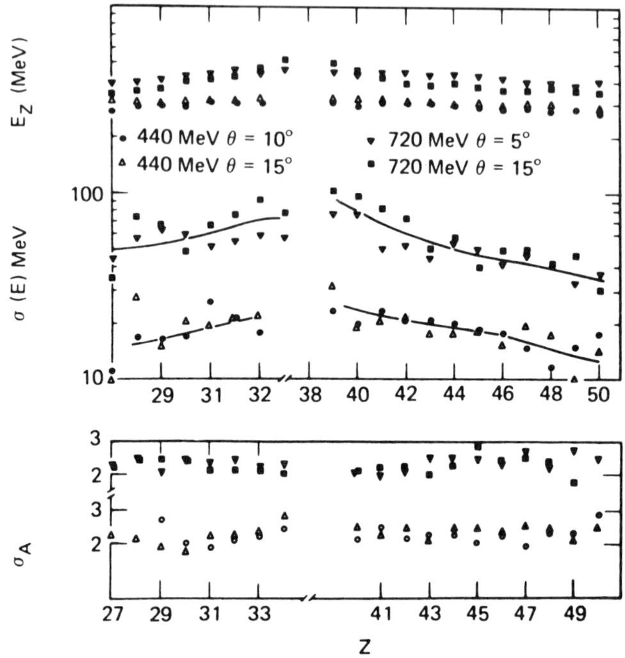

Figure 152. Average measured kinetic energies, E_Z, and standard deviation in energy, σ, and mass, σ_A, for fragments from the reaction $^{124}Sn + ^{86}Kr$ at $E_{Lab} = 440$ and 720 MeV at two angles each, plotted vs. atomic number Z of the detected fragment. For each Z value, the upper and lower pairs of data points correspond to $E_{Lab} = 720$ MeV and 440 MeV, respectively. The curves are drawn to guide the eye. From (Bla 79b).

Broglia and collaborators (Bro 74,76,76c,78c,78e, Esb 78, Das 82) in a semiclassical model, as discussed briefly in Section 2.3. This group has performed calculations for several systems, solving the equations for damped harmonic oscillators [cf. Equation (2.3.17)], describing the multipole surface vibrations, coupled to the equations of relative motion. In the model, this coupling is effected by the external driving force [see Equation (2.3.17)], deduced from the sum of nuclear proximity and multipole Coulomb interactions, for a given multipole (λ, μ):

$$F_{\lambda\mu} = \left[R_0 \frac{\partial V_{\text{Prox}}(s)}{\partial r} - \frac{3e^2 Z_P Z_T}{2\lambda + 1} \frac{R_0^\lambda}{r^{\lambda+1}} \right] Y_\mu^{\lambda *}(\hat{r}) \qquad (3.5.17)$$

Here, the radius parameter of each nucleus, R_0, is defined in Equation (2.3.16), r is the center separation distance, and s denote the surface separation depending on the relative angles of orientation $(\hat{r})$ of target and projectile surface multiples. The dissipative effect of nucleon exchange between the fragments is accounted for (Bro 78e) by adopting the "window dissipation" formula (2.6.25). The properties of the surface vibrations, including both low-lying and giant-resonance-type isoscalar modes, are calculated (Bro 78c) in random-phase approximation. A phenomenological nonlinear damping is adopted for low-energy multipole vibrations, whereas linear damping according to Equation (2.3.17) with $c_\lambda = 2B_\lambda \Gamma_\lambda / \hbar$ is assumed for the giant-resonance states. The latter resonances, lying between 8 and 35 MeV of excitation energy, are given a spreading width of $\Gamma_\lambda = 5$ MeV.

In Figure 153, results of such a calculation (Bro 78e) for the reaction $^{208}\text{Pb} + {}^{136}\text{Xe}$ at $E_{\text{Lab}} = 1130$ MeV are compared to results obtained (Sch 78g) from a phenomenological analysis of data for the similar reaction $^{209}\text{Bi} + {}^{136}\text{Xe}$ at $E_{\text{Lab}} = 1130$ MeV. Here, the final total kinetic energy E is plotted vs. impact parameter b. The solid curve represents the decrease in

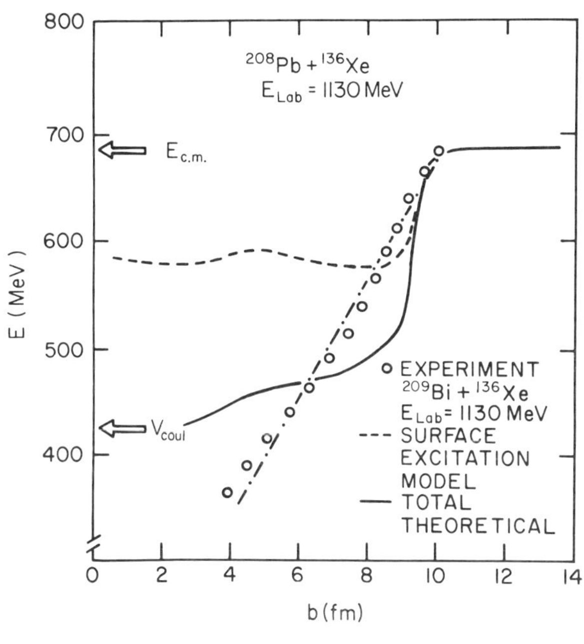

Figure 153. Final total kinetic energy E vs. impact parameter b for the reaction $^{208}\text{Pb} + {}^{136}\text{Xe}$ at $E_{\text{Lab}} = 1130$ MeV. The solid curve represents the results of calculations emphasizing the excitation of collective surface oscillations, as explained in the text. The dashed curve indicates the energy loss due to these excitations. The difference between solid and dashed curves is due to energy dissipation effected by nucleon exchange. The circles and the dashed-dotted line represent results of an l-wave decomposition of the experimental energy distribution for the similar reaction $^{209}\text{Bi} + {}^{136}\text{Xe}$ measured at the same bombarding energy.

energy due to both excitation of the surface modes and nucleon exchange-induced dissipation. The dashed curve indicates the contribution due to the collective mode alone, amounting to roughly 30% of the total energy loss at low impact parameters. As can be seen from Figure 153, the calculations provide a qualitative reproduction of the data. The model calculations yield an almost step-like dependence E(b) separating quasi-elastic and damped events by a large energy gap. The further slow variation of E with B corresponds to a narrow peak in the predicted damped energy distribution, located between E = 400 and 500 MeV. Here, indeed, a bump is observed in the experimental total kinetic energy spectrum (cf. Figure 147). Experimentally, however, this bump is much broader ($\Gamma_E \approx 220$ MeV) than predicted, leading to an experimental correlation between final energy and impact parameter that is almost linear, in contrast to the calculated sudden onset of collective excitations. In addition, the data for the highest energy losses, which indicate large static deformations in the exit channel, are not well reproduced by theory. Similar arguments appear to apply in the case of the ^{144}Sm + ^{84}Kr reaction, although a detailed comparison of data, obtained for this system by Mignerey *et al.* (Mig 84) at $E_{\text{Lab}} = 470$, 595, and 720 MeV, with model calculations (Bro 78e) has not been made. With regard to the earlier discussion of the high-energy structure observed in the energy spectra for lighter systems such as ^{40}Ca + ^{40}Ca (cf. Figure 143), it is an important feature of collective excitation processes described in the model of Broglia *et al.* that they lead to energy peaks in the strongly damped domain. Therefore, it is unlikely that such mechanisms can be made responsible for the structure in question. This is also the general conclusion of adiabatic-TDHF calculations performed by Tricoire *et al.* (Tri 81) for quadrupole excitation in ^{40}Ca + ^{40}Ca collisions, where the number of such phonons is predicted to be exceedingly small.

As noted previously, although the position of the bump in the energy distribution predicted by the collective model of Broglia *et al.* is approximately in the region where broad structure is observed experimentally, the theoretical peak is too narrow. However, since the model employs classical trajectories, neglecting quantal and statistical fluctuations, such an underestimate is expected and may not invalidate the microscopic mechanism on which this approach is based. Esbensen *et al.* (Esb 78) have considered the effect of quantal zero-point fluctuations of surface vibrations (cf. Sections 2.4 and 2.7) on the damped energy distributions. In calculations for the ^{208}Pb + ^{136}Xe reaction at $E_{\text{Lab}} = 1130$ MeV, these fluctuations are simulated by extending the calculation by Broglia *et al.* (Bro 78e) to a statistical ensemble of initial conditions characterized by dispersions of the oscillation amplitudes and corresponding momenta as given by these zero-point fluctuations. Results of this calculation for final kinetic energy spectra shown on the left of Figure 154 for several impact parameters ρ indicate

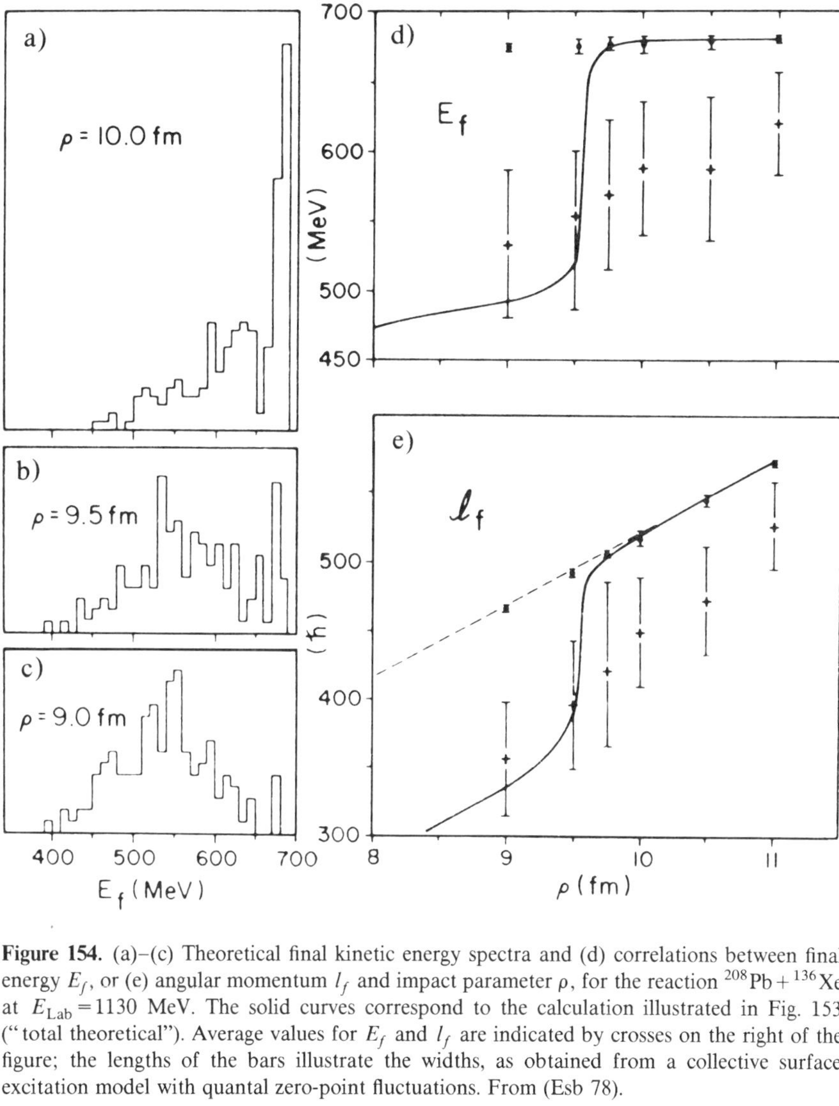

Figure 154. (a)–(c) Theoretical final kinetic energy spectra and (d) correlations between final energy E_f, or (e) angular momentum l_f and impact parameter ρ, for the reaction $^{208}Pb + ^{136}Xe$ at $E_{\text{Lab}} = 1130$ MeV. The solid curves correspond to the calculation illustrated in Fig. 153 ("total theoretical"). Average values for E_f and l_f are indicated by crosses on the right of the figure; the lengths of the bars illustrate the widths, as obtained from a collective surface excitation model with quantal zero-point fluctuations. From (Esb 78).

that, indeed, widths of the order of those experimentally observed are predicted by the model. On the right of this figure, the theoretical correlations between final energy, E_f, and angular momentum, l_f, and the impact parameter ρ are depicted, where the dispersions are indicated by the lengths of the bars. The solid curves represent the calculations neglecting these fluctuations. For each impact parameter, there are obviously two energy components predicted by the model, interpreted (Esb 78) to be due to the spread in effective impact parameters introduced by the zero-point fluctuations already for ρ values larger than the grazing-impact parameter. Although the appearance of the energy spectrum for a near-grazing collision, i.e., for $\rho = 10.0$ fm, is reminiscent of the experimental spectral shapes shown in Figure 141 for low-energy heavy-ion reactions, this latter feature is probably unrealistic and leads to theoretical Wilczyński diagrams that are in conflict with the observed angular focusing property (cf. Figure 74).

The question whether coherent modes of the nuclear response function, undoubtedly explored in other types of reactions, are important in damped collisions, is certainly one of the most interesting problems addressed by current experimental and theoretical studies. Comparisons such as discussed above neither strongly contradict nor support a certain role of an energy dissipation mechanism that proceeds via the excitation of collective nuclear surface modes. At present, one would tentatively conclude that these modes are probably more strongly damped than assumed by theory and, hence, are less probable to develop fully in damped reactions, as already alluded to in Section 2.3. In addition, the early opening of a neck between the reaction partners could destroy their structural individuality, such that coherent response modes of much lower frequency become more adequate for a description of the system. Clearly, this subject matter deserves more experimental and theoretical investigation.

Although it remains at present uncertain whether nuclear surface degrees of freedom are excited in damped heavy-ion reactions, the data clearly indicate that the intermediate dinuclear complex can attain significant static deformations in deeply penetrating collisions. If these deformations evolve slowly in time, they are able to store energy supplied by the relative motion. In addition, a smaller amount of energy is consumed by thermal shape fluctuations. Hence, the slow development of dynamical fragment deformations can emulate a dissipative process. Such a situation is considered in the work of Wolschin (Wol 79b) and Riedel *et al.* (Rie 79, 80). Assuming an overdamped motion, the time evolution of a deformation parameter ε common to both fragments is described (Wol 79b) in terms of a Fokker–Planck equation (2.7.14) with essentially constant, though l-dependent, transport coefficients. Then average deformation parameter and dispersion are determined by Equations (2.8.2) as functions of interaction time, which is connected to experimental observables via a classical reaction model or a parametrized deflection function (see Section 3.3).

In Figure 155, results of such a phenomenological transport model (Rie 79) are compared to experimental energy loss (ΔE) and integral fragment charge distributions, as well as to measured correlations between energy loss and second moments σ_Z^2 of the energy-differential Z distribution. Considered are the two reactions ^{166}Er + ^{86}Kr and ^{209}Bi + ^{136}Xe, for which data are shown for two bombarding energies each. Constant transport coefficients

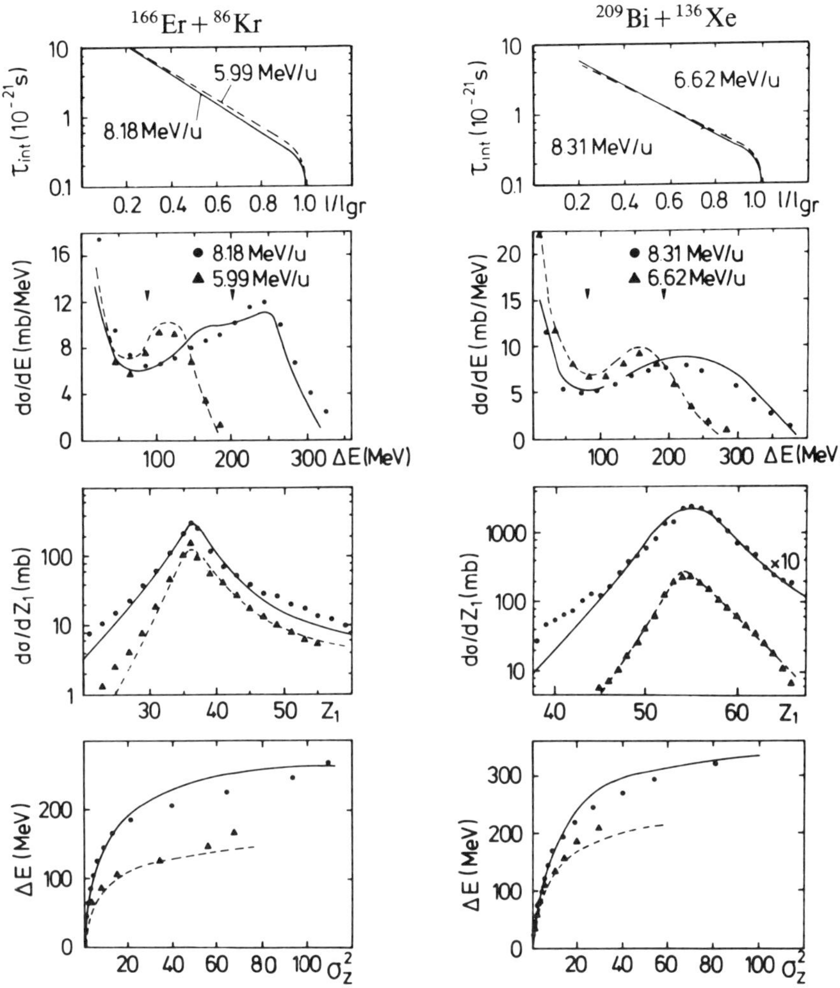

Figure 155. Comparison of experimental energy loss distributions, $d\sigma/d(\Delta E)$, total fragment charge distributions $d\sigma/dZ_1$, and correlations between energy loss and variance, σ_Z^2, of fragments from the reactions $^{166}Er + ^{86}Kr$ and $^{209}Bi + ^{136}Xe$ at two incident energies each, with the results (solid and dashed curves) of a phenomenological transport model. The upper left and right portions display the dependence of the interaction time on angular momentum l, relative to the grazing value l_{gr}, as deduced from a fit of a parametrized deflection function to experimental angular distributions. From (Rie 79).

and relaxation times for mass and charge asymmetry, angular momentum, radial kinetic energy, and fragment deformation are assumed in the model and extracted from a fit to the experimental data shown in Figure 155. However, statistical fluctuations in the deformation parameter are neglected. Obviously, a consistent description of experimental charge and energy distributions is obtained with the model for both reactions, using a constant $t_{\rm rel}(\varepsilon)=5\times10^{-21}$ s for the characteristic shape relaxation time, which is roughly an order of magnitude larger than estimated (Wol 78b) for the relaxation of the radial energy. Mass transport coefficients derived simultaneously exhibit the peculiarities already referred to in Section 3.4.

While the phenomenological model of Riedel *et al.* (Rie 79) takes no account of statistical fluctuations in the scission shapes of the dinuclear system, they are included in a similar, but simpler, model analysis neglecting mass exchange, as performed by Wolschin (Wol 79b) in the case of the ^{166}Er + ^{86}Kr reaction at $E_{\rm Lab}=5.99$ MeV per nucleon. Here, the shape relaxation time determines the position of the damped energy peak. A fit to the experimental spectral shape yields (Wol 79b) the relatively small diffusion coefficient for the deformation degree of freedom

$$D_{\varepsilon\varepsilon}\approx 7.2\times10^{17}\ \mathrm{s}^{-1} \tag{3.5.18}$$

corresponding to a contribution of only $\Gamma_E \lesssim 15$ MeV of statistical shape fluctuations to the width of the total kinetic energy distribution. As has already been noticed in Section 3.3, the effect of dynamical fragment deformations is most noticeable for small values of angular momentum, yielding increased energy losses and interaction times (Rie 79). Of course, the magnitude of the deduced diffusion coefficient [Equation (3.5.18)], describing statistical fluctuations in the dinuclear shapes, is dependent on the details of the reaction model employed in the analysis. In view of the expectation, however, that there are additional possible sources of fluctuations, e.g., due to quantal zero-point motion, one may tentatively conclude from the analyses discussed above that the contribution of statistical shape fluctuations to the dispersion observed in the total kinetic energy spectrum is only of the order of $\sim 10\%$ or less, while the main contribution is due to dynamical dissipation processes as well as to dynamic dispersions in dinuclear shapes, remotely analogous to the dynamic dispersions in scattering angle discussed in Sections 2.2 and 3.3 (see discussion of Figure 71).

A more complete treatment of nuclear shape deformations in damped reactions is possible based on phenomenological dynamical reaction models such as those of Deubler and Dietrich (Deu 75, 77b) and of Gross *et al.* (Gro 81), already referred to in Section 3.3. In these models, the dinuclear shapes are approximated by touching axially symmetric spheroids allowed to undergo damped collective vibrations, as described by classical equations of motion of the type of Equation (2.3.15). Statistical fluctuations are not considered in these models. As noted in Section 3.3, the theory of Deubler and Dietrich is unsuccessful in the reproduction of experimental energy-angle patterns of damped reaction fragments. It is, hence, not discussed here any further.

On the other hand, the agreement between experimental data for medium-weight systems and the friction model of Gross *et al.* is dramatically improved through the inclusion of spheroidal fragment deformations, as is demonstrated in Figure 78 for the $^{232}\mathrm{Th}+{}^{40}\mathrm{Ar}$ reaction. In these calculations, aligned spheroids with equal axis ratios α are considered. Liquid-drop driving forces for this degree of freedom and Coulomb interaction are taken from the work of Nix (Nix 64). A friction coefficient of the form

$$c_q = \frac{16}{\pi}\left(R_1^3 + R_2^3\right)\eta \tag{3.5.19}$$

is adopted, describing the damping of the vibrational modes. Here, the quantities R_i are the equivalent sharp radii of the spherical nuclei, and $\eta = 1\times10^{-22}$ MeV s fm^{-3} is a viscosity coefficient. In this model, large oblate nuclear deformations develop in the entrance channel, with a maximum axis ratio of $\alpha = 1.3$ in the case of an $l = 100$ trajectory of the system $^{209}\mathrm{Bi}+{}^{136}\mathrm{Xe}$ at $E_{\mathrm{Lab}} = 1130$ MeV (Gro 81). After passage of the classical turning point, oblate deformations regress and the fragments attain prolonged prolate shapes with $\alpha \gtrsim 0.6$. Figure 156 shows final values of the deformation parameter α and the energy ΔE_{def} stored in deformations as a function of angular momentum L, as predicted (Gro 81) by the model for the $^{209}\mathrm{Bi}+{}^{136}\mathrm{Xe}$ reaction. Although the decrease in final total kinetic energy due to the presence of deformation amounts to some 50 MeV for low l values, only approximately 20 MeV are eventually stored in the form of deformation energy. For somewhat lighter systems such as $^{232}\mathrm{Th}+{}^{40}\mathrm{Ar}$, the amount of deformation energy ($\Delta E_{\mathrm{def}} \lesssim 6$ MeV) is even less significant. Hence, the main effect of the deformations on the energy loss is attributed to a lowering of the Coulomb barrier in the exit channel and a prolonged action of the friction force between the fragment surfaces. Similar conclusions have been reached in model calculations by Abul-Magd *et al.* (Abu 78).

The strongly focused angular distribution observed for very heavy systems (cf. Section 3.3) is not well produced by the model of Gross *et al.* (Gro 81), predicting a more Coulomb-like backward dispersion of the cross section. It is currently not possible to identify the single element in the calculation most responsible for this shortcoming. A possible cause may, however, be suspected in the magnitude of oblate deformations in the

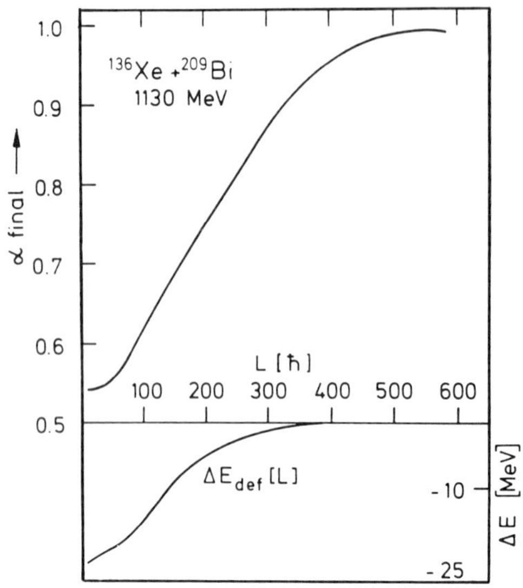

Figure 156. Common axis ratio α and deformation energy ΔE_{def} as functions of angular momentum L predicted by the dynamical model of Gross *et al.* for the scission configuration of dinuclear system produced in the reaction $^{209}\mathrm{Bi} + {}^{136}\mathrm{Xe}$ at $E_{\mathrm{Lab}} = 1130$ MeV. From (Gro 81).

entrance channel, which could have been overestimated by the model. For example, as discussed by Ko (Ko 79c), the liquid-drop model stiffness used to calculate the driving force for the deformation degree of freedom is too low, compared to what is obtained in microscopic calculations of collective nuclear states. Secondly, the damping parameter associated with vibrational modes applicable in the situation encountered in damped collisions is currently not known and may be considerably underestimated by Equation (3.5.19). Such effects would diminish the significance of deformations of the individual fragments in both entrance and exit channels, which, however, could be counterbalanced by the formation of a neck (cf. Figures 9 and 10) between the fragments.

A classical dynamical model for damped heavy-ion reactions has been proposed recently by Samaddar *et al.* (Sam 81), who account for neck formation between deformed fragments, stretching and mass exchange between the reaction partners. The classical model (Sam 79b) for the average system trajectories assumes spherical fragments in the entrance channel, interacting via Coulomb, nuclear proximity, and dissipative forces, given by the radial part of the "window formula" [cf. Equation (2.6.25)]. It is further assumed that the two nuclei form a neck and stick together upon reaching the critical distance $r = R_1 + R_2$, whereafter the system rotates rigidly and axially symmetric deformations of its constituents begin to develop. Stretching of the intermediate mononucleus occurs, until repulsive Coulomb and centrifugal forces exceed the nuclear tensile strength. The one-body "wall formula" [cf. Equation (2.6.19)] is employed to model a strong damping of the motion along the collective shape degrees of freedom, while collective inertias are calculated from the hydrodynamical model for irrotational flow (Sam 79b). Fluctuations in these collective variables are modeled in a Fokker–Planck approach [see Equation (2.7.18)], where the diffusion and friction tensors D_{ij} and c_{ij}, respectively, are connected via a thermodynamic temperature τ, i.e., $D_{ij}=c_{ij}\cdot\tau$.

The above model has been applied (Sam 82) to the reactions $^{197}Au+{}^{63}Cu$, $^{209}Bi+{}^{84}Kr$, and $^{209}Bi+{}^{136}Xe$ at laboratory energies of 433, 600, and 1130 MeV, respectively. In the calculations, an initial neck radius is chosen in order to fit the position of the maximum in the angular distribution of the differential reaction cross section. Model predictions (Sam 81) are compared in Figure 157 to experimental data (Sch 78g) on final total kinetic energy spectra for the reaction $^{209}Bi+{}^{136}Xe$ at $E_{\mathrm{Lab}}=1130$ MeV, for the various final projectile-like fragments indicated. The dashed curves in this figure result from the model under the assumption that all l values lead to the formation of an intermediate mononucleus with continuous interior. The solid curves represent calculations in which this is assumed to be the case only for angular momenta below a critical value, l_d. For $l>l_d$, a different reaction model (De 78, 78b, 78c) is used in these latter hybrid calculations. As is obvious from Figure 157, only a qualitative description of the data is achieved with the model. Significant deviations between theory and experiment result also for fragment Z and angular distributions. It is concluded (Sam 81) from these comparisons that statistical shape fluctuations and fluctuations in the scission times are too large, if a mononucleus with the assumed family of shapes is formed in damped reactions, except for low l values.

It is common to all classical and semiclassical calculations discussed in this section that statistical fluctuations are derived from the conditions existing on the average system trajectory at any given time. If, however, there is a limiting nuclear shape with maximum deformation, as indicated by the various sets of data presented previously, it presents (Aga 79b) a barrier which is impenetrable by classical trajectories. Such a barrier would, of course, restrict the development of statistical shape fluctuations. From all the above observations, it is plausible to conclude tentatively that statistical shape fluctuations of the intermediate system formed in damped reactions probably do not play a major role in determining the shape of the total kinetic energy spectrum but that the dynamical evolution of the average shape has a major influence on it and other reaction observables, such that its consideration is important for an understanding of the energy loss mechanism. Clearly, up to now neither theoretical nor experimental work has taken full advantage of the opportunity to employ damped reactions in a detailed investigation of the dynamics of large nuclear deformations at high spins and excitation energies. Problems associated with the nature of

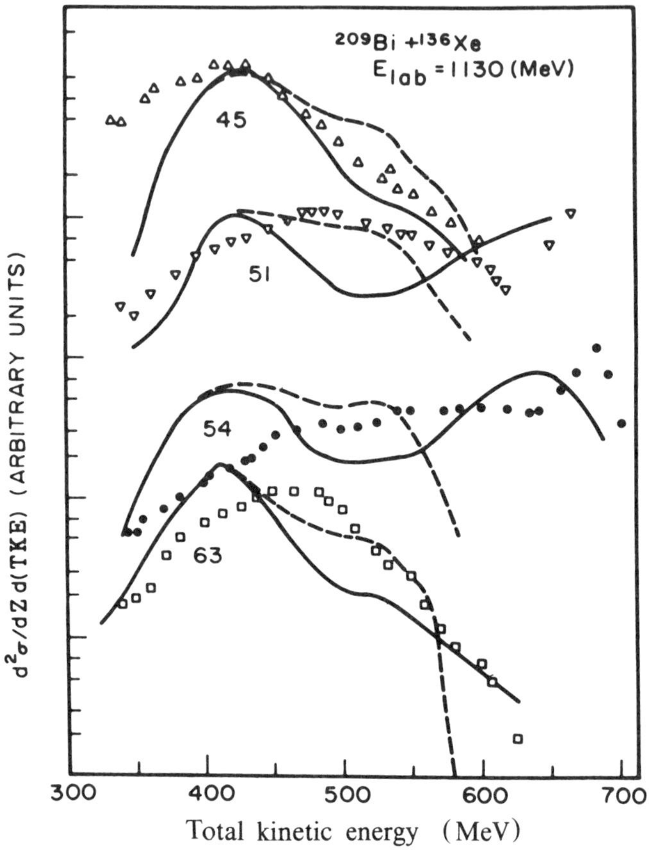

Figure 157. Total kinetic energy spectra for fragments from the reaction $^{209}Bi + ^{136}Xe$ at $E_{Lab} = 1130$ MeV with the Z values indicated. The symbols represent the data in Fig. 150. Dashed and solid curves are the results of model calculations accounting for statistical fluctuations in collective shape variables, as is explained in the text. From (Sam 81).

the collective inertia and the dominance of either one-body or two-body viscosity that can be addressed by such studies have consequences that reach beyond the limited scope of interest here.

Models for the microscopic origins of energy dissipation and its connection with other dissipative features in damped reactions have been discussed in Sections 2.6–2.8. Several applications of these models have already been illustrated in previous experimental sections. Basically, the microscopic mechanism of nucleon exchange and particle–hole excitation are the noncollective modes of energy dissipation to be considered, since there is no evidence for a generally dominant transfer of nucleon clusters in damped reactions (cf. Section 3.4). The questions to be answered then pertain to whether these processes can be understood in terms of independent-particle models for the interacting nuclei that may carry a signature of the individual nuclear structure of each reaction partner. It is also of considerable interest whether these mechanisms give rise to an adiabatic motion where the collision dynamics of a system that is initially far from thermodynamic equilibrium is, nevertheless, determined by its equilibrium properties. Dissipation can, under these circumstances, be expected to be effected by classical friction forces. If, however, diabatic excitation, memory effects, and quantal behavior of the intrinsic system are important in damped reactions, the process can have a more complex dissipative structure, which may only within a limited range of reaction conditions be simulated by the action of frictional forces.

As explained in Section 2.8, the microscopic transport model of Agassi, Ko, and Weidenmüller retains many of the quantal and nonequilibrium statistical features of inelastic excitations in damped collisions that cause dissipation of kinetic energy of relative motion. In actual numerical calculations, however, a weak-coupling approximation with a renormalization of parameters is used (Aga 79b). Then, an (isotropic) friction coefficient is obtained which has the property that it is not constant but decreases strongly with decreasing energy of relative motion. Einstein relations [cf. Equation (2.7.17)] are only valid in the limit of vanishing momentum transfer in a single elementary interaction. Typically, however, the associated diffusion coefficient may be (Aga 79b) by a factor of 2 larger than predicted by the Einstein relation. The exchange of protons between the reaction partners is included in a phenomenological way, but correlations between charge exchange and other reaction variables are not considered. Hence, energy dissipation is essentially due to inelastic excitation, with possible contributions due to nucleon exchange hidden in model parameters. An example of such a calculation has been presented already in Figure 79.

Figure 158 compares experimental laboratory energy spectra of K fragments produced in the reaction ^{232}Th$+^{40}$Ar at $E_{\mathrm{Lab}}=388$ MeV and various reaction angles with the predictions of the model (Aga 79b). One observes only a rather approximative reproduction of the data (Art 73b) by the calculations. For energy losses larger than ~ 150 MeV for this reaction, deformations of the fragments become important, which are not accounted for in the model. Hence, the disagreement between theory and experiment increases for small final kinetic energies.

Similar model calculations (Aga 78b), but for the very heavy system ^{209}Bi$+^{136}$Xe at $E_{\mathrm{Lab}}=1130$ MeV, are compared in Figure 159 to experimental results on the correlation between energy loss and impact parameter. The data obtained (Sch 78g) from an l-wave decomposition of the energy

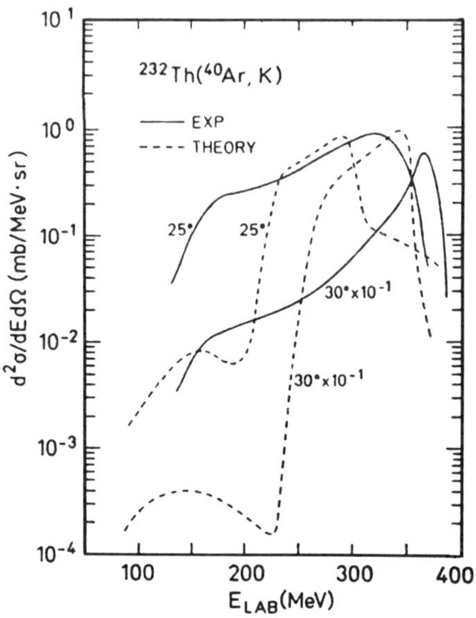

Figure 158. Laboratory kinetic energy spectra of K fragments produced in the reaction $^{232}Th + {}^{40}Ar$ at $E_{Lab} = 388$ MeV at two observation angles. Data (Art 73b) are indicated by the solid curves. Transport calculations are represented by dashed curves. From (Aga 79b).

loss spectrum, shown already in Figure 153 are represented by the solid line. Obviously, the model predictions represented by the dashed curve in Figure 159 give a proper account of the deduced relation between energy loss and l value, for a range of intermediate impact parameters b. Again, agreement cannot be expected for small values of b, because of the neglect of deformations. However, the predicted energy loss is somewhat too large at high l values, leading to a quasielastic peak which is approximately twice as broad as observed for this energy (cf. Figure 147). An improved description of the general form of the measured energy loss spectrum is obtained by Ko (Ko 79c), who included damped harmonic quadrupole shape oscillations in the model, using phenomenological inertia, stiffness, and damping parameters.

From the comparisons made in Figures 158 and 159 and in accord with remarks made in Section 3.3, it appears difficult at present to give a definite appreciation of the basic mechanism underlying the transport model of Agassi *et al.* In order to obtain a more decisive comparison of the predicted peculiar dissipative behavior of the model forces with experimental data, it seems desirable to perform model calculations for experimentally studied medium-weight and heavy systems at several bombarding energies each.

The microscopic friction model of Gross and collaborators, outlined in Section 2.6, assumes ph excitations induced by the moving edge of the single-particle potential to be the dominant source of kinetic-energy dissipation in damped heavy-ion collisions. The effectiveness of this mechanism

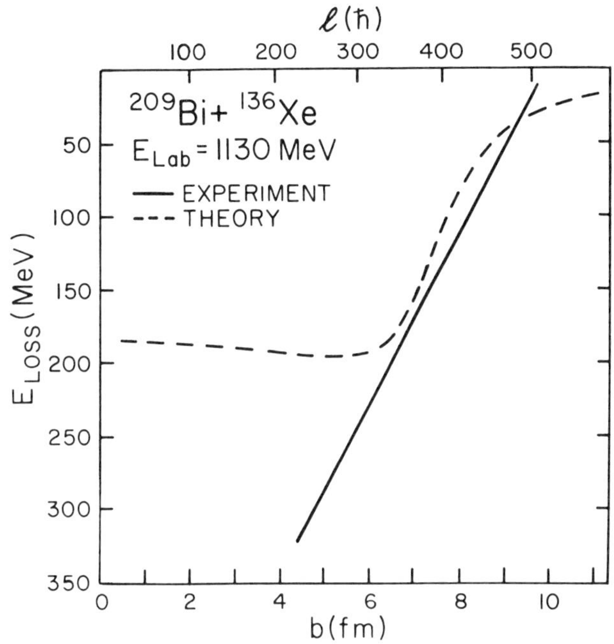

Figure 159. Correlation between average total kinetic energy loss and impact parameter b or l value, for the reaction $^{209}Bi + ^{136}Xe$ at $E_{Lab} = 1130$ MeV. The solid straight line represents the values deduced from experiment (see Fig. 153), the dashed curve is calculated from the transport model of Agassi *et al.* (Aga 78b).

depends crucially on the presence of high frequencies in the Fourier transform of the conservative force. Microscopic perturbation calculations (Gro 78) suggest a quadratic dependence of the resulting friction form factor on the gradient of the interaction potential [cf. Equation (2.6.17)], implying a relatively sudden (~ 10^{-22} s) stoppage of the relative motion at separation distances corresponding to relatively small matter overlaps between the interacting heavy ions (hard friction model). Utilizing a steep folding-type interaction potential (cf. Section 2.5), the expected relation between friction strength and conservative force is observed (Gro 78, 81) to be consistent with experimental Wilczyński diagrams of the damped reaction cross section, as exemplified in Figure 78. If, however, a shallower interaction potential such as the Krappe–Nix potential (cf. Section 2.5) is utilized in the classical dynamical calculations, the radial dependence of the friction form factors has to be taken (Gro 78) as proportional to the fourth power of the interaction force, in order to provide acceptable fits to the data. The energy loss occurs at smaller distances than in the former approach but is comparably sudden in both models. Evaluated at the respective turning points of a trajectory with an l value just above the critical angular momentum for fusion, both form factors have the same strength and gradient.

The above remarks illustrate the ambiguity of attempts to interpret experimental energy–angle correlations in terms of conservative and dissipative forces. Earlier discussion in this section has emphasized the additional complication due to dynamical deformations. The latter, however, have their greatest effect usually in near-central collisions. Hence, it may be of significance to an appreciation of the hard-friction model to note that calculations (Gro 78, 81, 81b) consistently underestimate the grazing angle and the scattering angle for weakly damped events by 5–10° in the c.m. system. Figure 160 presents an example for such a comparison between hard-friction model calculations (Gro 78) and experimental data (Wol 76) for the Wilczyński diagram of projectile-like fragments from the reaction $^{209}Bi + ^{84}Kr$ at $E_{Lab} = 714$ MeV. The experimental grazing or quarter-point angle is reported (Bir 82b) as $\theta_{1/4}(\text{c.m.}) \approx 50°$. The hard-friction model calculation by Gross and Kalinowski (Gro 78) represented by the solid and dashed curves in Figure 160 miss this angle by 6 or 7 degrees, depending on whether or not fragment deformations are taken into account.

Brosa and Gross (Bro 80f) have suggested using the features of experimental mass and charge distributions for weakly damped events, presumably associated with peripheral collisions and small matter overlap, to distinguish between the hard-friction model and other dissipation mechanisms. As is illustrated in Figure 116, the barriers of the combined single-

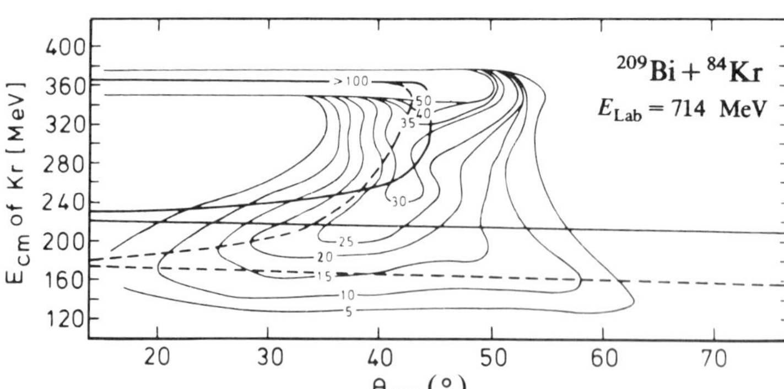

Figure 160. Wilczyński diagram $d^2\sigma/dEd\theta$ in units of mb(MeV rad)$^{-1}$ for the reaction $^{209}Bi + ^{84}Kr$ at $E_{Lab} = 714$ MeV plotted vs. c.m. energy and scattering angle of Kr-like fragments. The contour lines represent experimental data (Wol 76). The full curves reflects results from dynamical calculations with the hard-friction model of Gross *et al.*; the dashed curves result from similar calculations accounting for fragment deformations. From (Gro 78).

particle potential for neutrons and protons drop successively below the Fermi level, with decreasing separation distance of the reaction partners. It has been delineated already in Section 3.4 that this effect leads to a neutron-rich mass flow between projectile and target for a region of l values below the grazing angular momentum, that, in the case of a hard-friction model, can be rather broad. Figure 117 demonstrates a quite adequate modeling of the large ratios σ_N^2/α_Z^2 observed (Sch 81h) in Xe-induced reactions with such a theory (Gro 81b). The predictions are obtained (Gro 81b) by integrating a two-dimensional Fokker–Planck equation for the exchange of individual protons and neutrons along trajectories determined completely by the hard-friction model (Gro 78). The exchange of nucleons between the reaction partners is governed by transport coefficients that take account of the quantum-mechanical barrier penetration [cf. Equation (3.4.18)] and the Pauli blocking. However, energy dissipation due to the exchange processes themselves is not included and, therefore, the energy loss per exchanged nucleon is underestimated within the framework of the model. Accepting the basic dissipation mechanism underlying the hard-friction model, one is led to expect the model to predict somewhat larger variances of fragment mass and charge distributions at a given energy loss than experimentally observed.

In contrast to these expectations, however, Figure 161 (Gro 81b) illustrates for several ^{56}Fe-induced reactions a systematical drastic underestimation by the model of experimental isotopic mass variances in a broad range of energy losses. Moreover, the general shapes of experimental and theoretical correlations differ significantly. A similar observation holds for the comparison of theoretical and experimental isobaric charge variances for the same reactions. Taking account of complications arising from sequential particle emission should not alter these conclusions for the heavy targets, as is discussed in Section 3.4. A possible explanation for the discrepancies apparent from Figure 161 could lie in an overestimation of the range of the dissipative force by the hard-friction model that leads already to significant energy losses for separation distances where nucleon exchange is prohibited by substantial barriers of the single-particle potential. Suggestions offered by Gross *et al.* (Gro 81b) include the possibility of cluster transfer in damped reactions for which, however, no positive experimental evidence exists. Currently available data and calculations do not allow one to draw definitive conclusions as to the insignificance of particle–hole excitations as a dissipation mechanism in damped reactions. However, it seems unlikely that it constitutes the dominant damping process. Certain properties of particle emission mechanisms, to be discussed in Section 3.7, also disfavor the hard-friction model. It remains, nevertheless, remarkable how well this theory is able to describe the features of experimental energy-angle correlations and to predict absolute cross sections (Gro 78).

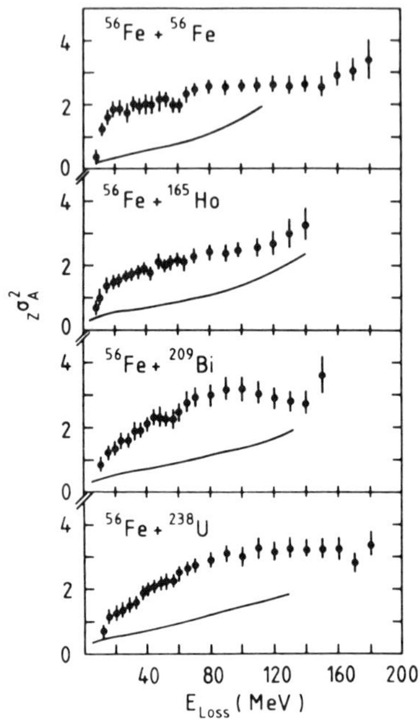

Figure 161. Isotopic mass variances for projectile-like fragments produced in ^{56}Fe-induced reactions at 8.3 MeV/nucleon. Experimental data are by Breuer *et al.* (Bre 81). The curves represent results of calculations using the hard-friction model (Gro 81b).

While classical dynamical friction models usually consider only the time evolution of an average system trajectory, yielding, for example, predictions of the development of the maximum cross section with angle and energy, account can be taken of the fluctuations arising in the final total kinetic energy through application of generalized fluctuation–dissipation relations (cf. Section 2.7). Berlanger *et al.* (Ber 78e) have calculated double-differential cross sections $d^2\sigma/d\theta_{\text{c.m.}}\, dE_{\text{c.m.}}$ for several systems employing the linear response theory of Hofmann and collaborators, in order to calculate the quantal and statistical dispersions associated with the average dissipated kinetic energy. The energy density potential (cf. Section 2.5) was used to calculate the conservative interaction force. Friction forces were assumed to scale with the fourth power of this force, adopting the friction proportionality constants proposed by Gross *et al.* (Gro 74b). In analogy to Equation (3.4.6), the double-differential cross section from the model can be written as (Ngô 80c) an integral over impact parameters b,

$$\frac{d^2\sigma}{dE\,d\theta} = \left(\frac{\mu}{2E}\right)^{1/2} \int 2\pi b\,db \lim_{t\to\infty} P(\theta, p_r, t) \tag{3.5.20a}$$

with the joint probability function

$$P(\theta, p_r, t) = \frac{1}{4\pi\sqrt{\Delta}} \exp\left[-\frac{(\theta-\theta_c)^2\chi_{rr} - 2(\theta-\theta_c)(p_r - p_{rc})\chi_{r\theta} + (p_r - p_{rc})^2\chi_{\theta\theta}}{4\Delta} \right] \tag{3.5.20b}$$

Here, $\chi_{rr} = \sigma_r^2/2$, $\chi_{\theta\theta} = \sigma_\theta^2/2$, and $\chi_{r\theta} = \sigma_{r\theta}/2$ are the time-dependent elements of the covariance tensor associated with the two-dimensional probability distribution P with respect to radial linear momentum p_r and angle θ. The quantity Δ stands for its determinant, μ is the reduced mass of the system, and E is its final total kinetic energy. The average classical quantities indicated by the subscript "c" are determined by the trajectory model, while the dispersions χ are calculated from the linear-response theory. The form of Equations (3.5.20) results from an assumption that changes in momentum and angle occur in a statistically independent fashion. Mass exchange and the development of dynamical deformations are not taken into account in this simplified model.

Model calculations for the 1560-MeV $^{208}Pb + ^{208}Pb$ reaction are illustrated in Figure 162 in the form of a Wilczyński diagram. For comparison, experimental data for the same reaction measured (Tan 80c) at the slightly higher incident energy of 1675 MeV are displayed in Figure 75. It should be noted that somewhat different energy quantities are plotted in the two figures, with Figure 75 using the total kinetic energy TKE* modified by the difference in entrance- and exit-channel Coulomb energies (see Section 3.4). This complicates a quantitative comparison of data and calculations.

The qualitative features of experimental and theoretical Wilczyński diagrams shown in Figures 75 and 162, respectively, appear to be rather similar. The larger theoretical grazing angle is expected from the lower bombarding energy assumed in the calculations, in comparison to the experiment. A backward-moving cross-section ridge observed experimentally is also predicted by the model, with average energies of the fully relaxed backward-angle component of about 650 MeV. As in the experimental data, the width of the final energy distribution increases with increasing angle and increasing average energy loss. However, it is quite obvious that the theory underestimates the widths of the experimental energy distributions consistently by factors of 2–3. This is true also of results for the $^{208}Pb + ^{132}Xe$ reaction obtained (Ber 78e) with the same model, if compared to experimental data displayed in Figure 74 for the similar $^{209}Bi + ^{136}Xe$ reaction.

Because of the lack of systematic studies using successive approximation within a given approach, it is impossible at present to identify the most

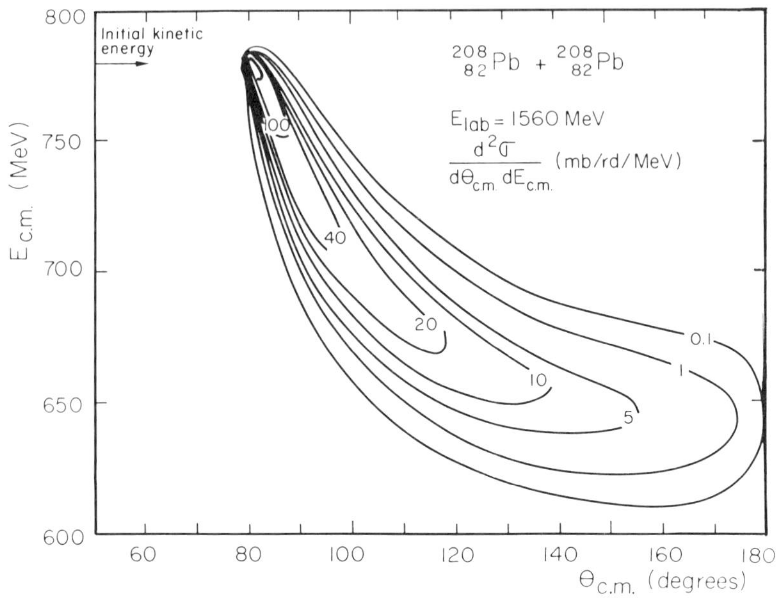

Figure 162. Contour diagram of the double-differential cross section $d^2\sigma/d\theta_{c.m.}\,dE_{c.m.}$ for the reaction $^{208}Pb + ^{208}Pb$. Results are given of a dynamical calculation employing a hard-friction model including dispersions. The conservative interaction force is derived from the energy-density formalism. From (Ber 78f).

important reason for the above quantitative disagreement between experiment and model prediction. It is, in particular, not obvious that this shortcoming can be attributed to the ansatz of the frictional force used, since the observed angular dependence of the dissipated energy is reasonably well described by the hard-friction model used in the calculations of Berlanger *et al.* (Ber 78e). Calculations (Ber 78f) performed by the same group employing the reaction model proposed by Siwek-Wilczyńska *et al.* (Siw 76), which accounts for fragment deformations in a phenomenological simulation, also yield only a qualitative similarity between theoretical cross section patterns for individual final charge fragmentations and experimental data. In these more complete calculations, charge exchange between the reaction partners is considered in a simplified fashion, using a charge diffusion coefficient proportional to the matter density overlap. Only average correlations between the charge asymmetry degree of freedom and other reaction coordinates are taken into account, as is suggested by general experimental observations [cf. Equation (3.4.7)]. One may then tentatively conclude that the dispersions in excitation energy or energy loss attributed, within the framework of the linear response theory, to a combination of thermal fluctuations and those due to the curvature of the effective static interaction potential [cf. Equations (2.7.19)–(2.7.22)] can at most contribute to the total observed widths but probably do not constitute the dominant source of the dispersion of system trajectories about the classical mean path. This conjecture complements an observation made in Section 3.4 with respect to the reluctance of mass and charge asymmetry in damped reactions to respond to the action of the static driving force. Although a consistent and detailed inclusion of correlations between energy loss and the development of dynamical deformations into the model calculations may alter somewhat the above conclusions, it appears more promising to inspect the influence of the dynamical properties of the microscopic transport processes on average dissipation and associated fluctuations in the total kinetic energy of relative motion.

It is quite interesting to note that the recent transport model of Gross and Hartmann (Gro 81b) is able to reproduce satisfactorily (cf. Figure 117) the energy loss dependence of experimental ratios of moments of the joint fragment (A, Z) probability distribution or of correlation parameters, at least for several low-energy reactions, but does not account well for the evolution of mass and charge variances with the amount of dissipated energy. Apparently, energy dissipation and mass exchange are more strongly coupled than assumed in this model.

The experimental data presented in Section 3.4 are clearly suggestive of a mass transport mechanism in damped reactions in which individual nucleons can be exchanged back and forth between the fragments a considerable number of times. On the other hand, in quasielastic and weakly damped reactions of lighter systems, definite correlations are observed between the maximum nucleon transfer cross section and an optimum reaction Q value, calculated [see Equations (3.5.7) and (3.5.8)] from the changes in ground-state Q values and interaction energies, as well as from the kinematical recoil associated with the successive transfer of nucleons. Depending on the incident c.m. energy, the relative contributions due to the different effects vary. In low-energy transfer reactions, the effects of ground-state Q value and static interaction force are most strongly felt, while the consideration of a recoil term becomes inevitable at higher bombarding energies. Although damped reactions are much more complex than direct few-nucleon transfer reactions, they are viewed to develop out of these relatively simple initial reaction stages. It appears, hence, interesting to try to discover whether the same basic dissipation mechanisms are at work both in quasielastic and damped reactions.

Early attempts to understand the experimentally observed strong correlations between energy loss and the widths of fragment charge distributions in terms of an exchange-induced dissipation mechanism are due to Huizenga and collaborators (Hui 76, 76c, 77, 77b, 82, Sch 77d, 77e, 78g, Wil 80, Wol

81). Details of the phenomenological models employed in their work have been described already in the previous section. Assuming nucleons at rest within the reacting fragments, in the event of the transfer of a nucleon, its share of the total relative momentum is dissipated. This model leads to the prediction of a strong correlation between energy loss and the variance σ_Z^2 of the fragment charge distribution, represented by Equation (3.4.12) with $\alpha = 1$. Here, the variance σ_Z^2 is taken as reflective of the total number N_{ex} of nucleons exchanged, which scales classically with this variance in different ways, depending on the correlations present in the joint fragment (A, Z) distribution (Bec 78d).

Such a simple nucleon exchange process generates a classical friction force $F = -k_{ex}u$ acting on the velocity u of relative motion. If, in addition to this process, other frictional mechanisms such as induced by particle–hole excitations operate concurrently, the nucleon exchange process provides at least a time scale (Sch 77d) for a study of these extraneous friction mechanisms. If the latter can be characterized by a friction constant k_{nex}, and if the elementary change in the relative kinetic energy E is small, the total energy loss associated on the average with a single exchanged nucleon is approximately given by

$$\delta E_{\text{Loss}} = -dE/dN_{ex} = (m/\mu)(1 + k_{nex}/k_{ex})E = (m/\mu)\alpha E \tag{3.5.21}$$

Under the additional assumption that the phenomenological parameter α is constant in a reaction, integration of Equation (3.5.21) yields Equation (3.4.12).

Fits to experimental correlations between energy loss E_{Loss} and fragment charge variance σ_Z^2 using Equation (3.4.12) are usually quite successful for energy losses not exceeding the kinetic energy initially available above the interaction barrier. This is illustrated in Figure 163 (Sch 80) for the $^{209}\text{Bi} + {}^{136}\text{Xe}$ reaction at two bombarding energies. Obviously, the shapes of the experimental correlations are well reproduced by the simple form of Equation (3.4.12). Only one of the parameters x or α entering this equation can be extracted from such a fit; the remaining parameter has to be estimated independently. It is not in general possible to derive a unique relation between N_{ex} and the variances σ_Z^2 or σ_A^2 and, hence, to obtain an estimate for the correlation parameter x. Fortunately, however, this is not necessary for the present purpose, provided the same relation holds approximately for all reactions compared. Assuming arbitrarily a value of $x = 2$, i.e., strictly correlated exchange of classical nucleons, one obtains the values $\alpha = 1.4$ for the $^{209}\text{Bi} + {}^{136}\text{Xe}$ reaction at 1130 MeV from the best fit indicated by the solid curve on the left of Figure 163. For the lower

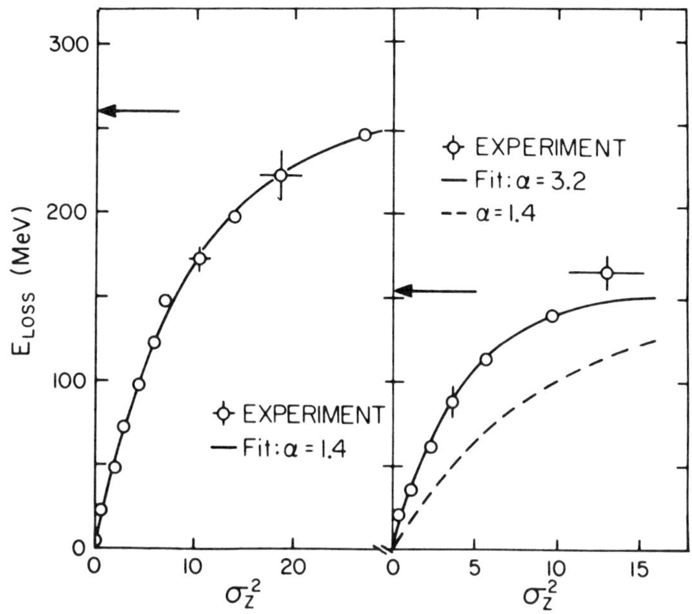

Figure 163. Correlation between energy loss and variance of the fragment-Z distribution for the reaction $^{209}Bi + ^{136}Xe$ at $E_{Lab} = 1130$ MeV (left) and 940 MeV (right). Arrows denote initially available kinetic energies. The curves represent predictions according to Equation (3.4.12) with the indicated α values and $x = 2$. From (Sch 80).

bombarding energy of 940 MeV, this α value predicts the dashed curve on the right of this figure. In order to obtain a fit to these lower-energy data, the parameter α has to be more than doubled in comparison to the 1130-MeV result. Hence, based on such a simple picture, one could be led to the conclusion that nucleon exchange as a dissipation mechanism loses efficiency with decreasing bombarding energy, as the ratio k_{ex}/k_{nex} drops from 2.5 to 0.45 when the bombarding energy is decreased from 1130 to 940 MeV. Similar trends are observed also for other reactions, as is demonstrated in Figure 164 (Sch 80), where α is plotted vs. the approach energy $(m/2)u^2 = (m/\mu)(E_{c.m.} - V_{Coul})$ per nucleon. The average trend suggested by the solid curve in Figure 164 predicts that α approaches unity for incident energies above the barrier in excess of ~ 4 MeV per nucleon, implying a vanishing contribution of nonexchange processes to energy dissipation in damped collisions. This trend is quite well established by systems with an initial projectile–target combination located near a local minimum in the potential energy surface [cf. Equation (2.5.30)], as calculated from the sum of liquid-drop model binding energies for two spherical nuclei. Systems with ^{56}Fe as a projectile are removed from such minima (see

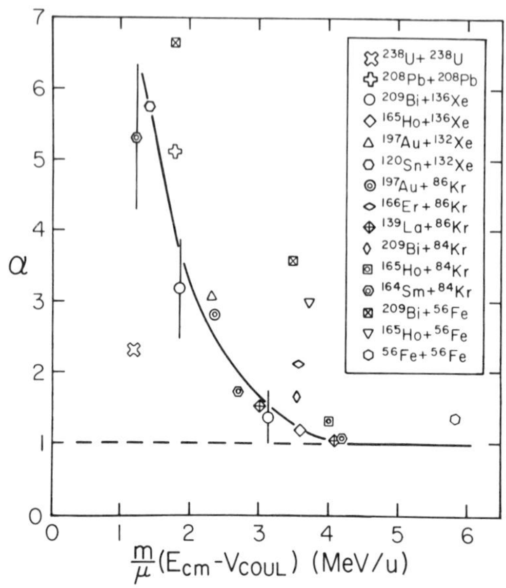

Figure 164. Values of the parameter α obtained from fits of Equation (3.4.12) to available data plotted vs. the initially available energy per nucleon above the Coulomb barrier. From (Sch 80).

Figure 22). They appear to show a significant deviation from the average behavior, which could be interpreted as an additional energy loss per exchanged nucleon, due to the associated change in ground-state Q values. Transport calculations based on Randrup's exchange model referred to in earlier sections and to be discussed further below take full account of these energy loss contributions, as can be seen from inspection of the model dissipation rate given by Equation (2.6.30). These calculations suggest a relatively small contribution of the variations in ground-state Q values or, more generally, of the static driving forces to the total energy dissipated in most reactions.

In contrast, Dakowski and collaborators (Dak 79, 80, Gob 81, Dak 82) have emphasized the importance of a structure energy DE, dependent on the excitation energy and the particular nuclear structure, that contributes to the energy dissipated in a nucleon exchange process. Instead of Equation (3.5.21), these workers use

$$\delta E^*_{\text{Loss}} = -\,dE/dN_{\text{ex}} = (m/\mu)E + \text{DE}(E^*_{\text{Loss}}) \qquad (3.5.22)$$

for the interpretation of correlations between fragment charge variances and the energy loss E^*_{Loss}, modified according to Equation (3.4.9) for changes in the Coulomb energies in entrance and exit channel. The total number of exchanged nucleons is assumed to be given by $N_{\text{ex}} = (A/Z)^2\sigma_Z^2$, where A and Z correspond to the global mass and atomic numbers, respectively, of the composite system. Obviously, Equation (3.5.22) does not contain a multiplicative parameter exhibiting the bombarding energy dependence depicted in Figure 164, hence a corresponding dependence must be observed in the deduced values of the quantity DE. The bombarding energy dependence expected for DE may, however, be rather weak. Comparison of Equations (3.5.21) and (3.5.22) suggests that for equivalent fits of the experimental data, the relation

$$\mathrm{DE} = (m/\mu)(\alpha - 1)E \tag{3.5.23}$$

should hold for the fit parameters DE and α. For a given incident bombarding energy, DE is expected to decrease with decreasing kinetic energy above the barrier or increasing energy loss, because of the apparent independence of α on E_{Loss}. This is borne out in the analyses of Dakowski *et al.*

In Figure 165 several sets of data are compiled for experimental correlations between variances of fragment charge distributions and energy loss (TKEL = E_{Loss}) or modified energy loss parameter (TKEL* = E^*_{Loss}). For the systems illustrated in this figure, the A/Z ratios of projectile and

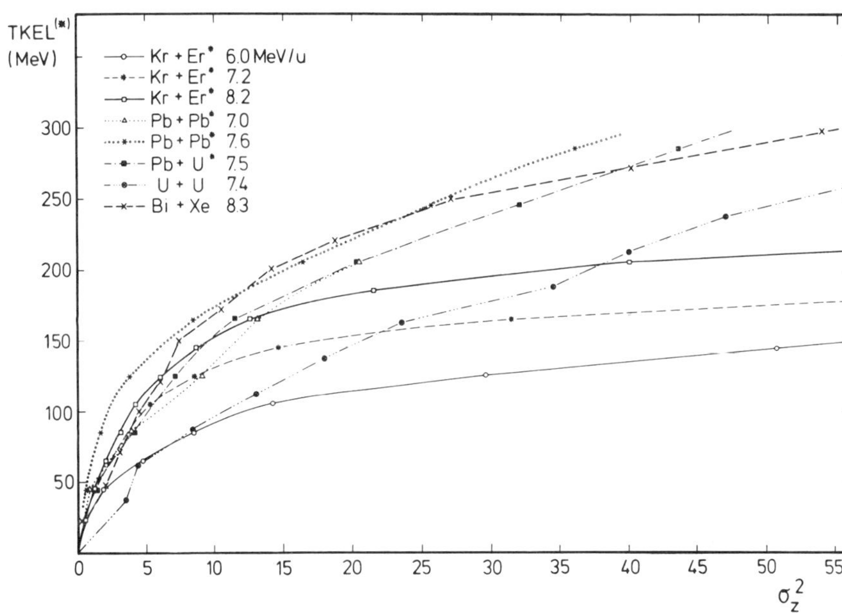

Figure 165. Experimental correlations between energy loss TKEL (= E_{Loss}) and variance σ_Z^2 of the fragment charge distribution for various heavy-ion reactions. Reactions, where a modified energy loss parameter TKEL* (= E^*_{Loss}) is plotted, are marked with an asterisk in the figure. From (Dak 80).

target are rather similar. The $^{208}Pb + ^{208}Pb$ and $^{166}Er + ^{86}Kr$ data included in Figure 165 were measured at several incident energies. The data were analyzed (Dak 80) by differentiating the curves drawn through the experimental points and subtracting the recoil energy term $RE = Em/\mu$. Results for the quantities RE and DE obtained from this procedure are exhibited in Figure 166. Part (a) of this figure displays the linear variation of RE with energy loss as straight lines, calculated for the $^{166}Er + ^{86}Kr$ system at various bombarding energies. The remainder DE of the experimental energy loss rate is shown as a histogram in each case. As expected from Equation (3.5.23), this extra "structure energy" DE needed to fit the data of Figure 165 decreases with increasing energy loss. The differential analysis of Dakowski *et al.* is subject to considerable systematical and statistical uncertainties reflected in the scatter of the derived values for DE. As can be seen from Figure 166a, the recoil term (RE) dominates for high incident

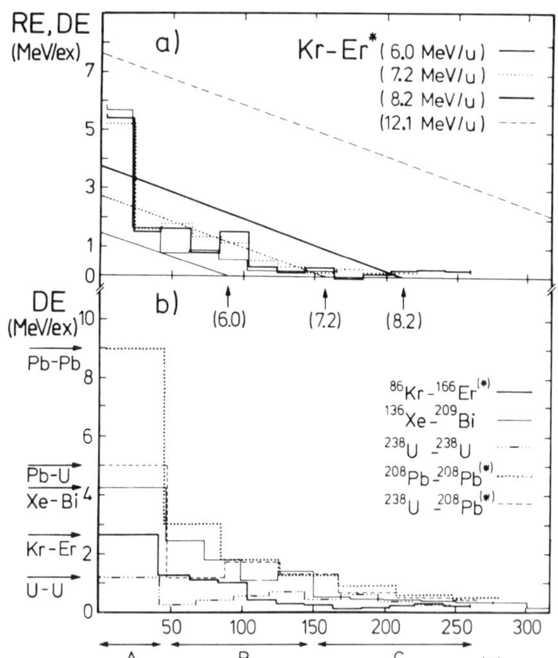

Figure 166. Average rate of energy loss per nucleon exchange (MeV/ex) as a function of TKEL*: (a) For the Kr–Er reaction at three bombarding energies. The histograms show the structure term DE. The straight lines visualize the recoil term ($RE = E \cdot m/\mu$) for different bombarding energies. (b) Comparison of the DE curves for the different systems. Arrows on the ordinate show the positions of DE_0 values normalized to 5–45 MeV bins in TKEL or in TKEL*. Regions A, B, and C discussed in text.

energies, while the structure term becomes relatively more important at lower bombarding energies. Within the accuracy of data and analysis, no bombarding energy dependence of DE can be discerned in the $^{166}Er + ^{86}Kr$ results shown in this figure. This is qualitatively what is expected from Equation (3.5.23) and the general energy dependence of α exhibited in Figure 164.

The dependence of the structure energy DE on the projectile–target combination is depicted in Figure 166b. Here, it is seen that among the systems considered in the analysis, at low energy losses the tightly bound $^{208}Pb + ^{208}Pb$ system yields the largest structure energy correction DE, which diminishes rapidly with increasing energy loss. Three energy loss regions are distinguished in Figure 166 by the letters A to C. The DE values obtained (Dak 80) for 5 MeV $\leq$ TKEL $\leq$ 45 MeV (A) are observed to correlate strongly with an effective Q value calculated (Dak 80) as an average of the changes in the ground-state Q values induced by the exchange of two protons or two neutrons between projectile and target. This finding is taken as strong indication for the structure dependence of DE in the region of low energy losses. In the region (B), corresponding to energy losses between 50 and 150 MeV, a strong decrease of DE is found in the analysis performed by Dakowski *et al.* (Dak 80). This feature is interpreted as being due to a melting of the nuclear shell structure at increased temperatures of the fragments. The region (C) of the highest energy losses is thought to be dominated by effects due to the increasing deformations of the constituents of the intermediate dinuclear system.

A more fundamental analysis of the influence of nuclear structure on the energy dissipation mechanism in damped reactions has been performed recently by Dakowski *et al.* (Dak 82), based on a random-walk model for nucleon exchange between fragments endowed with single-particle shell structure. In this approach, the energy dissipated per exchanged nucleon is written as

$$-\frac{dE}{dN_{\text{ex}}} = \frac{-dE/dt}{dN_{\text{ex}}/dt} = \frac{\Sigma_{i,j}(\varepsilon_i - \varepsilon_j)W_{ij}f(\varepsilon_j)}{\Sigma_{i,j}W_{ij}f(\varepsilon_j)[1-f(\varepsilon_i)]} \tag{3.5.24}$$

Here, the quantities ε_i denote the nuclear single-particle energies, W_{ij} are time-dependent transition probabilities expressed in terms of a Gaussian in the energy difference of single-particle states involved in a transition, with phenomenological overall strength and l-dependent width parameters (cf. Section 2.7). The functions f are single-particle occupation probabilities. Account is taken of the Pauli principle in nucleon exchange processes, as is obvious from the blocking factor in the denominator of Equation (3.5.24). The prime indicates summation over levels in different nuclei, while the unrestricted sum in the numerator of Equation (3.5.24) includes also inelastic particle–hole excitations. The above energy loss rate is evaluated within a phenomenological model (Rie 79) involving interaction and relaxation times for energy and angular momentum. Comparisons to experimental data are made by identifying the number of exchanged nucleons N_{ex} with the variance σ_A^2 of the fragment mass distribution or, as before, with the fragment charge variance σ_Z^2 scaled by the square of the global A/Z ratio.

Figure 167 illustrates for the two symmetric systems $^{208}Pb + ^{208}Pb$ and $^{238}U + ^{238}U$ that the model is able to produce perfect fits (solid curves) to experimental energy-loss–variance correlations, whereas the simple recoil calculation according to Equation (3.5.21) with $\alpha = 1$ significantly underestimates the energy loss for a given value of σ_Z^2. Fits of similar quality are obtained (Dak 82) for correlations between energy loss and mass variance σ_A^2 for a number of heavy and medium-weight systems, although the bombarding-energy dependence of $E_{Loss}(\sigma_Z^2)$ correlations is overestimated by the model in the case of the $^{139}La + ^{86}Kr$ reaction (Dak 82).

The quality of reproduction of experimental data by the above model, which incorporates elements of the diffusion theory developed by Nörenberg and collaborators (see Sections 2.7 and 2.8), is certainly impressive. However, it is not obvious what can actually be deduced unambiguously for the dissipation mechanism from this agreement. The calculations contain several adjustable parameters and are based on phenomenological procedures whose quantitative validity may be questioned. In some cases, for example, the agreement between calculations and experimental $E_{Loss}(\sigma_Z^2)$ relations could be destroyed if instead of the charge variance σ_Z^2, scaled with the square of the global A/Z ratio, the actual energy-loss-dependent (see Section 3.4) value of σ_A^2 is used for N_{ex}. Even the latter identification must be viewed with suspicion, as its validity depends drastically on the correlations present in the joint fragment (A, Z) distribution. It is, furthermore, significant to remember that statistical models for nucleon exchange such as employed in these calculations neglect important dynamical features of the transport mechanism. As has been illustrated in Section 3.4, fits obtained with such models to certain experimental correlations actually violate basic requirements of the theory, e.g., Einstein relations, while relations between other quantities remain totally unexplained. To these latter class of relations belong experimental drifts in mass and charge asymmetry that follow the static driving force only rather reluctantly with increasing energy loss contrasted to the strong correlations within the fragment (A, Z) probability distribution that this force can nevertheless induce. Many of these aspects are not addressed by the stochastic model of Dakowski *et al.* (Dak 82), making it impossible to give a final assessment of the validity of this approach.

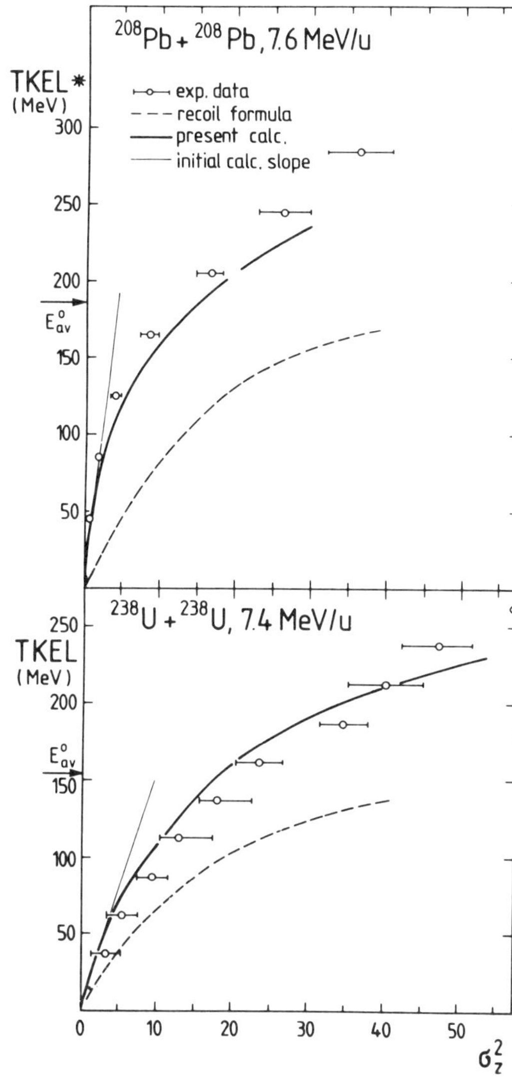

Figure 167. Correlations between energy loss and variance of fragment charge distributions for the reactions $^{208}Pb + ^{208}Pb$ and $^{238}U + ^{238}U$. For the latter reaction, data are somewhat uncertain because of important distortions due to the sequential fission process. Solid curves represent calculations with the stochastic model described in the text; dashed curves indicate contributions from the recoil model of energy dissipation. From (Dak 82).

The two phenomenological dissipation models discussed above treat the relative motion in a rather simplified fashion, leaving ample latitude for an interpretation of results on comparisons to experimental data. A more satisfactory, though still approximate, way to investigate the connection between energy dissipation and nucleon exchange consists in solving the Lagrange–Rayleigh Equations (2.3.1) for the average classical system trajectory in a multidimensional coordinate space. With the reaction model of Gross *et al.* (Gro 81b), an example has already been discussed in this section for a theory in which nucleon exchange during a heavy-ion collision is controlled by the dynamics of relative motion. In this model, however, the feedback effect of this exchange process on the relative motion is neglected. That these effects can be of dominant influence on the whole reaction course is, on the other hand, the principle on which the transport model proposed by Randrup (Ran 78, 78d, 79d) is based. This theory attributes all dissipative transport phenomena to the exchange of individual nucleons between the reaction fragments but neglects possible contributions to energy dissipation by inelastic particle–hole excitations. For example, the energy dissipation rate [cf. Equation (2.6.29)] contains both a contribution due to the driving force F_{AB}, governing the flow of mass and charge between the fragments, and a recoil term. The model makes, hence, an interesting connection between the optimum Q-value systematics observed [cf. Equations (3.5.7)–(3.5.9)] in few-nucleon transfer reactions and dissipative processes in damped reactions. The model driving force is derived from the balance of dissipative, conservative, and inertial forces and constitutes a dynamical force that reduces to the static driving force only in the limit of vanishing kinetic energy of relative motion. Track of the influence of nuclear structure on exchange and dissipation is kept within the model through the static part of the driving force calculated from the difference in Fermi levels as given by the liquid-drop binding energies, corrected or uncorrected for shell effects. The influence of the static driving potential, however, is relatively weak at kinetic energies of a few MeV per nucleon above the barrier. This is directly seen in experimental data on mass and charge exchange, as discussed in Section 3.4, and also qualitatively expected from the transport model.

While the model of Dakowski *et al.* discussed previously suggest that the bombarding-energy dependence of experimental $\sigma^2(E_{\text{Loss}})$ correlations, in excess to what is predicted by the classical recoil term in Equation (3.5.22), is due to an extraneous structure energy dissipated in an exchange process, the model of Randrup attributes this to a bombarding-energy dependence of the width of the region of single-particle levels around the Fermi surface that can actually participate in exchange processes. This behavior can, for peripheral collisions, be modeled (Sch 80) by an energy-dependent scaling factor α presented in Equation (3.4.13), except for a ratio of form factors, by which the classical recoil term has to be multiplied. The quantity α is found to be inversely proportional to the "effective temperature" τ^*, a flux average of the energy transported in a nucleon exchange process. This temperature parameter also enters the generalized Einstein relations of Equation (2.8.21) and, hence, determines in the model effectively the statistical dispersions of reaction observables about their average value. The variation of τ^* with energy loss and bombarding energy is illustrated in Figure 37.

As has been discussed in Section 2.8, the effective temperature parameter τ^* is proportional to the average exchange flux of nucleons, which is restricted by Pauli blocking most effectively at low relative energies. The variation of this restriction along a simple trajectory can be inferred from Figure 168, where the predicted time dependence of τ^* and that of the thermodynamic temperature τ is illustrated for the 465-MeV $^{165}Ho+^{56}Fe$ reaction and $l=180$. The quantity τ^* is obviously considerably larger than τ, especially early and late in the collision, where the relative velocity is high. For a window of constant size between the interacting fragments, one would then expect a significant fraction of the exchanged nucleons to be

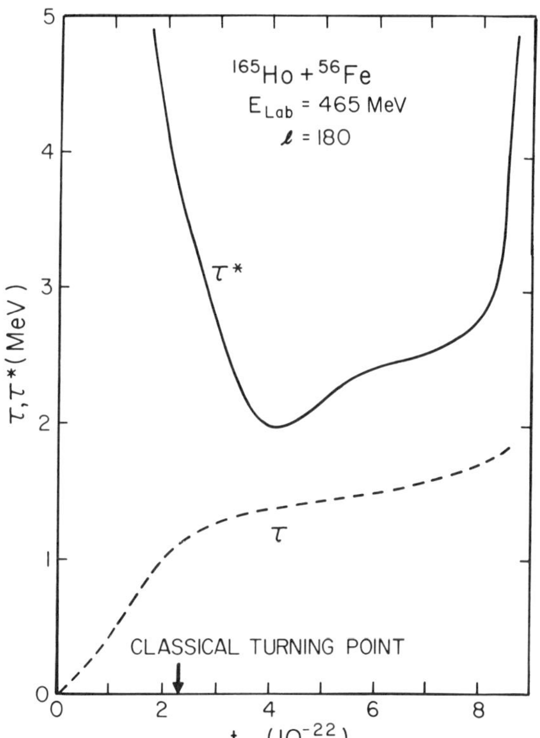

Figure 168. Illustration of the time dependence of effective temperature τ^* and thermodynamic temperature τ along an $l=180$ trajectory for the reaction $^{165}Ho+^{56}Fe$ at $E_{\rm Lab}=465$ MeV. The abscissa gives the interaction time in units of 10^{-22} s, counted from passage of the strong-absorption separation distance. The classical turning point is reached at a time indicated by the arrow.

transferred at high relative velocities, implying a large energy deposit per exchanged nucleon. Actually, however, the window transmission varies simultaneously, as given in the model by the differential flux form factor N' (see Section 2.8), and, hence, the effect is considerably smaller than expected from the variation of τ^* alone.

In the classical limit where Pauli blocking is neglected, all nucleons can contribute in exchange processes, to an extent permitted by the window size, as is given in the model by the integral flux form factor N (cf. Section 2.8). However, only those nucleons exchanged between the displaced Fermi surfaces actually lead to energy dissipation, an effect discussed in the context of the piston model (cf. Section 2.6). Hence, in this case, the energy dissipated per exchanged nucleon is relatively small. Figure 169 (Sch 80) illustrates the importance of including the Pauli blocking effect for an understanding of experimental $E_{\mathrm{Loss}}(\sigma_Z^2)$ correlations for two reactions. Calculations with the full transport model, details of which have been given previously, are represented by the full curves giving an adequate description of the data (circles). If Pauli blocking is neglected in the calculations, the broken curves result, indicating dramatically increased variances due to larger transfer rates for a given total kinetic energy loss. The theoretical correlations are predicted to become somewhat steeper than displayed in Figure 169 when account is taken of restrictions imposed on the proton exchange by the curvature of the underlying driving potential (Sch 81b),

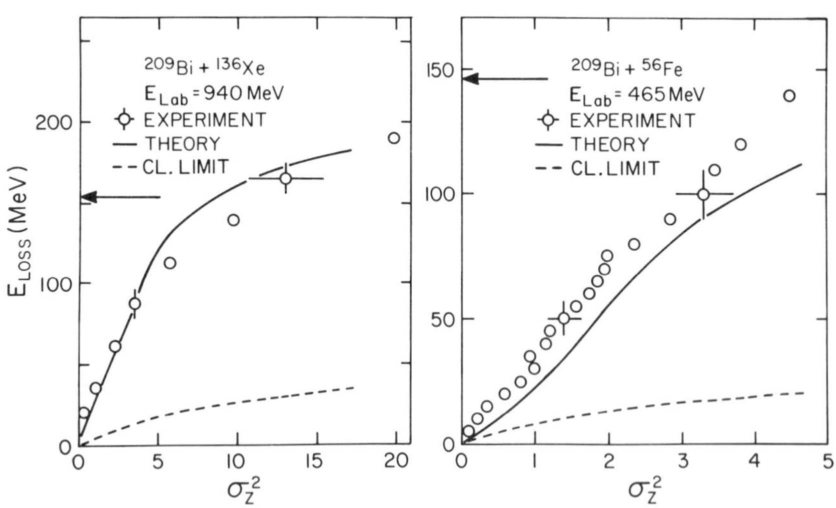

Figure 169. Comparison of transport model predictions for the correlation $E_{\mathrm{Loss}}(\sigma_Z^2)$ with data for the reactions ^{209}Bi + ^{136}Xe (left) and ^{209}Bi + ^{56}Fe (right). The dashed curves represent the classical limits of the full calculations (full curves), neglecting Pauli blocking. From (Sch 80).

which were neglected in the early calculations shown in this figure. For the reaction $^{144}Sm + ^{84}Kr$, associated changes in the predictions are illustrated in Figure 111. However, this complication does not alter the general conclusion drawn from comparisons such as shown in Figure 169 that the Pauli principle constitutes an important aspect to be considered in the interpretation of correlations of experimental observables with the energy loss.

The dynamical reaction model based on Randrup's transport theory is successful in describing the qualitative features of experimental correlations between fragment mass distributions and dissipative energy, although the agreement is in some cases only semiquantitative. The bombarding-energy dependence of these correlations observed experimentally is consistent with model predictions, although the latter appear to overestimate the effect. Furthermore, the inclusion of macroscopic correlations in the fragment (A, Z) distributions, induced by the curvature of the driving potential based on spherical fragments, does not lead to an improved description of all reaction aspects, as is exemplified in Figure 111. Both these deficiencies could be due to unrealistic idealizations made in the model that contains no free parameters, a possibility that has to be investigated further. On the other hand, experimental energy–angle correlations are well described. The sequence of Wilczyński diagrams for the reaction $^{139}La + ^{86}Kr$ at several bombarding energies (cf. Figure 76) providing a stringent test for any dissipation mechanism, is well described by the model in the domain of small deformations. This includes the quasielastic region, where the model gives an adequate reproduction of the grazing angle in most cases. As has been shown in previous sections, it can also describe the change in character of the average deflection function with the charge product of the projectile–target system. These arguments provide strong support to the importance of nucleon exchange as a mediator for energy dissipation in damped collisions. Further support will be discussed in later sections.

In contrast to the hard-friction model proposed by Gross and collaborators, the nucleon exchange mechanism leads to a gradual disposition of available kinetic energy of relative motion along the whole nuclear trajectory. Obviously, in order for this dissipation mechanism to become effective, the separation distance between the reaction partners has to become sufficiently small to produce at least a slight matter overlap and a sufficient transparency of the interface between the two heavy ions. This behavior can be observed in Figure 170a, where the theoretical time dependence of the dissipated energy above the barrier, E_{Diss}, is plotted for an $l = 190$ trajectory of the $^{165}Ho + ^{56}Fe$ system at $E_{\text{Lab}} = 462$ MeV. The turning point is passed approximately $(1\text{–}3) \times 10^{-22}$ s after the strong-absorption radius has been reached in the approach phase. During this time, only a fraction of the energy is dissipated that will eventually have been lost

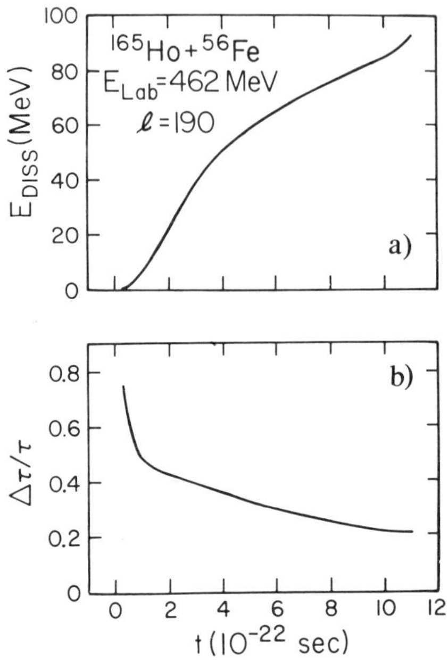

Figure 170. (a) Transport model prediction for the kinetic energy above the barrier dissipated (E_{Loss}) as a function of reaction time. The turning point is passed approximately at $(2–3)\times10^{-22}$ s after reaching the strong-absorption radius. (b) Time dependence of the difference $\Delta\tau$ in thermodynamic temperatures of the light and heavy reaction partners, predicted by the transport model. The quantity τ is the average nuclear temperature of the system.

in that collision, corresponding to an average dissipation rate in the entrance channel of $\sim10^{23}$ MeV s^{-1}. This rate decreases somewhat in the course of the collision, due to a decrease in relative kinetic energy, which is counterbalanced by the acceleration by the Coulomb force. The final turnup of the theoretical curve in Figure 170a is due to the snapping of the cylindrical neck assumed in the model to have formed between the spherical reaction partners.

From the variance in the fragment mass distribution predicted by the transport model, one estimates that roughly 10 nucleons have been exchanged during the collision time of the trajectory depicted in Figure 170a, or approximately one nucleon every 10^{-22} s. The majority of these nucleons are exchanged with relatively small excess velocities near and beyond the turning point when the relative motion is comparatively slow. Comparison of the results of these model calculations with experimental data for the ^{165}Ho + ^{56}Fe reaction (Hoo 82) suggests that this scenario of a rather gentle dissipation mechanism is compatible with observations.

This property of the one-body, exchange-induced dissipation mechanism is quite characteristic and leads to predictions that can be tested in experiments. If one adopts an intrinsic thermal relaxation time of $t_{\text{rel}} \approx 10^{-22}$ s, as suggested by Wong [(Won 79f), see Section 2.7], the intrinsic system is expected to remain rather close to its equilibrium state at all times, receiving only small distortions of the nucleon momentum distribution that dampen out largely in between two successive transfer events. Direct or preequilibrium particle emission from the intermediate system is not likely in such a situation, as will be referred to in Section 3.7. The reaction model presupposes, furthermore, that the identities of projectile- and target-like fragments are preserved during a collision to a large extent, despite the neck assumed to have formed between them. Experimental data presented in Section 3.4 appear to be in support of such a view, except for collisions associated with the highest energy losses and longest interaction times, where the picture of a mononuclear intermediate system seems more realistic. This, however, would lead to a temperature disparity between the constituents of an asymmetric projectile–target system, as has been explained in Section 2.7. Equations (2.7.33) and (2.7.34) represent a simple model (Ran 81b) for an intranuclear convection current that tends to balance this thermal asymmetry.

It turns out that the above heat convection mechanism can be rather effective in equalizing the fragment temperatures by the time that the reaction subsides. In Figure 170b, results are shown of the time dependence of the relative difference $\Delta\tau/\tau$ in fragment temperatures, predicted by integration of Equations (2.7.33) and (2.7.34) along the $l=190$ $^{165}Ho+{}^{56}Fe$ system trajectory referred to in Figure 170a. It can be seen that the initial large relative temperature gradient between the fragments is first rapidly destroyed, followed by a more gradual decay, until it stabilizes at $\sim 20\%$, in the final stages of this collision.

Results of such a heat transfer calculation are compared in Figure 171 to experimental data (Hil 79c) on the asymptotic nuclear temperatures of heavy and light collision partners produced in the reaction $^{165}Ho+{}^{56}Fe$ at $E_{Lab}=476$ MeV. Values for an effective temperature (Lec 59) τ characterizing the evaporation cascade of neutrons emitted from the primary hot fragments are plotted in Figure 171 vs. the total kinetic energy loss. Within experimental accuracy, the temperatures of the reaction partners are found (Hil 79c) to be equal, suggesting that thermal equilibrium has been attained for the intermediate system at breakup. The experimental temperatures are in good agreement with the full curve representing the Fermi gas temperature for full equilibration of the dissipated energy. However, the calculations represented by the dashed curves in this figure predict only small thermal disparities to survive at the end of a collision, which are not in gross violation of the data. Obviously, in order to be able to test these implications of the dissipation model more strictly, experiments with higher accuracy are needed. As seen from Figure 170, the chance to observe relatively large relative deviations from thermal equilibrium exists for partially damped reactions, where, however, the absolute temperatures are expected to be relatively small and difficult to measure accurately.

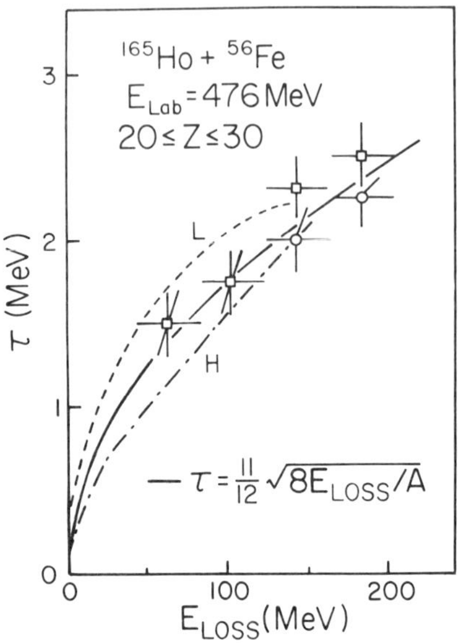

Figure 171. Asymptotic nuclear temperatures of light (L, circles) and heavy (H, squares) reaction partners produed in 476-MeV ^{165}Ho + ^{56}Fe collisions. The quantity τ, plotted vs. total kinetic energy loss, represents an effective temperature characterizing a neutron evaporation cascade. The open symbols are experimental results (Hil 79c). The solid curve represents a prediction by the Fermi gas model for a full equilibration of the dissipated energy. The dashed curves are the results of a heat convection calculation described in the text.

Further access to the interesting energy equilibration process can be gained experimentally by determining higher moments of the excitation energy distribution in each of the reaction partners, which is possible (Hil 79d, Sch 79f) through a measurement of corresponding moments of the multiplicity distribution of neutrons evaporated from the fragments. The statistical model predicts definite correlations between the various moments of the excitation energy distribution of a given reaction fragment and between the corresponding moments of two reaction partners in thermodynamic equilibrium, that can be tested by experiments. Morrissey and Moretto (Mor 81b) have recently discussed theoretical consequences of excitation energy equilibration on the number and energy spectrum of particles evaporated from damped reaction fragments. A heat transfer mechanism effecting a thermalization of the energy dissipated in damped collisions similar to the one discussed above has been considered by Moretto (Mor 83) emphasizing its relation to the dynamical driving force controlling the flux of mass and charge between the partners of a damped interaction.

The discussion of experimental data and theoretical model calculations presented in this section has emphasized a dominantly stochastic nature of the energy transformation process that can, at least in the rather limited range of incident energies of a few MeV per nucleon above the barrier, be well approximated by a classical friction force. There is no convincing evidence characteristic of a two-step dissipation mechanism in which kinetic energy is initially transformed into vibrational energy of collective doorway states that subsequently decay via the interaction with more complex intrinsic states. Although some energy is converted to deformation energy of the fragments, this amounts probably not to a significant fraction of the kinetic energy lost from the entrance channel. It also appears unlikely that the shape of the final energy spectrum reflects statistical fluctuations in the deformation of the intermediate system, although the rigidity of experimental end-point energies is indicative of critical values of the deformation parameters that cannot be surpassed in damped reactions.

Although the energy loss mechanism in damped reactions exhibits stochastic features, it is not well described in terms of a classical equilibrium energy diffusion process. Its dynamic features are the most important ones, masking the influence of changes in the static driving forces, except possibly for low bombarding energies. Fluctuations associated with the average dissipation rate are expected to reveal further characteristics of the energy loss mechanism, but these have so far received unduly little attention. Final preference to either hard or soft friction models cannot be given yet, but the close correlations observed between energy dissipation and mass transport suggest that the exchange of individual nucleons between the reaction partners plays an important, if not the dominant role for the dissipation of kinetic energy in damped reactions. Such a mechanism constitutes a relatively gentle interaction in which kinetic energy is dissipated in many small steps, never driving the intrinsic system very far from its thermodynamic energy equilibrium. Thermal disparity may still exist between the two reaction partners, but strong feedback mechanisms are expected to effect a fast decay of this disequilibrium state, the remnant of which may be difficult to detect experimentally. A final establishment of such an exchange-induced dissipation mechanism would close the gap between quasielastic few-nucleon transfer reactions and strongly damped collisions.

3.6. Transmutation of Angular Momentum

The discussion of transport phenomena in damped heavy-ion reactions presented in previous sections would remain incomplete, if the dissipation of orbital angular momentum and the generation of fragment spins were omitted from consideration. Not only should a consistent model view of the reaction mechanism be able to accommodate the observed transformation of angular momentum, but an investigation of the spin distribution of the reaction fragments can provide additional access to the properties of the transport processes that is not readily achieved with other methods. One is then interested in the dependence of these distributions on the projectile–target combination, bombarding energy, and kinetic energy loss, as well as in the correlations with the final fragmentation in mass and charge. Basic questions to be answered by such studies pertain to the success or failure of phenomenological statistical models to account for the fragment spin distributions and their intranuclear correlations, to possible contributions of collective spin-carrying nuclear excitation modes in angular momentum dissipation, and to the significance of macroscopic driving forces such as represented by spin-dependent potential energy surfaces. It is not only the magnitude of the spins acquired by the reaction fragments that is of interest in this context, but also their orientation, i.e., the population of magnetic substates described in terms of spin alignment or polarization. From the role of the energy loss parameter as an indicator of reaction time and of the peripherality of a heavy-ion collision, established in previous sections, a characteristic change in the fragment spin distributions with increasing energy loss is expected, because of increasingly smaller orbital angular momenta in the entrance channel.

Experimentally, the nuclear spins of excited primary reaction fragments can be inferred from a comparison of measured deexcitation patterns with the predictions of the statistical model (see, e.g., Wol 51, Bla 52, Hau 52, Eri 58, 59, 60, Tho 64, Gro 62, Hal 68, Mor 75c, 81c, Bac 78, Dos 81). These include the probability for fission and the cross sections, multiplicity distributions, and branching ratios of light particles and γ rays emitted in the statistical deexcitation cascade, as well as their angular correlations measured with respect to the fragment direction of flight. These quantities are thought to be determined basically by the product of the phase space volume accessible to the particular deexcitation channel and a statistical S-matrix element depending on the densities of states of the nuclei involved. In practice, an analysis of intensity and correlation patterns of nuclear decay products in terms of spin parameters, entering the predictions of the statistical model, faces serious difficulties because of unknown properties of the exotic nuclei produced in damped reactions, with unusual mass-to-charge ratios, high excitation energies, and possibly large spins and deformations. Investigations of such nuclei form an active field of nuclear research, part of which has been reviewed recently by Diamond and Stephens (Dia 80).

Attempts to derive information on the fraction of the initial orbital angular momentum l, dissipated in a damped reaction, from the balance of total kinetic energies and its variation with bombarding energy and charge fragmentation have already been discussed in Section 3.5. It has been demonstrated there that phenomenological analyses of experimental data on final fragment energies are ambiguous with respect to the amount of dissipated angular momentum, as can be inferred, for example, from the discussion of Equations (3.5.5) and (3.5.6). This work will not be reviewed here in any detail, but attention is focused on experimental studies of the decay of the primary reaction fragments, aiming at a determination of the magnitude and alignment of the spins acquired by the reaction products, and their implications for the damped reaction mechanism.

The dependence of the angular anisotropy of the cross section for fusion–fission reactions on the various components of the spin of the fissioning system has already been discussed in Section 3.3 [cf. Equations (3.3.2)–(3.3.4)]. The corresponding formalism can easily be expanded to include sequential emission processes originating from the highly excited primary damped reaction fragments. The spin quantization axis or the reference frame most appropriate for a theoretical analysis of data can experimentally be defined in coincidence experiments. In Figure 172, this principle is illustrated for the example of a damped reaction in which only the target-like primary fragment decays by sequential fission, but where the projectile-like fragment remains intact. By measuring the emission angle and momentum of the light fragment, both the orientation of the reaction plane and the momentum vector of the recoiling primary heavy fragment are determined. The Z direction, normal to the reaction plane, and the disintegration axis, whose orientation with respect to the recoil direction is established by measurement of one of the sequential reaction products, serve to define M_Z and K as the respective components of the spin J. In extension of Equation (3.3.2), one expects then classically an angular

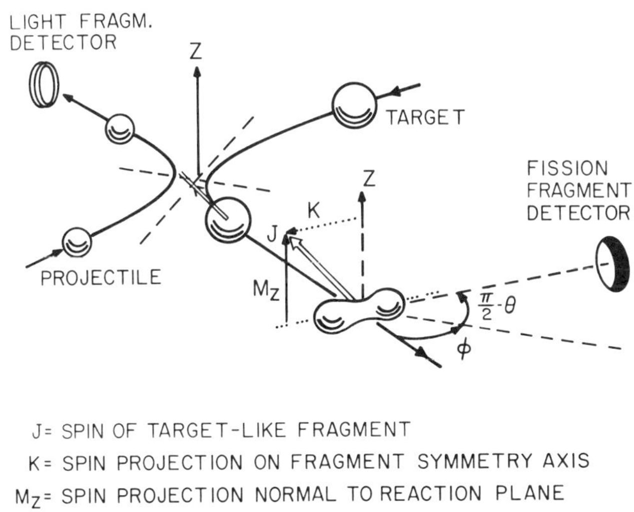

Figure 172. Schematics of a sequential-fission fragment angular correlation experiment. The quantity J is the spin of the fissioning primary fragment; M_Z and K are the projections on the reaction normal and the fission axis, respectively.

distribution of sequential reaction products such as given by

$$\frac{d\sigma(\theta)}{d\Omega} \propto \sum_{JM_ZK} P(J, M_Z)P(J, M_Z, K)\left|D^J_{M_ZK}(\theta)\right|^2 \tag{3.6.1}$$

The angular distribution depends on the probability distribution $P(J, M_Z)$, according to which primary fragments with a given spin J and projection M_Z are produced in the reaction, and the relative probability $P(J, M_Z, K)$ for a particular orientation of the fragment disintegration axis. In terms of these distributions, one can define both average polarization P_Z and alignment P_{ZZ} of the fragment spin as (Dos 81)

$$P_Z = \left\langle \frac{M_Z}{J} \right\rangle, \qquad P_{ZZ} = \left\langle \frac{3M_Z^2 - J(J+1)}{J(2J-1)} \right\rangle \tag{3.6.2}$$

Figure 173 illustrates in a schematical way how the magnitudes and signs of polarization and alignment can be used to distinguish between various spin distributions. The separation of the total spin probability distribution into two product terms is suggested by the two-step nature assumed for the process, where $P(J, M_Z)$ contains the desired physical information about the primary reaction. Once it has been established that the sequential reaction occurs after a time sufficiently long to allow a thermodynamic equilibrium between all degrees of freedom to be achieved by the recoiling fragment, the statistical model can be consulted to provide an approximate

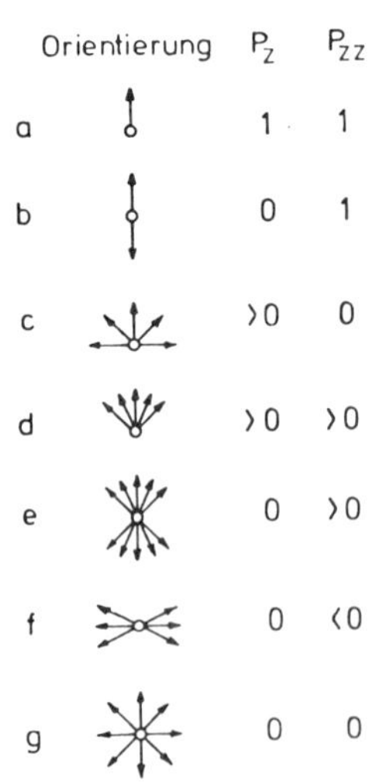

Figure 173. Schematic illustration of polarization P_Z and alignment P_{ZZ} for various spin distributions. The quantization axis is in the direction of the spin in the distribution (a). From (Tra 81d).

procedure to calculate the function $P(J, M_Z, K)$ for a given spin J and projection M_Z. This method has recently been applied by Back and Bjørnholm (Bac 78), as well as in other related work discussed below and in Section 3.8.

The statistical model predicts a population of states that is proportional to the density of intrinsic nuclear levels, in a simplified form expressed as

$$\rho(E^*, J, K) \propto (2J+1)\exp\left[\left(E^* - E_{\text{coll}}\right)/\tau\right] \tag{3.6.3}$$

for given total energy content E^* of the decaying nucleus at the thermodynamic temperature τ. The quantity E_{coll} contains all energies not contributing to intrinsic excitation, i.e., mainly those tied up in collective degrees of freedom, both in the form of kinetic or potential energies. Significant contributions to E_{coll} come from the total potential energy V and the rotational energy E_{rot}, although energy contents of other collective degrees of freedom can in principle be important also for the population of states, as considered in the statistical model of Moretto *et al.* (Mor 75c, 81c).

In a classical treatment, the rotational energy of a nuclear system can be divided into components according to the spin projections parallel and perpendicular to the disintegration axis, which is usually identified with the axis of a rotationally symmetric system, such that,

$$\begin{aligned} E_{\text{rot}}(J, K) &= \frac{\hbar^2}{2\mathscr{I}_{\parallel}}K^2 + \frac{\hbar^2}{2\mathscr{I}_{\perp}}(J^2 - K^2) \\ &= \frac{\hbar^2}{2}\left(\frac{1}{\mathscr{I}_{\parallel}} - \frac{1}{\mathscr{I}_{\perp}}\right)K^2 + \frac{\hbar^2}{2\mathscr{I}_{\perp}}J^2 \end{aligned} \tag{3.6.4}$$

Here $\mathscr{I}_{\parallel}$ and $\mathscr{I}_{\perp}$ are, in general J-dependent moments of inertia with respect to symmetry axis and perpendicular axis, respectively. Consequently, the model predicts a Gaussian distribution $P(K)$ for the population of K substates, with a variance K_0^2 depending on the above two moments of inertia and the nuclear temperature. This is the form usually employed in the analysis of fission or sequential-fission angular distributions with values of K_0^2 scaled from those derived in low-energy fission (Van 73) or calculated from a model for the shape of the system. Obviously, for a realistic calculation of $\mathscr{I}_{\parallel}$, information on the microscopic structure of the primary fragment under reaction conditions is required, since this moment of inertia is quantum-mechanically not associated with collective rotation but with the spin alignment of individual nucleons (Dia 80). Even in the classical regime of high spins and excitation energies, however, where one may expect the rotating-liquid-drop model (Coh 74) to provide valid predictions for the nuclear moments of inertia, experiments to be discussed in Section 3.8 seem to indicate a somewhat different behavior.

Another important factor entering the calculation of the equilibrium spin population is a proper estimate of the collective energies, for example, of the potential energy experienced by the decaying system. Although in general this potential is expected to vary during the decay, it is often assumed that knowledge of the exact energy is required only for a characteristic point on the system trajectory. This procedure appears to be justified if beyond such a point the decay is sufficiently rapid and the system inert enough to prevent any readjustment of the disintegration axis with respect to the spin. Such an assumption is made in the transition-state theory of fission (see Van 73) postulating that the spin distribution is frozen in at the saddle point of the potential energy surface. Hence, in this model, it is the height of the effective fission barrier that acquires significance in the estimate of the equilibrium population of states. Knowledge about the precise value of the fission barrier has so far remained fragmentary. Several studies (see, e.g., Bec 77, 78c, 78e) suggest a deviation of the J-dependent fission barriers from the predictions of the rotating-liquid-drop model (Coh 74).

Despite the above remarks emphasizing the model dependence of analyses of sequential-fission data in terms of fragment spin parameters, this method is relatively direct in that it considers a process that occurs in the initial stages of the deexcitation cascade, except for nuclei that may undergo multiple-chance fission. A similar observation is true for particle evaporation processes, notably those involving light or heavier charged particles, requiring sufficient amounts of excitation energy in the nucleus to overcome the barrier. Particle branching ratios depend (Eri 58, 59, 60) on both excitation energy E^* and spin J available at each stage of the evaporation cascade. The statistical model predicts

$$\Gamma_i(E^*, J) \propto (2S_i+1) \sum_{l_i=0}^{\infty} \sum_{I=|J-l_i|}^{J+l_i} (2I+1) \int_0^{E^*-E_I-B_i} d\varepsilon_i \, \rho(E^*-E_I-B_i-\varepsilon_i) T_i(l_i, \varepsilon_i) \quad (3.6.5)$$

for the decay width associated with a given particle i of intrinsic spin S_i and binding energy B_i. In Equation (3.6.5), the sums extend over all values of the relative angular momentum l_i of the particle and the spin I of the daughter nucleus that couple to the initial spin J. The density of states ρ is evaluated at the intrinsic excitation energy of this nucleus, i.e., from the initial total excitation energy E^*, the yrast energy E_I as well as kinetic

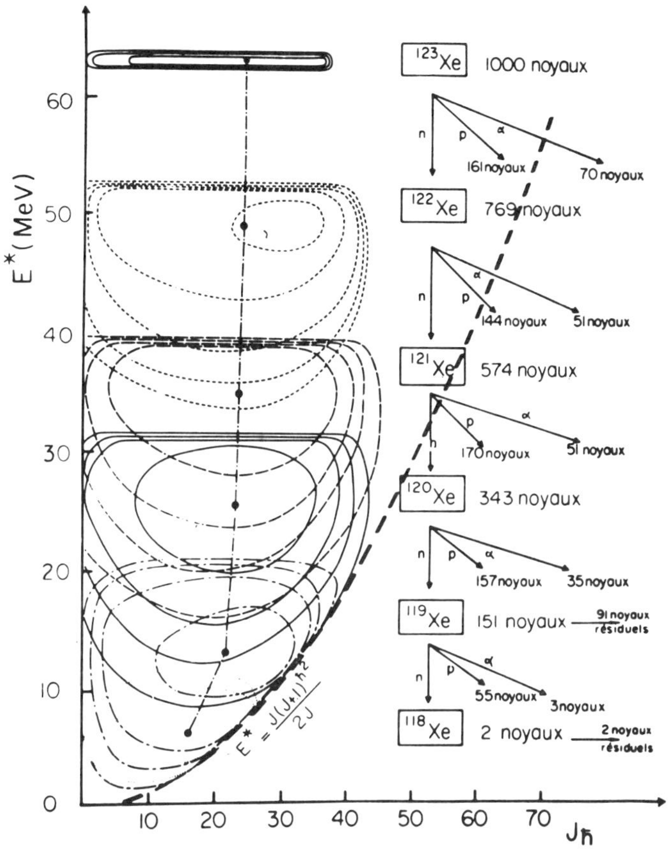

Figure 174. Schematic neutron evaporation cascade in the decay of an excited ^{123}Xe nucleus. Contour lines of the probability distribution are plotted for each step vs. excitation energy E^* and nuclear spin J. The dashed curve represents the yrast line calculated, using the rigid-body moment of inertia. The decay branching is illustrated on the right of the figure, for each step reached in the neutron cascade. From (Lef 78).

energy ε_i and binding energy B_i of the particle are subtracted. The quantity T_i in Equation (3.6.5) denotes the transmission probability for the particle.

A schematic representation of a deexcitation cascade starting from an excited ^{123}Xe nucleus is given in Figure 174 in the form of a contour diagram of the relative population probability plotted vs. excitation energy and spin of intermediate Xe isotopes. These are results of a computer simulation reported by Lefort *et al.* (Lef 78). The dashed curve in Figure 174 represents an estimate of the yrast line, the location of the level of minimum energy for a given spin. The closeness of the deexcitation path to it, and the relative magnitudes of the particle binding energies influence strongly the subsequent decay pattern. As an example, the varying particle branching ratios along a neutron cascade starting from ^{123}Xe are illustrated on the right-hand side of the figure.

Statistical model predictions reported by Beckerman and Blann (Bec 78) for the spin population $\sigma(A, Z, J)$ of representative evaporation residues produced in the fusion reaction $^{141}\mathrm{Pr}+{}^{35}\mathrm{Cl}$ at three bombarding energies are shown in Figure 175. A sharp-cutoff initial distribution $\sigma(\varepsilon, I)$ in angular momentum I is assumed at each bombarding energy ε. As can be seen from this figure, the spin distributions of the final evaporation residues, identified by the decay chains indicated at each curve, are generally bell-shaped with r.m.s. widths of 5–8$\hbar$ and slightly negative skewnesses. Fission is found in the calculations to deplete the cross section at high initial angular momenta, such that it dominates in the $^{141}\mathrm{Pr}+{}^{35}\mathrm{Cl}$ fusion reaction at $E_{\mathrm{Lab}}=185$ MeV. The distributions in spin not carried away in fission or

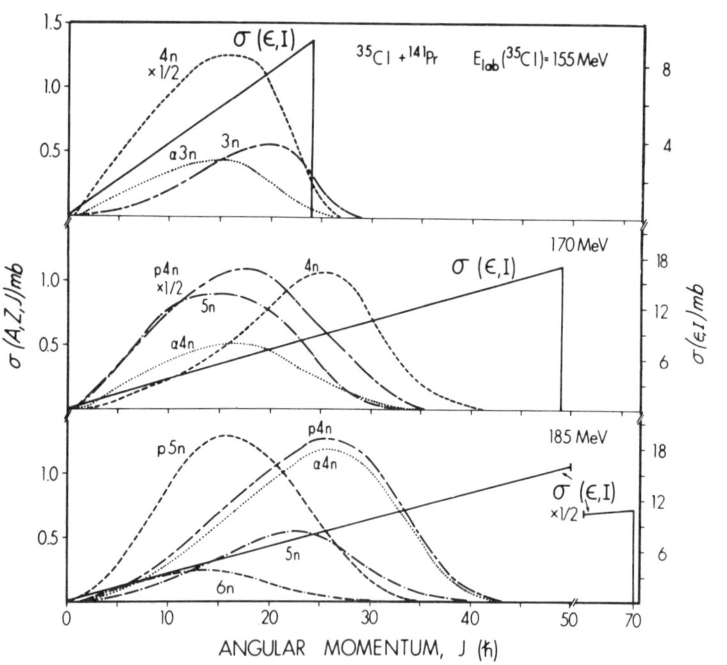

Figure 175. Predictions of spin distributions $\sigma(A, Z, J)$ for various evaporation residues (A, Z) produced in the fusion reaction $^{141}\mathrm{Pr}+{}^{35}\mathrm{Cl}$ at three bombarding energies. The evaporation model considers both fission and particle evaporation as decay modes. The final nuclides are characterized by the decay chains indicated at the curves. An initial sharp-cutoff spin distribution $\sigma(\varepsilon, I)$ assumed in the calculation is also illustrated in each part of the figure. From (Bec 78).

particle evaporation processes represent the starting distributions for the subsequent γ deexcitation cascades. When summed over all final isotopes, these distributions become relatively broad and vary only slightly with the width of the initial spin distribution. Final and initial spins are more strongly correlated with one another for particular decay channels, as can be inferred from Figure 175. This complicates the extraction of fragment spin values from measured cross sections of evaporated particles.

Complementary information on the spins J of the emitting fragment is contained in the evaporated particle-fragment angular correlations. In the rest frame of a fragment with a distribution of spins J aligned perpendicular to its direction of motion ($\theta_{\text{c.m.}} = 0^\circ$), the angular distribution of evaporated particles is expected (Eri 58, 60, Gal 74b) to be approximately given by

$$W(\theta_{\text{c.m.}}) \propto \begin{cases} 1 + \dfrac{1}{3} mR^2 \dfrac{\hbar^2 \langle J^2 \rangle}{2\mathscr{I}^2 \tau} P_2(\cos\theta_{\text{c.m.}}), & \dfrac{\hbar^2 Jl}{2\mathscr{I}} \lesssim \tau \qquad (3.6.6\text{a}) \\ 1/\sin\theta_{\text{c.m.}} & \dfrac{\hbar^2 Jl}{2\mathscr{I}} \gg \tau \qquad (3.6.6\text{b}) \end{cases}$$

Here, m and l are mass and angular momentum, respectively, of the evaporated particle, R is the nuclear radius, $\mathscr{I}$ its moment of inertia, and τ the nuclear temperature. The brackets in Equation (3.6.6a) indicate averaging over the initial spin distribution. In the limit of strong alignment between particle angular momentum l and nuclear spin J, a classical $(1/\sin\theta_{\text{c.m.}})$-type angular distribution represented by Equation (3.6.6b) results. In the limit of extremely weak coupling between l and J, when $\hbar^2 Jl/2\mathscr{I} \ll \tau$, the angular distribution of evaporated particles would be isotropic in the rest frame of the emitter, carrying no information about the fragment spin. This would be the case for spin-independent level densities or very high nuclear temperatures.

If the direction of the fragment spin is restricted to be perpendicular to the reaction plane selected by an experiment, the out-of-plane angular distribution of coincident evaporated particles is expected to be approximated by

$$W(\phi_{\text{c.m.}}) \propto 1 + \frac{1}{2}\left(\frac{\hbar^2}{2\mathscr{I}\tau}\right)^2 \langle J^2 \rangle \langle l^2 \rangle \cos^2\phi_{\text{c.m.}} \qquad (3.6.7)$$

where $\phi_{\text{c.m.}}$ is the out-of-plane angle of emission in the rest frame of the fragment. Again, the anisotropy vanishes in the limit of high temperature or spin-independent level densities. The mean squared angular momentum $\langle l^2 \rangle$ carried away by a particle of kinetic energy ε is estimated (Eri 58) to be

$$\langle l \rangle \approx mR^2(\varepsilon - V_c)/\hbar^2 \qquad (3.6.8)$$

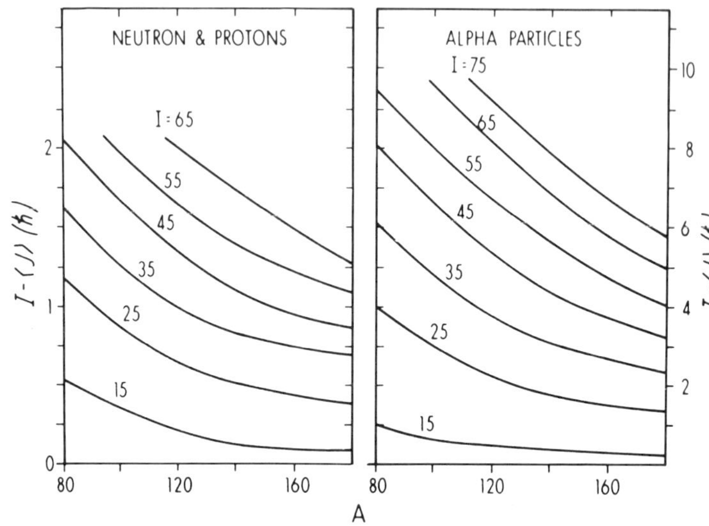

Figure 176. Average angular momentum, $\langle l \rangle = I - \langle J \rangle$, carried away by nucleons (left) and α particles (right) evaporated from a nucleus of mass number A and initial spin I. Results are shown of evaporation calculations for a fixed excitation energy of 90 MeV. Note the different ordinate scales. From (Bec 78).

Here, V_c is the Coulomb barrier for charged-particle evaporation. It is expected from Equation (3.6.8) that measurements of fast and heavy sequentially emitted particles are most sensitive to the magnitude and alignment of the fragment spin. In Figure 176, mean angular momenta $\langle l \rangle = I - \langle J \rangle$, where I is the initial and J the final spin, are plotted vs. the mass number A of a nucleus at a fixed excitation energy $E^* = 90$ MeV, for nucleons (left) and alpha particles (right). Shown are results (Bec 78) of evaporation calculations referred to earlier. It is concluded from these calculations that, on the average, neutrons and protons remove angular momenta between $0.1\hbar$ and $2\hbar$, while corresponding values for α particles range from $0.2\hbar$ to more than $10\hbar$ (Bec 78). The excitation energy dependence of these values is predicted (Bec 78) to be only slight, for a medium-weight nucleus such as ^{149}Tb.

If the spin of the excited fragment is polarized, evaporated nucleons or heavy particles with spin are expected to be polarized also according to the statistical model. The effect has been studied by Karnaukhov (Kar 78), who predicts sizable polarizations for neutrons evaporated from high-spin excited fragments. No experimental studies of nucleon polarization in damped reactions have so far been reported and, hence, this interesting method will not be discussed here any further.

While light particles emitted from the primary fragments provide a rather direct means to estimate the nuclear spins, results of methods involving the measurement of γ rays are more difficult to interpret and generally require calibration measurements in which nuclei similar to the damped reaction fragments of interest are produced at known spins, for example, in compound nucleus reactions. It has been established (see, e.g., Dia 80) experimentally that γ decay of evaporation residues sets in at excitation energies approximately one neutron binding energy above the yrast line, but does not compete significantly with particle evaporation at higher excitation energy. This entry limit line for γ decay is schematically illustrated in Figure 177. As is indicated by the various typical γ deexcitation paths in this figure, there are mainly two types of transitions to be distinguished. At relatively high excitation energies above the yrast line, the nucleus deexcites mainly by emission of "statistical" electrical dipole γ

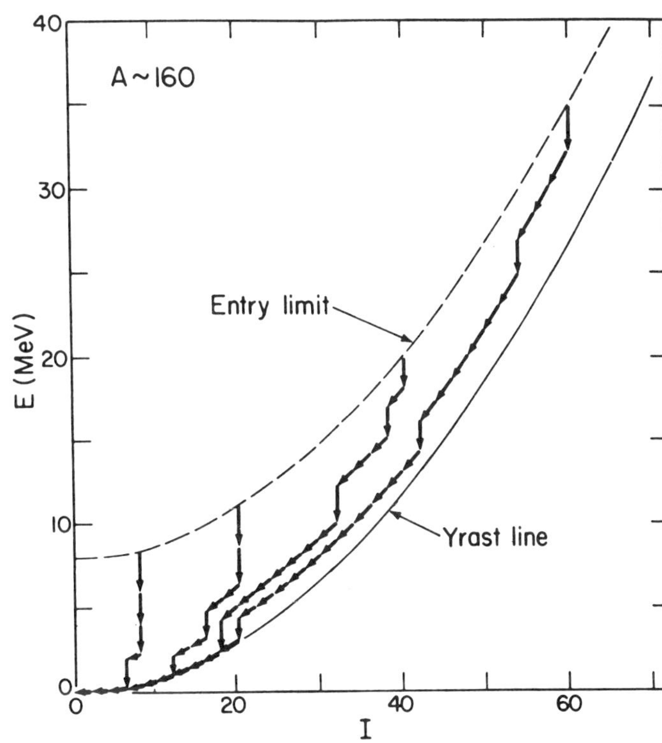

Figure 177. Schematics of typical γ-ray deexcitation paths shown in a plot of excitation energy E vs. nuclear spin I. The yrast line represents the minimum excitation energy of a level at spin I. The entry line drawn as a dashed curve marks the onset of γ-ray emission following the particle evaporation cascade as a function of spin. From (Dia 80).

radiation. The transition probabilities T_{if} for such γ rays depend mainly (Dia 80) on an average matrix element, $M_{if,}$ and the level density measured with respect to the yrast line E_I, i.e.,

$$T_{if} \propto E_\gamma^{2l+2}|M_{if}|^2\rho\left(E_f^* - E_I\right)/\rho\left(E_i^* - E_I\right) \tag{3.6.9}$$

Here, the first term gives the characteristic dependence of the transition probability for radiation of multipolarity l on the transition energy $E_\gamma = E_i^* - E_f^*$. It determines the shape of the γ spectrum at low γ energies, while the exponential level density factor controls the high-energy γ-ray spectrum.

Such statistical emission of γ rays is observed in thermal-neutron-induced fission occurring at very low spins of the fissioning system, as reviewed by Vandenbosch and Huizenga (Van 73). Figure 178 illustrates the

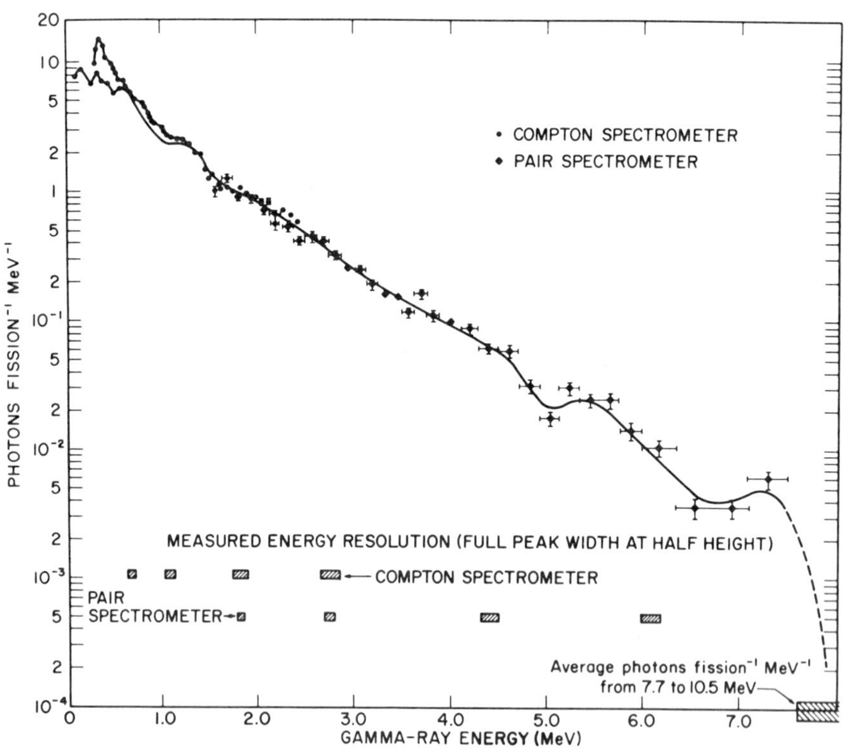

Figure 178. Energy spectrum of prompt γ rays from thermal-neutron-induced fission of ^{235}U. The spectrum represents a combination of two measurements. Experimental energy resolutions are indicated by horizontal bars. From (Mai 65).

energy spectrum of γ rays emitted in thermal neutron fission of ^{235}U (Mai 65). Approximately 7–10 γ rays are observed per low-energy actinide fission event. Their energy spectrum, exemplified in Figure 178, decreases in an approximately exponential fashion, except at energies of a few hundred keV. This is expected for statistical γ rays, according to Equation (3.6.9). Strutinsky (Str 60) has derived an approximate expression for the angular distribution of these γ rays,

$$W_l(\theta) \propto 1 + k_l\left(\hbar^2 J^2/\mathscr{I}\tau\right)^2 \sin^2\theta \tag{3.6.10}$$

measured with respect to the fission fragment direction of flight, where the values of k_l are $+1/8$, $-3/8$, and $-81/64$ for $l=1$ (dipole), $l=2$ (quadrupole), and $l=3$ (octupole), respectively. The general structure of this equation, due to the spin dependence of the level density, is similar to those of Equations (3.6.6)–(3.6.7) describing angular anisotropies of evaporated particles.

The characteristic exponential spectral shape observed in the fission γ-ray spectrum of Figure 178 is discovered also in the high-energy spectrum of γ rays emitted in heavy-ion compound-nucleus reactions. This is illustrated in Figure 179 for several fusion reactions investigated by

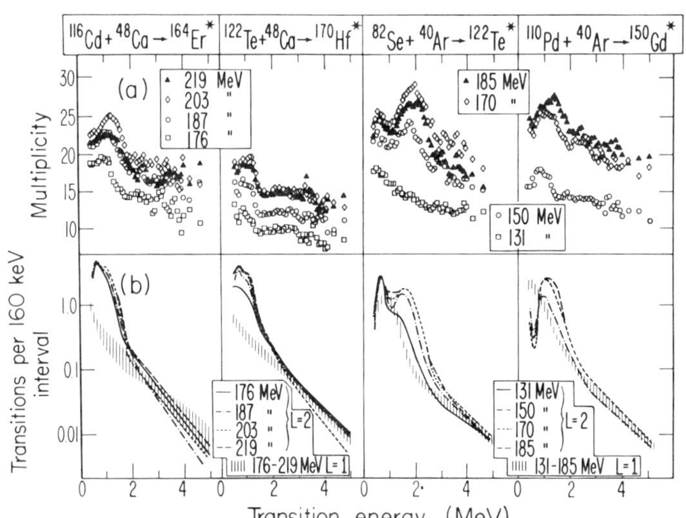

Figure 179. (a) γ-ray multiplicity spectra are shown for the four indicated systems at several bombarding energies. (b) The corresponding energy spectra are shown for stretched quadrupole and dipole components. The lines show the averaged quadrupole component for each bombarding energy, while the hatched area covers the full range of the dipole spectra for all bombarding energies listed. From (Del 78b).

Deleplanque *et al.* (Del 78b). These authors have measured angular correlations of continuum γ rays in coincidence with an array of NaI detectors selecting high-angular-momentum events. The γ-ray multipolarities are determined in this work from intensity ratios for various angle pairs. The lower part (b) of Figure 179 shows a decomposition of the total measured spectrum in $l=1$ and 2 components for several bombarding energies. The hatched band covers the full range of dipole γ-ray spectra observed at all energies. There is some evidence for a low-energy bump in the dipole spectra of some of the spectra shown in Figure 179, which has been associated (Del 78b) with the particular nuclear shell structure, but the most characteristic feature of such γ rays is their exponential energy spectrum. The multiplicity of statistical γ rays has been estimated (Sar 78, Hil 79, Dia 80) to amount to 3–4 in the mass region of $A \approx 160$. The energy spectra of quadrupole transitions, shown as various curves in Figure 179b, however, exhibit a well-developed bump, the so-called yrast bump, associated (Dia 80) with stretched E2 transitions proceeding parallel to the yrast line (see Figure 176). This bump usually moves up in γ energy, as the bombarding energy and, hence, the maximum angular momentum introduced into the system is increased. Figure 179 demonstrates, however, that this bombarding-energy dependence is a function of the particular nucleus considered.

In the top part of Figure 179, the variation of the multiplicity of γ rays with γ ray energy and bombarding energy is shown for four fusion reactions leading to lanthanide compound nuclei. Multiplicities as high as 30 γ rays per event have been measured (Del 78b) in these reactions. The definite dependence observed in these data on the bombarding energy suggests the suitability of the method of deriving information on the spin distribution as it exists at the entry line, from the average γ multiplicity $\langle M_\gamma \rangle$. In principle, this quantity can already be determined from the ratio of coincident (N_c) to singles (N_s) events in a $\gamma-\gamma$ coincidence experiment using two γ counters, since (Hag 75)

$$\frac{N_c}{N_s} = 1 - \left(1 - \frac{\Omega_1}{1-\Omega_2}\right)^{\langle M_\gamma - 1 \rangle} \tag{3.6.11}$$

where Ω_1 and Ω_2 are the total γ efficiencies of the coincidence detector and the trigger counter, respectively. Modern detection devices, however, use multcounter arrays with solid angles approaching 4π. The packing scheme of NaI crystals of one of such detection systems, the Heidelberg "crystal ball" (Met 83), is illustrated in Figure 180. In this setup, γ radiation emitted from the target region is observed by a spherical shell of 162 20-cm-thick NaI detectors. This allows one to measure the sum energy carried away by all γ rays and the determination of various moments of the γ-ray multiplic-

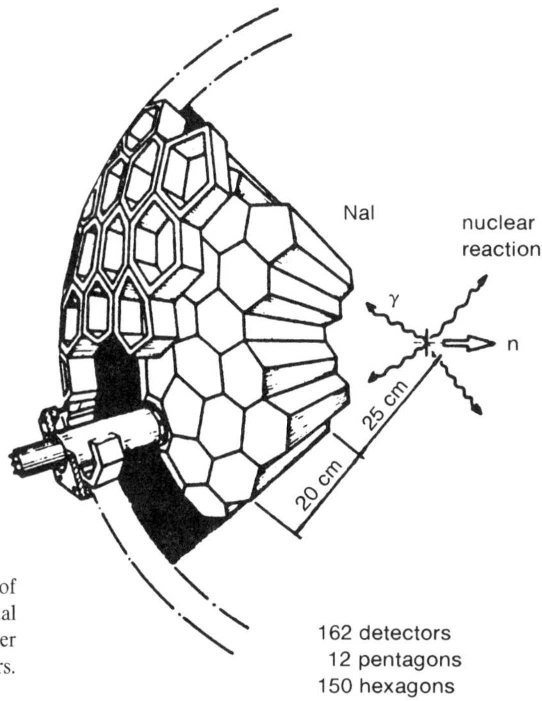

Figure 180. Schematic diagram of a section of the Heidelberg "crystal ball" γ-ray multiplicity counter consisting of 162 NaI detectors. From (Met 83).

ity distribution. If multipolarities of these γ rays are known, the moments of the multiplicity distribution reflect similar moments of the spin distribution at the entry line.

Measurements of various moments of the γ multiplicity distribution, which can be used for calibration purposes in investigations of damped reactions, have been performed by several groups for compound-nucleus reactions (Tjo 74, Hag 75, Mat 74, Ban 75, Fen 75, New 75,77, Tra 75, Sar 76, And 78, Geo 78,78b, Nam 79). Some of the data on average γ multiplicities are summarized (Nam 79) in Figure 181. In this figure, average γ multiplicities $\langle M_\gamma \rangle$ are plotted in a two-dimensional contour diagram vs. average initial angular momentum $\langle J \rangle$ and mass number of the nucleus. Values of $\langle J \rangle$ used in this plot are inferred from experimental fusion cross sections, where available, interpreted in the sharp cutoff approximation. Data are combined in this figure that correspond to different excitation energies of the compound nucleus and are, hence, not strictly comparable. Although more extensive and consistent data are clearly in demand, the results exhibited in Figure 181 emphasize already several interesting points.

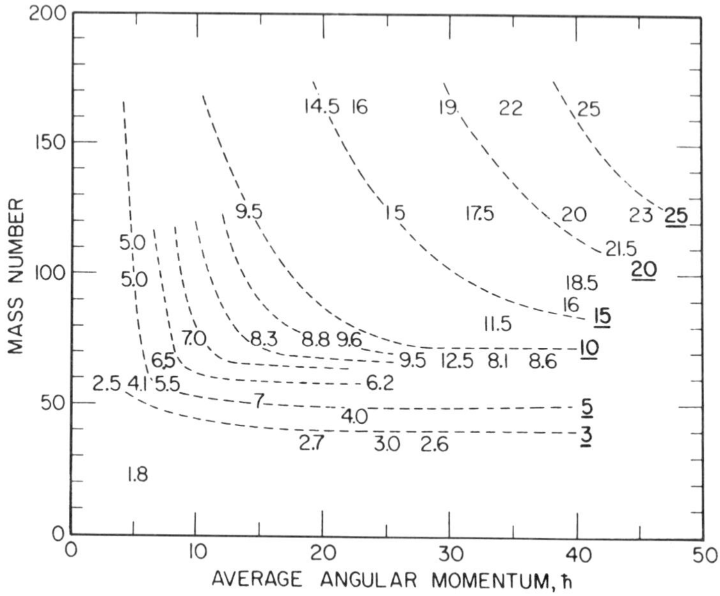

Figure 181. Contour plot of experimental average γ-ray multiplicity, $\langle M_\gamma \rangle$, as a function of mass and average spin, $\langle J \rangle$, of the compound nucleus produced in fusion reactions at various excitation energies. The dashed curves interpolate the data and correspond to the underlined multiplicity values given at the right of each curve. From (Nam 79).

For nuclear masses of $A = 50$–70, there is a broad region in spin, where the γ multiplicity does not provide sensitive information on the angular momentum of the emitter. Consequently, for damped reactions leading to asymmetric fragmentations with the higher fragment within the above mass range, a determination of the total γ multiplicity essentially measures the spin of the heavy reaction partner (Nam 79). For heavy nuclei, an approximately linear relation between average multiplicities (M_γ) and spins (J) is suggested by the roughly constant spacing of the interceptions of the contour lines in Figure 181 with parallels to the abscissa. A relation of the form

$$\langle J \rangle \approx 2(\langle M_\gamma \rangle - \delta) \tag{3.6.12}$$

is commonly employed in the literature, where $\delta \approx 2$–4 (Dia 80) is an effective number of statistical γ rays that, on the average, do not carry angular momentum away from the compound nucleus, and the factor 2 arises from the assumption of a dominantly stretched E2 character of the nonstatistical γ rays.

The breakup of the projectile before fusion and an emission of fast, preequilibrium particles has consequences for the suitability of the γ multiplicity as an indicator of the spin of the compound nucleus. This is most clearly seen in reactions induced by relatively light projectiles. For example, it is observed (Wur 80) in the $^{27}Al + ^{16}O$ reaction that $\langle M_\gamma \rangle$ increases with increasing bombarding energy, up to $E_{Lab} \approx 45$ MeV, reflecting increasing amounts of angular momentum introduced into the system. At higher bombarding energies, however, $\langle M_\gamma \rangle$ no longer increases but saturates at about $\langle M_\gamma \rangle = 6$, an effect attributed (Wur 80) to the emission of fast α particles, above a critical energy. Hence, above this energy, this reaction ceases to provide an experimental relation between $\langle M_\gamma \rangle$ and $\langle J \rangle$ that would be useful in the interpretation of multiplicity values observed for damped reaction fragments in the mass region $A \approx 43$. Similar preequilibrium effects have been reported by Sarantites *et al.* (Sar 78b) in the $^{158}Gd + ^{12}C$ reaction at $E_{lab} = 112$–165 MeV, corresponding to excitation energies between 94 and 143 MeV of the ^{170}Yb composite system. Here, it is observed that $\langle M_\gamma \rangle$ corresponding to different $(^{12}C, xn)$ or $(^{12}C, \alpha xn)$ channels increases at first with increasing excitation energy, then levels off and even decreases, at higher energies. These effects, which severely restrict the information content of experimental γ multiplicities in the case of reactions induced by lighter ions, are less significant for processes induced by heavier projectiles, although they may not be completely absent (cf. Section 3.7).

While γ multiplicity measurements may be used to determine average values of the spin remaining in heavy reaction residues, they do not provide direct information on polarization or alignment of the spin. This information can, however, be extracted from results of experiments determining the γ-ray circular polarization or, in special cases, from the anisotropy of the electrons from the subsequent beta decay of the fragment in its metastable ground state.

The circular polarization of γ rays can be defined as (Tra 81c, 81d)

$$P_\gamma = \frac{W(h=+1) - W(h=-1)}{W(h=+1) + W(h=-1)} \tag{3.6.13}$$

where $W(h)$ is the probability of finding a γ ray of given helicity h. For well-defined γ transitions, this quantity can be calculated (Ste 75). For stretched γ transitions, P_γ is a direct measure of the nuclear polarization P_Z (Tra 81d). For an admixture of nonstretched transitions, P_γ is smaller than P_Z by an attenuation factor, both because of a smaller polarization of the emitted γ rays and their depolarizing effect. Hence, the γ deexcitation patterns of the nuclei under study have to be well known, in order to interpret measured γ circular polarizations.

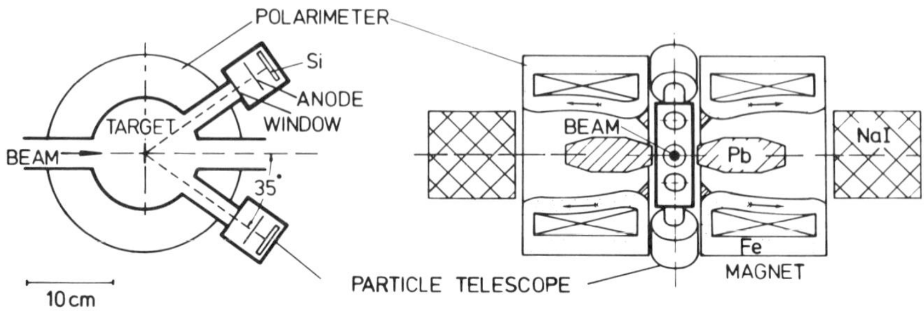

Figure 182. Schematical arrangement of γ polarimeters and heavy-ion telescopes. Two identical scattering polarimeters with NaI γ-ray detectors are placed on either side of the beam, while detectors for measuring projectile-like fragments in coincidence are located at a fixed angle of 35° above and below the beam line. From (Tra 81d).

Experimentally, the value of P_γ is inferred from count rate asymmetries induced by Compton scattering of polarized γ rays on the magnetized iron (Sch 58) of polarimeters that can be of a forward-scattering or transmission design. Both these types have been employed in the studies of Trautmann *et al.* (Tra 77b, 79, 80, 81b, 81c, 81d, Lau 81, Deb 81). Figure 182 shows a schematic drawing of the detector arrangement used by this group, where two γ polarimeters with NaI γ detectors are placed on the left and right of the beam and two heavy-ion telescopes are positioned at a fixed angle above and below. Such a double-symmetrical detector arrangement has the advantage that ratios of several coincidence count rates can be formed in which asymmetries in the apparatus cancel approximately. From such ratios, the product $P_\gamma \cdot A$ of the circular polarization and analyzing power can be determined, with the latter having low typical values of $A \approx 1\%–2\%$ (Tra 81d). Hence, such measurements require a very high statistical count rate accuracy.

Using the angular anisotropy of beta-decay electrons to deduce the spin of a reaction fragment is possible only for a few particular reaction channels, leading to one of the fragments that is beta-unstable in its ground state. Measurements of this kind have been performed by Sugimoto *et al.* (Sug 77, Tak 78, 80d, Ish 79) for the (^{14}N, ^{12}B) reaction on ^{100}Mo and ^{232}Th targets and the ^{197}Au(^{14}N, ^{12}B) reaction. The experimental apparatus employed by this group is schematically depicted in Figure 183. Light reaction fragments emerging from the target pass an Al absorber and are stopped in Pt or Mg foils, whereby a specific window in final kinetic energies is selected. ^{12}B nuclei are identified through their decay electrons of 13.4 MeV end-point energy, measured by plastic-scintillator counter telescopes placed above and below the stoppers. To reduce asymmetries in the detection system, the spins of the ^{12}B products are flipped alternatively up and down,

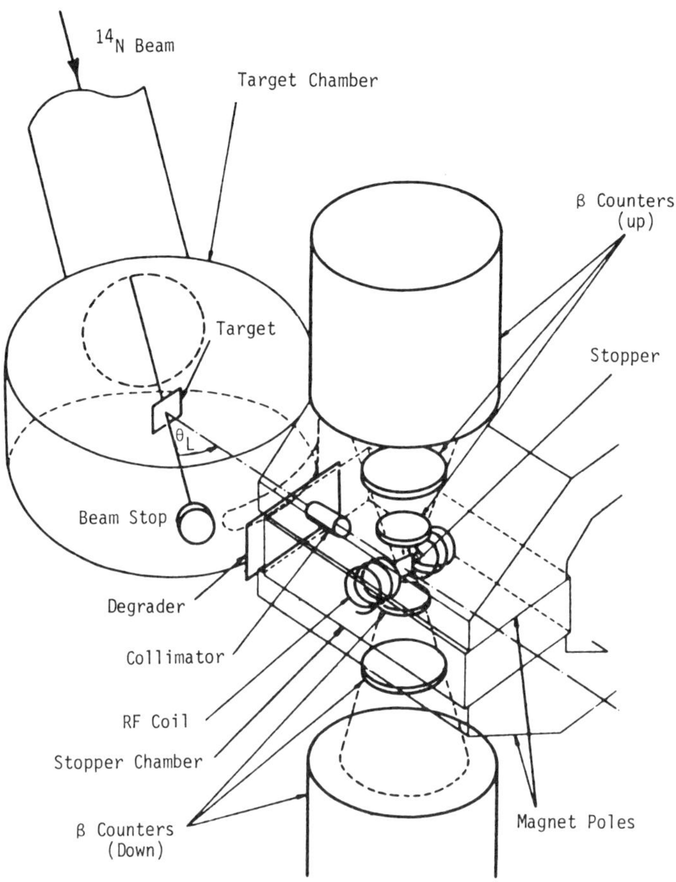

Figure 183. Schematical setup for a measurement of the spin polarization of ^{12}B reaction fragments via the anisotropy of beta-decay electrons. Fragments are degraded in energy and collected by stopper foils, selecting a kinetic energy window. Decay electrons are detected by plastic-scintillation counters above and below the stoppers. ^{12}B spins precessing in the vertical magnetic field can be flipped by application of rf power, in an NMR method. From (Sug 77).

using an NMR technique. Although this is an elegant experimental method, it is highly specific and provides a test of reaction conditions only rather indirectly, yielding information on the fragment spin at the very end of the sequential deexcitation cascade and for fragments that have undergone several multiscattering processes.

After the above outline of particular experimental methods for a determination of the spins of reaction fragments, the following will concentrate on the results of measurements and their interpretation in macroscopic and microscopic models of damped heavy-ion interactions. To obtain a basis for comparison, it appears useful to recall some characteristics of spin transfer or generation in quasielastic and fission processes, that form the two limits of the damped reaction mechanism.

Turning to the fission process first, one infers from the energy spectrum of promptly emitted γ rays, such as shown in Figure 178 for thermal-neutron-induced fission of ^{235}U, that these γ rays mainly result from statistical cascades. This would be expected for a low-energy fission process, where the parent nucleus is endowed with spin values close to zero, However, as summarized by Vandenbosch and Huizenga (Van 73), detailed inspection of the balance of excitation energies carried away by γ rays from the fission fragments are, nevertheless, produced with nonnegligible average spins of the order of $8\hbar$ each. If one assumes that after evaporative particle emission, a fission fragment remains on the average with an excitation energy equal to one-half the neutron binding energy, the total γ-ray energy for both fragments is expected to equal approximately the total neutron binding energy of 5–6 MeV, which is several MeV less than observed (Van 73). The existence of sizable amounts of spin in the primary fragments would explain this effect, because the mainly low-energy evaporation neutrons do not carry away much spin, such that γ emission may start to compete with neutron emission already at relatively high excitation energies of 7–9 MeV. Theoretical work cited in the above review (Van 73), indeed, supports such a picture. Hence, there is some evidence that fragment spin is generated by the dynamics of the fission process, although the total angular momentum of both fission fragments retains its original low value. For example, Nix and Swiatecki (Nix 65) have considered the effectiveness of several collective modes in producing fragment spins, including stretching, bending, wriggling, and twisting motion. In their semiclassical statistical approach, similar in spirit to what is discussed in the context of Equations (2.7.20)–(2.7.22), these authors use a harmonic expansion of the multidimensional liquid-drop model potential energy surface for touching spheroidal fission fragments. They then calculate the probability distribution [compare Equation (2.7.21)] for each macroscopic observable, assuming the model of a quantal multidimensional harmonic oscillator coupled to a heat bath of temperature τ. A similar but simplified approach has been followed recently by Moretto *et al.* (Mor 79c, 79d, 79e, 80) for heavy-ion reactions.

Theoretical predictions (Nix 65) are shown in Figure 184 for the probability distribution $P(L_1)$ of the magnitude L_1 of angular momentum of a single fission fragment at the fission saddle point, for a temperature of the compound nucleus of $\tau = 1.13$ MeV. The solid curve represents the prediction for no damping of the collective motion, while the dashed–dotted curve is calculated under the assumption of infinite viscosity. If the electrostatic torque exerted on the fragments is neglected, as probably realized at

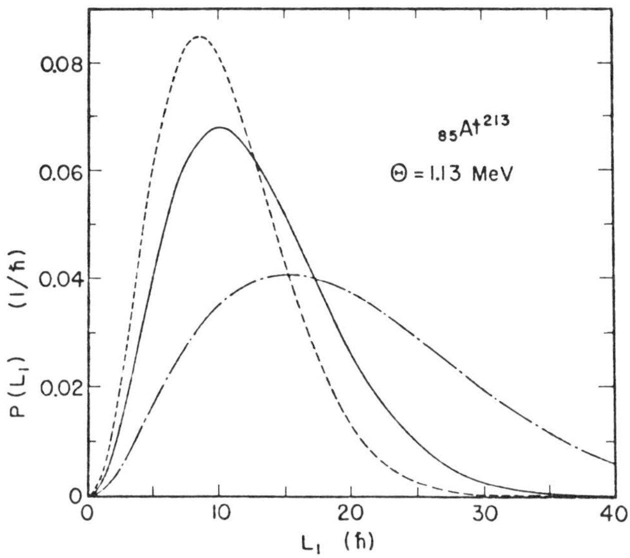

Figure 184. Theoretical spin probability distribution for one fission fragment, predicted from a semiclassical statistical model of collective motion at the fission saddle point. The solid curve represents results for nonviscous fragments, while the dot-dashed curve is calculated for infinitely viscous motion. The short-dashed curve results from a neglect of the torque exerted on the fragments due to the Coulomb interaction. From (Nix 65).

scission, the short-dashed curve results. Evidently, the most probable magnitude of angular momentum of a fission fragment at the instance of scission is predicted to be $L_1 \approx 8$–$9\hbar$. This value would be increased if the fragments retained their deformations to some extent during mutual acceleration in the Coulomb field.

Although such collective models reproduce qualitative features observed in the data, this agreement does not necessarily indicate a significant excitation of collective modes in fission. Discussion presented in Section 2 has emphasized the fact that rather different theoretical approaches can yield results of a similar mathematical structure, where the underlying physical phenomenon determines the properties of model parameters. For the case of low-energy fission processes, there are no detailed systematical studies that would allow a definitive identification of the origin of spin generation. Nevertheless, the above model presents an example of a mechanism that could be encountered in very strongly damped, central heavy-ion collisions.

In quasielastic heavy-ion reactions, on the other hand, there is usually enough orbital angular momentum present in the entrance channel to enforce or, at least, enhance particular transitions in accordance with angular-momentum selection rules. Similar to the dependence of the nucleon transfer cross section on the Q value, discussed in Section 3.5, there are effective kinematical conditions known to exist for the population of states with specific angular momentum. This behavior can be observed most dramatically in reactions induced in relatively light systems, where the abrupt change in the penetration function (cf. Figure 54) restricts the occurrence of quasielastic reactions to a very narrow band of angular momenta close to the grazing l value.

One of the many examples for angular-momentum selectivity in heavy-ion transfer reactions (see, e.g., Any 74, Hod 78, Gol 79) is given in Figure 185 (Any 74), illustrating the bombarding-energy dependence of the population probability of various ^{15}O-states in the reaction ^{12}C(^{12}C,^{9}Be)^{15}O. It is rather striking to observe in this figure how much the transitions to states of successively higher spins become favored at the expense of low-spin states, as the bombarding energy is increased. This feature cannot be explained (Any 74, Hod 78, Gol 79) entirely in terms of a change with bombarding energy of the structure of states in the donor nucleus that contribute to a given transfer. Hence, the extra angular momentum required to effect an

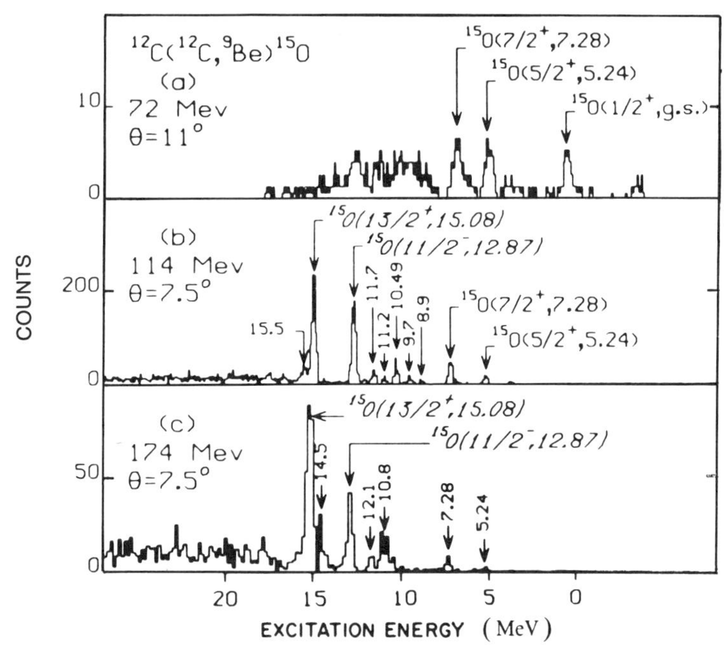

Figure 185. Illustration of the selective population of high-spin states in the ^{12}C(^{12}C,^{9}Be)^{15}O transfer reaction at three laboratory energies. From (Any 74).

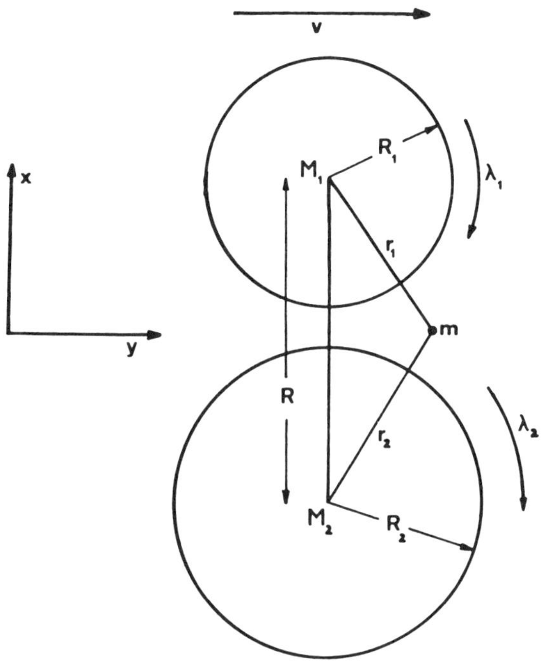

Figure 186. Schematics of a nucleon transfer process in a grazing collision. The quantities λ_1 and λ_2 are the z components of the orbital angular momentum of the nucleon before and after the transfer, respectively. From (Bri 72).

increased population of high-spin states must come out of the relative motion of the reaction partners. A classical relation between the effective Q value, $Q_{\text{eff},}$ on the total kinetic energy loss (cf. Section 3.5), and the amount of transformed orbital angular momentum associated with the transfer of a nucleon has been estimated by Brink (Bri 72). Figure 186 represents a schematical illustration of the transfer of a nucleon of mass m between touching spherical nuclei of masses M_i and radii R_i, at a separation distance R. Pictured is a grazing collision with only a tangential component of the relative velocity v. The quantities λ_1 and λ_2 are the z components of the orbital angular momenta of the nucleon in its initial and final state, respectively, while its spin is neglected. According to the model of Brink (Bri 72), maximum transfer probability is achieved, if the following kinematical conditions are fulfilled:

$$\Delta k = k_0 - (\lambda_1/R_1 + \lambda_2/R_2) \approx 0 \tag{3.6.14a}$$

$$\Delta L = \lambda_2 - \lambda_1 + \tfrac{1}{2}k_0(R_1 - R_2) + Q_{\text{eff}}R/(\hbar v) \approx 0 \tag{3.6.14b}$$

Here, $\hbar k_0 = mv$ is the linear momentum of the transferred nucleon due to the relative motion. The condition of Equation (3.6.14a) requires that the difference in angular velocities of the nucleon in initial and final orbits is matched to the tangential part of the nucleon velocity that is due to the relative motion. In other words, this condition implies that the y component of the velocity of the transferred nucleon is approximately conserved. The second condition in Equation (3.6.14b) reflects the conservation of the z component of the total angular momentum of the system. It requires that in the transfer process, the change, $\hbar(\lambda_1 - \lambda_2)$, in intrinsic spins is offset by an opposite change in orbital angular momentum, $\Delta L = \delta(\mu v R)$, of the system that receives contributions from a change in reduced mass, μ, touching distance, R, and relative velocity, v. The quantity Q_{eff} appears in Equation (3.6.14b), because it is in the model approximated by $Q_{\text{eff}} = \delta(\frac{1}{2}\mu v^2)$ and, therefore, related to the change in orbital angular momentum.

In the semiclassical evaluation of Anyas-Weiss *et al.* (Any 74), a transfer probability P is suggested that decreases in a Gaussian fashion with the magnitudes of Δk and ΔL, given by Equations (3.6.14), according to

$$P(\lambda_2, \lambda_1) = P_0(R) \left| Y_{\lambda_1}^{L_1}(\pi/2, 0) Y_{\lambda_2}^{L_2}(\pi/2, 0) \right|^2 \exp\left[-\left(\frac{R \Delta k}{\sigma_1} \right)^2 - \left(\frac{\Delta L}{\sigma_2} \right)^2 \right] \tag{3.6.15}$$

Here, the quantities $Y_\lambda^L(\theta, \phi)$ are spherical harmonics, the angular parts of the nucleon's wave functions. For the width parameters σ_1 and σ_2, Anyas-Weiss *et al.* (Any 74) suggest as rough estimates based on the uncertainty principle,

$$\sigma_1^2 \approx \pi^2, \qquad \sigma_2^2 \sim (2m\varepsilon)^{1/2} R/\hbar \tag{3.6.16}$$

where ε is some average of the binding energies of the particle in the two nuclei. Evidently, in this approximation, the width parameters are independent of bombarding energy. Such a simple semiclassical approach is successful in explaining, at least qualitatively, branching ratios in few-nucleon transfer reactions, such as in (^{12}C,^{10}B) and (^{12}C,^{10}C) processes (Any 74). However, as pointed out by Goldfarb and von Oertzen (Gol 79), who discuss more extended models of kinematic selectivity, the model should not be taken too literally in applications to quasielastic reactions induced by relatively light projectiles. Because of the small angular-momentum window contributing and rapidly varying penetration and deflection functions, these systems are actually rather close to the quantal limit. The underlying assumptions and simplifications made in the model are, however, much better fulfilled for heavy systems such that the mechanism provides an interesting chance to trace both the dissipation of kinetic energy and of angular momentum to the same microscopic transfer process.

In turning now to the discussion of experimental information about angular-momentun dissipation in heavy-ion reactions, first, results obtained from polarization studies will be reviewed that are of a more global importance for the understanding of the damped reaction mechanism.

Measurements of spin polarization in damped reactions employing the beta decay method described earlier have been performed only for reactions induced by light projectiles that lead to the beta-unstable ^{12}B projectile-like fragments in transfer processes involving only a few nucleons (Sug 77, Tak 78, 80d, Ish 79). With an apparatus similar to the one illustrated in Figure 183, Takahashi *et al.* (Tak 78) measured the spin polarization of ^{12}B fragments, produced in the ^{100}Mo(^{14}N, ^{12}B) reaction at bombarding energies of 90, 125, and 200 MeV, as a function of reaction Q value. In Figure 187, results of their experiment at $\theta_{\text{lab}} = 20°$ are shown. The ^{12}B Q-value spectra depicted in part (a) of this figure exhibit only a weakly damped component at the lowest bombarding energy, whereas a fully damped energy component develops at the higher incident energies. Values for the nuclear spin polarization P are depicted at the bottom of the figure. The sign of the polarization is taken positive when the spin vector is parallel to the vector product $\mathbf{k}_f \times \mathbf{k}_i$ of initial and final linear momenta of the relative motion, $\mathbf{k}_i$ and $\mathbf{k}_f$, respectively. This definition is opposite to what is used in the work on γ-ray circular polarization discussed further below.

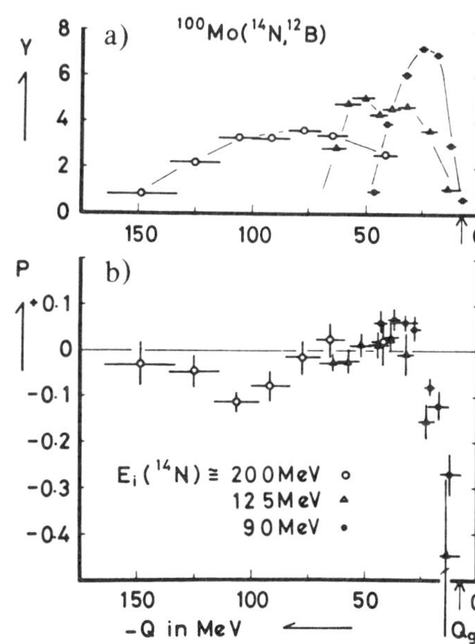

Figure 187. Q-value spectrum (a) and spin polarization (b) of ^{12}B fragments produced in the reaction ^{100}Mo + ^{14}N at $E_{\text{Lab}} = 90$, 125, and 200 MeV. From (Tak 78).

It is observed in Figure 187 that the polarization is large and negative in the quasielastic range of Q values between -7.5 and -25 MeV, with a magnitude strongly decreasing with increasing energy loss. For somewhat larger energy losses, between 25 and 50 MeV, small positive values of the polarization result. For much larger negative Q values, reached in the reaction at $E_{\text{Lab}} = 200$ MeV, the polarization becomes again negative but smaller than in the region of very low energy losses. The negative values for the polarization at high energy losses correspond to negative-angle scattering, as expected from Wilczyński's orbiting model (see Figure 56). Small energy losses, in the range between 25 and 50 MeV, are associated with positive-angle scattering, also as predicted by such a classical model. It is only the regime of energy losses smaller than 25 MeV, in the reaction ^{100}Mo(^{14}N, ^{12}B), that deviate from the expectations based on a classical trajectory model. This latter behavior is, however, successfully modeled (Uda 78) by the quantal direct-reaction theory of Tamura and collaborators (Tam 77, 77b), where it is shown to be due to a quantal interference of contributions from neighboring l waves. Hence, apart from a range of Q values, where classical models are evidently inapplicable, the data for the above relatively light system are in agreement with the classical orbiting picture.

Measurements of γ-ray circular polarization have been performed so far also only for a few reactions such as ^{27}Al + ^{16}O (Tra 80), ^{48}Ti + ^{16}O (Tra 81e), ^{58}Ni + ^{16}O (Lau 78), ^{62}Ni + ^{16}O (Lau 78), ^{nat}Ag + ^{40}Ar (Tra 77b), ^{nat}Ag + ^{86}Kr (Gra 78), and ^{197}Au + ^{86}Kr (Deb 78).

In Figure 188, experimental data (Tra 77b) on energy spectra of projectile-like fragments with $11 \leq Z \leq 21$ from the reaction ^{nat}Ag + ^{40}Ar at two bombarding energies are displayed at the top, while the bottom part shows values of the product of polarization and analyzing power for two regions of final fragment energies. The schematical illustrations of collision trajectories included in the figure give an interpretation of the sign of the polarization in terms of positive and negative deflection angles. As can be seen from the figure, in the region of strongly damped events (DI), the positive value of $P \cdot A$ indicates classically negative scattering angles, whereas weakly damped reactions (QE) are associated with positive scattering angles. This measurement supports the classical orbiting picture referred to in Section 3.3, for medium-weight and light systems. Table 4 gives a summary (Tra 81d) of measured γ-ray circular polarizations, P_γ, for the weakly damped (QE) and strongly damped (DI) reaction components. The systematic trend of the polarization with the projectile–target charge product, $Z_1 \cdot Z_2$, is indicated in Figure 189 (Tra 81d), for several reactions measured at laboratory energies of about 7 MeV per nucleon.

Although in these measurements Z values of projectile-like fragments are selected, as indicated in Table 4, the experimental average circular

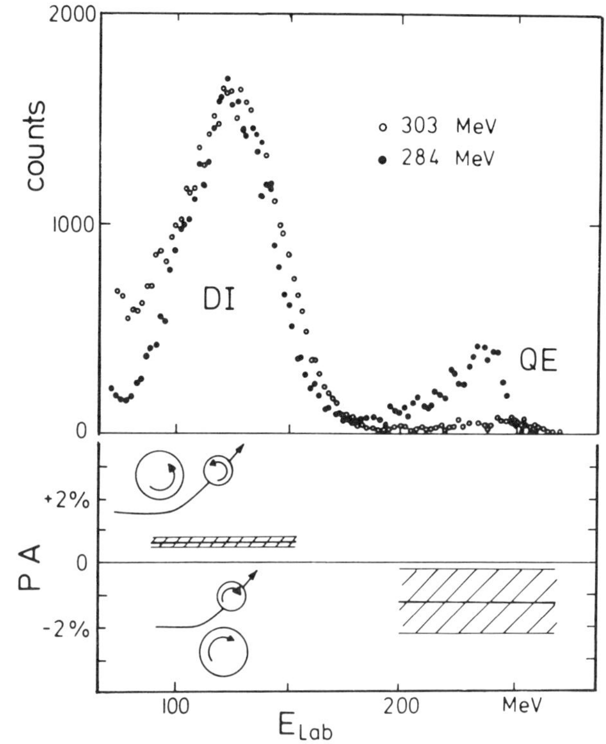

Figure 188. Energy spectrum of projectile-like fragments from the reaction $^{\mathrm{Nat}}Ag + ^{40}Ar$ at $E_{\mathrm{Lab}} = 303$ and 284 MeV (top), and count rate asymmetry $P \cdot A$ for quasielastic (QE) and strongly damped (DI) events (bottom). The schematical collision trajectories illustrate the connection between sense of rotation and sign of the circular polarization. From (Tra 77b).

polarizations P_γ contain contributions from both reaction partners, weighted with the corresponding γ-ray multiplicities. Furthermore, the effects of positive-angle and negative-angle collisions leading to similar final kinetic energies can partially cancel and result in small values of P_γ, dominated by the contribution from reactions with high γ multiplicities. The polarization is further reduced by nonstretched γ transitions that may be present in the deexcitation cascade. These are complications in the interpretation of small values of P_γ.

With these reservations in mind, one nevertheless discovers a systematic change in character of the γ radiation between values of the charge product of $Z_1 \cdot Z_2 = 1000$ and 3000. In agreement with the development of experimental energy–angle correlations with increasing Coulomb repulsion, the above polarization measurements suggest a nonorbiting behavior of the

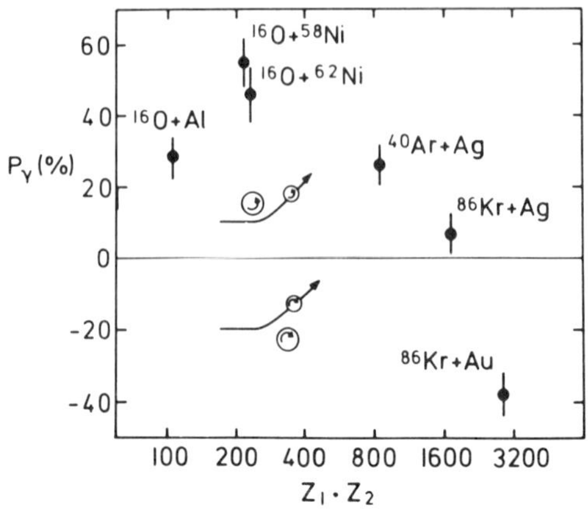

Figure 189. Average circular polarization, P_γ, of γ rays associated with strongly damped processes, plotted vs. charge product, $Z_1 \cdot Z_2$, of projectile and target. From (Tra 81d).

deflection function for the heaviest system studied, $^{197}Au + ^{86}Kr$, while the data indicate strong orbiting tendencies for lighter systems such as $^{nat}Ag + ^{40}Ar$. It is interesting to note that dynamical calculations (Bir 82b) based on Randrup's transport model discussed earlier predict orbiting for completely relaxed $^{209}Bi + ^{84}Kr$ collisions at $E_{lab} = 712$ MeV, but not for $E_{lab} = 600$ MeV. In contrast, the hard-friction model of Gross *et al.* (Gro 78) suggests orbiting for such collisions at either energy. For the $^{144}Sm + ^{84}Kr$ reaction (Mig 84), the former transport model predicts orbiting to set in at bombarding energies between 470 and 595 MeV. Results of similar model calculations are compared in Figure 76 to experimental data on the reaction $^{139}La + ^{86}Kr$ at $E_{lab} = 505$, 610, and 710 MeV, where negative-angle scattering is predicted for the very strongly damped collisions at all three incident energies. New experimental results on γ-ray circular polarization for this

Table 4. Average Circular Polarization P_γ of γ Rays Associated with Quasi-elastic (QE) and Strongly Damped (DI) Energy Components of Fragments with Charges Between Z_1 and Z_2[a]

Target	Projectile	P_γ(QE)	P_γ(DI)	Z_1	Z_2	Ref.
^{27}Al	^{16}O	—	0.27 ± 0.06	5	10	(Tra 80)
^{48}Ti	^{16}O	-0.23 ± 0.12	0.38 ± 0.02	5	10	(Tra 81e)
^{58}Ni	^{16}O	-0.27 ± 0.19	0.65 ± 0.09	5	10	(Lau 78)
^{62}Ni	^{16}O		0.53 ± 0.08	5	10	(Lau 78)
^{nat}Ag	^{40}Ar	-0.6 ± 0.5	0.36 ± 0.08	11	21	(Tra 77b)
^{nat}Ag	^{86}Kr	-0.54 ± 0.24	0.11 ± 0.09	33	39	(Gra 78)
^{197}Au	^{86}Kr	-0.2 ± 0.3	-0.44 ± 0.05	26	46	(Deb 78)

[a] From (Tra 81d).

system, reported by Schandera *et al.* (Sch 82d), appear to indicate a weak dominance of negative scattering angles for strongly damped events at a bombarding energy of 705 MeV. At a higher incident energy of 860 MeV, the observed polarizations are positive but consistent with zero. A definite trend is, however, observed for the polarization P_γ, measured for $^{139}La + ^{86}Kr$ at 705 MeV, as a function of final mass asymmetry. It indicates a dominating contribution of negative-angle scattering for more symmetric fragmentations, while events with a similar degree of energy damping but larger mass asymmetries are associated with negative γ polarizations and, hence, positive deflection angles. While the polarization measurements are qualitatively in agreement with what is expected from the transport model calculations employing proximity forces, these latter observations are taken (Sch 82d) to suggest that calculations of average system trajectories are inadequate in the region of very strongly damped collisions where fluctuations in deflection angle can be very large.

The polarization measurements discussed above have provided unique information on the character of the deflection function in damped reactions, as well as on fluctuations of system trajectories about an average path. As explained already in Section 3.3, the orbiting-like behavior of experimental energy–angle correlations for medium-weight systems appears to disfavor the existence of a strongly repulsive core in the nuclear interaction potential. It has, hence, been possible to derive general information about the balance of interaction forces in damped reactions from the polarization of the spins of the final products. For a more detailed study of the reaction mechanism, however, it is necessary to consider in addition the absolute magnitude and the alignment of the fragment spins generated in heavy-ion collisions. Pertinent experimental results and their implications are discussed in the following.

Although difficulties inherent in the interpretation of inclusive final fragment kinetic energy spectra in terms of rotational energies [cf. Equation (3.5.5)] have been emphasized in Section 3.5, results of such an analysis (Adl 80) are given in Figure 190. In the experiment of Adler *et al.* (Adl 80), the two pairs of very similar composite systems, ^{59}Cu and ^{60}Zn and ^{83}Y and ^{88}Zr, were produced in the reactions indicated in the figure. Plotted vs. atomic number of the primary fragments are the most probable values of initial orbital angular momentum, as derived from the most probable final fragment energies, using Equation (3.5.5) and the sticking limit. The center separations R entering this expression are calculated (Adl 80) from liquid-drop model saddle shapes, which are taken as representatives of the breakup shapes of the dinuclear system. The arrows in Figure 190 indicate critical angular momenta for fusion, L_c, estimated from a systematics of experimental fusion cross sections, and the grazing l value, L_g. One observes from Figure 190 definite trends of the deduced orbital angular momentum,

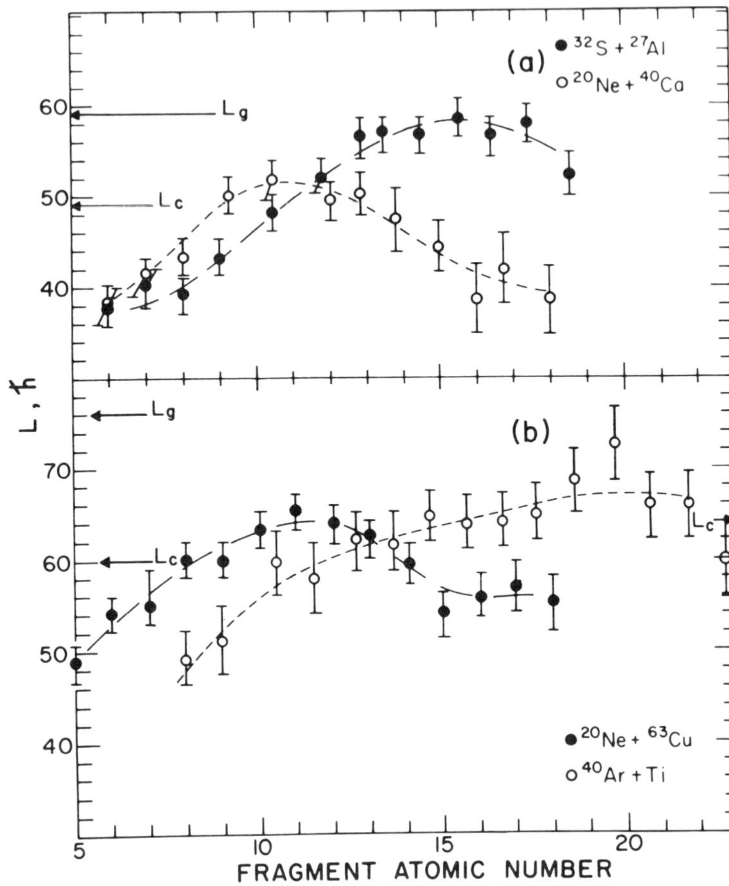

Figure 190. Most probable values of initial orbit angular momentum L as a function of fragment Z values for two cross bombardments leading to similar composite systems. The L values were derived from the final kinetic energy spectra under the assumption of attainment of the sticking limit and scission shapes of the dinuclear system suggested by the liquid-drop model. The grazing angular momentum, L_g, and the critical angular momentum for fusion, L_c, are indicated by arrows. They are matched for the reactions illustrated in part (a). In the lower part (b), L_g and L_c for the $^{63}Cu + ^{20}Ne$ reaction are indicated on the left, L_c for $Ti + ^{40}Ar$ is shown on the right. From (Adl 80).

remaining within the system, with fragment atomic number and characteristic differences in these relations, that appear to reflect the initial projectile–target asymmetry. Fragments with Z values close to that of the projectile correspond in each case to the highest l values, while those with atomic numbers far removed from the projectile Z result from lower partial waves. This is also what is expected from the discussion of the development of fragment A and Z distributions with energy loss, given in Section 3.5. The above analysis (Adl 80) assumes a complete dissipation of orbital angular momentum in excess of the sticking value [cf. Equation (2.3.11)] in these reactions. However, while an approach to the sticking situation by such light systems may constitute a realistic picture, other models of angular momentum dissipation and of dinuclear shapes could also have been employed in the analysis, leading possibly to different conclusions about angular momentum fractionation of heavy-ion processes with different final mass or charge asymmetries.

The main body of data on angular-momentum dissipation in damped reactions is provided by investigations of the γ decay of the excited reaction fragments. Measurements of moments of the γ-ray multiplicity distribution and the γ anisotropy have been reported for damped reactions induced by ^{16}O (Alb 75, Van 77), ^{20}Ne (Gla 77, Day 77, 80, Geo 78b, Nat 78), ^{40}Ar (Nat 78, Geo 79), ^{63}Cu (Ger 79, Ber 82), $^{84,86}Kr$ (Olm 78b, Ale 78, Reg 78, Chr 78, 80, Sob 81b), ^{136}Xe (Agu 79), and ^{165}Ho (Mor 81c, McD 82) projectiles.

In an early experiment, Albrecht *et al.* (Alb 75) bombarded Ni targets with a 6-MeV per nucleon ^{16}O-beam and studied the multiplicity of γ rays coincident with damped reaction fragments, with the aim of deducing the fragment spins and, hence, the energy loss due to tangential friction (cf. Section 2.3). Results of this low-efficiency experiment, using only three $3'' \times 3''$ NaI γ detectors, are displayed in Figure 191, where the average multiplicity M_γ is plotted vs. c.m. scattering angle, as a function of reaction

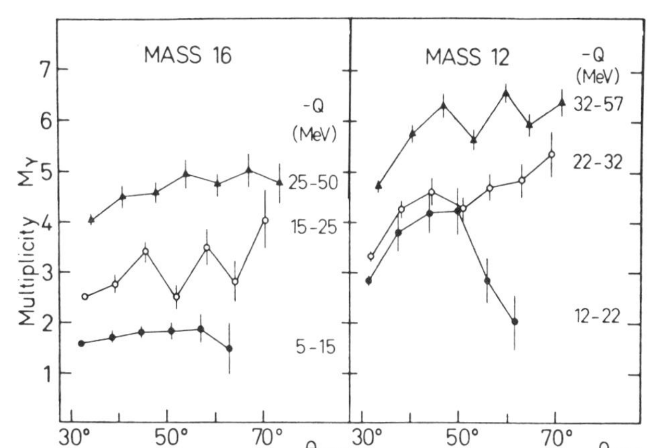

Figure 191. Average multiplicity, M_γ, of γ rays coincident with $A=16$ (left) and $A=12$ (right) fragments from the reaction $Ni + {}^{16}O$ at $E_{Lab} = 96$ MeV, plotted vs. c.m. scattering angle, $\theta_{c.m.}$. The various sets of connected data points correspond to the bins in reaction Q value indicated on the right of each set. From (Alb 75).

Q value and mass of the detected fragment. The γ-ray multiplicity was determined from the coincidence and singles counting rates [compare Equation (3.6.11)]. An averaging over the fragment–γ angular correlation was accomplished by summing the coincidence yields of all three detectors. It is observed from Figure 191 that increasing negative Q values or energy losses are associated with higher average γ multiplicities, although for the $A=12$ fragments corresponding to quasielastic collisions, already relatively large multiplicities result. This latter feature is not seen for the $A=16$-fragment γ multiplicities depicted on the left of the figure. Applying an approximate correction for the fragment spin carried away by evaporated particles, Albrecht *et al.* (Alb 75) derived characteristic spins of $J=15\hbar$ and $J=12\hbar$, for damped reaction events with light fragments of $A=16$ and $A=12$, respectively. These are the total spins, $J=J_1+J_2$, of both collision partners. Splitting the total Q value or energy loss into radial and tangential parts, and assuming a fixed interaction distance R, one obtains (Alb 75) classically for the tangential part of the dissipated energy

$$E_{\text{loss}}^{\tan}(R)=\left(l_i^2-l_f^2\right)/2\mathscr{I}(R) \tag{3.6.17}$$

where $J=l_i-l_f$ is the dissipated angular momentum, and $\mathscr{I}(R)=\mu R^2$ is the moment of inertia for orbital motion. Inserting into Equation (3.6.17) for l_i the critical angular momentum for fusion, estimated from empirical systematics (Nat 72), and for R the interaction radius, one obtains (Alb 75) an estimate for the lower limit of $E_{\text{loss}}^{\tan}$ for the damped reaction component. It is found in this work that tangential friction in damped collisions is substantial, making up on the average of 50%–60% of the total average energy loss experienced by the light $\mathrm{Ni}+{}^{16}\mathrm{O}$ system. Although the results of the above analysis are rather model dependent, it is interesting to note that Albrecht *et al.* (Alb 75) find their data consistent with an approach to a rolling situation (cf. Section 2.3), in which $l_f/l_i=5/7$ [Equation (2.3.9)] for l values close to the critical angular momentum for fusion.

For the somewhat lighter system ${}^{27}\mathrm{Al}+{}^{16}\mathrm{O}$ at $E_{\text{lab}}=100$ MeV, Van Bibber *et al.* (Van 77) suggest a complete dissipation of angular momentum in excess of the sticking value l_s [cf. Equation (2.3.11)], to occur for strongly damped collisions. In this work, yields of discrete γ transitions have been measured in coincidence with reaction fragments. The observed anisotropies in the fragment–γ angular correlations are indicative of a high degree of spin alignment, in this reaction. For example, an in-plane/out-of-plane anisotropy of larger than 2 is measured for 4.4 MeV γ rays corresponding to the $2^+\to0^+$ transition in ${}^{12}\mathrm{C}$ fragments. For perfect spin alignment perpendicular to the reaction plane, a value of 2.2 is expected (Van 77) for this ratio.

An array of 9 NaI detectors has been used by Dayras *et al.* (Day 80) in an extensive study of angular-momentum transfer in the damped reaction $^{63}Cu+{}^{20}Ne$ at $E_{lab}=166$ MeV. The first three moments of the multiplicity distribution and fragment–γ angular correlations have been determined in this work, yielding results that are somewhat at variance with the above studies emphasizing a classical frictional angular-momentum dissipation mechanism and implying a complete alignment of fragment spins perpendicular to the reaction plane.

In Figure 192, continuum γ-ray spectra are shown for the reaction $^{63}Cu+{}^{20}Ne$, measured (Day 80) in coincidence either with evaporation

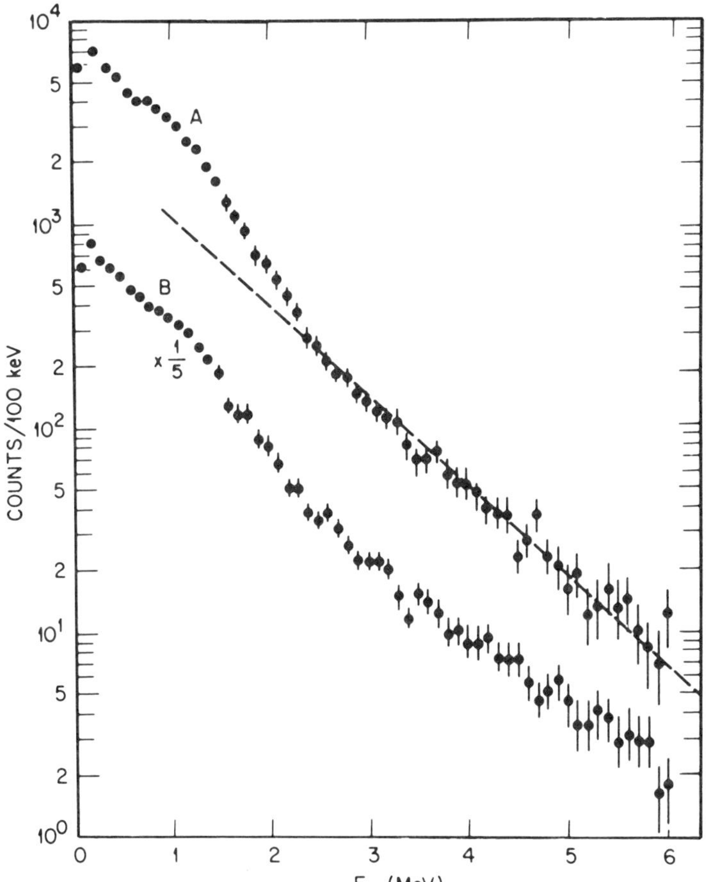

Figure 192. Energy spectra of γ rays in coincidence with evaporation residues (A) at $\theta_{lab}=20°$ and damped reaction fragments (B) at $\theta_{lab}=45°$ from the reaction $^{63}Cu+{}^{20}Ne$ at $E_{Lab}=166$ MeV. The dashed line emphasizes the enhanced yield of low-energy γ rays. From (Day 80).

residues (A) deflected to $\theta_{\mathrm{lab}} = 20°$ or with damped fragments (B) with $3 \leq Z \leq 16$ emitted at $\theta_{\mathrm{lab}} = 45°$. The general shapes of the two γ spectra are apparently rather similar, exhibiting an exponential tail towards high energies and an "yrast bump" in the region of 1–2 MeV, that is more pronounced in the spectrum (A) coincident with evaporation residues. These aspects of the γ spectra have been discussed earlier in this section. It is observed (Day 80) for this reaction that the γ multiplicity increases with increasing charge asymmetry of the final fragmentation, with a more noticeable Z dependence at the higher energy losses. For a given Z of the light fragment and given energy loss, the average multiplicity $\overline{M}_\gamma$ is not strongly dependent on the reaction angle. In Figure 193, the experimental dependence on Z is shown for $\overline{M}_\gamma$, the r.m.s. width σ_r of the multiplicity distribution, as well as of its skewness, $s_\gamma = \langle (M_\gamma - \overline{M}_\gamma)^3 \rangle / \sigma_\gamma^3$. These data correspond to a 30-MeV-wide energy bin, centered at the maximum of the damped component of reaction fragments, detected at $\theta_{\mathrm{lab}} = 45°$, some 27° backward of the grazing angle. Figure 193 shows an increase of $\overline{M}_\gamma$ with increasing asymmetry of the fragments, while the width of the γ multiplicity distribution exhibits only small fluctuations around an average value of $\sigma_\gamma \approx 4$. The skewness s_γ is seen to increase with increasing charge of the light fragment, from ~ -1.5 to ~ 0.

Several of these features, pertaining mainly to the average multiplicity $\overline{M}_\gamma$, have been interpreted by Dayras *et al.* (Day 80) in terms of a classical

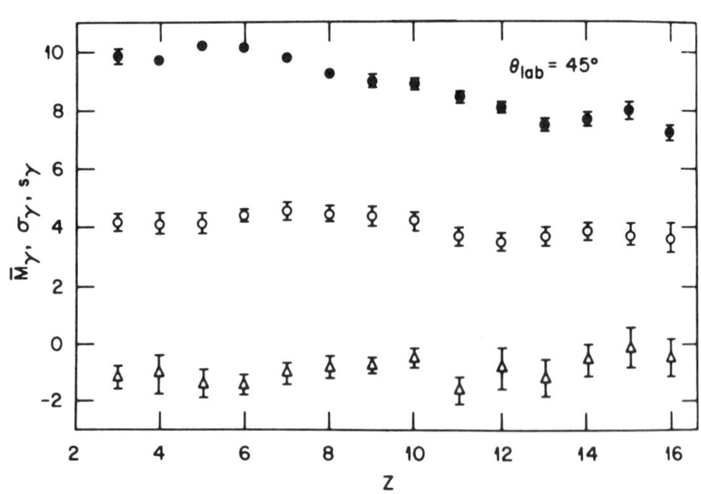

Figure 193. Average multiplicity $\overline{M}_\gamma$ (solid points), r.m.s. width (open circles), and skewness (triangles) of the γ multiplicity distribution for the $^{63}Cu + ^{20}Ne$ reaction at $E_{\mathrm{Lab}} = 166$ MeV, with damped reaction products of atomic number Z measured at $\theta_{\mathrm{lab}} = 45°$. The data are for a 30-MeV-wide energy bin centered at the maximum of the damped energy component. From (Day 80).

friction model. This, of course, requires a conversion of the measured γ multiplicity to the total average spin $\langle J\rangle = \langle |\mathbf{J}_1| + |\mathbf{J}_2|\rangle$ of both reaction partners. Instead of Equation (3.6.12), these authors use as a conversion formula

$$\langle J\rangle = \langle J_0\rangle + \tfrac{3}{16}\left(E^* - E_0\right) + 2\left(\overline{M}_\gamma - \delta\right) \tag{3.6.18}$$

based on available experimental information on the decay properties of compound nuclei in the mass region $A \approx 60$ of interest for this particular reaction, for a considerable range of excitation energies, E^*. In Equation (3.6.18), $\langle J_0\rangle \approx (11\text{–}12)\hbar$ and $E_0 = 80$ MeV (Day 80). A value of $\delta = 2$ was adopted in this work for the number of statistical γ rays. This equation accounts implicitly for the effect of particle evaporation on the fragment spin and, hence, yields values for the primary fragment spins.

The variation of the total fragment spin $\langle J\rangle$ with energy loss is depicted in Figure 194 for the above system, as a function of the charge of the light fragment and its detection angle. It can be seen from this figure that $\langle J\rangle$ increases with energy loss, up to about $E_{\text{loss}} = 80\text{–}90$ MeV, from where on it either remains constant or decreases slightly. It should be remembered from the discussion in previous sections that it appears that, for many systems, the energy loss is (almost linearly) related to the initial orbital angular momentum. Hence, implicitly, an approximate correlation between the final intrinsic spins of the fragments and the entrance-channel orbital angular momentum is pictured in Figure 194. This then suggests that, as the initial l value decreases, an increasing fraction of the initial angular momentum is converted into intrinsic fragment spin, for most of the energy loss range covered in the figure.

Performing an analysis of their data in terms of final rotational energies, similar to what is expressed by Equation (3.5.5), Dayras *et al.* (Day 80) conclude from the Z dependence of the total fragment spin that the system rotates rigidly, i.e., attains the sticking limit. A similar conclusion had been reached earlier by Glässel *et al.* (Gla 77) in a study of the $^{\text{nat}}\text{Ag} + {}^{20}\text{Ne}$ reaction. In the rolling limit, according to Equation (2.3.9), the final angular momentum is expected to scale with the initial l value, independent of the fragmentation. Such an interpretation, however, is based on strong model assumptions about a single separation distance R [cf. Equation (3.5.5)] that is equal for all fragmentations. Furthermore, as noted by the authors (Day 80), a relatively small angular anisotropy, observed for the γ rays, disfavors such a simplified classical-mechanical view of angular-momentum dissipation in damped reactions. The r.m.s. widths σ_γ of the γ multiplicity distribution shown for the $^{63}\text{Cu} + {}^{20}\text{Ne}$ reaction in Figure 193 are also larger (Day 80) than expected from the width of the l window, even if one assumes that for a given fragmentation all l waves between the

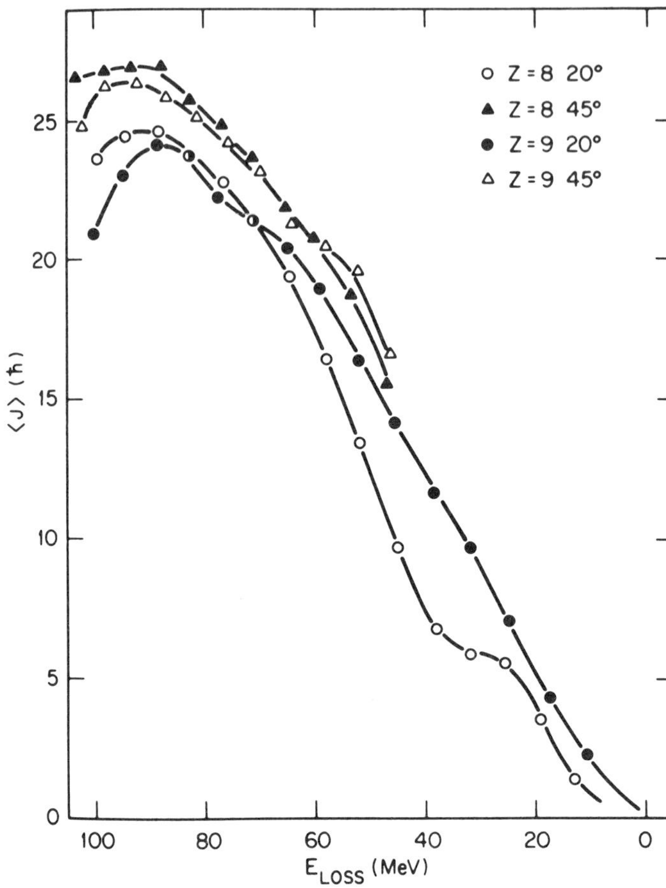

Figure 194. Transferred angular momentum $\langle J \rangle$ as a function of energy loss, for fragments with $Z = 8$ or 9 produced in the reaction $^{63}Cu + ^{20}Ne$ at $E_{Lab} = 166$ MeV. At a laboratory angle of 45°, only the fully relaxed energy component is present in the experimental data. From (Day 80).

grazing l and the critical l value for fusion contribute. Particle emission from the excited primary fragments and uncertainties in the conversion of γ multiplicity into total fragment spin cannot account (Day 80) for the large values of σ_γ and the small γ anisotropy. An interpretation of these observations in terms of an additional fluctuating component of the spin generated in the fragments will be discussed further below.

One may expect, as outlined in Sections 2.1 and 2.2, that classical models are more appropriate in the description of interactions in heavier systems, on which attention is focused in the following. Namboodiri *et al.* (Nam 79), for example, have studied angular momentum transfer for the medium-weight system $^{89}Y + ^{40}Ar$ at $E_{lab} = 237$ MeV employing similar techniques of measurement and analysis as just discussed. These authors suggest evidence in support of an approach to rigid rotation for the lowest l waves contributing to the damped-reaction component. Employing the equations

$$\langle M_\gamma \rangle \langle E_\gamma \rangle = \langle M_\gamma \rangle_H \langle E_\gamma \rangle_H + \langle M_\gamma \rangle_L \langle E_\gamma \rangle_L$$

$$\langle M_\gamma \rangle = \langle M_\gamma \rangle_H + \langle M_\gamma \rangle_L \tag{3.6.19}$$

they were able to decompose the measured total γ multiplicity $\langle M_\gamma \rangle$ into contributions from the heavy (H) and light (L) reaction partners. Here, $\langle E_\gamma \rangle$ is the measured average γ energy, and $\langle E_\gamma \rangle_H$ and $\langle E_\gamma \rangle_L$ are determined from compound-nucleus reaction systematics (Nam 79). In Figure 195, a comparison between deduced values, $\langle J \rangle_H$, of the spins of the target-like fragments and those deduced from the fragment kinetic energies is given. The curves are calculated for the indicated values of the initial l wave, under the assumption of a rigidly rotating, deformed dinuclear system, where the most probable final total kinetic energy is given by the sum of Coulomb repulsion and centrifugal energy evaluated at the breakup distance. The dissipated fraction of the initial l value is then assumed to divide between the fragments according to their moments of inertia, yielding $\langle J \rangle_H$ and $\langle J \rangle_L$ for each fragmentation. As can be seen from Figure 195, the data are well represented by the theoretical curve corresponding to rigid rotation of a system formed on the average in collisions with partial waves of $l \approx 70$, which is close to the critical angular momentum for fusion. No measurements of higher moments of the γ multiplicity distribution were made in these experiments, which would have provided an additional consistency check of the assumptions made in this classical phenomenological analysis, such as in the case discussed above.

While the l bands for the transitions from sliding to rolling and sticking are expected to be rather narrow for higher systems, owing to an abruptly changing penetration function (cf. Figure 54), these bands should be much broader for heavy systems, where the chance exists to amplify the different regions by choice of reaction parameters. Figure 196 illustrates results for the dependence of the γ multiplicity, M_γ, on the charge fragmentation, identified by the Z values, Z_3, of the detected light reaction products, for three ^{86}Kr-induced reactions on heavy targets. These data, measured by Aleonhard *et al.* (Ale 78) at $E_{lab} = 618$ MeV, are for the damped reaction component only, while the multiplicity for quasielastic events is observed (Ale 78) to increase approximately linearly with the

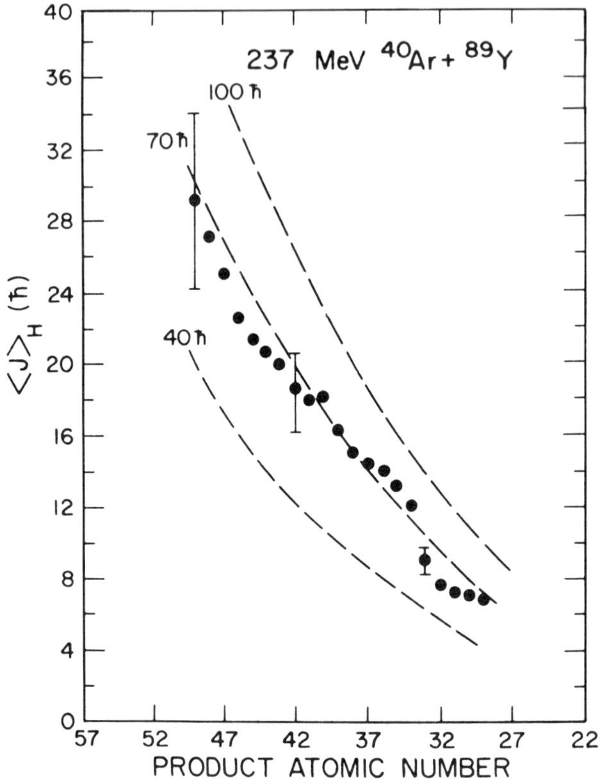

Figure 195. Comparison between experimental values $\langle J \rangle_H$ of the spins of heavy, target-like fragments produced in the reaction $^{89}Y + ^{40}Ar$, and values predicted from centrifugal energies of the final fragments for the indicated entrance channel angular momenta, assumed to be damped to their sticking limits. Projectile-like fragments were detected at 55°. From (Nam 79).

distance in atomic number, $|Z_3 - Z_p|$ of the light fragment (Z_3) from the projectile (Z_p). From the data displayed in Figure 196, one notices generally a slight increase of the multiplicity with increasing weight of the target. In addition, the variation of M_γ with Z_3 is larger for the heavy system $^{197}Au + ^{86}Kr$ than for $^{107,109}Ag + ^{86}Kr$. Furthermore, in the case of the $^{197}Au + ^{86}Kr$ reaction, the multiplicity M_γ is larger at forward than at backward angles. This is in contrast to results on the $Ag + ^{20}Ne$ reaction, reported by Glässel *et al.* (Gla 77), where the highest γ multiplicities correspond to backward-angle scattering.

The dashed and solid theoretical curves in Figure 196, corresponding to the rolling and sticking situations, respectively, have to be regarded as rough estimates only. A connection between γ multiplicity and total fragment spin is made by Aleonard *et al.* (Ale 78), assuming the predominance of γ

deexcitation by stretched E2 transitions [cf. Equation (3.6.12)], neglecting statistical γ rays. An average value of the entrance channel angular momentum of $\langle l \rangle = (2/3) l_{\max}$ is used in the calculation for all fragmentations, assuming a contribution from all l waves as given by a triangular sharp-cutoff distribution in differential cross section. The moments of inertia are taken in this work as those of spherical fragments.

Inspection of Figure 196 suggests that neither sticking nor rolling models fit the data well, which is particularly obvious for the $^{197}Au + {}^{86}Kr$ reaction. Except, possibly, in the case of $^{107,109}Ag + {}^{86}Kr$, the measurements

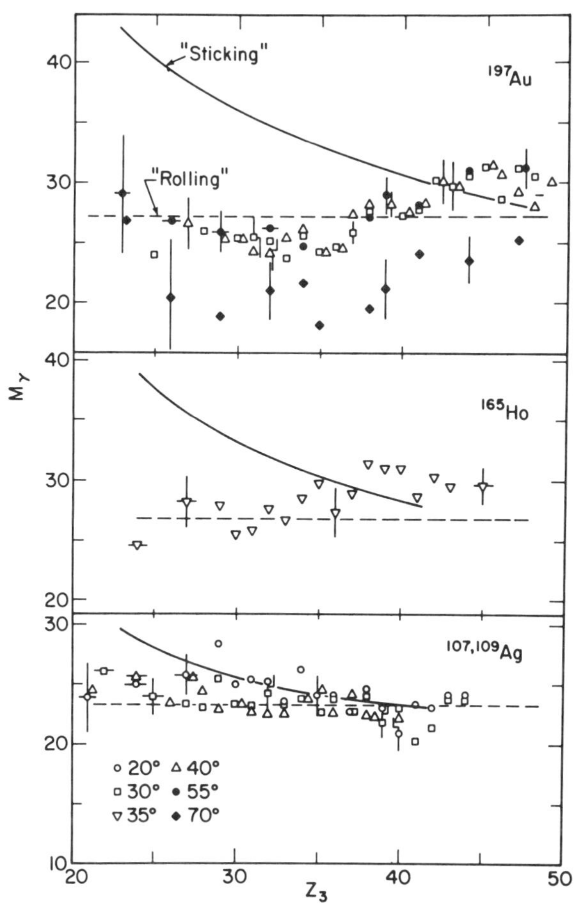

Figure 196. Average γ multiplicity, M_γ, vs. atomic number, Z_3, of light fragments, for 618-MeV ^{86}Kr-induced reactions on targets of $^{107,109}Ag$, ^{165}Ho, and ^{197}Au. The solid curves correspond to rigid rotation of two touching spheres, the dashed lines represent the rolling limit. From (Ale 78).

do not yield multiplicities that rise with increasing charge asymmetry, as required by the sticking model. The rolling situation is dismissed by Aleonard *et al.* (Ale 78), also because of a lack of physical justification for the presence of rolling friction in the interaction of heavy nuclei. In addition, it is found that the unexpected behavior of the γ multiplicity is not due to distortions in the γ multiplicity distribution induced by sequential fission of the heavy reaction partners. As a possible explanation for the effect, it is suggested (Ale 78) that highly asymmetric fragmentations are produced in near-central collisions, characterized by low l values and long interaction times. It is concluded by these workers that then the data may still be consistent with an attainment of rigid rotation of the intermediate system. This effect, termed (Ale 78) "'angular-momentum fractionation," is suggested by a characteristic change of the driving potential, V_Z, for charge exchange with decreasing value of l. This is illustrated in Figure 197,

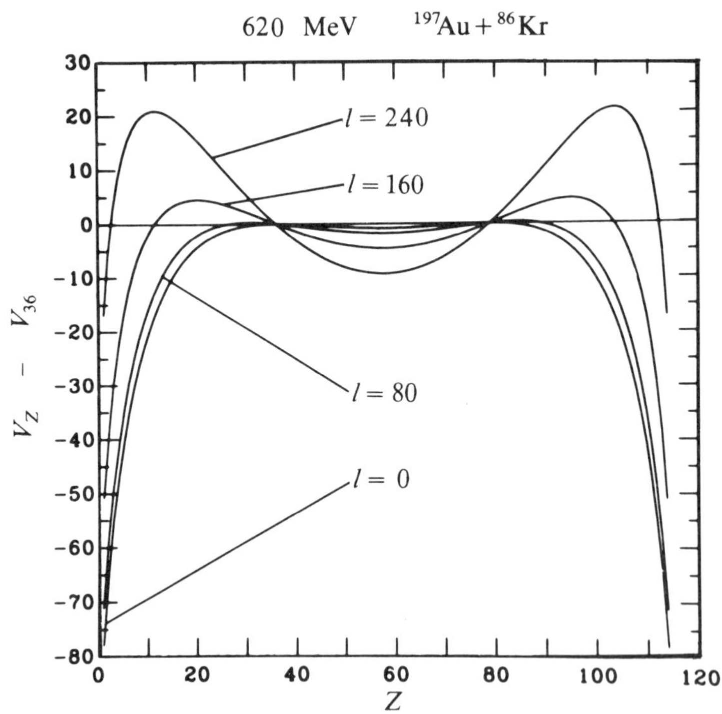

Figure 197. Driving potential, V_Z, for charge exchange in the reaction $^{197}Au + {}^{86}Kr$ plotted vs. Z of one of the fragments. The potential is calculated from liquid-drop, Coulomb, proximity, and centrifugal energies, for various values of angular momentum l. It is normalized to zero for the initial charge asymmetry (V_{36}). From (Mor 79d).

depicting this potential for the $^{197}Au + ^{86}Kr$ system, for various l values. The potential is normalized to zero at the initial charge asymmetry (V_{36}) and plotted vs. Z of the light fragment. It is calculated [cf. Equation (2.5.30)] from the sum of liquid-drop binding energies, Coulomb energy, nuclear proximity potential, and centrifugal energy. Obviously, the static driving force predicted from such a model is directed towards symmetry, for high l values, but favors a production of fragments with large charge asymmetries, in more central collisions. Although this consideration reflects an attractive and simple physical picture, it is probably not realistic. The discussion presented in Section 3.4 suggests that the role of the static driving force is not decisive for mass and charge exchange in damped reactions at the relatively high bombarding energies at which the experiments have been performed. Although it appears true that angular-momentum fractionation along the mass asymmetry coordinate occurs for initially rather asymmetric projectile–target combinations, in a sense that more asymmetric final fragmentations result from lower l waves, this is, at least, questionable for Kr-induced reactions. Here, data discussed in Section 3.4 indicate that fragmentations with asymmetries either much larger or smaller than the entrance-channel asymmetry sample lower l waves and longer interaction times.

A similar view is adopted by Gerschel *et al.* (Ger 79), who studied angular momentum transfer in the reactions $^{192}Au + ^{40}Ar$ at $E_{lab} = 227$ MeV and $^{197}Au + ^{63}Cu$ at $E_{lab} = 365$ and 443 MeV. Figure 198 illustrates their results on the dependence of the average γ multiplicity M_γ on the final fragment mass ratio M_H/M_L, for the $^{197}Au + ^{63}Cu$ reaction at 365 MeV. They show that, for each reaction angle, the γ multiplicity decreases with increasing final mass asymmetry. Again, from the above correlation between interaction time and distance of the final fragmentation from the initial asymmetry, one would have expected a decrease in γ multiplicity on either side of the injection point, $M_{Au}/M_{Cu} = 3.13$, which is clearly not the case. This situation is experimentally quite similar to that of the $^{197}Au + ^{86}Kr$ reaction discussed earlier. However, the calculated (Ger 79) driving potential for the $^{197}Au + ^{63}Cu$ fragmentations does not exhibit a characteristic change in character with varying l value, as seen for the $^{197}Au + ^{86}Kr$ reaction in Figure 197. The model driving force (Ger 79) is always directed towards symmetry and, hence, one expects the more symmetric fragmentations to sample the longer interaction times and lower l values. The smaller γ multiplicities predicted by this model for symmetric fragmentations with $M_H/M_L \approx 1$ are in clear disagreement with the data of Figure 198. The discrepancy seems to arise from the unfounded assumption that the dissipative forces, acting on the tangential motion, are strong enough to drive a heavy-ion system to a situation of rigid rotation, the sticking limit, regardless of the initial orbital angular momentum. The conclusion then that the

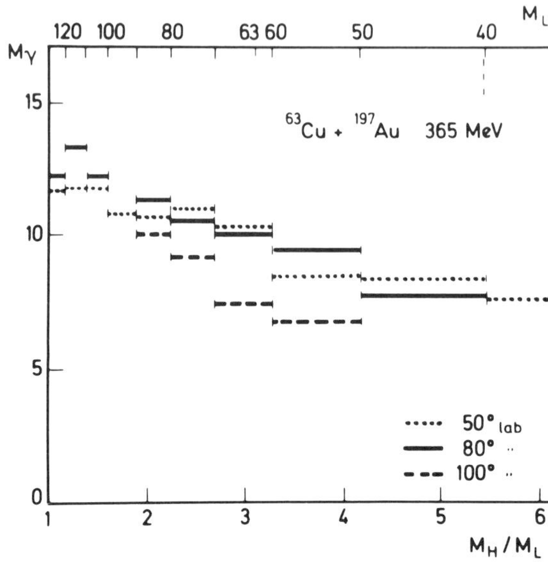

Figure 198. Average γ multiplicity, M_γ vs. mass ratio, M_H/M_L, of the final heavy (H) and light (L) fragments, produced in the 365-MeV ^{197}Au + ^{63}Cu reaction. Plotted are data at various laboratory scattering angles. Typical experimental uncertainties are of the order of 20%. The initial mass ratio (injection point) corresponds to $M_H/M_L = 3.13$. From (Ger 79).

sticking limit is in general not reached in damped reactions, except possibly for relatively low l values, provides the most plausible explanation for the observed properties of γ multiplicities.

As in the preceding experimental sections, also in the present discussion of angular-momentum transformation, it will prove most instructive to consider correlations of the experimental observables with the amount of initial total kinetic energy dissipated in a damped reaction, rather than those with reaction angle or final fragmentation. There is ample evidence for the importance of the energy loss parameter as an indicator of the reaction time and the complexity of the reaction, although this is not a sharply defined relation, as previously discussed data suggest a coexistence of different reaction types at large energy losses.

Experimental results reported by Olmi *et al.* (Olm 78b) for the reactions ^{120}Sn + ^{86}Kr and ^{166}Er + ^{86}Kr at 5.99 MeV per nucleon are displayed in Figure 199. A laboratory Wilczyński diagram for the former reaction, measured with a position-sensitive ionization chamber, is shown in part (a), in form of a scatter plot. The three regions, defined by the solid lines in this figure, correspond to elastic scattering (A), the partially damped region (B), and the fully relaxed component (C), the latter of which is attributed to

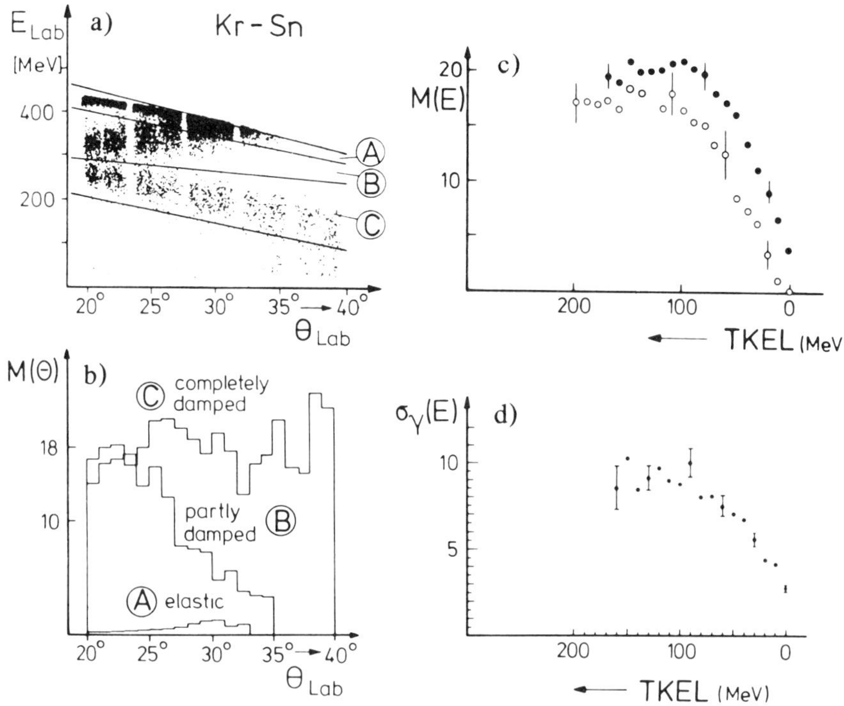

Figure 199. (a) Scatter plot of the Wilczyński diagram obtained with a position-sensitive ionization chamber. The discontinuities in the data along the abscissa are due to the window supports of the gas counter. (b) Average γ multiplicity M of the Kr–Sn reaction as a function of the laboratory scattering angle for the elastic (A), the partly damped (B), and the fully relaxed (C) components. (c) Dependence of M on the energy dissipated in the Sn + Kr (open circles) and Er + Kr reaction (solid dots). (d) r.m.s. width of the multiplicity distributions for the Er + Kr reaction as a function of energy loss (TKEL). From (Olm 78b).

negative-angle scattering. Part (b) of the figure depicts the development of the average γ multiplicity, M, with observation angle of the projectile-like fragments, for the above three regions. For the elastic branch, the multiplicity is small but increases with angle, as the domain of quasielastic processes is approached. A dramatic increase of M is observed for the partially damped events, increasing further with the degree of damping. In contrast, the fully damped events associated with region (C) exhibit a large γ multiplicity that does not change significantly any further, as angle or energy loss increases. As the correlations of main interest to the present discussion, Figures 199 (c) and (d) illustrate the dependences of average γ multiplicity, M, and the r.m.s. width, σ_γ, of the multiplicity distribution, on the magnitude of the total kinetic energy, TKEL, lost in a collision. The correlations between M and energy loss shown in Figure 199 for the ^{166}Er + ^{86}Kr (solid points) and ^{120}Sn + ^{86}Kr (open circles) reactions are rather similar to the energy loss dependence of the angular momentum $\langle J \rangle$ transferred in the reaction ^{63}Cu + ^{20}Ne, illustrated in Figure 194. With increasing energy loss, at first the average multiplicity, and presumably the total fragment spin, increase. However, for energy losses in excess of approximately 80–100 MeV, no further increase in M occurs. Hence, it appears that the multiplicity saturates and forms a plateau of ~100 MeV in width, when plotted vs. energy loss. A corresponding relation exists (Olm 78b) also for the correlation between $\langle J \rangle$ and the initial l value, if a transformation to l scale is made by relating the cross section found in a given energy loss bin to an angular-momentum bin. For the ^{166}Er + ^{86}Kr reaction, Olmi *et al.* (Olm 78b) report a total spin $\langle J \rangle$ of the fragments, that is approximately constant for initial l values between $l_i = 30$ and 110. This invalidates simple sticking or rolling models of angular-momentum dissipation, where the final spin is always proportional to the initial spin, although the proportionality factor for the former model depends on the mass fragmentation [cf. Equation (2.3.11)]

This peculiar behavior of the γ multiplicity is repeated by the r.m.s. widths σ_γ plotted vs. energy loss (TKEL) in Figure 199(d). This correlation exhibits a sharp discontinuity at small energy losses, followed by a steady rise with increasing energy loss, until a plateau is apparently reached at a similar energy loss where the average multiplicity saturates. If the γ deexcitation cascade were due predominantly to stretched E2 transitions, the saturation values of σ_γ would correspond to r.m.s. widths σ_J of the fragment total spin distributions in the range of $(8–10)\hbar$. Regardless of the procedure used to derive the spin from the γ multiplicity, it is interesting to note that the ratio $\sigma_J/\langle J \rangle$ is independent of energy loss or excitation energy for a significant range of energy losses, in this reaction. This observation is of immediate consequence for the validity of statistical models of spin transport in damped reactions, in analogy to similar relations for fragment mass and charge distributions discussed in Section 3.4. Higher moments of the γ multiplicity distribution have rarely been determined in heavy-ion reactions. For the reaction ^{144}Sm + ^{86}Kr, however, Christensen *et al.* (Chr 78) find skewnesses of the order of -0.6, implying asymmetric tails of the fragment spin distributions towards small angular momenta.

Returning to the unexplained dependence of the average γ multiplicity on the final mass or charge asymmetry, discussed in the context of Figures 196–198, the double-differential γ multiplicity $M_{\Delta E}(Z, E)$, measured by Olmi *et al.* (Olm 78b) for the ^{166}Er + ^{86}Kr reaction, is shown in Figure 200, plotted vs. fragment Z value, for adjacent, $\Delta E = 20$-MeV-wide bins in total kinetic energy loss. Apart from the general trend of the γ multiplicity with energy loss, already shown in Figure 199(c), one observes in Figure 200 a

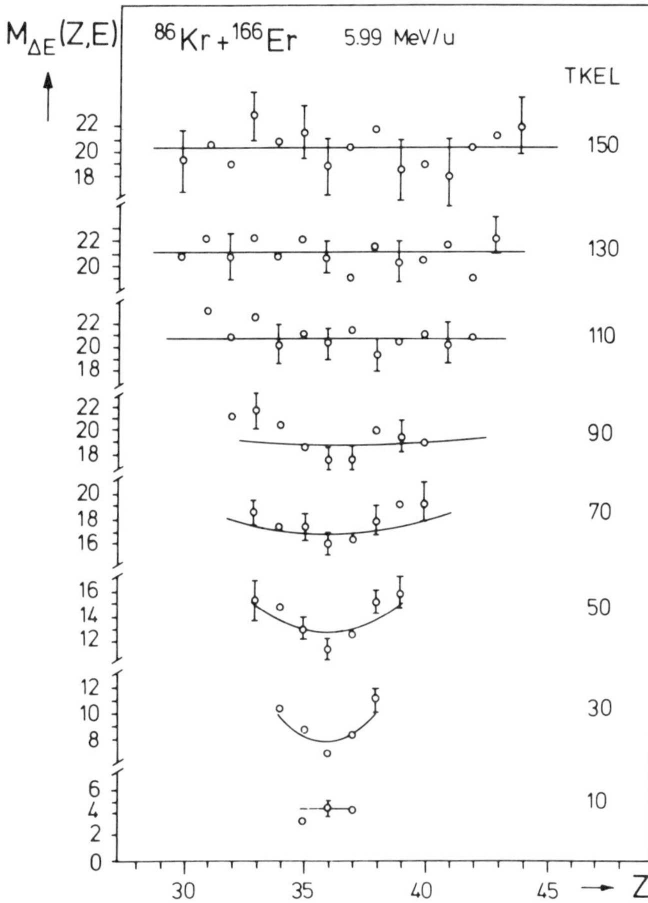

Figure 200. Average multiplicity of γ rays, plotted vs. fragment atomic number, for the reaction $^{166}Er + ^{86}Kr$. The various sets of data points correspond to adjacent bins in total kinetic energy loss (TKEL), each $\Delta E = 20$ MeV wide with a central energy given at the right. The curves represent calculations based on measured cross sections and demonstrate the independence of the multiplicity on the final fragmentation for a given sharp value of the energy loss. From (Olm 78b).

definite dependence of $M_{\Delta E}$ on the charge fragmentation, for relatively small and intermediate values of TKEL. No significant correlation with the fragment Z value is discernible at the higher energy losses. Olmi *et al.* (Olm 78b) ascribe the apparent Z dependence at smaller energy losses to the finite TKEL bin width ΔE, according to which the data have been sorted. Using experimental cross sections $d^2\sigma(Z, E)/dE\,dZ$ and assuming that the γ multiplicity depends only on the energy loss but is independent of Z, they calculate

$$M_{\Delta E}(Z,E)=\left[\int_E^{E+\Delta E} dE\, M(E)\frac{d^2\sigma}{dE\,dZ}\right]\left[\int_E^{E+\Delta E} dE\,\frac{d^2\sigma}{dE\,dZ}\right]^{-1} \tag{3.6.20}$$

The solid curves in Figure 200 representing these calculations are evidently in good agreement with the data suggesting that, at least in the case of the ^{166}Er + ^{86}Kr reaction, the Z dependence of the γ multiplicity is rather weak, once a particular energy loss is chosen. This finding illustrates, once again, the dominant role of the energy loss parameter in characterizing a damped reaction. Furthermore, the weak Z dependence of the energy-differential γ multiplicity for intermediate and high energy losses apparently contradicts the assumption that fragment spins are dominantly acquired through dissipation of orbital angular momentum in collisions leading to rigid rotation of the intermediate dinuclear complex and fully aligned (polarized) nuclear spins. Hence, at least for highly damped collisions, other mechanisms of spin generation have to be invoked.

Additional evidence for a different mechanism of fragment spin generation is obtained in experiments measuring the in-plane–out-of-plane angular anisotropy of deexcitation γ rays (Pui 80, Woz 80, Laz 81). Puigh *et al.* (Pui 80), for example, studied this anisotropy for discrete γ rays of known multipolarity emitted in coincidence with projectile-like fragments from the reaction ^{166}Er + ^{86}Kr at 600 MeV. In this experiment, the $6^+ \to 4^+$, $4^+ \to 2^+$, and $2^+ \to 0^+$ γ transitions for several residual nuclei were studied, employing Ge(Li) γ detectors in and out of the reaction plane, defined by a solid-state detector telescope measuring projectile-like fragments in coincidence with the γ rays. Figure 201 displays results (Pui 80) for the ratio of summed yields of the $6^+ \to 4^+$ and $4^+ \to 2^+$ transitions, taken at $\phi = 90°$ and 0° with respect to the reaction plane.

For comparison, the angular emission pattern of stretched E2 radiation with respect to the spin direction is given by

$$W(\phi)=\tfrac{5}{4}(1-\cos^4\phi) \tag{3.6.21}$$

In a classical picture of angular momentum dissipation, the resulting fragment spins would be aligned perpendicular to the reaction plane, in which case ϕ measures the angle with respect to the normal to this plane. The experimental anisotropies depicted in Figure 201 are significant, indicating considerable spin alignment of the final reaction fragments, but this alignment is not perfect. This could be due to a nonstretched E2 character of the γ radiation or be attributed to the depolarizing effect of particle evaporation and emission of statistical γ rays whose anisotropy is found (Pui 80) to be considerably smaller than that of the transitions illustrated in

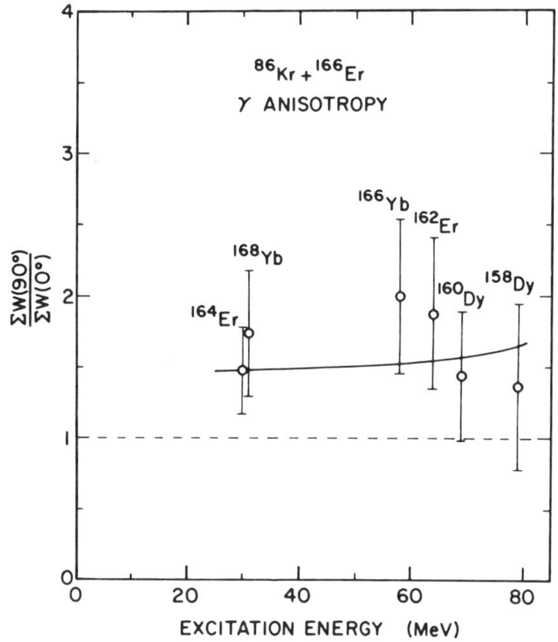

Figure 201. Out-of-plane anisotropy of γ rays from various residual fragments produced in the reaction $^{166}\mathrm{Er}+{}^{86}\mathrm{Kr}$ at $E_{\mathrm{lab}}=600$ MeV. The data correspond to summed yields of respective $6^+\rightarrow 4^+$ and $4^+\rightarrow 2^+$ transitions. The solid curve is the result of a simple model of spin transfer due to nucleon exchange. From (Pui 80).

Figure 201. The solid curve in Figure 201 represents a calculation using a simple model introduced by Vandenbosch (Van 79b) for spin transfer in peripheral heavy-ion collisions, due to stochastic nucleon exchange between the reaction partners. A small correction is made in these calculations for the deorientation effect produced in the early stages of the deexcitation cascade. The implications of the good agreement between model calculations and data will be discussed in detail further below.

The corresponding behavior of γ multiplicity and anisotropy with increasing energy loss is demonstrated in Figure 202. Here, average multiplicity, M_γ, and the in-plane/out-of-plane γ intensity ratio are plotted for continuum γ rays produced in the $^{100}\mathrm{Mo}+{}^{165}\mathrm{Ho}$ reaction, measured by Lazzarini *et al.* (Laz 81) at a c.m. bombarding energy of 450 MeV. Nonlinear angle and Q-value scales are given at the top of the figure. The abscissa is scaled in units of initial orbital angular momentum, deduced from a comparison of a trajectory calculation with the experimental average energy–angle relation. The experimental deflection function is of the double-rainbow type (cf. Section 3.3), noticeable in the inversion in the $\theta_{\mathrm{c.m.}}$

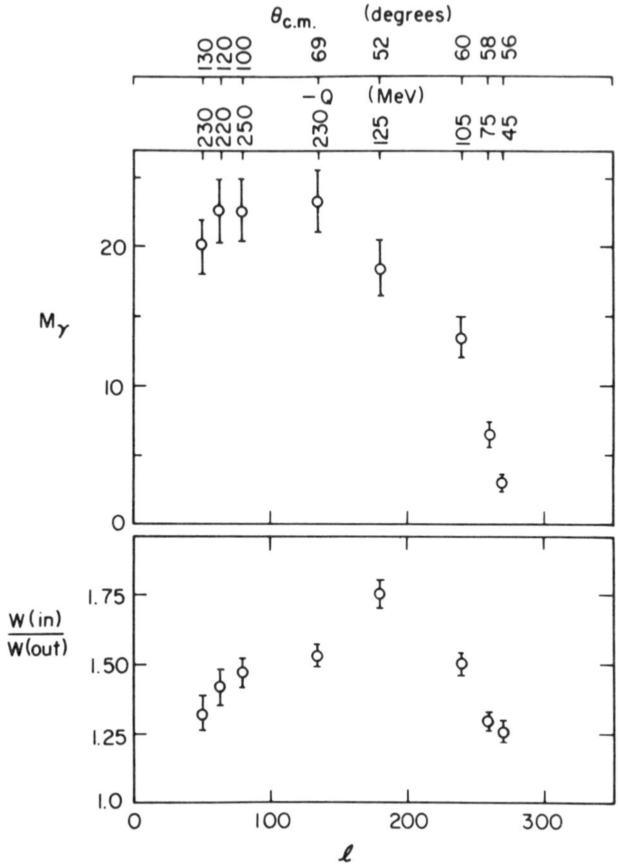

Figure 202. Average multiplicity, M_γ, and in-plane/out-of-plane anisotropy for γ rays emitted in the reaction ^{100}Mo + ^{165}Ho at $E_{c.m.} = 450$ MeV. The scales at the top of the figure represent the average angle–energy loss correlation defined by the experimental Wilczyński diagram. The bottom l scale is derived from a comparison to a dynamical trajectory calculation. From (Laz 81).

scale. The anisotropy data in Figure 202 are based on events collected with one NaI detector placed perpendicular to the reaction plane and one in-plane detector at $\theta = 45°$.

The dependence of M_γ on energy loss or l value seen in Figure 202 exhibits the saturation property familiar from the data in Figure 199. The γ anisotropy, shown in the bottom part of Figure 202, at first increases with increasing energy loss. This behavior reverses itself, however, for the larger energy losses corresponding to a backward-going cross section ridge in the experimental Wilczyński diagram. This latter domain is also characterized by large fluctuations in the final fragment charge fragmentation, indicative of a large number of nucleon exchange processes.

The data of Figure 202 clearly suggest that in strongly damped collisions, significant fragment spin components can be generated that have random directions and are, therefore, not due to dissipation of initial orbital angular momentum. Angular-momentum conservation requires that the random spin components of the two reaction partners be antiparallel, with no net polarization. Hence, the dominance of the random components at high energy losses could explain the small residual γ polarization, observed (Sch 82d) in the $^{139}La+^{86}Kr$ system, as has been discussed earlier in this section. However, this feature depends probably very sensitively on the system and the bombarding energy. Figure 189, for example, indicates significant γ polarizations, both for lighter and heavier projectile–target combinations.

Similar measurements have been made by Wozniak *et al.* (Woz 80) for the heavy system $^{165}Ho+^{165}Ho$, studied at 8.5 MeV per nucleon. The deduced energy loss dependence of fragment spins and γ anisotropy conforms with the observations made for lighter systems (cf. Figure 202). The above authors also discuss the influence of disturbing effects, such as neutron evaporation and the emission of statistical γ rays, on the γ anisotropy. They conclude that, for the system studied, these effects do not lead to significant distortions of the measured γ anisotropy.

It remains a fact, however, that the information on angular-momentum transfer and fragment spin generation that can be extracted from measurements of γ multiplicities and anisotropies is gained rather indirectly, relying on a precise knowledge of the character of the deexcitation cascade and the nature of the γ transitions. This knowledge is, so far, rather restricted. It, hence, appears desirable to inspect information from the earlier stages of the deexcitation cascade, which are dominated by the evaporation of mainly neutrons, protons, and α particles. For very heavy fragments, the sequential fission process can successfully compete with particle evaporation. Unfortunately, the number of experiments in which particle evaporation patterns are consulted to derive the spins of the primary damped reaction fragments is rather small. Experimental results have been reported for reactions induced by beams of ^{16}O (Alb 75, Ho 77, Wur 80), ^{40}Ar (Kuh 80, Bab 80, Log 80b), ^{56}Fe (Hil 79c), ^{84}Kr (Sob 81b), and ^{165}Ho (Sob 83). In interpreting these measurements, one has to appreciate the fact that there are several potential sources of evaporation particles, whose contributions overlap to a certain extent. As will be explained in further detail in Section 3.7, it is in general possible to select kinematical regions in which sequential emission from either the projectile-like or the target-like reaction fragment predominates. This is of considerable advantage as compared to measurements of continuum γ rays, where the contributions from the two reaction partners are indistinguishable.

The sensitivity of fragment–particle angular correlations to the magnitude and the alignment of the spin of the emitting nucleus has been referred to earlier in this section [cf. Equations (3.6.6)–(3.6.8)], in terms of the predictions of the statistical model. The implications of this model for the decay of highly excited compound nuclei are still being discussed in recent experimental and theoretical work (e.g., Bra 73, Gal 74b, 74c, 74d, Cat 80, Plo 82, Raj 82).

The emission of light charged particles in damped collisions of 280-MeV ^{40}Ar projectiles with ^{58}Ni target nuclei has been used as a probe on angular-momentum transfer by Babinet *et al.* (Bab 80), who have measured in- and out-of-plane angular correlations of protons and α particles in coincidence with light reaction fragments. Figure 203 illustrates typical out-of-plane correlations for α particles in the rest frames of the emitting fragments characterized by the atomic numbers given on the right of each distribution. The solid curves in this figure represent fits to the data using the formula (Bab 80).

$$W(\phi_{\text{c.m.}}) \propto \exp\left[\frac{\hbar^2\left(I+\frac{1}{2}\right)^2}{2\mathscr{I}\tau}\,\frac{\mu_\alpha R^2}{\mu_\alpha R^2+\mathscr{I}}\sin^2\phi_{\text{c.m.}}\right] \tag{3.6.22}$$

This equation is obtained from the statistical model of Ericson and Strutinsky (Eri 58, 59, 60) after integration over all possible angular momenta and energies of the evaporated particles and assuming a sharp-cutoff model for the transmission coefficients. The quantities entering Equation (3.6.22) have the usual notation; $\mu_\alpha R^2$ is the moment of inertia for the orbital motion of the α particle at the nuclear surface. The angle $\phi_{\text{c.m.}}$ is the out-of-plane angle, with $\phi_{\text{c.m.}} = 0°$ in plane. Deducing the nuclear temperature τ from the falloff of the particle energy spectrum and adopting a moment of inertia, $\mathscr{I}$, for the nucleus, it is possible to estimate its r.m.s. spin value, $I_{\text{r.m.s.}} = |I + \frac{1}{2}|$.

Assuming for the nuclear moment of inertia the rigid-body value [cf. Equation (2.3.6)], calculated with a nuclear radius of $R = 1.2A^{1/3}$ fm, Babinet *et al.* (Bab 80) arrive at the preevaporative total spins of both fragments shown in Figure 204. Included in this figure are also results of a γ multiplicity measurement evaluated under the usual assumption of a predominance of stretched E2 transitions in the γ cascade. The apparent discrepancy of $(8–12)\hbar$ between the two experiments is probably (Bab 80) due to the limitation of the γ multiplicity study in estimating preevaporative fragment spins.

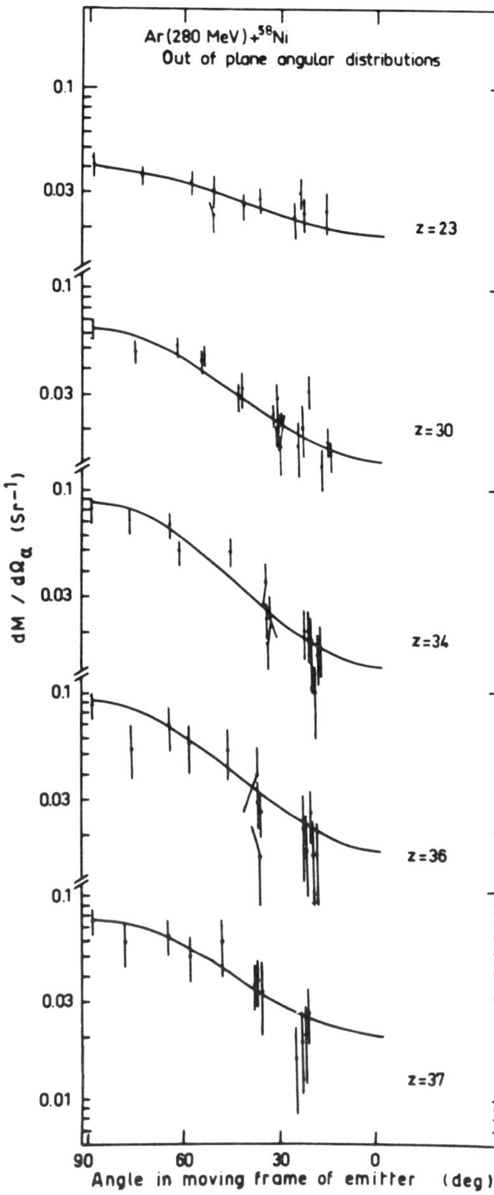

Figure 203. Out-of-plane angular correlation of α particles emitted sequentially from fragments produced in the reaction $^{58}Ni + {}^{40}Ar$ at $E_{Lab} = 280$ MeV. The fragment atomic numbers are indicated on the right of each distribution. Solid curves represent fits using Equation (3.6.22). From (Bab 80).

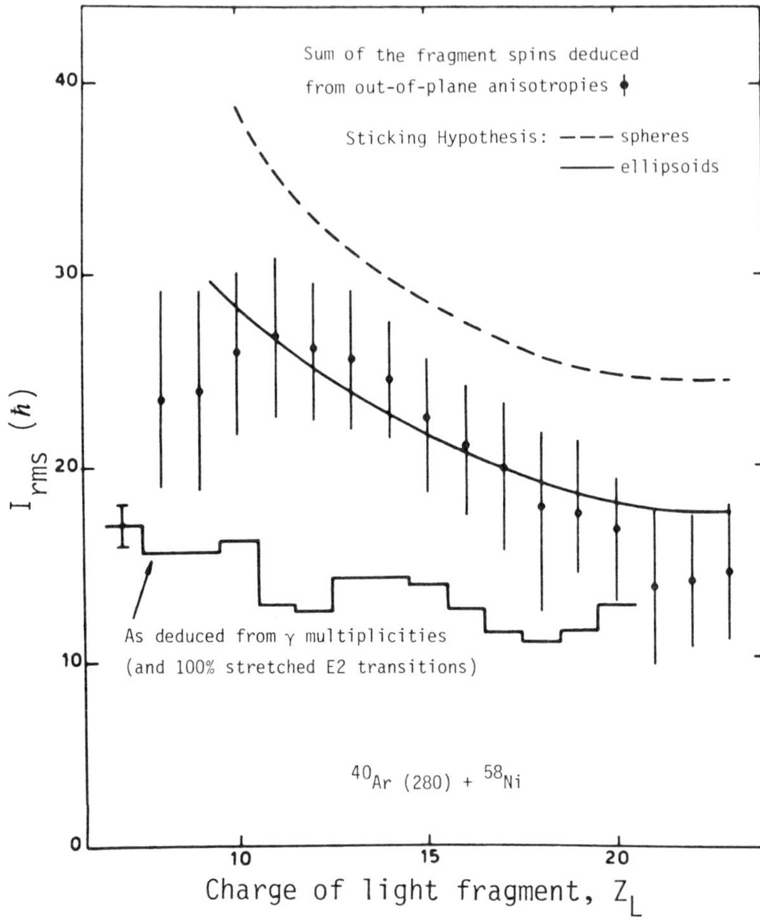

Figure 204. Variation of total intrinsic spin (r.m.s.) with charge asymmetry, indicated by atomic number of the light reaction partners, for the reaction $^{58}Ni + ^{40}Ar$ at $E_{Lab} = 280$ MeV. The data points are results of a measurement of α-particle out-of-plane anisotropy, while the histogram refers to γ multiplicities obtained for the same reaction. The curves represent predictions of the sticking hypothesis for spherical (dashed curve) or deformed (solid curve) fragments. From (Bab 80).

Model calculations performed by Babinet *et al.* (Bab 80), assuming that the sticking condition is reached in this relatively light system, are represented by the solid and dashed curves in Figure 204. As can be seen from the comparison, the experimental total fragment spins are significantly less than expected for rigid rotation of sticking spherical fragments (dashed curve). The solid curve, giving a much better account of the data, is obtained (Bab 80) for symmetrically deformed fragments in the sticking situation. In this latter case the distance of the fragment centers from each other, at breakup of the intermediate system, was assumed to be given by the final fragment kinetic energies. The individual deformation parameters were then taken to scale with the linear dimensions of the undisturbed fragments. While such a procedure may appear plausible to some extent, the agreement between data and the schematic model calculation shown in Figure 204 should not be considered to provide more than a consistency check, because it relies so evidently on stringent model assumptions. As the discussion of the final kinetic energy spectra in Section 3.5 has emphasized, there is actually no independent experimental evidence for a particular division of final fragment total kinetic energies into contributions due to radial kinetic energy, present at breakup of the system, centrifugal energy, Coulomb repulsion, and thermal fluctuations in various collective degrees of freedom.

Out-of-plane particle–fragment angular correlations have been discussed by Wurm (Wur 80) for the somewhat heavier system $^{93}Nb + ^{40}Ar$ at $E_{Lab} = 400$ MeV. Differential backward-angle multiplicities $dM_{\alpha}^{R}/d\Omega_{\alpha}^{R}$, and $dM_{p}^{R}/d\Omega_{p}^{R}$ for α particles and protons, respectively, from the recoiling target-like fragment, are plotted in Figure 205 vs. out-of-plane angle ϕ^{R}. The data are summed over the in-plane correlation angle, of which the anisotropy is observed to be independent. It is clear from these measurements that anisotropies for α particles are definitely larger than those for protons, a consequence of the larger mass of the α particle and its ability to carry off larger amounts of angular momentum [cf. Equations (3.6.7) and (3.6.8)]. The curves in Figure 205 represent statistical-model fits performed in the framework of Equation (3.6.22). However, instead of assuming a fixed average fragment spin, Wurm (Wur 80) describes an analysis in terms of a Gaussian spin distribution centered at J_0 and with a variance σ_J^2, by which the statistical misalignment of a fluctuating fragment spin component is

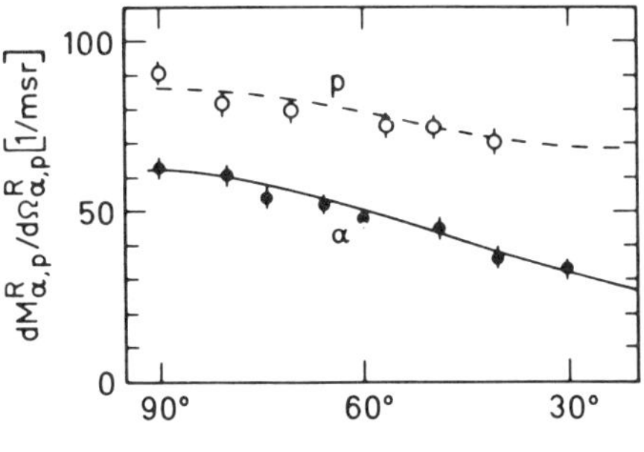

Figure 205. Out-of-plane angular dependence of the differential multiplicity $dM^{R}/d\Omega^{R}$ for protons (open circles) and α particles (solid dots) from the reaction $^{93}Nb + ^{40}Ar$ at $E_{Lab} = 400$ MeV. The angle ϕ^{R} is measured from the normal of the reaction plane. The curves represent statistical-model fits. From (Wur 80).

modeled. This fluctuating-spin distribution has the effect of damping the angular anisotropy, which is accounted for (Wur 80) by introducing a damping factor:

$$K = \left(1 + \sigma_J^2 \frac{\hbar^2}{\mathscr{I}T} \frac{\mu_\alpha R^2}{\mu_\alpha R^2 + \mathscr{I}}\right)^{-1} \tag{3.6.23}$$

into the argument of the exponential function in Equation (3.6.22). Correlation data such as shown in Figure 205 are not sufficient to determine two moments, J_0 and σ_J^2, of the fragment spin distribution. The region in the $J_0 - \sigma_J$ plane compatible with the experimental α anisotropies is depicted in Figure 206 as a hatched band. It is intersected by the line of constant spin, calculated from the sticking assumption at values of σ_J suggested by the transport theory of Nörenberg and collaborators (cf. Section 2.8). However, the sticking situation is not particularly endorsed by these experiments. Correspondingly, the uncertainty in the deduced $J_0 - \sigma_J$ values translates into a considerable error in the inferred alignment parameter P_{zz} [cf. Equation (3.6.2)]. For the above $^{93}Nb + ^{40}Ar$ reaction, as well as for the $^{58}Ni + ^{16}O$ reaction at $E_{Lab} = 96$ MeV, Wurm (Wur 80) quotes $0.65 < P_{zz} < 1$. These uncertainties can be reduced, in principle, by utilizing the spin dependence of particle branching ratios [cf. Equation (3.6.5)]. The technique has been discussed recently by Catchen *et al.* (Cat 80), reporting a rather drastic dependence of the multiplicity ratio M_α/M_p on the spin of an excited ^{86}Kr fragment. This implies, on the other hand, that data obtained in measurements of evaporated nucleons and α particles are not rigorously comparable to each other, as different regions of the spin distribution are sampled with different weights.

As a further example, experimental results on average spins of fragments from the reaction $^{nat}Ag + ^{84}Kr$ at $E_{Lab} = 664$ MeV are shown in

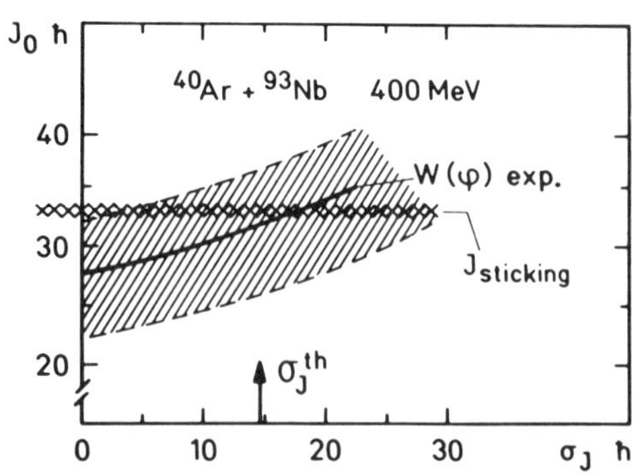

Figure 206. Combinations of average value, J_0, and r.m.s. width, σ_J, consistent with measured α anisotropies are given as a solid curve with estimated uncertainties represented by the hatched band. The cross-hatched band indicates the spin expected for the target-like fragment in the sticking limit. The quantity σ_J^{th} is suggested by a spin diffusion model. From (Wur 80).

Figure 207, as reported by Sobotka *et al.* (Sob 81b). These experiments combine γ multiplicity measurements, yielding an estimate of the total spin, $I_H + I_L$, of both fragments, with a determination of the out-of-plane anisotropy for α particles from the heavy reaction partner, providing information on its spin, I_H. Open circles in Figure 207 represent the γ multiplicity results, while the solid dots are data obtained for the heavy reaction partner from the α-particle anisotropies. In the upper part (a) of the figure, average

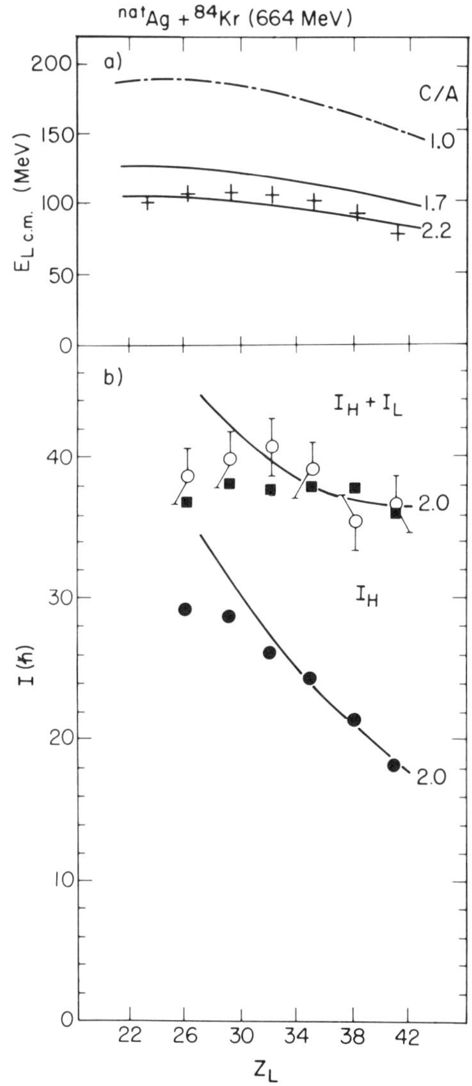

Figure 207. Center-of-mass energies and spin values plotted vs. atomic number of the light fragment, Z_L, for the $^{nat}Ag + ^{84}Kr$ reaction at $E_{Lab} = 664$ MeV. Crosses in (a) represent average final fragment energies in an approximately 100-MeV-wide energy bin. The solid dots in (b) are experimental spin values for the heavy reaction partner, derived from out-of-plane anisotropies of coincident α particles. The squares are the total spins obtained assuming the sticking hypothesis. Open circles represent γ multiplicity data. From (Sob 81b).

c.m. energies of the detected light reaction fragments are plotted vs. their atomic numbers, Z_L. It should be noted, however, that coincident reaction fragments in the whole strongly damped region were accepted in the experiment, corresponding to a width in c.m. energies in excess of 100 MeV. The curves in Figure 207 correspond to rigid rotation of the intermediate system thought to consist of two equally deformed spherical fragments with the indicated axis ratios and a tip separation distance of 1 fm.

The filled squares in Figure 207(b) are calculated values, derived from the data (solid dots) under the assumption of rigid rotation. They agree with the total spins obtained from the γ multiplicities, assuming stretched E2 transitions, following the emission of, on the average, three statistical γ rays per fragment. A correction of the order of 30% was also made (Sob 81b) in this conversion, to account for spin carried away by neutrons and α particles. In view of the substantial latitude available in the interpretation of the primary data, not too much significance can be attached to the agreement between data and calculations, as has already been stated for the other cases discussed previously.

To complement the experimental information on fragment spins acquired in damped heavy-ion collisions, investigations studying the sequential fission mode will be illustrated next. Obviously, such measurements are possible only for systems in which heavy fragments are produced. Except for highly fissile nuclei, the method samples regions of high excitation energy and spin and is affected by the multiple-chance fission process. Only a few experiments have so far been performed (Dye 77, 79b, Har 79, Raj 79, Leb 80, Pui 79), using this method, which is schematically illustrated in Figure 172.

In-plane and out-of-plane angular correlations of fission fragments from the reaction $^{238}U + ^{86}Kr$ at $E_{Lab} = 730$ MeV are depicted in Figure 208, as reported by Puigh *et al.* (Pui 79). The reaction plane was defined in the experiment by a solid-state detector telescope at $\theta_{Lab} = 35°$, measuring projectile-like fragments. Coincident fission fragments were detected in two single-element counters placed both in and out of the reaction plane. The angular distributions shown in Figure 208 are grouped from top to bottom according to increasing negative reaction Q value. They are sorted in 40-MeV-wide energy bins centered at the indicated values of $\overline{Q}$. The out-of-plane angle $\theta_{c.m.}$ is defined such that $\theta_{c.m.} = 0$ corresponds to detection of a fission fragment in a direction perpendicular to the reaction plane. The in-plane angle $\theta_{c.m.}$ is measured with respect to the recoil direction of the primary target-like reaction fragment, inferred from the detected projectile-like reaction partner.

As is evident from the data of Figure 208, the in-plane angular correlations are not very pronounced and almost isotropic. This is contrasted with the large anisotropies exhibited by the out-of-plane data. For

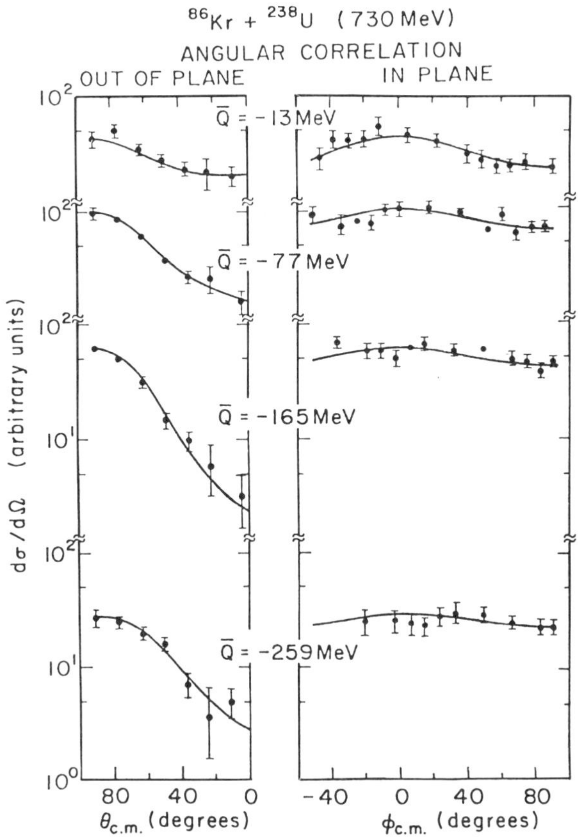

Figure 208. Semilogarithmic plot of in-plane and out-of-plane angular distributions of fission fragments from the reaction $^{238}U + ^{86}Kr$ at $E_{\text{Lab}} = 730$ MeV. The distributions are grouped according to 40-MeV-wide bins in reaction Q value with the indicated centroids, $\overline{Q}$. The curves represent fits to the data. From (Pui 79).

higher energy losses, these anisotropies are seen to be definitely larger than those for low energy losses. This behavior indicates immediately that angular-momentum transfer increases with increasing energy loss. Adopting a monotonic, descending relationship between entrance-channel angular momentum and energy loss, it also suggests that the spin acquired by the U-like primary reaction fragment is disproportional to the initial l value. This would exclude both the simple-minded sticking or rolling situations from being realized uniformly in the reaction, for which such a proportionality is predicted, although with different, mass-dependent proportionality factors.

The measured in-plane and out-of-plane angular correlations of fission fragments from the $^{238}U + ^{86}Kr$ reaction are fitted (Pui 79) simultaneously with a theoretical form given by Equation (3.6.1). A single average spin value, J, of the primary target-like fragment is assumed in this analysis. The distribution in spin projections (K) on the scission axis is taken to be a Gaussian with a variance K_0^2, extrapolated from low-energy fission systematics (Van 73), according to the variation in thermodynamic temperature. Misalignment of the fragment spin with respect to the normal to the reaction plane, chosen as the quantization axis, is modeled classically by assuming a random component of the spin distributed isotropically in the reaction plane. The first moments, J, and the alignments, P_{zz} [cf. Equation (3.6.2)], obtained from the fits shown in Figure 208 are plotted in Figure 209 vs. reaction Q value. The errors are obtained from the dependence of χ^2 on the fit value J after minimalization with respect to the random spin component.

In agreement with γ multiplicity experiments discussed earlier in this section (cf. Figures 199 and 202), the $^{238}U + ^{86}Kr$ data exhibited in Figure 209 indicate that kinetic energy and angular momentum are dissipated concurrently. As before, it is also seen in these data that the acquired fragment spin saturates at high energy losses, although the plateau is less

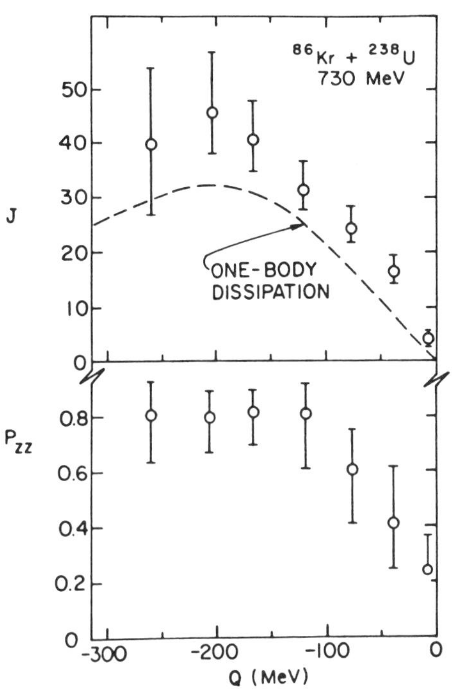

Figure 209. First moments, J, and alignment parameters, P_{zz}, of the spin distributions of primary U-like fragments, produced in the $^{238}U + ^{86}Kr$ reaction at $E_{Lab} = 730$ MeV, are plotted vs. reaction Q value. The data are obtained from angular correlations of sequential fission fragments. The dashed curve in the top part of the figure represents a simplified model calculation based on the one-body dissipation theory of Randrup. From (Pui 79).

well defined than a similar feature in the average γ multiplicity data shown in Figure 199. Quite surprisingly, the fragment spins are well aligned already at moderate energy losses, as indicated by the large values of $P_{zz} \approx 0.8$. The alignment is not destroyed even at the highest energy losses measured in the $^{238}U + ^{86}Kr$ experiment. This appears to be at variance with the observations made in γ-ray measurements (Pui 80, Laz 81) where the γ anisotropy decreases at high energy losses, for systems somewhat lighter than $^{238}U + ^{86}Kr$.

Kinematically complete experiments have been performed by V. Harrach *et al.* (Har 79), studying three-body events in the bombardments of ^{58}Ni and ^{90}Zr targets with beams of ^{208}Pb and ^{238}U at 7.5 MeV per nucleon. The experiments were done with a large-solid-angle counter array such as illustrated in Figure 48. Rather isotropic in-plane angular correlations were found in these measurements for the sequential fission fragments produced in all four reactions. This is consistent with r.m.s. angular widths of the out-of-plane distributions that are nearly independent (Har 79) of the in-plane-angle of fission fragments, measured with respect to the recoil direction of the primary fissioning fragment. One, hence, concludes (Har 79) that the spin components parallel to the reaction plane are distributed isotropically.

In Figure 210, experimental results (Har 79) are summarized for the dependence of various spin-related quantities on the total kinetic energy loss. The top part shows the fission probabilities, on the left for the ^{208}Pb-induced reactions, on the right for the ^{238}U-induced processes. From top to bottom then follow the relative fission probabilities, the r.m.s. widths of the out-of-plane distributions, and, finally, the first moments, $\langle I_z \rangle$, of the distributions over the fragment spin components perpendicular to the reaction plane. It is of relevance to note that sequential fission in the ^{208}Pb-induced reactions acquires significance only at considerable energy losses, whereas it dominates in the reactions $^{58}Ni + ^{238}U$ and $^{90}Zr + ^{238}U$, at all energy losses. One also recognizes a different behavior of the width of the out-of-plane correlation with increasing energy loss, in the reactions induced by ^{238}Pb and ^{238}U particles. The energy loss dependence of the heavy fragment spin component, $\langle I_z \rangle$, estimated from fits similar to those described previously, reflect this difference. In the reactions induced by the highly fissile ^{238}U, the aligned spin component is observed to increase steadily with increasing energy loss, until a plateau is reached at about 180 MeV. For energy losses smaller than that, the heavy fragments produced in the ^{208}Pb-induced reactions are hardly fissionable. Those, however, that actually undergo sequential fission, have acquired significant spins $\langle I_z \rangle$. These latter events correspond, therefore, to a nonrepresentative, selective sampling of the high-spin tails of the fragment spin distribution. For relatively low and intermediate energy losses, the values of $\langle I_z \rangle = (20\text{–}30)\hbar$

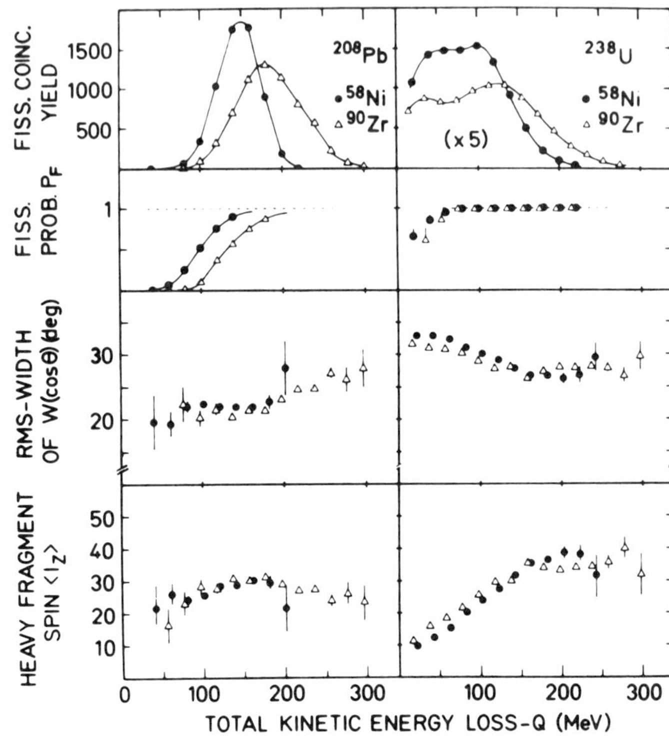

Figure 210. Sequential fission yield, fission probability, r.m.s. width of the out-of-plane fission angular correlation, and average Z component of the spin acquired by heavy fragments formed in reactions induced by ^{208}Pb (left) and ^{238}U (right) projectiles of 7.5 MeV per nucleon. From (Har 79).

observed in Figure 210 for Pb-like reaction fragments could then simply signify the minimum spins necessary to induce fission in such nuclei, at excitation energies of $E^* = 40$–100 MeV.

The spin-selectivity of the sequential fission decay channel has also been observed in the $^{209}\text{Bi} + {}^{86}\text{Kr}$ reaction investigated by Dyer *et al.* (Dye 79b). These authors report some evidence for a correlation between the magnitude of the aligned spin component of the primary Bi-like fragment and its atomic number or fissility. Data obtained in these measurements, integrated over all fragment Z values, are reproduced as circles in Figure 211, while triangles correspond to heavy fragments with $Z > 83$. This figure illustrates the correlation between average aligned spins, J_{Bi}, of the heavy fragments and the initial orbital angular momentum, l_i. The latter quantity is deduced from the measured cross section in adjacent bins of total kinetic energy loss, E_{loss}, assuming a monotonic dependence between l_i and E_{loss}.

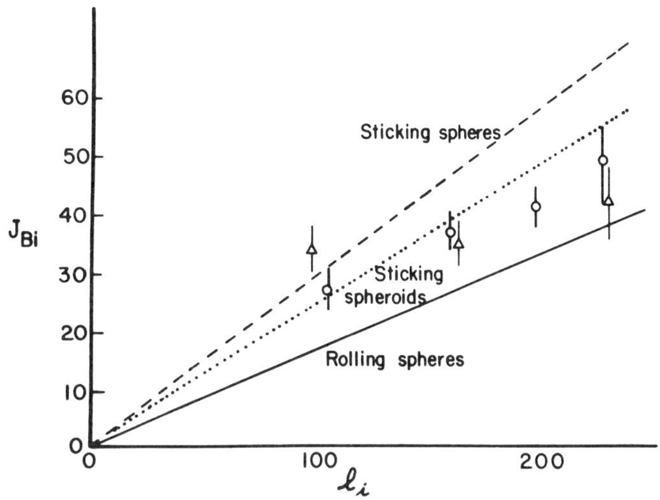

Figure 211. Average spin acquired by fissioning Bi-like fragments in the reaction $^{209}Bi + ^{86}Kr$ at $E_{Lab} = 610$ MeV, plotted vs. entrance-channel orbital angular momentum, as deduced from the cross section in various Q-value bins. The circles are based on Z-integrated data, while the triangles are for heavy fragments with charges larger than that of the target. Straight lines illustrate the relations expected for several simple classical models. From (Dye 79b).

Straight lines in Figure 211 represent $J_{Bi}(l_i)$-relations expected from various sticking and rolling models. The data are seen to be consistent with the assumption of sticking spheroids for all l values, with a deformation inferred from the most probable final energy of the damped reaction component. However, as pointed out by Dyer *et al.* (Dye 79b), the bias present in the data precludes any conclusion with respect to the attainment of sticking for large values of l_i, or low energy losses. Reservations such as this should be kept in mind when judging the comparisons between data and various model predictions discussed in the following part of this section.

It is evident from the previous discussion of experimental data that the present knowledge about angular-momentum transfer and spin generation in damped reactions is still rather incomplete, owing to a lack of systematical studies. The theoretical situation can be regarded with similar concern, since only a few detailed comparisons of model predictions to experiments have been made. Angular momentum transmutation is discussed in terms of classical phenomenological models (Bon 74b, 75, Gro 74, 74b, 78, 81, Deu 77b, Ban 80), of reaction theories based on collective excitations of surface vibrations (Bro 78c, 79e, Esb 78) or more exotic modes (Mor 79c, 79d, 79e, 80, 80d, 81c, Ger 79, Woz 80), and of stochastic transport models (Wol

78, 78b, 78c, Aga 77b, 78b, Hof 76c, Ngo 77, Ber 78f, Ger 79, Ran 78, 79d, 80b, 82, Fri 81). The hierarchy of the various theories and the basic physical principles they exploit have been outlined in Section 2.

In classical dynamical models, dissipation of orbital angular momentum is ascribed to frictional forces acting on the orbital motion. Since this is the only process considered, by which reaction fragments gain intrinsic spin, the predictions of such models can only be compared to the aligned component of the fragment spin. Most of these reaction theories assume a tensor of friction form factors which is diagonal in the radial and tangential components of the relative velocity. However, these friction models can differ drastically with respect to the radial variation of the friction form factors and the general anisotropy of the dissipative force. In Figure 27, for example, radial and tangential form factors suggested by the one-body window formula [Equation (2.6.25)] are compared to those of the hard-friction model proposed by Gross *et al.* [Gro 78, cf. Equation (2.6.17)]. Since the latter theory predicts a highly anisotropic friction force, with the radial component some 400 times stronger than its tangential counterpart, one expects a relatively small contribution of angular momentum dissipation to the total energy loss.

The predictions of the hard-friction model (Gro 78) for the 600-MeV $^{209}Bi + ^{84}Kr$ reaction are summarized in Figure 212. It is seen in this figure that the fraction ΔL of initial orbital angular momentum L dissipated in the reaction is suggested to be rather small, in general. For the smallest l values considered, the dissipated angular momentum barely approaches the rolling limit [cf. Equation (2.3.9)]. This is in contrast to the large values of $J_{Bi} = (30\text{–}50)\hbar$ found experimentally (Dye 79b) for the heavy reaction fragment alone. Since the experimental data are biased towards high spins and the deduced J_{Bi} values contain a random component, the calculations may, nevertheless, be consistent with experiment. Because of the high fissility of U-like fragments, reactions such as $^{238}U + ^{86}Kr$ (see Figure 209) or $^{90}Zr + ^{238}U$ (Figure 210) represent more suitable test cases. Calculations with the model are available (Ban 83, see Figure 213) for the $^{238}U + ^{86}Kr$ system, where it is found that the predictions significantly underestimate the experimental spin values. The generally small amounts of dissipated angular momentum, predicted (Gro 78, 81) by this theory for a range of systems, suggest that the large and highly aligned spins acquired by U-like fragments in the above reactions are probably not due to the dissipative processes assumed in the hard-friction model.

In the above model, as in most other approaches, orbital motion of two interacting heavy ions is damped only due to a tangential component of the adopted friction force. This force becomes ineffective when the sliding motion of interacting surface elements is completely damped, i.e., when the fragments roll on each other, while the dinuclear complex rotates. Classi-

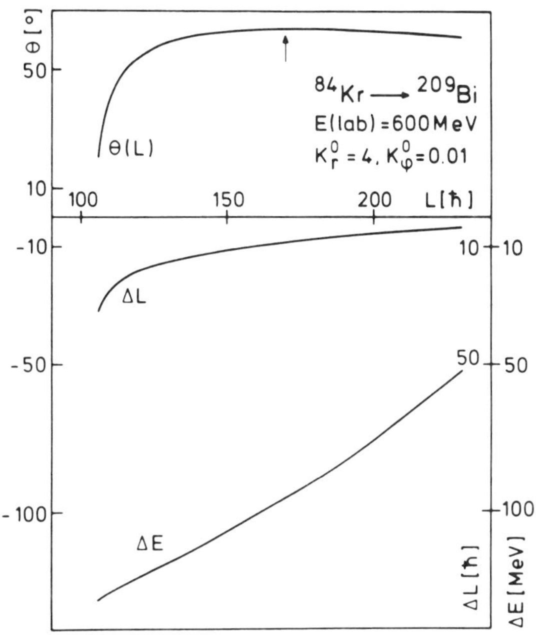

Figure 212. Theoretical deflection function $\theta(L)$, dissipated angular momentum, ΔL, and total energy loss, ΔE, plotted vs. initial orbital angular momentum, L, as predicted for the 600-MeV ^{209}Bi + ^{84}Kr reaction by the hard-friction model. From (Gro 78).

cally, further dissipation of orbital motion and, hence, of angular momentum is only possible through the action of rolling friction. This type of friction has been considered by Siwek-Wilczyńska and Wilczyński (Siw 76) and, more recently, by Bangert and Freiesleben (Ban 80, 83). In these models, classical equations of motion are solved that are similar to those in Equation (2.3.8), but modified by a rolling-friction term. The spin acquisition rate for each fragment i is given, for example (Ban 80), by

$$\frac{dI_i}{dt} = -c_t(r)r\frac{R_i}{R_1 + R_2}u_t + (-)^i c_{\text{roll}}(r)g^2(\dot{\theta}_1 - \dot{\theta}_2) \qquad (3.6.24)$$

Here, the radius and angle parameters are defined as in Section 2.3 (cf. Figure 6). The quantity u_t is the tangential velocity, and g stands for an effective radius of gyration (Tsa 74). The form factors $c_t(r)$ and $c_{\text{roll}}(r)$ describe the coefficients for sliding and rolling friction, respectively. In the work of Bangert *et al.* (Ban 80, 83), the radial dependence of the friction form factors is assumed to be given by the one-body window formula [cf. Equation (2.6.25)].

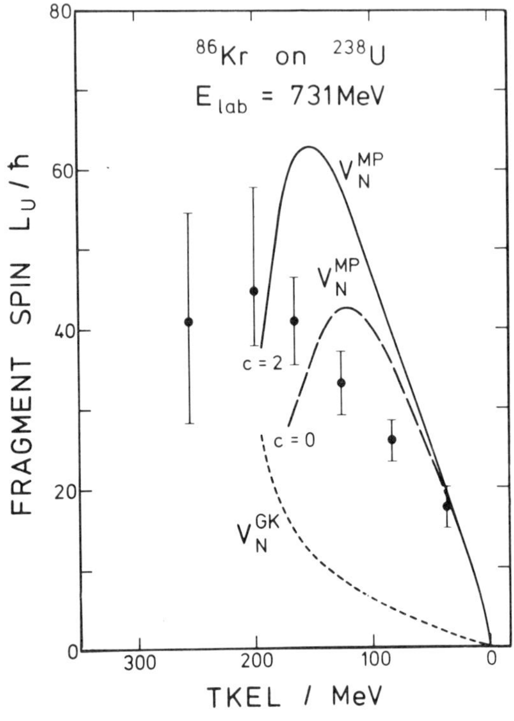

Figure 213. Spin L_u vs. total kinetic energy loss (TKEL), for U-like fragments produced in the reaction $^{238}U+{}^{86}Kr$. The solid dots are experimental data (Van 79b), while the dashed and solid curves correspond to values predicted by the trajectory calculation described in the text. The quantity c gives the ratio, c_{roll}/c_t, of rolling to tangential friction coefficients. For comparison, the predictions of the hard-friction model are included in this figure as a short-dashed curve. From (Ban 83).

Results obtained (Ban 83) from the above trajectory model are compared in Figure 213 to experimental data on the energy loss (TKEL) dependence of heavy fragment spins produced in the reaction $^{238}U+{}^{86}Kr$ discussed earlier. The long-dashed curve corresponds to $c=c_{roll}/c_t=0$ and, hence, to the neglect of rolling friction, whereas the solid curve represents results obtained under the assumption of strong rolling friction, $c=2$. Although there is qualitative agreement between data and calculations for low and intermediate energy losses, neither model reproduces the experiment quantitatively. While the discrepancy between experiment and calculation for energy losses higher than ~200 MeV could be attributed to the neglect of fragment deformations in the calculations, the overestimation of the fragment spins acquired at low values of TKEL may be more significant. This latter conclusion is reached because of the relatively small alignment (P_{zz}) parameters measured (Pui 79, see, Figure 209) in this energy domain, which suggest a dominant spin generation mechanism not due to dissipation of orbital angular momentum. Similar calculations (Ban 83) performed for the $^{209}Bi+{}^{86}Kr$ reaction yield a better quantitative agreement with experimental fragment spins, deduced from a measurement of sequential fission fragments. But, as stated earlier, these experimental data sample selectively the tail of the fragment spin distribution, while the calculation describes an average classical path. Hence, from such comparisons, it is not possible to derive arguments in support of rolling friction in damped collisions.

The predictions of the hard-friction model are also included in Figure 213 as a short-dashed curve. It is clear from this figure that these calculations cannot account for the fragment spin dependence on energy loss, even in the domain of energy losses associated with highly aligned spins. A similar conclusion is reached by Gerschel *et al.* (Ger 79) who studied angular momentum transfer in the reactions ^{197}Au + ^{40}Ar and ^{197}Au + ^{63}Cu.

Classical dynamical calculations of angular-momentum dissipation have also been performed (Deu 77b) with the model of Deubler and Dietrich (Deu 77), discussed in Section 2.3. This model accounts for fragment deformations and employs adjustable phenomenological friction parameters. As has been mentioned in previous sections, the interaction force used in the calculations is too repulsive for larger interpenetrations, leading to a deflection function not in accord with the experimental observations on γ-ray polarization. For heavy systems at low bombarding energies, such as ^{209}Bi + ^{84}Kr at $E_{\text{Lab}} = 600$ MeV, however, such deep interpenetrations are not probable. Hence, this reaction model gives a better account of the collision features, for such cases. In fact, for the ^{209}Bi + ^{84}Kr reaction at 600 MeV, Deubler *et al.* (Deu 77b) obtain a qualitative agreement between calculations and experiment on the angular position of the strongly damped component and its cross section. With an isotropic friction tensor, the model predicts losses in orbital angular momentum of the order of $30\hbar$, for strongly damped ^{209}Bi + ^{84}Kr collisions at $E_{\text{Lab}} = 600$ MeV. This result is qualitatively consistent with the data (Dye 79b), although there is no provision in the model for the generation of additional random spin components.

The model of collective doorway excitation in damped collisions (see Section 2.3) proposed by Broglia and collaborators (Bro 76, 78c, 78e, 79e, Esb 78) considers a microscopic mechanism for the dissipation of kinetic energy and angular momentum. The calculations, essentially a coupled-channels scheme, allow various channels to be fed for a particular impact parameter and produce, hence, fluctuations in the final observables for any given l wave. Model predictions for the final orbital angular momentum in the reaction ^{208}Pb + ^{136}Xe at $E_{\text{Lab}} = 1130$ MeV have already been shown in Figure 155. As seen in this figure from the solid curves describing a hybrid calculation (see Section 3.5), both final energy, E_f, and final angular momentum, l_f, decrease rather abruptly with decreasing impact parameter ρ. The dissipated angular momentum approaches in magnitude the sticking limit, as the kinetic energy loss approaches its predicted maximum. Although this appears to be a plausible scenario, it has been argued in Section

3.5 that this transition occurs too rapidly to be consistent with the data. The bars in Figure 155 represent the results of extended model calculations by Esbensen *et al.* (Esb 78), taking account of quantal zero-point fluctuations.

In this model calculation, it is observed (Esb 78) that, for a fixed impact parameter ρ, there is a correlated distribution of final energies and angular momenta. This correlation is linear and may be expressed as

$$l_f - \langle l_f \rangle = c_0 \left(E_f - \langle E_f \rangle \right) \tag{3.6.25}$$

where brackets indicate the average taken over the distribution. Neglecting zero-point fluctuations, the coefficient of proportionality is given (Esb 78) by

$$c_0 = \frac{\langle E \cdot l \rangle - \langle E \rangle \langle l \rangle}{\left\langle \left(E - \langle E \rangle \right)^2 \right\rangle} = \frac{\Sigma_{n\lambda\mu} \hbar \mu \langle E_{n\lambda\mu} \rangle}{\Sigma_{n\lambda\mu} \hbar \omega_{n\lambda} \langle E_{n\lambda\mu} \rangle} \tag{3.6.26}$$

Here, $\langle E_{n\lambda\mu} \rangle$ is the expectation value of the energy $E_{n\lambda\mu}$, absorbed by the collective mode of frequency $\omega_{n\lambda}$. The mode is specified by the number n of quanta, and its angular momentum λ. The quantity μ is the component of the spin perpendicular to the reaction plane. The value of the correlation parameter is estimated (Esb 78) empirically to $c_0 = 0.78\hbar/\mathrm{MeV}$, for a fixed impact parameter of $\rho = 10$ fm in the $^{208}\mathrm{Pb} + ^{136}\mathrm{Xe}$ reaction at $E_{\mathrm{Lab}} = 1130$ MeV. Its magnitude is partially due to the excitation of spin-carrying collective modes and partially due to nucleon exchange, accounted for in this hybrid model calculation. For other impact parameters in the range of 9–11 fm, the coefficient c_0 is empirically found (Esb 78) to vary only slightly between 0.74 and $0.80\hbar/\mathrm{MeV}$.

It is of interest in this context to remember the correlation between final total kinetic energy and average initial orbital angular momentum l_i, deduced (Wol 82) from experimental energy-loss–cross-section relations for the $^{209}\mathrm{Bi} + ^{136}\mathrm{Xe}$ reaction. This experimental correlation, depicted in Figure 149, yields a correlation coefficient of about $c = 1\hbar/\mathrm{MeV}$. This result is remarkably close to the above theoretical estimates. However, while the theoretical correlation is one between final energy and final angular momentum, the experimentally deduced correlation relates presumably final energy and initial angular momentum. Hence unless a constant fraction of the initial orbital angular momentum is dissipated for all l values, a linear relation between final orbital angular momentum and final energy appears to be inconsistent with the observations. There is actually no experimental data justifying such an assumption. Experimental measurements discussed earlier in this section seem to suggest, in fact, that the fraction of dissipated angular momentum increases with increasing energy loss.

As has been emphasized before, the experimental data on energy dissipation do not provide direct evidence either for or against the model of collective doorway state excitation. This is true also for the dissipation of angular momentum in damped reactions. In Figure 214, predictions of this collective model are shown as reported by Broglia *et al.* (Bro 79e) for the reaction $^{208}Pb + ^{86}Kr$ at $E_{Lab} = 610$ MeV. The results are obtained in a

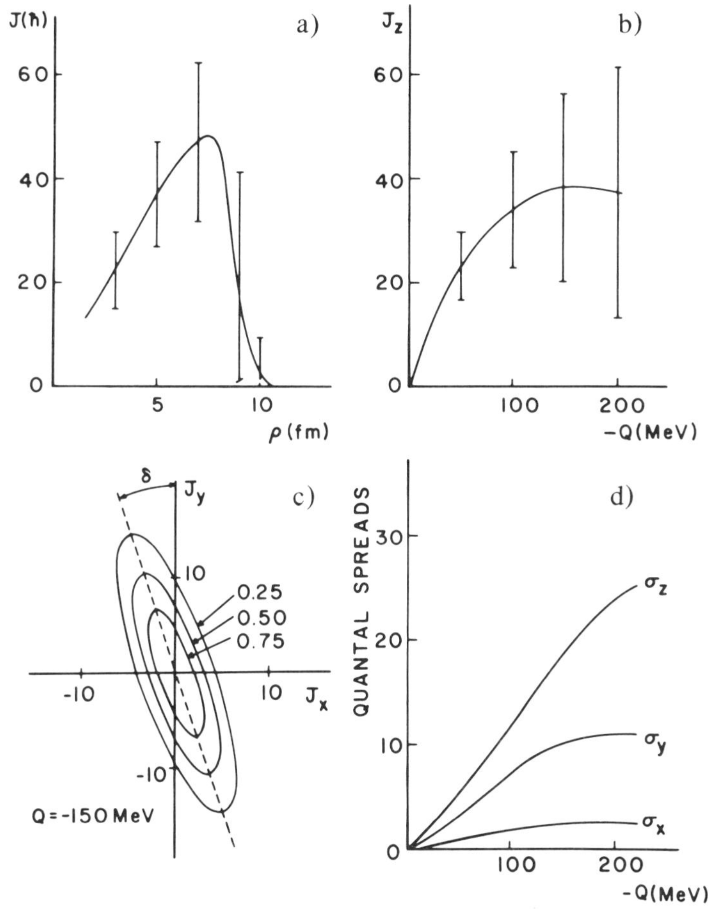

Figure 214. Spin transfer to the heavy fragment in the $^{208}Pb + ^{86}Kr$ at 610 MeV, predicted by the model of collective surface excitations. Part (a) illustrates the average fragment spin J (solid curve) and its fluctuation (vertical bars), as a function of impact parameter ρ. In part (b), the Q-value dependence of the aligned spin component J_z and its fluctuation is depicted. The r.m.s. widths of the spin distribution, due to quantal zero-point fluctuations are plotted vs. $-Q$ in part (d), where x, y, and z denote the heavy fragment recoil direction, a direction perpendicular to it in the reaction plane, and the normal to the reaction plane, respectively. In (c), the joint probability distribution $P(J_y, J_x)$ is given for a Q value of -150 MeV. From (Bro 79e).

hybrid calculation described previously, that takes some account of quantal zero-point fluctuations and of the effect of stochastic nucleon exchange. The upper left part of Figure 214 illustrates the predicted variation with impact parameter of the average spin value of Pb-like fragments (solid curve) and of the associated fluctuations (bars). In parts (b) and (d) of this figure, theoretical averages and fluctuative r.m.s. widths, respectively, of the component, J_z, perpendicular to the reaction plane are plotted vs. negative Q value or energy loss. Figure 214(c) illustrates the predicted probability distribution of spin components J_x and J_y of the heavy reaction partners, taken along the recoil direction (x) and a direction (y) perpendicular to it in the reaction plane. The dependence of the aligned component, J_z, on energy loss, observed (Dye 79b) in the similar reaction $^{209}\text{Bi} + {}^{86}\text{Kr}$ at $E_{\text{Lab}} = 610$ MeV, is in approximate agreement with the model predictions exhibited in Figure 214. However, a direct comparison between theory and experiment for this reaction is not possible, because of the spin selectivity of the experimental method discussed earlier. More stringent tests of the model are possible for reactions involving highly fissile targets, for which experimental data are already available. In addition, experiments in which spin–spin correlations of the two primary reaction fragments, as well as the joint probability distribution in various components of the heavy fragment's spin are measured, would allow a more differentiated judgement of the validity of this and other reaction theories.

A collective model of angular-momentum transfer and spin generation based on a physical idea quite different from the principles underlying the theory of Broglia *et al.* has been proposed by Moretto and collaborators (Mor 79, 80, 81c, Agu 79, McD 82). This model does not refer to a particular microscopic transmutation mechanism but assumes simply that, at the end of a damped collision, a thermodynamic equilibrium is reached for all collective and intrinsic degrees of freedom, except for the general shape of the intermediate system. In such an approach, the thermodynamic partition function determines the excitation probability for each degree of freedom. This is equivalent to a probability distribution in accord with the density of states, as described, in an approximate fashion, by Equation (3.6.3). As pointed out in Section 2.8, the equilibrium situation is approached at long times by a diffusion process if the motion along the collective degrees of freedom is overdamped.

A pictorial view of the classical tilting, wriggling and bending modes for two spherical nuclei in contact is given in Figure 215, as considered in the above work. Note that some of the modes, invoking rotation about a symmetry axis, have no quantum-mechanical analogs.

For the simple situation of statistical equilibrium between intrinsic rotation of the fragments and the orbital motion, the most probable spin values $\hbar s$ for the two reaction fragments can easily be calculated (Mor 80)

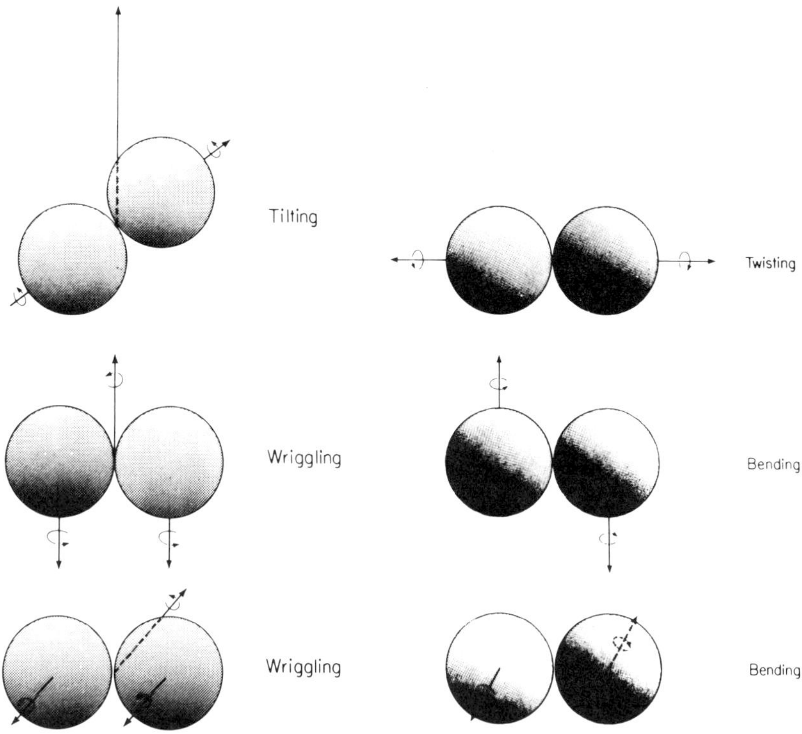

Figure 215. (a) A pictorial description of tilting and doubly degenerate wriggling modes for a two-equal-sphere model. The arrow originating at the point of tangency represents the orbital angular momentum, while the shorter arrows represent the individual fragment spins. (b) Twisting and bending modes with pairwise cancellation of fragment spins. From (Mor 81c).

for any given value $J = \hbar I$ of total angular momentum of the system. Assuming that all spins are parallel, the rotational energy can be written as (Mor 80)

$$E(s) = \frac{\hbar^2 (I - 2s)^2}{2\mu r^2} + \frac{2\hbar^2 s^2}{2\mathscr{I}} = \left(\frac{2}{\mu r^2} + \frac{1}{\mathscr{I}}\right)\hbar^2 s^2 - \frac{2\hbar^2 Is}{\mu r^2} + \frac{\hbar^2 I^2}{2\mu r^2} \tag{3.6.27a}$$

where the quantities have the usual notation. The partition function is then calculated to be

$$\mathscr{Z} \propto \int ds\, e^{-E(s)/\tau} = \left(\frac{\pi}{\hbar^2}\frac{\mu r^2 \mathscr{I}\tau}{2\mathscr{I} + \mu r^2}\right)^{1/2} \exp\left[-\frac{I^2}{2\tau(2\mathscr{I} + \mu r^2)/\hbar^2}\right] \tag{3.6.27b}$$

It can be used to calculate moments of the thermodynamic probability distribution for any observable. Moretto *et al.* (Mor 80) recover the sticking limit as the most probable situation for angular-momentum transfer at equilibrium in a collision. It is characterized by first and second moments of the spin distribution

$$\langle 2s \rangle = 2\frac{\mathscr{I}}{\mu r^2 + 2\mathscr{I}} I = \frac{2}{7} I$$

$$4\sigma_s^2 = 2\frac{\mathscr{I}\mu r^2}{\mu r^2 + 2\mathscr{I}} \tau = \frac{10}{7}\mathscr{I}\tau \tag{3.6.28}$$

respectively. As expected, the variance σ_s^2 in fragment spin varies proportionally to the temperature τ. The same simple formalism can be employed to calculate moments of the distribution over the spin moduli, $|s|$, measured in γ-ray multiplicity experiments. Even for a total angular momentum of $I = 0$, the average of $|s|$ remains nonzero and increases (Mor 80) with the square-root of the temperature. The distribution of spins carried by various modes of collective rotation (see Figure 215) in thermal equilibrium can be calculated by partitioning the energy in terms of the respective rotational energy components, employing classical moments of inertia associated with each rotational degree of freedom. A classical coupling scheme then allows the derivation of the total intrinsic spin distribution for a heavy-ion system in thermodynamic equilibrium, given (Mor 81c) for the three components s_x, s_y, and s_z in terms of a Gaussian,

$$P(\mathbf{s}) \propto \exp\left\{-\left[\frac{s_x^2}{2\sigma_x^2} + \frac{s_y^2}{2\sigma_y^2} + \frac{(s_z - \langle s_z \rangle)^2}{2\sigma_z^2}\right]\right\} \tag{3.6.29}$$

Here, the variances are proportional to the product of moment of inertia and temperature. This distribution is then folded with theoretical expressions for the anisotropy of fission fragments, or of light particles and γ rays, emitted from the primary reaction fragments, in order to obtain predictions that can be compared to experimental data. For the example of stretched quadrupole radiation, an angular distribution of γ rays of the form

$$\begin{aligned} W(\theta,\phi) &= \int d\mathbf{s}\, P(\mathbf{s}) W_{\mathrm{E}2}(\alpha(\theta,\phi)) \\ &= \tfrac{5}{4}(1-\cos^4\theta) - \tfrac{5}{2}\big[(3\sin^2\theta\cos^2\theta\cos^2\phi - \cos^4\theta)\sigma_x^2/\langle s_z^2\rangle \\ &\quad + (3\sin^2\theta\cos^2\theta\sin^2\phi - \cos^4\theta)\sigma_y^2/\langle s_z^2\rangle\big] \end{aligned} \tag{3.6.30}$$

has been derived by Moretto *et al.* (Mor 81c). In this equation, the quantity α is the angle of emission with respect to the spin direction, and θ and ϕ are the usual out-of-plane and in-plane angles.

The above formalism has been applied by McDonald *et al.* (McD 82) to the anisotropy of γ rays emitted in the reaction $^{165}Ho + ^{165}Ho$ at $E_{Lab} = 8.5$ MeV per nucleon. Figure 216 shows a comparison between experimental data and model calculations. The calculation proceeds iteratively, using experimental γ multiplicities and estimates of the numbers of stretched E2 and statistical γ transitions. A correction is employed to account for spin carried away by neutron evaporation and the associated misalignment of the fragment spin it produces. Several curves are shown in Figure 216, representing various stages of the analysis in terms of the above collective model. The dashed curve gives an interpretation of the experimental multiplicity data, assuming fully aligned fragment spins. Accepting the procedure employed to calculate fragment spins from the γ multiplicity, predicted anisotropies are obviously too large, as compared to the data. If the misalignment produced in the neutron evaporation cascade is accounted for, the dotted curve in Figure 216 results, which still largely overestimates the experimental anisotropies. It is concluded (McD 82) that significant

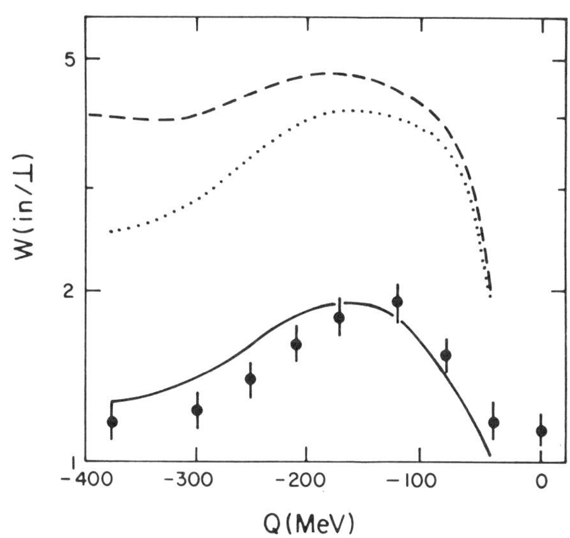

Figure 216. In- to out-of-plane anisotropy of γ rays ($E_\gamma > 0.6$ MeV) emitted in the reaction $^{165}Ho + ^{165}Ho$ at 8.5 MeV per nucleon, plotted vs. reaction Q value. The solid points are experimental data. The dashed curve represents anisotropies derived from the experimental multiplicity data, assuming complete spin alignment, whereas the dotted curve illustrates the depolarizing effect of neutron evaporation. The full calculation, taking account of equilibrium statistical spin fluctuations, results in the solid curve. From (McD 82).

primary fluctuations are present in the initial fragment spins. These can be accounted for by the model, as the agreement between data and the solid curve demonstrates. It should be noted, however, that there is considerable latitude in devising such a phenomenological analysis procedure. For example, in the calculation represented by the solid curve in Figure 216, it was assumed that the variance of the spin distribution narrows, as neutrons are evaporated from the primary reaction fragments. Calculations for compound-nuclear decay patterns, such as presented in Figure 174, suggest that this is probably the correct direction. These calculations, however, suggest also that the final spin distribution is a complicated convolution of components from different decay chains that may or may not reflect the shape of the initial distribution. Although the various curves in Figure 216 appear to demonstrate a large sensitivity of the data to the initial fragment spin distribution, alternative model descriptions are not excluded from such a comparison. Again, the situation suggests a more systematical study of reactions in which parameters of important consequences for model predictions are varied over a significant range. In view of the discussion of other reaction observables, given in previous sections, one may expect contributions from processes modeled in the above theory of Moretto *et al.* only in the domain of very high energy losses. For small or intermediate energy losses, there is no independent information that the reaction proceeds slowly enough that collective shape degrees of freedom have time to attain a thermal equilibrium. Even if such modes could be excited during a damped reaction, it is not self-evident that their effect could be observed in the spin distributions of the asymptotic fragments. There is no mechanism known that ensures both adiabatic motion along collective coordinates allowing the establishment of a thermal equilibrium, including these collective modes, and the freezing-in of the spin distribution of any point of the system trajectory except at the instant of breakup. At this point, however, the collective modes of excitation may have dampened out. A more realistic treatment of these modes would have to include consideration of the damping, inertial, and driving forces, in a dynamical model calculation.

Similar reservations may be held with respect to the model proposed by Friedman (Fri 81), already referred to in Section 2.7. This approach predicts angular-momentum transfer as given by the conditions of statistical equilibrium. It is not based on a particular microscopic reaction mechanism and postulates simply a population distribution of intrinsic states, according to the combined level density of both fragments to be achieved in any damped collision. This model predicts for the relation between the dissipated angular momentum, $\Delta l = l_i - l_f$, and the acquired nuclear temperature τ,

$$\langle \Delta l \rangle = \frac{2}{\sqrt{\pi}\,\hbar}\left[2(\mathscr{I}_1 + \mathscr{I}_2)\tau\right]^{1/2} \tag{3.6.31a}$$

A random spin component, perpendicular to the orbital angular momentum,

$$\langle \Delta \rangle = \frac{\pi}{4}\left[\frac{(\mathscr{I}_1/\mathscr{I}_2)^{1/2}}{1+\mathscr{I}_1+\mathscr{I}_2}\right]\langle \Delta l \rangle \tag{3.6.31b}$$

follows also in the context of the model, where the quantities $\mathscr{I}_i$ are the moments of inertia of the fragments. The total final spins of each fragment are obtained from adding aligned and random components quadratically, with a partition of the aligned part, $\langle \Delta l \rangle$, according to the moments of inertia, calculated for spherical fragments.

In Figure 217, predictions of the model of Friedman (Fri 81) are compared to experimental data on first moments, J_U, of U-like fragments produced in the reaction $^{238}U + ^{86}Kr$ at 8.5 MeV per nucleon (Pui 79). As seen from this comparison, there is reasonable agreement between calculations and experiment, for the whole range of energy losses studied. The model is also able to account for the large values of spin alignment observed in this reaction. Fits of a quality similar to that shown in Figure 217 are also accomplished with the model for γ multiplicity data on the reaction

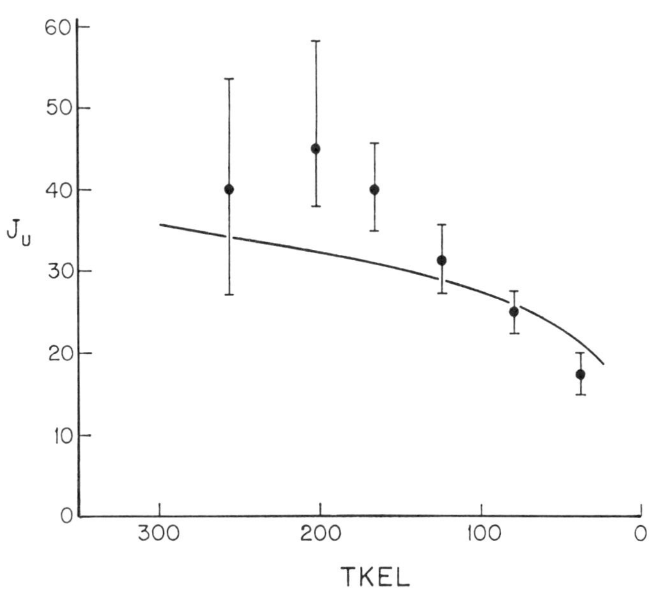

Figure 217. Comparison of experimental (Pui 79) average spin values transferred to U-like fragments in the reaction $^{238}U + ^{86}Kr$ at 8.5 MeV per nucleon to predictions (solid curve) by the equilibrium model of Friedman. From (Fri 81).

$^{166}Er + ^{86}Kr$ discussed earlier, depending somewhat on the adopted relation between spin and γ multiplicity.

Despite the good reproduction of the U + Kr data achieved with the equilibrium model of Friedman (Fri 81), it generally does not provide a consistent picture of the reaction mechanism. For example, dynamical considerations are not included in the model, where all l waves can contribute to all energy losses. Whereas experimentally the cross section vanishes at the maximum energy losses possible in a damped reaction (see Section 3.5), the model predicts maximum cross sections for these highest energy losses, similar to what is observed in long-lifetime fusion–fission reactions. As delineated in previous sections, distributions of other reaction variables indicate that statistical equilibrium is not attained in damped reactions, except for the highest energy losses.

The above statistical models of angular-momentum transfer in damped reactions assume essentially that the extreme long-time limit of stochastic transport processes be reached, in which there are no significant influences of the collision dynamics. Consequently, these models connect only those experimentally measureable parameters that enter the adopted expressions of the density of states, yielding predictions for only very few correlations that can be tested by experiment. Furthermore, a microscopic mechanism, by which some state of thermodynamic equilibrium could be reached for essentially all l waves, is not considered in these models. There are, obviously, a number of possibilities to define such an equilibrium state, depending on the degrees of freedom assumed to participate. The examples discussed above indicate that the experimental fragment spin distributions alone can be described both by models that include or neglect the excitation of spin-carrying collective modes. Apparently, the thermodynamic partition function [compare Equation (3.6.27)] can be modeled in different ways, leading to comparable results for the fragment spin distributions. Since only isolated cases have been calculated in these models, nothing can be inferred with respect to the validity of the somewhat different general trends predicted by each model. It is presently an open question whether or not the models would exhibit significant differences when applied more rigorously also to additional observables such as the angular distribution of the damped reaction fragments. From the discussion given early in this section and in Section 3.3, one might expect theoretical angular distributions that are similar to those of fusion–fission fragments, in a situation where collective shape degrees of freedom of the intermediate double-nuclear system are at thermal equilibrium. Such angular distributions are, in general, not observed in experiments, as illustrated in Section 3.3.

Less drastic assumptions are usually made in stochastic transport models that also consider possible microscopic mechanisms as carriers of mass, charge, energy, and spin. The performance of these theories in the reproduction of experimental data has already been exhibited repeatedly in previous sections. In the following, their predictions with respect to angular-momentum transmutation and fragment spin generation will be compared to the observations.

The transport theory developed by Agassi, Ko, and Weidenmüller (cf. Section 2.7) has been applied (Aga 78b, 79b) to calculate angular-momentum transfer in ^{40}Ar, ^{84}Kr, and ^{136}Xe-induced reactions. This theory results in a generalized Fokker–Planck equation (2.7.10). The actual calculations are performed with a simplified version of the model that is, in principle, capable of describing statistical and quantal fluctuations in a non-Markovian transport process. The simplifications include the neglect of certain correlations in the gain and loss terms in Equation (2.7.10), describing the quantum-statistical fluctuations, and of their off-shell behavior. In addition, a Gaussian form is adopted for the shape of the joint probability functions P_s and P_t. On the other hand, charge exchange between projectile and target is included in an approximate fashion, by introducing a discrete variable $x = (Z_1 - Z_2)/(Z_1 + Z_2)$ and by extending the statistical assumptions on the form factors in an analogous way also to their dependence on x. Averages and dispersions of observables are then defined in terms of a joint probability function, expressed in spherical coordinates (r, θ, ϕ), and depending on these variables, the associated momenta p_r, p_θ, p_ϕ, as well as on the charge asymmetry x. Consequently, the multidifferential cross section is a multivariate Gaussian in coordinates and momenta. Correlations are included in the model for radial and angular coordinates, as well as for the charge diffusion coefficient, which depends on the density overlap between the reaction partners and the relative momentum. A set of coupled differential equations is solved for average values and half-variances, Δ, of the probability distribution.

Results of the model calculation (Aga 79b) for ^{208}Pb + ^{84}Kr at $E_{\mathrm{Lab}} =$ 719 MeV are depicted in Figure 218. Here, average values of deflection angle, θ, energy loss, ΔE, and dissipated angular momentum, ΔL, are plotted vs. impact parameter b. The resulting energy–angle relation can be compared to the experimental Wilczyński diagram shown for the similar reaction ^{209}Bi + ^{84}Kr in Figure 161. Obviously, only a qualitative description of the data is possible with the model, for relatively large impact parameters. For smaller values of b, predicted reaction angles are too large. This is partly a consequence of the repulsive core of the proximity potential adopted in the model. As can be seen in Figure 218, the largest energy and angular-momentum losses occur for impact parameters close to the second (nuclear) rainbow. In this region, the model predicts an almost complete dissipation of orbital angular momentum, with about 10 units remaining in orbital motion. These values appear much larger than experimentally observed (cf. Figure 212) in similar reactions, as discussed previously. This

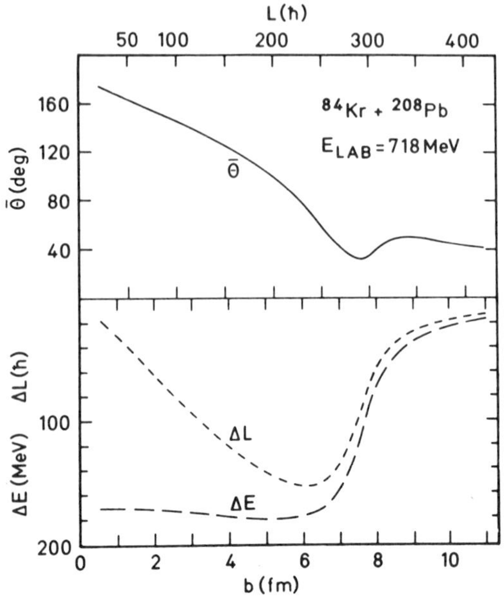

Figure 218. Mean deflection angle, $\bar{\theta}$, energy loss, ΔE, and angular-momentum loss, Δl, vs. impact parameter, b, predicted by the transport model of Agassi *et al.* for the reaction $^{208}Pb+^{84}Kr$ at E_{Lab} = 718 MeV. From (Aga 79b).

discrepancy could, in part, be due (Aga 79b) to the large angular-momentum cutoff parameters used in the model.

Fluctuations in transferred angular momentum are also predicted to increase with decreasing impact parameter. For the $^{208}Pb+^{84}Kr$ reaction at $E_{Lab} = 718$ MeV, the half-variance $\Delta_{p_\theta p_\theta}$ is expected, according to the model (Aga 79b), to decrease slightly again for impact parameters smaller than those in the rainbow region, where the average angular-momentum loss ΔL decreases also (cf. Figure 218). This almost saturation-like behavior is shown in Figure 219, where $\Delta_{p_\theta p_\theta}$ is plotted for this reaction vs. impact parameter, along with the values of other half-variances. The experimental data of Figure 212 show a qualitatively similar dependence on l as suggested by the theoretical variances alone, but the combination of average dissipated angular momentum and its fluctuative width, that is experimentally measurable, does not comply with the observations. It is interesting to note that the transport model of Agassi *et al.* (Aga 79b) predicts a drastically different dependence of angular momentum dissipation on impact parameter, when the bombarding energy is decreased from 718 to 494 MeV. In the low-energy case, theoretical average dissipated angular momentum and its variance both increase monotonically with decreasing impact parameter. Close to the orbiting impact parameter, this transport model predicts drastic increases both of average value and variance in transferred angular momentum. It would be interesting to test this predicted general trend in a series of experiments at different bombarding energies.

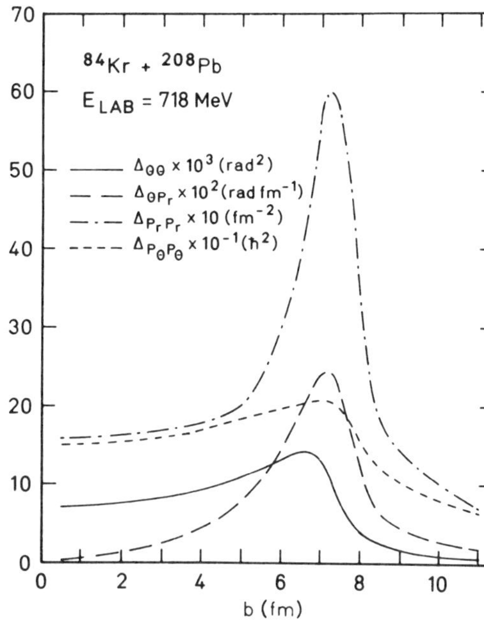

Figure 219. Theoretical half-variances plotted vs. impact parameter b, for the $^{208}Pb + ^{84}Kr$ reaction at $E_{Lab} = 718$ MeV, as predicted by the transport model of Agassi *et al.* The half-variance $\Delta_{p_\theta p_\theta}$ of the distribution in transferred angular momentum is represented by the short-dashed curve. From (Aga 79b).

Presently, it is not possible to conclude whether or not the model is able to reproduce at least the overall features of angular-momentum transmutation in damped reactions.

In contrast to the above microscopic transport theory, the linear response model of Hofmann, Siemens, and collaborators, outlined in Section 2.7, merely generates dispersive features of a damped reaction from knowledge of conservative and dissipative forces derived from other sources. Of course, such model forces can be processed in the formalism only if certain conditions are met by the system of two interacting heavy ions, as explained in Section 2.7. In practice, these conditions are essentially those of intrinsic thermal equilibrium. Because of its versatility, this transport model has been employed in a variety of settings, employing different friction and diffusion models, as described in previous sections. However, there are only a few attempts to describe angular-momentum dissipation within this framework.

Berlanger *et al.* (Ber 78f) apply the linear response model to calculate multidifferential cross sections for the 280-MeV $^{58}Ni + ^{40}Ar$ and 365-MeV $^{197}Au + ^{63}Cu$ reactions, using the conservative and dissipative forces suggested by Siwek-Wilczyńska *et al.* (Siw 76). This model assumes a phenomenological potential simulating deformation effects in the exit channel, a two-body friction force of the type given by Equation (2.6.9), and statistically uncorrelated motion along the mass asymmetry degree of freedom. By construction, the frictional force is isotropic, and hence, angular-momentum dissipation in this model contributes significantly to the total kinetic energy loss. The cross section as a function of different variables can be written, in a form similar to that of Equation (3.5.19), in terms of a reduced density distribution. As discussed in previous sections, the model is capable of describing general trends in mass, energy, and angular distributions of fragments from several reactions but fails in important details.

Results on average angular momenta $\langle \Delta l \rangle$ dissipated in the 365-MeV $^{197}Au + ^{63}Cu$ reaction, as predicted by this transport model, are shown in Figure 220 (Lef 79b) as a function of the fragment mass ratio M_H/M_L, for three reaction angles. Curves $b(\theta_{Lab} = 80°)$ and $c(\theta_{Lab} = 100°)$ can be directly compared to the γ-multiplicity data depicted in Figure 199. As can be seen from the comparison of the two figures, the general trend observed in the data of a γ multiplicity increasing with decreasing mass ratio, or

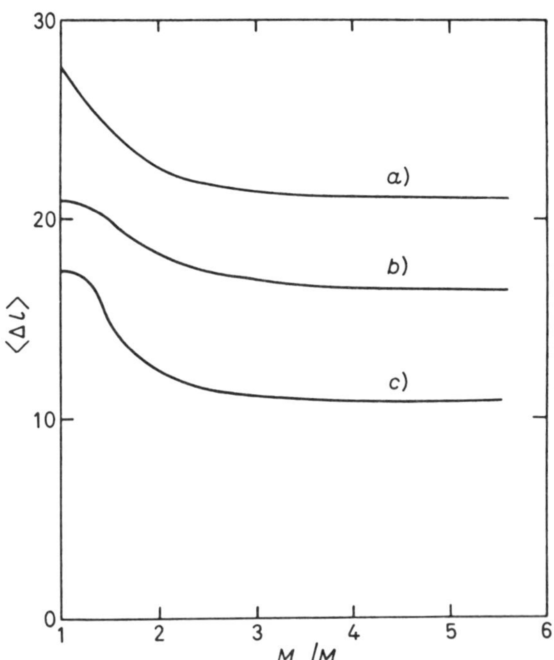

Figure 220. Average angular momentum, $\langle \Delta l \rangle$, transferred in the reaction $^{197}Au + ^{63}Cu$ at $E_{Lab} = 365$ MeV, vs. fragment mass ratio M_H/M_L, as predicted by the linear-response theory and two-body friction forces. The curves (a)–(c) represent calculations performed for the reaction angles $\theta_{Lab} = 50°(\theta_{c.m.} = 70°)$ 80°(100°), and 100°(130°), respectively. From (Lef 79b).

increasing symmetry, is reproduced by the calculations. Also, the predicted increase of the transferred angular momentum at backward angles is reflected in a corresponding behavior of the experimental γ multiplicities. However, a detailed comparison reveals that the amounts of dissipated angular momentum predicted by the model are systematically too small and the theoretical angular dependence too strong to be compatible with the values of $\langle \Delta l \rangle_{\text{exp}} = 25$, 25, and $20\hbar$, deduced experimentally (Ger 79) for $\theta_{\text{Lab}} = 50°$, 80°, and 100°, respectively. Including statistical fluctuations in the model would increase, to some extent, the predicted average magnitude of the dissipated angular momentum, the quantity to which γ-multiplicity measurements are sensitive. This, however, is not expected to remove the large discrepancy between theory and experiment. It is presently impossible to identify the effect most responsible for the disagreement. A plausible cause, however, may be seen in an inadequacy of the isotropic two-body friction force adopted in the model, which may be too weak. This is also suggested by the incapability of the model to account for the observed energy–angle cross section relations of damped events in the $^{197}\text{Au} + {}^{63}\text{Cu}$ reaction (Ber 78f). Unfortunately, experiments spanning a range of bombarding energies, as well as corresponding calculations, are lacking, which could be expected to elucidate the origin of the discrepancy between experiment and theory.

Early attempts to understand angular momentum dissipation in damped reactions as a diffusive mechanism are due to Wolschin and Nörenberg (Wol 78, 78b, 78c) who applied the equilibrium-transport theory of Nörenberg and collaborators, discussed in Sections 2.7 and 2.8. Using transport coefficients [cf. Equation (2.8.14)], evaluated within a single-particle model, mean values and variances of the spin distribution of reaction fragments are deduced. From the Fokker–Planck equation (2.7.14) governing angular-momentum dissipation, according to the model, one expects (Wol 78) an average z component of the spin of both fragments

$$\langle M \rangle = \frac{\mathscr{I}_1 + \mathscr{I}_2}{\mathscr{I}_{\text{rig}}} l \left[1 - \exp(-t/t_l)\right] \tag{3.6.32a}$$

Here, l is the initial orbital angular momentum, $\mathscr{I}_1$ and $\mathscr{I}_2$ are the moments of inertia of the two reaction partners, and $\mathscr{I}_{\text{rig}}$ is the total rigid-body moment of inertia of the intermediate dinuclear system. The quantity t_l is the spin relaxation time. Such a functional dependence is expected from the discussion in Section 2.8. Similarly, the model predicts

$$\sigma_M^2 = D_{MM} t_l \left[1 - \exp(-2t/t_l)\right] \tag{3.6.32b}$$

for the variance in the distribution of z components of the spin, where D_{MM}

is the angular-momentum diffusion coefficient. The spin relaxation time that results (Wol 78) from this diffusion model is

$$t_{\mathrm{rel}} = (\mathscr{I}_1 + \mathscr{I}_2)\frac{\mathscr{I}_{\mathrm{rel}}}{\mathscr{I}_{\mathrm{rig}}}\left(\frac{E_l^*}{g_1 + g_2}\right)^{1/2}\frac{1}{D_{MM}} \tag{3.6.32c}$$

In Equation (3.6.32c), the quantity $\mathscr{I}_{\mathrm{rel}}$ is the moment of inertia for orbital motion, E_l^* denotes the local excitation energy [compare Equation (2.8.10)], calculated under the assumption that the radial kinetic energy is completely dissipated at the interaction barrier. The effect of the single-particle model underlying the transport theory is seen in the appearance of the single-particle level density parameters g_1 and g_2 in Equation (3.6.32c), which also enter the angular-momentum diffusion coefficient D_{MM} [Equation (2.8.14)]. For sufficiently long interaction times $t_{\mathrm{int}} > t_l$, the average fragment spin approaches the sticking limit. The essential difference with respect to classical friction models consists in the predictions of large variances in the Gaussian spin distributions made by the transport model, which also includes stochastic mass and charge exchange between the reaction partners. Hence, the dependence of the total spin $\langle J \rangle = \langle |\mathbf{J}_1| + |\mathbf{J}_2| \rangle$ can be calculated as function of the charge, Z_1, of one of the fragments, according to

$$\langle J(Z_1)\rangle = 2\pi\lambda\!\!\!^{-2}\left(\frac{d\sigma}{dZ_1}\right)^{-1}\int dl\, l\langle J(Z_1, t_{\mathrm{int}}(l))\rangle P(Z_1, t_{\mathrm{int}}(l)) \tag{3.6.33}$$

In the evaluation of $\langle J \rangle$, it is assumed (Wol 78) that the z components and the spin variances divide between the fragments according to their moments of inertia, and that variances for all three spatial components are equal. In Equation (3.6.33), $t_{\mathrm{int}}(l)$ denotes the l-dependent interaction time, adjusted to fit the experimental fragment Z distribution with the theoretical expression for the probability $P(Z_1, t)$. As explained in Section 3.4, such a fit requires an adjustment of the theoretical mass transport coefficients, in violation of the classical dissipation–fluctuation theorem.

In Figure 221, results of the model calculation are compared to γ-multiplicity data obtained by Olmi *et al.* (Olm 78b). Plotted in this figure are experimental values of γ multiplicity and deduced total spin vs. the Z value of reaction fragments produced in the reaction $^{166}\mathrm{Er} + {}^{86}\mathrm{Kr}$ at $E_{\mathrm{Lab}} =$ 5.99 MeV per nucleon. The data are summed over all energy losses larger than 20 MeV. Two theoretical curves are shown in Figure 221, demonstrating the effect of including statistical fluctuations in the calculation of the first moments of the total spin distributions. The dashed curve, obtained by

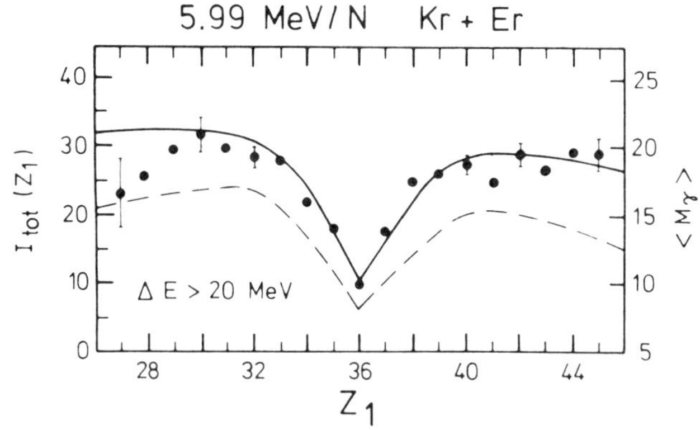

Figure 221. Experimental average γ multiplicities and deduced total spin values for fragments produced in the damped reaction ^{166}Er+^{86}Kr at 5.99 MeV per nucleon. The data, averaged over all energy losses larger than 20 MeV, are plotted vs. atomic number of one of the final fragments. The curves represent calculations with the diffusion model of Nörenberg and collaborators, including (solid curve) or neglecting (dashed curve) statistical fluctuations. From (Wol 78).

neglecting these fluctuations, obviously does not reproduce the data as well as is possible when fluctuations are taken into account (solid curve).

The general feature of data and calculations displayed in Figure 221 are understood in terms of the varying contributions of different fragmentations to the cross section with varying energy loss. For example, fragmentations close to the initial projectile–target combination are largely concentrated in the region of small or medium energy losses. The decrease of the total angular momentum for Z_1 values close to that of the projectile can then be interpreted as a reflection of the circumstance that small energy losses correspond to small angular-momentum losses, although the initial orbital angular momenta are largest for such events. Similarly, an increase with decreasing value of Z_1 of the average spin, suggested by the variation of Equation (3.6.23a) with the moments of inertia, is offset by decreasing values of initial l value sampled by more asymmetric fragmentations. This behavior is more clearly illustrated in correlations of total fragment spin with total kinetic energy loss. In fact, Wolschin and Nörenberg (Wol 78) are able to give a good account of the saturation behavior of the γ multiplicity measured (Olm 78b) for the ^{166}Er+^{86}Kr reaction (cf. Figure 200). This is possible because the decrease of the aligned component $\langle M \rangle$ with increasing energy loss or decreasing initial l value is counteracted by thermal spin fluctuations that build up at large energy losses.

In contrast to the above model that relates the individual average spins and associated variances of the two partners in a damped reaction to each other as suggested by a classical friction model, Li and Wolschin (Li 82) have treated these quantities as independent variables in a diffusion model, otherwise similar to the one discussed above. Here, the angular-momentum diffusion tensor is a 6×6 matrix with components $D^{ij}_{kk'}$, corresponding to the two fragments, i.e., k, $k'=1,2$, and three spatial directions, i, $j=1,2,4$. The earlier restriction (Wol 78) to equal fluctuations of the spin in all directions perpendicular to the initial orbital angular momentum is also relaxed in the extended version of the model (Li 82). However, spin–spin correlations on a microscopic level reflected by the mixed diffusion coefficients are neglected. Only average, macroscopic correlations imposed on the spin transport by the centrifugal part of the driving potential

$$U_l = \frac{J_1^2}{2\mathscr{I}_1} + \frac{J_2^2}{2\mathscr{I}_2} + \frac{(\mathbf{l} - \mathbf{J}_1 - \mathbf{J}_2)^2}{2\mathscr{I}_{\text{rel}}} \tag{3.6.34}$$

are retained in the model. A set of coupled linear differential equations for the spin variances with respect to time is then solved, where the connection between reaction time, l values, and reaction angle is modeled by a parametrized deflection function such as shown in Figure 66.

Theoretical dependences on the initial orbital angular momentum l of the mean absolute values of fragment spins and alignment are depicted in Figure 222 for the reaction $^{238}U+^{208}Pb$ at 8.5 MeV per nucleon, as

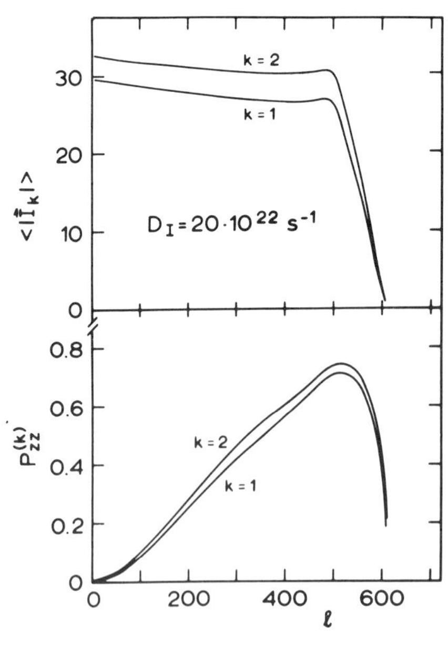

Figure 222. Angular-momentum dependence of average magnitudes and alignment of spins acquired by the fragments $k=1$ and 2 in the reaction $^{238}U+^{208}Pb$ at 8.5 MeV per nucleon, as predicted by the transport model of Li and Wolschin. The quantity D_I is an overall strength factor of the angular-momentum diffusion coefficients. From (Li 82).

predicted by the model. Although, according to the calculations, the mean values $\langle I_1 \rangle$ and $\langle I_2 \rangle$ of the fragment spins decrease with decreasing l value, after an initial rise for large l values, the absolute values $\langle |I_1| \rangle$ and $\langle |I_2| \rangle$ show the saturation behavior observed generally in experimental data. On the other hand, the alignment parameter P_{zz}, defined in Equation (3.6.2) decreases for smaller l values, after an initial rise at large initial angular momenta. This latter feature is not observed in experimental data on the $^{238}U + {}^{86}Kr$ reaction, illustrated in Figure 210, where large values of the alignment parameter are seen to persist at large energy losses, i.e., in the plateau region of the correlation $|\mathbf{I}(E_{loss})|$.

Comparisons of the model calculations by Li and Wolschin (Li 82) are shown in the middle part of Figure 223 for the reaction $^{154}Sm + {}^{86}Kr$ measured by Christensen *et al.* (Chr 80). The top and bottom parts of this figure depict the theoretical energy spectra and the variation of the spin variances with energy loss ΔE, respectively. As is inferred from Figure 223, the model gives a good account of the data in the strongly damped region; however, it overestimates the increase of the average total spin with energy loss, for low and intermediate values of ΔE. It is of significance for a judgement of the performance of the model that, in order to obtain such a description of the data, the overall magnitudes of the total spin diffusion coefficients have to be adjusted from values suggested by the microscopic theory (see Section 2.8) and to be made bombarding-energy dependent. The fit values reported by Li and Wolschim (Li 82) for these diffusion strengths are $D_I = 7 \times 10^{22}$ s^{-1} and 10×10^{22} s^{-1}, for the ^{154}Sm reaction at 5.7 and 7 MeV per nucleon, respectively. This shortcoming of the diffusion model is a general feature in violation of the Einstein relation required by the theory, noticed also for mass and charge exchange, as discussed in an earlier section.

A simple model for angular momentum transmutation in damped reactions has been considered by Vandenbosch (Van 79b), incorporating both the effects of relative motion of the reaction partners and of the intrinsic motion of individual nucleons that are assumed to generate fragment spins in an exchange mechanism. Hence, the underlying physical picture in this model is that of one-body, window-type nuclear friction discussed in Section 2.6. It is also reminiscent of optimum-Q-value considerations made in few-nucleon transfer reactions, such as leading to the relations of Equation (3.6.14) between energy loss and angular-momentum dissipation derived by Brink (Bri 72) and others. The essential difference between these early kinematical considerations and the one-body transport model lies in the fact that, in damped reactions, considerable mass overlaps between the reaction partners can be reached, such that the velocity direction of a transferred nucleon is not well defined, but reflects to a certain extent the intrinsic Fermi velocity distribution of the nucleons. In

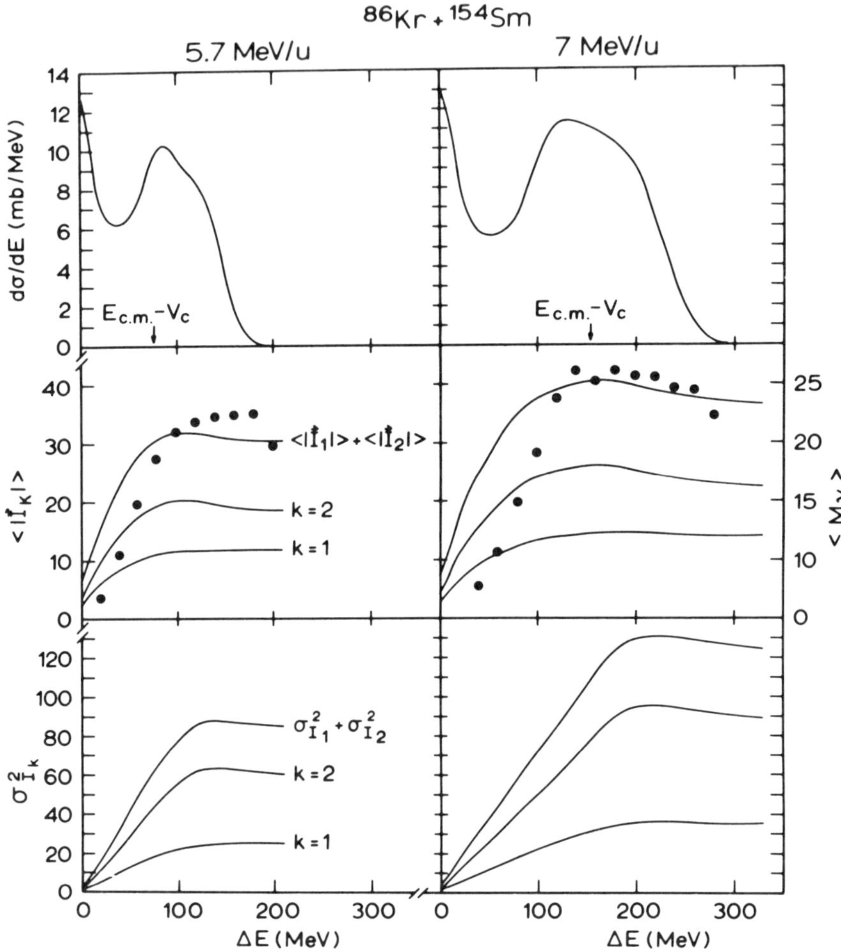

Figure 223. Energy spectra, mean absolute value and variance of the dissipated angular momentum in the reaction $^{154}Sm + ^{86}Kr$ at two bombarding energies. Experimental data on the variation of the mean spin with energy loss, ΔE, are represented by solid dots. From (Li 82).

essence, nucleon exchange processes then exert random torques on both reaction fragments, yielding random final fragment spins. Nevertheless, the average velocity mismatch of the Fermi velocity distributions of the two reaction fragments, due to their relative motion, leads to a component of the spins generated within the fragments that is aligned perpendicular to the reaction plane. Figure 224 gives a pictorial illustration of sample directions of spins acquired by a fragment in the exchange of one or ten nucleons. This figure (Van 79b) already explains qualitatively the dependence of the alignment parameter P_{zz} on energy loss (cf. Figure 209) observed experimentally. Whereas in few-nucleon exchange processes the randomly oriented spin, corresponding to the Fermi velocity of a transferred nucleon, masks the aligned spin component due to the relative nuclear velocity, the coherent addition of aligned components in many-nucleon exchanges can dominate the random spin that does not grow as fast.

In the simple picture suggested by Vandenbosch (Van 79b), nucleons are exchanged in the equatorial plane of two touching nuclei, at a radial distance R from the center of one of the fragments. The aligned component of this fragment's spin then increases with the number N of nucleon exchange events as (Van 79b)

$$j_a = NRm\langle u^2\rangle^{1/2} = NR\left(2m^2\overline{E}/\mu\right)^{1/2} \tag{3.6.35}$$

In Equation (3.6.35), u is the relative velocity of the two interacting heavy ions, μ is the reduced mass of the dinuclear system, and $\overline{E}$ denotes the average relative kinetic energy. The r.m.s. value of the spin component due

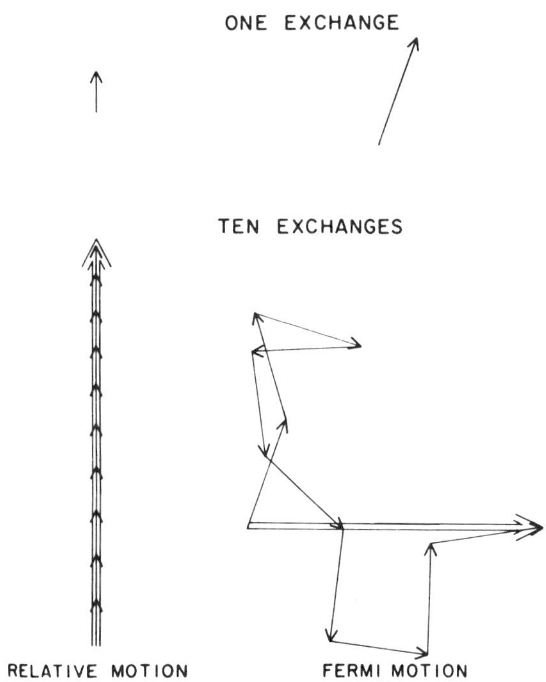

Figure 224. Illustration of the relative magnitudes and alignments of the components of fragment spins, due to the velocity of relative motion (left) and to the Fermi velocity of nucleons exchanged between the fragments. Thin arrows correspond to individual transfers, the double arrows indicate the spin resultant. From (Van 79b).

to the Fermi velocity of an exchanged nucleon, assumed to lie in the plane perpendicular to the line connecting the nuclear centers, is given by (Van 79b)

$$j_F(N=1) = R\left(\tfrac{2}{3}2mT_F\right)^{1/2} \tag{3.6.36}$$

where T_F is the Fermi kinetic energy. This random component grows with the square root of the number of nucleons exchanged, $N^{1/2}$, such that the mean square value, J, of the total spin acquired by a fragment in N exchanges has an average projection on the reaction normal of

$$\langle M^2\rangle = \langle J^2\rangle - \tfrac{1}{2}j_F^2(N) \tag{3.6.37}$$

From these relations, Vandenbosch (Van 79b) derives for the spin alignment parameter

$$P_{zz} = \frac{3}{2}\frac{\langle M^2\rangle}{\langle J^2\rangle} - \frac{3}{4}\frac{J_F'^2}{\langle J^2\rangle} \tag{3.6.38}$$

Obviously, for a comparison of this model to experiment, a connection has to be made between the number N of exchanged nucleons with reaction observables. The relation $N=(A/Z)\sigma_Z^2$ between N and the variance, σ_Z^2, of the fragment charge distribution is adopted for this purpose, in the above work, where A and Z are mass and atomic numbers, respectively, of the composite system. The uncertainties associated with such an identification have been outlined in Section 3.4. However, in a correlation between dissipated energy and dissipated angular momentum, such uncertainties are expected to cancel to some extent (cf. Section 3.4).

Results of calculations reported by Vandenbosch (Van 79b) for the reaction $^{238}U + {}^{86}Kr$ at $E_{Lab} = 730$ MeV are compared to experimental data in Figure 225. The data (Pui 79) have already been shown in Figure 209. It can be seen from Figure 225 that the above simple nucleon exchange model provides a remarkably good description of the Q-value dependence of both r.m.s. spin, J, and alignment parameter, P_{zz}. Model calculations represented by solid curves in Figure 225 are only shown for small and intermediate negative Q values, associated with peripheral collisions, as for more deeply interpenetrating collisions, idealizing model assumptions are expected (Van 79b) to break down. The dashed curves in Figure 225 are obtained, if a correction is made to the calculated relative velocity for the intrinsic rotation of the two fragments. Other approximations made in the simplified calculation are discussed by Vandenbosch (Van 79b) and found to be relatively unimportant. Hence, the agreement between experiment and

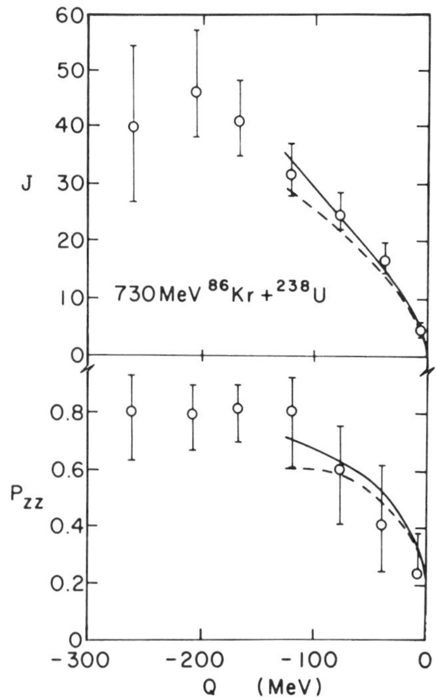

Figure 225. First moments of the magnitude, J, and the alignment, P_{zz}, of the spin distribution of U-like fragments, produced in the reaction $^{238}U + ^{86}Kr$ at $E_{Lab} = 730$ MeV. The quantity Q is the reaction Q value. The curves represent calculations based on a simple nucleon exchange model for angular-momentum dissipation. The dashed curve differs from the solid curve in that a correction is made for the intrinsic rotation of the fragments. From (Van 79b).

calculations observed in Figure 225 appears to add support to a basic damped-reaction mechanism, in which transport phenomena are mediated by the exchange of independent nucleons between the collision partners. However, further evidence is needed to reach a definite conclusion in this regard. As has been shown in previous sections, some of the experimental correlations between the widths of fragment Z (and A) distributions and the total kinetic energy loss can qualitatively be understood also in terms of similarly simple nucleon exchange models. However, deviations from the experimental behavior became obvious, once a more complete set of data became available. In order to test a frictional model more stringently, it appears necessary to confront it with data obtained for a range of bombarding energies, and for a variety of different systems. There is clearly a need for experiments filling in the gap that is now present.

A more general theory for transport induced by nucleon exchange has been developed by Randrup. Its performance in view of experimental data has been exhibited repeatedly in previous sections. Recently, the scope of this model has been enlarged (Ran 81b) to include spin transport in damped reactions. Elements of the theory discussed in Sections 2.5–2.8 indicate how macroscopic probability distributions in any experimental observable can be inferred, within the framework of the model, from the elementary transfer probabilities associated with the exchange of individual nucleons. In its essential microscopic part, the model makes predictions of transport coefficients applicable to two interacting nuclei at individual thermodynamic equilibrium, retaining important dynamical features neglected in other approaches. This latter feature manifests itself in the appearance of a dynamical quantity τ^*, an "effective temperature" (cf. Figure 37), in the generalized Einstein relations of Equation (2.8.21). The microscopic transport model can then be combined with a (generalized) Fokker–Planck equation, for example, those given by Equations (2.7.14) and (2.7.18) describing the time evolution of the average system trajectory and its dispersions, for a given set of nuclear shapes and associated potential energies. In practical applications of the model, dinuclear shapes of spherical fragments, connected by a cylindrical neck, as well as liquid-drop and proximity forces are adopted. The dynamical driving forces for the given coordinate are obtained from the partial derivatives of the Hamiltonians or of the Langrangian with respect to the conjugate momenta. As discussed in Section 2.8, the nonlinearities in the driving potential impose constrictions on the growth of dispersions, which approach, for long reaction times, saturation values characterizing a dynamical equilibrium. The latter, however, differs from the thermodynamic equilibrium to an extent determined by the degree to which the kinetic energy of relative motion is damped in a collision.

In the case of angular-momentum transmutation, the model spin diffusion tensor can be written as (Ran 81b)

$$\mathsf{D}_{J_iJ_i} = N\tau^*\left(a_i^2\mathsf{P}_\perp + \bar{c}^2\mathsf{I}\right), \qquad i = P, T$$

$$\mathsf{D}_{J_iJ_j} = N\tau^*\left(a_ia_j\mathsf{P}_\perp - \bar{c}^2\mathsf{I}\right), \qquad i \neq j = P, T \tag{3.6.39}$$

Here, N is the bulk one-way flux of nucleons, discussed in Section 2.8, $\mathsf{P}_\perp$ denotes a tensor projecting onto a plane perpendicular to the axis joining the nuclear centers, and I is the identity tensor. The quantities a_P and a_T entering Equation (3.6.39) are the imaginary "lever arms" associated with the torques exerted on target-like and projectile-like fragments, respectively, in the exchange of a nucleon. Their sum equals the center separation distance, i.e., $a_P + a_T = r$. The quantity $\bar{c}$ denotes an average displacement from the symmetry axis of the exchanged nucleons, as they pass the interface between the two interacting fragments. From Equation (3.6.39), the drift coefficients for the fragment spins J_P and J_T can be derived employing the generalized Einstein relations [cf. Equation (2.8.21)]. It is an attractive opportunity offered by the model that it relates mean values and fluctuations of various observables to one another, through the common underlying mechanism of nucleon exchange.

To give a detailed account of the somewhat involved formalism used (Ran 81b) to describe the generation of fragment spins in the rotating dinuclear system would exceed the scope of the present discussion. Hence, only the results of one calculation, performed by Randrup (Ran 81b) for the 610-MeV $^{209}Bi+^{86}Kr$ reaction, are illustrated in the following. Figure 226 displays the theoretical dependence of the mean spin, $\langle|\mathbf{S}_B|\rangle$, transferred to the target-like fragment, as well as of its component normal to the reaction plane, as a function of the initial orbital angular momentum, J. The dashed straight lines indicate the values corresponding to spin equilibrium (sticking condition) of two fragments touching with zero surface separation ($s=0$) or at a larger distance where, in the exit channel, the neck ruptures ($c=0$) according to the model. The model predictions for the dissipation of orbital angular momentum suggest that the sticking condition is not reached for high and intermediate l waves. For lower values of angular momentum, corresponding to fully damped events, the sticking condition is approached until the narrowing of the neck in the separation phase prevents a full spin

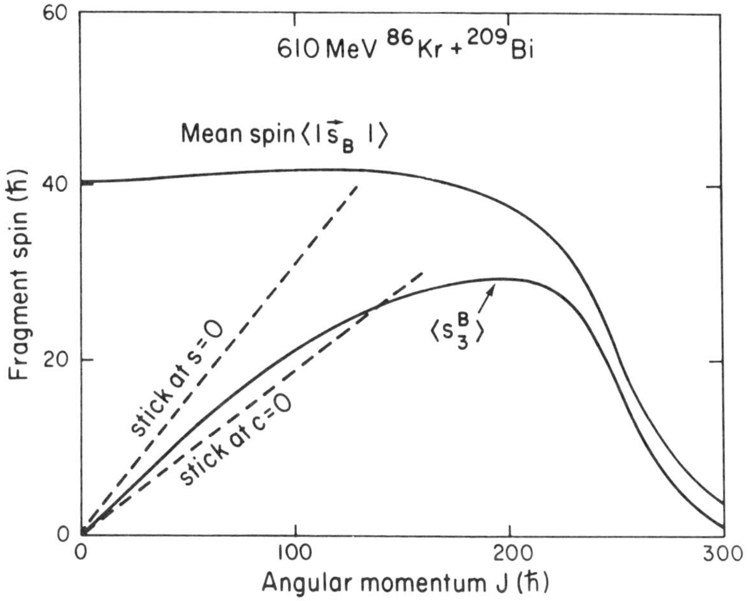

Figure 226. Average spin and projection normal to the reaction plane of target-like fragments in the reaction $^{209}Bi+^{86}Kr$ at $E_{\text{Lab}}=610$ MeV, as predicted by Randrup's transport model. The dashed lines illustrate the sticking condition for zero surface separation ($s=0$) and the separation ($c=0$) at neck rupture. The quantity J corresponds to the initial orbital angular momentum. From (Ran 81b).

equilibration and allows slippage to occur (Ran 81b). Due to the statistical dispersions, the average spin is always larger than the average projection on the reaction normal, an effect most evident for small l values associated with large excitation energies.

As expected from the previous discussion of the simple nucleon exchange model, both polarization and alignment of fragment spins increase at first drastically with decreasing total angular momentum or increasing numbers of exchanged nucleons. In the region of low and intermediate l values, however, both quantities decrease with decreasing angular momentum, where random spin components dominate to an increasing extent. The variation of polarization and alignment with orbital angular momentum, predicted by Randrup's model, is depicted in Figure 227. Also shown in Figure 227 is the theoretical in-plane spin asymmetry, defined in terms of the spin projections on the dinuclear symmetry axis and an in-plane direction perpendicular to it. This asymmetry is responsible for the in-plane anisotropy predicted for the products of the subsequent decay of the primary reaction fragments. Similar calculations have also been performed (Ran 81b) in the case of the ^{165}Ho + ^{165}Ho reaction at 1.4 GeV.

No attempt is made in the work cited above to make a detailed comparison between model predictions and experimental data, because it is felt (Ran 81b) that subsequent deexcitation mechanisms could result in significant distortions of the primary fragment spin distributions. Furthermore, in comparison to experimental results, note should be taken of the possibility that fluctuations in energy loss could be large for the more central collisions. The general dependence of the mean fragment spins on angular momentum, predicted by the transport model, appear to be in qualitative agreement with experimental spin–energy-loss correlations in the cases of the ^{209}Bi + ^{86}Kr and the ^{165}Ho + ^{165}Ho reactions discussed earlier. The feature of spin polarization and alignment predicted to decline with decreasing l value, however, is not readily discovered in the limited set of data currently available. A final judgment of this model with respect to angular momentum transmutation has to be deferred, therefore, until its more specific aspects have been tested experimentally. Measurements of spin–spin correlations in damped reactions may prove elucidating in this respect.

In summary of this section, it is concluded that processes of angular-momentum transmutation in damped nuclear reactions reveal interesting features of the microscopic reaction mechanism that are not yet fully understood in terms of various model views proposed in the literature. The data exhibit coherent as well as stochastic features of angular-momentum generation in the reaction fragments, which remember the direction of the initial orbital angular momentum to some extent, depending on its magnitude and the amount of kinetic energy damped in a collision. It has been possible to derive important information on the general character of the

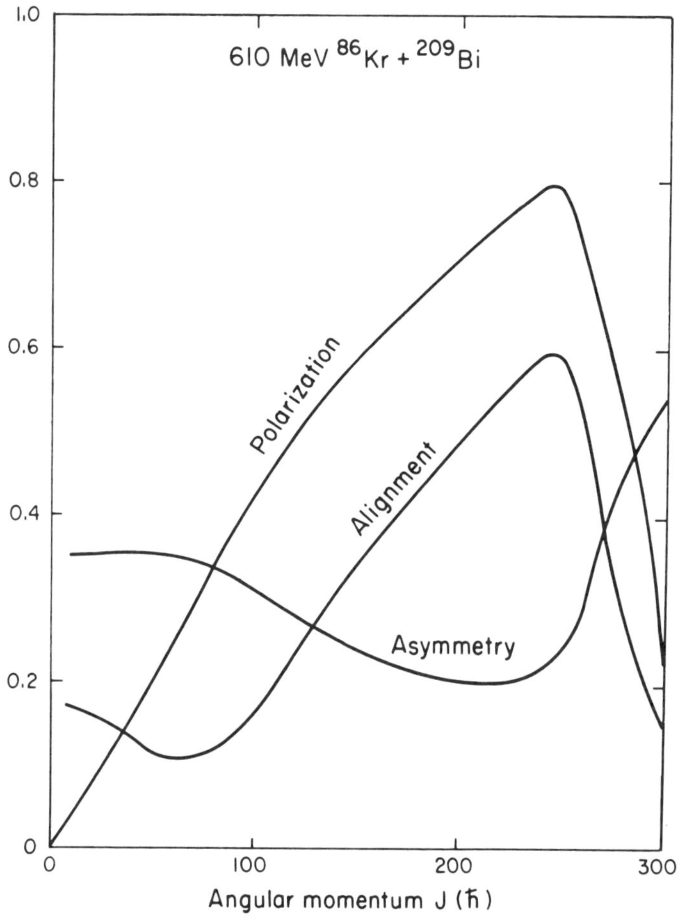

Figure 227. Polarization, alignment, and in-plane spin asymmetry parameters as a function of total angular momentum, predicted by Randrup's transport model for the heavy fragments produced in the 610-MeV $^{209}Bi + ^{86}Kr$ reaction. From (Ran 81b).

damped deflection function from studies of the nuclear spin polarization performed in a few cases.

However, although there have been a number of experiments that provide information on the spins of the fragments produced in damped collisions, much of this information appears to be rather indirect and strongly influenced by the method of analysis as well as by unknown details of the deexcitation cascade that deorients the spins of the primary reaction fragments. Other experiments contain a bias, making the deduced spin distributions nonrepresentative for the reaction, which complicates a direct comparison to most of the models. There is clearly a need for more thorough and detailed studies of the bombarding-energy dependence of angular momentum dissipation, preferably those that involve highly fissile systems or investigate particle evaporation from the early stages of the deexcitation cascade. Measurements of γ multiplicity distributions add complementary information on these dissipation processes, but calibration experiments are necessary to reduce the considerable latitude that presently exists in the interpretation of such data.

The comparison of the somewhat scattered data discussed in this section to theoretical models has not yet provided a consistent picture of the nature of angular momentum dissipation and fragment spin generation in damped collisions. Individual sets of data have been interpreted in terms of a variety of conflicting models. No model has been shown as yet to provide a satisfactory account of a considerable subset of experimental observations. This may not be too surprising in view of the fact that often important aspects of the nuclear collision dynamics are neglected in both collective and statistical reaction theories. Classical dynamical models, on the other hand, are mostly incomplete in their treatment of the stochastic reaction features.

It appears currently that there are mainly two sources producing fragment spins in damped reactions. One contributor may be seen in a friction-type dissipation mechanism that acts on the orbital motion of the two reaction partners, generating fragment spins that are aligned perpendicular to the reaction plane. However, neither classical sticking nor rolling conditions supply a satisfactory explanation of the data for all degrees of energy damping. There appears to be a definite correlation of the characteristics of the fragment spin distribution with the total amount of kinetic energy dissipated in a collision. Correlations with other variables seem to be secondary and of less importance. The second important spin generation mechanism exhibits stochastic features, gaining some dominance at high excitation energies of the intermediate system. However, in some experiments large values of the spin alignment parameters are observed to persist even at the highest energy losses measured in the experiments. With respect to the magnitude and alignment of the transferred angular momentum, the stochastic mechanism does not seem to reflect a situation where the intermediate system is at complete thermodynamic equilibrium. The consequences of the microscopic exchange-induced transport mechanism, favored by other dynamical reaction features discussed in earlier sections, appears to describe qualitatively some of the properties of angular-momentum transfer, but a definite conclusion about its validity in the description of these aspects has not been reached as yet. It is very obvious that further experimental and theoretical investigations are required to gain a consistent insight in the important question concerning the microscopic nature of angular momentum transmutation in damped reactions.

3.7. Simultaneous and Sequential Multiparticle Processes

It is one of the surprising characteristics of damped nuclear collisions that the binary reaction channel with only two massive final fragments exhausts the cross section almost entirely, despite an energy content of the interacting system sufficient to disintegrate the projectile to a large extent. As has been discussed in Section 3.2, the probability of observing a direct disintegration of either projectile or target in a damped heavy-ion collision is in general extremely small, for relative kinetic energies of a few MeV per nucleon above the interaction barrier. This feature is indicative of a relatively gentle interaction of projectile and target, in which the relative motion is slowed down in a succession of many small elementary steps, lacking significant high-frequency components in the interaction force that could induce disrupture of nuclear matter in the approach phase of a heavy-ion collision. Evidence for such a dominantly noncollective nature of the energy dissipation mechanism has been delineated in Section 3.5, as derived from final fragment total kinetic energy spectra. A basically sequential, multistep character of the reaction mechanism would not be expected to induce a collective nuclear response at elevated bombarding energies either, where coherence in the nucleonic motion is increasingly lost, due to a decreasing mean free path at the higher nuclear temperatures that gradually build up during a collision. However, the decay of the intermediate dinuclear system prior to its eventual split into massive, projectile-like and target-like fragments offers the opportunity to investigate important relaxation processes of nuclear matter at high excitation energies and spins, associated with extremely short lifetimes. Questions that are of interest in this respect pertain to the thermalization of intrinsic degrees of freedom, to the nuclear phase transition from a Fermi liquid to a Fermi gas, and to the eventual appearance of preequilibrium decay, signifying a vanishing capacity of nuclear matter to sustain momentum and energy gains introduced by the impact of nucleons received from the collision partner. These processes are expected to result in, not necessarily equilibrium-statistical decay products, emitted during and after the damped collision.

While the above simultaneous emission processes are currently very little understood and form an active topic of research, sequential multiparticle processes, such as sequential fission or light particle evaporation from the excited primary reaction fragments, are experimentally already somewhat better documented. An understanding of sequential fragment decay is essential for an interpretation of results of experiments that measure only quantities related to the asymptotic secondary reaction fragments. More significant for damped-reaction studies is, however, the sensitivity of the deexcitation patterns to the reaction conditions that prevailed immediately prior to the breakup of the intermediate system. Sections 3.5 and 3.6 have presented various examples for studies, in which experimental branching ratios, angular and energy distributions of sequentially emitted fission fragments, light particles and γ rays are interpreted in terms of statistical models, in order to derive information on the damped-reaction mechanism. The following discussion intends to broaden the scope of consideration of simultaneous and sequential multibody processes to include the mechanism of production and to draw conclusions from the degree to which they compete successfully with the two-body damped collision process.

The most basic task in experiments studying simultaneous and sequential emission of light or heavy particles in damped reactions consists in a discrimination of contributions to the cross section from the different possible sources of emission. These sources include the projectile and target nuclei in the entrance channel, the intermediate dinuclear system at various stages of evolution, and the final reaction fragments during and after acceleration in the Coulomb field. The most general case, where all sources contribute significantly, presents a most difficult task that can be met only to some extent even by the most sophisticated exclusive measurements. It is helpful, however, to notice that the emission patterns of particles are characteristic of the emission source and the stage of the reaction. It is this feature that, in fact, renders the measurement of associated particle emission in damped reactions a powerful investigative tool.

Among the properties of particle emission patterns, that allow one to disentangle the contributions from different sources to some extent are the symmetries of particle angular correlations and energy spectra that reflect the velocity and direction of flight, as well as the spin, of the emitter. In the case of charged-particle emission, which is hindered by the Coulomb barrier, a shift in energy of the particle spectrum is expected that is characteristic of the charge Z of the emitting source. Furthermore, in a truly asymptotic sequential decay of one of the reaction fragments, quantities such as Q value for the decay and particle angular correlations, in the proper rest frame of the emitter, should be largely independent of the history of the reaction partner. Overall angular correlations are, however, induced when the reaction plane is selected through the measurement of one primary fragment, in particular, when a preferred spin direction of the fragments is thereby also defined in the procedure. The effect has been discussed in the previous section.

On the other hand, strong correlations in angle and energy of emitted particles are expected if the decay of a fragment occurs when there is still significant nuclear or Coulomb interaction between the two primary reaction partners, through what may be termed "proximity scattering." For heavy-ion reactions, this effect has been discovered recently by Harrach *et al.* (Har 82) and Glässel *et al.* (Gla 82) in kinematically complete measurements of sequential fission induced in the reactions $^{122}Sn + ^{129}Xe$, $^{166}Er + ^{84}Kr$, and $^{90}Zr + ^{84}Kr$, at $E_{Lab} = 12.4–12.5$ MeV per nucleon. Figure

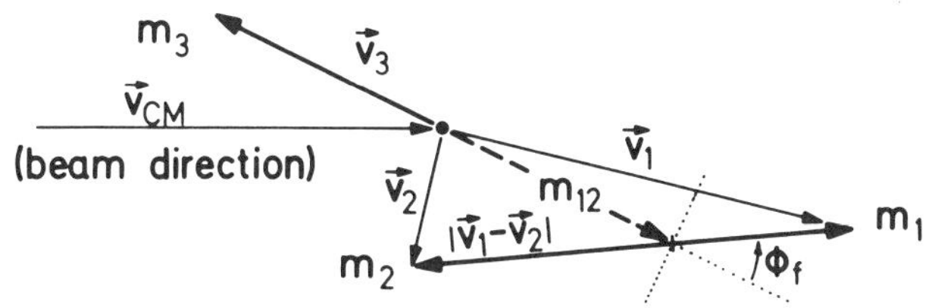

Figure 228. Velocity diagram of a damped reaction leading to primary fragments of masses m_{12} and m_3, followed by sequential fission of fragment m_{12}. Shown is the projection of the initial reaction plane. The azimuthal angle ϕ_f is measured with respect to the recoil direction of fragment m_{12}. From (Gla 82).

228 (Gla 82) illustrates the kinematics of such a three-body event, where a symmetric primary fragment m_3 ($A_3 \approx 125$) is scattered to $\theta_{c.m.} = 160°$ and the forward-going reaction partner m_{12} sequentially fissions in flight. The orientation of the fission axis is described in terms of a polar angle, θ_f, measured from the normal to the plane of the primary reaction, as defined by the directions of the beam and of the velocity $\mathbf{v}_3$ of particle m_3. Its azimuthal angle, ϕ_f, is counted from the direction of flight of the fissioning particle m_{12}.

Results of the measurement (Gla 82) on the $^{122}Sn + ^{129}Xe$ reaction are depicted in Figure 229. The top part of this figure displays the dependence of the average total relative velocity, $\langle|\mathbf{v}_1 - \mathbf{v}_2|\rangle$, of the sequential-fission products on the aximuthal angle ϕ_f. The data points are seen to oscillate about the dashed horizontal line representing the relative velocity expected from the systematics of the total fission decay energy. This average relative velocity is found (Har 82) to be essentially independent of the properties of the first, damped-reaction step, establishing the sequential character of the fission process. However, the relative velocity is smallest for small aximuth angles, corresponding to a situation where the velocity vectors $\mathbf{v}_3$, $\mathbf{v}_1$, and $\mathbf{v}_2$ are nearly collinear.

The solid curves in Figure 229 are the results of a Coulomb trajectory calculation following the time evolution of sequential fission events, starting at scission in the first reaction step. From the magnitude of the proximity effect observed in Figure 229a, the trajectory calculation suggests a very short time of $t_{ss} \approx 1 \times 10^{-21}$ s between breakup of the intermediate dinuclear system and the sequential scission act.

The angular distribution of the fission fragments with respect to the recoil direction of the fissioning particle is shown in Figure 229b. The broad maximum at $\phi_f \approx 30°$ indicates (Gla 82) that memory of the separation direction in the damped reaction is retained to some extent. The data are clearly at variance with the above Coulomb trajectory calculations (solid

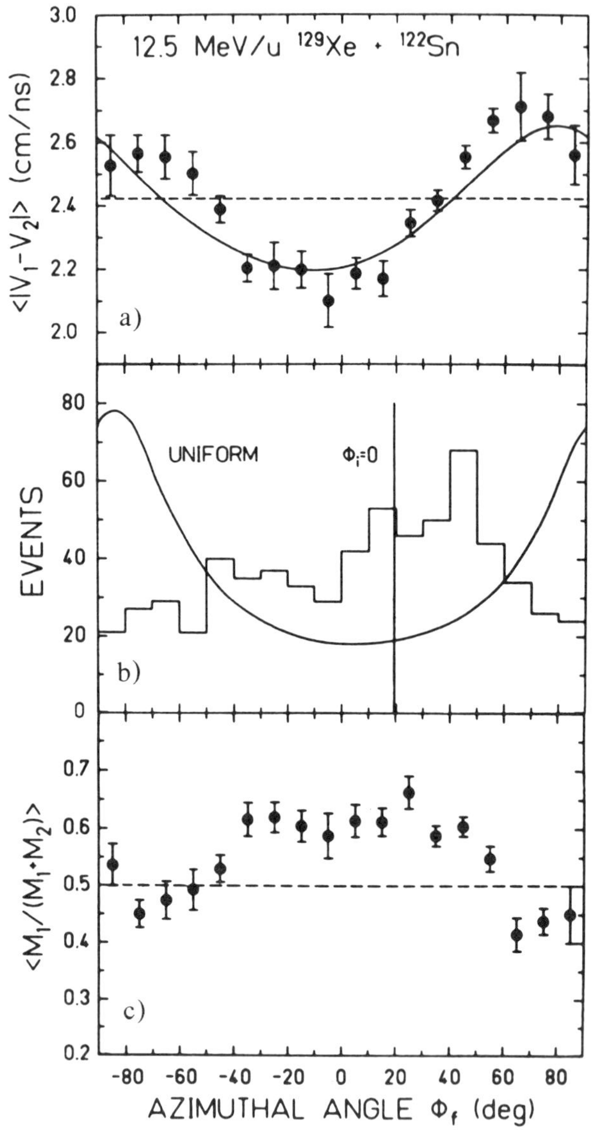

Figure 229. (a) Average relative velocity $\langle |\mathbf{v}_1 - \mathbf{v}_2| \rangle$, (b) yield, and (c) average mass fraction $\langle m_1/(m_1 + m_2) \rangle$ of the fission fragment pair vs. the azimuthal angle ϕ_f. The data are integrated over all polar angles θ_f, but represent a selection in $E_{k,\text{tot}}$ (400–450 MeV), m_{12} (110–130 u), and $\theta_{\text{c.m.}}$ (0°–40°). The dashed line in (a) is the ϕ_f-averaged value of $\langle |\mathbf{v}_1 - \mathbf{v}_2| \rangle$. The solid lines in (a) and (b) are the result of a model trajectory calculation. From (Gla 82).

curve in Figure 229b) if a uniform distribution of initial orientations ϕ_i of the sequential-fission axis is assumed.

In Figure 229c, the average mass fraction $\langle m_1/(m_1+m_2)\rangle$ of the sequential fission fragments is displayed, where m_1 is the fragment emitted at forward angles, in the frame of the fissioning particle. Obviously, heavier fission fragments are emitted close to the recoil direction, away from the flight direction of the primary reaction partner m_3. On the whole, the fission fragment mass distribution is asymmetric. With decreasing total kinetic energy loss, i.e., decreasing interaction time, the velocity modulation seen in Figure 229a disappears (Gla 82), whereas azimuthal anisotropies and average mass fractions approximately persist. As expected, the probability for three-body events also decreases drastically at the smaller-energy losses. However, the sequential fission probability, approaching 10% of the probability for binary events at high energy losses, is found (Har 82) to be an order of magnitude larger than expected from statistical model calculations or fission systematics.

The above findings indicate that the interaction times in very strongly damped collisions are not significantly shorter than scission times and that fission of a damped reaction fragment can be triggered and driven by the nuclear interaction with its collision partner. This appears to be true for damped reactions at bombarding energies in excess of 10 MeV per nucleon.

A similar effect is observed with 12.1-MeV per nucleon ^{86}Kr-induced reactions on targets of ^{166}Er and ^{89}Y by Olmi *et al.* (Olm 80), where, however, the projectile-like fragment is found to break up. Figure 230

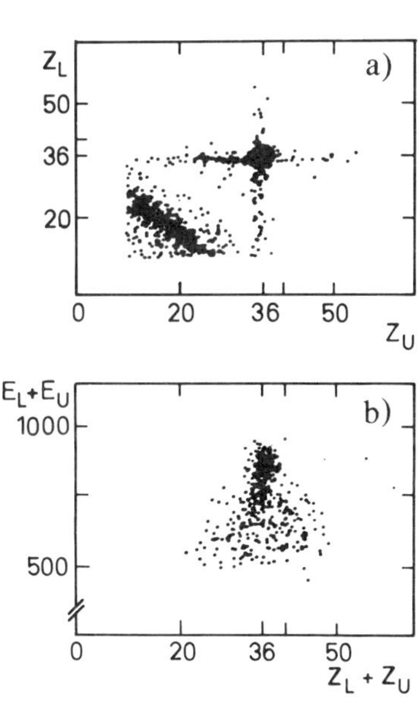

Figure 230. Probability scatter diagrams for sequential decay products of projectile-like fragments produced in the reaction $^{89}Y+^{86}Kr$ at 12.1 MeV per nucleon. Part (a) displays the correlation between the atomic numbers Z_L and Z_U of two light-fragment remnants measured in coincidence. Horizontal and vertical ridges correspond to accidental coincidences with elastically scattered projectiles. Part (b) illustrates the evolution of the summed charge, $Z_L + Z_U$, with decreasing total kinetic energy, $E_L + E_U$, of both projectile remnants. From (Olm 80).

displays experimental results for the $^{89}Y + ^{86}Kr$ reaction, where two remnants of the projectile-like fragment are measured in coincidence at forward angles. Shown are scatter diagrams, where the density of points is proportional to the cross section, plotted vs. the atomic numbers, Z_L and Z_U, of two coincident fragments (top), or vs. total final kinetic energy $E_L + E_U$ and total charge $Z_L + Z_U$ (bottom). Horizontal and vertical intensity ridges in Figure 230a are due to random coincidences between elastically scattered projectiles. However, a cross section ridge corresponding to a constant total charge $Z_L + Z_U \approx 36$ is clearly discernible in the lower left corner of Figure 230a, corresponding to an estimated total breakup probability of projectile-like fragments of the order of 4% (Olm 80). Although this constitutes only a small contribution to the cross section and is not observed at all (Olm 80) at the lower bombarding energy of 8.2 MeV per nucleon, it presents an unexpectedly high fission probability of a medium-weight nucleus. Hence, these findings could indicate the onset of new reaction mechanisms at elevated bombarding energies, characterized by nonequilibrium, nonadia-

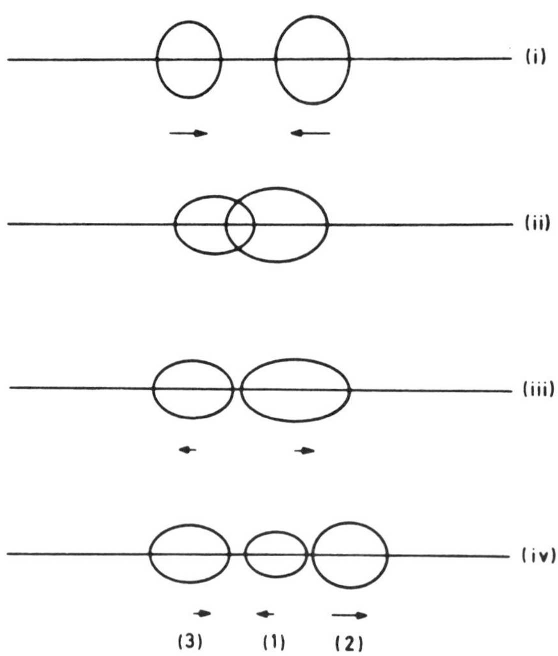

Figure 231. Schematic illustration of an instantaneous three-body reaction, where one of the damped products fissions while in nuclear contact with its collision partners. Interaction time proceeds from top to bottom. From (Deu 76).

batic nuclear excitation and decay modes. Further investigations are required to obtain a clearer experimental picture of such reaction modes.

Very little theoretical work has been done so far that pertains to the above fission process of a damped reaction fragment, while still in nuclear contact with its collision partner. An early attempt to describe such instantaneous fission processes in a classical model is due to Deubler and Dietrich (Deu 76, 78). Figure 231 gives a schematical representation of the reaction sequence suggested by these authors from their trajectory model that includes spheroidal deformations of the interacting heavy ions (cf. Section 2.3). In this model, the strong nuclear attraction produces large deformations of the nuclei at late stages of the damped reaction, with deformation energies that can reach values comparable to the fission barrier of a highly fissile target nucleus. A strong influence of the external interaction field exerted on the target nucleus by its collision partner is predicted for the instantaneous fission process. According to the model, it determines essentially the time after which fission occurs and the most probable mass asymmetry of the fission fragments. The change in "deformation" energy E_{Def} at scission, due to the presence of the interaction partner, is schematically pictured in Figure 232 (Deu 76). Here, the shape of the fissioning ^{236}U target nucleus at scission is replaced by tangent spheres of volume ratio λ,

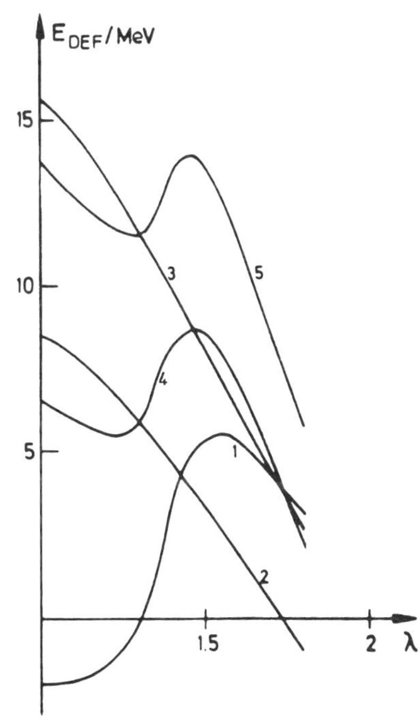

Figure 232. Deformation energy E_{Def} of ^{236}U target nucleus at its scission point as a function of the volume ratio λ of the two fission fragments, assumed to be spherical (curve 1). Representative Coulomb energies due to the presence of Kr (curve 2) or Xe (curve 3) projectiles as third particles, and total deformation potential for Kr (curve 4) and Xe (curve 5) projectiles are also plotted vs. λ. From (Deu 76).

that are coaxially situated with respect to spherical Kr or Xe projectiles. Curve 1 in Figure 232 depicts the dependence of this deformation energy for an isolated ^{236}U target nucleus at an excitation energy of 100 MeV on the fission fragment volume ratio. Curves 2 and 3 illustrate the external interaction potential for a Kr and a Xe projectile, respectively, as the third fragment. The potential is calculated as a sum of Coulomb and energy-density nuclear potential (cf. Section 2.5). In Figure 232, the total deformation potential is shown by curves 4 and 5, for Kr and Xe reaction partners, respectively. It can be seen from this figure that the minimum in the deformation potential is filled in and shifted to asymmetries $\lambda > 1$, due to the external field. This effect suggests a preferentially asymmetric mass split in the instantaneous fission process, as, indeed, observed in the experiments discussed earlier. It would also lead to an increased fission probability, although quantitative comparisons between experiments and realistic dynamical models have yet to be made. At present, it appears that no definite conclusion can be reached with respect to the equilibrium or nonequilibrium nature of the instantaneous fission process, if the equilibrium point is defined as the minimum in the total deformation potential. However, as discussed previously, nonnegligible breakup probabilities of Kr-like fragments from the 12.1-MeV per nucleon ^{89}Y + ^{86}Kr reaction could possibly indicate the onset of a nonequilibrium fission process at elevated bombarding energies. It appears to be a worthwhile endeavor to expand both experimental and theoretical investigations of this process and its implications on the damped reaction mechanism.

In experiments at bombarding energies below 10 MeV per nucleon, where the fission decay channel of excited primary fragments is studied, the above proximity scattering process is not observed. In the following, evidence is discussed that indicates that such lower-energy reactions, involving heavy targets or projectiles, lead frequently to sequential fission of heavy primary fragments, occurring well outside the interaction region. Wilcke *et al.* (Wil 80), for example, report a 30% probability for sequential fission of target-like fragments produced in the reaction ^{209}Bi + ^{136}Xe at $E_{\text{Lab}} = 940$ MeV. For such heavy fragments, fission and neutron evaporation are the dominant decay channels. If fission and neutron emission are both statistical deexcitation processes, the interesting branching ratio Γ_n/Γ_f can be estimated from the density of available final states. For neutron and, more generally, light-particle evaporation, one of the dominant parameters determining the decay width Γ [see Equation (3.6.5)] is the level density of the daughter nucleus, characterized by a level density parameter (see, e.g., Boh 69) $a = A/8$. Here, A is the mass number of the residual nucleus. In equilibrium-statistical fission theory, reviewed by Vandenbosch and Huizenga (Van 73), the density of transition states present at saddle-point nuclear deformations is regarded to be the important quantity governing the fission probability. In this theory, the transition state is associated with a constriction in available phase space that directs the probability flow in an irreversible fashion. Adopting the Bohr–Wheeler formula (cf. Van 73) for Γ_f and Fermi-gas expressions (cf. Boh 69) for the level densities, one obtains approximately for the neutron evaporation-to-fission branching ratio at a fixed excitation energy E^*

$$\left(\frac{\Gamma_n}{\Gamma_f}\right)_{E^*} = \frac{4A^{2/3}a_f(E^* - B_n)}{k_0 a\left[2a_f^{1/2}(E^* - E_f)^{1/2} - 1\right]} \times \exp\left[2a^{1/2}(E^* - B_n)^{1/2} - 2a_f^{1/2}(E^* - E_f)^{1/2}\right] \quad (3.7.1)$$

Here, a_f is the level density parameter for the transition states at the fission saddle point, E_f denotes the fission barrier, B_n is the neutron binding energy, and the parameter k_0 is given by $k_0 = \hbar^2/(2mr_0^2) \approx 9.8$ MeV. Equation (3.7.1) is valid only rather approximately, since spin dependences of the level densities are neglected in this formulation. Wilcke *et al.* (Wil 80) have parametrized the dependence of the branching ratio Γ_n/Γ_f on excitation energy, E^*, and spin, J, of the fissioning nucleus of charge Z in terms of a critical energy $E^*_{cr}(Z, J)$. At this excitation energy, $(\Gamma_n/\Gamma_f)_{E^*_{cr}} = 1$. Fits to experimental data and extrapolations using Equation (3.7.1) suggest (Wil 80)

$$E^*_{cr}(Z, J) = \begin{cases} E^*_{cr}(Z,0)\left[1 - J^2/6400\right], & J \le 80\hbar \\ 0, & J > 80\hbar \end{cases} \quad (3.7.2a)$$

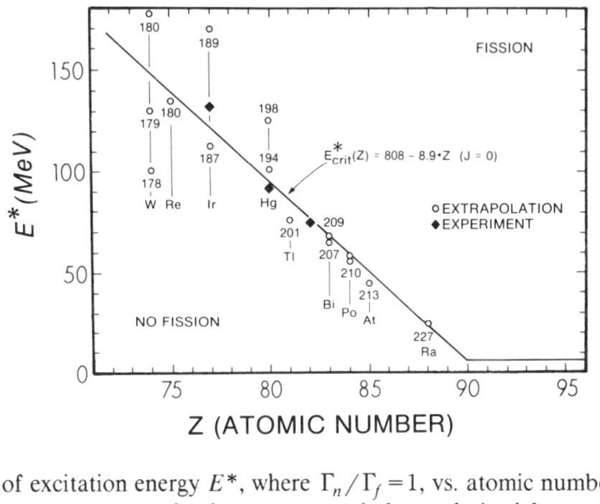

Figure 233. Plot of excitation energy E^*, where $\Gamma_n/\Gamma_f = 1$, vs. atomic number Z of a nucleus. Filled symbols represent measured values; open symbols are derived from extrapolation using Equation (3.7.1). From (Wil 80).

with

$$E_{cr}^{*}(Z,0) = \begin{cases} 808 - 8.9 \cdot Z \text{ MeV}, & Z < 91 \\ 6 \text{ MeV}, & Z \geq 91 \end{cases} \tag{3.7.2b}$$

Figure 233 depicts the dependence of the critical excitation energy $E_{cr}^{*}(Z,0)$ on the charge Z of the fissioning nucleus. Obviously, the parametrization in Equation (3.7.2) provides an adequate description of measured (filled symbols) branching ratios and of extrapolated values (open symbols). The above systematics can, of course, be transformed into one relating a critical Z value to a given pair of spin and excitation energy.

The actual fission probability of damped reaction fragments has been measured for several systems, as referred to already in Section 3.6. A further example is presented in Figure 234 for the $^{197}Au + ^{86}Kr$ reaction at 618 MeV

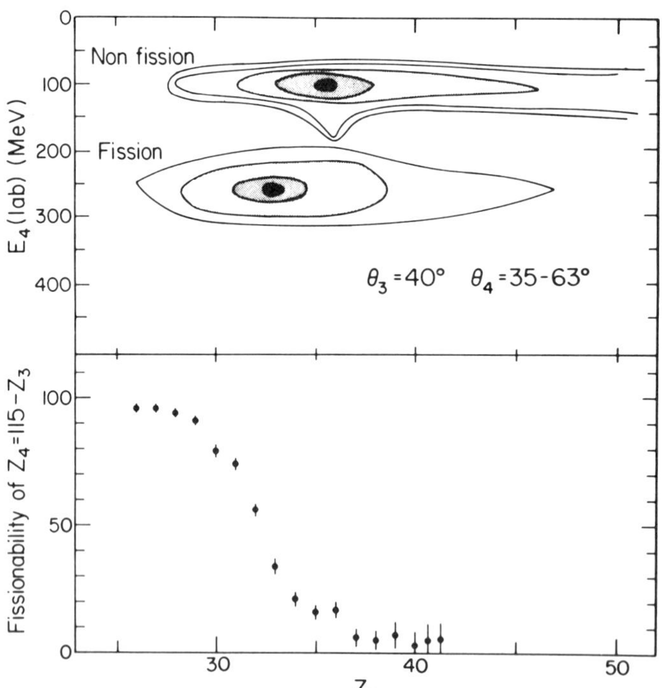

Figure 234. Energy–Z distribution (top) and fissionability (bottom) of fragments produced in the reaction $^{197}Au + ^{86}Kr$ at $E_{Lab} = 618$ MeV. The data are results of a coincidence experiment, where the charge, Z_3, and the energy, E_3, of the light reaction partner are measured in coincidence with the heavy reaction partner (Z_4, E_4) or one of its fission fragments. From (Woz 78b).

(Woz 78b). In the top part of the figure, isocontour lines of the cross section are plotted vs. atomic number, Z_3, of a projectile-like fragment, detected in a gas $\Delta E - E$ telescope, and the energy, E_4, of the coincident heavy reaction partner or of one of its fission fragments. A clear separation of damped and sequential-fission events is apparent at a detection angle of $\theta_3 = 40°$. The bottom part of Figure 234 depicts the fissionability of the heavy fragment, plotted vs. atomic number Z_3 of its collision partner. The fission probability is rather small for Z_3 values around 40, but rises rapidly and approaches 100% for $Z_3 < 30$ ($Z_4 > 85$). The actual shape of the correlation between fissionability and atomic number displayed in Figure 234 reflects the distribution over spin and excitation energy acquired by the heavy fragment in the damped reaction. It may, therefore, be used to test these reaction conditions.

Very large fission probabilities, exceeding 80%, have been reported by Wozniak *et al.* (Woz 78b) for target-like fragments produced in the ^{197}Au + ^{86}Kr reaction. This is to be contrasted with a value of only $\approx 10\%$ (Woz 78b) measured for the light-ion reaction ^{197}Au + 130 MeV ^{4}He, yielding a similar excitation energy as in the ^{197}Au + ^{86}Kr reaction at high energy losses. This fission enhancement following damped reactions has been qualitatively explained in terms of angular-momentum transfer processes (see Section 3.6), endowing the heavy primary fragments with considerable spins. Hence, from large fragment fission probabilities alone, it is not possible to draw conclusions on the presence of nonequilibrium features of the sequential fission process. Proton pickup by the heavy partner in a damped reaction, of course, also enhances the sequential fission probability of such fragments. This effect has been observed, for example, by Russo *et al.* (Rus 77) in the case of the 979-MeV ^{197}Au + ^{136}Xe reaction, where the sequential-fission Z distribution is found to be centered at Z values larger than one half the charge of Au. Fission following transfer or inelastic scattering has also been studied for other heavy targets (Fre 75, Hab 77, Fra 79c, Lei 79, Mat 79, Raj 79, She 79, Leb 80).

The sequential character of fragment fission following damped collisions is apparent from the correlation of the kinetic energy of the fission fragments with the recoil angle of the fissioning nucleus. This is exemplified in Figure 235 displaying the angular dependence of the mean energy vs. detection angle of sequential-fission products from the reaction ^{197}Au + ^{136}Xe at 979 MeV. Data are averaged over fragments with Z values of 36, 37, and 38, where the sequential-fission component is well separated in energy from the damped-reaction events. The curve drawn through the data points in Figure 235 serves to guide the eye. The overall angular distribution of projectile-like damped fragments peaks near to the grazing angle (38°) with a corresponding recoil angle of 50° for the heavy primary reaction partner. The peak in the energy of the fission fragments observed in Figure 235 just

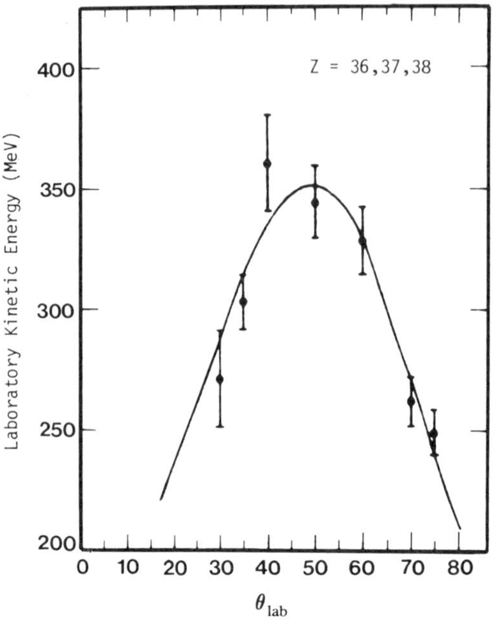

Figure 235. Mean laboratory energy of sequential-fission fragments of Z values between 36 and 38, produced in the reaction $^{197}Au + ^{136}Xe$ at $E_{Lab} = 979$ MeV, plotted vs. detection angle of a sequential fission fragment. The curve through the data points serves to guide the eye. From (Rus 77).

near this recoil angle indicates that fission of the heavy reaction partner occurs well after the recoil direction is firmly established in the exit channel, i.e., after acceleration of the primary reaction fragments in the Coulomb field. This separation in time scales between strongly damped collisions and sequential fission at bombarding energies below 10 MeV per nucleon, as well as the nearly symmetric sequential-fission mass and charge distributions, imply that the interaction force is not effective enough during the short collision time to predeform the heavy reaction fragment into its saddle shape.

Of far greater importance as a reaction channel than sequential fission is the emission of light particles, such as neutrons, protons, or α particles in damped reactions. For certain systems, heavier remnants of one of the reaction partners, usually of the lighter, projectile-like fragment, can also be observed. The instantaneous fission process discussed earlier represents an example for such a mechanism. It bears some resemblance to the breakup mechanism of light ions, studied particularly well for deuterons (see, e.g., Ser 47, Joh 70, Raw 74, Bau 76, 79c, 80, Aus 78, Ama 79) as projectiles. More recently, however, breakup has been investigated for heavier ions as well, both experimentally (Ost 76, Ina 77, 79, Bho 78, 79, Zol 78, Wu 79, Yam 79, Siw 79b, 79c, Wil 79b, 79d, 80b, 82, San 79b, Dri 81, Tak 81, Hom 82) and theoretically (Uda 79, 79b, 80, Bau 79b, 79c, 79d, 80, 80b, Hus 79d, 81, Pai 81).

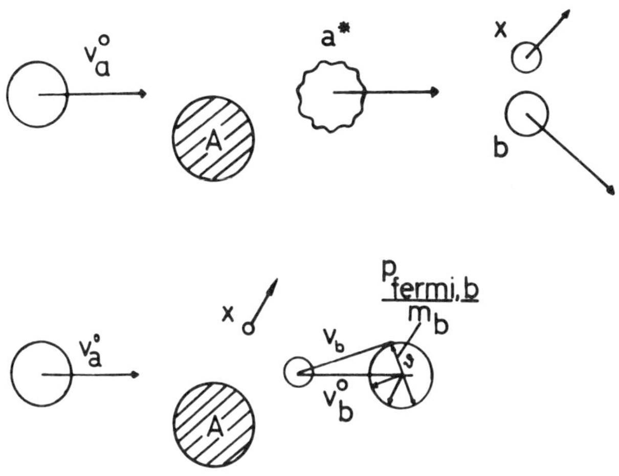

Figure 236. Schematical illustration of projectile breakup mechanisms. Elastic breakup is pictured at the top, a spectator process is depicted at the bottom. This part also includes a velocity vector diagram involving the Fermi velocity $p_{\text{fermi},b}/m_b$ of the constituent b in the projectile a. From (Bau 79d).

Several types of breakup of a fragment into two remnants can be distinguished. In elastic breakup, both remnants emerge in their ground states, as does the reaction partner. This process is schematically illustrated in the top part of Figure 236 (Bau 79d), depicting a composite projectile $a = b + x$, excited into the continuum via interactions with the target A, followed by the subsequent decay into components b and x. Another possibility, shown in the lower part of Figure 236, consists in an interaction of only constituent x with the target, while the component b acts as a spectator. If effects of binding and acceleration in the Coulomb field can be neglected, the spectator will move essentially with an average velocity equal to that in the entrance channel, i.e., $v_b^0 = v_a^0$. However, the Fermi motion of the component b within the initial nucleus a, characterized by an average Fermi momentum, $\mathbf{p}_{\text{fermi},b}$, will lead to a distribution in final velocities $\mathbf{v}_b$ of the ejectile b, corresponding to energies

$$E_b(\theta) = \tfrac{1}{2} m_b \mathbf{v}_b^2 = \tfrac{1}{2} m_b \left(\mathbf{v}_b^0 + \mathbf{p}_{\text{fermi},b}/m_b\right)^2$$

$$\approx \frac{m_b}{m_a} E_a + v_b^0 p_{\text{fermi},b} \cos\theta \tag{3.7.3}$$

The approximate equality in Equation (3.7.3) is valid for $m_b v_b^0 \gg p_{\text{fermi},b}$, and θ denotes the angle between $\mathbf{v}_b^0$ and $\mathbf{p}_{\text{fermi},b}$.

In the more general case of inelastic breakup, one or both constituents of the projectile can interact with the target in entrance and/or exit channels, leading to more complicated energy relations. It is, therefore, of importance (Hus 79d) to consider the local momentum balance as it exists at the time of breakup. Furthermore, initial- or final-state interactions following breakup can lead to the occurrence of incomplete fusion (Siw 79b, 79c, Wil 79b, 79d, 80b, 82) or of damped reaction processes involving one or both of the breakup fragments.

Breakup reactions induced by lighter ions appear to be understood theoretically (e.g., Ser 47, Bau 79d, Uda 79) in terms of direct peripheral processes, localized in l space close to the grazing angular momentum. Baur *et al.* (Bau 79d) introduce the parametrization

$$T_l^{\text{bu}}(a,b)=\beta(E_a,E_{\text{bind}})e^{-(b-b_0)^2/(\Delta R)^2} \tag{3.7.4}$$

for the transmission coefficient for breakup of the projectile, a, at incident energy E_a, associated with an ejectile, b, with binding energy E_{bind} in the projectile. In Equation (3.7.4), b_0 is the central value of the impact parameter window of width ΔR contributing to the breakup process, and the quantity β is a strength factor that increases with bombarding energy, until it attains a saturation value. The total breakup cross section, obtained (Bau 79d) by summation over all impact parameters,

$$\sigma_{\text{tot}}^{\text{bu}}(a,b)=2\pi\beta\int_0^{\infty} db\, b e^{-(b-b_0)^2/(\Delta R)^2}\approx 2\pi^{3/2}\beta b_0\,\Delta R \tag{3.7.5}$$

is approximately proportional to $b_0=r_0(A^{1/3}+a^{1/3})$, where A and a are the mass numbers of target and projectile, respectively, with $r_0\approx 1.2$ fm (Bau 79d). For light projectiles and heavy targets, the central impact parameter depends only on the target, i.e., $b_0\approx r_0A^{1/3}$, whereas the width parameter, ΔR, characterizes the projectile. Hence, the cross section in Equation (3.7.5) factorizes into a factor describing the projectile and one depending only on the target geometry.

The scaling law of the breakup cross section with the radius of the target, $R_A=r_0A^{1/3}$, is illustrated in Figure 237, where the angular distribution of ^{3}He particles from the 140-MeV α-breakup reaction on a variety of targets is shown as a function of laboratory angle. As is obvious from the figure, the cross sections for different targets follow the same universal behavior when divided by $A^{1/3}$ of the target. The theoretical curve in Figure 237 is based on a particular distribution of Fermi momenta of ^{3}He within the incident α particle (Wu 78), which is reflected both in the energy [cf. Equation (3.7.3)] and the angular distribution of breakup ejectiles.

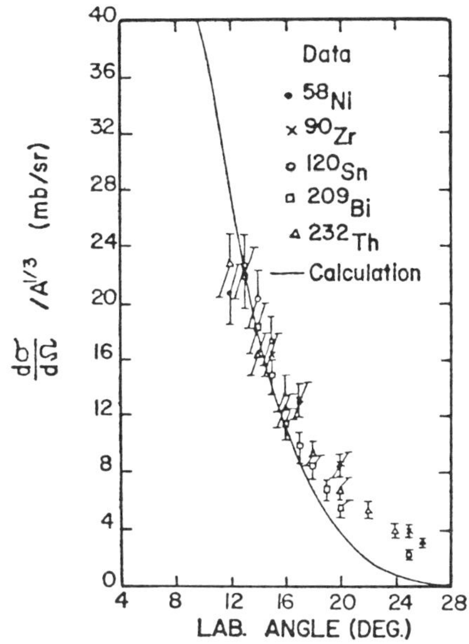

Figure 237. Differential $(\alpha, {}^3\text{He})$ breakup cross section divided by $A^{1/3}$ for $E_\alpha = 140$ MeV and the indicated targets. The curve represents a theoretical angular distribution normalized to data at $\theta_{\text{lab}}({}^3\text{He}) = 13°$. From (Wu 78).

Consideration of breakup processes associated with damped reactions is of significance for understanding heavy-ion collisions, because the presence or absence of sizable breakup or incomplete fusion reactions yields an indication of the rapidity of energy dissipation processes and the strength of conservative interaction forces, as well as of the relative importance of collective features in damped reactions. It has already been delineated that breakup-type reactions for relatively heavy ions appear with small, but noticeable strength only at bombarding energies in excess of 10 MeV per nucleon. It is interesting to discover whether there is a strong bombarding-energy dependence of the breakup probability also for lighter heavy ions, whose cluster structure is expected to render such projectiles more amenable to a direct breakup.

Results of an early experiment by Britt and Quinton (Bri 61) on breakup–α-particle emission in the reaction ${}^{197}\text{Au} + {}^{12}\text{C}$ at $E_{\text{lab}} = 126$ MeV are shown in Figure 238, depicting a bell-shaped c.m. α-energy spectrum for $\theta_{\text{c.m.}} = 43°$. This distribution peaks at an energy reduced from that corresponding to the beam velocity. The authors have studied the emission of protons, as well as that of α particles, for ${}^{197}\text{Au}$ and ${}^{209}\text{Bi}$ targets and 7.1–10.5-MeV per nucleon ${}^{12}\text{C}$, ${}^{14}\text{N}$, and ${}^{16}\text{O}$ projectiles. The laboratory angular distributions of α particles emitted in several of these reactions are displayed in Figure 239. One can distinguish the characteristics of two different emission mechanisms in these angular distributions. At backward angles, the angular distribution is relatively flat with an average cross section depending visibly on both the target mass and the projectile energy. These α particles have energy spectra characteristic of statistical evaporation from the composite system. Superimposed on the distribution of statistically emitted particles, a strongly forward-peaked component of α particles appears in Figure 239. As can be observed in this figure, there is no noticeable dependence on target mass of either the absolute magnitude of the differential α cross section or the shape of the angular distribution at forward angles. Although the total α emission probability decreases drastically with decreasing bombarding energy, the relative contributions of forward-peaked and evaporated particles remain approximately invariant. However, the width of the forward-peaked α component increases with decreasing incident energy, due to a lesser angular focusing of the α particles at lower projectile energies. These features are found (Bri 61) consistent with a direct breakup of projectiles that are relatively unstable against α-particle emission. Results obtained by Britt and Quinton (Bri 61) for proton emission in the $^{209}Bi + ^{16}O$ reaction indicate that, although evaporative cross sections are larger for protons than for α particles, the reverse is true for the direct particles emitted at forward angles. This feature suggests a strong influence of the individual nuclear structure on the emission mechanism for lighter nuclei, which may not be representative of general properties of reactions involving heavier ions.

The particular shape of the breakup α-particle energy spectrum depicted in Figure 238 can be interpreted (Bri 61) in terms of an intrinsic

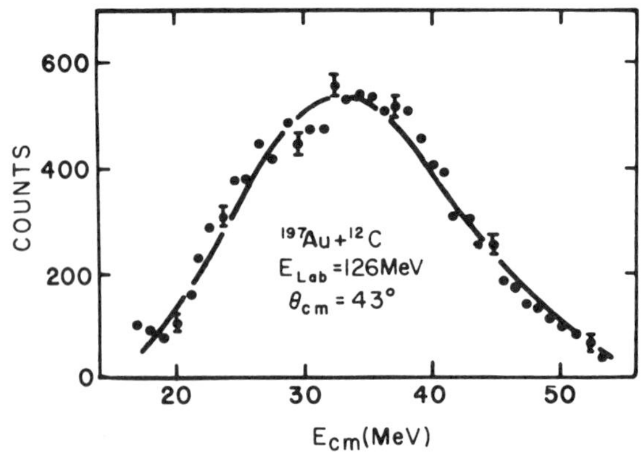

Figure 238. Center-of-mass energy spectrum of breakup α particles emitted in the reaction $^{197}Au + ^{12}C$ at $E_{Lab} = 126$ MeV and $\theta_{c.m.} = 43°$. The curve drawn through the data points represents a theoretical fit. From (Bri 61).

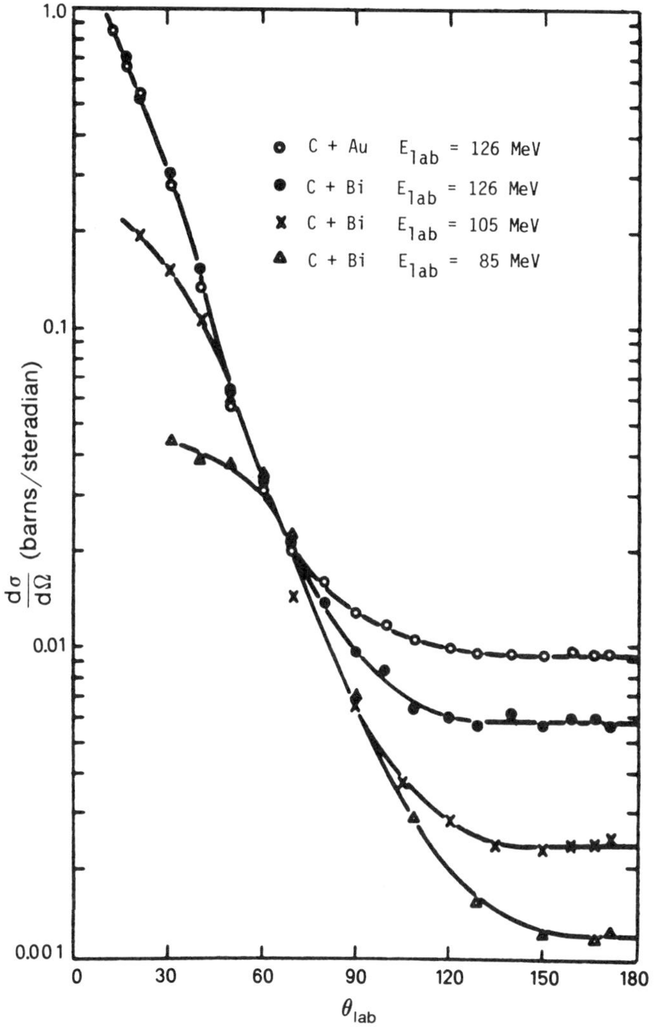

Figure 239. The laboratory angular distributions for alpha particles emitted in carbon-induced reactions on targets of Au and Bi. Individual points are estimated to be accurate to about 10%. From (Bri 61).

momentum distribution of the form

$$P(p) \sim p^2 \exp(-p^2/p_0^2) \tag{3.7.6}$$

of the α particles within the projectile, as illustrated by the solid curve in this figure. Here, the characteristic α particle momentum p_0 corresponds to a relatively low value of 0.25 MeV per nucleon. In order to obtain the fit shown in Figure 238, it was also necessary (Bri 61) to assume that the projectile had lost some 25 MeV of kinetic energy prior to breakup. However, as pointed out by Britt *et al.* (Bri 61), such a simple analysis probably does not do justice to the intricacies of the breakup process.

Angular and energy correlations between heavy ejectiles and α particles have also been measured by Sanderson *et al.* (San 79b) for 10.5 MeV per nucleon ^{14}N reactions on the relatively light targets ^{12}C, ^{27}Al, and ^{58}Ni. Experimental results from these measurements are shown in Figure 240. Here, the differential α multiplicity is plotted vs. angle, for the various heavier ejectiles indicated at the curves. The angles of the ejectiles detected in coincidence are marked by arrows on the abscissa. For the ^{58}Ni target and, approximately, also for the ^{27}Al target, the angular correlations of α particles are symmetric about the beam axis, with a full width of about 40°. There is no systematical dependence on the heavy ejectile of these angular correlations discernible in the data of Figure 240. Similarly, there is no appreciable correlation with ejectile type or its energy of the α-particle energy spectra. In fact, most of the angular distributions in Figure 240 are well approximated by the solid curves representing experimental singles distributions. Sanderson *et al.* (San 79b) are able to parametrize their coincidence data with a factorized form

$$\frac{d^4\sigma(E_{\mathrm{HI}}, \theta_{\mathrm{HI}}, E_\alpha, \theta_\alpha)}{dE_{\mathrm{HI}}\, d\Omega_{\mathrm{HI}}\, dE_\alpha\, d\Omega_\alpha} \propto \frac{d^2\sigma(E_{\mathrm{HI}}, \theta_{\mathrm{HI}})}{dE_{\mathrm{HI}}\, d\Omega_{\mathrm{HI}}} \cdot \frac{d^2\sigma(E_\alpha, \theta_\alpha)}{dE_\alpha\, d\Omega_\alpha} \tag{3.7.7}$$

where the product terms correspond to the singles cross sections for the heavy ejectiles (HI) and the α particles. This factorization is in accord with a fast breakup mechanism, where an α particle is produced first, followed by subsequent interactions of the projectile remnant with the target nucleus. Integral multiplicities of α particles are found to be quite large, i.e., 0.4–0.8 per event for ^{58}Ni and ^{27}Al, and 0.8–1.6 per event for the ^{12}C target.

Sequential emission of α particles from either the projectile-like or the target-like fragment in a damped reaction is claimed to be unlikely by Bhowmik *et al.* (Bho 78) in the case of $^{12}\mathrm{C}+{}^{11}\mathrm{B}$ and $^{58}\mathrm{Ni}+{}^{11}\mathrm{B}$ reactions at $E_{\mathrm{Lab}} = 116$ MeV, and for the $^{58}\mathrm{Ni}+{}^{14}\mathrm{N}$ reaction at $E_{\mathrm{Lab}} = 148$ MeV (Bho 79). As in the case of the ^{14}N-induced reactions discussed above, large α

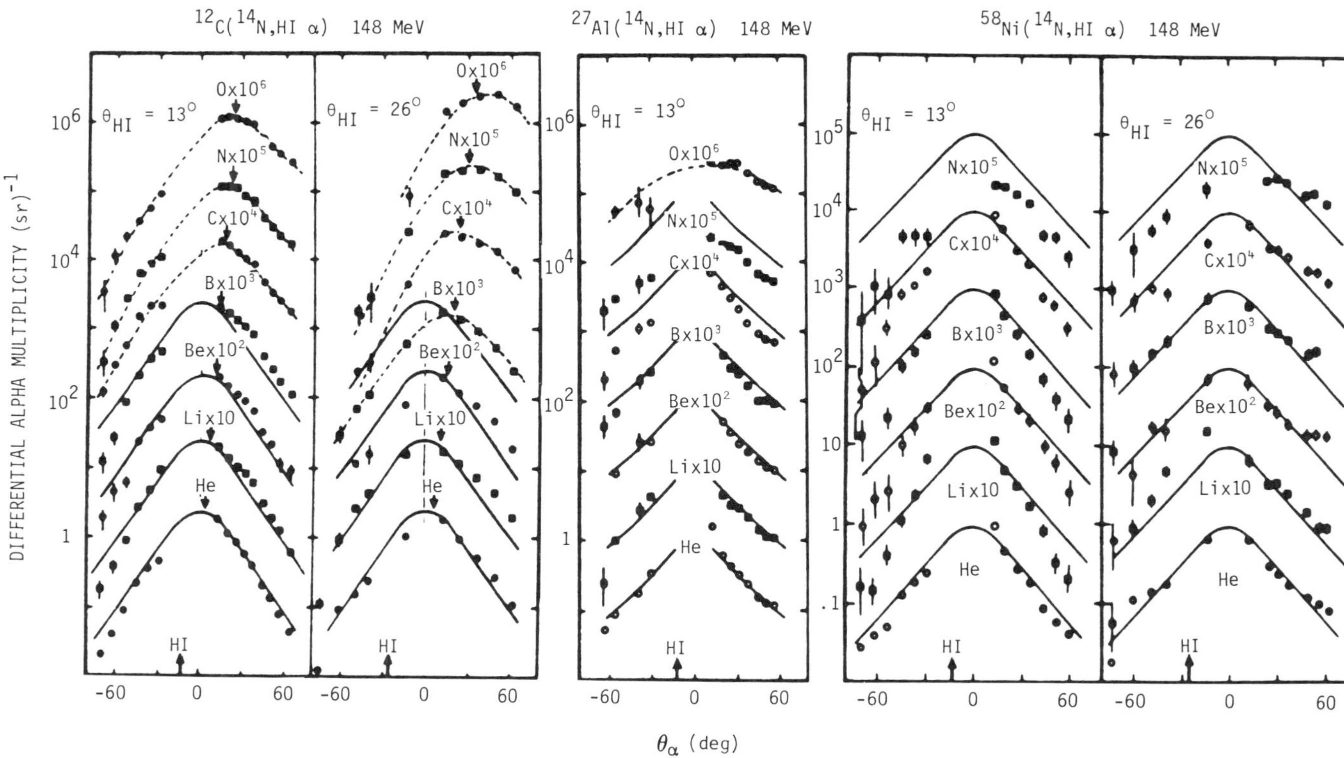

Figure 240. Differential α multiplicities plotted vs. angle, for ^{14}N-induced reactions on ^{12}C, ^{27}Al, and ^{58}Ni targets. The heavy ejectiles measured in coincidence are indicated at the distributions. The ejectile detection angles are marked by arrows (HI) on the abscissa. Solid curves drawn through the data points represent the α angular distributions measured in singles. From (San 79b).

multiplicities, of the order of 0.5 per fragment, are observed also in the ^{11}B-induced reactions. For these latter reactions (Bho 78), an analysis of three-body Q values is made, in order to obtain information on the α emission mechanism. The dependence of various average energy values on the α emission angle is depicted in Figure 241 (Bho 78), on the left for the

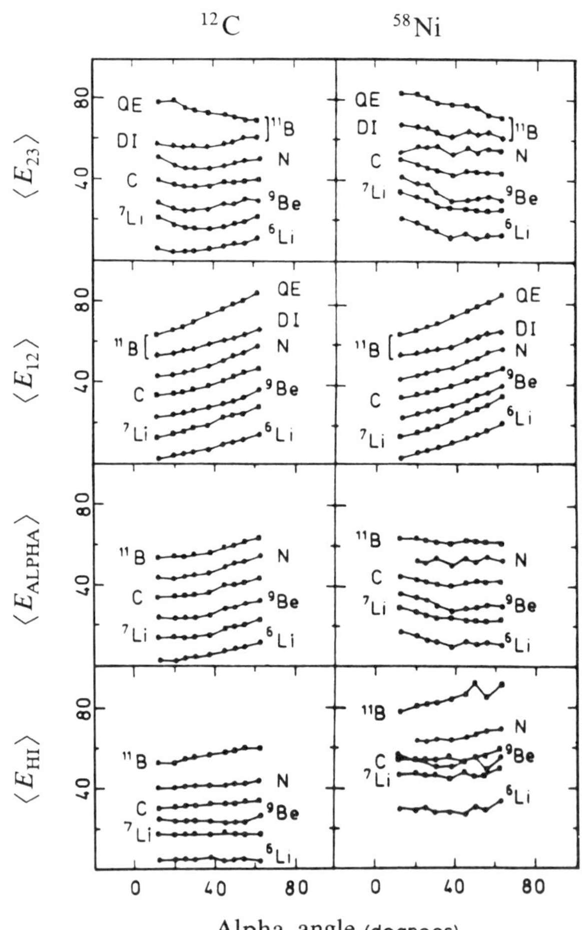

Figure 241. Average values of ejectile and α particle energies, as well as of 3-body Q values E_{12} and E_{23}, plotted vs. α-particle detection angle in the reactions $^{12}C+^{11}B$ (left) and $^{58}Ni+^{11}B$ (right) at $E_{Lab}=116$ MeV. Different data sets are marked by the detected ejectile; sets for adjacent ejectiles are displaced by 10 MeV. For ^{11}B ejectiles, data points are shown separately for quasielastic (QE, $Q>-30$ MeV) and damped (DI, $Q<-30$ MeV) events. From (Bho 78).

^{12}C target, and on the right for the ^{58}Ni target. Data points belonging to adjacent ejectiles are shifted by 10 MeV for purposes of presentation. In Figure 241, the quantities $\langle E_{HI} \rangle$ and $\langle E_{ALPHA} \rangle$ are the average energies of the ejectiles and the α particles, respectively. The variable $\langle E_{12} \rangle$ is the average total kinetic energy of the α-ejectile system in its own rest frame, and $\langle E_{23} \rangle$ denotes the similar quantity evaluated for the combined system of α particle and heavy reaction partner. For a sequential emission of α particles from the projectile or target-like reaction fragments, $\langle E_{21} \rangle$ or $\langle E_{23} \rangle$, respectively, would be independent of the α emission angle. Neither situation is, obviously, met by the experimental results of Figure 241, apparently in contradiction to the assumption of sequential α emission from the asymptotic damped reaction fragments. Table 5, collecting available information on light-particle emission associated with damped reactions, gives further examples of reactions in which nonstatistical particle emission patterns have been reported.

Conclusions drawn in some of the experiments listed in Table 5 were recently put in perspective by Young *et al.* (You 80), who caution that anisotropic emission of high-energy α particles in reactions induced by relatively light heavy ions such as ^{16}O may, nevertheless, be consistent with sequential emission from the primary reaction fragments. These authors have studied α emission in the reaction $^{93}Nb + ^{16}O$ at $E_{Lab} = 204$ MeV. In Figure 242, results of experiments and Monte Carlo simulations of α particle emission are displayed, as reported by Young *et al.* (You 80). The figure shows α–^{12}C coincidence data from the $^{93}Nb + ^{16}O$ experiment with a heavy-ion telescope placed at $\theta_{HI} = 21°$ (arrow). The top part depicts the differential α multiplicity as a function of α scattering angle; the lower parts give the angular dependences of the average three-body Q values, where E_{12} corresponds to the relative kinetic energy in the α–^{12}C system and E_{23} to that in the α–^{93}Nb system. Obviously, the angular distribution of α particles is not centered around the ^{12}C detection angle, nor are the three-body Q values independent of the α emission angle. The solid curves in Figure 242 are the results of a Monte Carlo evaporation simulation, where the primary reaction fragments are assumed to evaporate statistically α particles in flight. If the primary projectile-like reaction fragment is assumed to be scattered to an angle close to the detection angle of $-21°$, the dashed curve in Figure 242 results, which is in clear violation of the data. However, the angular distribution of the primary fragments is strongly forward-peaked for such reactions (cf. Section 3.3), causing a preferential weighting of forward-emitted fragments that are deflected into the detector by the recoil imparted on them by an evaporated α particle. Taking account of the effect of the primary angular distribution yields the solid curves in Figure 242, which provide a much better reproduction of the data. Similar calculations can also qualitatively explain (You 80) experimental results on other ^{16}O-induced reactions.

Table 5. Light-Particle Emission in Damped Reactions

Target	Projectile	E_{Lab} (MeV)	$(E_{\text{c.m.}} - V_B)/\mu$ (MeV/nucleon)	Light particle	Comment	Ref.
^{197}Au	^{63}Cu	365, 400	0.56, 1.12	n	a	Pet 77, Tam 79b
^{197}Au	^{40}Ar	240	1.17	n	b	Dak 80b
^{197}Au	^{132}Xe	990	1.95	n	a	Gou 78, Gou 80
^{181}Ta	^{165}Ho	1354	2.21	α	c	Sob 83
^{27}Al	^{16}O	65	2.40	α	d	Har 77
^{27}Al	^{16}O	65	2.40	α	e	Tsa 81
^{166}Er	^{86}Kr	675, 602, 402	3.19, 2.34, 0.02	n	a	Eya 78, Eya 80
^{165}Ho	^{136}Xe	1130	3.29	n	a	Sch 79f
^{58,64}Ni	^{16}O	96	3.41	n	f	Gem 80c
^{197}Au	^{86}Kr	724	3.42	α	g	Mil 78
^{165}Ho	^{56}Fe	465	3.59	n	a	Hil 79c
^{165}Ho	^{56}Fe	465	3.59	α, p	c	Kos 83
Ag	^{84}Kr	664	3.81	α	c	Sob 81, Sob 81b
^{58}Ni	^{40}Ar	280	4.13	α, p	c	Bab 80
^{27}Al	^{14}N	100, 70	5.55, 3.40	α	c	Bil 79
^{58,60,64}Ni	^{13}C	105	5.75	α, p	h	Ost 76
^{197}Au	^{32}S	373	6.34	α	i	Gam 78, Gam 79
^{58}Ni	^{16}O	144, 86	6.38, 3.38	α	j	Ho 83, Ho 77
^{93}Nb	^{40}Ar	400	6.66	α, p	c	Kuh 80
^{166}Er	^{86}Kr	1020	7.20	n	f	Tse 81
^{58}Ni	^{14}N	148	8.01	α	j	Bho 79
^{93}Nb	^{16}O	204	9.45	α	k	You 80
^{93}Nb	^{16}O	204	9.45	b	f	Gav 81
Cu	^{20}Ne	252	9.89	p	k	Sch 81c
Ti	^{16}O	310	17.20	α	j	Ho 80

[a] Neutron evaporation from fully accelerated fragments.
[b] Report that the deformation energy at scission is a large part of the total excitation energy, especially for the light fragment.
[c] Dominant sources of α particles in the damped reaction process are the two fully accelerated fragments.
[d] Preequilibrium α-particle emission is a major fragment-decay mode following (^{16}O,^{12}C) reaction. See, however, next comment, e.
[e] Evaporation of α's from an intermediate nucleus ^{31}P* accounts for the majority of the C–α coincidence events.
[f] Component of preequilibrium neutrons from damped reaction fragments.
[g] Although the dominant sources of α particles are the two fully accelerated fragments, some yield of α particles in the direction perpendicular to the fragment separation axis.
[h] Some evidence that at least part of the damped reaction events are due to many-particle reactions.
[i] Coincident α particles arise from quasielastic break up of the projectile-like reaction fragment.
[j] Some α emission processes are proposed in addition to sequential emission from the fragments.
[k] Angular and energy distributions of emitted charged particles are consistent with evaporation from fully accelerated fragments if thermal fluctuations in the division of the excitation energy between the two fragments are included.

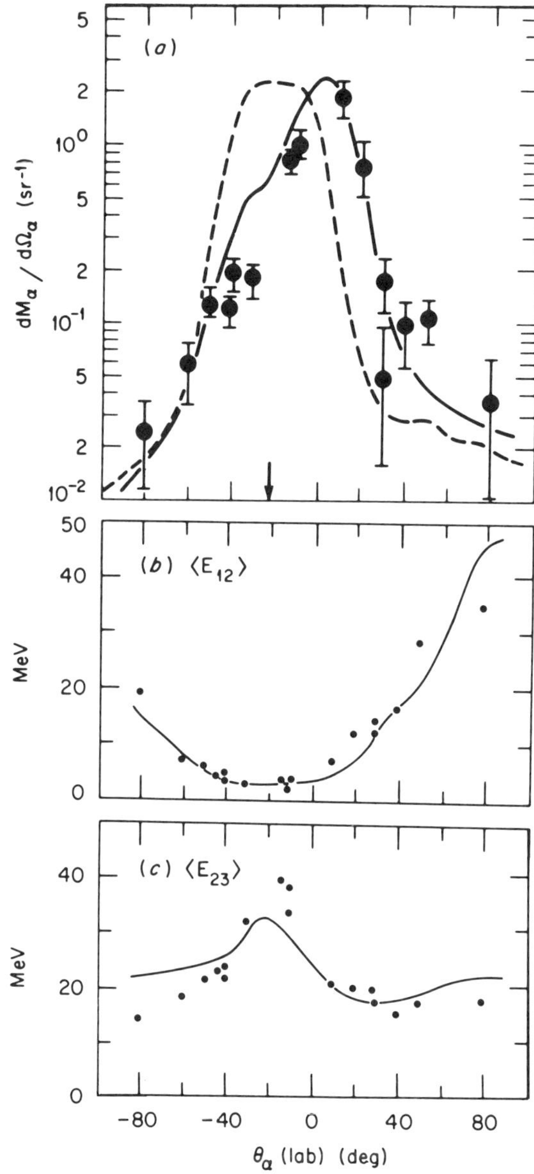

Figure 242. Alpha particles in coincidence with ^{12}C fragments detected at $\theta_{HI} = -21°$ from the reaction $^{93}Nb + ^{16}O$ at $E_{Lab} = 204$ MeV. (a) Differential α-multiplicity plotted vs. laboratory emission angle, (b) average α-^{12}C Q value, and (c) average α-^{93}Nb Q value vs. α emission angle. The dashed curve in (a) represents a simulation of sequential α emission from the asymptotic fragments, assuming a primary reaction angle close to $\theta_{HI} = -21°$. The solid curves result from accounting for the deflection of the primary fragments due to the recoil of the emitted α particle. From (You 80).

Another important effect to be considered in the interpretation of fast particle emission patterns is due to thermal fluctuations in the division of the energy loss between the partners of a damped reaction. This has been pointed out by Schmitt *et al.* (Sch 81c), who studied proton emission in the reaction Cu + ^{20}Ne at 12.6 MeV per nucleon. In this work, forward-peaked angular distributions of high-energy protons were observed both in singles and in coincidence with a fragment, corresponding to 1–2.4 times the velocity of the beam particles. Inclusive and exclusive proton spectra were found to have a similar shape, a fact often assumed to be indicative of a direct breakup-like emission process. Nevertheless, Schmitt *et al.* (Sch 81c) were able to reproduce these data by calculations based on a sequential proton evaporation process following a damped reaction which leads to a Gaussian distribution in excitation energies of the damped fragments that has a considerable width. The vector addition of the velocity of the projectile and of the relative velocity of the evaporated proton in the projectile frame of reference leads, in this model, to the high-energy proton component observed experimentally. Their angular distribution is also found (Sch 81c) to be approximately consistent with the model assumptions. The question of thermal fluctuations in the fragment excitation energy distribution will be addressed further below.

However, while the work of Young *et al.* (You 80) and Schmitt *et al.* (Sch 81c) point to potentially important complications in the interpretation of α-particle emission patterns, these effects are found (Ho 82) insufficient to account satisfactorily for energy and angular correlations of α particles in coincidence with damped reaction fragments from the ^{58}Ni + ^{16}O reaction at 6 and 9 MeV per nucleon. In these experiments, excess intensities of high-energy α particles are observed (Ho 83) that are not easily reconcilable with the above model of sequential emission from thermally equilibrated primary reaction fragments. Although the dominant production mechanism for these particles is presently not firmly established, the breakup mechanism may contribute to these emission processes.

Projectile breakup processes appear also to be present in reactions induced by the slightly heavier projectiles ^{20}Ne and ^{32}S. For example, Takada *et al.* (Tak 81) report on the observation of projectile-breakup and breakup-fusion in the reaction ^{40}Ca + ^{20}Ne at $E_{\rm Lab} = 260$ MeV. In Figure 243, results of this experiment, where ^{16}O ejectiles and α particles have been measured in coincidence, are represented in terms of an energy correlation plot, where the ^{16}O energy is plotted vs. α energy for each measured event. Figure 244 allows an identification of the kinematical regions in the experimental distribution shown in Figure 243. The heavy straight lines in both figures represent loci of constant three-body Q value, Q_3', corrected for the 4.73-MeV of minimum energy necessary for a disintegration of ^{20}Ne into ^{16}O and an α particle. The dashed curves in Figure 244 indicate constant

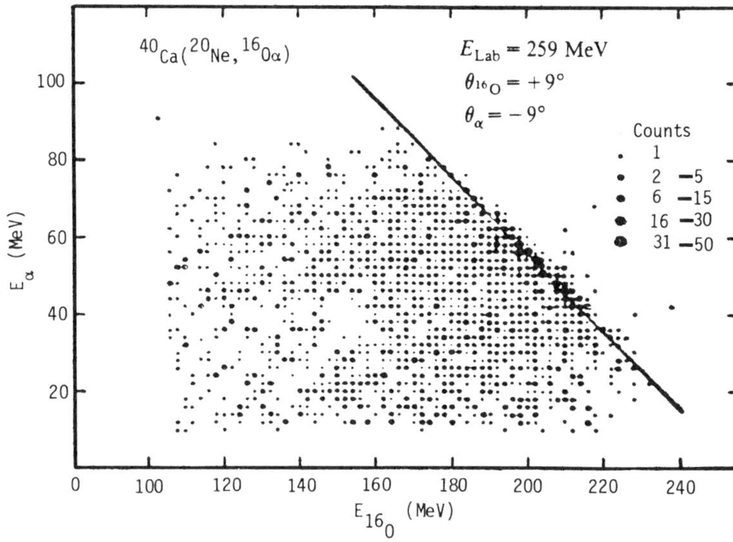

Figure 243. Two-dimensional energy correlation plot of coincident ^{16}O ejectiles and α particles from the reaction $^{40}Ca + ^{20}Ne$ at $E_{Lab} = 259$ MeV. The size of the dots reflects the experimental intensity of counts. The heavy-ion telescope was placed at $\theta_{^{16}O} = +9°$, symmetrically to the α telescope ($\theta_\alpha = -9°$). The solid line shows the kinematical locus for zero three-body Q-value. From (Tak 81).

excitation energies of the ^{20}Ne system, whereas the dotted curves give the loci of constant scattering angle of ^{20}Ne, θ_{Ne}, which is the center of mass of the outgoing ^{16}O and α particles.

A clustering of events about the line of $Q_3' = 0$ is observed in Figure 243, signifying a relatively high probability for elastic or nearly elastic breakup with high ^{16}O kinetic energies. However, there are also a considerable number of more inelastic events, corresponding to larger excitation energies of the ^{20}Ne system. It is noted by Takada *et al.* (Tak 81) that the α–^{16}O coincidence cross section is considerably smaller than expected from the ^{16}O singles production cross section corresponding to the high-energy, breakup part of the ^{16}O-energy spectrum, believed (Tak 81) to be distinguishable from a broader, lower-energy distribution of transfer events. Hence, if it can be assumed that the high-energy ^{16}O ejectiles are dominantly due to direct projectile breakup, the associated α particles must have been captured by the ^{40}Ca-target nuclei. This process, termed "breakup fusion" by Takada *et al.* (Tak 81), is similar to "incomplete fusion" or "massive transfer" processes identified by other investigators of lighter-ion-induced reactions quoted earlier.

The above experimental results were interpreted in terms of the direct-breakup model of Udagawa *et al.* (Uda 79, 80), who evaluate the DWBA

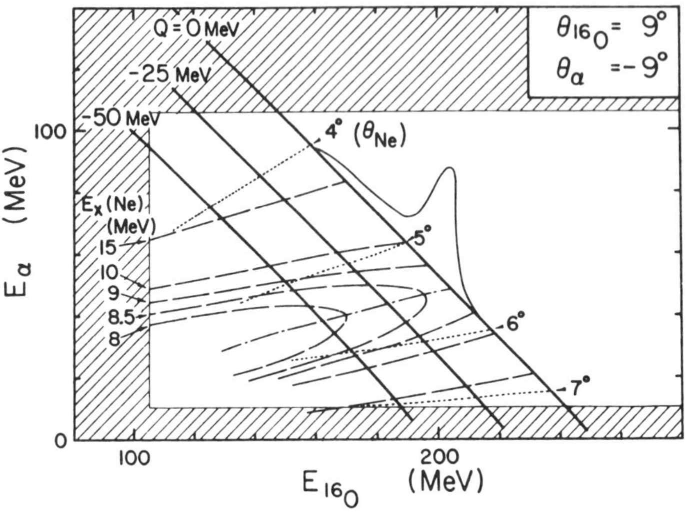

Figure 244. Schematics of kinematical regions for the experimental distribution shown in Fig. 243. The heavy straight lines indicate loci of constant three-body Q value. The dashed curves outline regions of constant intrinsic excitation energy, E_x(Ne), of the projectile. The scattering angle, θ_{Ne}, of the outgoing ^{20}Ne can be determined from the dotted curves. The hatched area in the diagram signifies the region excluded by the detector thresholds. From (Tak 81).

amplitude

$$T = \left\langle \chi_{\mathrm{Ne}}^{(-)} \chi_{\alpha}^{(-)} \middle| U_{\mathrm{OCa}} + U_{\alpha\mathrm{Ca}} \middle| \phi_{\alpha} \chi_{\mathrm{Ne}}^{(+)} \right\rangle \tag{3.7.8}$$

Here $\chi_{\mathrm{Ne}}^{(\pm)}$ are the distorted waves of ^{20}Ne with respect to the target (Ca), with the usual notation for the boundary conditions. The functions ϕ_{α} and $\chi_{\alpha}^{(-)}$ are the bound-state and continuum wave functions, respectively, of the α particle with respect to the ^{16}O projectile remnant. The interactions U_{OCa} and $U_{\alpha\mathrm{Ca}}$ are, in the actual calculations, approximated by real potentials of the Woods–Saxon type. The effect of subsequent absorption of breakup α particles by the target nuclei is, in the model, accounted for by renormalizing the amplitudes in the partial-wave analysis of the transition matrix element in Equation (3.7.8) by factors involving the penetration probability of α particles, in relation to the S-matrix element for α scattering. This theoretical description provides a good representation (Tak 81) of singles and coincidence ^{16}O cross sections in the above ^{40}Ca + ^{20}Ne reaction, if model parameters are allowed to be adjusted slightly. Figure 245 demonstrates the agreement between experimental and theoretical singles energy spectra of ^{16}O and α particles from the ^{40}Ca + ^{16}O reaction. Hence, from

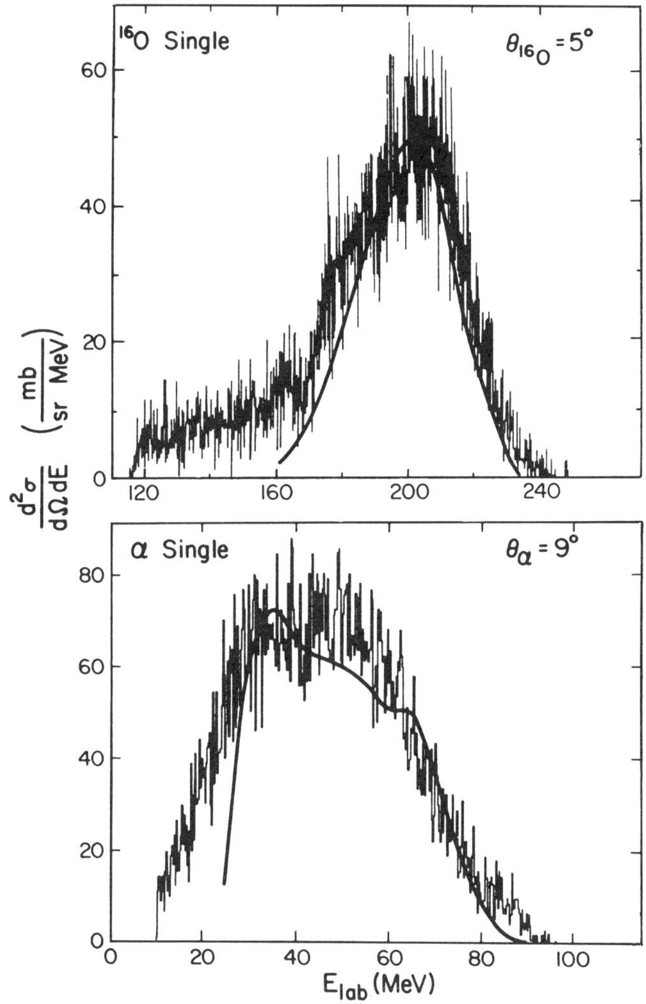

Figure 245. Comparison of singles spectra of ^{16}O and α particles from the reaction $^{40}Ca + (259$ MeV) ^{20}Ne to theoretical results accounting for breakup fusion. The solid curves show the calculated spectra normalized to the experimental spectra. From (Tak 81).

this work (Tak 81) evidence is suggested to be present in support of a direct projectile breakup reaction mechanism, for this system.

However, recent exclusive, high-resolution experiments performed by Homeyer *et al.* (Hom 82) on the dissociation of ^{20}Ne in the reaction $^{197}Au + ^{20}Ne$ at 290 MeV provide evidence conflicting with the above one-step direct reaction mechanism (Uda 79, 80). The bombarding energy corresponds approximately to 170 MeV above the interaction barrier. Figure 246 reproduces some of the experimental results obtained in the measurements. Part (a) depicts correlation plots between α energy $E_{\alpha\,\mathrm{Lab}}$ and the energy E_{PL} of the coincident projectile-like fragments ^{16}O (top) and ^{15}N (bottom). Parts (b) and (c) show the excitation energy distributions of the corresponding target-like and projectile-like fragments, respectively.

From the energy correlation diagrams in Figure 246a, it is discovered that coincident events are preferentially associated with discrete relative energies E_{rel} of ejectile and α particle. These discrete energies correspond (Hom 82) to particular excited states in the ejectile–α systems ^{20}Ne* and ^{19}F*. Figure 246b indicates that, in the (^{20}Ne, ^{16}Oα) reaction, the target-like reaction partner remains with a high ($\cong 50\%$) probability in its ground state (line at $E^*_{\mathrm{TL}} = 0$). With about equal probability, a Au fragment can be produced with a broad excitation distribution, centered around 20 MeV. These latter processes are the only ones of significance in the (^{20}Ne, ^{15}Nα) reaction, involving a transformation of the target nucleus. The excitation energy spectra of the coincident projectile-like fragments displayed in Figure 246c are characterized by a pronounced line structure corresponding to excited states in ^{20}Ne and ^{19}F, respectively. Several of these states are also found (Hom 82) to be selectively excited in direct inelastic scattering or proton pickup reactions on ^{20}Ne. The Ne and F fragments are identified (Hom 82) as the sources of the emitted α particles as inferred from the directional correlation of the α-particle velocities with the fragment direction of flight.

Monte Carlo simulation calculations, assuming sequential emission of α particles from projectile-like primary fragments and adopting experimental fragment angular and excitation energy distributions, performed by Homeyer *et al.* (Hom 82) are found in good overall agreement with experimental angular correlations of the coincident α particles. A small unexplained α cross-section component, observed on the side of the beam opposite to the heavy ejectile detection angle, is taken (Hom 82) as the only possible indication of uncorrelated [cf. Equation (3.7.7)] breakup or early α emission processes occurring prior to the formation of the primary damped reaction fragments. Hence, these results corroborate an earlier study of the reaction ^{27}Al + ^{16}O performed by Tsang *et al.* (Tsa 81), where it was found that α-particle emission resulted mainly from the decay of an intermediate ^{31}P* nucleus produced in a first reaction step.

It would be interesting to learn whether the sequential decay mechanism observed (Hom 82) in the above ^{197}Au + ^{20}Ne reaction is the dominant source of fast α particles also in other reactions induced by relatively light ions, where high-resolution measurements resolving individual states are possible. Although the ^{197}Au + ^{20}Ne reaction was studied at a relatively high energy and may, hence, not be easily compared to the lower-energy damped

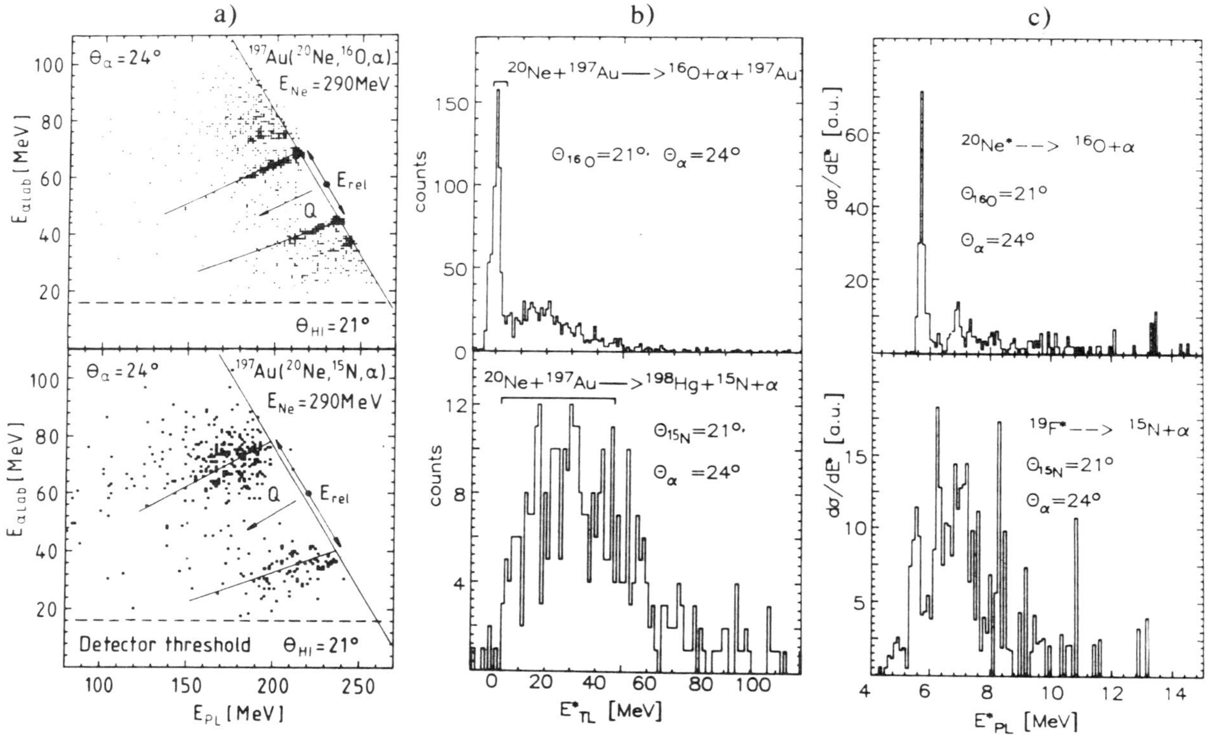

Figure 246. Experimental spectra of correlated fragments from the 290-MeV ^{197}Au(^{20}Ne, ^{16}Oα) (top) and ^{197}Au(^{20}Ne, ^{15}Nα) (bottom) reactions. Part (a) gives the distribution of correlated laboratory energies of projectile-like (PL) ejectile and α particle. Parts (b) and (c) display the excitation energy (E^*) spectra of target-like (TL) and projectile-like fragments. The arrows in part (a) indicate the direction of increasing negative Q values; E_{rel} denotes the relative kinetic energy in the ejectile-α particle frame. From (Hom 82).

reactions of interest here, a significant contribution of fast particles from direct breakup processes in damped reactions seems unlikely. Such a behavior may, at least, be expected for peripheral heavy-ion collisions, where reaction times are sufficiently short. For more central collisions, however, it is conceivable that breakup of the projectile may occur even during the relatively long reaction time. Consistency with the findings discussed above would suggest that such processes are initiated only after substantial amounts of relative kinetic energy have been dissipated. Due to the close proximity of all particles reached in near-central collisions, a variety of subsequent processes may occur. These include proximity scattering, discussed earlier in this section, as well as fusion of one or both breakup fragments with the reaction partner.

It is such a scenario that would allow to bridge the gap between incomplete fusion (or massive-transfer) reactions, on the one hand, and damped, fusion and fusion–fission-like reactions, on the other hand. A partially or completely damped reaction stage would then usually precede the subsequent fusion or scattering processes. In particular, fusion involving only one of the projectile remnants would occur to the extent to which it is allowed by the reaction conditions prevailing after dissipation of kinetic energy and initial angular momentum in the entrance channel. This appears to justify the following brief discussion of incomplete fusion in the context of this section.

Incomplete fusion reactions between heavy ions, associated with the emission of fast particles, have been studied by a number of groups (e.g., Ina 77, 79, Zol 78, Yam 79, Siw 79b, 79c, Wil 79b, 79d, 80b, 82). Wilczyński and collaborators, for example, have investigated ^{12}C- and ^{14}N-induced reactions on rare-earth targets, at bombarding energies between 7.5 and ~17 MeV per nucleon. Energy spectra, absolute cross sections, and average γ-ray multiplicities for incomplete-fusion reactions were measured in these coincidence experiments. It is found (Wil 82) that reaction channels leading to α particles as ejectiles play a dominant role in reactions such as $^{159}\mathrm{Tb}+{}^{14}\mathrm{N}$, although other reactions, in which smaller fractions of the projectile are captured by the target nucleus, contribute also significantly to the total cross section.

Following Bondorf *et al.* (Bon 71), Wilczyński *et al.* (Wil 82) assume that the probability $P(i)$ for a binary division i of the intermediate dinuclear system is given by the statistical level density of the heavy reaction partner alone, i.e.,

$$P(i) \propto \exp\{[Q_{gg}(i) - Q_c(i)]/\tau\} \tag{3.7.9}$$

Here, the quantities Q_{gg} and Q_c denote the ground-state Q value for the fragmentation i and the change in Coulomb energy due to charge transfer, respectively, and τ stands for an effective nuclear temperature of the target nucleus. The limited validity of a similar formulation for the description of damped-reaction mass and charge distributions in terms of the thermodynamic temperature has been outlined in Section 3.4 [cf. Equation (3.4.2)]. In addition, Wilczyński *et al.* (Wil 82) assume that Equation (3.7.9) determines also the probability for complete fusion of the system. Furthermore, the model postulates that transfer of mass to the target occurs only when the orbital angular momentum of the captured fragment is smaller than a critical angular momentum l_{cr}, calculated from the liquid-drop surface tension. Specifically, this angular momentum is given by

$$\left(l_{cr}+\tfrac{1}{2}\right)^2=\frac{\mu_{12}(C_1+C_2)^3}{\hbar^2}\left[4\pi\gamma\frac{C_1C_2}{C_1+C_2}-\frac{e^2Z_1Z_2}{(C_1+C_2)^2}\right] \tag{3.7.10}$$

Here, μ_{12} denotes the reduced mass of target (1) and captured fragment (2), and the quantities C_i stand for the half-density radii defined by Equations (2.5.3) and (2.5.5). The expression (3.7.10) is based on the assumption that the total interaction force between target nucleus and captured fragment at a separation distance of $r=C_1+C_2$ should be attractive, in order to effect the capture. The capture cross section is then written as

$$\sigma(i)=\pi\lambda\!\!\!^{-2}\sum_{l=0}^{l_{max}}(2l+1)T_l(i)P(i)\Big/\sum_j T_l(j)P(j) \tag{3.7.11}$$

where T_l is a smooth-cutoff transmission coefficient

$$T_l(i)=\left[1+\exp\left(\frac{l-l_{cr}(i)}{\Delta}\right)\right]^{-1} \tag{3.7.12}$$

depending on the critical angular momentum, l_{cr}, given in Equation (3.7.10) and a width parameter, Δ. This latter quantity, along with the effective temperature τ and a Coulomb radius R_c, used to calculate $Q_c(i)$ in Equation (3.7.9), constitute the free parameters of the phenomenological model. It is important to realize that the form of Equation (3.7.11) implies that no dissipation of initial orbital angular momentum occurs prior to fusion.

Figure 247 demonstrates the quality of model fits to the various partial cross sections observed (Wil 82) in the reaction $^{159}Tb+{}^{14}N$ at $E_{Lab}=140$ MeV. Values of the fit parameters are $\tau=3.5$ MeV, $R_c/(A_1^{1/3}+A_2^{1/3})=1.5$ fm, and $\Delta=1.7\hbar$. The overall reproduction of the data shown in Figure 247 is satisfactory, although several noticeable discrepancies exist. In particular,

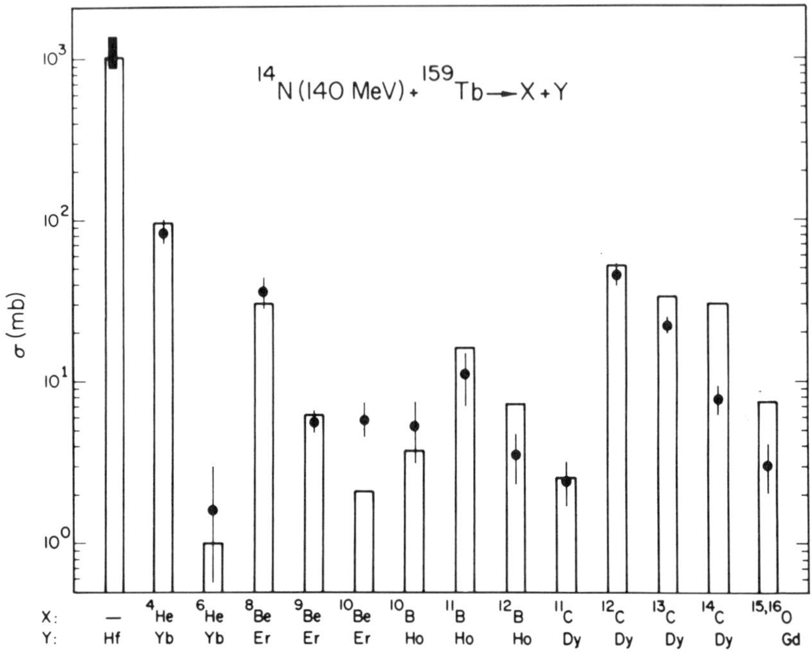

Figure 247. Integrated cross sections for various channels in the reaction ^{159}Tb$+^{14}$N at $E_{\rm Lab}=140$ MeV, indicated on the abscissa by the final products X and Y. Circles represent experimental data, while the bars are the results of a fit with the incomplete-fusion (sum rule) model proposed by Wilczyński *et al.* From (Wil 82).

the α-particle production cross section in relation to neighboring reaction channels is well described by the model.

As noted earlier, the above model for incomplete fusion, also called the "sum rule model," does not account for dynamical contributions to the Q values discussed in Section 3.5. It also neglects the excitation energy of the projectile-like reaction partner in the energy balance, which, however, can be significant, as data such as shown in Figure 246 suggest. For these reasons, it is presently not obvious how to interpret the overall agreement between model and data displayed in Figure 247, although the extracted parameters appear (Wil 82) to have reasonable values.

Dynamical effects of angular-momentum dissipation are considered in a study of 282-MeV ^{20}Ne-induced reactions on targets ranging from ^{165}Ho to ^{238}U performed by Huizenga *et al.* (Hui 83). These authors employ experimental information on the total linear momentum of either evaporation residues or fission fragments produced in a reaction to infer whether complete or incomplete fusion of projectile and target nuclei have occurred.

The sizes of the windows in initial angular momentum contributing to various complete and incomplete fusion processes were deduced from the experimental relative cross sections for events with linear-momentum transfers indicative of the emission of up to four α particles emitted prior to or simultaneously with the fusion of target and projectile remnant. The limiting values, l_{FLMT}, of initial angular momentum leading to complete fusion turned out (Hui 83) to be consistently larger than predicted by the condition that the total effective interaction potential for target and projectile should exhibit a minimum (pocket). Employing the one-dimensional nuclear proximity potential (cf. Section 2.5), the corresponding maximum l_{POCKET} can be calculated for the different projectile–target systems.

In Figure 248 (Hui 83), the deduced ratios, $l_{\text{FLMT}}/l_{\text{POCKET}}$, are plotted, for six different reactions, as a function of incident energy per nucleon

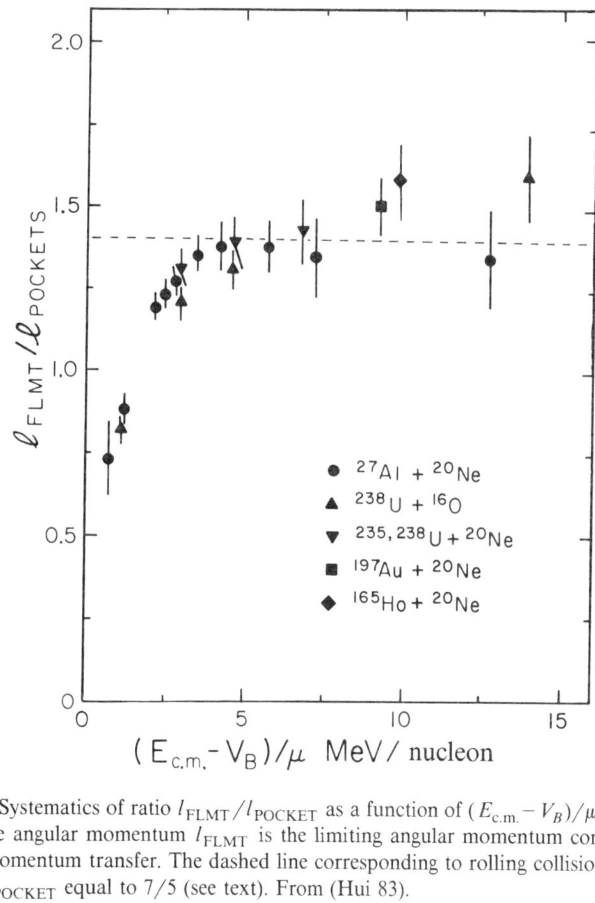

Figure 248. Systematics of ratio $l_{\text{FLMT}}/l_{\text{POCKET}}$ as a function of $(E_{\text{c.m.}} - V_B)/\mu$ in energy per nucleon. The angular momentum l_{FLMT} is the limiting angular momentum corresponding to full linear momentum transfer. The dashed line corresponding to rolling collisions has a value of $l_{\text{FLMT}}/l_{\text{POCKET}}$ equal to 7/5 (see text). From (Hui 83).

above the fusion barrier. The dashed horizontal line in this figure represents a ratio 7/5, corresponding to the rolling limit [cf. Equation (2.3.9)] attained by the projectile–target system. As can be seen from the experimental data points in Figure 248, the critical angular momenta l_{FLMT} deduced from experiment increases with increasing bombarding energy, until, at about 4 MeV per nucleon above the barrier, the rolling limit is reached. Above this energy, no further increase in limiting l value for complete fusion occurs. The additional l waves are found (Hui 83) to be associated with an incomplete fusion–fission reaction of the target with only part of the projectile, i.e., with simultaneous emission of light particles.

If the above model interpretation is correct, one would expect that also for a band of l waves adjacent to the fusion-like angular-momentum window, emission of light particles would occur only after substantial damping has taken place. Such a process would be indicated by experimental systematics similar to the one demonstrated in Figure 248, but where the bombarding-energy dependence of the limiting angular momentum $l_{MT}(i)$, corresponding to the momentum transfer induced by the fusion of projectile fragment i with the target, is plotted in relation the last l value, $l_{POCKET}(i)$, for which the interaction potential exhibits a minimum. Unfortunately, no such systematical studies are available yet.

In summary of the previous discussion, one can conclude that there is substantial evidence for the emission of light particles during or after a nuclear reaction induced by a relatively light heavy ion. Sequential breakup (see top of Figure 236) appears to be a likely candidate for such an emission mechanism, that can be found, in an analogous form, also in incomplete fusion reactions. Such an emission mechanism, mainly associated with α particles, attains increasing importance at energies of the order of or above 10 MeV per nucleon, but has been observed (e.g., Bri 61, Ho 77, Tsa 81) also at incident energies of 6–7 MeV per nucleon.

While the above process may be of significance only for damped reactions involving nuclei whose structure renders them relatively unstable against breakup into particular fragmentations, it may not constitute a general characteristic of damped reactions. However, there are a number of possible particle emission processes that are related to particular dissipation and equilibration mechanisms that have been proposed in the literature, and are, hence, of potentially higher significance for an understanding of the microscopic features of the damped reaction mechanism. These emission processes include the so-called Fermi jet emission (Bon 76b, 79, 79c, 80, Gro 77b, Rob 79, Moo 81), a mechanism closely related to the one-body picture of nuclear interactions, the possible ejection (Gro 77b) of particles into the continuum, due to a sudden stoppage of the relative motion, as predicted by the hard-friction model (Gro 78), and preequilibrium emission of fast particles from localized zones of hot nuclear matter (Bet 38, Got 77, 79,

Moo 81), possibly created in a damped heavy-ion collision. The latter mechanism is a special example of more general preequilibrium emission processes that are commonly observed in nucleon- and light-ion-induced reactions, as reviewed by Blann (Bla 75,76). Their presence or absence in damped reactions is indicative of the rapidity of the energy dissipation process, as can be inferred from the dependence of particle emission times on the temperature of the emitter, illustrated in Figure 32. Finally, there will be emission of particles from the last stage of a heavy-ion collision, when excited primary fragments decay in flight with deexcitation patterns characteristic of the excitation energy and spin distributions, acquired by the fragments at the end of the damped interaction. Possibly, this decay could occur for fragments that are in simple modes of excitation, such as collective surface vibrations (cf. Sections 2.3 and 3.5).

The phenomenon of Fermi jets predicted first by Bondorf (Bon 76b) is schematically outlined in Figure 249 (Sch 79f). It is due to the exchange of nucleons between projectile and target during the approach phase, when the relative ion–ion velocity is high and a relatively small window between the reaction partners has already opened. Owing to the vector addition of momenta illustrated in Figure 249, an exchanged nucleon with intrinsic Fermi momentum $\mathbf{p}$ and kinetic energy ε will have a momentum $\mathbf{q} = \mathbf{p} + m\mathbf{u}$ in the recipient nucleus, which is augmented, due to the relative ion–ion velocity $\mathbf{u}$. Depending on the relative directions of $\mathbf{p}$ and $\mathbf{u}$, the energy

$$E \approx \varepsilon + \mathbf{p} \cdot \mathbf{u} \tag{3.7.13}$$

of the nucleon may be sufficiently high to render its emission energetically possible. Kinematical conditions on the bombarding energy E_{Lab} and

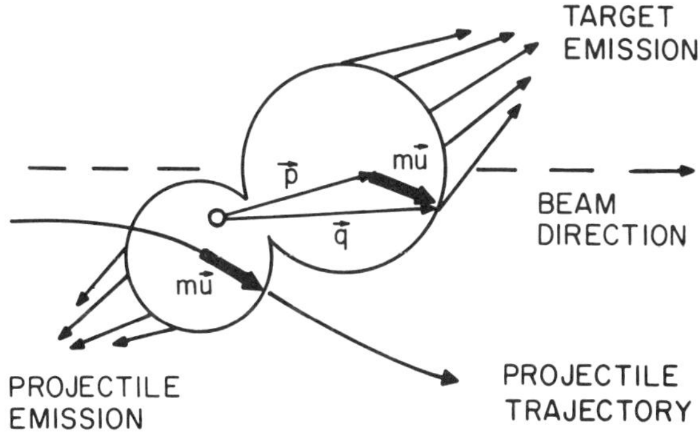

Figure 249. Pictorial view of the Fermi-jet emission mechanism. The quantity $\mathbf{p}$ is the intrinsic momentum of nucleons of mass m; $\mathbf{u}$ denotes the velocity of relative motion. The direction of propagation changes at the nuclear surface because of refraction. From (Sch 79f).

angular momentum l permitting Fermi jets energetically are given by

$$E_{\text{Lab}} \geq \frac{A_P + A_T}{A_T}\left[V_{\text{Coul}}(\overline{R}) + \mu(\sqrt{V} - \sqrt{\varepsilon})^2\right]^{1/2}$$

$$l \leq \overline{R}\left\{2\mu\left[E_{\text{c.m.}} - V_{\text{Coul}}(\overline{R}) - \mu(\sqrt{V} - \sqrt{\varepsilon})^2\right]\right\}^{1/2} \qquad (3.7.14)$$

as derived by Robel (Rob 79). In Equations (3.7.14), A_P and A_T are the mass numbers of projectile and target, μ is the reduced mass, and $V_{\text{Coul}}(\overline{R})$ is the Coulomb interaction potential evaluated at a characteristic separation distance $\overline{R}$. The quantity V is the potential barrier for the emission of the Fermi-jet particle. Emission can in principle occur, with 180° symmetry, from the target or the projectile, in the frame of reference rotating with the dinuclear system.

There are, obviously, several effects that limit the Fermi-jet emission probability. First of all, whenever the relative velocity **u** is large, the window between the collision partners is small, prohibiting an effective exchange of particles. Secondly, there is nucleon–nucleon scattering in the recipient nucleus, due to the residual interaction, randomizing the motion of the transferred nucleons. The direct emission of these nucleons is further diminished by refraction at the nuclear surface. These effects have been taken into account in an approximative fashion by Robel (Rob 79), who combines the Fermi jet model with a trajectory calculation based on one-body proximity forces.

Figure 250 shows results of calculations with this Fermi-jet model (Rob 79) for the $^{64}\text{Cu} + {}^{20}\text{Ne}$ reaction at $E_{\text{Lab}} = 252$ MeV and an initial angular momentum of $l = 82$. In the figure, isomultiplicity contour lines $d^2N/d\Omega_p\, dE_p$ for proton emission from the target are depicted in a polar plot vs. emission angle and proton laboratory energy. Note that the multiplicity is given in the relatively small units of 10^{-4}/(MeV sr). The arrow in Figure 250 indicates the asymptotic projectile scattering angle predicted by the trajectory calculation. Several characteristics are quite obvious from the distribution depicted in Figure 250. First, Fermi-jet proton energies can be much larger than those corresponding to beam velocities, due to the amplifying effect of the intrinsic Fermi motion of the nucleons [cf. Equation (3.7.13)]. Secondly, these protons are predicted to be emitted into a narrow angular range, which may allow their experimental distinction from particles emitted in other processes, despite the low overall intensity predicted (Rob 79) for Fermi-jet emission.

For reasons of energetics, the Fermi-jet emission mechanism can be expected to be of significance only for the high-momentum components of the intrinsic nucleon distribution. The calculations reproduced in Figure 250

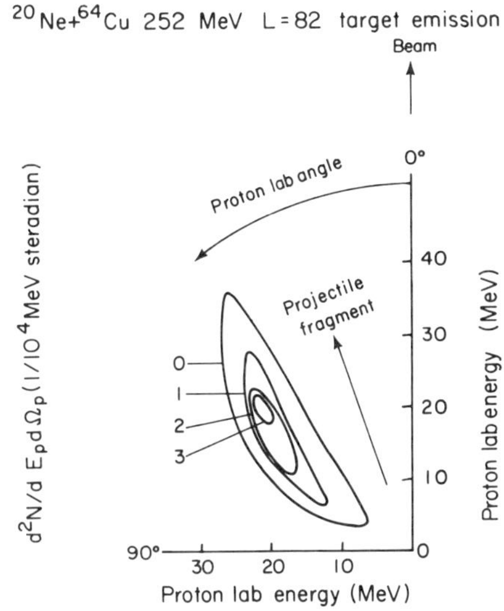

Figure 250. Fermi-jet emission of protons predicted for the reaction $^{64}Cu + ^{20}Ne$ at $E_{\rm Lab} = 252$ MeV and initial orbital angular momentum $l = 82\hbar$. Depicted are contour lines of the theoretical double-differential multiplicity $d^2N/dE_p\,d\Omega_p$ in units 10^{-4}/(MeV sr) in a polar plot where the distance from the origin indicates the laboratory proton energy. The arrow indicates the asymptotic projectile scattering angle predicted from a trajectory calculation. From (Rob 79).

are based on a Fermi-type momentum distribution of uncorrelated nucleons, which underestimates just these high-momentum components of the Fermi motion in real nuclei. A Fermi-type momentum distribution is given by

$$f(p) = \left[1 + \exp\left(\frac{p^2 - p_F^2}{2m\tau}\right)\right]^{-1} \tag{3.7.15}$$

where p_F is the Fermi momentum, m the nucleon mass, and τ the nuclear temperature. This distribution overemphasizes the low-momentum components but decreases exponentially towards high momenta at $p \approx p_F$, with a steep gradient determined by the generally small value of the nuclear temperature. In contrast, the momentum distribution of nucleons in a harmonic oscillator well is given by a more smoothly varying Gaussian dependence with a width parameter that is significantly larger than that describing the falloff of the Fermi distribution [Equation (3.7.15)]. As discussed, e.g., by Zabolitzky *et al.* (Zab 78), such a Gaussian momentum distribution of nucleons within a nucleus may be realistic for momenta $k \leq 2$ fm^{-1}, corresponding to instantaneous kinetic energies of approximately 80 MeV inside the nucleus. These authors suggest a probability for higher momenta that exceeds the Gaussian distribution by orders of magnitude, an effect due to nucleon–nucleon correlations. Although such high-momentum components are probably not tested in a low-energy heavy-ion collision, using a more realistic momentum distribution of uncorrelated nucleons could lead to predictions of enhanced emission and significantly different energy and angular distributions of the emitted particles. This uncertainty complicates an unambiguous experimental identification of Fermi jet emission patterns.

Calculations of instantaneous emission of neutrons and protons in heavy-ion collisions have also been performed by Bondorf *et al.* (Bon 79, 80) in their "PEP" (promptly emitted particles) model, which is rather similar to the above Fermi-jet model. However, it accounts also for second-chance emission of nucleons having suffered nucleon–nucleon collisions (two-body PEPs). The total cross section for (one-body) PEP emission can be written (Bon 79, 80) as

$$\sigma_t = 2\pi \int_0^{b_{\max}} db\, b \int_{t_{\min}}^{t_{\max}} dt\, P_{\text{PEP}}(t, b) \tag{3.7.16a}$$

with a differential emission probability

$$P_{\text{PEP}}(t, b) = \int_A dA \int_0^{v_F} d^3\mathbf{v}_a\, \rho(\mathbf{v}_a)\, n(\mathbf{v}_b)\, e^{-s/\lambda} \tag{3.7.16b}$$

that depends implicitly on impact parameter, b, and interaction time, t, through the time evolution of the system trajectory. In Equation (3.7.16a), $b_{\max}$ is the grazing-impact parameter, the quantity $t_{\min}$ is the time at which the interface between the interacting collision partners becomes transparent to nucleon exchange. At time $t_{\max}$, the relative ion–ion velocity $\mathbf{u}$ has diminished by so much, due to conservative and dissipative forces, that the velocity $\mathbf{v}_b = \mathbf{v}_a + \mathbf{u}$ of a nucleon transferred from nucleus A to nucleus B is no longer sufficient to overcome the barrier. In Equation (3.7.16b), the function $\rho(\mathbf{v}_a)$ denotes the zero-temperature Fermi velocity distribution of the nucleons in the donor nucleus A extending up to the Fermi velocity v_F, and $n(\mathbf{v}_b)$ stands for the one-sided proximity flux of nucleons. Absorption within the recipient nucleus is taken into account by the exponential factor in Equation (3.7.16b). Here, the nucleon track length s is comparable to the nuclear diameter, since the nucleonic mean free path λ is relatively long (compare Figure 25). The mean free path is approximately connected to the imaginary part W of the optical potential by the relation $\lambda = \hbar v_b / 2W$.

Both of the above models predict very small cross sections for the prompt, or instantaneous, emission of light particles in damped collisions.

The best chance to observe such particles experimentally appears to be provided (Bon 80) in studies of very asymmetric projectile–target combinations with high relative velocities **u** in the c.m. system, because of the resulting distinct angular anisotropy of particle cross section and energy. This effect is due to the vector addition of particle and fragment velocities. If $\mathbf{v}_b$ is the velocity of a transferred nucleon inside the recipient nucleus B, its velocity $\mathbf{v}_b'$ after emission from this nucleus is reduced from $\mathbf{v}_b$ because of the potential barrier. In the overall center-of-mass system, the particle velocity $(\mathbf{v}_b')_{\text{c.m.}}$ is further reduced by the c.m. velocity $(\mathbf{v}_B)_{\text{c.m.}}$ of the emitter, as given by (Bon 80)

$$(\mathbf{v}_b')_{\text{c.m.}} = \mathbf{v}_b' + (\mathbf{v}_B)_{\text{c.m.}} \tag{3.7.17}$$

because the emitting nucleus has generally a velocity component antiparallel to $\mathbf{v}_b'$. The reduction is, obviously, smallest for a heavy nucleus B, since in the c.m. system, its velocity $(\mathbf{v}_B)_{\text{c.m.}}$ is inversely proportional to its mass. The forward–backward asymmetry of the c.m. energy spectra of PEP or Fermi-jet neutrons emitted in a collision of a light projectile with a heavy target is schematically depicted in Figure 251 (Bon 80). This figure displays for a forward and a backward angle the Fermi jet component (dotted region) on which an evaporation component (hatched region) is superimposed. In the backward hemisphere, where projectile Fermi jet emission is predicted to contribute also, this component is largely masked by the dominant evaporation spectrum, due to the relatively high c.m. velocity of the projectile. In the forward region, on the other hand, a tail of high-energy neutrons associated with the instantaneous Fermi jet emission mechanism extends in c.m. energy, $E^n_{\text{c.m.}}$, significantly above the evaporation component.

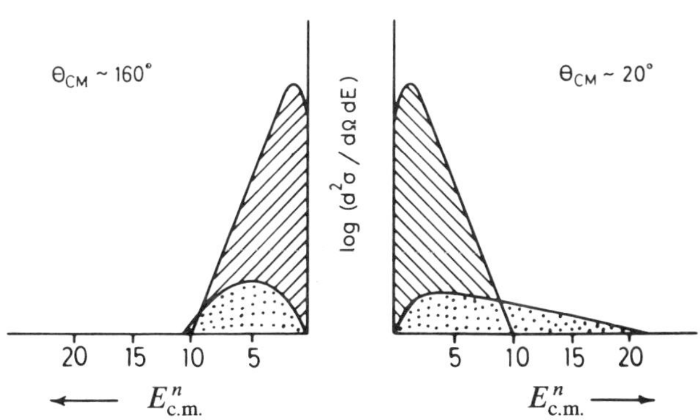

Figure 251. Schematical illustration of the c.m. energy spectrum of Fermi-jet (dotted area) and statistically evaporated (hatched area) neutrons produced in a reaction of a light projectile with a heavy target. From (Bon 80).

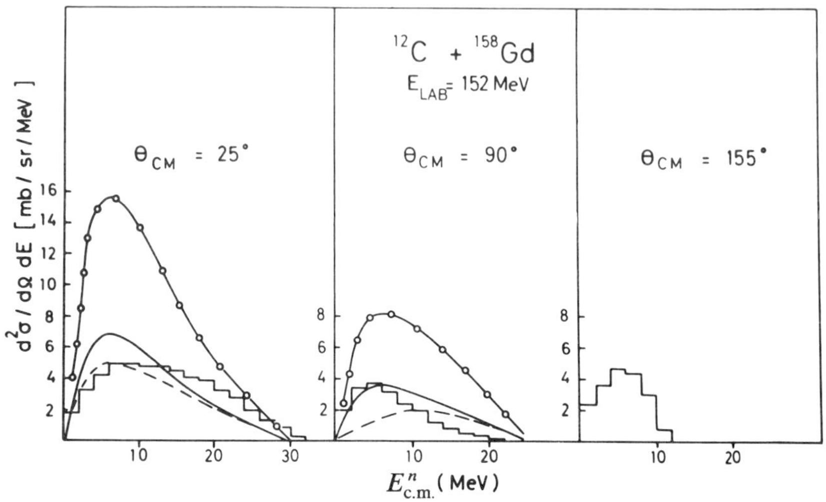

Figure 252. Experimental high-energy components (solid curves) of neutrons emitted in the fusion-evaporation reaction $^{158}Gd + ^{12}C$ at $E_{Lab} = 152$ MeV and predictions (histograms) by the PEP model. The differential neutron cross section is plotted vs. neutron energy, $E^n_{c.m.}$, in the c.m. system, for three angles $\theta_{c.m.}$. The dashed and intermediate solid curves represent two different decompositions of the experimental neutron spectrum, which also contains an evaporative component. The curves connecting circles result from the solid curves by scaling the experimental spectrum of neutrons associated with evaporation residues up by the ratio of total fusion to evaporation residue cross section. From (Bon 80).

The above model of Bondorf *et al.* (Bon 79, 80) has been applied to neutron emission from the fusion reaction $^{158}Gd + ^{12}C$, investigated experimentally at $E_{Lab} = 152$ MeV by Westerberg *et al.* (Wes 78d, Sar 78b). In these experiments, neutron energy spectra are observed that exhibit a high-energy component, in addition to a lower-energy evaporation spectrum. In Figure 252, the experimental high-energy, nonevaporative c.m. neutron energy distributions are shown as solid curves, for three neutron emission angles. These spectra decay at high energies approximately in an exponential fashion, i.e.,

$$dN/dE^n_{c.m.} \propto \exp(-E^n_{c.m.}/E^n_0) \qquad (3.7.18)$$

characterized by a decay constant $E^n_0 \approx 6$ MeV. Curves in Figure 252 with open circles are obtained (Bon 80) from the solid experimental curves by scaling with the ratio of the measured fusion and evaporation residue cross sections. The dashed curves are also results of an analysis of the experimental total neutron spectra using, however, a slightly increased nuclear temperature for a description of the Maxwilliam evaporation spectrum. Hence, the circles and the dashed curves illustrate the latitude in possible decompositions of the data in terms of equilibrium and nonequilibrium neutron components.

The histograms in Figure 252 represent the predictions of the model of Bondorf *et al.* (Bon 80) for one-body and two-body PEP-neutrons. Although the theoretical spectrum for 25° is consistent with observations, within the large margins of error, at 90° the same calculation underestimates considerably the high-energy portion of the experimental spectrum, indicating a broader than predicted angular distribution of energetic particles emitted in the reaction. No data are available for $\theta_{\text{c.m.}} = 155°$, where the model predicts still a sizable neutron cross section associated with a broad bump in the theoretical angular correlation, due to backward projectile emission of PEP neutrons (cf. Figure 249).

The spectra of Fermi-jet neutrons and protons differ from each other, of course, owing to the presence of the additional Coulomb barrier affecting the emission of charged particles. The energy-dependent barrier transmission for emission of charged particles of asymptotic energy E and angular momentum l, can approximately be calculated employing the Hill–Wheeler formula (Hil 53),

$$P_l(E) = \left[1 + \exp\left(\frac{2\pi}{\hbar\omega}\left[E_B(l) - E\right]\right)\right]^{-1} \tag{3.7.19}$$

Here, $E_B(l)$ is the barrier height and ω is its curvature, in analogy to the definition in Equation (3.4.18). Although the above transmission coefficient influences the overall shape of the charged-particle energy spectrum, its main effect is a low-energy cutoff near $E = E_B$ and, hence, generally a sizable reduction in cross section for charged-particle emission. Furthermore, refraction at the nuclear surface will be larger for charged particles as compared to the effect on neutrons, leading to a broadening of angular correlation functions.

In view of the significance of one-body nucleon exchange for the relaxation of mass and charge asymmetry as well as for the dissipation of relative kinetic energy in damped reactions, suggested by data discussed in previous sections, it appears important to establish the conditions under which Fermi-jet or PEP emission mechanisms are operative in these reactions. So far, there is no unambiguous experimental evidence for their presence, although a potential contribution of these processes in certain reactions cannot be ruled out (see, e.g., Ger 82). Theoretical models, on the other hand, are also still rather crude and incorporate unjustified assumptions on which the predictions depend dramatically. The insufficiency of Fermi-gas estimates for a description of the higher-momentum components of the nucleonic momentum distribution has already been mentioned.

Furthermore, the use of geometric optics to model trajectories and scattering processes of low-energy nucleons exchanged between the reaction partners seems problematic. Recent quantum-mechanical calculations by Brosa and Krone (Bro 81e), within a schematical model of particle emission, illustrate the possible importance of quantal fluctuations for an explanation of the broad angular distributions of high-energy particles emitted in damped reactions above bombarding energies of 10 MeV per nucleon.

In addition to the above Fermi-jet particle emission mechanism, Gross *et al.* (Gro 77b) discuss qualitatively the possibility that the moving wall of the single-particle potential in a heavy-ion collision can excite a few nucleons or an α particle to unbound states ("piston model"). These states can subsequently decay via the emission of nucleons or α particles with velocities comparable to or even higher than the projectile velocity. It appears essential (Gro 77b) for the validity of the "hard-friction" model of nuclear dissipation that a large fraction of the initial relative kinetic energy is transferred to single-particle modes within a short pulse ($\sim 10^{-22}$ s). In contrast to the Fermi-jet mechanism, the strong radial friction force assumed in the model should be capable of inducing fast-particle emission with significant probability, already for partially damped, peripheral collisions. Figure 253 depicts typical system trajectories for l values between the grazing l, L_{gr}, and the critical l value for fusion L_{cr}, predicted by the hard-friction model for the reaction ^{232}Th$+^{40}$Ar at $E_{\text{Lab}} = 379$ MeV. This reaction exhibits a characteristic orbiting-type deflection function, discussed in earlier sections. As is indicated in Figure 253 by arrows, labeled in correspondence to the heavy-ion trajectories shown, emission angles for fast α particles, relative to the heavy-ion scattering angle, depend on the character of the trajectory. Whereas for partially damped collisions α emission is predicted to occur dominantly on the side of the beam opposite to the heavy-ion detection angle, α particles from strongly damped, orbiting-type collisions are expected on the same side of the beam as the projectile-like reaction fragments. Except for near-central collisions, the emission patterns of highly energetic charged particles in coincidence with damped reaction fragments, are, according to these qualitative considerations, asymmetric with respect to the beam direction.

Thus far, no experimental data on particle angular correlations have been reported in support of the predictions of the above piston model for fast-particle emission in damped reactions. Experiments performed by Gelbke *et al.* (Gel 77b) on angular correlations between fast α particles and projectile-like fragments in ^{16}O-induced reactions on ^{208}Pb and ^{197}Au targets at $E_{\text{Lab}} = 140$ and 310 MeV yield, in fact, results that appear to be inconsistent with such an emission mechanism. The results of these measurements are interpreted (Gel 77b) in terms of a fast, nonsequential emission process, where α particles associated with partially damped reac-

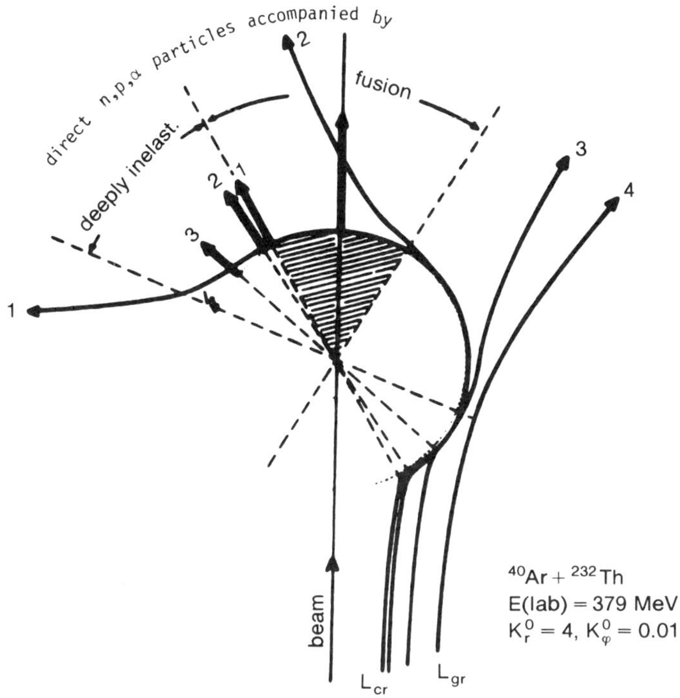

Figure 253. Correlations between deflection angles of damped reaction fragments (trajectories) from the reaction $^{232}Th + 379$-MeV ^{40}Ar and estimated emission angles (heavy arrows) of α particles produced by the strong radial friction force in the hard-friction model. From (Gro 77b).

tions are observed on the same side of the beam as the heavy ejectiles, whereas for strongly damped, orbiting-type collisions, α-particle emission occurs on the side opposite to the fragment detection angle. In view of the earlier discussion in this section, it appears difficult, however, to draw conclusions about the details of the α-particle production mechanism, before simulation calculations, including the effects of α-particle recoil and of statistical fluctuations become available for the reactions studied by Gelbke *et al.* (Gel 77b).

In reality, sharp angular correlations between instantaneously emitted particles and the heavy reaction fragments cannot be expected. The dispersing effect of nuclear refraction has already been discussed as one of the contributors to the broadening of particle–fragment angular-correlation functions. A similar effect is due to statistical, dynamical, and quantal dispersions of the system trajectories about an average classical path, as discussed in Section 2.2. The two latter types of dispersion have been considered by Dünnweber *et al.* (Dun 80) in a calculation based on Strutinsky's phenomenological model of heavy-ion angular distributions (cf. Section 2.2).

The expressions given in Equations (2.2.8) and (2.2.9) for the angular distribution of the coherent reaction cross section can be approximated by a sum of the two Gaussian cross section terms

$$\frac{d\sigma^{(\pm)}}{d\Omega_{\mathrm{HI}}} \propto \frac{1}{\sin\theta_{\mathrm{HI}}} \exp\left[-\frac{(\theta_{\mathrm{HI}} \pm \theta_0)^2}{\xi^2}\right] \tag{3.7.20}$$

where θ_0 is the average deflection angle, associated with the classical deflection function [compare Equation (2.2.7)]. As discussed in Section 2.2, the angular dispersion parameter ξ contains contributions from dynamical and quantal fluctuations. The two terms in Equation (3.7.20) can classically be interpreted as originating from interactions of the projectile with opposite sides of the target nucleus. Figure 254 gives a schematic illustration of realistically broad angular distributions (solid curves), associated with average strongly damped trajectories of both types. The dashed curves in this figure indicate the angular distributions $\sigma_\alpha^{(\pm)}$ assumed in the calculation for α particles emitted early in the reaction, labeled according to the associated heavy-ion trajectories. The mean deflection angles of these α particles correspond (Dun 80) to continuations of the projectile Coulomb trajectories. Assuming, for simplicity, a factorization of the coincidence cross section, the calculations predict for the shape of the α-particle angular correlations distributions of the type shown in Figure 255 (Dun 80) for the reaction Th(Ar, Sα) at $E_{\mathrm{Lab}} = 400$ MeV, and an α-particle energy of 40 MeV.

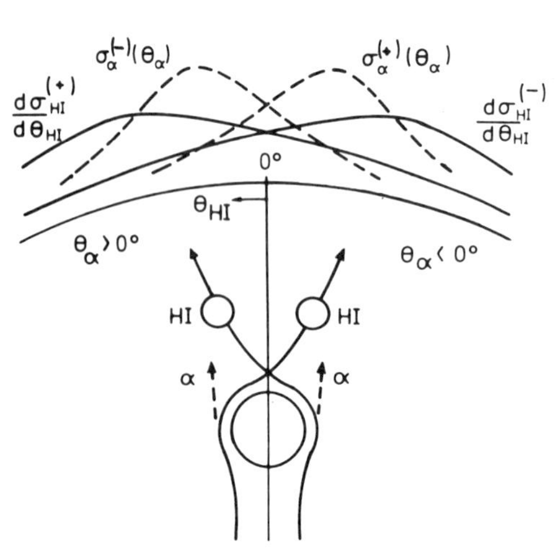

Figure 254. Schematic illustration of light charged-particle emission at an early stage of a damped reaction. Average heavy-ion trajectories and their angular spread $d\sigma^{(\pm)}/d\theta_{\mathrm{HI}}$ are depicted for orbiting-type collisions with cross sections $\sigma_\alpha^{(\pm)}$ of associated α particles concentrated on the side of the beam opposite to the heavy-ion detection angle. From (Dun 80).

Depicted are α-particle correlation functions for two windows of reaction Q values. The experimental dependences of cross section and average scattering angle on the Q value have been used to determine average values of the entrance-channel angular momentum and of the angular dispersion parameter ξ in an analysis of the type described in Section 3.3. As observed from Figure 255, for partially damped, nonorbiting collisions, the model predicts broad angular distributions of α particles, centered close to the fragment detection angle θ_{HI}. Because of the relatively small dispersion ξ for partially damped events (cf. Section 3.3), the contribution of α particles, associated with average deflection to the opposite side of the beam, is small, resulting only in a shoulder in the angular distribution. This situation is changed for strongly damped, orbiting-type collisions. Here, the angular correlation exhibits clearly peaks on both sides of the beam, as seen in the lower part of Figure 255.

Although the model calculation discussed above makes reference to a particular, breakup-like emission mechanism of α particles, it serves as an illustration of the sizable widths of particle–fragment angular correlations that can realistically be expected in experiments. It also emphasizes the importance of examining experimentally the dependence of such correlations on the energy loss or reaction Q value. The variation of the correlation function with Q value or energy loss, demonstrated in Figure 255, is characteristic for the particular emission mechanism. For example, for the mechanism suggested by Gross *et al.* (Gro 77b), outlined further above, one

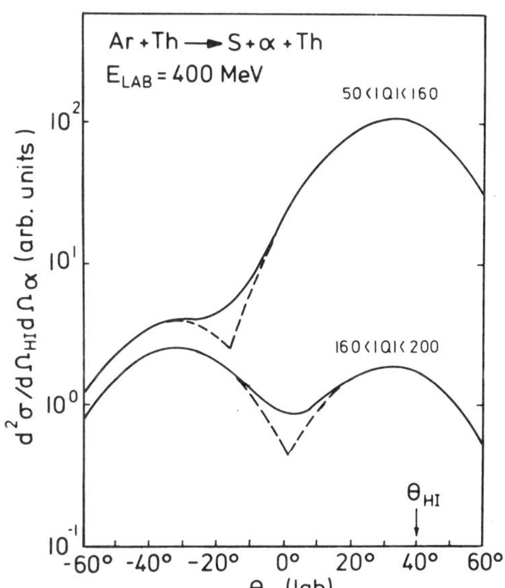

Figure 255. Theoretical α-particle angular correlations for the reaction Th + Ar at $E_{Lab} = 400$ MeV and two regions of reaction Q values. Plotted in this figure is the double-differential cross section vs. α-particle detection angle. The scattering angle of the projectile-like fragment is indicated by an arrow. The dashed curves indicate the separate contributions from either side of the beam, while solid curves represent their superpositions. From (Dun 80).

might expect angular correlations which are qualitatively similar to those of Figure 255 but reflected about $\theta_\alpha = 0°$. Unfortunately, however, for this latter model (Gro 77b), no quantitative calculations of particle cross sections and angular correlations are available.

Correlations between α particles and projectile-like fragments have been measured by Gamp *et al.* (Gam 78, 79) in the reaction $^{197}Au + ^{32}S$ at $E_{Lab} = 373$ MeV, corresponding to 6.3 MeV per nucleon above the interaction barrier. The fast α particles detected in singles in this experiment have features similar to that of breakup particles discussed earlier in this section. The singles α-particle angular distribution is strongly forward-peaked, with a large integrated cross section of approximately 800 mb (Gam 78). However, only 5% of the damped reaction fragments, mainly ^{30}Si products, are found (Gam 78) to be in coincidence with α particles of energies greater than 25 MeV. This finding excludes projectile breakup as a major source of fast α particles in the Au + S reaction.

Average kinetic energies of the α particles emitted in this reaction are found to be ~10 MeV per nucleon, i.e., slightly smaller than the corresponding value (12 MeV per nucleon) for the projectile. In Figure 256 (Gam 78), angular distributions of the coincidence cross section $d^2\sigma/d\Omega_{HI}\,d\Omega_\alpha$ are plotted vs. laboratory α detection angle $\theta_{\alpha\,Lab}$, for several Z values of the projectile-like fragment. The top part of this figure corresponds to the higher-energy, partially damped fragment component, whereas correlations for the completely damped, orbiting-type component are displayed in the lower portion of Figure 256. The fragment detection angle of $\theta_{Lab\,HI} = 35°$ is indicated by an arrow in Figure 256. It is quite surprising that most of the coincident α particles are observed (Gam 78) to be associated with the partially damped fragment component and not with completely relaxed events.

As can be inferred from Figure 256, the α correlation function for partially damped fragments are peaked at $\theta_{\alpha\,Lab} \approx 15°$, well inside the heavy-ion detection angle. This is qualitatively what is expected according to the theoretical distributions, shown for a somewhat heavier system at the top of Figure 255. Although no quantitative comparison between these calculations and the $^{197}Au + ^{32}S$ data is possible, it appears, however, that for this system the experimental angular correlations for fragments with relatively high Z values are somewhat narrower than what the predictions for fragment breakup in the Th + Ar reaction suggest. For angles removed from the maximum of the correlation function in Figure 256, data are of poor statistics, but there is no definite indication of a factorization of the coincidence cross section into a product of terms describing α emission and damped reaction separately. The angular distribution appears to broaden significantly with increasing energy loss, as is most obvious from a comparison of the data for the strongly damped events, shown in the bottom

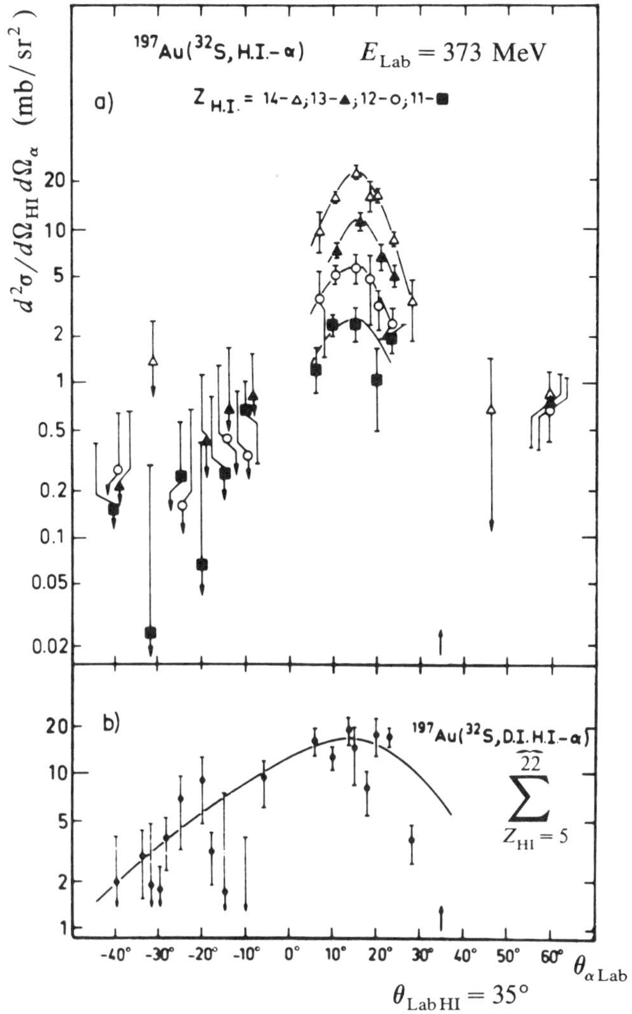

Figure 256. Angular correlation of α particles produced in coincidence with projectile-like fragments in the reaction $^{197}Au + ^{32}S$ at $E_{Lab} = 373$ MeV. The top part shows the double-differential cross section $d^2\sigma(\theta_{\alpha Lab})/d\Omega_{HI}\, d\Omega_{\alpha Lab}$ for partially damped fragments with $Z = 11$, 12, 13, and 14. The lower part corresponds to completely relaxed events, summed over fragment Z values from $Z = 5$ to 22. The position of the coincident-fragment detector ($\theta_{LabHI} = 35°$) is indicated by an arrow. The curves are drawn to guide the eye. From (Gam 78).

portion of Figure 256, with those for partially damped Au + S collisions at the top. Similarly, the widths of the out-of-plane α-particle distributions are broader (Gam 79) for strongly damped Au + S collisions than for partially damped events. For the latter class of reactions, the out-of-plane distribution broadens with decreasing charge of the detected projectile-like fragment, i.e., increasing charge exchange between the reaction partners.

Certainly, the reaction $^{197}Au + {}^{32}S$ at $E_{Lab} = 373$ MeV, studied by Gamp *et al.* (Gam 78, 79), does not exhibit the characteristics of a fragment breakup process in an obvious fashion. Calculations with the Fermi jet or PEP models are, unfortunately, not available, nor has the piston model referred to earlier been applied in detail to this system. Since, however, these models can be expected to predict particle cross sections in coincidence with partially damped fragments to peak on the side of the beam opposite to the fragment detection angle, there appears to be no definite experimental evidence in support of a significant contribution of these emission mechanisms in the partially damped Au + S reactions at 12 MeV per nucleon considered here. In the case of the strongly damped, orbiting component, however, neither mechanism can be excluded, in the absence of detailed model calculations.

The fact that in the above $^{197}Au + {}^{32}S$ reaction, no α particles are observed with ^{32}S ejectiles, leads Gamp *et al.* (Gam 78) to conclude that these α particles are emitted from the projectile or the projectile-like fragment, in a partially damped collision. Most of the coincident α particles are found (Gam 79) to be correlated with ^{30}Si ejectiles, suggesting α emission after pickup of two neutrons by the ^{32}S projectile. The particular emission mechanism proposed initially by these authors (Gam 78, 79) is illustrated in Figure 257, depicting in part (a) the three-body (Au, Si, α) initial condition for a Coulomb trajectory calculation. Part (b) displays the predicted coincidence cross section, as well as the correlation between laboratory α scattering angle (θ_α) and the initial emission angle θ_{em} in the frame of the ^{28}Si nucleus. If all emission angles $0 \leq \theta_{em} \leq 360°$ were allowed in this model, a differential α-particle coincidence cross section peaking at 15° and 60°, respectively, results from the calculation, for an ejectile detection angle of $\theta_{HI} = 35°$. However, only a peak at $\theta_{\alpha Lab} = 15°$ is observed experimentally (cf. Figure 256). The absence of the second cross-section peak at 60° then appears to indicate that the α particle is produced in a region of the projectile facing the target nucleus, but mainly with forward emission angles $0° \leq \theta_{em} \leq 115°$ (Gam 78).

The mechanism proposed implies (Gam 78) that the dissociation of a ^{34}S projectilelike fragment occurs on a fast time scale, during the damped interaction, in a process similar to inelastic breakup. Information about the out-of-plane component of the initial velocity of the α particle can be obtained from the width of its out-of-plane angular correlation, assuming an

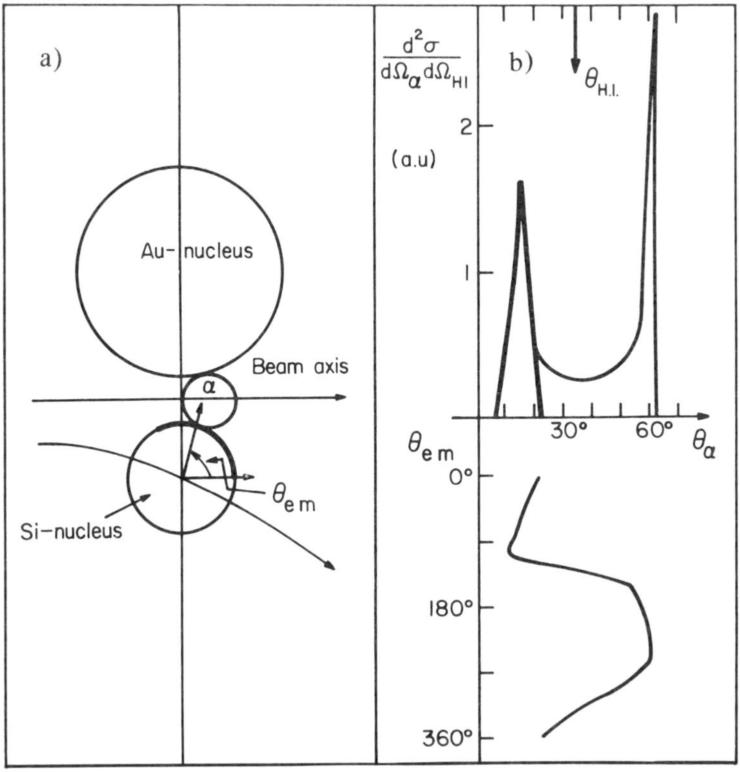

Figure 257. Model for fast α-particle emission in the reaction ^{197}Au + ^{32}S at $E_{\mathrm{Lab}} = 373$ MeV. Initial conditions for the Coulomb trajectory calculation are illustrated in part (a), with an initial emission angle θ_{cm} of the α particle in the frame of the projectile remnant. Part (b) shows the calculated correlation between θ_{cm} and the α-particle laboratory scattering angle θ_α (bottom) and the corresponding differential coincidence cross section. Restriction of θ_{cm} to the segment emphasized in part (a) corresponds to the forward peak of the correlation function shown in (b). The fragment detection angle is indicated by an arrow. From (Gam 78).

initial configuration such as depicted in Figure 257. This is so, because an α particle will receive the Coulomb acceleration by the combined target–ejectile system only to the extent that its velocity is larger than the velocity of separation of the two heavy fragments. The calculations of Gamp *et al.* (Gam 78, 79) then allow one to derive a mean-square relative velocity $\langle v_\alpha^2 \rangle^{1/2}$ between α particle and Si core. From the experimental width $\Delta\phi = 24°$ for the out-of-plane α correlation corresponding to $Z_{\mathrm{HI}} = 14$, one obtains (Gam 79) $\langle v_\alpha^2 \rangle^{1/2} = 0.35$ cm/ns. This corresponds to an energy of only 55 keV per nucleon of the α particle inside the S nucleus, prior to breakup. This energy of relative motion of the α cluster appears to be surprisingly small, although theoretical estimates for this quantity are unavailable. However, the interpretation of the data is highly model dependent. An adoption of initial conditions rather different from those in Figure 257 could well be expected to influence the conclusions on emission mechanism and initial α-particle velocities reached from a comparison of trajectory calculations to experimental data.

This is borne out, in fact, in interpretations of more recent experiments, where Gamp *et al.* (Gam 81, 82) have studied coincident α-particle emission in the damped reaction ^{197}Au + ^{36}Ar at $E_{\text{Lab}} = 380$ MeV. Qualitatively, the α-particle angular correlations have an appearance similar to the corresponding correlations for the ^{197}Au + ^{32}S reaction shown in Figure 256. In Figure 258, experimental α coincidence data are shown for partially damped ^{197}Au + ^{36}Ar collisions, associated with S ejectiles. Here, the Galilean-invariant cross section

$$\frac{1}{p_\alpha}\frac{d^3\sigma}{dE_\alpha\, d\Omega_\alpha\, d\Omega_{\text{HI}}} \propto \frac{1}{v_\alpha^2}\frac{d^3\sigma}{dv_\alpha\, d\Omega_\alpha\, d\Omega_{\text{HI}}} \tag{3.7.21}$$

of coincident α particles is plotted vs. the component, $v_{\|}$ and $v_{\perp}$, of the α-particle velocity v_α, parallel and perpendicular, respectively, to the beam direction. Strictly, the expression of Equation (3.7.21) represents the invariant cross section only for a fixed energy and emission angle of the heavy ion. Then, the specific shape of the invariant cross section pattern remains the same in reference frames connected to one another by Galilean transformations. Particular symmetries of these patterns, for example, an isotropic angular distribution of the cross section in a particular reference frame, can be used to identify the probable emission source of the particles and its velocity vector.

The circle labeled "1" in Figure 258, centered at the tip of the laboratory velocity vector $\mathbf{v}_S$ of a S ejectile, describes the locus of all laboratory velocity vectors of α particles, emitted with minimum (barrier) energy and isotropically, from the ejectile traveling with its asymptotic velocity. A corresponding circle (2) is drawn in the figure for emission from the target-like fragment. As can be inferred from Figure 258, there is no significant emission of α particles from the heavy reaction partner in a partially damped ^{197}Au + ^{36}Ar reaction, although it presumably obtains the major portion of the dissipated energy. Most of the particles exhibit a definite correlation with the circle corresponding to emission from the projectile-like fragment. In addition, a weak α component is visible on the side of the beam opposite to the fragment detection angle.

From a comparison of the cross section patterns depicted in Figure 258 and an evaporation calculation, it can be concluded (Gam 81) that the

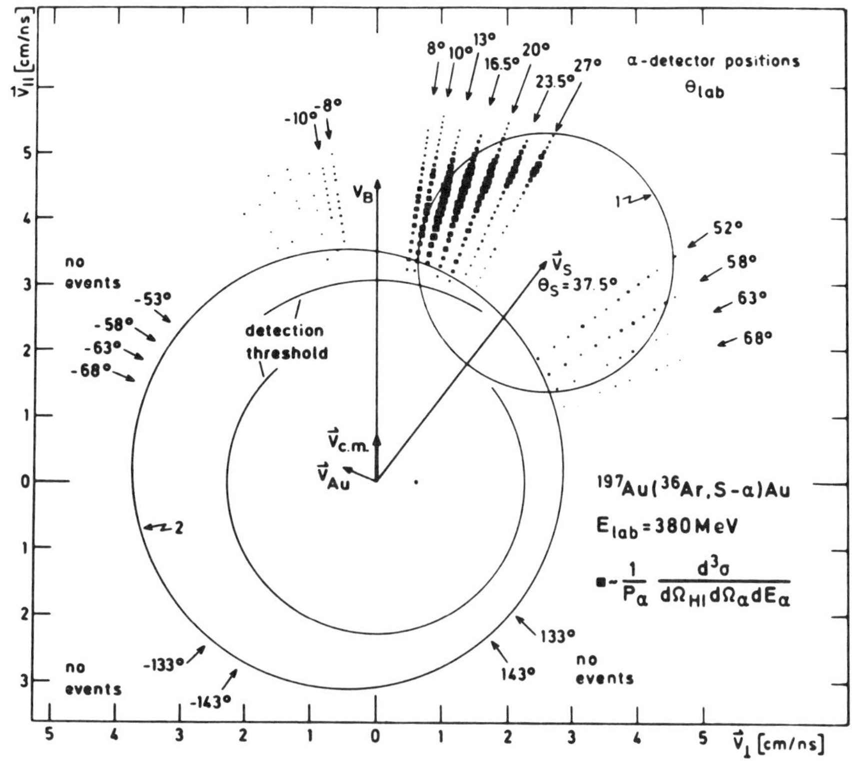

Figure 258. Galilean-invariant cross section (dots) of α particles emitted in the partially damped reaction $^{197}Au + ^{36}Ar$ at $E_{Lab} = 380$ MeV, plotted vs. α-particle velocity components parallel ($v_{\|}$) and perpendicular ($v_{\perp}$) to the beam velocity (v_B). Circle 1, centered at the tip of the velocity vector $\mathbf{v}_S$ of a detected S ejectile, describes the locus of velocity vectors of α particles with barrier energy emitted isotropically from the asymptotic fragment. Circle 2 corresponds to emission from a target-like fragment. The size of the dots represents the magnitude of the coincident cross section. Experimental α-particle detection thresholds are indicated by circle segments. From (Gam 81).

majority of α particles are emitted sequentially from the projectile-like reaction product. The distortion of the experimental cross section distribution from an expected circular shape is understood (Gam 81) to be due to the deflection that the secondary fragments suffer from the recoiling α particles. The strong forward peaking of the angular distribution of the primary reaction fragments is responsible for the apparent asymmetry of the coincident cross section about the ejectile detection angle. It is noteworthy that, except for the small cross section component on the opposite side of the beam, no other source of α particles is discovered within the experimental accuracy. Although a considerable hindrance of α-particle emission, due to the Coulomb barrier of a heavy nucleus, can be expected, it is, nevertheless, surprising that no charged particles are found in this reaction to be associated with target-like reaction fragments. A more detailed discussion of this effect would, however, require data on other possible decay channels.

In view of the above results, it appears likely that the α particles emitted in the partially damped reaction ^{197}Au + ^{32}S at 12 MeV per nucleon, discussed earlier, are also due to the process observed in the ^{197}Au + ^{36}Ar reaction. This finding emphasizes again that considerable care has to be exercised in interpreting multiparticle cross-section patterns, since statistical evaporation processes may emulate to a certain extent the properties of more exotic particle emission processes.

While the α-particle spectra and angular correlations for partially damped ^{197}Au + ^{36}Ar collisions at 380 MeV find a satisfactory explanation in terms of dominant statistical emission from the asymptotic projectile-like fragment (Gam 81), strongly damped interactions of the same system are probably associated with additional α-particle emission mechanisms. As can be seen from Figure 259 (Gam 82), showing the energy-integrated angular correlation of α particles in coincidence with strongly damped fragments from the ^{197}Au + ^{36}Ar reaction at $E_{\text{Lab}} = 380$ MeV, the α-particle component at negative angles, developed only weakly in partially damped collisions (cf. Figure 258), is now comparable in intensity to the sequential evaporative component at positive angles. This bimodal angular distribution for events with energy losses between 150 and 200 MeV is reminiscent of the theoretical correlation for strongly damped events displayed, for a different system,

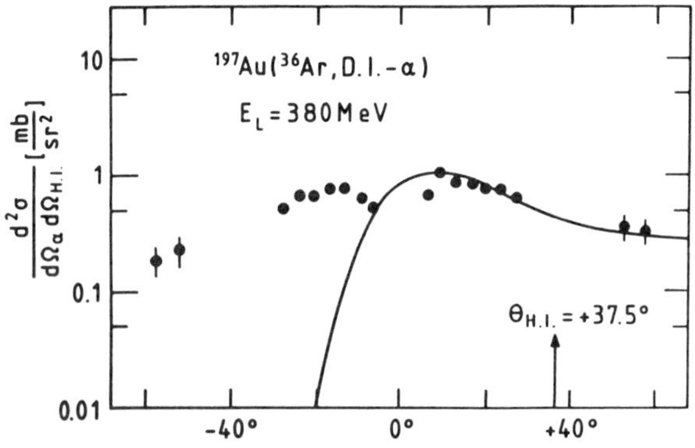

Figure 259. Angular correlation of α particles in coincidence with strongly damped fragments from the reaction ^{197}Au + ^{36}Ar at $E_{\text{Lab}} = 380$ MeV, plotted vs. laboratory α-particle emission angle θ_α. The detection angle of the coincident fragment is indicated by an arrow. The solid curve represents a Monte Carlo calculation of α-particle evaporation from the asymptotic projectile-like fragment. From (Gam 82).

in Figure 255. At α emission angles larger than those included in Figure 259, the cross section is found (Gam 82) to drop significantly. No coincident events are observed at a backward α-particle angle of $\theta_\alpha = 140°$, precluding significant contributions of particle evaporation from the target-like fragment which should yield almost equal intensity at forward and backward angles, because of the small velocity (0.5 cm/ns) of the heavy fragment in the laboratory system.

The solid curve in Figure 259 represents a Monte Carlo calculation of α-particle evaporation from the asymptotic projectile-like fragment, whose remnant is detected in the experiment at $\theta_{\mathrm{HI}} = 37.5°$. The theoretical curve has been normalized at $\theta_\alpha = 20°$ and accounts for an α-particle multiplicity of $M_\alpha = 0.35$ (Gam 82). However, this value for the integrated α-particle multiplicity is unexpectedly high. Although considerable uncertainty exists with respect to the N/Z ratio of the projectile-like fragment, its excitation energy and spin distributions, statistical model calculations (Gam 82), evaluating decay widths expressed in terms of nuclear level densities [cf. Equation (3.6.5)], predict multiplicities only of the order of $M_\alpha \sim 10^{-2}$. In the absence of more complete experimental data, including measurements of fragment masses and of other decay products, it remains somewhat uncertain what implications these discrepancies may present for an understanding of the damped-reaction mechanism.

In any case, the above calculations for statistical evaporation of α particles from the final fragments, produced in $^{197}\mathrm{Au} + {}^{36}\mathrm{Ar}$ collisions, cannot account for the component of α particles observed (cf. Figure 259) on the side of the beam opposite to the detection angle of the projectile-like fragment. Rather, this type of angular correlation is suggestive of a fast mechanism of α-particle production such as illustrated in Figure 254. Here, an α particle is emitted from the projectile-like fragment relatively early in a damped interaction, that eventually leads to an orbiting collision. Comparison of experimental α-particle angular correlations with results of trajectory calculations (Gam 82) with the "hard-friction model" (Gro 78) indicate emission times that range from 10^{-22} to 10^{-21} s. According to the model, α-particle emission takes place during the damped interaction, after some dissipation of relative kinetic energy. With increasing interaction time and increasing energy loss, the laboratory α-particle energy decreases, as the velocity of the emitter is slowed down. Simultaneously, the theoretical angular correlation function becomes broader and, eventually, extends to the other side of the beam. From the experimental data, however, it is quite impossible to disentangle unambiguously the components associated with instantaneous and sequential emission and to determine the character of the angular correlation of the fast α-particle component. More complete experimental investigations of various heavy-ion systems, covering a range of bombarding energies, appear to be necessary to elucidate this poorly understood mechanism of instantaneous α-particle emission.

A mechanism of fast-particle emission in damped reactions, conceptually different from the ones considered previously, has been introduced by Gottschalk *et al.* (Got 77, 79), who consider the preequilibrium decay of a spatially localized hot zone (Bet 38) of nuclear matter produced in a heavy-ion collision. Similar mechanisms have been discussed subsequently also by other workers (Nom 78, Gar 80, Mor 80c).

The possibility of producing a "hot spot" in damped reactions has so far not been rigorously analyzed in theoretical models. However, such a classical entity must correspond to a rather complicated superposition of single-particle and hole states, which is unlikely to be created in near-adiabatic heavy-ion collisions, damped by relatively gentle one-body friction forces. Hence, an experimental discovery of localized hot zones in the dinuclear system formed in damped collisions or in its breakup products would definitely associate the reaction mechanism with concepts such as diabaticity, strong coupling scheme (cf. Section 2.7), and strong, non-friction-like energy dissipation.

In the work of Gottschalk *et al.* (Got 77, 79) these aspects of hot-spot formation are not considered. The proposed model simply postulates the occurrence of local heating in damped reactions and follows the decay of a hot spot in terms of the heat diffusion model of Weiner *et al.* (Wei 77c), already discussed in Section 2.7. The solutions of the corresponding heat diffusion equation [Equation (2.7.31)] describe the time evolution of the temperature fields $\tau(\mathbf{r}, t)$ of the two interacting nuclei. Although this is not an inherent requirement of the model approach, the numerical evaluation of the temperature fields and of particle emission starts at the time ($t = 0$) of breakup of the intermediate dinuclear system, whose constituents are assumed to follow subsequently Coulomb trajectories. The model, hence, treats only sequential particle emission from the accelerating fragments. Emission probabilities of particles of type i, with binding energy S_i and kinetic energy ε, are calculated (Got 79) as

$$P_i(\varepsilon, \theta, t) = c_i \sigma_{\text{inv}}^i(\varepsilon, B_i)\, \varepsilon \exp\left[-\frac{\varepsilon + S_i}{\tau(R, \theta, t)}\right] \tag{3.7.22}$$

where θ is the emission angle in the rest frame of the emitting nucleus of radius R. The quantity σ_{inv}^i is the cross section for the inverse (capture) reaction, depending on the particle energy and the barrier B_i, as well as on the properties of the capturing nucleus. Equation (3.7.22) differs from familiar expressions for particle evaporation in the statistical model only by the replacement of the unique equilibrium temperature by the angle- and time-dependent values of the temperature field at the surfaces of the reaction fragments. It is a trivial transformation of Equation (3.7.22) to calculate particle energy spectra and angular distributions in other frames of reference, once particular conditions of the reaction fragments at breakup are adopted.

It is the nature of the above thermal relaxation mechanism that the decay in time of the temperature gradient assumed to exist at $t = 0$ is not characterized by a unique relaxation time, i.e., a purely exponential decay, but a whole distribution of such times. One can, however, roughly distinguish (Wei 77c) an early time, in which this decay occurs very fast, followed by a slower mode of approach to the equilibrium nuclear temperature. Consequently, the model predicts a component of fast particles emitted at the earliest stages of the thermal relaxation process, when the reaction partners are still in close proximity. A slower component of particles is emitted over a longer period of time, during the acceleration phase of the reaction fragments. According to the model (Got 79), this latter emission component becomes the dominant one for times larger than typically one-quarter of the asymptotic thermal relaxation time (cf. Section 2.7), which itself is a function of the equilibrium temperature. Estimates (Got 79) for the time scale on which the fast decay mode dominates yield values of the order of 10^{-22} s for a nucleus such as Ni and several tens of MeV of excitation energy. The emission patterns of light particles emitted during such short time periods after scission of the intermediate system are, hence, expected to be strongly influenced by the proximity of the primary reaction fragments.

Results of calculations performed by Gottschalk and Weström (Got 79), employing the above model, are displayed in Figures 260 and 261 for the emission of α particles and neutrons, respectively, in the reaction $^{58}Ni + {}^{16}O$ at $E_{Lab} = 96$ MeV. They represent a large number ($\approx 2 \times 10^4$) of three-body trajectories determined by the Coulomb repulsion, in the case of charged particles. Absorption of a light particle is assumed in the calculation, whenever its trajectory intersects the mass density distribution of one of the heavy reaction fragments. The calculations start from somewhat unrealistic initial conditions assumed to prevail at scission time of the intermediate complex. These conditions correspond to a $^{58}Ni + {}^{16}O$ grazing collision leading to a sticking situation. The scission radius is taken to be given by the sum of the radii of the two nuclei. A hot spot is assumed to have been formed on the surface of the Ni-like reaction fragment in the region of overlap with its collision partner.

In Figure 260 (Got 79), the time-integrated in-plane angular correlations and energy spectra of α particles from the $^{58}Ni + {}^{16}O$ reaction are exhibited for the fast (solid curves) and slow (dashed curves) decay modes. The center of the figure displays a contour plot of the energy–angle correlations of the α particles. The top part of Figure 260 shows the angular correlation of the α particles, measured with respect to the scattering angle

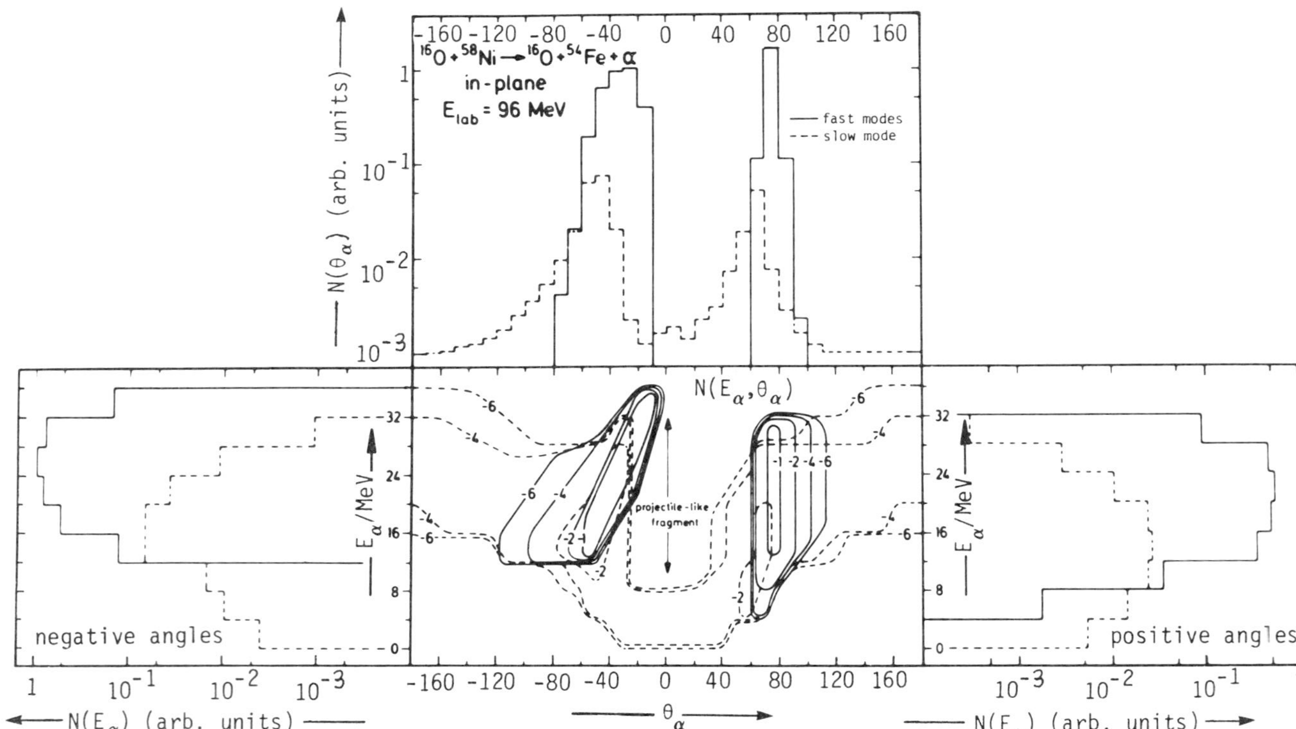

Figure 260. Theoretical c.m. angular and energy correlations of α particles emitted from a hot spot assumed to be produced in the damped reaction $^{58}Ni + ^{16}O$ at $E_{Lab} = 96$ MeV. A peripheral damped collision is considered, leading to rigid rotation of the dinuclear complex and negative scattering angles of the projectile-like fragment. Emission of α particles is assumed to occur after breakup of the system, when a hot spot created on the surface of the Ni-like fragment decays in time and space with a fast and a slow characteristic decay component. From (Got 79).

($\theta_\alpha = 0$) of the projectile-like fragment, while the outer parts represent the particle energy spectra for positive and negative emission angles θ_α. Positive angles θ_α correspond to the side of the beam direction, where the projectile makes initial contact with the target.

As is observed in Figure 260, the energy–angle correlations and their projections are characteristically different for α particles emitted in the fast and slow decay modes. The angular distributions of the α particles are asymmetric with respect to $\theta_\alpha = 0$ and exhibit a depression about the scattering direction of the projectile, more pronounced for particles emitted on the fast time scale. This effect is, in the model, due to absorption by the projectile. On both sides of this central angular region, the angular distribution of the fast-mode α particles decreases exponentially, while the distribution decreases more slowly for particles emitted in the slow mode. The spectra of the former particles are concentrated at higher energies than those of the slow-mode α particles. Marked differences exist between distributions for positive and negative angles. These features are related (Got 79) to the angular gradient of the temperature field $\tau(R, \theta, t)$ and the intrinsic rotation of the emitter.

The shadow effects clearly observed in the theoretical angular and energy distributions of "hot-spot α particles" shown in Figure 260, are also present in the corresponding distributions of neutrons and protons from the same source, as illustrated in Figure 261. However, due to the absence of Coulomb focusing, in the case of neutrons, these features are expected to be less well defined. For the same reason, the relative contribution of the fast mode should be smaller for neutrons than for charged particles. This is due (Got 79) to a higher absorption probability for neutrons, which are not deflected away from the projectile-like charged particles.

The hot-spot model calculations discussed above only illustrate the effects to be expected for nonequilibrium particle emission from localized regions on the primary reaction fragments. They are not realistic enough to warrant a detailed comparison to experimental data. However, features observed by Ho *et al.* (Ho 77, 83) in the experimental angular correlations of α particles in coincidence with oxygen and carbon fragments produced in the $^{58}Ni + ^{16}O$ reaction at 96 MeV are claimed (Ho 77, 83) to resemble qualitatively the features predicted from such an emission mechanism, although it does not represent the only alternative. The experimental data (Ho 77, 83) are displayed in Figure 262, together with simulation calculations (curves) based on sequential α-particle evaporation from the asymptotic fragments at thermal equilibrium. The structure seen in the experimental differential multiplicity, in an angular region of $\pm 40°$ around the fragment detection angle, can be understood quite well in terms of sequential evaporation from the asymptotic projectile-like fragment. It corresponds to a multiplicity of $M_\alpha(L) = 0.14 \pm 0.04$ (Ho 83). Assuming the

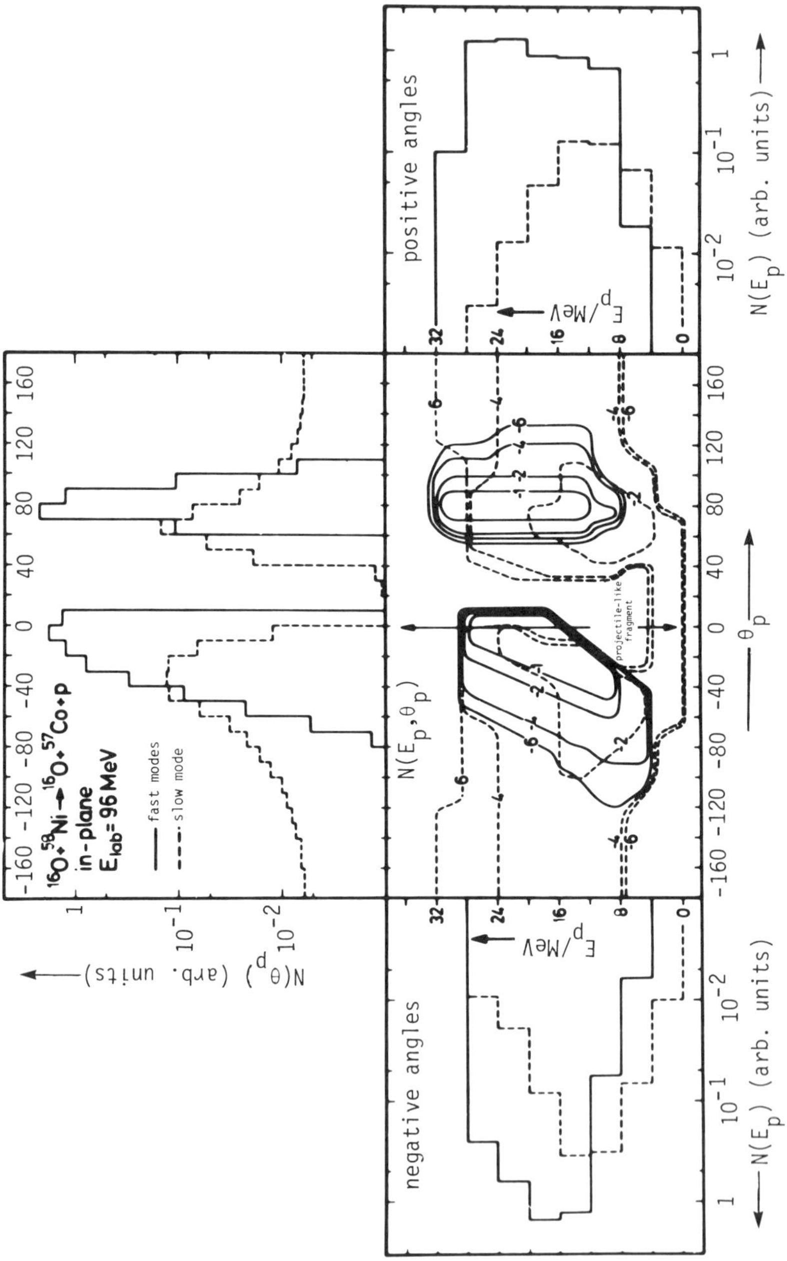

$^{16}O + ^{58}Ni \to ^{16}O + ^{57}Co + p$
in-plane
$E_{lab} = 96$ MeV
fast modes
slow mode
$N(\theta_p)$ (arb. units)
$N(E_p, \theta_p)$
projectile-like fragment
θ_p
E_p/MeV
positive angles
negative angles
$N(E_p)$ (arb. units)

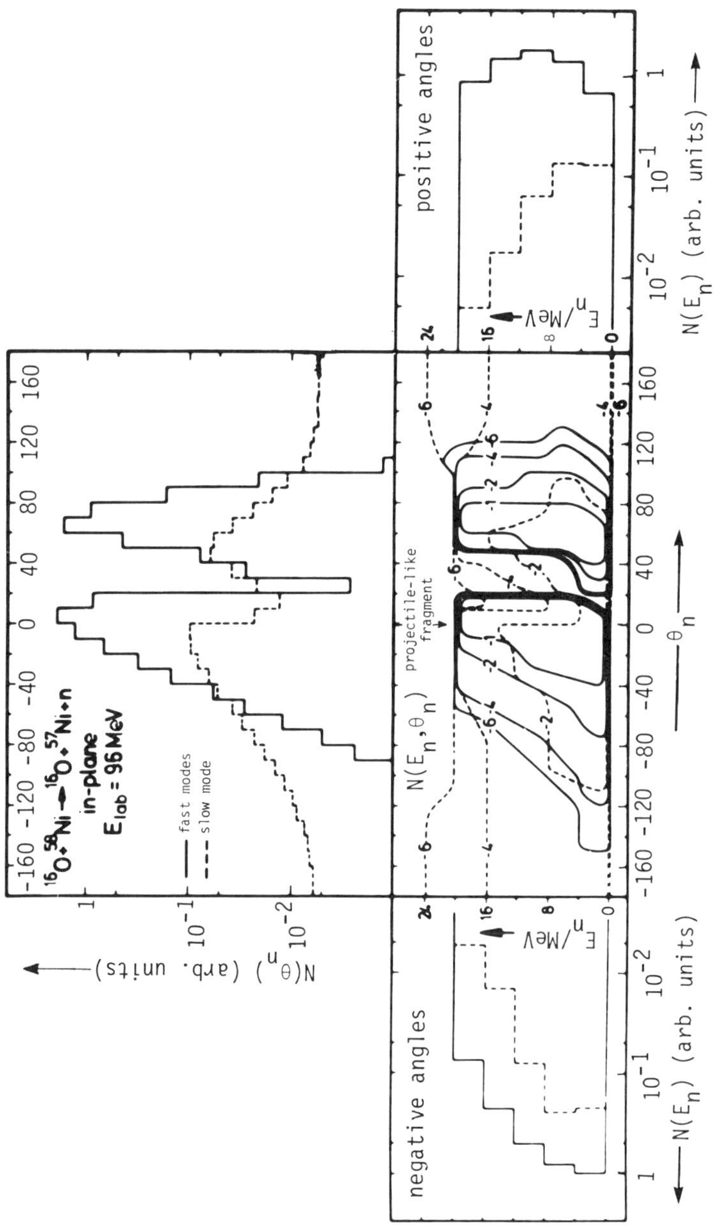

Figure 261. Same as Fig. 260, but for hot-spot emission of neutrons and protons. From (Got 79).

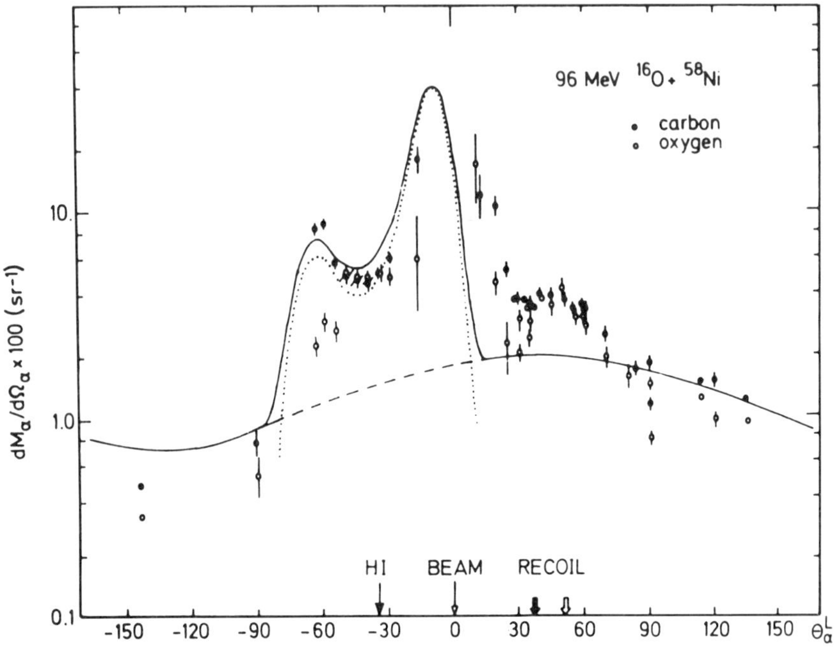

Figure 262. In-plane angular distribution of α particles emitted in the reaction $^{58}Ni + {}^{16}O$ at $E_{Lab} = 96$ MeV. The open and solid symbols correspond to strongly damped O and C fragments, measured in coincidence. Arrows indicate the emission angles of their reaction partners. The α-particle scattering angle θ_α^L is counted from the beam direction in the laboratory frame. Curves represent evaporation calculations for projectile emission (dotted curve) and target emission (dashed and solid curve); the solid curve gives the sum of all evaporative α-particle components. From (Ho 83).

same mechanism for the decay of the target-like fragment, however, leads to a very broad angular distribution of α particles, represented in Figure 262 by the smoothly oscillating solid and dashed curves. This component corresponds to $M_\alpha(H) = 0.10 \pm 0.03$. The structure concentrated around the recoil angles of the target-like fragments and a rise of the multiplicity towards forward angles are not reproduced by the calculation. Qualitatively similar experimental and theoretical distributions are found (Ho 83) at the higher bombarding energy of $E_{Lab} = 147$ MeV.

It is noteworthy to observe that for the $^{58}Ni + {}^{16}O$ reaction, α-particle multiplicities associated with evaporation from the light and heavy reaction fragments at $E_{Lab} = 147$ MeV, $M_\alpha(L) = 0.04 \pm 0.01$ and $M_\alpha(H) = 0.53 \pm 0.16$, respectively, estimated by Ho *et al.* (Ho 83) from their data, are rather different from the ones measured for $E_{Lab} = 96$ MeV. While the α multiplicity associated with the light reaction fragment decreases considerably with increasing bombarding energy, the multiplicity corresponding to emission from the target increases.

The evaporation calculations represented by curves in Figure 262 are rather extensive and account for the strong variation of the angular distribution of the primary fragments, the α-recoil effect, and the finite width of the fragment excitation energy distribution. The inability of these calculations to reproduce the experimental data can, therefore, with some justification be considered (Ho 83) indicative of an additional α-particle emission mechanism that results in an anisotropic angular distribution. The multiplicity M_α^{ne} of the excess α particles, believed (Ho 83) to originate from a nonstatistical, nonequilibrium emission process, increases strongly with increasing bombarding energy. This feature is illustrated in Figure 263 (Ho 83) displaying experimental values of this excess multiplicity for several reactions as a function of the relative velocity v' of projectile and target nuclei at the Coulomb barrier. The barrier is evaluated by Ho *et al.* (Ho 83) as

$$V_{\text{Coul}} = e^2 Z_P Z_T / R, \qquad R = \left[1.22\left(A_P^{1/3} + A_T^{1/3}\right) + 2.0\right] \text{ fm} \quad (3.7.23)$$

Although the number of data points included in Figure 263 is limited, an exponentially increasing trend of the nonequilibrium multiplicity M_α^{ne} with increasing relative velocity is suggestive.

According to the data displayed in Figure 263, nonequilibrium α-particle emission for relatively light and medium-weight systems is present

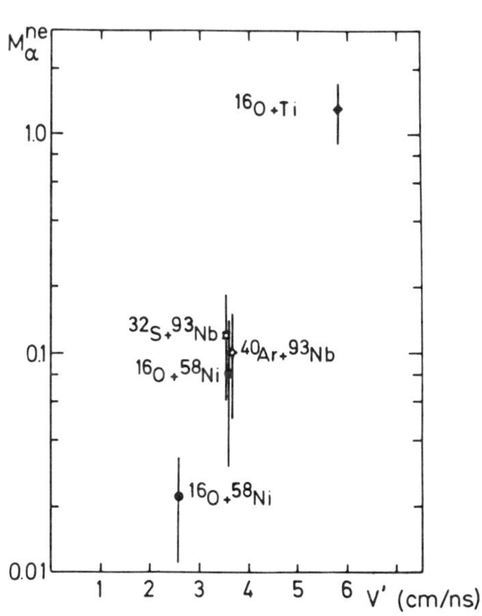

Figure 263. Multiplicity M_α^{ne} of nonequilibrium α particles plotted vs. relative velocity of projectile and target above the Coulomb barrier. The quantity M_α^{ne} is derived from comparison of experimental data with evaporation calculations. From (Ho 83).

already for relative velocities corresponding to energies of 3–4 MeV per nucleon above the barrier, where M_α^{ne} amounts to a few percent. In most experiments included in Table 5, however, nonequilibrium particle components appear to be more clearly visible at bombarding energies in excess of 5–6 MeV per nucleon above the barrier.

In contrast to the model (Got 77,79) of preequilibrium emission of particles from the reaction fragments discussed above, Garpman *et al.* (Gar 80) consider nucleon emission from hot zones of the intermediate composite system produced at early stages of a central heavy-ion collision. Figure 264 illustrates the geometry assumed in the model for a stationary spherical hot zone, comprised of all A_P nucleons of the projectile and A'_T nucleons of the target. It is further assumed that the hot zone is characterized by an equilibrium thermodynamic temperature τ_S, while the surrounding nuclear matter is cold. The temperature of the hot spot is calculated (Gar 80) within the Fermi-gas model as

$$\tau_S = \frac{2}{\pi}\left[\frac{E_F(\rho_S)}{A_P + A'_T}\right]^{1/2}\left\{\frac{A_P A'_T}{A_P + A'_T}T_{\text{inc}} - \frac{3}{5}(A_P + A'_T)[E_F(\rho_S) - E_F(\rho_0)]\right\}^{1/2} \tag{3.7.24}$$

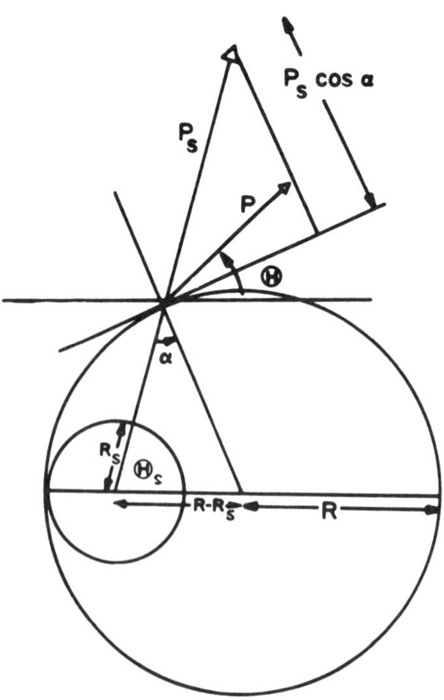

Figure 264. Geometry of nucleon emission from a spherical hot zone of radius R_S, formed in the composite nucleus produced in a central heavy-ion collision. At the nuclear surface, the tangential component, $p_S \sin\alpha$, of the nucleon's momentum is continuous, while the radial component is reduced, due to the action of the potential. The angles θ and θ_S are the polar angles of the particle trajectory with respect to the beam direction, outside and inside the nucleus, respectively. From (Gar 80).

In Equation (3.7.24), $T_{\rm inc}$ denotes the incident energy per nucleon and the quantity $E_F(\rho)$ is the Fermi energy,

$$E_F(\rho) = (2\pi^2\hbar^2/m)[3\rho/(4\pi g)]^{2/3} \tag{3.7.25}$$

which depends on the mass density ρ, the nucleonic mass m, and the weight factor $g(=4)$. The quantities ρ_S and ρ_0 in Equation (3.7.24) are the density of the hot zone and the normal nuclear matter density, respectively. The density ρ_S of the hot zone and its radius, R_S, are free model parameters.

Employing the notation in Figure 264, nucleons of average velocity $\langle v_S \rangle$, emitted from the hot zone, will reach the edge of the composite nucleus within very short times, of the order $\Delta t \approx (R - R_S)/\langle v_S \rangle$. Here, refraction will occur, such that only particles with sufficiently large angles of incidence, α, corresponding to radial kinetic energies larger than the potential well depth V, will be emitted. According to the model (Gar 80), the differential multiplicity of nucleons emitted from the composite nucleus is given by

$$\frac{d^2N}{p_S^2 dp_S\, d\Omega_S} = \frac{g4\pi R_S^2(R-R_S)v_S}{h^3\langle v_S \rangle} f(\varepsilon_S, \tau_S)\Theta\left[p_S - (2mV)^{1/2}/\cos\alpha\right] \tag{3.7.26}$$

Here, the quantities ε_S, p_S, and v_S are the kinetic energy, linear momentum, and velocity, respectively, of a nucleon inside the composite nucleus. The function Θ is Heaviside's step function describing total reflection of nucleons with insufficiently large radial velocity components. The relation between the differential multiplicities expressed in terms of kinematical quantities of the nucleons inside the emitting nucleus and in the overall c.m. frame is given by (Gar 80)

$$\frac{d^2N}{p^2dp\,d\Omega} = \frac{d^2N}{p_S^2\, dp_S\, d\Omega_S}\frac{p_S^2 \sin\theta_S}{p^2 \sin\theta}\mathscr{J}\left[\frac{(p_S, \theta_S)}{(p, \theta)}\right] \tag{3.7.27}$$

in a Galilean-invariant form. Here, the angles θ_S and θ are the angles inside and outside, respectively, of the nucleus, formed between the classical particle trajectory and the beam direction (cf. Figure 264). The quantity $\mathscr{J}$ is the Jacobian for this transformation.

Experimental data and model fits are presented in Figure 265 (Gar 80) for neutron emission from the fusion-evaporation reaction ^{154}Sm + ^{16}O at $E_{\rm Lab} = 153$ MeV. In this figure, the experimental double-differential cross section $d^2\sigma/d\Omega\, dE$ for neutrons, in coincidence with known yrast γ rays from particular evaporation residue nuclei, is plotted vs. c.m. emission angle, for laboratory neutron energies of 10.5 MeV (circles), 13 MeV (squares) and 17 MeV (triangles). It is quite obvious from Figure 265 that the experimental angular distributions become steeper, i.e., more forward-peaked, with increasing neutron energy. The dashed-dotted, dashed and solid curves in Figure 265 represent corresponding results, obtained (Gar 80) with the above hot-spot emission mechanism, under the assumption of a large radius, $R_S = 0.7R$, of the hot zone, taken to be at equilibrium matter density ($\rho_S = \rho_0$). The parameters were adjusted to obtain a fit to the experimental angular distribution at $\theta = 20°$ and for a neutron energy of 17 MeV. The calculations give a reasonable account of the data at forward

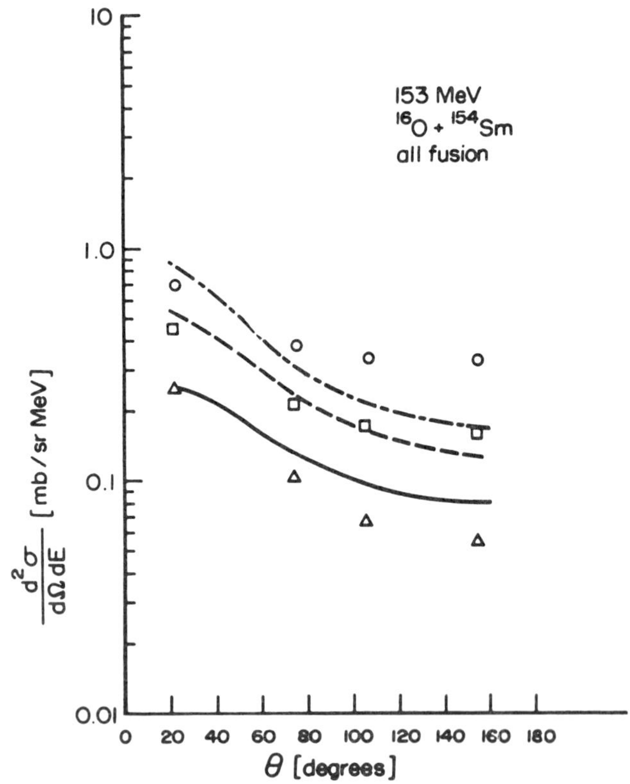

Figure 265. Double-differential cross section of neutrons emitted in the fusion reaction $^{154}Sm + ^{16}O$ at $E_{Lab} = 153$ MeV, plotted vs. c.m. emission angle θ. The circles, squares, and triangles correspond to laboratory neutron energies of 10.5, 13, and 17 MeV, respectively. The dashed-dotted, dashed, and solid curves represent the corresponding hot-spot model calculations, normalized to the data at $\theta = 20°$ and a neutron energy of 17 MeV. From (Gar 80).

angles, an agreement claimed (Gar 80) to be in general support of the underlying physical picture of a hot sphere surrounded by relatively cold matter. However, it is apparent from Figure 265 that the theoretical angular distributions do not vary as strongly with neutron energy as the experimental data. At backward angles, theoretical cross sections are observed to be systematically smaller and larger than the experimental values for neutron energies of 10.5 and 17 MeV, respectively.

Calculations with the above hot-spot model, including the absorption of the primary nucleons, due to their finite mean free path λ, have not been reported on. However, one can argue qualitatively that inclusion of absorption would not significantly improve the reproduction of experimental data by the model. As is illustrated in Figure 25, the mean free path λ of nucleons in nuclear matter generally decreases with increasing nucleon energy ε. Hence, forward emission of fast nucleons is expected to be suppressed, due to absorption along the forward-going nucleon trajectories that correspond to small angles θ_S (cf. Figure 264) and relatively long distances to the nuclear edge. At backward angles, this absorption process is less effective, given the geometry displayed in Figure 264. The expected consequence of absorption is then a broadening of the theoretical angular distributions for the high-energy nucleons, which is uncharacteristic of the experimental data.

The hot-spot model calculations (Gar 80) discussed above are performed for central collisions, leading to fusion or fusion-like reactions, and may therefore not be entirely characteristic of the corresponding mechanism in peripheral damped reactions. Nevertheless, as indicated by Garpman *et al.* (Gar 80), one expects only a weak impact parameter dependence of the hot-spot emission pattern. Hence, the results of the calculations presented in Figure 256 may be taken as qualitative model predictions of hot-spot neutron emission patterns associated with very strongly damped, near-central collisions.

As an intuitively more appropriate geometry for a hot zone assumed to be produced in a medium-energy (20 MeV per nucleon) damped collision, Morison *et al.* (Mor 80c) and Mooney *et al.* (Moo 81) take the juxtaposed spherical caps on projectile and target nuclei. Kinetic energy of relative motion is assumed to be converted into intrinsic excitation energy via two successively operating mechanisms. The one-body, exchange-induced dissipation mechanism (cf. Sections 2.6 and 3.5), leading to a uniform heating of both interaction partners, is for near-central collisions of spherical nuclei not effective enough to account for the entire amount of dissipated energy observed in experiments (see Section 3.5). The deficit is, in the model, attributed to the effect of neck formation and other changes in shape of the intermediate dinucleus, developing into a mononuclear system. It is assumed (Mor 80c, Moo 81) that this extra energy is transformed into heat of the spherical caps of the reaction partners at contact. The number of nucleons among which this energy is distributed and, hence, the temperature of the hot zones depends on the volumes adopted for them in the model. Nucleon emission is assumed to occur from these hot zones, on a time scale of 2×10^{-22} s, almost one order of magnitude faster than typical reaction times.

Figure 266 displays experimental data on energy spectra of protons emitted in the $^{197}Au+^{16}O$ reaction at $E_{\rm Lab}=315$ MeV reported by Symons *et al.* (Sym 80). They are compared to the predictions of the above model (Mor 80c) represented by thin solid curves. Obviously, a qualitative fit of the

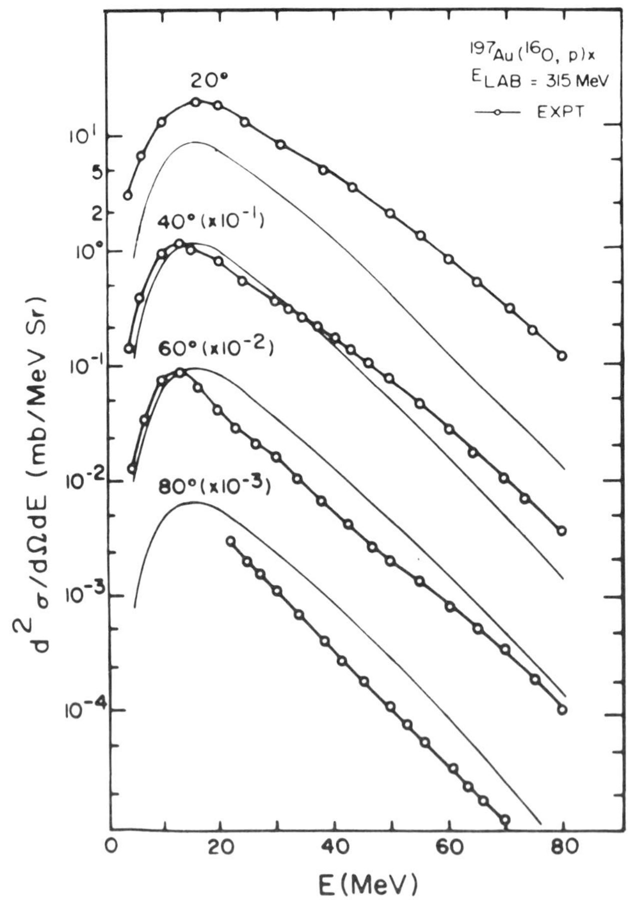

Figure 266. Proton energy spectra at various detection angles for the reaction $^{197}Au+^{16}O$ at $E_{\rm Lab}=315$ MeV (Sym 80). The thin solid curves represent calculations performed with a hot-spot model. From (Mor 80c).

experimental data is possible with the model, assuming a thickness of 2.5 fm for the hot zones. Similar to the model for hot-spot emission in central collisions discussed earlier (cf. Figure 265), the experimental angular distributions cannot be explained satisfactorily by the above calculations. The inclusion of instantaneous emission of PEP particles (Moo 81) in the model can only partially remedy this insufficiency.

Further freedom in the description of experimental particle spectra and angular distributions in terms of phenomenological models can be obtained by allowing the emitting source to move with a velocity intermediate between the center-of-mass velocity and that of the projectile. For example, in Figure 267, a comparison is made between inclusive experimental spectra (Awe 81) of protons from the ^{197}Au + ^{16}O reaction for three incident energies and fits (solid curves), based on such a "moving-source" model. The experimental spectra illustrated in Figure 267 are smooth distributions with a width increasing strongly with increasing bombarding energy. At backward angles, the high-energy spectral components are seen to be much less important than at forward angles. As is demonstrated by the solid curves in this figure, the experimental energy and angular dependences can be well described in terms of an isotropic Maxwellian energy distribution (Awe 81),

$$\frac{d^2\sigma}{dE\,d\Omega} \propto (E - E_c)^{1/2} \exp\left\{ -\left[E - E_c + E_S - 2E_S^{1/2}(E - E_c)^{1/2}\cos\theta \right] / \tau \right\} \tag{3.7.28}$$

in the rest frame of the source, moving in the beam direction with a kinetic energy per nucleon E_S, which is an adjustable parameter. Particle energy and emission angle, E and θ, respectively, are given in the laboratory system. The quantity E_c is the kinetic energy gained by the Coulomb repulsion from the target, and τ denotes the temperature parameter.

As pointed out by Awes *et al.* (Awe 81), the good quality of the fits obtained with the moving-source model should not be regarded as evidence for a physical picture in which particles are emitted from a hot gas of nucleons separated from the target. The parametrization of the data merely implies a successful decomposition of the particle velocities into one component, v, in beam direction and another, randomly oriented, velocity component. The extracted best-fit parallel velocities v, indicated in Figure 267, are found (Awe 81) to be sensitive to the angular range included in the fits. They are roughly of the order of one-half of the relative ion–ion velocity in a grazing collision, but do not reflect the considerable variation of this energy in the experiments illustrated in Figure 267. By the same token, the temperature parameter τ (T in Figure 267) is not necessarily equivalent to

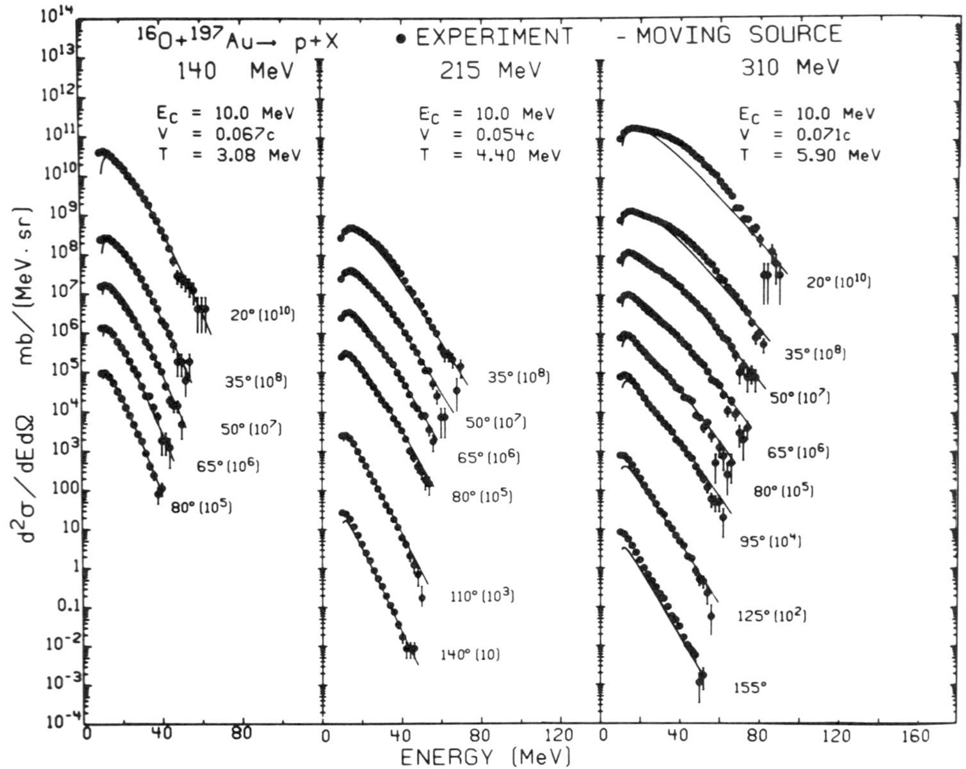

Figure 267. Inclusive energy spectra of protons from the reaction $^{197}Au + ^{16}O$ at $E_{Lab} = 140$, 215, and 310 MeV, for several emission angles. Experimental data are indicated by dots, whereas the solid curves represent calculations using a moving-source model. Model parameters given in the figure correspond to Coulomb repulsion between the target and a proton (E_c), the velocity of the hypothetical source, v, and the temperature parameter, T, of a Maxwellian energy spectrum in the rest frame of the emitter. From (Awe 81).

an equilibrium thermodynamic temperature. The deduced values appear to be generally consistent with the temperature of a Fermi gas and exhibit an appropriate variation with the bombarding energy. However, systematical differences between temperature parameters deduced from deuteron, triton, and proton spectra remain unexplained (Awe 81) in such a thermal model.

Although, as mentioned previously, a physical justification for the formation of localized hot zones in damped reactions has yet to be provided, the lack of success in describing fast-particle emission for intermediate bombarding energies in terms of hot-spot models renders this mechanism rather unlikely to occur at incident energies of less than 10–15 MeV per nucleon. Although some experiments (see Table 5) on charged-particle and neutron emission at energies of a few MeV per nucleon above the barrier have yielded results that appear difficult to be understood in terms of equilibrium-statistical evaporation processes occurring in the final reaction fragments, no hard experimental evidence is currently available in support of exotic reaction mechanisms that could produce localized zones of equilibrated hot nuclear matter, in low-energy heavy-ion collisions. Even with regard to more conventional nonequilibrium emission processes, care should be exercised in the interpretation of experimental results. The analysis of an experiment by Young *et al.* (You 80) on α-particle production in the reaction $^{93}Nb+{}^{16}O$ at 204 MeV, discussed earlier (cf. Figure 242), can be regarded as an example of a reaction where fast particles with signatures commonly associated with preequilibrium emission are found to be consistent with sequential evaporation from the asymptotic reaction fragments.

It appears reasonable to expect, however, that preequilibrium nuclear decay, well studied in the case of nucleon- and light-ion-induced reactions (Bla 75), will eventually, at sufficiently high incident energies, assume a role also in reactions involving heavy ions. Such an expectation can be based on the importance of nucleon exchange processes in damped reactions, for which experimental evidence has been discussed in previous sections. If the damped reaction mechanism could be understood entirely in terms of a succession of individual nucleon transfers between projectile and target, each of the transferred nucleons could be expected to initiate an equilibration cascade. Depending on whether the weak- or the strong-coupling limit (cf. Section 2.7) is more characteristic of a damped nuclear collision, the individual elementary cascades would either proceed to develop independently of one another or be strongly influenced by the history of concurrent cascades, in both reaction partners.

Following the pioneering work of Griffin (Gri 66, 67), various models of preequilibrium decay have been developed, as reviewed by Blann (Bla 75). The basic physical ideas have also been used in recent work (Aga 75, Man 75, 76, Obl 78, Akk 79, Mac 79, 80, Bla 79, 81, Kal 79, Mad 80, Gad 81) to devise more sophisticated models yielding, for example, predictions of angular distributions and of complex-particle emission during the equilibration cascade. In Figure 268, an elementary cascade is pictured (Bla 75) in terms of Blann's exciton model. An incident particle of energy E above the Fermi surface, ε_F, undergoes a collision with a nucleon below the Fermi surface, generating $\Delta n = 2$ excitons, one particle and one hole. Such collisions are the result of the residual interactions that vanish in the extreme one-body limit, where all nucleons move independently in the mean field of the nucleus. The preferred direction, in which the equilibrium cascade proceeds, is that of increasing density of states of the nucleus, indicated by long arrows in Figure 268. The lower part of the figure exhibits examples for exciton states with one particle in the continuum, which may be emitted from the nucleus prior to attainment of equilibrium.

In the master equation exciton model employed by Blann and coworkers (Bla 75), the time evolution of the population probability $P(n, t)$

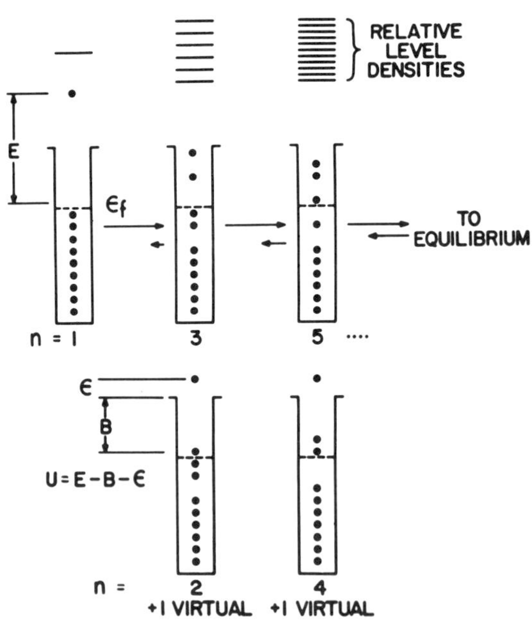

Figure 268. Schematics of a preequilibrium cascade pictured in the exciton model. An initial particle at energy E above the Fermi surface, ε_F, initiates an avalanche of secondary particles and holes (excitons). The quantity n denotes the number of excitons in each stage of the cascade. The length of the horizontal arrows illustrates the preferred direction of transitions. The lower part of the figure gives examples for exciton states with a nucleon in the continuum. From (Bla 75).

for n-exciton states is followed by solving coupled master equations of the form

$$\frac{dP(n,t)}{dt} = P(n-2,t)\langle W_{+}^{n-2}(E)\rangle + P(n+2,t)\langle W_{-}^{n+2}(E)\rangle - P(n,t)[\langle W_{+}^{n}(E)\rangle + \langle W_{-}^{n}(E)\rangle] \tag{3.7.29}$$

These equations connect the n-exciton states with $(n \pm 2)$ exciton states via excitation-energy (E) dependent transition probabilities $W_{\pm}^{i}$ leading to an increased $(+)$ or decreased $(-)$ number of excitons. Estimates of the transition rates can be obtained, for example, from the imaginary part, W, of the optical potential or from the free nucleon–nucleon scattering cross section as discussed in Section 2.6. Preequilibrium particle emission is possible in the model at any stage of the cascade, depending on the relative abundance of exciton states with particles in the continuum. The number of particles with channel energy ε, emitted from an n-exciton state with excitation energy E, leaving the residual nucleus at excitation energy U in a state with $(p-1)$ excited particles can approximately be calculated (Bla 75) as

$$\frac{dP_n(\varepsilon)}{d\varepsilon} = \frac{2S+1}{\pi^2\hbar^3}\frac{m\sigma(\varepsilon)}{gE}\varepsilon\left(\frac{U}{E}\right)^{n-2} p(n-1)t_n \tag{3.7.30}$$

Here, the term U^{n-2} is characteristic of the density of final $(n-1)$-exciton levels. The quantity $(2S+1)$ denotes the spin degeneracy of the emitted particle of mass m, g stands for the single-particle level density parameter, and t_n characterizes the mean lifetime of an n-exciton state. The latter quantity is often used as an adjustable parameter. An additional model is required for the calculation of the cross section, $\sigma(\varepsilon)$, for the inverse capture reaction. Summation of Equation (3.7.30) over all numbers n of excitons from initial value, n_0, up to the equilibrium value $\bar{n}$, yields the total preequilibrium particle spectrum. Angular distributions of these particles are not treated within the model. Physically, the initial exciton number n_0 should be characteristic of the projectile. In practice, adjustments of this parameter are found necessary to fit experimental data.

An illustration of the time development of the equilibration cascade, predicted by the above model [cf. Equation (3.7.30)], is given in Figure 269 (Bla 75) for a nucleus in the region $A \approx 100$, excited by a nucleon to 24 (left) or 96 (right) MeV, with an initial exciton number $n_0 = 3$. The top portion of the figure shows the exciton distribution $P(n,t)$ as a function of time, which is indicated at the curves in arbitrary units. One observes an initially rather rapid increase of average value and width of the distribution function, followed by smaller changes at later times, when the distribution approaches the stationary equilibrium. The middle part of the figure exhibits the instantaneous spectral emission rates of light particles. The energy spectra of particles emitted from early stages of the cascade are relatively flat and extend to energies close to that of the projectile. The energy spectra of particles emitted at late times are characteristic of Maxwellian evaporation processes. Time-integrated theoretical particle spectra, depicted at the bottom of Figure 269, illustrate the division of the theoretical spectrum into components from fast and slow emission modes, a division that becomes more apparent at the higher excitation energy of 96 MeV. It is quite instructive to observe in Figure 269 the successive distribution of the excitation energy among an increasing number of degrees of freedom. The drastic dependence of the instantaneous particle emission spectra on the elapsed time also indicates a strong sensitivity of the shape of the integral particle energy spectrum on the initial number of excitons carrying the excitation energy. In this aspect, heavy-ion reactions may or may not differ drastically from light-ion or nucleon-induced reactions at similar excitation energies. For example, from a damped nuclear reaction proceeding in many small steps, one may qualitatively expect preequilibrium particle spectra comparable to those induced by nucleons at similar incident energies per nucleon, despite the higher excitation energies attained in heavy-ion reactions.

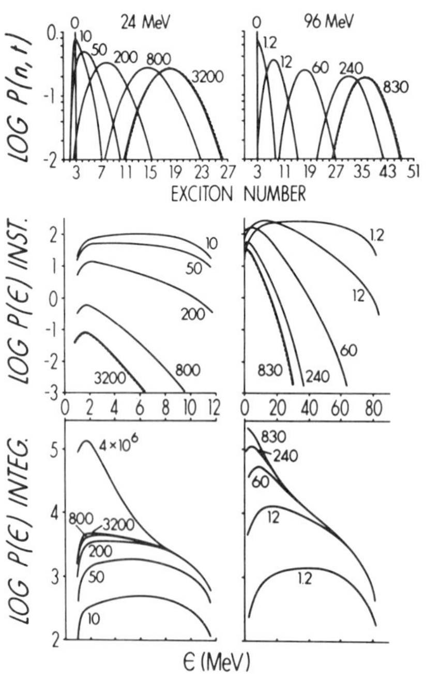

Figure 269. Preequilibrium cascade induced by a nucleon in a nucleus of $A \approx 100$, excited to 24 MeV (left) or 96 MeV (right). The calculations are made in the master-equation approach. The top part shows the time evolution of the exciton number distribution $P(n, t)$, given in arbitrary time units. The central portion illustrates the predicted instantaneous particle emission rates. Their time-integrated values are plotted at the bottom vs. particle energy ε and for different times elapsed after the projectile impact. From (Bla 75).

As mentioned earlier, one anticipates a rather complex convolution of individual equilibration cascades induced in a heavy-ion collision, where nucleon exchange between the reaction partners may occur continuously during the entire reaction time. To account approximately for this effect, Blann (Bla 81) has modified the master equation approach of Harp, Miller, and Berne (Har 68) to preequilibrium particle emission. In this approach, the intrinsic system is described in terms of a two-component Fermi gas of neutrons and protons. The equivalent coupled equations for a one-component Fermi gas can be written as (Bla 81)

$$\begin{aligned}\frac{d}{dt}(g_i f_i) = & \sum_{j,k,l} W_{kl,ij}(g_i g_j g_k g_l) f_k[1-f_i] f_l[1-f_j] \\ & - \sum_{j,k,l} W_{ij,kl}(g_i g_j g_k g_l) f_i[1-f_k] f_j[1-f_l] \\ & - W_{i,n}(g_i g_n) f_i + \frac{d}{dt}(g_i f_i)_{fus} \end{aligned} \tag{3.7.31}$$

In Equation (3.7.31), the indices $i-l$ characterize intervals in the single-particle energy spectrum of a nucleus, g the average number of single-particle levels in each energy interval, and f stands for the average occupation probability. The quantities $W_{ij,kl}$ denote the transition probabilities of nucleons in the initial energy bins i and j to scatter into bins k and l, evaluated from the free nucleon scattering cross section. The fractional occupation numbers $(1-f)$ account for the Pauli blocking effect. Hence, the first two terms in Equation (3.7.31) describe a two-body nucleon–nucleon collision process $(k, l) \rightarrow (i, j)$, while the third term gives the rate of ejection of a nucleon into the continuum (n). The last term in Equation (3.7.31) models the increase in the number of particles in the energy interval i due to fusion of part of the projectile with the target. Its time dependence, i.e., the variation in time of the partition of the dissipated energy over nucleonic degrees of freedom, is a function of the damped reaction mechanism, on which the equilibration cascade and the associated particle emission process depend crucially. The dependence of the preequilibrium particle emission spectra and probabilities on entrance channel conditions as well as on the amounts of dissipated energy can, hence, yield information about the reaction mechanism.

The above preequilibrium model has been applied (Bla 81) to central collisions of the system $^{197}Au + ^{16}O$ at 320 MeV. In the calculation, it is assumed that initially the A_p projectile nucleons partition all the excitation energy, i.e., the initial exciton number relative to the target is assumed to be close to $n_0 = 16$. As the projectile approaches the target, with a velocity estimated from the center-of-mass energy above the Coulomb barrier, nucleons of type i enter the target nucleus with a rate $n_i(t)$ determined by the geometrical overlap of spherical nuclei. They are, in the model, assumed to populate target exciton states in the excitation energy interval $[U, U+\Delta U]$ according to a distribution function

$$N_i(U)\Delta U = n_i(t)\left[U^{n_0-1} - (U-\Delta U)^{n_0-1}\right]/E^{n_0-1} \tag{3.7.32}$$

adopted from analyses of light-particle-induced preequilibrium reactions. In Equation (3.7.32), the quantity $N_i\Delta U$ represents the number of nucleons in the excitation energy interval, and E is the total excitation energy of the composite system. The exponents (n_0-1) reflect the power-law dependence on the excitation energy of the initial density of exciton states in the projectile. This distribution could be significantly altered (Bla 81) by nucleon transfer to the projectile or conversion of kinetic energy of relative motion into collective forms of excitation energy. These processes would either increase the initial exciton number n_0 or decrease the intrinsic excitation energy, which would lead to a softening of the energy spectra of preequilibrium particles, as is illustrated in Figure 269.

Results of the above model calculations are compared in Figure 270 to inclusive data on proton spectra in the ^{197}Au + ^{16}O reaction at $E_{\text{Lab}} = 320$ MeV. The uppermost spectrum, representing the data, is arbitrarily normalized with respect to the theoretical curves showing the development of the preequilibrium spectrum with time, as indicated at the curves. The effect of an increased exciton number of $n_0 = 19$ is illustrated in Figure 270 by an additional theoretical spectrum.

As can be seen from the comparison made in Figure 270, the above simple preequilibrium model is able to account for the general shape of the experimental proton spectrum, although the experimental high-energy slope is somewhat steeper than predicted. A similar discrepancy has also been found by Gavron *et al.* (Gav 83) for C- and Ne-induced fusion reactions at 4–11 MeV per nucleon leading to the ^{170}Yd nucleus. In appreciating the quality of reproduction of the shape of the experimental spectrum in Figure 270 by the calculations, it should be stated that the experiments did not distinguish between fusion, incomplete fusion, damped, and other reactions types. The model calculations, on the other hand, do not yield particle angular distributions and can, hence, only be compared to angle-integrated data. Nevertheless, the model offers an alternative to the physical picture of a stationary or moving hot zone at thermodynamic equilibrium. It is seen from the development of the theoretical spectra in time that high-energy particles can be emitted during the equilibration cascade up to times of the order of a few times 10^{-22} s, whereafter the spectra develop a low-energy evaporation component.

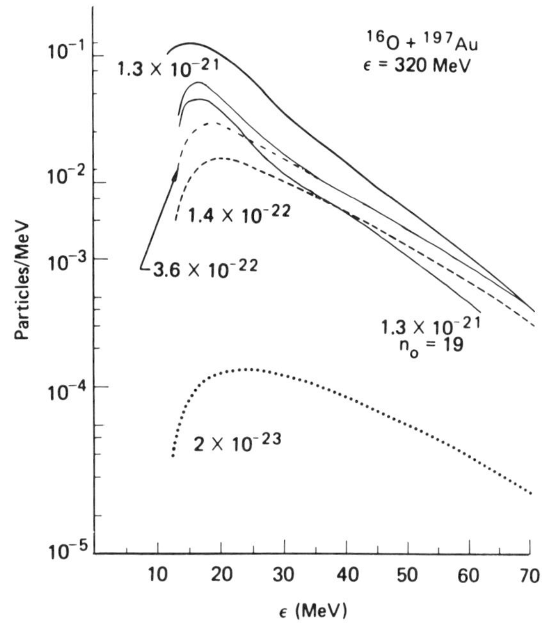

Figure 270. Experimental spectrum of protons emitted in the reaction $^{197}Au + ^{16}O$ at 320 MeV (heavy solid curve). Dashed, dotted, and thin solid curves represent predictions of a preequilibrium model, integrated up to the times (in units of s) indicated at the curves. The lower thin curve is a result of a calculation with an increased initial exciton number of $n_0 = 19$. The experimental spectrum is arbitrarily normalized with respect to the theoretical ones. From (Bla 81).

The preequilibrium particle emission model employed by Blann (Bla 81) predicts a strong dependence of both spectral shape and cross section of preequilibrium particles on the incident energy. For a bombarding energy of 20 MeV per nucleon, multiplicities of approximately 1 and 0.5, for preequilibrium emission of neutrons and protons, respectively, are estimated (Bla 81), decreasing to about half these values at a lower energy of ~ 15 MeV per nucleon. Approximately half the precompound nucleons are predicted to be emitted prior to a complete fusion of projectile and target. Such particles could then also be observed in damped reactions at elevated bombarding energies. Qualitatively, such a behavior has been observed by Hilscher *et al.* (Hil 83) in fusion and damped reactions induced by ^{20}Ne projectiles at incident energies of 6.5 to 15.5 MeV per nucleon above the barrier. It is a noteworthy result of the theoretical studies by Blann (Bla 81) that equilibration times are predicted to decrease with increasing incident energy, although the probability for preequilibrium emission increases in this direction.

There are too few sets of coincidence data available for preequilibrium particle spectra and angular correlations to allow a stringent, systematical test of the basic ideas of this emission process. Similarly, viable models should provide information on angular correlations of particles emitted at the various stages of the reaction and account for kinematical effects produced in the emission process. As pointed out earlier, a neglect of such influences may lead to deceptive results of analyses of experimental data. Some of the shortcomings of current models for heavy-ion-induced preequilibrium emission are avoided in the preequilibrium models of Mantzouranis *et al.* (Man 75,76), Akkermans (Akk 79), and Mädler *et al.* (Mad 80), who account for the angular distribution of fast preequilibrium particles either by considering the angular dependence of the scattering cross section for intranuclear nucleon–nucleon collisions or by including the total linear momentum (Mad 80) of the "exciton gas" in a model for the statistical partition function. Unfortunately, these models have so far not been employed in a description of particle emission in damped reactions, although a satisfactory explanation of the angular distribution of fast neutrons emitted in the ^{165}Ho$+^{20}$Ne fusion reaction at ~ 15 and ~ 20 MeV per nucleon is achieved (Hil 83) in the framework of the model of Mantzouranis *et al.* (Man 75,76).

It is clearly desirable to develop more fully the existing models of preequilibrium decay for applications in damped nuclear reactions, since they have been successful in describing the particle emission processes in light-ion reactions. Such theories model the time-dependent cooling-off process of an excited nucleus, which depends crucially on the complexity of the initial excitation process (cf. Figure 269). This initial excitation mechanism, however, not accessible to such models of nuclear deexcitation, is characteristic of the damped interactions between complex nuclei. Much-debated theories of fast-particle emission associated with these interaction models have been discussed previously. The most recent development of the diabatic reaction model by Nörenberg and collaborators (Nör 83) should be included, which describes possible modes of particle excitation into the continuum. A meaningful test of particle emission mechanisms proposed for damped and other heavy-ion reactions must, it seems, rely on a proper treatment of preequilibrium decay.

Most experiments searching for features associated with preequilibrium decay are not accompanied by detailed theoretical analyses in terms of appropriate models. The most characteristic property of such a process appears to be the two-component energy spectrum of associated particles and its angular correlation with the beam direction, rather than with the scattering angle of a coincident reaction fragment. More indirect inferences of preequilibrium properties, for example, from experimental excitation functions and particle and γ-ray multiplicities, should be viewed with caution, as illustrated by Blann (Bla 79). Relying on experimental information on selected reaction channels may introduce a bias that complicates the interpretation of data, as exemplified also in the previous section for the case of analyses of heavy-ion-induced fission of nonfissile targets in terms of angular-momentum transfer.

Preequilibrium-type energy spectra of particles emitted in the fusion reaction $^{158}Gd + ^{12}C$ at 152 MeV have been measured by Westerberg *et al.* (Wes 78d, Sar 78, 78b), in coincidence with evaporation residues identified by characteristic γ-ray transitions. Figure 271 illustrates some of their experimental α-particle spectra observed in coincidence with characteristic transitions in the final nuclei. All spectra exhibit an exponential tail towards high energies with a slope that decreases drastically at forward angles, i.e., high-energy ($E_\alpha > 25$ MeV) α particles are predominantly emitted at forward angles, while the angular distribution of lower-energy α particles is symmetric around 90°, indicative of compound-nucleus evaporation. The angular correlations of the high-energy α particles emitted in the ($^{12}C, \alpha 7n$) to ($^{12}C, \alpha 10n$) channels are observed (Wes 78d) to exhibit the same shape, suggesting that in these channels α emission precedes the neutron evaporation cascade. It appears consistent with this picture that the α-particle energy spectra shift to higher energies with decreasing number x of neutrons in a reaction channel ($^{12}C, \alpha xn$), as expected from the energy balance. Westerberg *et al.* (Wes 78d) conclude also from their coincident γ-ray data that high-energy α particles are emitted from high-spin initial states and populate high-spin states in the product nuclei. In contrast, low-energy α-particle emission is associated (Wes 78d) with a relatively broad spin region in the final nuclei.

The above authors (Wes 78d, Sar 78, 78b) have also studied particle emission in the reaction $^{150}Nd + ^{20}Ne$ at 175 MeV producing the same composite nucleus ^{170}Yb at an excitation energy comparable to that reached in the $^{158}Gd + ^{12}C$ reaction. However, preequilibrium effects are hardly, if at all, visible in the former reaction. This clearly suggests that the excitation energy is not the important parameter determining the occurrence of preequilibrium particle emission. A difference between the reactions $^{150}Nd + ^{20}Ne$ and $^{158}Gd + ^{12}C$ exists with respect to the relative velocity, which is approximately 20% larger for the latter reaction than for the former. Such a correlation between the appearance of nonequilibrium statistical emission and the relative velocity has been illustrated already earlier (cf. Figure 263).

Whereas the evaporative spectral component is somewhat suppressed for charged particles, due to the Coulomb barrier, the energy spectrum of neutrons emitted in an equilibration cascade is more clearly of the two-component character expected from precompound models (cf. Figure 269). Such a shape has, indeed, been observed for spectra of neutrons emitted in the reaction $^{158}Gd + ^{12}C$ discussed above, while for the 175-MeV ^{20}Ne-induced

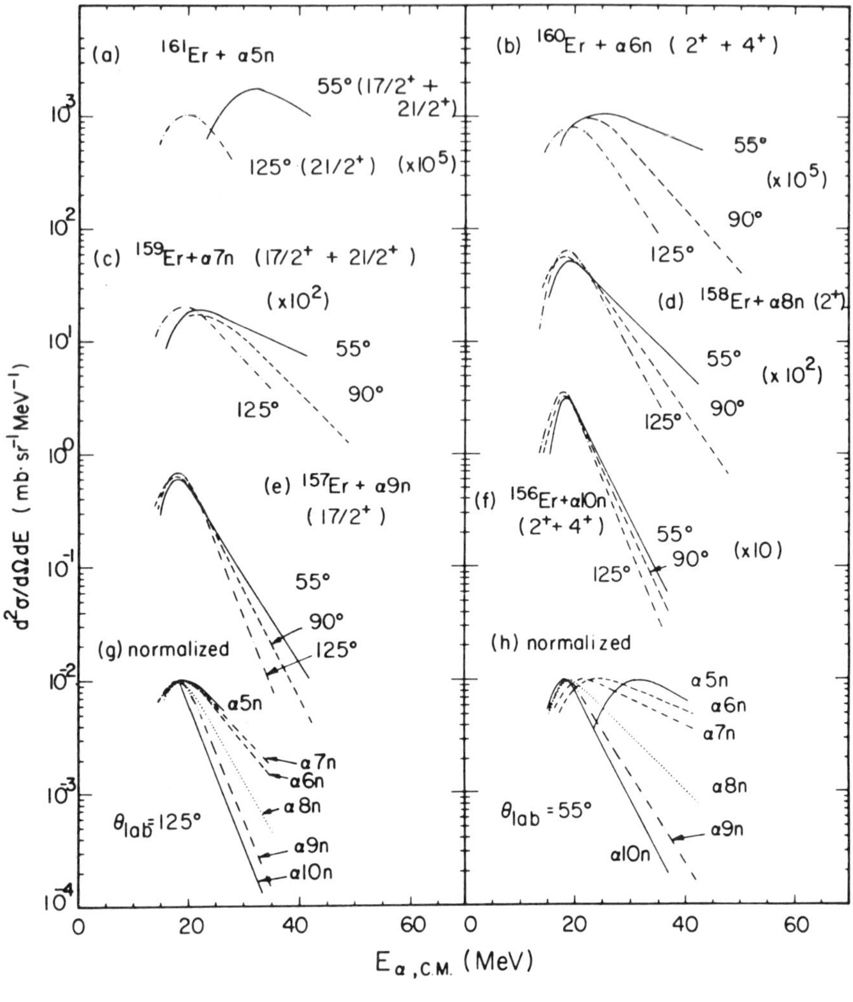

Figure 271. Center-of-mass energy spectra of α particles emitted in the reactions ^{158}Gd(^{12}C, $\alpha xn\gamma$)$^{166-x}$Er at $E_{\rm Lab} = 152$ MeV, for various emission angles. The curves represent smoothed experimental data. In parts (g) and (h) they are normalized to each other for easier comparison. Final states observed in the evaporation residue are indicated by spin and parity values given in brackets. From (Wes 78d).

reaction on ^{150}Nd (Wes 78d) only an equilibrium component is visible. However, preequilibrium effects reappear in the neutron spectra of ^{20}Ne-induced fusion reactions on ^{165}Ho (Hil 81) at incident energies of about 11 MeV per nucleon. Figure 272 exhibits the angular and energy dependences of the neutron multiplicity associated with the production of evaporation residues, as measured by Hilscher *et al.* (Hil 81, Hol 83) for this reaction at

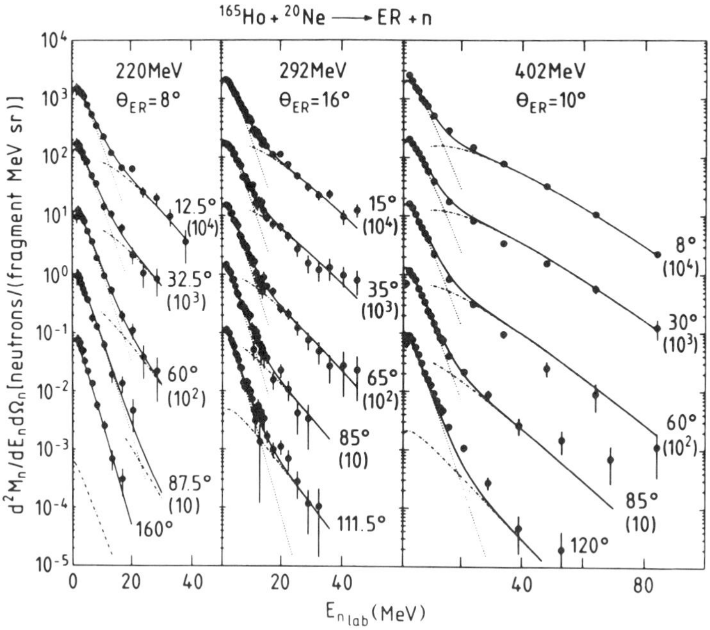

Figure 272. Differential multiplicities of neutrons emitted in the fusion-evaporation reaction $^{165}\mathrm{Ho}+{}^{20}\mathrm{Ne}$ at 11, 14.6, and 20.1 MeV per nucleon, plotted vs. laboratory energy for several detection angles. The quantity θ_{ER} denotes the detection angle of the coincident evaporation residues. The dashed and dashed-dotted curves represent the two components of a theoretical fit function [cf. Equation (3.7.32)]. From (Hol 83).

$E_{\mathrm{Lab}} = 220$, 292, and 402 MeV. While the hard, preequilibrium component is rather pronounced at the higher bombarding energies, already at 11 MeV per nucleon there is a clear excess of high-energy neutrons at forward angles, as compared to a theoretical evaporation spectrum. The curves in Figure 272 correspond to two-component fits using the theoretical differential multiplicity (Hol 83):

$$\frac{d^2M_n(\theta_n)}{dE_n\,d\Omega_n} = \sum_{i=1}^{2} \frac{M_i\sqrt{E_n}}{2(\pi\tau_i)^{3/2}} \exp\left\{-\left[E_n - 2(\varepsilon_i E_n)^{1/2}\cos(\theta_i-\theta_n)+\varepsilon_i\right]\Big/\tau_i\right\} \tag{3.7.33}$$

In this equation, E_n is the neutron kinetic energy in the laboratory system, and θ_n the emission angle. The quantities ε_i, τ_i, and θ_i denote the energy per nucleon, the temperature, and the angle, respectively, of the emitting neutron source. Hence, the fit function of Equation (3.7.33) corresponds to a parametrization of the spectra in terms of emission from hot, moving sources, although this physical picture is not adopted for the high-energy component of the neutron spectra. Good fits to the data are possible with the exciton-precompound model discussed earlier in this section, if the initial exciton number n_0 is adjusted.

The multiplicities of the high-energy neutron component extracted from the fits to the ^{165}Ho$+^{20}$Ne data vary from 0.4 to 2.3 for the bombarding-energy range from 220 to 402 MeV. This corresponds to preequilibrium fractions of 5% to 15% of the total multiplicity of neutrons emitted per fusion-type event in the ^{165}Ho$+^{20}$Ne reaction (Hol 83). The experimental energy spectra depicted in Figure 272, corresponding to temperature parameters τ_i [cf. Equation (3.7.33)] between 2.4 and 2.8 MeV are found (Hol 83) to be in disagreement with the predictions of Fermi-jet and PEP models described earlier in this section, in particular at angles larger than 35°. On the other hand, reasonable agreement is obtained between data and exciton-model calculations, if the initial exciton number n_0 is treated as a free parameter. Fits (Hol 83) require an increase with bombarding energy of the n_0 value from 20 to 28.

The character of the experimental spectra of neutrons emitted in fusion–fission events in the ^{165}Ho$+^{20}$Ne reaction is very similar to that of neutrons measured in coincidence with evaporation residues (Hol 83). However, in the former case, at least four emission processes have to be considered as possible contributors to the total particle cross section. These include, as before, preequilibrium and equilibrium emission from the composite nucleus prior to fission, as well as evaporation from the excited fission fragments. Fits to energy spectra of neutrons in coincidence with fission fragments from the ^{165}Ho$+^{20}$Ne reaction have been performed by Holub *et al.* (Hol 83), employing a fit formula analogous to that of Equation (3.7.33), but for four sources i. While the Maxwellian evaporation spectra in Equation (3.7.33) are consistent with the statistical model only in the approximation of a constant-temperature nuclear level density, they describe with similar accuracy the mean energy spectrum of all neutrons emitted in an evaporation cascade (LeC 59, Lan 64). In this case, the nuclear temperature τ of the product nucleus, related to its intrinsic excitation energy E^* by $E^* = a\tau^2$ and the level density parameter $a \approx A/8$, has to be replaced by an effective temperature parameter (LeC 59, Lan 64)

$$\tau \approx \tfrac{11}{12}(E^*/a)^{1/2} \tag{3.7.34}$$

given in terms of the temperature of the first emitter in the cascade. Similarly, the factor $\sqrt{E}$ in Equation (3.7.33) is to be replaced by $E^{5/11}$.

From such analyses of the $^{165}Ho+^{20}Ne$ neutron distributions, it is found that approximately 6 neutrons are statistically evaporated from the intermediate system at a temperature of about 2.6–3 MeV, independent of the bombarding energy. The energy balance is largely made up by preequilibrium neutrons, whose multiplicities and mean energies increase with increasing incident energy, as well as by neutrons emitted from the final fission fragments. In Figure 273, the correlation (Hol 83) between the total preequilibrium multiplicity $M_{n\mathrm{PE}}$ per fusion-like event, deduced from data on a variety of reactions, and the center-of-mass energy per nucleon above the Coulomb barrier V_{CB} is shown. The straight lines represent fits to the $^{165}Ho+^{20}Ne$ data on coincidences with evaporation residues (dashed line) and with fission fragments (solid line). A somewhat different correlation of nonequilibrium α-particle multiplicities with relative incident velocities has already been illustrated in Figure 263 (Hol 83). Although the data points collected in Figure 273 show considerable scatter, there appears to be a roughly linear increase of the preequilibrium neutron multiplicity with the relative kinetic energy per nucleon, i.e., with the square of the relative velocity. Since the mean energy of the preequilibrium neutrons is found also to increase with bombarding energy, the composite nucleus is left with decreasing fractions of the total balance of the excitation energy and of the linear momentum per nucleon. Such a behavior demonstrates a growing inability of nuclear matter to sustain the high amounts of energy and momentum transfer introduced by the impact of an energetic projectile on a target nucleus. The observed insensitivity to incident energy of the temperature of the intermediate system prior to scission, also reflected in a constant prescission neutron multiplicity reported by Gavron *et al.* (Gav 80c) for a weakly fissile system, is a possible consequence of this property.

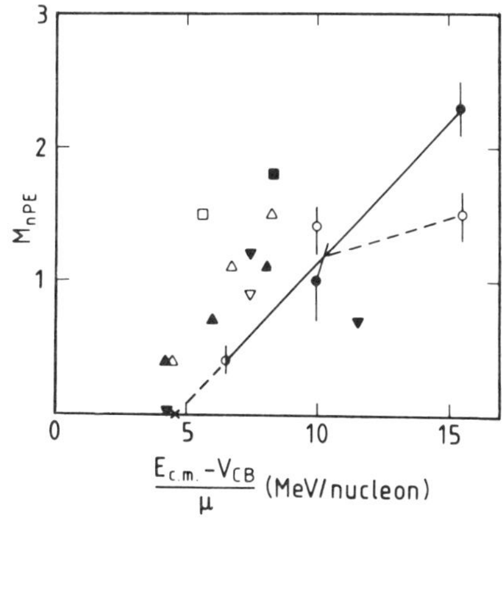

Figure 273. Mean multiplicity of preequilibrium neutrons in coincidence with fusion-like events plotted vs. the relative incident energy per nucleon above the Coulomb barrier. The data are for a number of different reactions. Solid dots correspond to the $^{165}Ho+^{20}Ne$ fusion–fission reaction. They are fitted by the solid straight line. Open circles and dashed lines are for $^{165}Ho+^{20}Ne$ fusion–evaporation process. From (Hol 83).

Average multiplicity and energy of particles emitted from the intermediate nucleus prior to scission can be used to obtain an order-of-magnitude estimate for the lifetime of such a nucleus, assuming that equilibration times are negligible (cf. Section 2.7). Using the empirical formula given by Stokstad (Sto 77c), which relates particle multiplicity, M_p, nuclear temperature of the emitter, τ, and average time for particle emission, t_p,

$$t_p \approx \tfrac{1}{2} M_p \exp(13\ \text{MeV}/\tau) \times 10^{-22}\ \text{s} \tag{3.7.35}$$

yields (Hol 83) evaporation times of the order of (2–4)$\times 10^{-20}$ s. These times have to be considered as rather long and, if confirmed in further studies, suggest a highly overdamped scission process, compatible (Neg 78) with strong one-body damping of the collective motion in the exit channel. In any case, these observations, together with further evidence discussed below, suggest interesting opportunities to study the dynamics of nuclear fusion and disintegration in general.

The mechanism of particle emission prior to scission is well established also in fission induced by neutrons and light ions (see, e.g., Che 70, Van 73, Fra 75, 78h), both at low and high excitation energies. For example, the multiplicity of prescission neutrons reported by Fraenkel *et al.* (Fra 75) for 16-MeV proton-induced fission of ^{226}Ra is 0.5, increasing to values of ~ 3 for fission of actinides induced by 45-MeV α particles. These prescission particles were distinguished from those evaporated from the accelerated fission fragments on the basis of their velocity and angular distributions, assuming an isotropic in-plane emission probability in the rest frames of the corresponding emission sources. Prescission proton and α particles are also observed (see Van 73) in low-energy fission, although with comparatively low probabilities of 10^{-5} and $\sim 10^{-3}$ per fission event, respectively. Particularly conspicuous are the long-range α particles in fission, that, because of their preferential emission in a direction perpendicular to the fission axis, have been associated (see Van 73) with the rupture of the neck, assumed to have been formed between the nascent fragments. The excitation-energy dependence of the production probability of such α particles is only known experimentally for a rather limited range of excitation energies, and very little is known about the dynamics of their production, making it difficult to arrive at a meaningful prediction of the occurrence of this mechanism in heavy-ion reactions.

The competition between particle evaporation and fission in heavy-ion reactions at lower energies than those discussed previously has been investigated by Miller and collaborators (Mil 78, 78b, Raj 79, 82, Del 79, Cat 80, Riv 82), who have studied ^{12}C, ^{19}F, ^{20}Ne, ^{40}Ar, and ^{86}Kr-induced reactions on a variety of medium-weight and heavy target nuclei. Various combinations of entrance-channel properties are chosen in these experiments to produce particular composite nuclei with different spins and excitation energies. For example, in experiments of Miller *et al.* (Mil 78b), the composite nucleus ^{194}Hg was produced in ^{182}W + ^{12}C and ^{175}Lu + ^{19}F reactions at 98 and 142 MeV, and inclusive measurements were made of evaporation residues, fission fragments, and of light charged particles. (In general, charged-particle data obtained in inclusive experiments should be taken with considerable caution, because of the possibly large distortion of the experimental results introduced by even miniscale ^{12}C and ^{16}O contaminations and deposits on the target.) The conclusions drawn from these inclusive experiments, corroborated by later exclusive measurements, performed by the same group, emphasize the importance of prescission emission of protons and α particles, successfully competing with fission at higher energies. Inclusive particle angular distributions for the reaction ^{164}Dy + ^{40}Ar (Del 79) are forward-peaked, with an additional component varying very little with angle, in the backward region, attributed to direct and evaporative emission processes, respectively. It is observed (Del 79) that the cross section for the direct, forward-peaked component of protons and α particles is large, (0.2–1) b, and relatively independent of the atomic number of the composite system produced in ^{40}Ar-induced reactions. In contrast, the evaporative component decreases drastically with the charge Z_{CN} of the composite nucleus, in the range $52 \le Z_{CN} \le 84$, increasing again in the domain $84 \le Z_{CN} \le 97$. These features of particle emission, found (Del 79) incompatible with statistical evaporation processes, are, hence, attributed (Del 79) to properties of the early reaction dynamics, i.e., ejection of protons and α particles during the initial contact of projectile and target.

These conclusions have been subsequently corroborated in experiments performed by Rajagopalan *et al.* (Raj 82), who produced the composite nucleus ^{194}Hg in twelve reactions, at excitation energies of 57–195 MeV and angular momenta of 25–142$\hbar$. Similar results are obtained in inclusive measurements of all of these reactions, with respect to a strong forward peaking of high-energy protons and α particles, and a flat angular distribution at backward angles. Energy spectra and angular distributions of protons and α particles in the backward hemisphere are used to study charged-particle evaporation from the intermediate composite system. It is concluded (Raj 82) from this work that in the reactions ^{182}W + ^{12}C, ^{175}Lu + ^{19}F, ^{174}Yb + ^{20}Ne, and ^{154}Sm + ^{40}Ar leading to the composite nucleus ^{194}Hg at excitation energies of 57, 80, 100, 142, and 195 MeV, most protons and α particles are evaporated from the composite nucleus prior to fission. No positive evidence of sequential charged-particle emission from fission fragments is reported (Raj 82) from these measurements. The fission branching ratio itself is observed to become smaller for excitation energies larger than 100 MeV, in comparison to evaporation-residue production, while proton and α decay widths increase (Raj 82). Some of the results reported from these experiments appear to be quite plausible. However, it is difficult to conceive of circumstances in which highly excited fission fragments are unable to evaporate charged particles. Quite likely, the limited sensitivity of the inclusive, backward-angle measurements to details of the particle emission patterns introduces ambiguities in the interpretation of data that are difficult to resolve.

Representative results of an experiment on proton and α-particle emission in the reaction ^{197}Au + ^{40}Ar at $E_{\text{Lab}} = 340$ MeV obtained by Logan *et al.* (Log 80) are depicted in Figure 274. In this figure, inclusive α-particle

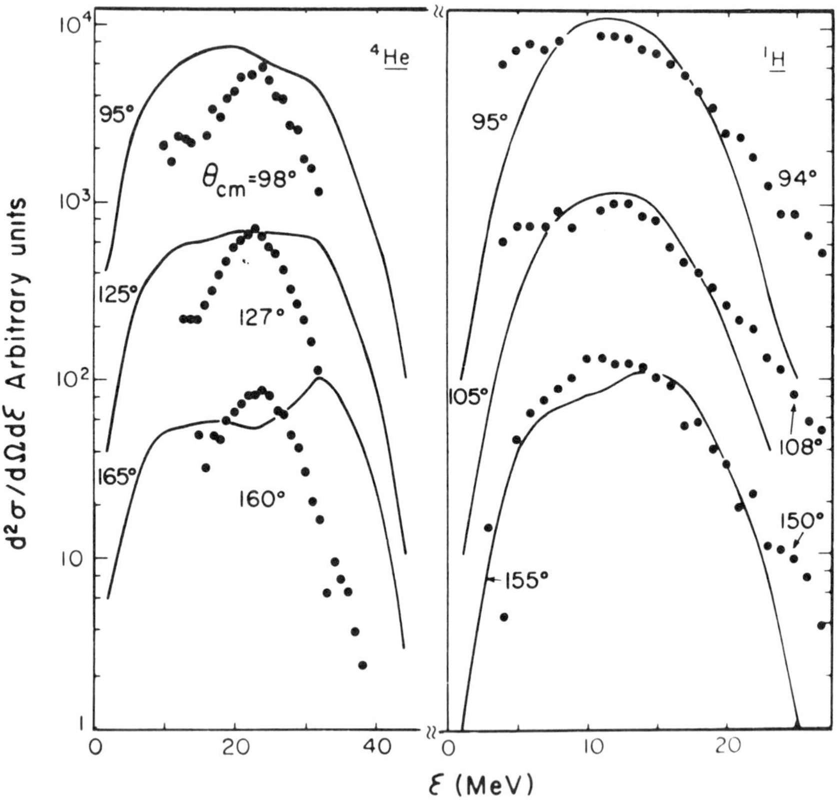

Figure 274. Inclusive energy spectra of α particles (left) and protons (right) emitted in the bombardment of a ^{197}Au target with a 340-MeV ^{40}Ar beam, for various emission angles. The solid curves represent Monte Carlo calculations assuming particle evaporation from fully accelerated fission fragments produced in the ^{197}Au + ^{40}Ar fusion–fission reaction. The fission fragment angular distribution adopted in the model calculations is of the $1/\sin\theta$ type, while isotropic emission in the rest frames of the emitters is assumed for the charged particles. From (Log 80).

(left) and proton (right) energy spectra are shown, as measured at various backward emission angles. The curves included in Figure 274 represent emission patterns predicted by Monte Carlo simulation calculations modeling statistical evaporation from fully accelerated fission fragments produced in the reaction. A symmetric Gaussian mass distribution with a standard deviation of $\sigma = 30$ mass units is assumed in the calculations. Fission is reported (Log 80) to be the main reaction process associated with the emission of charged particles in the reaction $^{197}\text{Au} + {}^{40}\text{Ar}$. This conclusion is drawn from a comparison of the angular distributions of protons and α particles, measured as singles distributions and in coincidence with heavy reaction fragments. Only $\sim 10\%$ of the coincident events are found (Log 80) to be in coincidence with damped reaction fragments.

As observed from Figure 274, the calculations for the α-particle spectra from fission fragments yield distributions that are much wider than the experimental ones. This disagreement leads Logan *et al.* (Log 80) to the conclusion that the α particles observed in the inclusive measurements are dominantly not due to emission from the fully accelerated fission fragments produced in the $^{197}\text{Au} + {}^{40}\text{Ar}$ reaction. In the case of the proton spectra shown in Figure 274, a better reproduction of the data is obtained with the Monte Carlo calculations. Assuming that a majority of the charged particles are produced in the $^{197}\text{Au} + {}^{40}\text{Ar}$ fusion–fission reaction, it is conjectured (Log 80) that they are due to an emission process preceding fission of the intermediate system. Of course, such inclusive measurements do not provide unambiguous evidence for a particular emission mechanism.

Exclusive experiments on light charged particles emitted in reactions between 330-MeV ^{40}Ar projectiles and targets of ^{116}Sn, ^{154}Sm, and ^{197}Au have been performed recently by Rivet *et al.* (Riv 82). In these measurements, fission fragments were detected at laboratory angles of 60° and 320° in coincidence with protons or α particles measured at a backward in-plane angle of 120° and out-of-plane angles of 0°, 27°, and 320°. Comparison of Monte Carlo calculations with experimental particle energy spectra indicates (Riv 82) that kinematical shifts in the high-energy wings, expected from the calculations modeling sequential evaporation from the accelerated fission fragments, are not as evident in the data. Only fractions of $< 20\%$ of the protons and $< 30\%$ of all α particles observed at backward angles are reported (Riv 82) to be compatible with such a sequential evaporation process. The conclusions then arrived at by Rivet *et al.* (Riv 82) are that probably more than 80% of the protons and at least 60% of all α particles emitted in the above ^{40}Ar-induced reactions are decay products of the intermediate system.

The data obtained in the above experiments have been analyzed (Riv 82) in terms of a model for statistical evaporation from a thermally equilibrated composite system, in order to deduce properties of this system.

For example, the energy spectrum

$$\frac{dM}{dE} \propto (E - E_B)\exp(-E/\tau) \tag{3.7.36}$$

for charged particle emission from a spin-zero nucleus depends both on the barrier E_B for evaporation and the temperature τ of the residual nucleus, corresponding to an average energy of $\langle E \rangle \approx E_B + \tau$. For nuclei with spin J, particle evaporation spectra are generally broader with an average energy increased (Riv 82) to

$$\langle E \rangle_J = E_B + \left(1 + \frac{\mathscr{I}}{\mathscr{I} + \mu R^2}\right)\tau + \frac{\hbar^2 (J + 1/2)^2}{2} \frac{\mu R^2}{(\mathscr{I} + \mu R^2)^2} \tag{3.7.37}$$

for emission in a plane perpendicular to **J**. In Equation (3.7.37), $\mathscr{I}$ is the moment of inertia of the emitter, and μR^2 denotes the moment of inertia for orbital motion of the emitted particle, evaluated at the disintegration radius. The parameters in Equation (3.7.37) involving moments of inertia and the nuclear spin also enter model expressions for the out-of-plane anisotropy, as discussed in the previous section, and may be determined from fits to experimental anisotropies. Analyses of charged-particle angular and energy distributions for the ^{40}Ar-induced reactions referred to above, have been performed by Rivet *et al.* (Riv 82) in terms of the statistical model, yielding estimates for r.m.s. spin values and barrier energies E_B for the emitting composite nucleus. Values deduced for the emission barriers (Riv 82) appear to be significantly (20%–30%) smaller than estimated (McM 80, Vaz 81) from the systematics of fusion cross sections.

While measurements and analyses of light charged-particle emission in ^{40}Ar-induced fusion–fission reactions, discussed above, have been interpreted in terms of overwhelming contributions due to statistical emission from a highly deformed intermediate system, particle emission patterns from damped reactions induced by ^{40}Ar, or heavier projectiles, as well as from fusion–fission reactions of heavier nuclei convey a contrasting picture, as discussed previously and further illustrated below.

Light-particle emission in damped ^{58}Ni + ^{40}Ar reactions have been investigated by Babinet *et al.* (Bab 80) in exclusive measurements of in-plane and out-of-plane angular correlations of protons and α particles, detected in coincidence with damped fragments emitted at a forward angle of 30° in the laboratory frame. Figure 275 depicts some of their results on

Galilean-invariant [cf. Equation (3.7.31)] cross sections for α particles in coincidence with fragments of charges $Z=10$ (top, left), $Z=16$ (bottom, left), and $Z=23$ (center, right). Here, the differential invariant α-particle emission cross section is plotted vs. the α-particle velocity components parallel ($v_{\parallel}$) and perpendicular ($v_{\perp}$) to the beam direction. Each row of dots in this figure corresponds to one measurement at a given lab angle. The dot size represents a measure of the magnitude of the cross section. The vectors $\mathbf{v}_Z$ and $\mathbf{v}_{Z\,\text{compl.}}$ included in the figure correspond to the velocity vectors of the detected fragment and its reaction partner. The velocity vector $\mathbf{v}_{\text{c.m.}}$ of the center-of-mass system in the laboratory frame is also indicated. Thin circles illustrate the detection thresholds.

Close inspection of the diagrams in Figure 275 shows (Bab 80) that the cross-section patterns lack the symmetry with respect to the beam direction required for dominant emission from the intermediate composite nucleus. In fact, the maxima of the invariant cross sections at the various laboratory detection angles define two ridges approximated by dashed circles in the figure, centered at the tips of the velocity vectors of projectile-like and target-like reaction fragments. As discussed earlier in this section, the loci of maximum cross section associated with sequential emission from the asymptotic reaction fragments are expected to outline an ellipsoidal curve, due to distortions by the recoil effect on projectile-like fragments, initially strongly focused to forward angles. For the very asymmetric fragmentation ($Z=10$), intensity patterns associated with possible emission from the composite system are kinematically not well enough separated from those for emission from the heavy reaction partners. From the more symmetric fragmentations, however, the cross section diagrams are uncharacteristic of significant contributions of α-particle emission from the composite nucleus.

In support of dominant sequential α decay of asymptotic damped reaction fragments is the observation (Bab 80) of a strong correlation of experimental α-particle energy spectra with the atomic number Z of the emitting fragment. The barrier energies for α-particle and proton emission, E_B^α and E_B^P, respectively, deduced from experimental data on a variety of compound-nucleus reactions can well be parametrized by (Bab 80)

$$E_B^\alpha(Z) = (0.1875Z + 1.4)\ \text{MeV}$$

$$E_B^P(\text{Z}) = (0.106Z - 1.7)\ \text{MeV} \tag{3.7.38}$$

The most probable particle energy, for a spectrum of the simple form of Equation (3.7.36) is given by $E(\max) = E_B + \tau$, and, hence, the observed

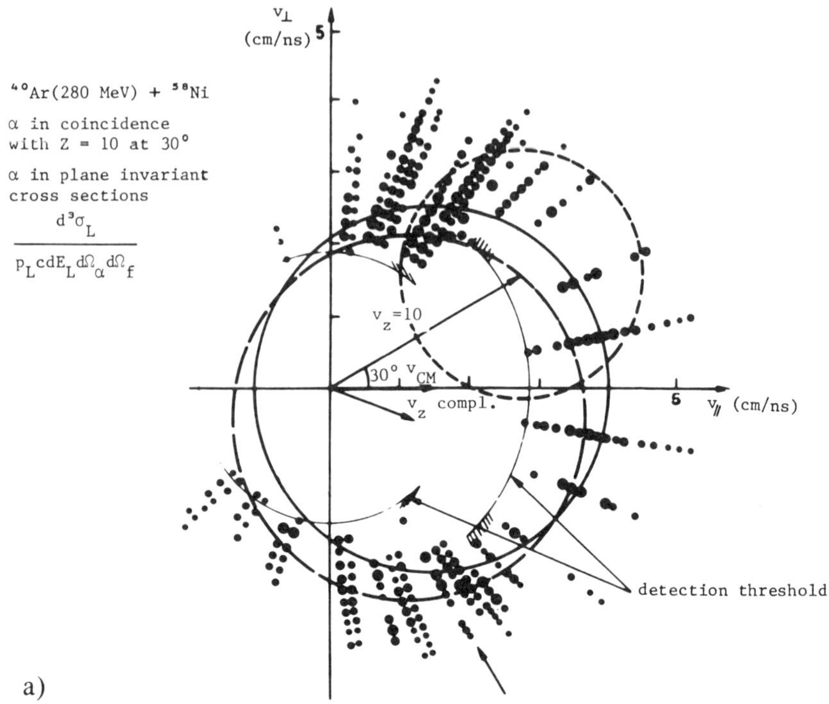

^{40}Ar(280 MeV) + ^{58}Ni
α in coincidence with Z = 10 at 30°
α in plane invariant cross sections
$\frac{d^3\sigma_L}{p_L c dE_L d\Omega_\alpha d\Omega_f}$
$v_\perp$ (cm/ns)
5
$v_z = 10$
30°
v_{CM}
v_z compl.
5
$v_{/\!/}$ (cm/ns)
detection thresholds

a)

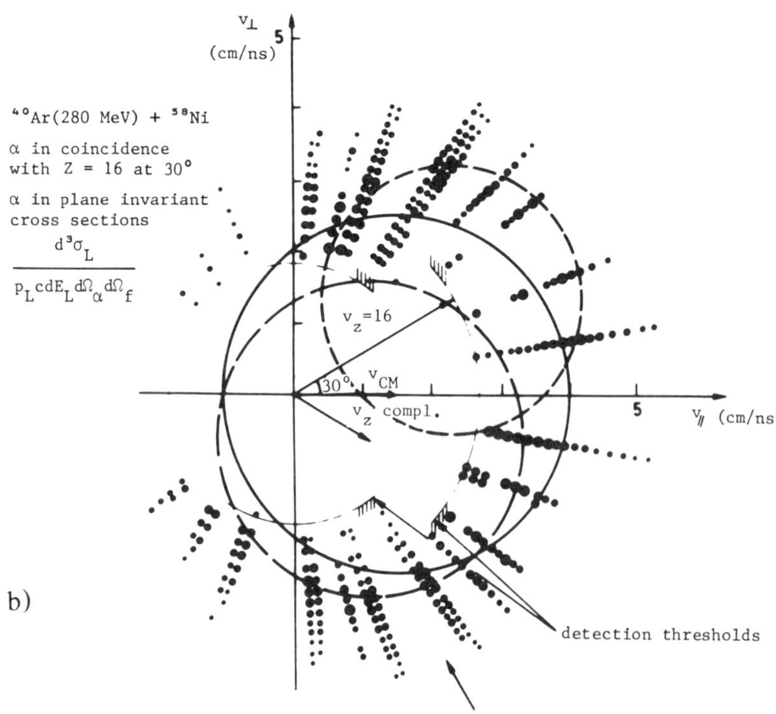

^{40}Ar(280 MeV) + ^{58}Ni
α in coincidence with Z = 16 at 30°
α in plane invariant cross sections
$\frac{d^3\sigma_L}{p_L c dE_L d\Omega_\alpha d\Omega_f}$
$v_\perp$ (cm/ns)
5
$v_z = 16$
30°
v_{CM}
v_z compl.
5
$v_{/\!/}$ (cm/ns)
detection thresholds

b)

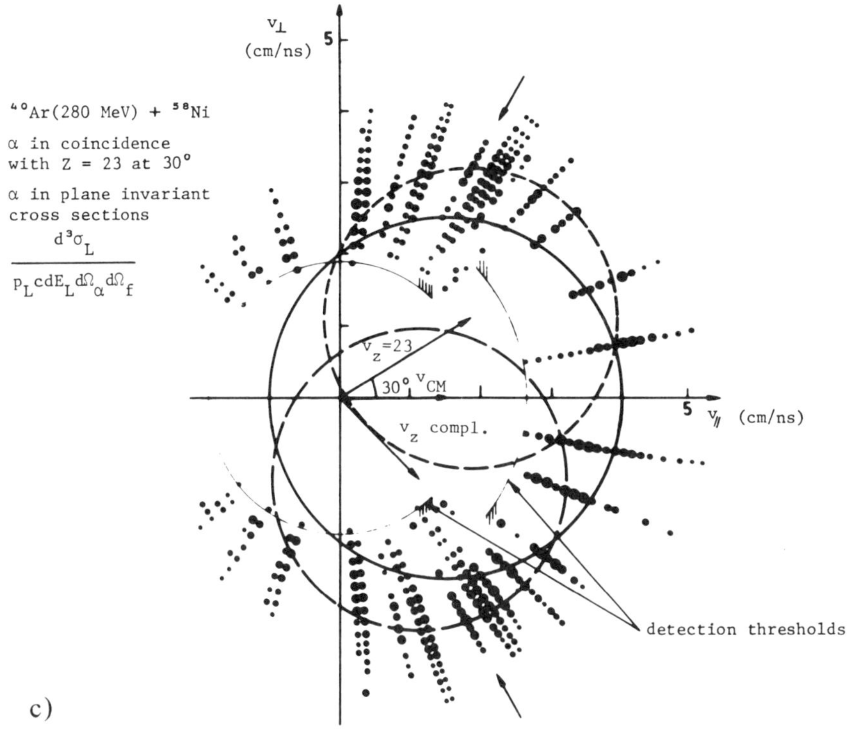

Figure 275. Galilean-invariant cross section of coincident α particles emitted in the damped reaction $^{58}Ni + ^{40}Ar$ at $E_{Lab} = 280$ MeV. Different fragment Z values are indicated at each diagram. The dashed circles centered at the tips of the velocity vectors of the coincident fragment and its reaction partner outline regions of maximum α-particle cross section. From (Bab 80).

correspondence of the most probable kinetic energies of α particles, measured in the above $^{58}Ni + ^{40}Ar$ reaction, to the barrier energies of Equation (3.7.38) can be regarded as evidence in support of dominant sequential emission from the asymptotic fragments.

Evidence against dominant particle emission from the intermediate system formed in damped reactions can also be constructed from an experimental study of charged-particle emission in the $^{40}Ca + ^{40}Ca$ reaction at 10 MeV per nucleon performed by Roynette *et al.* (Roy 81). Some of their experimental results are reproduced in Figure 276, showing Galilean-invariant cross sections of protons (top) and α particles (bottom), measured in coincidence with partially damped Ca and Ar fragments, respectively,

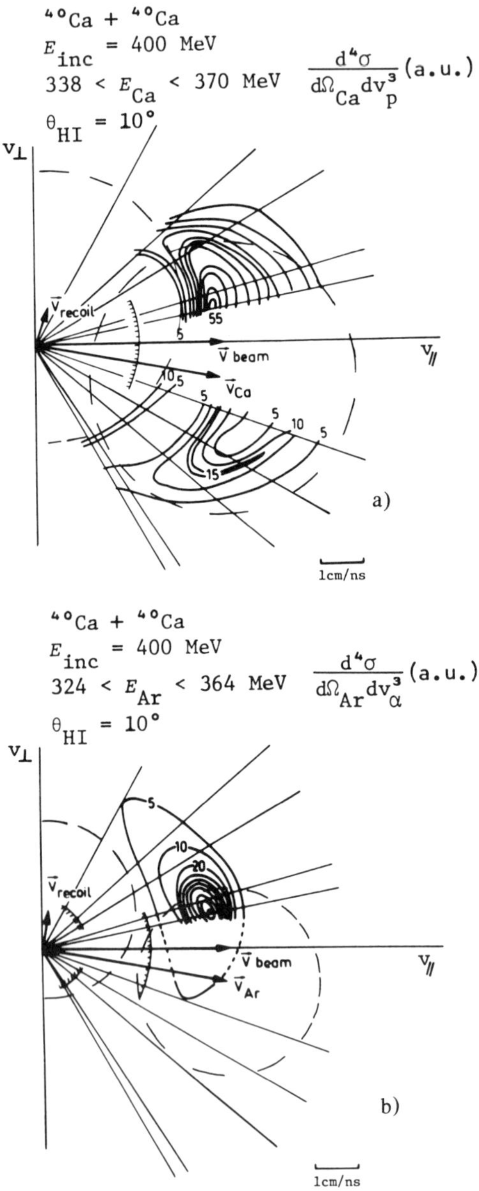

Figure 276. Contour plot of the Galilean-invariant cross section for protons (top) and α particles (bottom) emitted in the 10-MeV per nucleon reaction $^{40}Ca + ^{40}Ca$ in coincidence with partially damped Ca and Ar fragments, respectively. Velocity vectors for beam particles and the two reaction partners are indicated by arrows. The dashed circles outline the cross section features expected for sequentially emitted particles. Straight lines indicate laboratory detection angles of light charged particles. From (Roy 81).

corresponding to a region in the fragment laboratory energy spectrum exhibiting bump structure (cf. Section 3.5). Straight lines in Figure 276 indicate angles at which coincidence measurements were made. As before, circles are drawn in this figure about the tips of the velocity vectors of the reaction fragments illustrating emission patterns expected from sequential evaporation. Experimental thresholds are also indicated in Figure 276.

The asymmetry of both, α-particle and proton cross-section patterns about the beam direction is obvious from Figure 276, precluding significant contributions from composite-system emission of charged particles in the $^{40}Ca + ^{40}Ca$ reaction in the forward angular range. Relatively low particle energies are also inconsistent with such an emission mechanism, as is the lack of α-particle cross section on the side of the beam at which the coincident reaction fragment is detected in the experiment. Calculations (Roy 81) assuming sequential particle evaporation from the reaction fragments also do not provide a satisfactory description of the experimental angular correlations, although they reproduce some qualitative features of the experimental cross-section diagrams for α particles. Measurements of coincident charged particles emitted into the backward hemisphere have not been made, unfortunately. However, they would most probably not invalidate the present conclusions with respect to the absence of composite-system emission in partially damped $^{40}Ca + ^{40}Ca$ collisions. Angular and energy correlations of protons and α particles emitted in strongly damped reactions of the same system $^{40}Ca + ^{40}Ca$ follow more strictly the expectations (Roy 81) based on sequential emission of charged particles from the asymptotic reaction fragments. No quantitative analyses of the data have been reported so far, however.

The reaction $^{40}Ca + ^{40}Ca$, along with those of other medium-weight symmetric systems, has been discussed already in Section 3.5 in the context of possible collective doorway excitations in damped heavy-ion collisions. Structure in the laboratory energy spectra of partially damped reaction fragments (see Figure 143) has been attributed (Fra 77, 80, 80f) to such collective excitation mechanisms, while a contrasting view (Hil 79b) associates it with the effects of particle emission. In the coincidence experiments of Roynette *et al.* (Roy 81), such structure is observed (cf. Figure 144) in the energy spectrum of Ca and K fragments in coincidence with protons, but not with α particles. It is, however, observed in the spectrum of Ar fragments in coincidence with α particles. Hence, a correlation is observed between the occurrence of structure in the fragment energy spectra and the associated-particle emission probabilities, which depends on the primary fragment species. Since sequential evaporation of particles from excited primary fragments, as expected for the decay of giant resonance states, currently does not appear to present a viable explanation of data, the question of the origin of the observed structure in the fragment energy spectra remains an interesting question. It appears worthwhile to confirm the peculiarities of this reaction by different experimental methods and complete the set of experimental data available for partially damped ^{40}Ca + ^{40}Ca collisions.

A dominantly sequential decay by particle emission in strongly damped ^{165}Ho + ^{56}Fe collisions at $E_{\text{Lab}} = 465$ MeV is demonstrated in Figure 277 by the Galilean-invariant differential cross section of α particles measured by Kosky *et al.* (Kos 83) in coincidence with projectile-like fragments detected at a laboratory angle of $\theta_{\text{HI}} = -22°$. The data are integrated over fragment Z values of smaller than $Z = 32$ and energy losses of larger than $E_{\text{Loss}} = 100$ MeV. Hence, very strongly damped collisions are emphasized in the velocity contour diagram constructed of $\sim 2 \times 10^5$ coincidence events collected in plane in all four quadrants. Experimental cross sections are represented by heavy curves in Figure 277, whereas light line segments indicate regions of

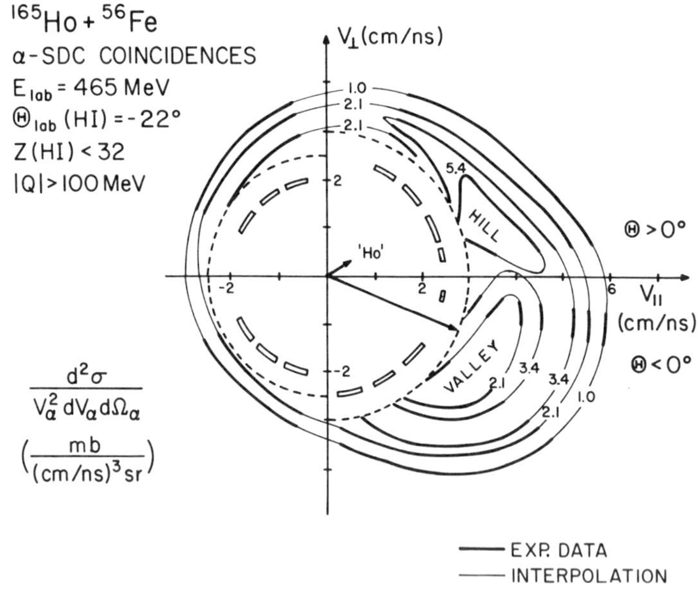

Figure 277. Contour diagram of the Galilean-invariant cross section for α particles emitted in coincidence with strongly damped reaction fragments produced in the reaction ^{165}Ho + ^{56}Fe at $E_{\text{Lab}} = 465$ MeV. The heavy lines represent experimental data; thin curves are the result of interpolations. Arrows indicate the velocity vectors of the detected projectile-like fragment and of its collision partner. Detection thresholds are marked by dashed semicircles. Open ring segments indicate acceptance angles of large-solid-angle charged-particle detectors. From (Kos 83).

interpolation of experimental data. Average velocity vectors of projectile-like and target-like reaction fragments are given by the arrows in this figure; experimental thresholds are denoted by dashed circles. Average α-particle multiplicities per projectile-like fragment are small, of the order of $\langle M_\alpha \rangle \approx 0.2$.

From the topology of the asymmetric pattern of the Galilean-invariant α-particle cross section, depicted in Figure 277, it is immediately obvious that isotropic emission from the intermediate system cannot be a dominant contributor to α-particle production in this damped reaction. The cross section contours are definitely stretched in the direction of flight of the projectile-like fragment. One observes a well-pronounced valley in the cross section, centered around the fragment detection angle, as expected from the circular shape (compare Figure 275) of the cross-section ridge associated with sequential evaporation from the asymptotic fragment. This emission pattern superimposes with a similar pattern of α particles emitted from the heavy reaction partner, producing a maximum in the first quadrant near 45°, where the two ridges intersect. Comparison of the data displayed in Figure 277 with Monte Carlo evaporation calculations lead to the conclusion (Kos 83) that sequential α-particle evaporation from the asymptotic fragments has to be considered the major production mechanism of α particles in the strongly damped reaction $^{165}\mathrm{Ho}+{}^{56}\mathrm{Fe}$. However, the angular correlations of α particles from this reaction appear to be generally somewhat less anisotropic than expected from purely isotropic emission in the rest frames of the accelerated fragments such that some contribution to the cross section, isotropic in the center-of-mass frame of reference, is accommodated by the data. Hence, an indication of α-particle emission from the composite system may have also been observed in the data for the $^{165}\mathrm{Ho}+{}^{56}\mathrm{Fe}$ reaction. In addition, there may be a component of high-energy α particles at forward angles on the side of the beam opposite to the fragment detection angle, reminiscent of a similar finding made by Gamp *et al.* (Gam 81, 82) in the reaction $^{197}\mathrm{Au}+{}^{36}\mathrm{Ar}$ (cf. Figures 258 and 259).

Similarly detailed measurements of associated α-particle emission have also been made (Wil 83) for the fusion–fission processes occurring in the above $^{165}\mathrm{Ho}+{}^{56}\mathrm{Fe}$ reaction. The cross-section diagram of α particles measured in coincidence with fission fragments bears resemblance to that for the damped process shown in Figure 277. Although the asymmetry about the beam direction is somewhat less pronounced than for α particles associated with damped $^{165}\mathrm{Ho}+{}^{56}\mathrm{Fe}$ collisions, the hill-and-valley pattern correlated with the fission fragment direction of flight (cf. Figure 277) is clearly visible also in this case, indicating dominant sequential evaporation from the asymptotic fission fragments. The latitude for the presence of additional emission mechanisms, associated with isotropic or anisotropic angular distributions in the center-of-mass frame, is considerably larger in the fusion–fission reaction than for damped ^{165}Ho + ^{56}Fe collisions. However, it appears unlikely that contributions from composite-system emission at the high levels claimed for the ^{40}Ar-induced reactions discussed earlier would be compatible with the experimental data.

Experimental data on charged-particle emission in damped reactions induced in very heavy systems are extremely rare. An exclusive measurement of α-particle emission from the reaction ^{181}Ta + ^{165}Ho at $E_{\text{Lab}} = 1354$ MeV has been performed recently by Sobotka *et al.* (Sob 83). Figure 278 displays a summary of their data in the form of a velocity diagram of the r.m.s. velocities (open circles) extracted from the coincident laboratory α-particle energy spectra. The arrows labeled "P" and "T" indicate the velocity vectors of projectile-like and target-like reaction fragments, enclosing the beam velocity vector. The large solid circles outline the loci of α-particle velocities expected for isotropic emission in the three possible rest frames. For target-like fragments, the loci of α-particle velocities for a 10% reduction in the expected average emission energy is illustrated by a circle segment. The lower part of Figure 278 depicts velocity diagrams for different bins in reaction Q value (open triangles, circles, and squares).

The close correspondence, observed in Figure 278, of the measured mean velocities of α particles emitted in the ^{181}Ta + ^{165}Ho reaction with the circle centered at the tip of the velocity vector of the target-like fragment (T) suggests strongly that these particles are emitted sequentially from strongly damped Ta-like reaction fragments. At one detection angle (solid dots), a contribution of α particles from the projectile-like fragment is also seen. The spectra of α particles in the rest frame of the target-like fragment are observed (Sob 83) to exhibit an evaporation-type shape [cf. Equation (3.7.36)] which is independent of detection angle, with the possible exception of the most forward angle. Peak energies correspond to generally ~ 18 MeV. At the most forward angle of 30° covered in the measurements, both differential particle multiplicity and mean energy are somewhat larger than measured for more backward angles. Although a small admixture of emission from the composite system is not ruled out (Sob 83) in this experiment, the possible level of contribution from such processes must be considered insignificant for such a heavy system.

Although the multiplicity of α particles emitted in damped and fusion–fission-like collisions varies with reaction Q value and charge of the emitter and can reach (Kos 83) integrated values of $\langle M_\alpha \rangle \sim 1$ per fragment, these emission processes constitute relatively minor decay channels of excited primary nuclei produced in such reactions. One must, hence, be aware of the possible bias introduced in the interpretation of heavy-ion reaction mechanisms, if the discussion is limited to such processes. Reasons for the overwhelming dominance of neutron emission over other decay channels may be seen in the very high excitation energies above the yrast

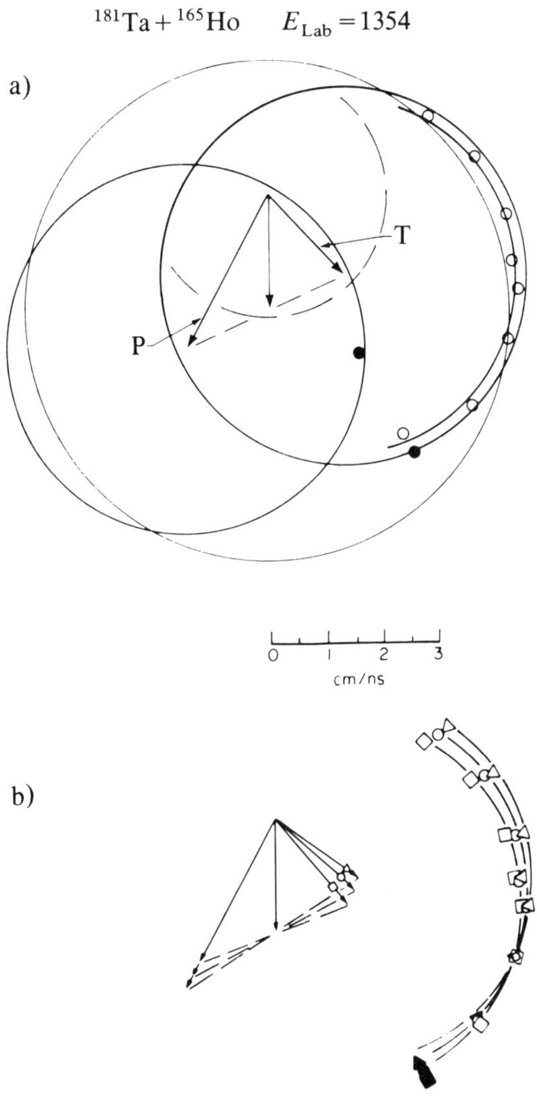

Figure 278. (a) Velocity diagrams of r.m.s. velocities of α particles emitted in coincidence with strongly damped reaction fragments from the reaction $^{181}Ta + ^{165}Ho$ at $E_{Lab} = 1354$ MeV. Large circles centered at the tips of the velocity vectors of the beam, the projectile-like fragment (P) or of the target-like fragment (T) illustrate the loci of α-particle velocity expected for isotropic emission in the three possible rest frames. Open and filled small circles represent measured values of α-particle velocities. (b) Velocity diagrams for three different bins in energy loss. Triangles indicate r.m.s. velocities of α particles measured in coincidence with fragments associated with smaller energy losses. Circle segments are drawn to guide the eye. The three pairs of corresponding fragment velocity vectors are given by the arrows. From (Sob 83).

line, attained usually in heavy-ion reactions, and the absence of emission barriers. As inferred from statistical model calculation, such as illustrated in Figures 174–176, α-particle emission is expected to become a significant deexcitation channel only for excitation-energy-spin domains close to the yrast line, while protons compete successfully with neutron evaporation already somewhat earlier in a deexcitation cascade. Hence, measurements of different decay channels provide useful complementary information on the dissipation and equilibration mechanisms operating in damped and fusion–fission-like reactions.

In Figure 279, experimental results relating to neutron emission in the reaction ^{165}Ho + ^{56}Fe at $E_{\text{Lab}} = 476$ MeV are displayed, as measured by Hilscher *et al.* (Hil 79c). The top part (c) of this figure shows the energy-loss spectrum of projectile-like damped reaction fragments detected at a laboratory angle of $\theta = 20°$ in coincidence with neutrons. The bottom portion (a) of Figure 279 illustrates the dramatic increase of the multiplicities, M_n, of neutrons emitted from the light (circles) and heavy (squares) reaction partners, while the center part (b) depicts the variation of the multiplicity ratio with increasing values of kinetic energy loss (TKE-LOSS). From these data, one infers that of the order of ~ 12 neutrons can be emitted in a strongly damped reaction such as ^{165}Ho + ^{56}Fe. The curves include statistical evaporation calculations discussed further below.

Neutron emission in damped nuclear reactions at energies below 10–12 MeV per nucleon have been studied by a number of groups (Pet 77, Gou 78, 80, Eya 78, 80, Hil 79c, 81, Tam 79, Dak 80b, Gem 80c, Tse 81, Hol 83) for reactions induced by projectiles ranging from ^{16}O to ^{132}Xe and a variety of targets. The emphasis in these investigations is usually placed on a study of the emission mechanism and information on the distribution of the energy dissipated in a damped collision among the reaction partners. As has already been noted in Section 3.5, the latter information is of crucial importance for an understanding of the dissipation and relaxation mechanism as well as for the validity of fundamental assumptions of reaction theories.

Identification of the dominant emission sources for neutron production in damped and fusion–fission-like reactions has to rely, of course, on the correlations of particle energies and emission angles with the velocities of the emitters. Experimental problems associated with such an identification are somewhat different from those encountered in inclusive charged particle measurements. Neutron emission is not hindered by a barrier of a height characteristic of the emitter. Hence, Galilean-invariant plots of neutron cross sections are less structured than those corresponding to charged particles. On the other hand, the usually quite low average neutron energies, for statistical emission from a nucleus at temperature τ simply given (cf., e.g., Hil 79c) by $\langle E_n \rangle = (3/2)\tau$, ensure an efficient kinematical focusing of

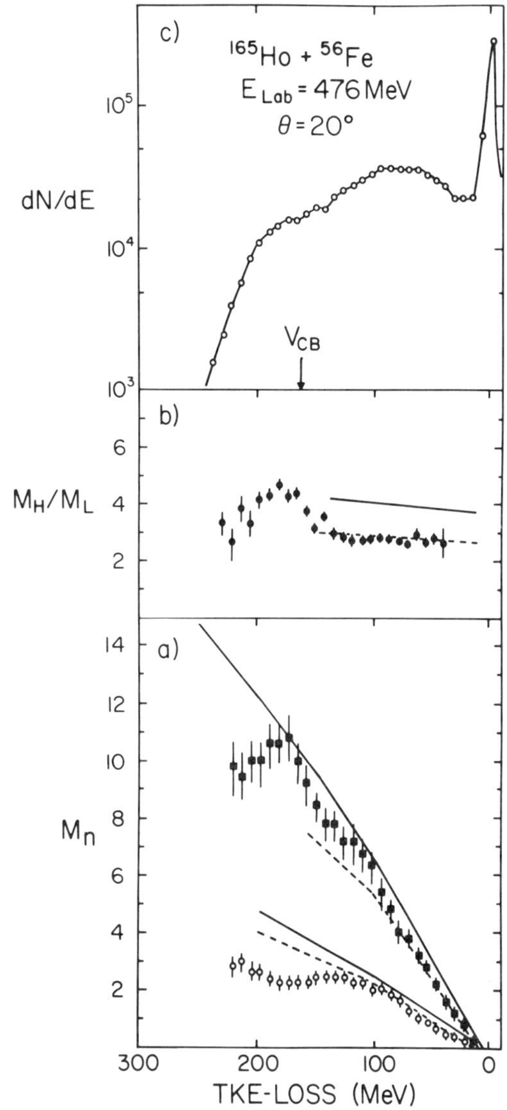

Figure 279. (a) Multiplicities of neutrons emitted from heavy (squares) and light (circles) partners in damped collisions of the ^{165}Ho + ^{56}Fe system at $E_{\mathrm{Lab}} = 476$ MeV, plotted vs. the total kinetic energy loss in the reaction. Curves represent results of statistical evaporation calculations. (b) Energy-loss dependence of the ratio of the multiplicities of neutrons from the heavy (M_H) and the light (M_L) reaction partners. (c) Z-integrated energy loss spectrum of projectile-like fragments detected at $\theta_{\mathrm{Lab}} = 20°$. The arrow (V_{CB}) marks the energy loss corresponding to complete damping of the relative kinetic energy down to the Coulomb barrier for spherical fragments. From (Hil 79c).

the majority of neutrons with low rest-frame energies into a relatively narrow cone centered around the direction of flight of the emitting nucleus.

This focusing effect is illustrated in Figure 280 (Hil 79d, Sch 79f) for statistical neutron evaporation from final fragments in the reaction ^{165}Ho $+ ^{136}$Xe at $E_{\rm Lab} = 1130$ MeV. The figure depicts laboratory intensity patterns of neutrons, at the bottom in the form of a two-dimensional contour plot vs. neutron energy and angle, with the total angular distribution shown in the top portion. Solid curves correspond to projectile emission, dashed curves to evaporation from the accelerated target-like fragment. In the calculation of the double-differential evaporation cross section $d^2\sigma/dE_n\,d\Omega$, complete conversion of the dissipated kinetic energy into intrinsic heat, establishment of thermal equilibration, and isotropic emission in the rest frames of the fully accelerated fragments were assumed. As can be seen from Figure 280, neutrons from the light, fast-moving fragment are generally higher in energy and confined to a narrower angular range than those from the slow-moving heavy fragment. Obviously, there are kinematical regions allowing a relatively unambiguous identification of the emitting fragment. A possible contribution of neutron emission from the intermediate system would fill in the gaps between the two components illustrated in Figure 280 around zero degrees.

Neutron emission in the reaction ^{197}Au $+ ^{63}$Cu at $E_{\rm Lab} = 400$ MeV has been measured by Tamain *et al.* (Tam 79, 79b) in an exclusive experiment in

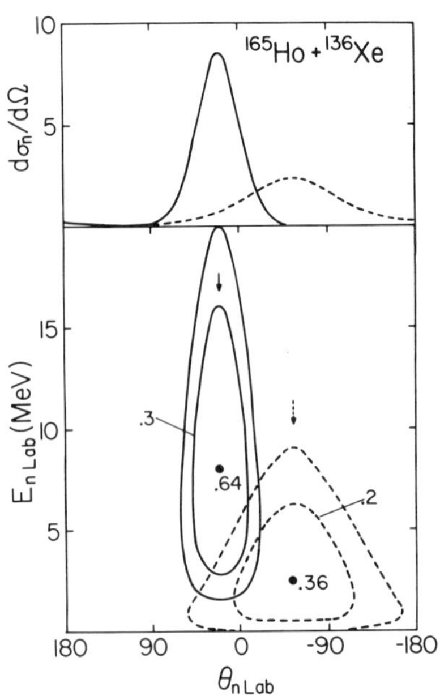

Figure 280. Theoretical angular distribution (top) and angle-energy correlation (bottom) of neutrons evaporated in flight from the light (solid curves) and heavy (dashed curves) fragments produced in the reaction ^{165}Ho $+ ^{136}$Xe at $E_{\rm Lab} = 1130$ MeV. The bottom portion depicts a two-dimensional contour diagram of the double-differential neutron cross section $d^2\sigma/dE_n\,d\Omega_n$, where the neutron emission angle is counted from the beam direction. The emission angles of the fragments are indicated by arrows. From (Hil 79d).

which damped reaction fragments were detected, at two laboratory angles, in coincidence with neutrons counted with nine independent plastic-scintillation detectors. The kinetic energy spectra of neutrons, transformed into the rest frames of projectile- and target-like fragments, exhibited a Maxwellian shape, corresponding to equal temperatures τ of the two reaction partners (compare Figure 171). This observation suggests that thermal equilibrium is attained in the damped ^{197}Au + ^{63}Cu reaction. One should note, however, that this result represents average information, weighted heavily at the high neutron multiplicities that are associated with very strongly damped processes (cf. Figure 279) and long interaction times.

The experimental results on neutron multiplicities in the ^{197}Au + ^{63}Cu reaction convey a similar message (Tam 79b). Here, the ratio ν_1/ν_2 of neutrons emitted from the projectile-like (M_1) and target-like (M_2) reaction fragments is plotted vs. the fragment mass ratio M_1/M_2. Within the experimental uncertainties, the data in Figure 281 are seen to follow a straight line. Very approximately, the average neutron multiplicity $\langle M_i \rangle$ (or ν_i) can be calculated from the intrinsic excitation energy E_i^*, the average thermal and binding energy, $\langle E_n \rangle \approx \langle 1.5\tau + B_n \rangle$, carried away by a neutron, and from the decay widths Γ [see Equation (3.6.5)] for competing emission processes, as

$$\langle M_i \rangle \approx \frac{E_i^*}{\langle E_n \rangle_i} \frac{(\Gamma_n)_i}{(\Gamma_{\text{tot}})_i} \tag{3.7.39}$$

Here, $(\Gamma_n)_i$ and $(\Gamma_{\text{tot}})_i$ are the partial neutron width and the total decay width, respectively, of nucleus i. If the neutron binding energy B_n and the branching ratio $(\Gamma_n/\Gamma_{\text{tot}})_i$ do not vary significantly with the particular fragment, and if the nuclear temperature can be calculated from the Fermi-gas formula, $\tau \cong (8E^*/A)^{1/2}$, then

$$\frac{\langle M_1 \rangle}{\langle M_2 \rangle} \approx \frac{E_1^*}{\langle B_n + 1.5\tau \rangle_1} \frac{\langle B_n + 1.5\tau \rangle_2}{E_2^*} \approx \frac{E_1^*}{E_2^*} \approx \frac{A_1}{A_2} \frac{\tau_1^2}{\tau_2^2} \tag{3.7.40}$$

In such a simple model, a multiplicity ratio $\langle M_1 \rangle / \langle M_2 \rangle$ proportional to the fragment mass ratio A_1/A_2, such as observed in Figure 281, then implies a constant ratio τ_1/τ_2 of the emitting fragments. The results for the ^{197}Au + ^{63}Cu reaction, reported by Tamain *et al.* (Tam 79b), appear to suggest $\tau_1 \approx \tau_2$ and, hence, the achievement of thermal equilibrium in this reaction prior to breakup of the intermediate system. For a fixed mass ratio of the emitting fragments, the neutron multiplicity ratio is reported (Tam 79b) to

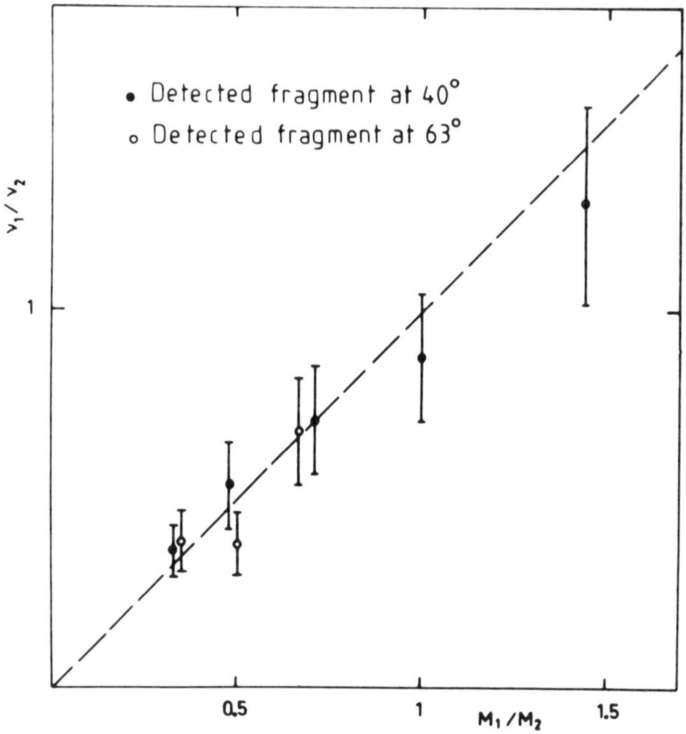

Figure 281. Neutron emission in the reaction ^{197}Au + ^{63}Cu at 400 MeV, where coincident damped reaction fragments are detected at 40° and 63°. Plotted is the ratio of the numbers, ν_1 and ν_2, of neutrons emitted by the detected fragment (M_1) and its reaction partner (M_2), respectively, vs. the fragment mass ratio, M_1/M_2. From (Tam 79b).

exhibit no noticeable variation with the amount of kinetic energy dissipated in the reaction. This behavior, however, is different from the observations made (Hil 79c), both for rather weakly and for very strongly (cf. Figure 279) damped ^{165}Ho + ^{56}Fe reactions, and may, hence, be peculiar of the ^{197}Au + ^{63}Cu reaction. In general, one would not expect a constant neutron multiplicity ratio, given by the mass ratio of the emitters, because of the variation of mass and charge asymmetries with increasing energy loss and the associated variation of binding energies and branching ratios.

In the above inclusive experiments on the ^{197}Au + ^{63}Cu reaction, covering an angular range from 15° to 105°, no indication of the presence of additional emission mechanisms, e.g., those involving the intermediate system, or of the fast emission processes discussed earlier in this section, has been reported (Tam 79b). An upper limit of 1 neutron per event is placed

(Tam 79b) on the possibly undetected contribution of such mechanisms, due to the limits of experimental precision.

The bombarding-energy dependence of neutron emission in the reaction $^{166}Er + ^{86}Kr$ has been investigated by Eyal *et al.* (Eya 78, 80), measuring neutrons in coincidence with damped reaction fragments, for incident energies of 5.7, 7.0, and 7.9 MeV per nucleon. The two complementary fragments from a collision were detected in these experiments with a pair of position-sensitive avalanche counters placed in the forward hemisphere. Neutron detectors were positioned at forward and backward angles, as well as at the detection angles of the reaction fragments, and out of the reaction plane.

It is found (Eya 80) in the $^{166}Er + ^{86}Kr$ experiments that the measured neutron angular correlations and energy distributions are fully compatible with statistical evaporation from the asymptotic reaction fragments, isotropic in the respective rest frames. Results of the measurements are represented in Figure 282, where the average neutron multiplicity, measured

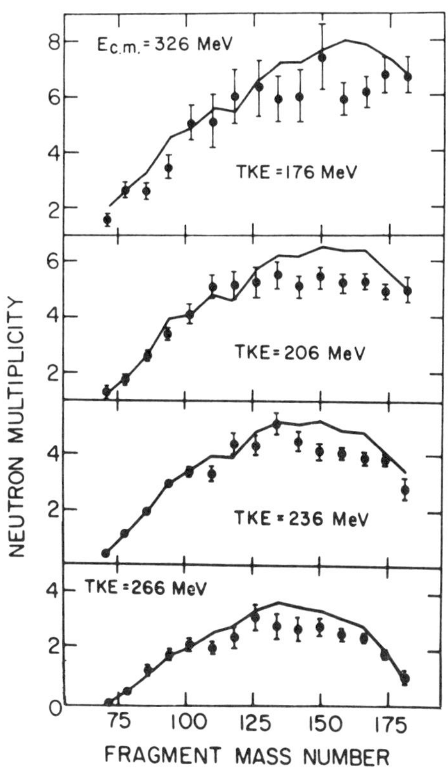

Figure 282. Multiplicity of neutrons emitted in the reaction $^{166}Er + ^{86}Kr$ at $E_{Lab} = 496$ MeV. Average neutron multiplicity is plotted vs. fragment mass number for various 30-MeV-wide bins in total kinetic energy, whose centroids are indicated at the distributions. The curves represent results of statistical evaporation calculations. From (Eya 80).

at an incident energy of $E_{Lab} = 496$ MeV, is plotted vs. mass number of the emitting fragment, for the different values of the final total kinetic energy stated at the distributions. For a given TKE value or energy loss, the neutron multiplicity is observed first to increase, until this dependence becomes rather weak for the larger fragment masses. The simple proportionality to the mass of the emitting fragment of the average neutron multiplicity, implied by the approximate formula of Equation (3.7.40), is not reflected in the data. It is seen from Figure 282 that the multiplicity for a given fragment increases with increasing energy loss. However, this energy-loss dependence is a function of the mass of the fragment, being relatively weak for mass numbers around $A = 125$.

Curves drawn through the data in Figure 282 represent Monte Carlo calculations (Gav 80), assuming evaporation from the accelerated fragments. Input parameters used in these calculations include fragment mass-to-charge ratios, as determined by an independent experiment (Eya 79) and fragment spin values derived from the experimental neutron out-of-plane anisotropies. It is further assumed that the total kinetic energy loss in a collision, corrected for the ground-state Q value, is completely converted into excitation energy of the fragments, and divided between them according to their mass ratio.

As can be inferred from the comparison in Figure 282, these evaporation calculations provide a satisfactory reproduction of the data. They show (Eya 80) that the deexcitation of all fragments is strongly dominated by neutron emission, allowing for an average charge of < 0.4 units per fragment to be evaporated. The fission probability is predicted to be essentially negligible, even for the heaviest fragments produced in the $^{166}Er + ^{86}Ke$ reaction. No positive evidence is reported (Eya 80) in this work for either an emission component due to the intermediate system, or for high-energy neutrons, in excess of the evaporation spectrum, that could be attributed to preequilibrium or Fermi jet emission mechanisms. Upper limits that can be placed on possible contributions of such processes are of the order of a few percent. However, no theoretical model calculations are available for these reactions to give estimates of emission probabilities that could be compared to the experimental upper limits.

Many properties of the reaction $^{165}Ho + ^{56}Fe$ at about 8 MeV per nucleon have been studied in considerable detail, as referred to previously. These investigations include measurements of multiplicities, energies, and angular correlations of neutrons, emitted in coincidence with fragments from fusion–fission-like and damped reactions. Results from these experiments on angular correlations of neutrons are exhibited in Figure 283 for coincident fission fragments, detected at 23° (arrow), and in Figure 284 for projectile-like damped reaction fragments. These figures illustrate the transition from more central to more peripheral reactions, as reflected in the neutron angular correlations with the direction of flight of coincident fragments. In both angular distributions, the main feature is a cross-section peak, centered around the fragment detection angle. This peak is rather narrow in the case of damped collisions (Fig. 284) and becomes broader for neutrons in coincidence with fission fragments. A less efficient angular focusing of sequentially emitted nuetrons is, of course, expected for the slower fission fragments, as compared to Fe-like damped reaction fragments that have about 5–7 MeV per nucleon of kinetic energy in the laboratory frame. Another structure is discernible in the angular correlations of Figures 283 and 284, on the side of the beam opposite to that on which the coincident fragment is detected. It, too, is less well defined in the case of fission fragments. The inserts in both figures illustrate the out-of-plane correlations measured at an in-plane angle of $\theta_n = 25°$. The out-of-plane angular anisotropy of neutrons is generally small and of little sensitivity to

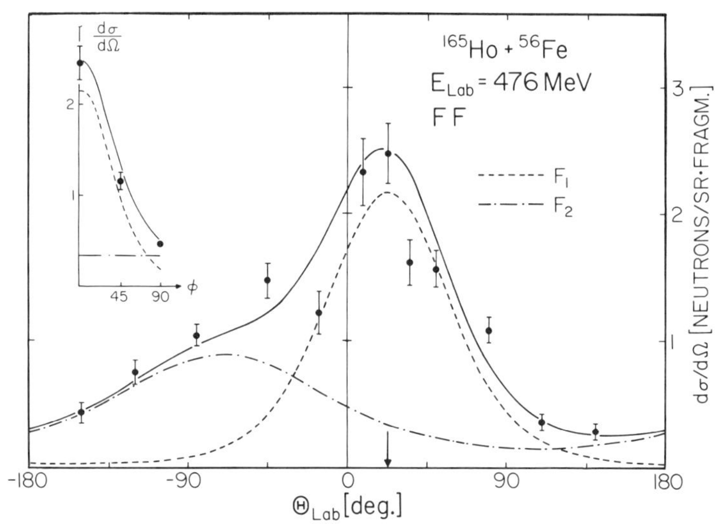

Figure 283. Angular correlation of neutrons from the reaction $^{165}Ho + ^{56}Fe$ at $E_{Lab} = 476$ MeV in coincidence with fission fragments ($30 \leq Z \leq 55$) recorded at $\theta_{Lab} = 23°$ (arrow). The differential neutron cross section $d\sigma/d\Omega$ is plotted vs. neutron detection angle measured with respect to the beam direction. The dashed and dashed-dotted curves represent results of evaporation calculations assuming sequential neutron evaporation from the accelerated detected (F_1) and undetected (F_2) fragments, respectively. The solid curve is the sum of the two individual contributions. The inset shows the out-of-plane correlation, for an in-plane neutron angle of $\theta_n = 25°$. From (Hil 79c).

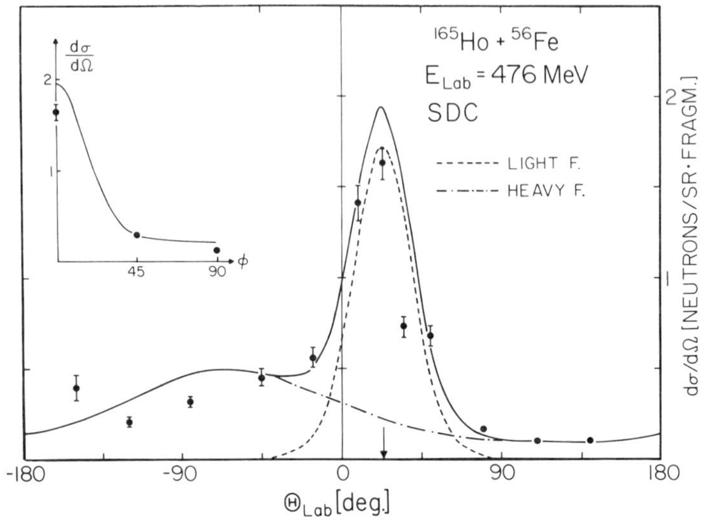

Figure 284. Same as Fig. 283, but for neutrons in coincidence with damped reaction fragments, integrated over all energy losses and fragment Z values between 10 and 29. From (Hil 79c).

the spin alignment of the emitter. This is a consequence of the small amounts of angular momentum that a neutron can carry away [see Equations (3.6.7) and (3.6.8)].

Plotting the angular dependences of the average neutron kinetic energy in the laboratory system results in strong correlations with the fragments' directions of flight. Highest neutron energies are observed (Hil 79c) close to the emission angle of the detected fragment, whereas the component attributed to evaporation from the slow, undetected fragment shows a rather weak angular dependence of the neutron energy. At all angles, the neutron spectra are well described by Maxwellian-type distributions (see above) in the rest frames of the emitters. The curves included in Figures 283 and 284 represent theoretical angular distributions resulting (Hil 79c) from integration over the neutron energy spectra in the respective rest frames. The dashed curves indicate the evaporation component from the detected fragment; the dashed-dotted curves represent contributions from evaporation from the undetected fragments. The sum of both components, represented by the solid curve in both figures, is observed to give a satisfactory account of the experimental data on energy spectra and absolute multiplicities at all measured angles. There is no definite indication of yet another source of neutrons in either fusion–fission or damped collisions of the $^{165}Ho+^{56}Fe$ system.

Hilscher *et al.* (Hil 79c) performed detailed statistical evaporation calculations, in order to describe the strong dependence of the neutron multiplicities with total kinetic energy loss (cf. Figure 279) and with the fragment atomic number in damped collisions. In the $^{165}Ho + ^{56}Fe$ experiments, in agreement with almost all other neutron measurements, the energy spectra of neutrons in the rest frames of the emitting fragments indicate (Hil 79c) that both fragments had attained thermal equilibrium at the end of the damped interaction. This is illustrated in Figure 171 for the energy loss dependence of the temperatures of the heavy and the light reaction partners, following the curve predicted from an equilibrated Fermi gas. According to estimates oriented at phenomenological reaction models, energy equilibration times of the order of 10^{-22}–10^{-21} s are suggested. It then appears justified to conclude that the total energy dissipated in a collision is equilibrated before the breakup of the intermediate system and, hence, shared between the reaction fragments in proportion to their numbers of nucleons. Adopting a plausible scheme of angular-momentum dissipation, in relation to the measured total energy loss, estimates for the A/Z ratios of the primary reaction fragments, prior to evaporation, are derived. It is concluded (Hil 79c) from this analysis that the data at a representative energy loss of ~ 100 MeV in the damped reaction are consistent with an average fragment A/Z ratio of 2.3, suggested by the minimum ground-state Q_{gg} values. Fragment mass-to-charge ratios close to the global A/Z ratio (2.38) are found to be still consistent with the data. Later measurements reported by Breuer *et al.* (Bre 79, 83), discussed in Section 3.4, demonstrate, however, a definite variation with energy loss of the charge density of projectile-like fragments, ranging from that of the projectile ($A/Z = 2.15$), at low energy losses, to the composite-system value for fully relaxed events. This additional information eliminates the charge density as a free parameter and allows the measured neutron multiplicities to be analyzed in terms of energy and angular-momentum dissipation mechanisms. Fundamental properties of these mechanisms are still largely unknown, as emphasized in Sections 3.5 and 3.6.

It is well known and common practice to deduce information on the excitation energy distribution of a nucleus from spectra and multiplicity distributions of sequentially emitted particles. Definite dependences exist (see, e.g., Hil 79d, Sch 79f) between corresponding moments of the distributions in excitation energy and in the number of neutrons emitted from an excited nucleus. These correlations can be amplified (Sch 79c) by considering relations between appropriated quantities associated with the two reaction partners of a damped reaction, for example, their charge densities or excitation energies. An illustration of this principle is given in Figure 285 (Hui 82b). Here, the experimental energy loss dependence of the ratio $\langle M_H \rangle / \langle M_L \rangle$ of the mean multiplicities of neutrons from the light and heavy fragments in damped $^{165}Ho + ^{56}Fe$ collisions is represented by circles, as reported by Hilscher *et al.* (Hil 79c). Also plotted, as solid and dashed curves, are results of simple neutron evaporation calculations (Sch 79f) for fragments with excitation energy distributions of finite r.m.s. width σ_E^*. It turns out (Sch 79f) that the magnitude of the multiplicity ratio, in the region of slow variation with energy loss, is sensitive to the mass-to-charge ratio and the relative average excitation energies of the reaction partners. For relatively low total kinetic energy losses, however, a strong energy loss dependence of the multiplicity ratio is observed in the data. This is a threshold phenomenon (Sch 79f), which can be modeled by the calculations only by adjusting the average division of energy between the reaction partners and/or the widths of the correlated excitation energy distributions of the fragments. In Figure 285, the sensitivity of the initial slope to the r.m.s. width of the excitation function is illustrated. The dependence on the first moment of the energy partition function has been discussed previously (Sch 79f). Although detailed analyses of the energy-loss dependences of fragment multiplicity correlations in damped reactions have not been made as yet, the comparison of calculations with data presented in Figure 285 indicates that the energy dissipated in partially damped $^{165}Ho + ^{56}Fe$ collisions is not distributed between the fragments in the statistical way suggested by a population of states in the fragments in accordance with accepted level density formulas. This, however, is just the effect predicted by transport models of energy dissipation and heat convection in damped reactions, discussed in Sections 2.7 [cf. Equations (2.7.33) and (2.7.34)] and 3.5.

A further pursuit of these important problems is, obviously, required. Apparently, the equivalence of temperatures of light and heavy reaction

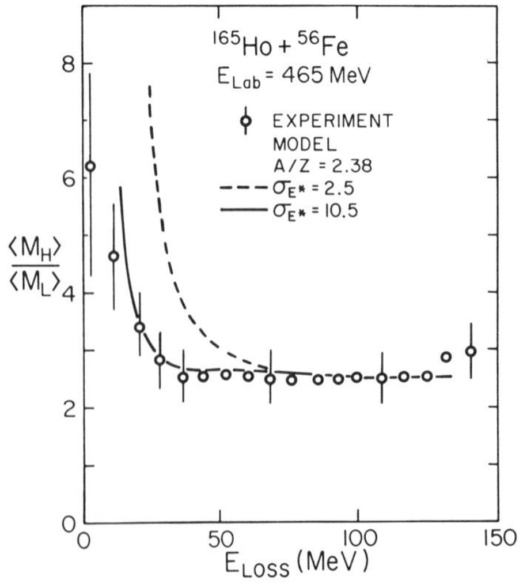

Figure 285. Ratio of average multiplicities of neutrons emitted from the heavy (M_H) and the light (M_L) reaction fragments produced in the reaction $^{165}Ho + ^{56}Fe$ at E_{Lab} = 465 MeV. Experimental data for a slightly higher incident energy are given by circles. Solid and dashed curves are the results of evaporation calculations, assuming an average division of the excitation energy according to the nucleon numbers in the fragments but different r.m.s. widths σ_E^* of the correlated excitation energy distributions. From (Hui 82b).

partners is not always a very good measure of the stage of equilibration of the dinuclear system at breakup. This is so, because the decay products of the reaction fragments, if emitted after the acceleration stage, are sensitive to the energetics of each fragment after it has fully equilibrated all degrees of freedom. Considering for example, only the effects of deformation of nascent fragments with equal temperatures τ, as specified by a deformation energy E^D, the asymptotic fragment temperatures τ_i reflect the energy balance

$$E_i^* = a\tau_i^2 \approx E_i^D + a\tau^2 \tag{3.7.41}$$

They can, therefore, be very different from the temperature τ at the instant of breakup. Only then will the original temperatures τ be maintained, when the total deformation energy, residing in a dinuclear system prior to breakup, is distributed among the final fragments according to their masses. In view of the large deformations that a dinuclear system can attain in strongly damped collisions (cf. Section 3.5), the experimental observations of equal fragment temperatures are quite surprising.

A slight disparity observed in measured asymptotic fragment temperatures, as well as deviations of the neutron multiplicity ratio from the expected mass dependence [cf. Equation (3.7.40)], has led Dakowski *et al.* (Dak 80b) to conclude that evidence exists, in the reaction $^{197}Au + {}^{40}Ar$ at 240 MeV, for an unbalanced division of the deformation energy at scission. According to their interpretation, the light fragment receives about one-half of the total deformation energy E^D, present in the dinuclear system at breakup.

The single piece of experimental evidence discussed above is, however, not very strong. In Figure 286, experimental neutron energy spectra for the light (circles) and the heavy (squares) fragments from the $^{197}Au + {}^{40}Ar$ reaction are displayed in their respective rest frames. The data are, obviously, not very accurate, but fits (Dak 80b) suggest definitely a higher temperature for the light reaction fragment, as compared to that of the heavy one. This magnitude of a temperature disparity is also expected from the heat transport mechanism referred to earlier (cf. Sections 2.7 and 3.5). Calculations predict average temperatures of $\tau = 1.15$ and 1.30 MeV for the heavy and the light reaction fragments, respectively, produced in the $^{197}Au + {}^{40}Ar$ reaction. The corresponding theoretical evaporation spectra are represented in Figure 286 by the dashed and solid curves, respectively. While these theoretical predictions may not correspond to the best possible fits to the data, they provide certainly an adequate reproduction of the experimental results. In view of the quality of the data and the crudeness of the calculations, these inferences should not be taken too literally. They stress a valid argument, however, referring to the influence of many different

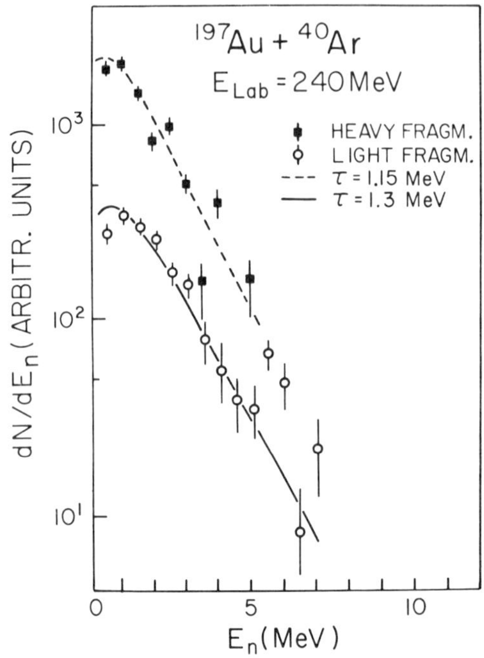

Figure 286. Experimental energy spectra of neutrons emitted from the heavy (squares) and the light (circles) fragments, produced in 240-MeV $^{197}Au + ^{40}Ar$ collisions. The spectra are plotted in the respective rest frames of the emitters. Dashed and solid curves represent theoretical spectra predicted by calculations modeling the heat transport mechanism due to nucleon exchange. From (Dak 80b).

effects on the temperatures of the asymptotic fragments. These quantities correspond to the first moments of the excitation energy distributions, and it seems inescapable to determine experimentally higher moments of this distribution, in order to disentangle the various effects that tend to mask one of the most important features of the energy dissipation and equilibration processes operating in damped nuclear reactions.

Finite distributions in the excitation energy, acquired by the fragments in damped nuclear reactions, have recently also been considered by Morrissey *et al.* (Mor 81b), who study the excitation energy equilibrium between two reaction partners, in a statistical model. They derive estimates for the ratio $\langle E_1^* \rangle / \langle E_2^* \rangle$ of mean excitation energies of two reaction fragments, and for the r.m.s. width σ_E^* of the correlated excitation energy distribution, given by

$$\frac{\langle E_1^* \rangle}{\langle E_2^* \rangle} = \frac{A_1}{A_2}, \qquad \sigma_E^* = 2\tau^3 \frac{A_1 A_2}{A_1 + A_2} \tag{3.7.42}$$

Here, A_1 and A_2 are the mass numbers of the two fragments, proportional to the corresponding level density parameters, and τ is the average temperature of both fragments.

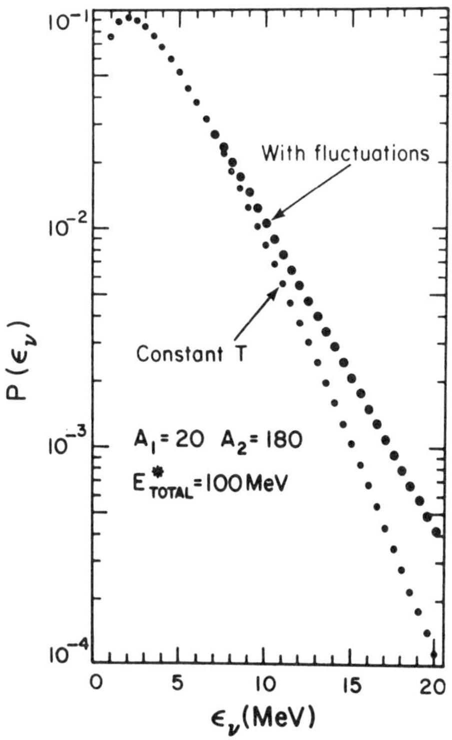

Figure 287. Effect on a neutron evaporation spectrum of a fluctuative width of the excitation energy distribution of the emitting nucleus. The theoretical spectra correspond to a nucleus of $A = 20$, emitting neutrons after having been in thermal equilibrium with an $A = 180$ reaction partner, sharing the total excitation energy E^*_{total}. The distribution labeled "constant T" results from a sharp division of the excitation energy, while fluctuations of this division, allowed according to the combined level density, lead to the broader spectrum. From (Mor 81b).

An immediate consequence of an excitation energy distribution of finite width is a broadening of evaporated particle energy spectra and of the multiplicity distributions. The former effect is illustrated in Figure 287 for the case of a spectrum of neutrons, evaporated from a fragment with mass $A = 20$ after establishment of thermal equilibrium with an $A = 180$ reaction partner, for a total excitation energy of $E^*_{\text{total}} = 100$ MeV. While the broadening effect is predicted to be small for symmetric systems, it is quite noticeable for an asymmetric mass split, such as illustrated in Figure 287.

Much more obvious than the broadening effect observed in the theoretical evaporation spectra displayed in Figure 287 are correlations between the multiplicity distributions of neutrons from two correlated fragments. An average consequence of this effect has already been discussed in the context of Figure 285. In Figure 288, theoretical results are displayed for the contour plot of the correlations between the multiplicities, ν_1 and ν_2, of neutrons emitted from correlated, symmetric fragments at a total excitation energy of 100 MeV, as modeled (Mor 81b) in Monte Carlo simulations of the evaporation process. The roughly symmetric correlation plot shown in Figure 288(a) has been calculated for the assumption of a fixed symmetric

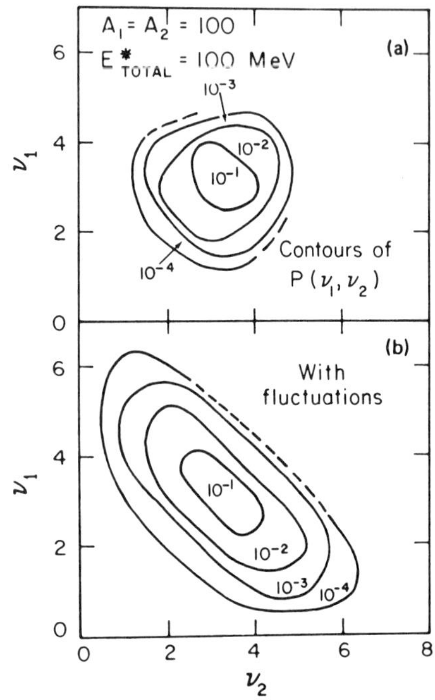

Figure 288. Correlations of multiplicities of neutrons, ν_1 and ν_2, emitted from correlated symmetric fragments at a total excitation energy of $E^*_{\text{total}} = 100$ MeV. Shown are contour lines for the product multiplicities $P(\nu_1, \nu_2)$ as predicted from a Monte Carlo simulation. In part (a) the division of the total excitation energy is assumed to be fixed. Part (b) displays a similar correlation diagram, however, with the inclusion of finite fluctuations in the partition of the total excitation energy between the reaction partners. From (Mor 81b).

division of the total energy. It is to be contrasted to the diagram in part (b) of the figure, corresponding to anticorrelated fragment excitation energy distributions of finite width. Not surprisingly, the multiplicity correlation diagram of Figure 288(b) reflects the anticorrelation in the fragment excitation energies.

It is this effect of thermal fluctuations that has been mentioned already in the earlier discussion of high-energy particle spectra carrying apparent insignias of preequilibrium emission. For example, the hard emission spectrum of protons produced in the $^{\text{nat}}\text{Cu} + {}^{20}\text{Ne}$ reaction at 12.6 MeV per nucleon has been well explained by Schmitt *et al.* (Sch 81c) in terms of statistical fluctuations in the partition of the excitation energy.

It appears quite possible that high-energy particles, observed in other heavy-ion-induced reactions discussed in this section, usually attributed to fast, nonequilibrium emission mechanisms, could be explained in terms of fluctuations and asymmetries in the excitation energy distributions. These aspects of the damped collision dynamics have so far received very little experimental and theoretical attention, which seems highly inappropriate in view of the fundamental importance of dissipation and equilibration mechanisms in damped reactions.

3.8. Transition from Damped to Fusion Reactions

As illustrated by the schematic Figures 1 and 39, the collision of two heavy ions leads to a variety of reaction mechanisms. The damped or deep inelastic reaction process has opened new insights into the dynamical properties of nuclear matter. During the short time that the two nuclei are in contact, several degrees of freedom are strongly relaxed. For example, several hundreds of MeV of kinetic energy are dissipated, charge equilibrium is attained to some extent, nucleon exchange occurs on a massive scale, angular momentum is transferred from orbital to intrinsic rotation, and deformation degrees of freedom are excited. Hence, the reaction products that emerge from damped reactions are considerably altered from the projectile and target nuclei. However, the mass asymmetry degree of freedom is only weakly relaxed during damped collisions. The average masses of the products are the same, or nearly the same, as those of the original nuclei. In the mass asymmetry degree of freedom, the identities of the target and projectile nuclei are not lost in the collision. This is illustrated in Figure 96 for the $^{209}Bi+^{136}Xe$ reaction at three different bombarding energies. Although the charge distributions of the reaction products have a sizable width, one observes that the centroids of the charge distributions are centered very near to $_{54}Xe$ and $_{83}Bi$.

Opposed to damped reactions, the compound nucleus reaction (see Figure 1) is characterized by the additional relaxation of the mass asymmetry degree of freedom. The decay modes of the compound nucleus, particle evaporation, and/or fission, are completely decoupled from the entrance channel. Hence, it is not possible to learn how mass relaxation is initiated in the entrance channel from measurements of the compound nucleus (Bjø 82). However, one can study mass relaxation by measurements of the probability for mass symmetrization as a function of entrance channel conditions (Bjø 82). This is illustrated in Figure 95, where the symmetric charge (mass) component is decomposed into an asymmetric damped reaction component and a symmetric component attributed to a fusion–fission-like process. The asymmetric $^{56}Fe+^{132}Xe$ reaction on the right gives a large cross section of symmetric events, whereas the $^{159}Tb+^{86}Kr$ reaction shown on the left has only a small contribution of symmetric events (compare with the $^{209}Bi+$ ^{136}Xe reaction shown in Figure 96). Asymmetric entrance channel systems are best suited for the above type of analysis. However, even with near-symmetric systems it is possible to distinguish fusion from damped collisions as illustrated in Figure 95 for the Ag+Kr reaction.

A number of reactions have been studied (Bir 79, Lef 81, Hui 81) in recent years where mass symmetrization occurs without compound nucleus formation. This is concluded from the fact that the cross section for events under the symmetric mass peak exceeds that corresponding to partial waves up to the limiting rotating liquid-drop l value, l_{RLDM}, where the fission barrier vanishes (Coh 74). Without a fission barrier a completely equilibrated compound nucleus is not likely to be formed. A simple sharp cutoff model formulation of the fusion cross section is given by

$$\sigma_{\mathrm{fus}} = \pi \lambda\!\!\!^{-2} l_f (l_f + 1) \tag{3.8.1}$$

where l_f is the maximum angular momentum that leads to fusion. The definition of fusion (Bir 79) used here is more inclusive than compound nucleus formation [see Figure 39 where fusion includes the compound nucleus (CN) and fusion-like (FL) events]. All events under the broad symmetric mass peak are termed fusion and are included in the cross section given by Equation (3.8.1). The ratio

$$\sigma_{\mathrm{fus}}(l > l_{\mathrm{RLDM}})/\sigma_{\mathrm{fus}}(l \le l_{\mathrm{RLDM}}) = [l_f(l_f+1)/l_{\mathrm{RLDM}}(l_{\mathrm{RLDM}}+1)] - 1 \tag{3.8.2}$$

gives a measure of noncompound nucleus fusion and is known to approach values as large as 10 for some systems as illustrated in Table 6.

Table 6. Fusion-Fission Properties of Heavy Systems[a]

System	Reaction: Target Projectile	$(Z^2/A)_{\mathrm{eff}}$	Ref.	$E_{\mathrm{c.m.}}$ (MeV)	l_f ($\hbar$)	l_{RLDM} ($\hbar$)	σ_{fus} (mb)	$\sigma_{\mathrm{fus}}(l > l_{\mathrm{RLDM}})/\sigma_{\mathrm{fus}}(l \le l_{\mathrm{RLDM}})$ [see equation (3.8.2)]
$^{162}_{68}\mathrm{Er}$	$^{30}\mathrm{Si} + {}^{132}\mathrm{Xe}$	23.31	(Oes 80)	200	97	87	1285	0.24
$^{234}_{94}\mathrm{Pu}$	$^{26}\mathrm{Mg} + {}^{208}\mathrm{Pb}$	25.23	(Gob 81)	190	95	72	1350	0.74
$^{149}_{65}\mathrm{Tb}$	$^{109}\mathrm{Ag} + {}^{40}\mathrm{Ar}$	25.27	(Bri 76)	247	103	83	975	0.54
$^{161}_{69}\mathrm{Tm}$	$^{121}\mathrm{Sb} + {}^{40}\mathrm{Ar}$	25.95	(Gau 75)	223	107	82	1130	0.70
$^{151}_{67}\mathrm{Ho}$	$^{116}\mathrm{Sn} + {}^{35}\mathrm{Cl}$	26.16	(Sik 80)	130	89	78	695	0.30
$^{205}_{85}\mathrm{At}$	$^{165}\mathrm{Ho} + {}^{40}\mathrm{Ar}$	28.88	(Tam 75)	242	126	79	1350	1.53
$^{237}_{97}\mathrm{Bk}$	$^{197}\mathrm{Au} + {}^{40}\mathrm{Ar}$	30.94	(Riv 82)	277	144	65	1480	3.87
$^{192}_{80}\mathrm{Hg}$	$^{56}\mathrm{Fe} + {}^{136}\mathrm{Xe}$	31.83	(Heu 78)	225	123	82	1118	1.24
$^{256}_{102}\mathrm{No}$	$^{48}\mathrm{Ca} + {}^{208}\mathrm{Pb}$	31.87	(Gob 81)	317	144	59	1114	4.90
$^{278}_{110}\mathrm{X}$	$^{238}\mathrm{U} + {}^{40}\mathrm{Ar}$	32.50	(Tam 75)	257	117	40	1030	7.42
$^{258}_{104}\mathrm{X}$	$^{50}\mathrm{Ti} + {}^{208}\mathrm{Pb}$	34.40	(Gob 81)	323	136	52	933	5.76
$^{231}_{93}\mathrm{Np}$	$^{165}\mathrm{Ho} + {}^{56}\mathrm{Fe}$	35.66	(Hoo 82)	345	123	73	720	1.82
$^{260}_{106}\mathrm{X}$	$^{52}\mathrm{Cr} + {}^{208}\mathrm{Pb}$	36.86	(Gob 81)	336	127	45	766	6.85
$^{266}_{108}\mathrm{X}$	$^{58}\mathrm{Fe} + {}^{208}\mathrm{Pb}$	37.96	(Gob 81)	365	131	41	683	9.04
$^{265}_{109}\mathrm{X}$	$^{209}\mathrm{Bi} + {}^{56}\mathrm{Fe}$	38.95	(Vio 80b)	377	127	36	640	11.20
$^{272}_{110}\mathrm{X}$	$^{64}\mathrm{Ni} + {}^{208}\mathrm{Pb}$	39.04	(Gob 81)	396	136	36	634	12.99

[a] From (Hui 81).

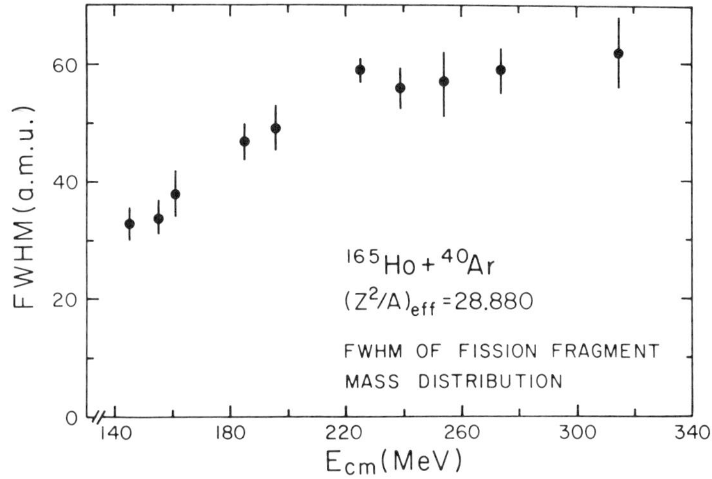

Figure 289. Width of the fission fragment mass distribution (FWHM) as a function of center-of-mass bombarding energy. Data from (Bor 81b).

The symmetric mass peak for reactions where $l > l_{RLDM}$ has been reported (Heu 78, Bor 81b) to be broader than that expected on the basis of the liquid-drop theory of nuclear fission (Nix 69). The width of the fission fragment mass distribution (FWHM) for the $^{165}Ho + ^{40}Ar$ reaction (Bor 81b) is plotted as a function of the center-of-mass bombarding energy in Figure 289. At $E_{c.m.} < 175$ MeV, $l_f < l_{RLDM}$ for this reaction. Whether or not the increasing FWHM of the mass distributions of the fission fragments for $E_{c.m.} > 175$ MeV is due to dynamical effects which are unique to the fusion-like (FL) process where $l > l_{RLDM}$ or simply a result of the combined effect of other factors like temperature and angular momentum is presently an open question. However, in view of the fact that the potential energy surface for symmetric fission of ^{205}At is extremely shallow, it is interesting to investigate the possible influence of angular momentum on the variance of the mass distribution of fragments from the fusion–fission process. A simple model is adopted where the variance in the mass distribution is given by

$$\sigma^2 = \tau / c \tag{3.8.3}$$

as expected for a system in thermodynamical equilibrium. Here, τ is the nuclear temperature and c denotes the stiffness (curvature) coefficient of the underlying potential energy surface. Further assuming that c is given by

$$c = c' - b\langle l^2 \rangle \tag{3.8.4}$$

and $\tau = (E^*/a)^{1/2}$, the width (FWHM) of the mass distribution can be written as

$$\Gamma(\text{FWHM}) = 2.355(E^*)^{1/4}/(a^{1/2}c' - a^{1/2}b\langle l^2\rangle)^{1/2} \qquad (3.8.5)$$

where a is the level density parameter, b is a constant, and c' is the stiffness coefficient for $l = 0$. Rearrangement of Equation (3.8.5) in the form

$$5.55(E^*)^{1/2}/\Gamma^2 = a^{1/2}c' - a^{1/2}b\langle l^2\rangle \qquad (3.8.6)$$

allows a direct comparison with the data shown in Figure 289. In Figure 290 the quantity $5.55(E^*)^{1/2}/\Gamma^2$ is plotted for these data as a function of $\langle l^2\rangle$, where $\langle l^2\rangle$ is assumed to be given by $\frac{1}{2}l_f^2$. In the application of Equation (3.8.6), a choice has to be made about the appropriate excitation energy E^*. Since the mass distribution is probably established somewhat prior to or near scission, the excitation energy at scission is used, $E^* = E_{\text{c.m.}} + Q - E_{\text{ff}}$, where $E_{\text{ff}} \approx 150$ MeV is the final total kinetic energy of fission fragments given by fission systematics (Van 73). As can be seen from Figure 290, this

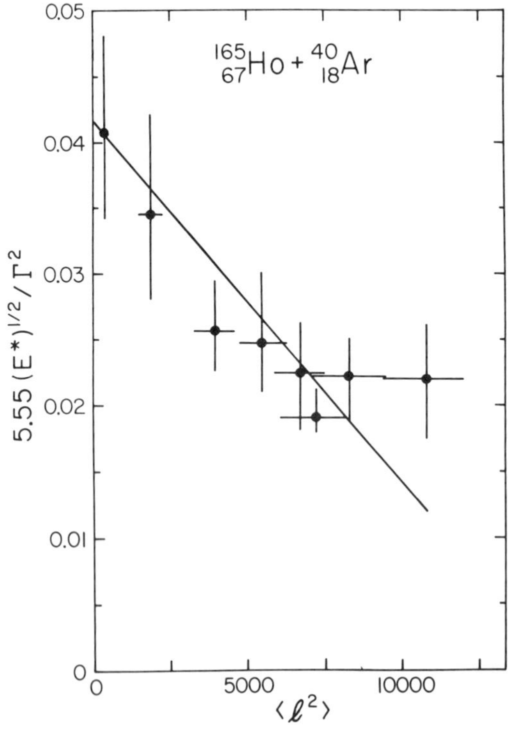

Figure 290. Dependence of the width (FWHM) of the mass distribution of fission fragments from the ^{165}Ho + ^{40}Ar reaction on excitation energy and angular momentum (see text).

simple empirical model gives a qualitative description of the data, suggesting that the presence of a fission barrier may not be an important factor in the determination of the width of the fission fragment mass distribution. This conclusion has been reached also by Faber (Fab 81) in a more sophisticated model calculation. The increase in the FWHM of the mass distribution as described above is a result of the potential energy surface along the mass asymmetry coordinate becoming more shallow with increasing angular momentum. In contrast to this explanation, Grégoire *et al.* (Gre 81b) have suggested a strong increase in the FWHM of the fission fragment mass distribution at l_{RLDM}.

The kinetic energy distributions of the symmetric fusion–fission-like events and the asymmetric damped reaction events differ from each other and give supplementary evidence in the identification of the two types of events. As illustrated in Figure 148, the kinetic energy distribution of damped or deep inelastic events terminates at the energy calculated for asymmetric fission fragments (Wol 82d) from systematics. In contrast to this energy spectrum, the symmetric fission events have an energy spectrum with a centroid at the energy predicted by systematics that coincides with the energy where the spectrum for damped events terminates. These results are consistent with the trajectories of the symmetric fission events penetrating inside the conditional saddle, whereas the damped events do not. Such a conditional saddle is shown at the bottom of Figure 21 and defined as a saddle point in a multidimensional plot of the potential energy at frozen mass asymmetry.

Interesting information on nuclear saddle deformations may be obtained also from fission fragment angular distribution measurements. For compound nucleus reactions it is thought that K (projection of the total angular momentum on the nuclear symmetry axis) may change continuously during the time evolution of the fusion–fission process as the system moves between the compound and saddle configurations. However, transition state theory assumes that K is frozen in at the fission saddle point and is not further altered by the Coriolis forces or thermal fluctuations on the path to scission. For low angular momenta and excitation energies in excess of the fission barrier, it has been shown that the transition state deformation as a function of the fissility parameter x, where $x = (Z^2/A)/50.883\{1 - 1.7826[(N - Z)/A]^2\}$, follows approximately the nonrotating liquid-drop saddle-point shapes (Rei 66).

The K distribution of the transition states is usually approximated by a Gaussian with a variance $K_0^2(I)$ given by (Hal 58, Van 73)

$$K_0^2(I) = \mathscr{I}_{\mathrm{eff}}(I)T/\hbar^2 \tag{3.8.7}$$

where T is the nuclear temperature of the intermediate system at the saddle-point deformation. The effective moment of inertia, which is angular momentum dependent and defined by,

$$\mathscr{I}_{\mathrm{eff}}(I)=\mathscr{I}_{\parallel}(I)\mathscr{I}_{\perp}(I)/\left[\mathscr{I}_{\perp}(I)-\mathscr{I}_{\parallel}(I)\right] \tag{3.8.8}$$

determines the magnitude of $K_0^2(I)$, where $\mathscr{I}_{\parallel}(\mathrm{I})$ and $\mathscr{I}_{\perp}(I)$ are the moments of inertia parallel and perpendicular to the nuclear symmetry axis, respectively. Hence, for a particular fissioning system with a given saddle shape, and corresponding value of K_0^2, the fission fragment angular distribution (for $M=0$) is given by (Hui 69, Van 73)

$$W(\theta)\propto\sum_{I_{\min}}^{I_{\max}}(2I+1)T_I\left\{\frac{\sum_{K=-I}^{I}\frac{1}{2}(2I+1)\left|D_{M=0,K}^{I}(\theta)\right|^2\exp\left[-\frac{K^2}{2K_0^2(I)}\right]}{\sum_{K=-I}^{I}\exp\left[-\frac{K^2}{2K_0^2(I)}\right]}\right\} \tag{3.8.9}$$

where the transmission coefficients are written as T_I, since $l=I$ when $M=0$. The $D_{M,K}^{I}(\theta)$ functions are defined in reference (Van 73). In the case of low energy helium-ion induced fission (Rei 66), values of I only up to approximately 20 were populated and in this limited range of I values, $\mathscr{I}_{\mathrm{eff}}(I)$ is approximately equal to $\mathscr{I}_{\mathrm{eff}}(I=0)$.

In the rotating-liquid-drop model (Coh 74), at a given value of the fissility parameter x, the fission barrier B_f decreases with increasing I until it becomes zero at I_{RLDM}. Correspondingly, the rotating-liquid-drop model (RLDM) predicts that $\mathscr{I}_{\mathrm{sph}}/\mathscr{I}_{\mathrm{eff}}(I)$ decreases, i.e., that the saddle-point configuration becomes more compact, with increasing I. For $I>I_{\mathrm{RLDM}}$, the RLDM does not make a prediction about the shape of the intermediate system. Hence, for very heavy ion induced fission where the angular momentum values approach and exceed I_{RLDM}, it is not possible to calculate theoretical angular distributions with the formalism described by Equations (3.8.7) to (3.8.9) without making *ad hoc* assumptions about $\mathscr{I}_{\mathrm{eff}}(I)$ for $I>I_{\mathrm{RLDM}}$ (Bac 81, Ros 83, Les 83, Tsa 83, Tsa 83b). In view of the fact that transition-state theory is not applicable for $I>I_{\mathrm{RLDM}}$, it has been proposed (Ros 84) to interpret fragment angular distributions from heavy ion induced fission with a completely different formalism called the statistical scission model (SSM).

Some years ago Ericson (Eri 60) proposed a statistical model to explain fission fragment angular distributions. Although never tested, this model assumes fission to be analogous to complex particle emission and the process is regarded to be entirely governed by available phase space. The statistical model is postulated to be valid when the fission barrier is small (or vanishes) and the excitation energy is large, as is the case for very heavy ion induced fission.

The statistical partition of the initial angular momentum into orbital angular momentum l and channel spin S of the two fragments, where $\mathbf{l}+\mathbf{S}=\mathbf{I}$, leads to the following expression for the angular distribution (Ros 84)

$$W(\theta) \propto \sum_{I,l,S,m} (2I+1)T_I\{(2l+1)T_l|\langle Sml0|Im\rangle|^2\rho(S)/\Gamma(I)\} \times \left(\frac{2I+1}{2}\right)|D^I_{M=0,m}(\theta)|^2 \tag{3.8.10}$$

where $\{(2l+1)T_l|\langle Sml0|Im\rangle|^2\rho(S)/\Gamma(I)\}$ is the probability of state I to fission with relative angular momentum l between the two final fragments, S is the channel spin (the vector sum of the two fragment spins) with projection m onto the emission axis ($m=\mathbf{n}\cdot\mathbf{S}$) and $\Gamma(I)$ is the total fission width for state I. The quantity $[(2I+1)|^2]D^I_{M=0,m}(\theta)|^2$ is the angular distribution function. The density of states $\rho(S)$ with projection m on the fission axis is assumed to be given by

$$\rho(S) \propto (2S+1)\exp\left[-(S+\tfrac{1}{2})^2 2S_0^2\right]. \tag{3.8.11}$$

The relationship between I, l, S and m is schematically illustrated in Figure 291. Although the equation for fission fragment angular distributions proposed by Bond (Bon 84), appears to be similar to Equation (3.8.10), it mixes the two very different transition state and statistical scission models. Multi-

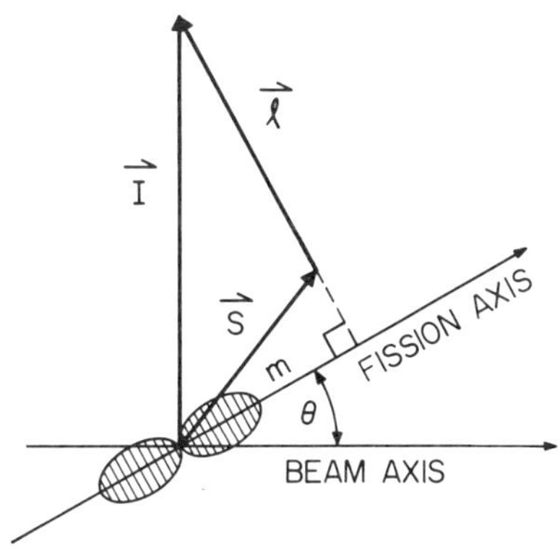

Figure 291. Schematic illustration of the relationship between, **I**, **l**, **S** and m for the statistical scission model. It is assumed that $M=0$. The combined channel spin S of the two primary fission fragments has a projection $m(m=\mathbf{n}\cdot\mathbf{S})$ along the fission fragment emission axis **n**. From (Ros 84).

plication of phase space factors at two different stages during the time sequence of the evolving system is generally not allowed.

Equation (3.8.10) may be rearranged to give a form similar to that of Equation (3.8.9). Summing over l and S gives an equation that well represents the results obtained with Equation (3.8.10)

$$W(\theta) \propto \sum_{I_{min}}^{I_{max}} (2I+1)T_I \frac{\sum_{m=-I}^{I} \left(\frac{2I+1}{2}\right) |D_{M=0,m}(\theta)|^2 \exp(-m^2/2S_0^2)}{\sum_{m=-I}^{I} \exp(-m^2/2S_0^2)} \tag{3.8.12}$$

The values of S_0^2 are estimated in the SSM by $S_0^2 = \sigma_1^2 + \sigma_2^2$, where the spin cutoff factors associated with the level densities of the two fission fragments are given by $\sigma_i^2 = (1/5)M_i(c^2 + a^2)T/\hbar^2$. The quantities c and a are the principal one-half axes of a spheroid, which is the assumed shape of each primary fission fragment.

Theoretical fission fragment angular distributions calculated with the statistical scission model are compared to experimental data in Figure 292. Excellent agreement between theory and experiment is obtained for both the $^{238}U + ^{40}Ar$ and $^{208}Pb + ^{32}S$ reactions. In these calculations, $c/a = 1.85$ and the nuclear temperature is calculated by $T = [\frac{1}{2}(E_{cm} + Q - E_{Coul} - E_{Def} - E_{Rot})/(LDP)]^{1/2}$ where LDP is the level density parameter of each of the symmetric primary fragments and E_{Rot} is the rotational energy of the scission configuration. The sum of the Coulomb (E_{Coul}) and deformation (E_{Def}) energies is the energy stored in potential energy at the instant of scission, where $E_{Coul}(\text{ MeV}) = 1.44(Z/2)^2/2c$ and Z is the charge of the composite system. In all calculations the liquid drop (Coh 74) value of $r_0 = 1.2249$ fm is used. The SSM gives an excellent description of fragment angular distributions for heavy ion induced fission for cases where the transition state model fails, even when *ad hoc* assumptions are included for large I values.

For light and intermediate heavy-ion systems with $Z_P \cdot A_T < 1500$ and $(E - V_B)/\mu \leq 2$ MeV per nucleon, fusion cross sections are remarkably well reproduced by a simple classical dynamical model (Bir 79) based on the one-dimensional proximity nuclear potential (Blo 77) and the one-body dissipation concept (Ran 78, Ran 80b, Swi 80) using present knowledge about nuclear radii (Mye 73, Blo 77). This indicates that experimental fusion cross sections are consistent with the simple model assumption of spherical nuclei in the approach phase where the only requirement for fusion is a net attractive balance of nuclear, Coulomb, and centrifugal forces at some radius.

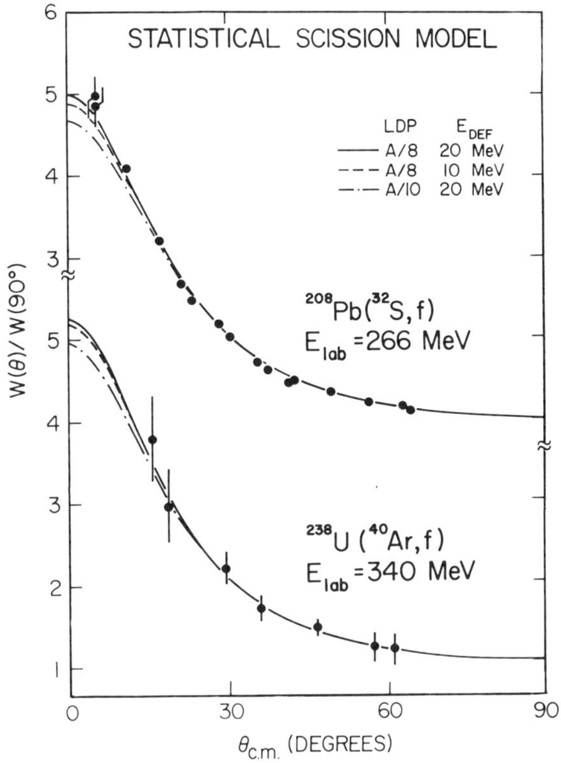

Figure 292. Experimental fission fragment angular distribution (solid data) for the $^{208}Pb(^{32}S, f)$ (Tsa 83b) and $^{238}U(^{40}Ar, f)$ (Les 83) reactions are compared with theoretical results (lines) calculated with the statistical scission model. The theoretical ratio, W(1°)/W(90°), calculated with Equation (3.8.10) is one percent smaller than that calculated with Equation (3.8.12). From (Ros 84).

The simple one-dimensional model is known, however, to fail for very heavy systems like, for example, the $^{165}Ho + {}^{84}Kr$ and $^{209}Bi + {}^{84}Kr$ reactions (Bir 79). For these reactions there is no reasonable way to adjust the parameters of the calculation, including the radii and forces, to remedy the large discrepancy between data and the predictions of the one-dimensional model. Recently, Swiatecki (Swi 81c, Swi 82b) has introduced a dynamical theory [Section 2.3, Equation (2.3.19)] that explains a wide variety of fusion data (San 81, Hui 81, Boc 82, Sik 82, Bla 82b, Wes 82), including those for the heaviest systems. This model focuses on three degrees of freedom: mass asymmetry, fragment separation, and neck size. In addition, the theory utilizes a macroscopic (liquid-drop) potential energy (Blo 77, Swi 81c), a macroscopic dissipation function (Ran 80b, Swi 80), and a simplified treatment of the inertial force. Two-dimensional contour maps of the sum of Coulomb and nuclear proximity potential vs. ratio of the neck radius to projectile radius and center separation are shown in Figure 21. For shapes with a pronounced neck (termed the mononuclear regime), the inertial forces are neglected since they are expected to be small compared to the conservative and one-body dissipative forces. On the other hand, for shapes with well-defined necks (termed the binary regime), the binary inertia is assumed to be that of the reduced mass for the relative motion of the mass centers.

The "Swiatecki model" identifies three important configurations in the dynamical evolution of a nucleus–nucleus collision (Bjø 82b). These are (A) the contact configuration (where the growth of the neck between two spherical nuclei at contact becomes energetically favorable), (B) the configuration of conditional equilibrium (a saddle-point in a multidimensional plot of the potential energy for a frozen mass asymmetry), and (C) the configuration of unconditional equilibrium (the fission saddle-point shape). The type (B) and type (C) configurations are identical for mass-symmetric systems (Nix 77), however, they are different for asymmetric systems. These configurations allow one to divide (at least conceptually) nucleus–nucleus reactions into four more or less distinct categories as shown in Figure 293. These reaction categories are similar to the categories outlined previously in Figures 1 and 39. However, in Figure 293 the reaction categories are associated with three threshold energies. These threshold energies are (A) the energy needed to make nuclear contact, (B) the energy needed to overcome the conditional saddle (extra push energy), and (C) the energy needed to overcome the unconditional saddle (extra-extra push energy).

For heavier systems, or for lighter systems with sufficient angular momentum, an extra amount of radial energy above the threshold for condition (A) is required to induce fusion. This extra energy E_x is necessary to balance the conservative force due to the growth of the neck, which tends to deflect the system trajectory away from the fusion path. It is parametrized in the following way:

$$E_x = \begin{cases} 0 \quad \text{for } (Z^2/A)_{\text{eff}} + (f \cdot l/l_{\text{ch}})^2 \leq (Z^2/A)_{\text{eff thr}} \\ K\left[(Z^2/A)_{\text{eff}} + (f \cdot l/l_{\text{ch}})^2 - (Z^2/A)_{\text{eff thr}}\right]^2 + \cdots \\ \quad \text{for } (Z^2/A)_{\text{eff}} + (f \cdot l/l_{\text{ch}})^2 > (Z^2/A)_{\text{eff thr}} \end{cases} \tag{3.8.13}$$

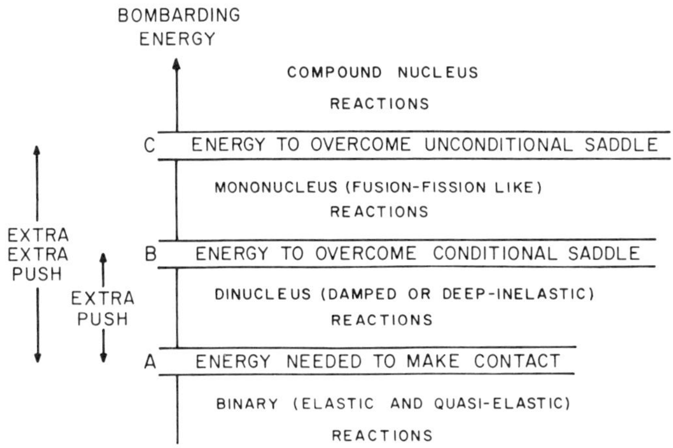

Figure 293. Schematic illustration of the relation between the three critical energies and the four types of nuclear reactions shown in Fig. 39. This figure is appropriate when the three milestone configurations exist and are distinct requiring the two kinds of extra push for moving from configuration (A) to (B) and (A) to (C). In some situations the critical energies (A), (B), and (C) may merge (pairwise or all together) squeezing out the regimes corresponding to dinucleus and/or mononucleus reactions. In other situations one or both of the upper boundaries [(B) and (C)] may dissolve, making the adjoining regions merge into continuously graduated reaction types. After (Bjø 82b).

where

$$(Z^2/A)_{\text{eff}} = 4Z_1Z_2/A_1^{1/3}A_2^{1/3}(A_1^{1/3}+A_2^{1/3}) \tag{3.8.14}$$

$$(Z^2/A)_{\text{eff thr}} = b\{1-1.7826[(N_1-Z_1+N_2-Z_2)/(A_1+A_2)]^2\} \tag{3.8.15}$$

$$l_{\text{ch}}^2 = \begin{cases} (e^2mr_0/4\hbar^2)\left[A_1^{4/3}A_2^{4/3}(A_1^{1/3}+A_2^{1/3})^2/(A_1+A_2)\right] \\ 0.01055A_1^{4/3}A_2^{4/3}(A_1^{1/3}+A_2^{1/3})^2/(A_1+A_2) \end{cases} \tag{3.8.16}$$

and

$$K = \begin{cases} (32/2025)(3/\pi)^{2/3}(e^4m/\hbar^2)a^2 \\ \qquad \times\left[A_1^{1/3}A_2^{1/3}(A_1^{1/3}+A_2^{1/3})^2/(A_1+A_2)\right] \text{ MeV} \\ 7.601\times10^{-4}a^2\left[A_1^{1/3}A_2^{1/3}(A_1^{1/3}+A_2^{1/3})^2/(A_1+A_2)\right] \text{ MeV} \end{cases} \tag{3.8.17}$$

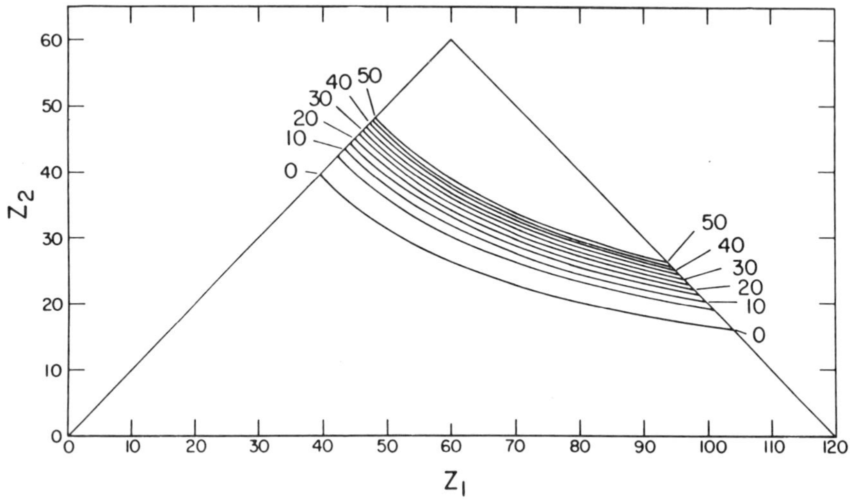

Figure 294. Contour lines of the "extra push" energy as calculated with Equation (3.8.13) when $a=12\pm2$ and $b=35.6\pm1.0$. From (Bjø 82b).

The magnitude of the extra push energy E_x given by Equation (3.8.13) is l dependent. This expression assumes an equivalent effect of the Coulomb and centrifugal forces on the fusion of two nuclei at contact in the entrance channel. An important feature of Equation (3.8.13) is that all parameters are fixed, except a, b, and f. The slope and threshold coefficients have been determined (Bjø 82b) empirically to be $a=12\pm2$ and $b=35.6\pm1.0$, respectively.

The contour lines of the "extra push" energy for $l=0$ according to Equation (3.8.13), with the above values of the parameters a and b are plotted in Figure 294. In the construction of Figure 294, each nucleus is assumed to be on the valley of stability such that $N-Z=0.4A^2/(200+A)$. From this plot one sees that the threshold energy for fusion increases rapidly as Z_1Z_2 increases. For the $^{165}_{67}Ho+^{84}_{36}Kr$ reaction to be discussed later, the "extra push" energy required for fusion is 50 MeV.

Three dynamical trajectories (Swi 81c) are illustrated in Figure 295 for a central collision ($l=0$) between two nuclei with $A=104$. The trajectories correspond to energies of 0, 3.6, and 10 MeV, respectively, above the interaction barrier. The light dashed curves represent some of the equipotential lines which delineate the saddle point whose energy is 0.033 in units of the surface energy $4\pi R^2\gamma$. For the present total system with $Z\approx82$ and $A\approx208$, this corresponds to a barrier of some 19 MeV. The approach of the two nuclei is on a trajectory along the ρ axis, where $\rho=r/(R_P+R_T)$. At the point $\rho=1.0$, the two spheres touch, and the lowest curve shows the subsequent dynamical evolution [see Equation (2.3.19)] in the case where

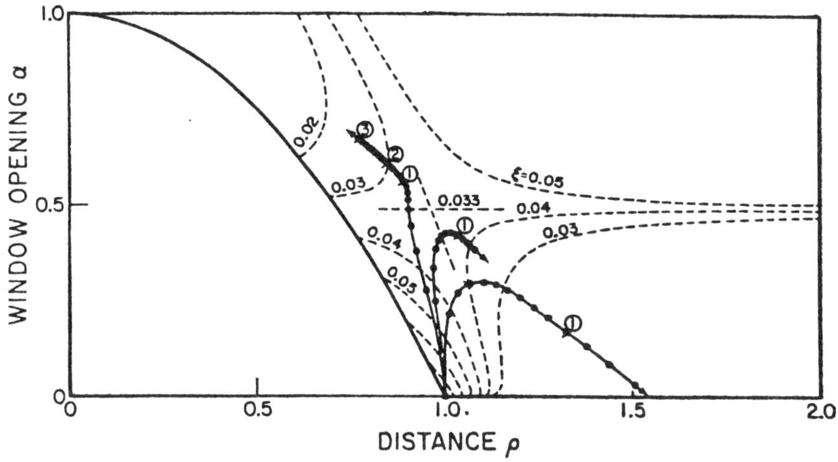

Figure 295. Example of a dynamical study of symmetric collisions between two medium-weight nuclei ($A_1 = A_2 \approx 104$, $Z_1 = Z_2 \approx 41$). The total system is $^{208}_{82}Pb$. The ordinate α and absicca ρ are defined in Section 3.2. The dashed lines are equipotentials, with energy expressed in units of the surface energy of a single sphere (≈ 583 MeV) and with reference to the energy of the spherical configurations. The location of the saddle is indicated by the crossing equipotentials. The two lower trajectories with energies 0 and 3.6 MeV above the interaction barrier fail to fuse and reseparate after dissipating some energy. The upper trajectory, with 10 MeV above the barrier, fuses. Dots are spaced at 1.2×10^{-22} s and circled dots indicate the lapse of 1.2×10^{-21} s. Adapted from (Swi 81c).

there is zero energy in excess of the barrier. Initially there is a rapid growth of the neck [α is defined in Section 2.3 and Figure 10b], then a leveling out and subsequent decrease in the neck size.

The dots on the trajectory mark equal intervals of time of 1.2×10^{-22} s. The whole reaction time for this trajectory from contact to snapping is about 16×10^{-22} s. Furthermore, the trajectory remains completely outside the saddle point. Even so, some energy has been dissipated during the neck growth and its collapse, and the two fragments would have less energy at infinity than that corresponding to the interaction barrier, which is what they started with.

The intermediate trajectory, with 3.6 MeV above the barrier at injection, also fails to induce capture and contributes to the damped or deep-inelastic reaction process. The uppermost trajectory with 10 MeV is easily captured and enters the mononuclear regime. This illustrates the two distinct classes of trajectories: trapping and nontrapping ones, leading in the present case to a compound nucleus and dinucleus, respectively.

In Equation (3.8.13) the quantity f represents the fraction of the total angular momentum that remains orbital after contact and allows for the fact that some of the initial angular momentum is converted into intrinsic fragment spin. In the present calculations between spherical nuclei, three types of collisions are assumed: sliding, rolling, and sticking collisions. For spherical nuclei, these correspond, respectively, to f values of 1, 5/7, and $(A_1^{1/3}+A_2^{1/3})^2/\{(A_1^{1/3}+A_2^{1/3})^2+\frac{2}{5}(A_1+A_2)(A_1^{2/3}A_2^{-1}+A_2^{2/3}A_1^{-1})\}$. With these values of a, b, and f it is possible to evaluate the energy shift as a function of the orbital angular momentum l. The fusion cross section is then given by (Hui 81)

$$\sigma_{\text{fus}}\left[l_f, E+E_x(l_f)\right]=\left\{E/\left[E+E_x(l_f)\right]\right\}\sigma'_{\text{fus}}(l_f, E) \qquad (3.8.18)$$

where $\sigma'_{\text{fus}}(l_f, E)$ is the fusion cross section calculated with the one-dimensional classical dynamical model (Bir 79) at a center-of-mass energy E, and l_f is the maximum model angular momentum that leads to fusion at energy E.

An alternate and simpler procedure, not requiring a classical trajectory calculation, has been suggested by Swiatecki (Swi 82b). If a finite amount of radial energy (extra push) E_x is needed to induce fusion, then one can write

$$\sigma E/\pi R_B^2 = E - V_B - E_x = E_{\tan} = l^2\hbar^2/2\mu R^2 \qquad (3.8.19)$$

where

$$l^2\hbar^2 = 2A_1A_2 m R_B^2 E_{\tan}/(A_1+A_2) \qquad (3.8.20)$$

The "extra push" energy can now be written as

$$E_x = (\alpha_x + \beta_x E_{\tan})^2 \qquad (3.8.21)$$

where

$$\alpha_x = K^{1/2}\left[\left(\frac{Z^2}{A}\right)_{\text{eff}} - \left(\frac{Z^2}{A}\right)_{\text{eff thr}}\right] \text{MeV}^{1/2} \qquad (3.8.22)$$

and

$$\beta_x = 8K^{1/2}(r_0/e^2)f^2/(A_1^{1/3}\cdot A_2^{1/3})\ \text{MeV}^{-1/2} \qquad (3.8.23)$$

The quantities K, $(Z^2/A)_{\text{eff}}$, and $(Z^2/A)_{\text{eff thr}}$ are defined by Equations (3.8.17), (3.8.14), and (3.8.15), respectively, and $(e^2/r_0)=1.1755$ MeV. In this procedure the fusion cross section is given by

$$\sigma_{\text{fus}} = \frac{\pi R_B^2}{E}\left\{-\left(\frac{\alpha_x\beta_x+1/2}{\beta_x^2}\right)+\left[\left(\frac{\alpha_x\beta_x+1/2}{\beta_x^2}\right)^2-\left(\frac{\alpha_x^2+V_B-E}{\beta_x^2}\right)\right]^{1/2}\right\} \qquad (3.8.24)$$

where $E > V_B - (\alpha_x/\beta_x)$. Equation (3.8.24) gives explicitly the energy dependence of the fusion cross section for systems that require an additional amount of radial energy above the interaction barrier V_B to induce fusion. When no "extra push" energy is required, the classical fusion cross section is given by the standard formula

$$\sigma = \frac{\pi R_B^2}{E}(E - V_B) \tag{3.8.25}$$

The values of the radial position R_B and height of the barrier V_B are determined with the proximity nuclear potential [see, for example Table 1 in (Bir 79)].

The experimental fusion excitation functions for three targets, ^{26}Mg, ^{50}Ti, and ^{64}Ni, bombarded with ^{208}Pb projectiles are shown in Figure 296 along with theoretical values based on Equation (3.8.24). Similar data for a ^{165}Ho target bombarded with three different projectiles, ^{40}Ar, ^{56}Fe, and ^{84}Kr, are shown in Figure 297. In both of these figures, the dot-dashed, dashed, and dotted theoretical curves correspond to sticking, rolling, and sliding collisions, respectively. The solid lines in each figure are calculated with Equation (3.8.25) and represent the energy dependence of the cross section for "solid contact" collisions. For the ^{26}Mg$+^{208}$Pb and ^{165}Ho$+^{40}$Ar reactions, where $(Z^2/A)_{\text{eff}} < 30$, the experimental fusion cross sections follow the predictions of Equation (3.8.25) at the lower energies, but deviate from the simple formula exceedingly with increasing energy, i.e., with increasing mean angular momentum. In examing a number of fusion excitation functions with $(Z^2/A)_{\text{eff}} < 30$, including some of very light systems, the fusion cross sections appear to start deviating from Equation (3.8.25) at values of $(E - V_B)/\mu \geq (2.0 \pm 0.3)$ MeV per nucleon. In the one-dimensional model this limitation in the fusion cross section is associated with the disappearance of attractive pockets in the effective interaction potential (Bir 79). However, in the Swiatecki model at this energy per nucleon an "extra push" energy becomes necessary when $(Z^2/A)_{\text{eff}} + (f \cdot l/l_{\text{ch}})^2$ begins to exceed $(Z^2/A)_{\text{eff thr}}$. The above observation may be useful in renormalizing the l_{ch} paramcters for reactions between light heavy ions, where the procedure as outlined here does not give a good fit to the data.

As $(Z^2/A)_{\text{eff}}$ increases, an "extra push" energy is required already at threshold ($l = 0$) when $(Z^2/A)_{\text{eff}}$ exceeds $(Z^2/A)_{\text{eff thr}}$. This is true for the ^{50}Ti$+^{208}$Pb and ^{165}Ho$+^{56}$Fe reactions where the theoretical "extra push" energies at threshold [see Equation (3.8.13)] are very small but finite and equal to 1.0 and 1.5 MeV, respectively, in good agreement with the experimental data. For the heaviest systems in Figures 296 and 297, the ^{64}Ni$+^{208}$Pb and ^{165}Ho$+^{84}$Kr reactions, the theoretical "extra push" energies are 31.0 and 50.2 MeV, respectively, at threshold. Finite fusion cross

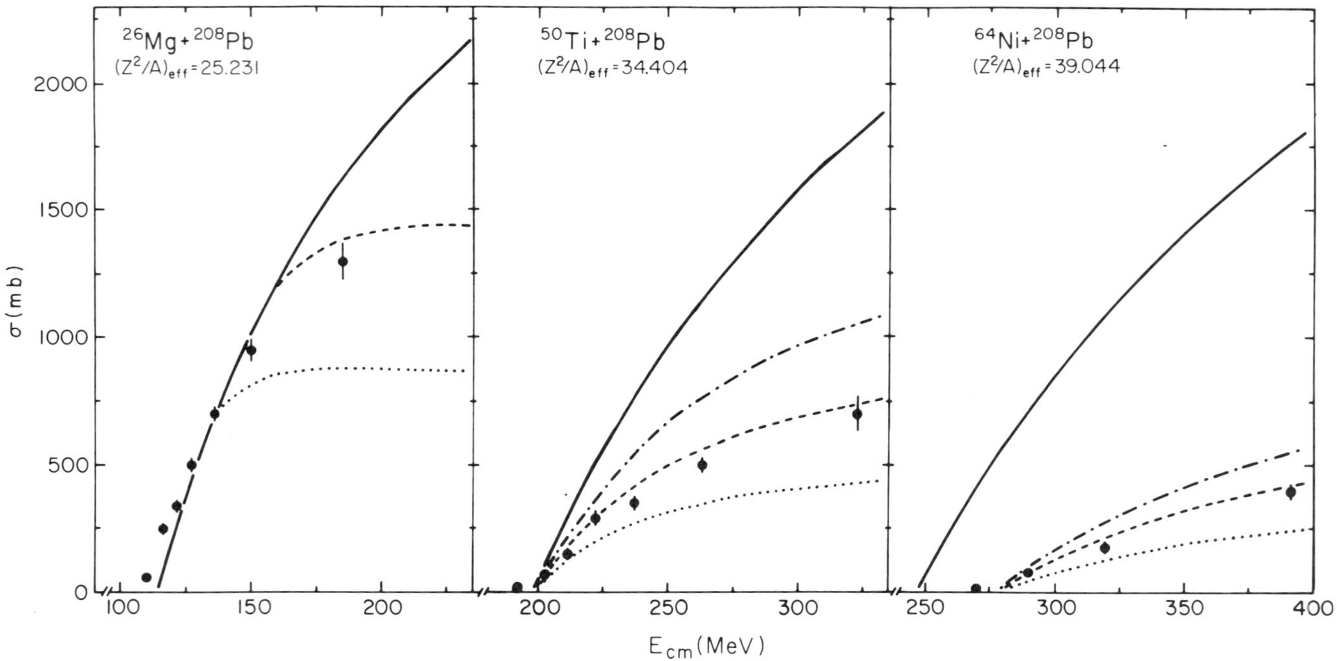

Figure 296. Fusion excitation functions for targets of ^{26}Mg, ^{50}Ti, and ^{64}Ni bombarded with ^{208}Pb projectiles. The dot-dashed, dashed, and dotted theoretical curves [Equation (3.8.24)] correspond to sticking, rolling, and sliding collisions, respectively. The solid curves are calculated with Equation (3.8.25). The data are from (Boc 82).

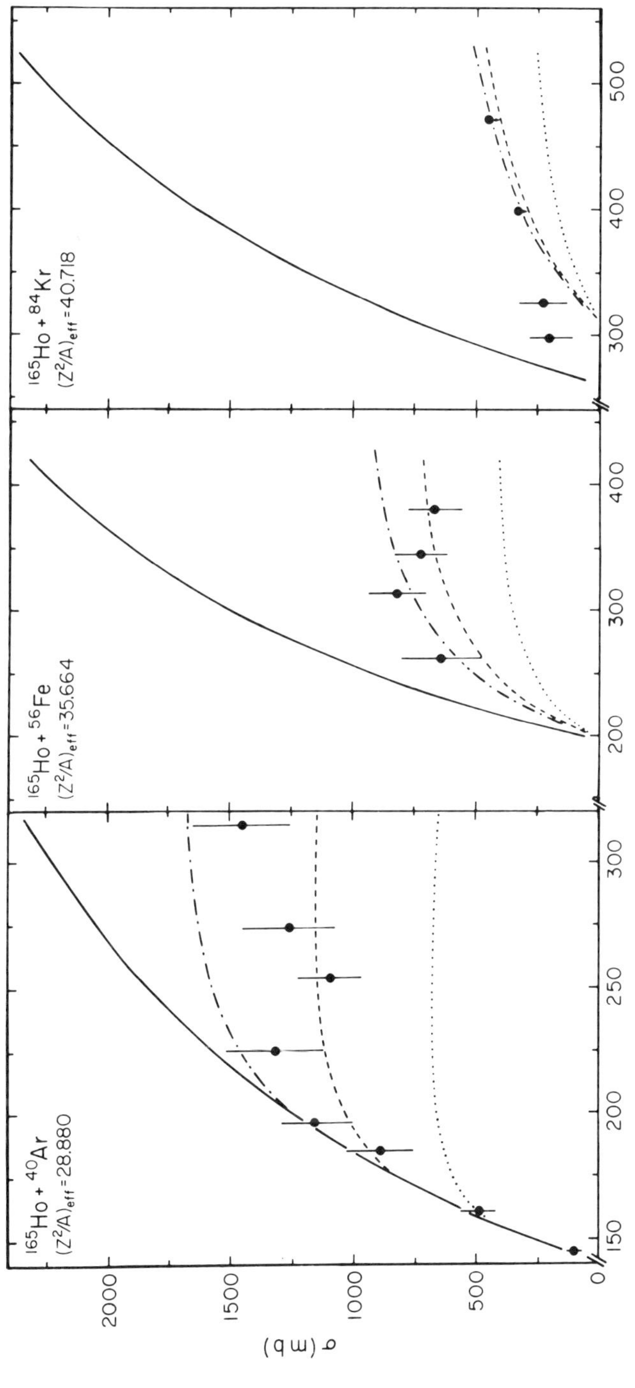

Figure 297. Fusion excitation functions for a ^{165}Ho target bombarded with ^{40}Ar, ^{56}Fe, and ^{84}Kr. For explanation of curves see caption of Fig. 296. The data are from (Bor 81b, Bir 83, Pet 75, Bir 83b).

sections were reported (Pet 75) for the ^{165}Ho + ^{84}Kr reaction at the two lowest bombarding energies of 492 and 525 MeV. However, the results of these early and difficult measurements should probably be considered as upper limits because of possible contributions from strongly damped collisions. Hence, not too much significance can presently be attributed to the apparent discrepancies between data and calculations for this reaction near the fusion threshold.

The six fusion excitation functions shown in Figures 296 and 297 are consistent with the "extra push" model of Swiatecki. With the empirical values of the constants a and b listed previously and used in the present calculations, the data indicate that a sizable fraction of the orbital angular momentum is converted into intrinsic spin at contact. As can be seen in these figures, all of the theoretical fusion excitation functions based on rolling collisions ($f = 5/7$) are in reasonable agreement with the experimental data. This is in accord with earlier analyses of fusion excitation functions of lighter systems with one-dimensional models (Bir 79, Bas 74).

The "extra-extra push" energy E_{xx}, illustrated in Figure 298, has been parametrized (Bjø 82b) in an analogous way to the "extra push" energy E_x defined in Equation (3.8.13) except $(Z^2/A)_{\text{eff}}$ is replaced by $(Z^2/A)_m$,

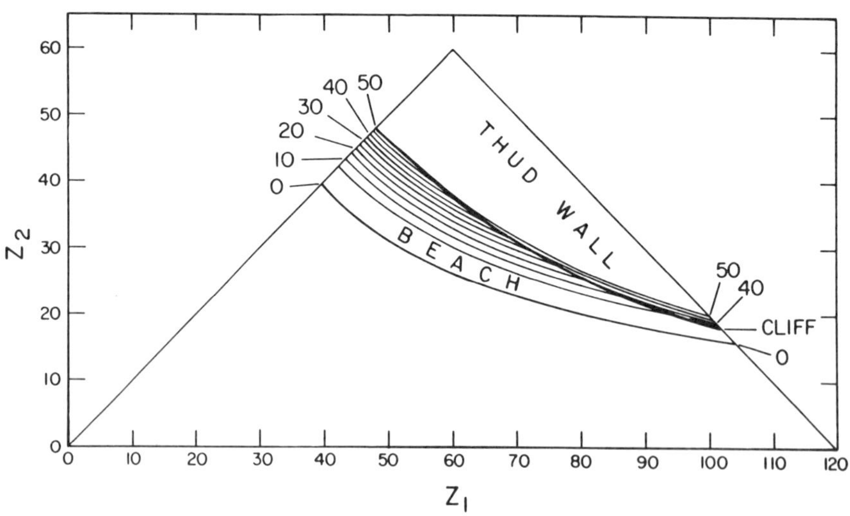

Figure 298. Contour lines of equal "extra-extra push" energy. In the region below the cliff the contour lines are identical to the "extra push" energy in Fig. 294. Above the cliff Equation (3.8.27) was used to calculate E_{xx} (see text). In the region of the "Thud Wall," the "extra-extra push" energy rises to very high values. From (Bjø 82b).

where

$$(Z^2/A)_m = \left[\left(\frac{Z^2}{A}\right)\left(\frac{Z^2}{A}\right)_{\text{eff}}\right]^{1/2} - k\left(\frac{Z^2}{A}\right)\left\{1 - \left[\left(\frac{Z^2}{Z}\right)_{\text{eff}} \Big/ \left(\frac{Z^2}{A}\right)\right]^{1/2}\right\}^2 \tag{3.8.26}$$

The "extra-extra push" energy for central collisions ($l = 0$) is now given by

$$E_{xx} = \begin{cases} E_x & \text{for } (Z^2/A)_m \leq 0.84(Z^2/A)_{\text{crit}} \\ K\left[(Z^2/A)_m - (Z^2/A)_{\text{eff thr}}\right]^2 & \text{for } (Z^2/A)_m > 0.84(Z^2/A)_{\text{crit}} \end{cases} \tag{3.8.27}$$

where

$$(Z^2/A)_{\text{crit}} = 50.833\left\{1 - 1.7826\left[(N_1 - Z_1 + N_2 - Z_2)/(A_1 + A_2)\right]^2\right\} \tag{3.8.28}$$

Contour lines of the "extra-extra push" energy (Bjø 82b) are plotted in Figure 298. As in the construction of Figure 294, it is assumed that each nucleus is on the valley of stability. In the region below the cliff the contour lines are identical with the "extra push" energy of Figure 294. Above the cliff, Equation (3.8.27) was used with $k = 1$ and $(Z^2/A)_{\text{cl}} = 0.84(Z^2/A)_{\text{crit}}$. Since neither k nor $(Z^2/A)_{\text{cl}}$ have been determined experimentally, the location and height of the cliff are for illustrative purposes only (Bjø 82b).

The compound nucleus cross section may be estimated from Equation (3.8.24) by replacing α_x and β_x by α_{xx} and β_{xx}, respectively, where (Bjø 82b)

$$\alpha_{xx} = K^{1/2}\left\{\left(\frac{Z^2}{A}\right)_m - \left(\frac{Z^2}{A}\right)_{\text{eff thr}}\right\} \text{MeV}^{1/2} \tag{3.8.29}$$

$$\beta_{xx} = 8K^{1/2}(r_0/e^2) f_m^2 / (A_1^{1/3} + A_2^{1/3}) \text{ MeV}^{-1/2} \tag{3.8.30}$$

All of the parameters in Equations (3.8.29) and (3.8.30) have been defined previously, except f_m, which is given (Bjø 82b) by

$$f_m = [\eta(x)]^{1/4} f^{1/2} (25/24) A_1^{1/3} A_2^{1/3} (A_1^{1/3} \cdot A_2^{1/3})^{1/2} (A_1 + A_2)^{-5/6} \tag{3.8.31}$$

Values of $\eta(x)$ as a function of the fissility parameter $x = (Z^2/A)/(Z^2/A)_{\text{crit}}$ are tabulated in Table II of (Bjø 82b). These theoretical formulas have been used recently, along with other information, to estimate the production of superheavy nuclei from different very-heavy-ion reactions (Bjø 82b).

4. Conclusions

It is apparent from the discussion of heavy-ion reaction theories and experimental investigations, presented in this volume, that the discovery of damped nuclear reactions has offered unique opportunities to nuclear science and access to properties of the response of nuclear matter heretofore unknown. These opportunities continue to present challenges, both to experiment and to theory. The impressive scientific activity in this area of nuclear research, manifesting itself in a rapidly growing number of publications, has made it impossible to include all relevant new developments in this review. Hence, it can represent at most an intermediate account of the properties of damped nuclear interactions and of their interpretation.

From the survey of theoretical approaches discussed to heavy-ion collision dynamics presented in Section 2 and their confrontation to experimental results in Section 3, it is obvious that the goal of providing a realistic and self-contained model of the microscopic mechanisms and of the associated simple modes of excitation in heavy-ion reactions has not yet been achieved. The challenging difficulties in extrapolating the processes known from light-ion-induced reactions into the very complex and extreme situations encountered in damped and fusion–fission-like reactions have already required the introduction of theoretical methods that are new to nuclear science. Although much has been learnt about these reactions, progress has not yet culminated in a unified theory of heavy-ion reaction dynamics.

In the model approaches discussed in previous sections, approximations are often made on the basis of rather crude estimates of fundamental reaction aspects such as, for example, relevant time scales for relaxation of intrinsic and collective degrees of freedom, transition strengths and couplings between different reaction modes. Internal consistency is frequently sacrificed to simplicity and tractability without a clear understanding of all of the important consequences. It also appears unsatisfactory when fundamental model quantities are adjusted to fit experimental data, a procedure that does not allow one to draw definite conclusions on the validity of the microscopic mechanism proposed as an explanation of the reaction phenomenology. Several models represent merely schemes that relate selected reaction phenomena such as dissipation and statistical fluctuations under certain boundary conditions, but do not, by themselves, offer significant insight into the collision dynamics. Other theories are designed to describe only isolated effects, relying on auxiliary models that are based on conflicting sets of assumptions. Furthermore, drastic successive approximations can lead to indistinguishable results from models derived from rather different physical pictures, eluding a stringent test by experiment, unless elaborate microscopic calculations of model quantities are provided also. Simplifying concepts such as that of a well-defined average system trajectory restrict the describable reaction modes to an extent that makes a thorough investigation of their validity highly desirable.

The richness of heavy-ion reaction dynamics is apparent from the current lack of a consistent understanding of the relative significance of the different microscopic processes evolving in such a collision. Mutually exclusive views of the reaction mechanism are still being vividly discussed in the literature. On the other hand, it is encouraging to observe the progress being made with a further diversification and completion of several of the most successful reaction models suggesting more specific experimental tests which can, eventually, lead to the development of a unified theory of heavy-ion reactions.

Experimental investigations can be viewed with much of the same criticism that is addressed to developments of reaction models. Deficiencies in both types of work are, of course, to some extent unavoidable, due to the overwhelming complexity of the damped reaction features. Although important first insights have been gained in the relatively simple, earlier experiments, the multitude of simultaneous and sequential processes occurring in damped reactions lead to rather convoluted final experimental distributions of observables, emphasizing the increased necessity for exclusive, multiparameter experiments. Examples have been presented in Section 3 for the discovery of new reaction modes in sophisticated coincidence experiments. There is also a growing need for more systematical experimental studies measuring as many correlations as possible. This would reduce the need to make ill-justified assumptions on which the interpretation of data may depend drastically. There are only very few cases in which a reaction has been studied rather completely and at more than one bombarding energy. Bombarding-energy systematics of damped reaction features are definitely of importance for stringent tests of frictional models and transport theories in general. For example, it has as yet not been possible to distinguish experimentally between different proposed sets of conservative, dissipative, and inertial interactions. The possible coexistence of coherent and stochastic reaction modes in dissipation and mass exchange processes remains to present an interesting problem, challenging experimenters to perform more systematic and accurate measurements, as well as careful investigations of the sequential decay of the primary reaction fragments that can, in certain cases, completely mask the reaction properties of interest.

Experimental work, employing the characteristics of measured deexcitation patterns, in order to derive information on the primary damped reaction mechanism, relies, of course, heavily on the availability of appropriate calibration measurements for the systems under study. Widely different prescriptions used in the inference of fragment spins from experimental γ-ray multiplicities can be taken as indication of the lack of proper calibration procedures. Furthermore, while the particle emission mechanisms constitute an important aspect of damped and fusion–fission-like reactions and are of considerable interest in their own right, one probably has to be quite aware of the possibility that different emission products sample different regions in the space of observables, such that restrictive consideration of one particular type may lead to a bias in the interpretation of experimental results.

In summary, however, one must conclude that an impressive body of information has been collected over the past decade on the exotic mechanisms operating in damped reactions between complex nuclei. A dominantly stochastic nature of the underlying physical processes appears currently to provide the best approximation to the phenomenology of damped reactions, a view that could be modified, as new experimental data become available. Such stochastic reaction mechanisms, however, are not easily understood in the framework of equilibrium-statistical theory. Undoubtedly, the transport mechanisms, observed at work in damped heavy-ion collisions, exhibit dynamical features that appear to be uncharacteristic of adiabatic motion of nuclear matter close to equilibrium. It can be anticipated that further progress in experimental and theoretical investigations of damped nuclear interactions, and their transitions to fusion–fission-like processes, will lead to a better understanding of the fundamental relaxation processes occurring in nuclear matter. This will also be of significance to the interpretation of new reaction phenomena that may appear in the range of intermediate bombarding energies.

Acknowledgments

This work was supported by the United States Department of Energy.

The authors are indebted to many of their colleagues in the field for useful and clarifying discussions. Particularly, the help and support received from the present and former members of the University of Rochester heavy-ion research group, J. R. Birkelund, M. A. Butler, A. D. Dougan, D. Hilscher, J. P. Kosky, H. H. Rossner, L. Schmieder, W. W. Wilcke, and H. J. Wollersheim, are deeply appreciated. A fruitful collaboration with J. Randrup on aspects of relaxation processes in damped reactions is gratefully acknowledged. He is thanked also for a careful reading of parts of this manuscript and helpful comments and suggestions. We are grateful to O. Zohni for extensive and illuminating discussions on several theoretical aspects discussed in this paper. His comments and suggestions on part of the manuscript have been very useful. We would also like to express our appreciation of Ms. K. P. Matsubara for her dedication in the typing of the manuscript. Some of the figures in the paper have been prepared by Ms. E. Korn. Most of all, we would like to thank our wives, Chojy and Dolly, for their encouragement and support.

References

(Abu 78) A. Y. Abul-Magd, M. H. Simbel, and H. A. Weidenmüller, *Z. Phys.* **A285**:41 (1978).

(Adl 80) L. Adler, P. Gonthier, J. H. K. Ho, A. Khodai, M. N. Namboodiri, J. B. Natowitz, and S. Simon, *Phys. Rev. Lett.* **45**:696 (1980).

(Aga 75) D. Agassi and H. A. Weidenmüller, *Phys. Lett.* **56B**:305 (1975).

(Aga 76) S. Agarwal, H. Gauvin, D. Guerreau, H. Jungclas, Y. Le Beyec, M. Lefort, and X. Tarrago, *Z. Phys.* **A278**:265 (1976).

(Aga 77) S. Agarwal, J. Galin, B. Gatty, D. Guerreau, M. Lefort, X. Tarrago, R. Babinet, B. Cauvin, J. Girard, and H. Nifenecker, *Nucl. Phys.* **A293**:230 (1977).

(Aga 77b) D. Agassi, C. M. Ko, and H. A. Weidenmüller, *Ann. Phys. (N. Y.)* **107**:140 (1977).

(Aga 78) D. Agassi, H. A. Weidenmüller and C. M. Ko, *Phys. Lett.* **73B**:284 (1978).

(Aga 78b) D. Agassi, C. M. Ko, and H. A. Weidenmüller, *Phys. Rev. C* **18**:223 (1978).

(Aga 79) D. Agassi, *Nucl. Phys.* **A330**:125 (1979).

(Aga 79b) D. Agassi, C. M. Ko, and H. A. Weidenmüller, *Ann. Phys. (N. Y.)* **117**:407 (1979).

(Aga 80) S. Agarwal, *Z. Phys.* **A297**:41 (1980).

(Aga 80b) S. Agarwal, J. Galin, B. Gatty, D. Guerreau, M. Lefort, X. Tarrago, R. Babinet, and J. Girard, *Z. Phys.* **A296**:287 (1980).

(Agu 79) P. Aguer, R. P. Schmitt, G. J. Wozniak, D. Habs, R. M. Diamond, C. Ellegaard, D. L. Hillis, C. C. Hsu, G. J. Mathews, L. G. Moretto, G. U. Rattazzi, C. P. Roulet, and F. S. Stephens, *Phys. Rev. Lett.* **43**:1778 (1979).

(Akk 79) J. M. Akkermans, *Phys. Lett.* **82B**:20 (1979).

(Alb 75) R. Albrecht, W. Dünnweber, G. Graw, H. Ho, S. G. Steadman, and J. P. Wurm, *Phys. Rev. Lett.* **34**:1400 (1975).

(Alb 75b) K. Albrecht, *Phys. Lett.* **56B**:127 (1975).

(Alb 77) K. Albrecht and W. Stocker, *Nucl. Phys.* **A278**:95 (1977).

(Alb 79) K. Albrecht, unpublished report (1979).

(Ale 78) M. M. Aleonard, G. J. Wozniak, P. Glassel, M. A. Deleplanque, R. M. Diamond, L. G. Moretto, R. P. Schmitt, and F. S. Stephens, *Phys. Rev. Lett.* **40**:622 (1978).

(Alh 78) Y. Alhassid and R. D. Levine, *Phys. Rev. A* **18**:89 (1978).

(Alh 79) Y. Alhassid and R. D. Levine, *Phys. Rev. C* **20**:1775 (1979).

(Alh 79b) Y. Alhassid, R. D. Levine, J. S. Karp, and S. G. Steadman, *Phys. Rev. C* **20**:1789 (1979).

(Alh 80d) Y. Alhassid, R. D. Levine, J. S. Karp, and S. G. Steadman, *Proc. Int. Symp. on Continuum Spectra*, held in San Antonio, Texas, December, 1979 (T.

Tamura, J. B. Natowitz, and D. H. Youngblood, eds.), Harwood Academic Publishers, CHUR, London, New York (1980), Vol. 2, p. 179.

(Ama 79) H. Amakawa, S. Yamaji, A. Mori, and K. Yazaki, *Phys. Lett.* **82B**:13 (1979).

(Ame 80) A. Ameaume, G. Bizard, J. L. Laville, and M. Louvel, P. Colombani, N. Frascaria, A. Gamp, J. P. Garron, J. C. Roynette, and C. Stephan, *Z. Phys.* **A294**:159 (1980).

(And 78) O. Andersen, R. Bauer, G. B. Hagemann, M. L. Halbert, B. Herskind, M. Neiman, H. Oeschler, and H. Ryde, *Nucl. Phys.* **A295**:163 (1978).

(Any 74) N. Anyas-Weiss, J. C. Cornell, P. S. Fisher, P. N. Hudson, A. Menchaca-Rocha, D. J. Millener, A. D. Panagiotou, D. K. Scott, D. Strottman, D. M. Brink, B. Buck, P. J. Ellis, and T. Engeland, *Phys. Rep.* **12**:201 (1974).

(Arm 76) P. Armbruster, *J. Phys. Colloq. C5* **37**:161 (1976).

(Arn 76) M. Arnould and W. M. Howard, *Nucl. Phys.* **A274**:295 (1976).

(Art 70) A. G. Artukh, G. F. Gridnev, V. L. Mikheev, J. Wilczyński, and V. V. Volvo, *Proc. Int. Conf. on Nuclear Reactions Induced by Heavy Ions*, Heidelberg (1969).

(Art 70c) A. G. Artukh, V. V. Avdeichikov, G. F. Gridnev, V. L. Mikheev, V. V. Volkov, and J. Wilczyński, *Phys. Lett.* **31B**:129 (1969).

(Art 71) A. G. Artukh, V. V. Avdeichikov, J. Ero, G. F. Gridnev, V. L. Mikheev, V. V. Volkov, and J. Wilczyński, *Nucl. Phys.* **A160**:511 (1971).

(Art 71b) A. G. Artukh, V. V. Avdeichikov, G. F. Gridnev, V. L. Milkheev, V. V. Volkov, and J. Wilczyński, *Nucl. Phys.* **A176**:284 (1971).

(Art 71c) A. G. Artukh, V. V. Avdeichikov, G. F. Gridnev, V. L. Mikheev, V. V. Volkov, and J. Wilczyński, *Nucl. Phys.* **A168**:321 (1971).

(Art 73) A. G. Artukh, G. F. Gridnev, V. L. Mikheev, V. V. Volkov, and J. Wilczyński, *Nucl. Phys.* **A211**:299 (1973).

(Art 73b) A. G. Artukh, G. F. Gridnev, V. L. Mikheev, V. V. Volkov, and J. Wilczyński, *Nucl. Phys.* **A215**:91 (1973).

(Art 73c) A. G. Artyukh, J. Wilczyński, V. V. Volkov, G. F. Gridnev, and V. L. Mikheev, *Sov. J. Nucl. Phys.* **17**:586 (1973).

(Art 77) A. G. Artukh, G. F. Gridnev, V. L. Mikheev, and V. V. Volkov, *Nucl. Phys.* **A283**:350 (1977).

(Art 78) A. G. Artyukh, V. V. Volkov, E. Gerlik, R. Gerstenberger, G. F. Gridnev, V. B. Zlokazov, A. N. Mezentsev, V. L. Mikheev, and T. S. Salamatina, *Sov. J. Nucl. Phys.* **27**:14 (1978).

(Art 78b) A. G. Artyukh, V. V. Volkov, E. Gierlik, G. F. Gridnev, A. N. Mezentsev, and V. L. Mikheev, *Sov. J. Nucl. Phys.* **28**:313 (1978).

(Art 79) A. G. Artukh, G. F. Gridnev, A. N. Mezentsev, V. L. Mikheev, A. Popescu, D. G. Popescu, and V. V. Volkov, *Nukleonika* **24**:191 (1979).

(Aue 75) N. Auerbach and A. Yeverechyahu, *Ann. Phys.* (*N. Y.*) **95**:35 (1975).

(Aug 79) P. Auger, T. H. Chiang, J. Galin, B. Gatty, D. Guerreau, E. Nolte, J. Pouthas, X. Tarrago, and J. Girard, *Z. Phys.* **A289**:255 (1979).

(Aug 79b) P. Auger, T. H. Chiang, J. Galin, B. Gatty, J. Girard, D. Guerreau, E. Nolte, J. Pouthas, and X. Tarrago, *Proceedings of Hirschegg Workshop VII*, (1979), p. 39.

(Aus 78) N. Austern, C. M. Vincent, and J. P. Farrell, Jr., *Ann. Phys.* (*N. Y.*) **114**:93 (1978).

(Avi 78) Y. Avishai, *Phys. Lett.* **76B**:5 (1978).

(Awe 79) T. C. Awes, C. K. Gelbke, B. B. Back, A. C. Mignerey, K. L. Wolf, P. Dyer, H. Breuer, and V. E. Viola, Jr., *Phys. Lett.* **87B**:43 (1979).

(Awe 80) T. C. Awes, C. K. Gelbke, G. Poggi, B. B. Back, B. Glagola, H. Breuer, V. E. Viola, Jr., and T. J. M. Symons, *Phys. Rev. Lett.* **45**:513 (1980).

(Awe 80b) T. C. Awes, B. B. Back, H. Breuer, C. K. Gelbke, B. Glagola, G. Poggi, and V. E. Viola, *Proc. Int. Conf., Berkeley*, LBL-11118 (1980), p. 548.

(Awe 81) T. C. Awes, G. Poggi, C. K. Gelbke, B. B. Back, B. G. Glagola, H. Breuer, and V. E. Viola, *Phys. Rev. C* **24**:89 (1981).

(Awe 82) T. C. Awes, S. Saini, G. Poggi, C. K. Gelbke, D. Cha, R. Legrain, and G. D. Westfall, *Phys. Rev. C.* **25**:2361 (1982).

(Ayi 76) S. Ayik, B. Schürmann, and W. Nörenberg, *Z. Phys.* **A277**:299 (1976).

(Ayi 76b) S. Ayik, *Phys. Lett.* **63B**:22 (1976).

(Ayi 76c) S. Ayik, B. Schürmann, and W. Nörenberg, *Z. Phys.* **A279**:145 (1976).

(Ayi 78) S. Ayik, G. Wolschin, and W. Nörenberg, *Z. Phys.* **A286**:271 (1978).

(Ayi 78b) S. Ayik and W. Nörenberg, *Z. Phys.* **A288**:401 (1978).

(Ayi 79) S. Ayik, *Z. Phys.* **A292**:257 (1979).

(Ayi 80) S. Ayik and W. Nörenberg, *Z. Phys.* **A297**:55 (1980).

(Ayi 80b) S. Ayik, *Z. Phys.* **A298**:83 (1980).

(Bab 76) R. Babinet, L. G. Moretto, J. Galin, R. Jared, J. Moulton, and S. G. Thompson, *Nucl. Phys.* **A258**:172 (1976).

(Bab 78) R. Babinet, B. Cauvin, J. Girard, H. Nifenecker, G. Gatty, D. Guerreau, M. Lefort, and X. Tarrago, *Nucl. Phys.* **A296**:160 (1978).

(Bab 80) R. Babinet, B. Cauvin, J. Girard, J. M. Alexander, T. H. Chiang, J. Galin, B. Gatty, D. Guerreau, and X. Tarrago, *Z. Phys.* **A295**:153 (1980).

(Bac 78) B. B. Back and S. Bjørnholm, *Nucl. Phys.* **A302**:343 (1978).

(Bac 80b) B. B. Back, K. L. Wolf, A. C. Mignerey, C. K. Gelbke, T. C. Awes, H. Breuer, V. E. Viola, Jr., and P. Dyer, *Phys. Rev. C* **22**:1927 (1980).

(Bac 81) B. B. Back, H.-G. Clerc, R. R. Betts, B. G. Glagola, and B. D. Wilkins, *Phys. Rev. Lett.* **46**:1068 (1981).

(Bal 77) A. J. Baltz, *Phys. Rev. Lett.* **38**:1197 (1977).

(Bal 78) A. J. Baltz, S. K. Kauffmann, N. K. Glendenning, and K. Pruess, *Phys. Rev. Lett.* **40**:20 (1978).

(Ban 75) M. V. Banaschik, R. S. Simon, P. Colombani, D. P. Soroka, F. S. Stephens, and R. M. Diamond, *Phys. Rev. Lett.* **34**:892 (1975).

(Ban 80) D. Bangert and H. Freiesleben, *Nucl. Phys.* **A340**:205 (1980).

(Ban 80b) B. Banerjee, N. K. Glendenning, and M. Gyulassy, Lawrence Berkeley Laboratory Report, LBL-10979 (1980), unpublished.

(Ban 83) P. Bandyopadhyay and J. Mahalanobis, *Phys. Rev. C* **27**:628 (1983).

(Bar 75) J. Barrette, P. Braun-Munzinger, C. K. Gelbke, E. Grosse, H. L. Harney, J. Kuzminski, I. Tserruya, and T. Walcher, *Z. Phys.* **A274**:121 (1975).

(Bar 77) J. Barrette, P. Braun-Munzinger, C. K. Gelbke, H. E. Wegner, B. Zeidman, A. Gamp, H. L. Harney, and T. Walcher, *Nucl. Phys.* **A279**:125 (1977).

(Bar 77b) J. Barrette and P. Braun-Munzinger, *Nucl. Phys.* **A287**:195 (1977).

(Bar 78) B. R. Barrett, S. Shlomo, and H. A. Weidenmüller, *Phy. Rev. C* **17**:544 (1978).

(Bar 78d) J. Barrette, P. Braun-Munzinger, C. K. Gelbke, H. L. Harney, H. E. Wegner, and B. Zeidman, *Nucl. Phys.* **A299**:147 (1978).

(Bar 80d) J. Bartel, and H. Feldmeier, *Z. Phys.* **A297**:333 (1980).

(Bas 74) R. Bass, *Nucl. Phys.* **A231**:45 (1974).

(Bas 76) R. Bass, *Proc. European Conf. Nucl. Phys. with Heavy Ions*, Caen, 1976, p. 147.

(Bas 80) R. Bass, *Nuclear Reactions with Heavy Ions*, Springer-Verlag, (1980).

(Bau 76) G. Baur and D. Trautmann, *Phys. Rep.* **25**:293 (1976).

(Bau 79b) G. Baur, M. Pauli, F. Rosel, and D. Trautmann, *Nucl. Phys.* **A315**:241 (1979).

(Bau 79c) G. Baur, F. Rosel, D. Trautmann, and R. Shyam, *Proc. Symp. on Deep-Inelastic and Fusion Reactions with Heavy Ions*, Springer-Verlag, Berlin (1979), Vol. 117, p. 268.

(Bau 79d) G. Baur, F. Rosel, D. Trautmann, and R. Shyam, *Proc. Int. Symp. on Continuum Spectra*, held in San Antonio, Texas, December, 1979 (T. Tamura, J. B. Natowitz, and D. H. Youngblood, eds.) Harwood Academic Publishers CHUR, London, New York (1980), Vol. 2, p. 131.

(Bau 80) G. Baur, R. Shyam, F. Rosel, and D. Trautmann, *Phys. Rev. C* **21**:2668 (1980).

(Bau 80b) G. Baur, R. Shyam, F. Rosel, and D. Trautmann, *Proc. Int. Conf., Berkeley*, LBL-11118 (1980), p. 490.

(Bec 73) R. Beck and D. H. E. Gross, *Phys. Lett.* **47B**:143 (1973).

(Bec 76) F. Beck, *Phys. Lett.* **62B**:385 (1976).

(Bec 77) M. Beckerman and M. Blann, *Phys. Rev. Lett.* **38**:272 (1977).

(Bec 78) F. Beck, K.-H. Müller, and H. S. Köhler, *Phys. Rev. Lett.* **40**:837 (1978).

(Bec 78b) F. Beck, J. Błocki, M. Dworzecka, and G. Wolschin, *Phys. Lett.* **76B**:35 (1978).

(Bec 78c) M. Beckerman and M. Blann, *Phys. Rev. C* **17**:1615 (1978).

(Bec 78d) F. Beck, M. Dworzecka, and H. Feldmeier, *Z. Phys.* **A289**:113 (1978).

(Bec 78e) M. Beckerman, *Phys. Lett.* **78B**:17 (1978).

(Bec 80) F. D. Becchetti, K. T. Hecht, J. Janecke, D. Overway, and G. Kekelis, *Phys. Rev. C* **21**:444 (1980).

(Ber 75) G. F. Bertsch, *Phys. Rev. Lett.* **34**:697 (1975).

(Ber 77) M. Berlanger, F. Hanappe, C. Ngô, J. Péter, F. Plasil, and B. Tamain, *Nucl. Phys.* **A276**:347 (1977).

(Ber 78b) G. Bertsch and D. Mundinger, *Phys. Rev. C* **17**:1646 (1978).

(Ber 78d) M. Berlanger, C. Ngô, P. Grange, J. Richert, and H. Hofmann, *Z. Phys.* **A284**:61 (1978).

(Ber 78e) M. Berlanger, P. Grange, H. Hofmann, C. Ngô, and J. Richert, *Phys. Rev. C* **17**:1495 (1978).

(Ber 78f) M. Berlanger, P. Grange, H. Hofmann, C. Ngô, and J. Richert, *Z. Phys.* **A286**:207 (1978).

(Ber 78g) G. Bertsch, *Z. Phys.* **A289**:103 (1978).

(Ber 79c) M. Berlanger, A. Gobbi, F. Hanappe, U. Lynen, C. Ngô, A. Olmi, H. Sann, H. Stelzer, H. Richel, and M. F. Rivet, *Z. Phys.* **A291**:133 (1979).

(Ber 82) G. F. Bertsch, MSUCL-385 preprint, October (1982).

(Bet 38) H. A. Bethe, *Phys. Rev.* **53**:675 (1938).

(Bet 79c) R. R. Betts, *Nucl. Instrum. Methods* **162**:531 (1979).

(Bet 81) R. R. Betts, H.-G. Clerc, B. B. Back, I. Ahmad, K. L. Wolf, and B. G. Glagola, *Phys. Rev. Lett.* **46**:313 (1981).

(Bho 78) R. K. Bhowmik, E. C. Pollacco, N. E. Sanderson, J. B. A. England, and G. C. Morrison, *Phys. Lett.* **80B**:41 (1978).

(Bho 79) R. K. Bhowmik, E. C. Pollacco, N. E. Sanderson, J. B. A. England, and G. C. Morrison, *Phys. Rev. Lett.* **43**:619 (1979).

(Bil 78) R. Billerey, C. Cerruti, A. Chevarier, N. Chevarier, and A. Demeyer, *Z. Phys.* **A284**:319 (1978).

(Bil 79) R. Billerey, C. Cerruti, A. Chevarier, N. Chevarier, B. Cheynis, and A. Demeyer, *Z. Phys.* **A292**:293 (1979).

(Bil 80b) R. Billerey, C. Cerruti, A. Chevarier, N. Chevarier, B. Cheynis, A. Demeyer, and O. Zohni, *Z. Phys.* **A297**:317 (1980).

(Bim 74) R. Bimbot, D. Gardes, R. L. Hahn, Y. De Moras, and M. F. Rivet, *Nucl. Phys.* **A228**:85 (1974).

(Bim 75) R. Bimbot, D. Gardes, R. R. Hahn, Y. De Moras, and M. F. Rivet, *Nucl. Phys.* **A248**:377 (1975).

(Bim 79) R. Bimbot, M. F. Rivet. J. Girard, R. Lucas, and J. Poitou, *Nucl. Phys.* **A318**:215 (1979).

(Bir 76) J. R. Birkelund, J. R. Huizenga, H. Freiesleben, K. L. Wolf, J. P. Unik, and V. E. Viola, Jr., *Phys. Rev. C* **13**:133 (1976).

(Bir 77) J. R. Birkelund, J. R. Huizenga, *Proc. Symp. on Heavy Ion Elastic Scattering*, Rochester, University of Rochester, (R. Devries, ed.) (1977), p. 210.

(Bir 78) J. R. Birkelund and J. R. Huizenga, *Phys. Rev. C* **17**:126 (1978).

(Bir 78b) J. R. Birkelund, J. R. Huizenga, J. N. De, and D. Sperber, *Phys. Rev. Lett.* **40**:1123 (1978).

(Bir 79) J. R. Birkelund, L. E. Tubbs, J. R. Huizenga, J. N. De, and D. Sperber, *Phys. Rep.* **56**:107 (1979).

(Bir 82) T. S. Biro and J. Zimanyi, *Phys. Lett.* **113B**:6 (1982).

(Bir 82b) J. R. Birkelund, H. Freiesleben, J. R. Huizenga, W. U. Schröder, W. W. Wilcke, K. L. Wolf, J. P. Unik, and V. E. Viola, Jr., *Phys. Rev. C* **26**:1984 (1982).

(Bir 83) J. R. Birkelund, A. D. Hoover, J. R. Huizenga, W. U. Schröder, W. W. Wilcke, *Phys. Rev. C* **27**:882 (1983).

(Bir 83b) J. R. Birkelund and J. R. Huizenga, *Ann. Rev. Nucl. Part. Sci.* **33**:265 (1983).

(Bis 78) J. Bisplinghoff, P. David, M. Blann, W. Scobel, T. Mayer-Kuckuk, J. Ernst, and A. Mignerey, *Phys. Rev. C* **17**:177 (1978).

(Bjø 82) S. Bjørnholm, *Comments Nucl. Part. Phys.* **11**:9 (1982).

(Bjø 82b) S. Bjørnholm and W. J. Swiatecki, *Nucl. Phys.* **A391**:471 (1982).

(Bla 52) J. M. Blatt and V. F. Weiskopf, *Nuclear Physics*, Wiley, New York (1952).

(Bla 54) J. S. Blair, *Phys. Rev.* **95**:1218 (1954).

(Bla 61) H. M. Blann, *Phys. Rev.* **123**:1356 (1961).

(Bla 75) M. Blann, *Ann. Rev. Nucl. Sci.* **25**:123 (1975).

(Bla 76) M. Blann, A. Mignerey, and W. Scobel, *Nukleonika* **21**:335 (1976).

(Bla 79) M. Blann, R. L. Ferguson, *Phys. Rev. C* **19**:289 (1979).

(Bla 79b) M. Blann, R. H. Stokes, B. Erkkila, H. C. Britt, P. D. Goldstone, R. L. Ferguson, F. Plasil, and H. H. Gutbrod, *Phys. Rev. C* **19**:1288 (1979).

(Bla 81) M. Blann, *Phys. Rev. C* **23**:205 (1981).

(Bla 82) M. Blann, D. Akers, T. A. Komoto, F. S. Dietrich, L. F. Hansen, J. G. Woodworth, W. Scobel, J. Bisplinghof, B. Sikora, F. Plasil, and R. L. Ferguson, *Phys. Rev. C* **26**:1471 (1982).

(Bla 82b) M. Blann and D. Akers, *Phys. Rev. C* **26**:465 (1982).

(Bło 77) J. Błocki, J. Randrup, W. J. Swiatecki, and C. F. Tsang, *Ann. Phys. (N. Y.)* **105**:427 (1977).

(Bło 78) J. Błocki, Y. Boneh, J. R. Nix, J. Randrup, M. Robel, A. J. Sierk, and W. J. Swiatecki, *Ann. Phys. (N. Y.)* **113**:330 (1978).

(Bło 80) J. Błocki, F. Beck, M. Dworzecka, and H. Feldmeier, *Proc. Int. Workshop on Gross Properties of Nuclei and Nuclear Excitations*, VIII (1980), p. 157.

(Bło 80b) J. Błocki, F. Beck, and M. Dworzecka, ORO 5126-84 Univ. of Maryland Technical Report No. 80-029 (1980).

(Bło 80c) J. Błocki, M. Dworzecka, F. Beck, and H. Feldemeier, Report IKDA 80/9 Darmstadt (1980).

(Bło 81b) J. Błocki and W. J. Swiatecki, *Ann. Phys. (N. Y.)* **132**:53 (1981).

(Boc 82) R. Bock, Y. T. Chu, M. Dakowski, A. Gobbi, E. Grosse, A. Olmi, H. Sann, D. Schwalm, U. Lynen, W. Muller, S. Bjørnholm, H. Esbensen, W. Wolfli, and E. Morenzoni, *Nucl. Phys.* **A388**:334 (1982).

(Boh 39) N. Bohr and J. Archibald, *Phys. Rev.* **56**:426 (1939).

(Boh 56) A. Bohr, *Proc. U. N. International Conference on the Peaceful Uses of Atomic Energy*, Paper P/911, U. N. (N. Y.) (1956), Vol. 2, p. 151.

(Boh 69) A. Bohr and B. R. Mottelson, *Nuclear Structure*, W. A. Benjamin, Inc., New York (1969), Vol. 1.

(Boh 75) A. Bohr and B. R. Mottelson, *Nuclear Structure*, W. A. Benjamin, Inc., New

York (1975), Vol. 2.

(Boh 80) W. Bohne, K. Grabisch, H. Morgenstern, and W. Stoffler, Int. Workshop on Heavy Ion Fusion Reactions, Bad Honnef, 1980.

(Bon 71) J. P. Bondorf, F. Dickmann, D. H. E. Gross, and P. J. Siemens, *J. Phys., Colloq. C6* **32**:145 (1971).

(Bon 73) J. P. Bondorf and W. Nörenberg, *Phys. Lett.* **44B**:487 (1973).

(Bon 74) J. P. Bondorf, *Proc. Int. Conf. on Reactions Between Complex Nuclei*, North-Holland, Amsterdam (1974), p. 383.

(Bon 74b) J. P. Bondorf, M. I. Sobel, and D. Sperber, *Phys. Rep.* **15**:83 (1974).

(Bon 75) J. P. Bondorf, J. R. Huizenga, M. I. Sobel, and D. Sperber, *Phys. Rev. C* **11**:1265 (1975).

(Bon 75b) J. P. Bondorf, Lecture Notes, Varenna, 1975.

(Bon 76) Y. Boneh, J. P. Błocki, and W. D. Myers, *Phys. Lett.* **63B**:265 (1976).

(Bon 76b) J. P. Bondorf, *J. Phys. Colloq. C5* **37**:195 (1976).

(Bon 76h) P. D. Bond, C. Chasman, and A. Z. Schwarzschild, *Proc. Symposium Macroscopic Features of Heavy-Ion Collisions*, Argonne, 1976, Report ANL/PHY-76-2, Vol. I, p. 475.

(Bon 78b) P. Bonche, B. Grammaticos, and S. Koonin, *Phys. Rev. C* **17**:1700 (1978).

(Bon 79) J. P. Bondorf, J. N. De, A. O. T. Karvinen, G. Fai, and B. Jakobsson, *Phys. Lett.* **84B**:162 (1979).

(Bon 79c) Y. Boneh, A. Gavron, and S. Wald, *Proceedings of Hirschegg Workshop VII*, (1979), p. 170.

(Bon 80) J. P. Bondorf, J. N. De, G. Fai, A. O. T. Karvinen, B. Jakobsson, and J. Randrup, *Nucl. Phys.* **A333**:285 (1980).

(Bon 84) P. D. Bond, *Phys. Rev. Lett.* **52**:414 (1984).

(Bor 74) B. Borderie, F. Hanappe, C. Ngô, J. Péter, and B. Tamain, *Nucl. Phys.* **A220**:93 (1974).

(Bor 81b) B. Borderie, M. Berlanger, D. Gardes, F. Hanappe, L. Nowicki, J. Peter, B. Tamain, S. Agarwal, J. Girard, C. Gregoire, J. Matuszek, and C. Ngô, *Z. Phys.* **A299**:263 (1981).

(Bra 72) M. Brack, J. Damgaard, A. S. Jensen, H. C. Pauli, V. M. Strutinsky, C. Y. Wong, *Rev. Mod. Phys.* **44**:320 (1972).

(Bra 73) W. J. Braithwaite and J. G. Cramer, *Phys. Rev. Lett.* **31**:401 (1973).

(Bra 76) P. Braun-Munzinger, C. K. Gelbke, J. Barrette, B. Zeidman, M. J. Le Vine, A. Gamp, H. L. Harney, and T. Walcher, *Phys. Rev. Lett.* **36**:849 (1976).

(Bra 76b) P. Braun-Munzinger, T. M. Cormier, and C. K. Gelbke, *Phys. Rev. Lett.* **37**:1582 (1976).

(Bra 78) P. Braun-Munzinger and J. Barrette, *Nucl. Phys.* **A299**:161 (1978).

(Bra 78c) K. Braune, Diplomarbeit MPI, Heidelberg (1978), unpublished.

(Bre 56b) G. Breit and M. E. Ebel, *Phys. Rev.* **103**:679 (1956).

(Bre 79) H. Breuer, B. G. Glagola, V. E. Viola, K. L. Wolf, A. C. Mignerey, J. R. Birkelund, D. Hilscher, A. D. Hoover, J. R. Huizenga, W. U. Schröder, and W. W. Wilcke, *Phys. Rev. Lett.* **43**:191 (1979).

(Bre 80) H. Breuer, K. L. Wolf, B. G. Glagola, K. K. Kwiatkowski, A. C. Mignerey, V. E. Viola, W. W. Wilcke, W. U. Schröder, J. R. Huizenga, D. Hilscher, and J. Birkelund, *Phys. Rev. C* **22**:2454 (1980).

(Bre 81) H. Breuer, A. C. Mignerey, V. E. Viola, K. L. Wolf, J. R. Birkelund, D. Hilscher, J. R. Huizenga, W. U. Schröder, and W. W. Wilcke, *Proc. of the IXth International Workshop on Gross Properties of Nuclei and Nuclear Excitations*, Hirschegg, Austria (1981), p. 162.

(Bre 83) H. Breuer, A. C. Mignerey, V. E. Viola, K. L. Wolf, J. R. Birkelund, D.

Hilscher, J. R. Huizenga, W. U. Schröder, and W. W. Wilcke, *Phys. Rev. C* **28**:1080 (1983).
(Bri 60) H. C. Britt and A. R. Quinton, *Phys. Rev.* **120**:1768 (1960).
(Bri 61) H. C. Britt and A. R. Quinton, *Phys. Rev.* **124**:877 (1961).
(Bri 72) D. M. Brink, *Phys. Lett.* **40B**:37 (1972).
(Bri 74) D. M. Brink and N. Rowley, *Nucl Phys.* **A219**:79 (1974).
(Bri 75) D. M. Brink and F. Stancu, *Nucl. Phys.* **A243**:175 (1975).
(Bri 76) H. C. Britt, B. H. Erkkila, R. H. Stokes, H. H. Gutbrod, F. Plasil, R. L. Ferguson, and M. Blann, *Phys. Rev. C* **13**:1483 (1976).
(Bri 78) D. M. Brink and F. Stancu, *Nucl. Phys.* **A299**:321 (1978).
(Bri 79) D. M. Brink, J. Neto, and H. A. Weidenmüller, *Phys. Lett.* **80B**:170 (1979).
(Bri 82) H. C. Britt, B. H. Erkkila, A. Gavron, Y. Patin, R. H. Stokes, M. P. Webb, P. R. Christensen, O. Hansen, S. Pontoppidan, F. Videbaek, R. L. Ferguson, F. Plasil, G. R. Young, and J. Randrup, *Phys. Rev. C* **26**:1999 (1982).
(Bro 72) R. A. Broglia and A. Winther, *Phys. Rep.* **4**:153 (1972.)
(Bro 72b) R. A. Broglia, S. Landowne, and A. Winther, *Phys. Lett.* **40B**:293 (1972).
(Bro 72c) R. A. Broglia and A. Winther, *Nucl. Phys.* **A182**:112 (1972).
(Bro 74) R. A. Broglia, C. H. Dasso, and A. Winther, *Phys. Lett.* **53B**:301 (1974).
(Bro 76) R. A. Broglia, C. H. Dasso, and A. Winther, *Phys. Lett.* **61B**:113 (1976).
(Bro 76c) R. A. Broglia, C. H. Dasso, G. Pollarolo, and A. Winther, *Proc. Symp. in Hakone*, Japan (1977), p. 251.
(Bro 78) U. Brosa and H. J. Krappe, *Z. Phys.* **A284**:65 (1978).
(Bro 78c) R. A. Broglia, O. Civitarese, C. H. Dasso, and A. Winther, *Phys. Lett.* **73B**:405 (1978).
(Bro 78e) R. A. Broglia, C. H. Dasso, G. Pollarola, and A. Winther, *Phys. Rev. Lett.* **41**:25 (1978).
(Bro 79) U. Brosa and H. J. Krappe, *Nukleonika* **24**:389 (1979).
(Bro 79b) U. Brosa, *Z. Phy.* **A292**:385 (1979).
(Bro 79e) R. A. Broglia, G. Pollarolo, C. H. Dasso, and T. Døssing, *Phys. Rev. Lett.* **43**:1649 (1979).
(Bro 80b) U. Brosa and D. H. E. Gross, *Z. Phys.* **A294**:217 (1980).
(Bro 80c) U. Brosa, *Z. Phys.* **A298**:77 (1980).
(Bro 80f) U. Brosa and D. H. E. Gross, *Z. Phys.* **A298**:91 (1980).
(Bro 81e) U. Brosa and S. Grossman, Fourteenth Summer School On Nuclear Physics, Mikolajki, Poland (1981).
(Bru 68) K. A. Brueckner, J. R. Buchler, and M. M. Kelly, *Phys. Rev.* **173**:944 (1968).
(Buc 79) B. Buck, L. C. Biedenharn, and R. Y. Cusson, *Nucl. Phys.* **A317**:205 (1979).
(Bus 57) U. L. Businaro and S. Gallone, *Nuovo Cimento* **5**:315 (1957).
(Bus 79) F. Busch, M. J. Canty, W. Pfeffer, B. Kohlmeyer, W. Schafer, W. F. W. Schneider, H. Freiesleben, and F. Pühlhofer, *Z. Phys.* **A290**:167 (1979).
(But 79) P. J. A. Buttle and L. J. B. Goldfarb, *Nucl. Phys.* **A176**:299 (1971).
(Cab 75b) C. Cabot, C. Ngô, J. Péter, and B. Tamain, *Nucl. Phys.* **A244**:134 (1975).
(Cab 75c) C. Cabot, H. Gauvin, Y. Le Beyec, and M. Lefort, *J. Phys. (Paris)* **36**:L289 (1975).
(Cas 79) W. Cassing and H. Friedrich, *Z. Phys.* **A291**:339 (1979).
(Cat 80) G. L. Catchen, M. Kaplan, J. M. Alexander, and M. F. Rivet, *Phys. Rev. C* **21**:940 (1980).
(Cau 78) B. Cauvin, R. P. Schmitt, G. J. Wozniak, P. Glässel, P. Russo, R. C. Jared, J. B. Moulton, and L. G. Moretto, *Nucl. Phys.* **A294**:225 (1978).
(Cau 78b) B. Cauvin, R. C. Jared, P. Russo, R. P. Schmitt, R. Babinett, and L. G. Moretto, *Nucl. Phys.* **A301**:511 (1978).

(Cha 43) S. Chandrasekhar, *Rev. Mod. Phys.* **15**:1 (1943).

(Che 70) E. Cheifetz, Z. Fraenkel, J. Galin, M. Lefort, J. Peter, and X. Tarrago, *Phys. Rev. C* **2**:256 (1970).

(Che 81) J. Cheng-Lie, P. R. Christensen, O. Hansen, S. Pontoppidan, F. Videbaek, D. Schull, and S. Wen-Qing, *Phys. Rev. Lett.* **47**:1039 (1981).

(Chi 79) T. H. Chiang, D. Guerreau, P. Auger, J. Galin, B. Gatty, X. Tarrago, and J. Girard, *Phys. Rev. C* **20**:1408 (1979).

(Chr 76) P. R. Christensen and A. Winther, *Phys. Lett.* **65B**:19 (1976).

(Chr 78) P. R. Christensen, F. Folkmann, O. Hansen, O. Nathan, N. Trautner, F. Videbaek, S. Y. Van Der Werf, H. C. Britt, R. P. Chestnut, H. Freiesleben, and F. Pühlhofer, *Phys. Rev. Lett.* **40**:1245 (1978).

(Chr 80) P. R. Christensen, F. Folkmann, O. Hansen, O. Nathan, N. Trautner, F. Videbaek, S. Y. Vander Werf, H. C. Britt, R. P. Chestnut, H. Freiesleben, and F. Pühlhofer, *Nucl. Phys.* **A349**:217 (1980).

(Clo 78) M. R. Clover and O. Zohni, *Phys. Lett.* **72B**:427 (1978).

(Cof 78) J. P. Coffin, P. Engelstein, A. Gallmann, B. Heusch, P. Wagner, and H. E. Wegner, *Phys. Rev. C* **17**:1607 (1978).

(Coh 74) S. Cohen, F. Plasil, and W. J. Swiatecki, *Ann. Phys. (N. Y.)* **82**:557 (1974).

(Col 74) P. Colombani, J. C. Jacmart, N. Poffe, M. Riou, C. Stephan, P. P. Singh, and A. Weidinger, *Phys. Lett* **48B**:315 (1974).

(Col 74b) P. Colombani, *Int. School of Phys. Enrico Fermi LXII Course* (1974).

(Col 74c) P. Colombani, *J. Phys. (Paris) Colloq. C5* **35**:75 (1974).

(Col 75) P. Colombani, N. Frascaria, J. C. Jacmart, M. Riou, C. Stephan, H. Doubre, N. Poffe, and J. C. Roynette, *Phys. Lett.* **55B**:45 (1975).

(Col 80) A. J. Cole, N. Longequeue, J. Menet, J. J. Lucas, R. Ost, and J. B. Viano, *Nucl. Phys.* **A341**:284 (1980).

(Col 80b) M. T. Collins, and J. J. Griffin, *Nucl. Phys.* **A348**:63 (1980).

(Con 78) M. Conjeaud, S. Gary, S. Harar, and J. P. Wielecko, *Nucl. Phys.* **A309**:515 (1978).

(Cor 74) T. M. Cormier, R. S. Galik, E. R. Cosman, and A. J. Lazzarini, *Nucl. Instrum. Methods* **119**:145 (1974).

(Cor 76) T. M. Cormier, A. J. Lazzarini, M. A. Neuhausen, A. Sperduto, K. Van Bibber, F. Videbaek, G. Young, E. B. Blum, L. Herreid, and W. Thoms, *Phys. Rev. C* **13**:682 (1976).

(Cor 76b) T. M. Cormier, *Proc. Symposium Macroscopic Features of Heavy-Ion Collisions*, Argonne, 1976, Report ANL/PHY-76-2, p. 153.

(Cor 76c) T. M. Cormier, E. R. Cosman, A. J. Lassarini, J. D. Garrett, and H. E. Wegner, *Phys. Rev. C* **14**:334 (1976).

(Cor 77) T. M. Cormier, E. R. Cosman, A. J. Lazzarini, H. E. Wegner, J. D. Garrett, and F. Pühlhofer, *Phys. Rev. C* **15**:654 (1977).

(Cor 77c) T. M. Cormier, P. Braun-Munzinger, P. M. Cormier, J. W. Harris, and L. L. Lee, Jr., *Phys. Rev. C* **16**:215 (1977).

(Cro 80) B. J. B. Crowley, *Phys. Rep.* **57**:47 (1980).

(Dak 79) M. Dakowski, P. Doll, A. Gobbi, U. Lynen, A. Olmi, G. Rudolf, H. Sann, and R. Bock, *Proc. Int. Symp. on Continuum Spectra*, held in San Antonio, Texas, December, 1979 (T. Tamura, J. B. Natowitz, and D. H. Youngblood, eds.), Harwood Academic Publishers CHUR, London, New York (1980), Vol. 2, p. 1.

(Dak 80) M. Dakowski, P. Doll, A. Gobbi, G. Rudolf, H. Sann, R. Bock, U. Lynen, and A. Olmi, *Phys. Lett.* **90B**:379 (1980).

(Dak 80b) M. Dakowski, R. Chechik, H. Fuchs, F. Hanappe, B. Lucas, C. Mazur, M. Morjean, J. Péter, M. Ribrag, C. Signarbieux, and B. Tamain, *Z. Phys.*

A294:289 (1980).

(Dak 82) M. Dakowski, A. Gobbi, and W. Nörenberg, *Nucl. Phys.* **A378**:189 (1982).

(Das 82) C. H. Dasso, T. S. Dumitrescu, and A. Vitturi, *Nucl. Phys.* **A378**:100 (1982).

(Dav 75) R. H. Davis, *International Symposium on Cluster Structure of Nuclei and Transfer Reactions Induced by Heavy Ions*, IKS-IPCR, Tokyo (1975).

(Dav 76) K. T. R. Davies, A. J. Sierk, and J. R. Nix, *Phys. Rev. C* **13**:2385 (1976).

(Dav 77b) K. T. R. Davies, R. A. Managan, J. R. Nix, and A. J. Sierk, *Phys. Rev. C* **16**:1890 (1977).

(Day 77) R. Dayras, Z. E. Switkowski, and S. E. Woosley, *Nucl. Phys.* **A279**:70 (1977).

(Day 80) R. A. Dayras, R. G. Stokstad, D. C. Hensley, M. L. Halbert, D. G. Sarantites, L. Westerberg, and J. H. Barker, *Phys. Rev. C* **22**:1485 (1980).

(De 76) J. N. De, D. H. E. Gross, and H. Kalinowski, *Z. Phys.* **A277**:385 (1976).

(De 77) J. N. De, *Phys. Lett.* **66B**:315 (1977).

(De 78) J. N. De and D. Sperber, *Phys. Lett.* **72B**:293 (1978).

(De 78b) J. N. De, S. I. A. Garpman, D. Sperber, and K. Tam, *Phys. Lett.* **76B**:39 (1978).

(De 78c) J. N. De, S. I. A. Garpman, A. Sherman, D. Sperber, and K. Tam, *Phys. Lett.* **78B**:13 (1978).

(De 83) J. N. De and S. S. Kapoor, *Phys. Rev. C* **27**:1328.

(Deb 78) J. Deboer, C. Lauterbach, W. Dünnweber, G. Graw, H. Hering, H. Puchta, W. Trautmann, and U. Lynen, *Nuclear Interactions*, Canberra, 1977, (B. A. Robson, ed.) Berlin (1978), p. 478.

(Del 78b) M. A. Deleplanque, Th. Byrski, R. M. Diamond, H. Hubel, F. S. Stephens, B. Herskind, and R. Bauer, *Phys. Rev. Lett.* **41**:1105 (1978).

(Del 79) H. Delagrange, D. Logan, M. F. Rivet, M. Rajagopalan, J. M. Alexander, M. S. Zisman, M. Kaplan, and J. W. Ball, *Phys. Rev. Lett.* **43**:1490 (1979).

(Deu 75) H. H. Deubler and K. Dietrich, *Phys. Lett.* **56B**:241 (1975).

(Deu 75b) H. H. Deubler and T. Fliessbach, *Nucl. Phys.* **A238**:409 (1975).

(Deu 76) H. H. Deubler and K. Dietrich, *Phys. Lett.* **62B**:369 (1976).

(Deu 77) H. H. Deubler and K. Dietrich, *Phys. Lett.* **69B**:137 (1977).

(Deu 77b) H. H. Deubler and K. Dietrich, *Nucl. Phys.* **A277**:493 (1977).

(Deu 78) H. H. Deubler, K. Lekkas, P. Sperr, and K. Dietrich, *Z. Phys.* **A284**:237 (1978).

(Dev 75) R. M. Devries and M. R. Clover, *Nucl. Phys.* **A243**:528 (1975).

(Dia 80) R. M. Diamond and F. S. Stephens, *Ann. Rev. Nucl. Part. Sci.* **30**:85 (1980).

(Die 73) K. Dietrich and K. Hara, *Nucl. Phys.* **A211**:349 (1973).

(Die 76) K. Dietrich, *Nukleonika* **21**:89 (1976).

(Die 77) K. Dietrich and C. Leclercq-Willain, *Ann. Phys. (N. Y.)* **109**:41 (1977).

(Die 78) K. Dietrich and M. Zielinska-Pfabe, preprint, Technische Universität München (1978).

(Die 80) M. Diebel, D. Glas, U. Mosel, and H. Chandra, *Nucl. Phys.* **A333**:253 (1980).

(Die 81b) K. Dietrich, *Nucl. Phys.* **A354**:319C (1981).

(Die 81c) K. Dietrich and Ch. Leclerq-Willain, *Nucl. Phys.* **A359**:201 (1981).

(Don 79) R. Donangelo, L. F. Canto, and M. S. Hussein, *Nucl. Phys.* **A320**:422 (1979).

(Dor 81) C. O. Dorso and E. S. Hernandez, *Nucl. Phys.* **A372**:215 (1981).

(Døs 81) T. Døssing, *Nucl. Phys* **A357**:488 (1981).

(Dou 78) H. Doubre, A. Gamp, J. C. Jacmart, N. Poffe, J. C. Roynette, and J. Wilczyński, *Phys. Lett.* **73B**:135 (1978).

(Dou 79) G. Doukellis, G. Hlawatsch, B. Kolb, J. B. Natowitz, G. Rosner, B. Sedelmeyer, and T. Walcher, *Proceedings of Hirschegg Workshop VII*, (1978), p. 185.

(Dov 75) C. B. Dover and J. P. Vary, *Proc. Class. Quant. Mech. Aspects Heavy Ion Collisions*, Springer-Verlag, Berlin (1975), Vol. 33, p. 1.

(Dri 81) J. Van Driel, S. Gonggrijp, R. V. F. Janssens, R. H. Siemssen, K. Siwek-Wilczyńska, and J. Wilczyński, *Phys. Lett.* **98B**:351 (1981).

(Dun 80) W. Dünnweber and K. M. Hartmann, *Phys. Rev. Lett.* **44**:729 (1980).

(Dye 77) P. Dyer, R. J. Puigh, R. Vandenbosch, T. D. Thomas, and M. S. Zisman, *Phys. Rev. Lett.* **39**:392 (1977).

(Dye 79) P. Dyer, T. C. Awes, C. K. Gelbke, B. B. Back, A. Mignerey, K. L. Wolf, H. Breuer, V. E. Viola, Jr., and W. G. Meyer, *Phys. Rev. Lett.* **42**:560 (1979).

(Dye 79b) P. Dyer, R. J. Puigh, R. Vandenbosch, T. D. Thomas, M. S. Zisman, and L. Nunnelley, *Nucl. Phys.* **A322**:205 (1979).

(Dye 80) P. Dyer, M. P. Webb, R. J. Puigh, R. Vandenbosch, T. D. Thomas, and M. S. Zisman, *Phys. Rev. C* **22**:1509 (1980).

(Eck 77) J. S. Eck and W. J. Thompson, *Am. J. Phys.* **45**:161 (1977).

(Egg 76) R. Eggers, M. N. Namboodiri, P. Gonthier, K. Geoffroy, and J. B. Natowitz, *Phys. Rev. Lett.* **37**:324 (1976).

(Eri 58) T. Ericson and V. Strutinski, *Nucl. Phys.* **8**:284 (1958).

(Eri 59) T. Ericson and V. Strutinsky, *Nucl. Phys.* **9**:689 (1959).

(Eri 60) T. Ericson, *Adv. Phys.* **9**:425 (1960).

(Esb 78) H. Esbensen, A. Winther, R. A. Broglia, and C. H. Dasso, *Phys. Rev. Lett.* **41**:296 (1978).

(Ess 79) H. Essel, K. Hartel, P. Kienle, H. J. Körner, K. E. Rehm, P. Sperr, and W. Wagner, *Phys. Lett.* **81B**:161 (1979).

(Ess 79c) H. Essel, K. Hartel, W. Henning, P. Kienle, H. J. Körner, K. E. Rehm, P. Sperr, W. Wagner, and H. Spieler, *Z. Phys.* **A289**:265 (1979).

(Est 80) R. A. Esterlund, W. Westmeier, A. M. Habbestad Watzig, U. Reus, W. Watzig, and P. Patzelt, Proc. Int. Conf., Berkeley, LBL-11118 (1980), p. 489.

(Eya 78) Y. Eyal, A. Gavron, I. Tserruya, Z. Fraenkel, Y. Eisen, S. Wald, R. Bass, G. R. Gould, G. Kreyling, R. Renfordt, K. Selzer, R. Zitzmann, A Gobbi, U. Lynen, H. Stelzer, I. Rode, and R. Bock, *Phys. Rev. Lett.* **41**:625 (1978).

(Eya 79) Y. Eyal, G. Rudolf, I. Rode, and H. Stelzer, *Phys. Rev. Lett.* **42**:826 (1979).

(Eya 80) Y. Eyal, A. Gavron, I. Tserruya, Z. Fraenkel, Y. Eisen, S. Wald, R. Bass, C. R. Gould, G. Kreyling, R. Renfordt, K. Selzer, R. Zitzmann, A. Gobbi, U. Lynen, H. Stelzer, I. Rode, and R. Bock, *Phys. Rev. C* **21**:1377 (1980).

(Eya 80b) Y. Eyal, *Phys. Rev. C* **23**:2745 (1981).

(Fab 79b) M. Faber, M. Ploszajczak, and A. Faessler, *Nucl. Phys.* **A326**:129 (1979).

(Fab 80) M. E. Faber, *Z. Phys.* **A297**:277 (1980).

(Fab 81) M. E. Faber, *Phys. Rev. C* **24**:1047 (1981).

(Fel 80) H. Feldmeier, *Nukleonika* **25**:171 (1980).

(Fen 75) M. Fenzl and O. W. B. Schult, *Z. Phys.* **272**:207 (1975).

(Fer 78) B. Fernandez, C. Gaarde, J. S. Larsen, S. Pontoppidan, and F. Vidbaek, *Nucl. Phys.* **A306**:259 (1978).

(Fer 80) R. L. Ferguson, A. Gavron, F. E. Obenshain, F. Plasil, R. L. Robinson, D. Shapira, A. H. Snell, and G. R. Young, Workshop on Nuclear Dynamics, LBL-10688 (1980), p. 21.

(Fey 81) R. P. Feynman, *Quantum Statistical Mechanics*, Frontiers in Physics, L. P. Kadanoff and G. Baym, eds. Benjamin/Cummins, London, Amsterdam, Don Mills, Ontario, Sydney, Tokyo (1981).

(Fin 74) H. J. Fink, W. Greiner, R. K. Gupta, S. Liran, H. J. Maruhn, W. Scheid, and O. Zohni, *Proc. Int. Conf. on Reactions Between Complex Nuclei*, North-Holland, Amsterdam (1974), p. 21.

(Fin 74b) H. J. Fink, J. Maruhn, W. Scheid, and W. Greiner, *Z. Phys.* **268**:321 (1974).

(Fle 74b) A. Fleury and J. M. Alexander, *Ann. Rev. Nucl. Sci.* **24**:279 (1974).

(Flo 78) H. Flocard, S. E. Koonin, and M. S. Weiss, *Phys. Rev. C* **17**:1682 (1978).

(Flo 78b) H. Flocard and M. S. Weiss, *Phys. Rev. C* **18**:573 (1978).

(Fra 63) W. E. Frahn and R. H. Venter, *Ann. Phys. (N. Y.)* **24**:243 (1963).

(Fra 64) W. E. Frahn and R. H. Venter, *Nucl. Phys.* **59**:651 (1964).

(Fra 66) W. E. Frahn, *Nucl. Phys.* **75**:577 (1966).

(Fra 69) W. E. Frahn and M. A. Sharaf, *Nucl. Phys.* **A133**:593 (1979).

(Fra 71) W. E. Frahn, *Phys. Rev. Lett.* **26**:568 (1971).

(Fra 72) W. E. Frahn, *Ann. Phys. (N. Y.)* **72**:524 (1972).

(Fra 74) W. E. Frahn, *Proc. Sym. on Classical and Quantum Mechanical Aspects of Heavy-Ion Collisions*, Springer-Verlag, Berlin (1974), Vol. 33, p. 102.

(Fra 75) Z. Fraenkel and I. Mayk, *Phys. Rev. C* **12**:1809 (1975).

(Fra 76c) W. E. Frahn and D. M. E. Gross, *Ann. Phys. (N. Y.)* **101**:520 (1976).

(Fra 77) N. Frascaria, C. Stephan, P. Colombani, J. P. Garron, J. C. Jacmart, M. Riou, and L. Tassan-Got, *Phys. Rev. Lett.* **39**:918 (1977).

(Fra 78) V. Franco and G. K. Varma, *Phys. Rev. C* **18**:349 (1978).

(Fra 78b) W. E. Frahn, *Nucl. Phys.* **A302**:281 (1978).

(Fra 78c) W. E. Frahn, *Nucl. Phys.* **A302**:301 (1978).

(Fra 78h) C. B. Franklyn, C. Hofmeyer, and D. W. Mingay, *Phys. Lett.* **78B**:564 (1978).

(Fra 79c) G. Franz, J. V. Kratz, W. Bruchle, H. Folger, and B. Haefner, *Z. Phys.* **A291**:167 (1979).

(Fra 80) N. Frascaria, P. Colombani, A. Gamp, J. P. Garron, M. Riou, J. C. Roynette, C. Stephan, A. Ameaume, C. Bizard, J. L. Laville, and M. Louvel, *Z. Phys.* **A294**:167 (1980).

(Fra 80b) W. E. Frahn and M. S. Hussein, *Phys. Lett.* **90B**:358 (1980).

(Fra 80f) N. Frascaria, J. C. Roynette, Y. Blumenfeld, J. C. Jacmart, A. Gamp, H. Fuchs, and W. Von Oertzen, *Proc. Int. Conf., Berkeley*, LBL-11118 (1980), p. 599.

(Fre 75) H. Freiesleben, G. T. Rizzo, and J. R. Huizenga, *Phys. Rev. C* **12**:42 (1975).

(Fre 79b) H. Freiesleben, K. D. Hildenbrand, F. Pühlhofer, W. F. W. Schneider, R. Bock, D. V. Harrach, and H. J. Specht, *Z. Phys.* **A292**:171 (1979).

(Fri 81) W. A. Friedman, *Phys. Lett.* **98B**:21 (1981).

(Fro 76) H. Fröhlich, P. Duck, W. Galster, W. Treu, H. Voit, H. Witt, W. Kuhn, and S. M. Lee, *Phys. Lett.* **64B**:408 (1976).

(Fro 78) P. Fröbrich and P. Kaufmann, *Nucl. Phys.* **A307**:297 (1978).

(Fuk 80) T. Fukuda, M. Tanaka, H. Ogata, I. Miura, M. Inoue, T. Yamagata, S. Nakayama, H. Kamitsubo, and M. Ishihara, *Proc. Int. Conf., Berkeley*, LBL-11118 (1980), p. 480.

(Gad 81) E. Gadioli and E. Gadioli Erba, *Z. Phys.* **A299**:1 (1981).

(Gah 77) U. Gahde and W. Stocker, *Nucl. Phys.* **A278**:177 (1977).

(Gal 70) J. Galin, D. Guerreau, M. Lefort, J. Péter, X. Tarrago, and R. Basile, *Nucl. Phys.* **A159**:461 (1970).

(Gal 74b) J. Galin, B. Gatty, D. Guerreau, C. Rousset, U. C. Schlotthauer-Voos, and X. Tarrago, *Phys. Rev. C* **9**:1113 (1974).

(Gal 74c) J. Galin, B. Gatty, D. Guerreau, C. Rousset, U. C. Schlotthauer-Voos, and X. Tarrago, *Phys. Rev. C* **9**:1126 (1974).

(Gal 74d) J. Galin, B. Gatty, D. Guerreau, U. C. Schlotthauer-Voos, and X. Tarrago, *Phys. Rev. C* **10**:638 (1974).

(Gal 75) J. Galin, L. G. Moretto, R. Babinet, R. Schmitt, R. Jared, and S. G. Thompson, *Nucl. Phys.* **A255**:472 (1975).

(Gal 76) J. Galin, *J. Phys. (Paris) Colloq. C5* **37**:83 (1976).

(Gal 76b) J. Galin, B. Gatty, D. Guerreau, M. Lefort, X. Tarrago, R. Babinet B. Cauvin,

J. Girard, and H. Nifenecker, *Z. Phys.* **A278**:347 (1976).

(Gal 77) J. Galin, B. Gatty, D. Guerreau, M. Lefort, X. Tarrago, S. Agarwal, R. Babinet, B. Cauvin, J. Girard, and H. Nifenecker, *Z. Phys.* **A283**:173 (1977).

(Gam 78) A. Gamp, J. C. Jacmart, N. Poffe, H. Doubre, J. C. Roynette, and J. Wilczyński, *Phys. Lett.* **74B**:215 (1978).

(Gam 79) A. Gamp, H. L. Harney, J. C. Roynette, E. Plagnol, H. Fuchs, H. Doubre, J. C. Jacmart, and N. Poffe, *Z.Phys.* **A291**:347 (1979).

(Gam 81) A. Gamp, M. Gurgel, M. R. Clover, C. Egelhaaf, H. Fuchs, B. Gebauer, H. Homeyer, and J. C. Jacmart, *Z. Phys.* **A300**:63 (1981).

(Gam 82) A. Gamp, M. Burgel, M. R. Clover, C. Egelhaaf, H. Fuchs, B. Gebauer, H. Homeyer, W. Raunch, and J. C. Jacmart, *Z. Phys.* **A309**:49 (1982).

(Gar 75) J. D. Garrett, H. E. Wegner, T. M. Cormier, E. R. Cosman, and A. J. Lazzarini, *Phys. Rev. C* **12**:481 (1975).

(Gar 80) S. I. A. Garpman, D. Sperber, and M. Zielinska-Pfabe, *Phys. Lett.* **90B**:53 (1980).

(Gar 80c) D. Gardes, R. Bimbot, J. Maison, and M. F. Rivet, *Phys. Rev. C* **21**:2447 (1980).

(Gat 74) B. Gatty, D. Guerreau, M. Lefort, J. Pouthas, X. Tarrago, J. Galin, B. Cauvin, J. Girard, and H. Nifenecker, *J. Phys. Lett.* **35**:117 (1974).

(Gat 75) B. Gatty, D. Guerreau, M. Lefort, X. Tarrago, J. Galin, B. Cauvin, J. Girard, and H. Nifenecker, *Nucl. Phys.* **A253**:511 (1975).

(Gat 75b) B. Gatty, D. Guerreau, M. Lefort, J. Pouthas, X. Tarrago, J. Galin, B. Cauvin, J. Girard, and H. Nifenecker, *Z. Phys.* **A273**:65 (1975).

(Gau 75) H. Gauvin, D. Guerreau, Y. Le Beyec, M. Lefort, F. Plasil, and X. Tarrago, *Phys. Lett.* **58B**:163 (1975).

(Gav 80) A. Gavron, *Phys. Rev. C* **21**:230 (1980).

(Gav 80c) A. Gavron, J. R. Beene, R. L. Ferguson, F. E. Obenshain, F. Plasil, G. R. Young, G. A. Petitt, K. A. Geoffroy, M. Jaaskelainen, J. G. Sarantites, and C. F. Maguire, *Phys. Rev. C* **24**:2048 (1981).

(Gav 81) A. Gavron, R. L. Ferguson, F. E. Obenshain, F. Plasil, G. R. Young, G. A. Petitt, K. G. Young, D. G. Sarantites, and C. F. Maguire, *Phys. Rev. Lett.* **46**:8 (1981).

(Gav 83) A. Gavron, J. R. Beene, B. Cheynis, R. L. Ferguson, F. E. Obenshain, F. Plasil, G. R. Young, G. A. Petitt, C. F. Maguire, D. G. Sarantites, M. Jaaskelainen, and K. Geoffroy-Young, *Phys. Rev. C* **27**:450 (1983).

(Gel 71) C. K. Gelbke, K. D. Hildenbrand, and R. Bock, *Nucl. Instrum. Methods* **95**:397 (1971).

(Gel 76) C. K. Gelbke, P. Braun-Munzinger, J. Barrette, B. Zeidman, M. J. Levine, A. Gamp, H. L. Harney, and Th. Walcher, *Nucl. Phys.* **A269**:460 (1976).

(Gel 76b) C. K. Gelbke, M. Buenerd, D. L. Hendrie, J. Mahoney, M. C. Mermaz, C. Olmer, and D. K. Scott, *Phys. Lett.* **65**:227 (1976).

(Gel 77) C. K. Gelbke, D. K. Scott, M. Bini, D. L. Hendrie, J. L. Laville, J. Mahoney, M. C. Mermaz, and C. Olmer, *Phys. Lett.* **70B**:415 (1977).

(Gel 77b) C. K. Gelbke, M. Bini, C. Olmer, D. L. Hendrie, J. L. Laville, J. Mahoney, M. C. Mermaz, D. K. Scott, and H. H. Wieman, *Phys. Lett.* **71B**:8 (1977).

(Gel 78) C. K. Gelbke, C. Olmer, M. Buenerd, D. L Hendrie, J. Mahoney, M. C. Mermaz, and D. K. Scott, *Phys. Rep.* **42**:311 (1978).

(Gem 80c) H. Gemmeke, P. Netter, A. Richter, L. Lassen, S. Lewandowski, W. Lucking, and R. Schreck, *Phys. Lett.* **97B**:213 (1980).

(Geo 78) K. A. Geoffroy, J. B. Natowitz, R. C. Eggers, and M. N. Namboodiri, *Nucl.*

Phys. **A302**:310 (1978).
(Geo 78b) K. A. Geoffroy, J. B. Natowitz, R. Eggers, P. Kasiraj, and M. N. Namboodiri, *Nucl. Phys.* **A302**:333 (1978).
(Geo 79) K. A. Geoffroy, D. G. Sarantites, M. L. Halbert, D. C. Hensley, R. A. Dayras, and J. H. Barke, *Phys. Rev. Lett.* **43**:1303 (1979).
(Ger 79) C. Gerschel, M. A. Deleplanque, M. Ishihara, C. Ingo, N. Perrin, J. Péter, B. Tamain, L. Vallentin, D. Paya, Y. Sugiyama, M. Berlanger, and F. Hanappe, *Nucl. Phys.* **A317**:473 (1979).
(Ger 82) C. Gerschel, *Nucl. Phys.* **A387**:297c (1982).
(Gho 81) G. Ghosh and R. W. Hasse, *Nucl. Phys.* **A372**:227 (1981).
(Gla 74) D. Glas and U. Mosel, *Phys. Rev. C* **10**:2620 (1974).
(Gla 74b) D. Glas and U. Mosel, *Phys. Lett.* **49B**:301 (1974).
(Gla 75) D. Glas and U. Mosel, *Nucl. Phys.* **A237**:429 (1975).
(Gla 76) D. Glas and U. Mosel, *Nucl. Phys.* **A264**:268 (1976).
(Gla 77) P. Glässel, R. S. Simon, R. M. Diamond, R. C. Jared, I. Y. Lee, L. G. Moretto, J. O. Newton, R. Schmitt, and F. S. Stephens, *Phys. Rev. Lett.* **38**:331 (1977).
(Gla 79) P. Glässel, D. V. Harrach, Y. Civelekoglu, R. Manner, H. J. Specht, J. B. Wilhelmy, H. Freiesleben, and K. D. Hildenbrand, *Phys. Rev. Lett.* **43**:1483 (1979).
(Gla 81) P. Glässel, D. V. Harrach, L. Grodzins, S. S. Kapoor, and H. J. Specht, Proc. 19TH Int. Winter Meeting on Nuclear Physics, Bormio, Italy (1981).
(Gla 82) P. Glässel, D. Harrach, L. Grodzins, and H. J. Specht, *Phys. Rev. Lett.* **48**:1089 (1982).
(Gle 75) N. K. Glendenning, *Rev. Mod. Phys.* **47**:659 (1975).
(Gle 77) N. K. Glendenning and G. Wolschin, *Nucl. Phys.* **A281**:486 (1977).
(Gnu 81) C. Gnucci and F. Stancu, *Nucl. Phys.* **A366**:520 (1981).
(Gob 80) A. Gobbi, and W. Nörenberg, *Heavy Ion Collisions* (R. Bock, ed.), North-Holland, Amsterdam (1980), Vol. 2, p. 128.
(Gob 81) A. Gobbi, *Nucl. Phys.* **A354**:337C (1981).
(Gol 48) M. Goldhaber and E. Teller, *Phys. Rev.* **74**:1046 (1948).
(Gol 79) L. B. J. Goldfarb, and W. von Oertzen, *Heavy Ion Collisions* (R. Bock, ed.) North-Holland, Amsterdam (1979), Vol. 1, p. 215.
(Gom 79) J. Gomez del Campo, R. G. Stokstad, J. A. Biggerstaff, R. A. Dayras, A. H. Snell, and P. H. Stelson, *Phys. Rev. C* **19**:2170 (1979).
(Gom 79b) J. Gomez del Campo, R. A. Dayras, J. A. Biggerstaff, D. Shapira, A. H. Snell, P. H. Stelson, and R. G. Stokstad, *Phys. Rev. Lett.* **43**:26 (1979).
(Gom 80) J. Gomez del Campo, K. Geoffroy, R. A. Dayras, and R. G. Stokstad, *Proc. Int. Conf., Berkeley*, LBL-11118 (1980), p. 497.
(Gon 80) P. Gonthier. H. Ho, M. N. Namboodiri, L. Adler, J. B. Natowitz, S. Simon, K. Hagel, R. Terry, and A. Khodai, *Phys. Rev. Lett.* **44**:1387 (1980).
(Gor 60) G. E. Gordon, A. E. Larsh, and T. Sikkeland, *Phys. Rev.* **118**:1610 (1960).
(Gor 60b) G. E. Gordon, A. E. Larsh, T. Sikkeland, and G. T. Seaborg, *Phys. Rev.* **120**:1341 (1960).
(Got 77) P.-A. Gottschalk and M. Weström, *Phys. Rev. Lett.* **39**:1250 (1977).
(Got 79) P.-A. Gottschalk and M. Weström, *Nucl. Phys.* **A314**:232 (1979).
(Got 79b) P.-A. Gottschalk, P. Vater, H.-J. Becker, R. Brandt, G. Grawert, G. Fiedler, R. Haag, and T. Rautenberg, *Phys. Rev. Lett.* **42**:359 (1979).
(Gou 78) C. R. Gould, R. Bass, J. V. Czarnecki, V. Hartmann, K. Stelzer, R. Zitzmann, and Y. Eyal, *Z. Phys.* **A284**:353 (1978).
(Gou 80) C. R. Gould, R. Bass, J. V. Czarnecki, V. Hartmann, K. Stelzer, R. Zitzmann,

and Y. Eyal, *Z. Phys.* **A294**:323 (1980).
(Gra 78) G. Graw. J. Deboer, W. Dünnweber, C. Lauterbach, II. Puchta, W. Trautmann, and U. Lynen, *J. Phys. Soc. Jpn., Suppl.* **44**:357 (1978).
(Gré 79) C. Grégoire, R. Ballini, B. Cauvin, J. Fabbro, J. Girard, M. C. Lemaire, C. Mazur, H. Nifenecker, J. C. Peng, and M. Ribrag, *Proc. Hirschegg Workshop VII,* (1979), p. 98.
(Gré 79b) C. Grégoire, R. Lucas, C. Ngô, and H. Hofmann, preprint, Céntres d'Etudes Nucléaires de Saclay (1979).
(Gré 80) C. Grégoire, H. Hofmann, and C. Ngô, Proceedings of Hirschegg Workshop VIII (1980), p. 129.
(Gré 80b) C. Grégoire, H. Hofmann, and C. Ngô, presented at XVIII Int. Winter Meeting on Nuclear Physics, Bormio (1980).
(Gré 81) C. Grégoire, C. Ngô, and B. Remaud, *Phys. Lett.* **99B**:17 (1981).
(Gré 81b) C. Grégoire, R. Lucas, C. Ngô, B. Schürmann, and H. Ngô, *Nucl. Phys.* **A361**:443 (1981).
(Gri 59) J. J. Griffin, *Phys. Rev.* **116**:107 (1959).
(Gri 66) J. J. Griffin, *Phys. Rev. Lett.* **17**:478 (1966).
(Gri 67) J. J. Griffin, *Phys. Lett.* **24B**:5 (1967).
(Gri 70) G. F. Gridnev, V. V. Volkov, and J. Wilczyński, *Nucl. Phys.* **A142**:385 (1970).
(Gri 81) J. J. Griffin, Y. Boneh, K. K. Kan, and M. Dworzecka, *Nucl. Phys.* **A369**:181 (1981).
(Gro 62) J. R. Grover, *Phys. Rev.* **127**:2142 (1962).
(Gro 67d) J. R. Grover and J. Gilat, *Phys. Rev.* **157**:802 (1967).
(Gro 73) D. H. E. Gross and H. Kalinowski, Munich Conference (1973), Vol. 1, p. 394.
(Gro 74) D. H. E. Gross and H. Kalinowski, *Phys. Lett.* **48B**:302 (1974).
(Gro 74b) D. H. E. Gross, H. Kalinowski, and J. N. De, *Proc. Symp. on Classical and Quantum Mechanical Aspects of Heavy-Ion Collisions*, Springer-Verlag, Berlin (1974), Vol. 33, p. 194.
(Gro 75) E. Gross, J. De Boer, R. M. Diamond, F. S. Stephens, and P. Tjom, *Phys. Rev. Lett.* **35**:565 (1975).
(Gro 75b) D. H. E. Gross, *Nucl. Phys.* **A240**:472 (1975).
(Gro 76) D. H. E. Gross, *Nucl. Phys.* **A260**:333 (1976).
(Gro 77) D. H. E. Gross, *Phys. Lett.* **68B**:412 (1977).
(Gro 77b) D. H. E. Gross and J. Wilczyński, *Phys. Lett.* **67B**:1 (1977).
(Gro 77c) D. H. E. Gross and H. Kalinowski preprint, HMI-Berlin (1977).
(Gro 77d) D. H. E. Gross and H. Kalinowski, Invited Talk, 4th Session D'Etudes Biennale De Physique Nucleaire, La Toussuire (1977).
(Gro 78) D. H. E. Gross and H. Kalinowski, *Phys. Rep.* **45**:175 (1978).
(Gro 79) D. H. E. Gross, *Proc. Symp. on Deep-Inelastic and Fusion Reactions with Heavy Ions*, Springer-Verlag, Berlin (1979), Vol. 117, p. 81.
(Gro 79b) D. H. E. Gross, *Z. Phys.* **A291**:145 (1979).
(Gro 80) S. Grossmann, *Z. Phys.* **A296**:251 (1980).
(Gro 80d) D. H. E. Gross, preprint 335, HMI-Berlin (1980).
(Gro 81) D. H. E. Gross, R. C. Nayak, and L. Satpathy, *Z. Phys.* **A299**:63 (1981).
(Gro 81b) D. H. E. Gross and K. M. Hartmann, *Phys. Rev. C* **24**:2526 (1981).
(Gue 80) D. Guerrau, J. Galin, B. Gatty, X. Tarrago, J. Girard, R. Lucas, and C. Ngô, *Z. Phys.* **A295**:105 (1980).
(Gup 77) R. K. Gupta, *Z. Phys.* **A281**:159 (1977).
(Gup 77b) R. K. Gupta, C. Parvulescu, A. Sandulescu, and W. Greiner, *Z. Phys.* **A283**:217 (1977).
(Gup 77c) R. K. Gupta, A. Sandulescu, and W. Greiner, *Phys. Lett.* **67B**:257 (1977).
(Gup 78) R. K. Gupta, C. Pirvulescu, A. Sandulescu, and W. Greiner, *Sov. J. Nucl.*

Phys. **28**:160 (1978).
(Gup 79) R. K. Gupta, preprint, Punjab University, Chandigarh, India (1979).
(Gup 80) R. K. Gupta, *Phys. Rev. C* **21**:1278 (1980).
(Hab 77) D. Habs, V. Metag, J. Schukraft, H. J. Specht, C. O. Wene, and K. D. Hildenbrand, *Z. Phys.* **A283**:261 (1977).
(Hag 75) G. B. Hagemann, R. Broda, B. Herskind, M. Ishihara, S. Ogaza, and H. Ryde, *Nucl. Phys.* **A245**:166 (1975).
(Hah 81) J. Hahn, K. Depta, A. Sandulescu, M. Uhlig, K. H. Ziegenhain, J. A. Maruhn, and W. Greiner, *Proc. of 3rd Adriatic Europhysics Study Conference on Dynamics of Heavy-Ion Collisions*, Hvar, Yugoslavia (N. Andro, R. A. Ricci, and W. Greiner, eds.), North-Holland, Amsterdam (1981), p.169.
(Hak 75) H. Haken, *Rev. Mod. Phys.* **47**:67 (1975).
(Hak 78) H. Haken, *Synergetics*, Springer-Verlag, Berlin (1978).
(Hal 58) I. Halpern and V. M. Strutinski, *Proc. 2nd. U. N. Conference on Peaceful Uses of Atomic Energy*, Paper D/1513 U. N. (Geneva) (1958), Vol. 15, p. 408.
(Hal 68) I. Halpern, B. J. Shepherd, and C. F. Williamson, *Phys. Rev.* **169**:805 (1968).
(Han 74b) F. Hanappe, M. Lefort, C. Ngô, J. Péter, and B. Tamain, *Phys. Rev. Lett.* **32**:738 (1974).
(Han 74c) F. Hanappe, C. Ngô, J. Péter, and B. Tamain, *Rochester Conf. on Physics and Chemistry of Fission*, (1973), p. 289.
(Har 68) G. D. Harp, J. M. Miller, and B. J. Berne, *Phys. Rev.* **165**:1166 (1968).
(Har 77) J. W. Harris, T. M. Cormier, D. F. Geesaman, L. L. Lee, Jr., R. L. McGrath, and J. P. Wurm, *Phys. Rev. Lett.* **38**:1460 (1977).
(Har 78) K. M. Hartmann, *Nucl. Phys.* **A305**:279 (1978).
(Har 79) D. V. Harrach, P. Glässel, Y. Civelekoglu, R. Manner, and H. J. Specht, *Phys. Rev. Lett.* **42**:1728 (1979).
(Har 79b) D. V. Harrach, P. Glässel, Y. Civelekoglu, R. Manner, H. J. Specht, J. B. Wilhelmy, H. Freiesleben, and K. D. Hildenbrand, *Proc. Int. Sym. on Physics and Chemistry of Fission*, IAEA-SM-241 (1979).
(Har 80) K. M. Hartmann, *Z. Phys.* **A294**:65 (1980).
(Har 82) D. V. Harrach, P. Glässel, L. Grodzins, S. S. Kapoor, and H. J. Specht, *Phys. Rev. Lett.* **48**:1093 (1982).
(Has 71b) R. W. Hasse, *Ann. Phys. (N. Y.)* **68**:377 (1971).
(Has 75) R. W. Hasse, *Ann. Phys. (N. Y.)* **93**:68 (1975).
(Has 78) R. W. Hasse, *Rep. Prog. Phys.* **41**:1027 (1978).
(Has 79b) R. W. Hasse, *Phys. Lett.* **85B**:197 (1979).
(Has 80) R. W. Hasse and K. Hahn, *Nukleonika* **25**:133 (1980).
(Hau 52) W. Hauser and H. Feshbach, *Phys. Rev.* **87**:366 (1952).
(Hen 78) W. Henning, Y. Eisen, H. J. Körner, D. G. Kovar, J. P. Schiffer, S. Vigdor, and B. Zeidman, *Phys. Rev. C* **17**:2245 (1978).
(Her 78b) A. Hertz, H. Essel, H. J. Körner, K. E. Rehm, and P. Sperr, *Phys. Rev. C* **18**:2780 (1978).
(Her 80) E. S. Hernandez and S. A. Moszkowski, *Phys. Rev. C* **21**:929 (1980).
(Her 81) E. S. Hernandez, W. D. Myers, J. Randrup and B. Remaud, *Nucl. Phys.* **A361**:483 (1981).
(Heu 78) B. Heusch, C. Volant, H. Freiesleben, R. P. Chestnut, K. D. Hildenbrand, F. Pühlhofer, and W. F. W. Schneider, *Z. Phys.* **A288**:391 (1978).
(Heu 81) B. Heusch, C. Beck, J. P. Coffin, R. M. Freeman, A. Gallmann, F. Haas, F. Rami, P. Wagner, and D. E. Alburger, *Phys. Rev. C* **23**:1527 (1981).
(Hil 53) D. L. Hill and J. A. Wheeler, *Phys. Rev.* **89**:1102 (1953).
(Hil 66) E. Hilf and G. Sussmann, *Phys. Lett.* **21**:654 (1966).
(Hil 77) K. D. Hildenbrand, H. Freiesleben, F. Pühlofer, W. F. W. Schneider, R. Bock,

D. V. Harrach, and H. J. Specht, *Phys. Rev. Lett.* **39**:1065 (1977).

(Hil 79) D. L. Hillis, J. D. Garrett, O. Christensen, B. Fernandez, G. B. Hagemann, B. Herskind, B. B. Back, and F. Folkmann, *Nucl. Phys.* **A325**:216 (1979).

(Hil 79b) D. Hilscher, J. R. Birkelund, A. D. Hoover, W. U. Schröder, W. W. Wilcke, J. R. Huizenga, A. C. Mignerey, K. L. Wolf, H. F. Breuer, and V. E. Viola, *Phys. Rev. C* **20**:556 (1979).

(Hil 79c) D. Hilscher, J. R. Birkelund, A. D. Hoover, W. U. Schröder, W. W. Wilcke, J. R. Huizenga, A. C. Mignerey, K. L. Wolf, H. F. Breuer, and V. E. Viola, Jr., *Phys. Rev. C* **20**:576 (1979).

(Hil 79d) D. Hilscher, J. R. Birkelund, A. D. Hoover, W. U. Schröder, W. W. Wilcke, J. R. Huizenga, A. C. Mignerey, K. L. Wolf, H. F. Breuer, and V. E. Viola, Jr., *Proc. Symp. on Deep-Inelastic and Fusion Reactions with Heavy Ions* (W. Von Oertzen, ed.), Springer-Verlag, Berlin (1980), Vol. 117, p. 100.

(Hil 81) D. Hilscher, E. Holub, U. Jahnke, H. Orf, and H. Rossner, *Proc. of Third Adriatic Europhysics Study Conf. on Dynamics of Heavy-Ion Collisions* (N. Cindro, R. A. Ricci, and W. Greiner, eds.), North-Holland, Amsterdam (1981), p. 225.

(Hil 83) D. Hilscher, E. Holub, G. Ingold, U. Jahnke, H. Orf, and H. Rossner, Fifteenth Summer School on Nuclear Physics, Mikolajki, Poland (1983).

(Ho 77) H. Ho, R. Albrecht, W. Dünnweber, G. Graw, S. G. Steadman, J. P. Wurm, D. Disdier, V. Rauch, and F. Scheibling, *Z. Phys.* **A283**:235 (1977).

(Ho 80) H. Ho, P. Gonthier, M. N. Namboodiri, J. B. Natowitz, L. Adler, S. Simon, K. Hagel, R. Terry, and A. Khodai, *Phys. Lett.* **96B**:51 (1980).

(Ho 82) H. Ho, G. Y. Fan, P. L. Gonthier, W. Kuhn, A. Pfoh, L. Schad, R. Wolski, J. P. Wurm, J. C. Adloff, D. Disdier, V. Rauch, and F. Scheibling, preprint, Max Planck Institute, Heidelberg, 1982.

(Ho 83) H. Ho, P. L. Gonthier, G. Y. Fan, W. Kuhn, A. Pfoh, L. Schad, R. Wolski, J. P. Wurm, J. C. Adloff, D. Disdier, A. Kamili, V. Rauch, G. Rudolf, F. Scheibling, and A. Strazzeri, *Phys. Rev. C* **27**:584 (1983).

(Hod 78) P. E. Hodgson, *Nuclear Heavy-Ion Reactions* Clarendon Press, Oxford (1978).

(Hof 75) H. Hofmann and P. J. Siemens, *Phys. Lett.* **58B**:417 (1975).

(Hof 76b) H. Hofmann and P. J. Siemens, *Nucl. Phys.* **A257**:165 (1976).

(Hof 76c) H. Hofmann and C. Ngô, *Phys. Lett.* **65B**:97 (1976).

(Hof 77) H. Hofmann and C. Ngô, *Proc. Fall Creek Falls*, Tenn. Conf-770602 (1977), p. 259.

(Hof 77b) H. Hofmann, Inv. Paper, Int. Symp. Nucl. Coll. and Microsc. Descr. Yugoslavia (1977).

(Hof 77c) H. Hofmann and P. J. Siemens, *Nucl. Phys.* **A275**:464 (1977).

(Hof 79) H. Hofmann, *Proc. Sym. on Deep-Inelastic and Fusion Reactions with Heavy Ions* (W. von Oertzen, ed.), Springer-Verlag, Berlin (1979), Vol. 117, p. 64.

(Hof 79b) H. Hofmann, C. Grégoire, R. Lucas, and C. Ngô, *Z. Phys.* **A293**:229 (1979).

(Hol 83) E. Holub, D. Hilscher, G. Ingold, U. Jahnke, H. Orf, and H. Rossner, *Phys. Rev. C* **28**:252 (1983).

(Hom 79) H. Homeyer, C. Egelhaaf, H. Fuchs, A. Gamp, H. G. Bohlen, and H. Kluge, *Proc. of Sym. on Heavy Ion Physics from 10 to 200 MeV/amu*, Report BNL-51115 (1979), Vol. 2, p. 769.

(Hom 79b) H. Homeyer, H. G. Bohlen, Ch. Egelhaaf, H. Fuchs, A. Gamp, and H. Kluge, *Proc. Sym. on Deep-Inelastic and Fusion Reactions with Heavy Ions*, Springer-Verlag, Berlin (1979), Vol. 117, p. 231.

(Hom 82) H. Homeyer, M. Burgel, M. Clover, C. Ehelhaaf, H. Fuchs, A. Gamp, D. Kovar, and W. Rauch, *Phys. Rev. C* **26**:1335 (1982).

(Hoo 81) A. D. Hoover, Univ. of Rochester Ph.D. thesis, 1981, unpublished.

(Hoo 82) A. D. Hoover, J. R. Birkelund, D. Hilscher, W. U. Schröder, W. W. Wilcke, J. R. Huizenga, V. E. Viola, Jr., K. L. Wolf, H. Breuer, and A. C. Mignerey, *Phys. Rev. C* **25**:256 (1982).

(Hor 78) D. Horn, H. A. Enge, A. Sperduto, and A. Graue, *Phys. Rev. C* **17**:118 (1978).

(Hub 80b) F. Hubert, A. Fleury, R. Bimbot, and D. Gardes, *Ann. Phys. (Paris) (Suppl.)* **5**:1 (1980).

(Hui 69) J. R. Huizenga, A. N. Behkami, and L. G. Moretto, *Phys. Rev.* **177**:1826 (1969).

(Hui 72) J. R. Huizenga and L. G. Moretto, *Ann. Rev. Nucl. Sci.* **22**:427 (1972).

(Hui 75) J. R. Huizenga, *Nukleonika* **20**:291 (1975).

(Hui 76) J. R. Huizenga, *Accounts Chem. Res.* **9**:325 (1976).

(Hui 76b) J. R. Huizenga, J. R. Birkelund, and W. Johnson, *Proc. Symposium Macroscopic Features of Heavy-Ion Collisions*, Argonne, 1976, Report ANL/PHY-76-2, p. 1.

(Hui 76c) J. R. Huizenga, J. R. Birkelund, W. U. Schröder, K. L. Wolf, and V. E. Viola, Jr., *Phys. Rev. Lett.* **37**:885 (1976).

(Hui 77) J. R. Huizenga, *Comments Nucl. Part. Phys.* **7**:17 (1977).

(Hui 77b) J. R. Huizenga, W. U. Schröder, J. R. Birkelund, and W. Wilcke, *Proc. Fall Creek Falls*, Tenn. Conf-770602 (1977), p. 179.

(Hui 79b) J. R. Huizenga, *Proc. Sym. on Deep-Inelastic and Fusion Reactions with Heavy Ions*, Springer-Verlag, Berlin (1979), Vol. 117, p. 1.

(Hui 81) J. R. Huizenga, J. R. Birkelund, W. U. Schröder, W. W. Wilcke, and H. J. Wollersheim, *Dynamics of Heavy-Ion Collisions* North-Holland, Amsterdam (1981), p. 15.

(Hui 81b) J. R. Huizenga, J. R. Birkelund, W. U. Schröder, W. W. Wilcke, and H. J. Wollersheim, *Proc. Nucl. Phys. Workshop*, I. C. T. P., Trieste, Italy (October, 1981) Nuclear Physics, (C. H. Dasso, R. A. Broglia, A. Winther, eds.), North-Holland, Amsterdam (1982), p. 583.

(Hui 82) J. R. Huizenga and J. R. Birkelund, University of Rochester report UR-NSRL 261 (1982), unpublished.

(Hui 82b) J. R. Huizenga, W. U. Schröder, J. R. Birkelund, and W. W. Wilcke, *Nucl. Phys.* **A387**:257C (1982).

(Hui 83) A. Huie, D. Beavis, S. Y. Fung. W. Gorn, D. Keane, J. J. Lu, R. T. Poe, B. C. Shen, and G. Vandalen, *Phys. Rev. C* **27**:439 (1983).

(Hus 79d) M. S. Hussein and K. W. McVoy, *Proc. Int. Sym. on Continuum Spectra*, held in San Antonio, Texas, December, 1979 (T. Tamura, J. B. Natowitz, and D. H. Youngblood, eds.) Harwood Academic Publishers, CHUR, London, New York (1980), Vol. 2, p. 187.

(Hus 81) M. S. Hussein, K. W. McVoy, and D. Saloner, *Phys. Lett.* **98B**:162 (1981).

(Imm 75) J. D. Immele, Kit-Keung Kan, and J. J. Griffin, *Nucl. Phys.* **A241**:47 (1975).

(Ina 77) T. Inamura, M. Ishihara, T. Fukuda, T. Shimoda, and H. Hiruta, *Phys. Lett.* **68B**:51 (1977).

(Ina 79) T. Inamura, T. Kojima, T. Nomura, T. Sugitate, and H. Utsunomiya, *Phys. Lett.* **84B**:71 (1979).

(Ing 56) D. R. Inglis, *Phys. Rev.* **103**:1786 (1956).

(Ish 79) M. Ishihara, T. Shimoda, H. Frohlich, H. Kamitsubo, K. Nagatani, T. Udagawa, and T. Tamura, *Phys. Rev. Lett.* **43**:111 (1979).

(Jac 75) J. C. Jacmart, P. Colombani, H. Doubre, N. Frascaria, N. Poffe, M. Riou, J. C. Roynette, C. Stephan, and A. Weidinger, *Nucl. Phys.* **A242**:175 (1975).

(Joh 70) R. C. Johnson and P. J. R. Soper, *Phys. Rev. C* **1**:976 (1970).

(Joh 77) P. J. Johansen, P. J. Siemens, A. S. Jensen, and H. Hofmann, *Nucl. Phys.* **A288**:152 (1977).

(Jor 79) W. J. Jordan, J. V. Maher, and J. C. Peng, *Phys. Lett.* **87B**:38 (1979).

(Kal 77c) R. Kalpakchieva, Yu. Ts. Oganessian, Yu. E. Penionzhkevich, H. Sodan, and B. A. Gvozdev, *Phys. Lett.* **69B**:287 (1977).

(Kal 77d) R. Kalpakchieva, Yu. Ts. Oganessian, Yu. E. Penionzhkevich, and H. Sodan, *Z. Phys.* **A283**:253 (1977).

(Kal 79) C. Kalbach, *Phys. Rev. C* **19**:1547 (1979).

(Kal 79b) R. Kalpakchieva, V. G. Kartavenko, Yu. Ts. Oganessian, Yu. E. Penionzhkevich, and H. Sodan, *Nukleonika* **24**:417 (1979).

(Kan 48) E. Kanai, *Prog. Theor. Phys.* **3**:440 (1948).

(Kan 74) K. K. Kan and J. J. Griffin, *Phys. Lett.* **50B**:241 (1974).

(Kar 78) V. A. Karnaukhov, *Nucl. Phys.* **A294**:246 (1978).

(Kar 80b) H. J. Karwowski, Indiana University, Cyclotron Facility report 80/1, thesis (1979).

(Kau 59) R. Kaufmann and R. Wolfgang, *Phys. Rev. Lett.* **3**:232 (1959).

(Kau 61) R. Kaufmann and R. Wolfgang, *Phys. Rev.* **121**:192 (1961).

(Kau 61b) R. Kaufmann and R. Wolfgang, *Phys. Rev.* **121**:206 (1961).

(Kau 74) S. B. Kaufman, E. P. Steinberg, B. D. Wilkins, J. Unik, A. J. Gorski, and M. J. Fluss, *Nucl. Instrum. Methods* **115**:47 (1974).

(Ker 57) D. D. Kerlee, J. S. Blair, and G. W. Farwell, *Phys. Rev.* **107**:1343 (1957).

(Kin 53) A. Kind, and G. Patergnani, *Nuovo Cimento* **10**:1375 (1953).

(Kle 73) M. Kleber and R. Beck, *Phys. Lett.* **43B**:98 (1973).

(Kno 74) J. Knoll and R. Schaeffer, *Phys. Lett.* **52B**:131 (1974).

(Kno 76) J. Knoll and R. Schaeffer, *Ann. Phys. (N. Y.)* **97**:307 (1976).

(Kno 77) J. Knoll and R. Schaeffer, *Phys. Rep.* **31**:159 (1977).

(Ko 76) C. M. Ko, H. J. Pirner, and H. A. Weidenmüller, *Phys. Lett.* **62B**:248 (1976).

(Ko 78b) C. M. Ko, G. F. Bertsch, and D. Cha, *Phys. Lett.* **77B**:174 (1978).

(Ko 79) C. M. Ko, *Phys. Rev. C* **19**:2417 (1979).

(Ko 79b) C. M. Ko, *Ann. Phys. (N. Y.)* **117**:237 (1979).

(Ko 79c) C. M. Ko, *Phys. Lett.* **81B**:299 (1979).

(Koe 75) T. Koeling and R. A. Malfliet, *Phys. Rep.* **22**:181 (1975).

(Koh 79) H. S. Köhler and H. Flocard, *Nucl. Phys.* **A323**:189 (1979).

(Koh 80) H. S. Köhler, *Nucl. Phys.* **A343**:315 (1980).

(Koo 76) S. E. Koonin, *Phys. Lett.* **61B**:227 (1976).

(Koo 76b) S. E. Koonin and J. R. Nix, *Phys. Rev. C* **13**:209 (1976).

(Koo 77) S. E. Koonin, R. L. Hatch, and J. Randrup, *Nucl. Phys.* **A283**:87 (1977).

(Koo 77b) S. E. Koonin and J. Randrup, *Nucl. Phys.* **A289**:475 (1977).

(Koo 79b) S. E. Koonin, preprint Map-5, California Institute of Technology (1979).

(Kos 72) M. D. Kostin, *J. Chem. Phys.* **57**:3589 (1972).

(Kos 75) M. D. Kostin, *J. Stat. Phys.* **12**:145 (1975).

(Kos 83) J. P. Kosky, W. W. Wilcke, J. R. Birkelund, M. A. Butler, A. D. Dougan, J. R. Huizenga, W. U. Schröder, H. J. Wollersheim, and D. Hilscher, *Phys. Lett.* **133B**:153 (1983).

(Kov 79) D. G. Kovar, D. F. Geesaman, T. H. Braid, Y. Eisen, W. Henning, T. R. Ophel, M. Paul, K. E. Rehm, S. J. Sanders, P. Sperr, J. P. Schiffer, S. L. Tabor, S. Vigdor, B. Zeidman, and F. W. Prosser, Jr., *Phys. Rev. C* **20**:1305 (1979).

(Kow 71) L. Kowlaski, A. Zebelman, A. Kandil, and J. M. Miller, *Phys. Rev. C* **3**:1370 (1971).

(Kra 74) H. J. Krappe and J. R. Nix, *Proc. Third Int. Atomic Energy Agency Symp. on Physics and Chemistry of Fission*, Int. Atomic Energy Agency, Vienna, 1:159 (1974).

(Kra 74c) J. V. Kratz, A. E. Norris, and G. T. Seaborg, *Phys. Rev. Lett.* **33**:502 (1974).

(Kra 76) H. J. Krappe, *Ann. Phys.* (*N. Y.*) **99**:142 (1976).

(Kra 76b) H. J. Krappe, *Nucl. Phys.* **A269**:493 (1976).

(Kra 76c) J. V. Kratz, J. O. Liljenzin, A. E. Norris, and G. T. Seaborg, *Phys. Rev. C* **13**:2347 (1976).

(Kra 77) J. V. Kratz, H. Ahrens, W. Bogl, W. Bruchle, G. Franz, M. Schädel, I. Warnecke, G. Wirth, G. Klein, and M. Weis, *Phys. Rev. Lett.* **39**:984 (1977).

(Kra 79) H. J. Krappe, J. R. Nix, and A. J. Sierk, *Phys. Rev. Lett.* **42**:215 (1979).

(Kra 79b) H. J. Krappe, J. R. Nix, and A. J. Sierk, *Phys. Rev. C* **20**:992 (1979).

(Kra 79d) J. V. Kratz, W. Bruchle, G. Franz, M. Schädel, I. Warnecke, G. Wirth, and M. Weis, *Nucl. Phys.* **A332**:477 (1979).

(Kra 80) J. V. Kratz, W. Bruchle, H. Gäggeler, M. Schädel, K. Summerer, and G. Wirth, *Z. Phys.* **A296**:141 (1980).

(Kra 81) J. V. Kratz, J. Poitou, W. Bruchle, H. Gäggeler, M. Schädel, G. Wirth, and R. Lucas, *Nucl. Phys.* **A357**:437 (1981).

(Kuh 80) W. Kuhn, R. Albrecht, H. Damjantschitsch, H. Ho, R. M. Ronningen, J. Slemmer, and J. P. Wurm, I. Rode, and F. Scheibling, *Z. Phys.* **A298**:95 (1980).

(Kup 74) W. A. Küpper, G. Wegmann, and E. R. Hilf, *Ann. Phys.* (*N. Y.*) **88**:454 (1974).

(Lan 37) R. E. Langer, *Phys. Rev.* **51**:669 (1937).

(Lan 54) J. M. B. Lang and K. J. Le Couteur, *Proc. Phys. Soc.* **A67**:586 (1954).

(Lan 64) D. W. Lang, *Nucl. Phys.* **53**:113 (1964).

(Lan 74) St. Landowne, *Nukleonika* **19**:2 (1974).

(Lau 78) C. Lauterbach, W. Dünnweber, G. Graw, W. Hering, H. Puchta, and W. Trautmann, *Phys. Rev. Lett.* **41**:1774 (1978).

(Lau 81) C. Lauterbach, Univ. of Munich Ph. D. thesis (1981).

(Laz 81) A. Lazzarini, V. Metag, A. G. Seamster, R. Vandenbosch, and R. Loveman, *Phys. Rev. Lett.* **46**:988 (1981).

(Leb 80) C. Le Brun, J. F. Lecolley, F. Lefebvres, M. L'Haridon, J. P. Patry, J. C. Steckmeyer, and R. Chechik, Proc. Int. Conf., Berkeley, LBL-11118 (1980), p. 538.

(Lec 59) K. J. Lecouteur and D. W. Lang, *Nucl. Phys.* **13**:321 (1959).

(Lec 81) Ch. Leclercq-Willain, M. Baus-Baghdikian, and K. Dietrich, *Nucl. Phys.* **A359**:237 (1981).

(Led 77) T. Ledergerber, H. C. Pauli, and Y. Yariv, *Nucl. Phys.* **A280**:241 (1977).

(Lee 81) S. M. Lee, Y. Higashi, Y. Nagashima, S. Hanashima, M. Sato, H. Yamaguchi, M. Yamanouchi, and T. Matsuse, *Phys. Lett.* **98B**:418 (1981).

(Lef 70) M. Lefort, R. Basile, J. Galin, D. Guerreau, J. Péter, and X. Tarrago, *Nuclear Reactions Induced by Heavy Ions*, North-Holland, Amsterdam (1970).

(Lef 72) M. Lefort, C. Ngô, J. Péter, and B. Tamain, *Nucl. Phys.* **A197**:485 (1972).

(Lef 73) M. Lefort, C. Ngô, J. Péter, and B. Tamain, *Nucl. Phys.* **A216**:166 (1973).

(Lef 74c) M. Lefort, *Phys. Scr.* **10A**:101 (1974).

(Lef 75b) M. Lefort, *Phys. Rev. C* **12**:686 (1975).

(Lef 76) M. Lefort, *Nukleonika* **21**:111 (1976).

(Lef 78) M. Lefort and Ch. Ngô, *Ann. Phys.* (*Paris*) **3**:5 (1978).

(Lef 79) M. Lefort, *Proc. Sym. on Deep-Inelastic and Fusion Reactions with Heavy Ions*,

Springer-Verlag, Berlin (1979), Vol. 117, p. 25.

(Lef 79b) M. Lefort and C. Ngô, *Riv. Nuovo Cimento* **2**:1 (1979).

(Lef 79c) M. Lefort, Lectures at Enrico Fermi Int. School, Varenna (1979).

(Lef 81) M. Lefort, *Z. Phys.* **A299**:47 (1981).

(Lei 79) J. R. Leigh, R. M. Diamond, A. Johnston, J. O. Newton, and S. H. Sie, *Phys. Rev. Lett.* **42**:153 (1979).

(Lej 81) A. Lejeune and J. Richert, *Z. Phys.* **A299**:273 (1981).

(Les 83) K. T. Lesko, S. Gil, A. Lazzarini, V. Metag, A. G. Seamster, and R. Vandenbosch, *Phys. Rev. C* **27**:2999 (1983).

(Lev 74) R. D. Levine and R. B. Bernstein, *Acc. Chem. Res.* **7**:393 (1974).

(Lev 78) R. D. Levine, S. G. Steadman, J. S. Karp, and Y. Alhassid, *Phys. Rev. Lett.* **41**:1537 (1978).

(Lev 78b) R. D. Levine, *Ann. Rev. Phys. Chem*, **29**:59 (1978).

(Lev 80) R. D. Levine and S. G. Steadman, *Phys. Lett.* **97B**:177 (1980).

(Li 82) J. Q. Li and G. Wolschin, *Phys. Rev. C* **27**:590 (1983).

(Loc 82) D. K. Lock, R. Vandenbosch, and A. Lazzarini, *Nucl. Phys.* **A384**:241 (1982).

(Log 80) D. Logan, M. Rajagopalan, M. S. Zisman, J. M. Alexander, M. Kaplan, and L. Kowalski, *Phys. Rev. C* **22**:104 (1980).

(Log 80b) D. Logan, H. Delagrange, M. F. Rivet, M. Rajagopalan, J. M. Alexander, M. Kaplan, M. S. Zisman, and E. Duek, *Phys. Rev. C* **22**:1080 (1980).

(Lov 77) W. G. Love, T. Terasawa, and G. R. Satchler, *Phys. Rev. Lett.* **39**:6 (1977).

(Lov 77b) W. Loveland, R. J. Otto, D. J. Morrissey, and G. T. Seaborg, *Phys. Rev. Lett.* **39**:320 (1977).

(Luc 76) R. Lucas, H. Nifenecker, J. Poitou, B. Tamain, and J. Péter, Caen, *Proc. Europ. Conf. Nucl. Phys. with Heavy Ions*, (1976), p. 138.

(Luc 77) R. Lucas, J. Poitou, N. Nifenecker, J. Péter, and B. Tamain, *Z. Phys.* **A283**:257 (1977).

(Luc 79b) R. Lucas, J. Poitou, J. V. Kratz, and G. Wirth, *Z. Phys.* **A290**:327 (1979).

(Lyn 79) U. Lynen, *Proc. of Sym. on Heavy Ion Physics from 10 to 200 MeV/amu*, report BNL-51115 (1979), Vol. 1, p. 221.

(Mac 79) H. Machner, *Phys. Lett.* **86B**:129 (1979).

(Mac 80) H. Machner, *Phys. Rev. C* **21**:2695 (1980).

(Mäd 80) P. Mädler and R. Reif, *Nucl. Phys.* **A337**:445 (1980).

(Mad 80b) G. Madurga, M. Lozano, and A. Jadraque, *Phys. Lett.* **95B**:358 (1980).

(Mai 65) H. Maier-Lebnitz, P. Armbruster, and H. J. Specht, *Proc. IAEA Symp. Phys. Chem. Fission*, Salzburg, IAEA, Vienna (1965), Vol. 2, p. 113.

(Man 75) G. Mantzouranis, D. Agassi, and H. A. Weidenmüller, *Phys. Lett.* **57B**:220 (1975).

(Man 76) G. Mantzouranis, *Phys. Lett.* **63B**:25 (1976).

(Man 77) G. Mantzouranis and H. C. Pauli, *Z. Phys.* **A281**:165 (1977).

(Man 80) G. Mantzouranis and H. C. Pauli, *Phys. Rev. C* **22**:1550 (1980).

(Mar 72) J. A. Maruhn and W. Greiner, *Z. Phys.* **251**:431 (1972).

(Mar 80) J. A. Maruhn, W. Greiner, and W. Scheid, *Heavy Ion Collisions* (R. Bock, ed.), North-Holland, Amsterdam (1980), Vol 2, p. 399.

(Mat 74) E. Der Mateosian, O. C. Kistner, and A. W. Sunyar, *Phys. Rev. Lett.* **33**:596 (1974).

(Mat 77) G. J. Mathews, G. J. Wozniak, R. P. Schmitt, and L. G. Moretto, *Z. Phys.* **A283**:247 (1977).

(Mat 79) G. J. Mathews, L. G. Sobotka, G. J. Wozniak, R. Regimbart, R. P. Schmitt, G. U. Rattazzi, and L. G. Moretto, *Z. Phys.* **A290**:407 (1979).

(Mat 80b) H. Matsuzaki, C. Ishii, T. Honda, and T. Tsukamoto, *Prog. Theor. Phys.* **63**:2119 (1980).

(Mat 80d) G. J. Mathews, L. G. Sobotka, G. J. Wozniak, R. P. Schmitt, R. Regimbart, G. U. Rattazzi, M. Lutolf, P. Biegeleisen, and L. G. Moretto, preprint, Lawrence Berkeley Laboratory (1980).

(Mcd 82) R. J. McDonald, A. J. Pacheco, G. J. Wozniak, H. H. Bolotin, L. G. Moretto, C. Schuck, S. Shih, R. M. Diamond, and F. S. Stephens, *Nucl. Phys.* **A373**:54 (1982).

(Mcm 80) M. A. McMahan and J. M. Alexander, *Phys. Rev. C* **21**:1261 (1980).

(Mer 81) A. C. Merchant and W. Nörenberg, *Z. Phys.* **A308**:315 (1982).

(Mes 65) A. Messiah, *Quantum Mechanics*, North-Holland, Amsterdam (1965).

(Met 83) V. Metag, D. Habs, K. Helmer, U. v. Helmolt, K. W. Heyng, B. Kolb, D. Pelte, D. Schwalm, W. Hennrerici, H. J. Hennrich, Q. Himmele, E. Jaeschke, R. Repnow, W. Wahl, R. S. Simon, and R. Albrecht, Proc. Symp. Detectors in Heavy-Ion Reaction, Berlin 1982, *Lecture Notes Phys.* **178**:163 (1983).

(Mey 76) W. G. Meyer, R. G. Clark, V. E. Viola, Jr., R. G. Sextro, and A. M. Zebelman, *Z. Phys.* **A277**:141 (1976).

(Mig 80) A. C. Mignerey, V. E. Viola, Jr., H. Breuer, K. L. Wolf, B. G. Glagola, J. R. Birkelund, D. Hilscher, J. R. Huizenga, W. U. Schröder, and W. W. Wilcke, *Phys. Rev. Lett.* **45**:509 (1980).

(Mig 80b) A. C. Mignerey, K. L. Wolf, H. Breuer, B. G. Glagola, V. E. Viola, Jr., J. R. Birkelund, D. Hilscher, J. R. Huizenga, W. U. Schröder and W. W. Wilcke, *Proceedings of the International Conference on Nuclear Physics*, Lawrence Berkeley Laboratory report LBL-11118 (1980), p. 611.

(Mig 82) A. C. Mignerey, A. Gokmen, H. Breuer, K. L. Wolf, R. R. Betts, C. Davids, B. G. Glagola, and V. E. Viola, Workshop on Nucl. Dynamics, Conf-820222 LBL14138 UC-34C (1982).

(Mig 84) A. C. Mignerey, K. L. Wolf, D. G. Raich, V. E. Viola, Jr., J. R. Birkelund, W. U. Schröder, and J. R. Huizenga, *Phys. Rev. C* **29**:158 (1984).

(Mik 76) T. Mikumo, I. Kohno, K. Katori, T. Motobayashi, S. Nakajima, M. Yoshie, and H. Kamitsubo, *Phys. Rev. C* **14**:1458 (1976).

(Mil 78) J. M. Miller, G. L. Catchen, D. Logan, M. Rajagopalan, J. M. Alexander, M. Kaplan, and M. S. Zisman, *Phys. Rev. Lett.* **40**:100 (1978).

(Mil 78b) J. M. Miller, D. Logan, G. L. Catchen, M. Rajagopalan, J. M. Alexander, M. Kaplan, J. W. Ball, M. S. Zisman, and L. Kowalski, *Phys. Rev. Lett.* **40**:1074 (1978).

(Min 72) M. M. Minor, A. E. Salwin, V. E. Viola, Jr., R. B. Theus, and R. O. Bondelid, *Nucl. Instrum. Methods* **99**:63 (1972).

(Moh 80) K. Möhring and U. Smilansky, *Nucl. Phys.* **A338**:227 (1980).

(Mol 77) P. Møller and J. R. Nix, *Nucl. Phys.* **A281**:354 (1977).

(Moo 81) P. Mooney, W. W. Morison, S. K. Samaddar, D. Sperber, and M. Zielinska-Pfabe, *Phys. Lett.* **98B**:240 (1981).

(Mor 65) H. Mori, *Prog. Theor. Phys.* **33**:423 (1965).

(Mor 65b) H. Mori, *Prog. Theor. Phys.* **34**:399 (1965).

(Mor 72b) L. G. Moretto, *Nucl. Phys.* **A182**:641 (1972).

(Mor 75) L. G. Moretto and J. S. Sventek, *Phys. Lett.* **58B**:26 (1975).

(Mor 75b) L. G. Moretto, R. P. Babinet, J. Galin, and S. G. Thompson, *Phys. Lett.* **58B**:31 (1975).

(Mor 75c) L. G. Moretto, *Nucl. Phys.* **A247**:211 (1975).

(Mor 75d) L. G. Moretto, S. K. Kataria, R. C. Jared, R. Schmitt, and S. G. Thompson, *Nucl. Phys.* **A255**:491 (1975).

(Mor 76) L. G. Moretto and J. S. Sventek, *Proc. Symposium Macroscopic Features of Heavy-Ion Collisions*, Argonne, 1976, Report ANL/PHY-76-2, Vol. I, p. 235.

(Mor 76b) L. G. Moretto and R. Schmitt, *J. Phys. (Paris) Colloq. C5* **37**:109 (1976).

(Mor 76c) L. G. Moretto, B. Cauvin, P. Glässel, R. Jared, P. Russo, J. Sventek, and G. Wozniak, *Phys. Rev. Lett.* **36**:1069 (1976).

(Mor 76d) L. G. Moretto, J. Galin, R. Babinet, Z. Fraenkel, R. Schmitt, R. Jared, and S. G. Thompson, *Nucl. Phys.* **A259**:173 (1976).

(Mor 77) L. G. Moretto, Proc. Symp. in Hakone, Japan (1977), p. 179.

(Mor 78b) D. J. Morrissey, W. Loveland, R. J. Otto, and G. T. Seaborg, *Phys. Lett.* **74B**:35 (1978).

(Mor 79) L. G. Moretto, J. Sventek, and G. Mantzouranis, *Phys. Rev. Lett.* **42**:563 (1979).

(Mor 79c) L. G. Moretto, Proc. Varenna Conference, Italy (1979) Lawrence Berkeley Laboratory report LBL-9130 (1979).

(Mor 79d) L. G. Moretto, Lawrence Berkeley Laboratory report LBL-8936 (1979).

(Mor 79e) L. G. Moretto, Lawrence Berkeley Laboratory report LBL-8613 (1979).

(Mor 80) L. G. Moretto and R. P. Schmitt, *Phys. Rev. C* **21**:204 (1980).

(Mor 80c) W. W. Morison, S. K. Samaddar, D. Sperber, and M. Zielinska-Pfabe, *Phys. Lett.* **93B**:379 (1980).

(Mor 80d) L. G. Moretto, XIII Masurian School, Mikolajki, Poland (1980).

(Mor 80e) L. G. Moretto, C. R. Albiston, and G. Mantzouranis, *Phys. Rev. Lett.* **44**:924 (1980).

(Mor 81b) D. J. Morrissey and L. G. Moretto, *Phys. Rev. C* **23**:1835 (1981).

(Mor 81c) L. C. Moretto, S. K. Blau, and A. J. Pacheco, *Nucl. Phys.* **A364**:125 (1981).

(Mor 83) H. Morgenstern, W. Bohne, K. Grabisch, H. Lehr, and W. Stoffler, *Z. Phys.* **A313**:39 (1983).

(Mos 76) U. Mosel, *Proc. Symposium Macroscopic Features of Heavy-Ion Collisions*, Argonne, 1976, Report ANL/PHY-76-2, p. 341.

(Mou 78) J. B. Moulton, J. E. Stephenson, R. P. Schmitt, and G. J. Wozniak, *Nucl. Instrum. Methods* **157**:325 (1978).

(Muk 79c) S. Mukamel, *J. Chem. Phys.* **71**:2012 (1979).

(Muk 79d) S. Mukamel, *J. Chem. Phys.* **70**:5834 (1979).

(Muk 81) S. Mukamel, U. Smilansky, D. H. E. Gross, K. Möhring, and M. I. Sobel, *Nucl. Phys.* **A366**:339 (1981).

(Mun 80) L. Munchow, A. Pfitzner, and H. Schulz, *Z. Phys.* **A296**:55 (1980).

(Mye 66) W. D. Myers and W. Swiatecki, *Nucl. Phys.* **81**:1 (1966).

(Mye 67) W. D. Myers and W. J. Swiatecki, Proceedings of the Lysekil Symposium (1966), *Ark. Fys.* **36**:343 (1967).

(Mye 69) W. D. Myers, *Phys. Lett.* **30B**:451 (1969).

(Mye 70) W. D. Myers, *Nucl. Phys.* **A145**:387 (1979).

(Mye 73) W. D. Myers, *Nucl. Phys.* **A204**:465 (1973).

(Mye 74) W. D. Myers and W. J. Swiatecki, *Ann. Phys.* (*N. Y.*) **84**:186 (1974).

(Mye 76) W. D. Myers, *Nukleonika* **21**:3 (1976).

(Mye 77b) W. D. Myers, *Droplet Model of Atomic Nucleus*, IFI/Plenum Press, New York (1977).

(Mye 80) W. D. Myers and W. J. Swiatecki, *Nucl. Phys.* **A336**:267 (1980).

(Nam 75b) M. N. Namboodiri, J. B. Natowitz, E. T. Chulick, K. Das, and L. Webb, *Nucl. Phys.* **A252**:163 (1975).

(Nam 76) M. N. Namboodiri, E. T. Chulick, and J. B. Natowitz, *Nucl. Phys.* **A263**:491 (1976).

(Nam 79) M. N. Namboodiri, J. B. Natowitz, P. Kasiraj, R. Eggers, L. Adler, P. Gonthier, C. Cerruti, and S. Simon, *Phys. Rev. C* **20**:982 (1979).

(Nat 70) J. B. Natowitz, *Phys. Rev. C* **1**:623 (1970).

(Nat 72) J. B. Natowitz, E. T. Chulick, and M. N. Namboodiri, *Phys. Rev. C* **6**:2133 (1972).

(Nat 73) J. B. Natowitz, E. T. Chulick, and M. N. Namboodiri, *Phys. Rev. Lett.* **31**:643 (1973).

(Nat 76) J. B. Natowitz, M. N. Namboodiri, and E. T. Chulick *Phys. Rev. C* **13**:171 (1976).

(Nat 77) J. B. Natowitz, M. N. Namboodiri, R. Eggers, P. Gonthier, K. Geoffroy, R. Hanus, C. Towsley, and K. Das, *Nucl. Phys.* **A277**:477 (1977).

(Nat 77b) J. B. Natowitz, Proc. Fall Creek Falls, Tenn. Conf-770602 (1977), p. 115.

(Nat 78) J. B. Natowitz, M. N. Namboodiri, P. Kasiraj, R. Eggers, L. Adler, P. Gonthier, C. Cerruti, and T. Alleman, *Phys. Rev. Lett.* **40**:751 (1978).

(Neg 78) J. W. Negele, S. E. Koonin, P. Møller, J. R. Nix, and A. J. Sierk, *Phys. Rev. C* **17**:1098 (1978).

(Neg 81) J. E. Negele and K. Yazaki, *Phys. Rev. Lett.* **47**:71 (1981).

(Nem 80) M. C. Nemes, and H. A. Weidenmüller, *Phys. Rev. C* **24**:450 (1981).

(New 75) J. O. Newton, J. C. Lisle, G. D. Dracoulis, J. R. Leigh, and D. C. Weisser, *Phys. Rev. Lett.* **34**:99 (1975).

(New 77) J. O. Newton, I. Y. Lee, R. S. Simon, M. M. Aleonard, Y. El Masri, F. S. Stephens, and R. M. Diamond, *Phys. Rev. Lett.* **38**:810 (1977).

(Ngô 75) C. Ngô, thesis, Univ. Paris-Sud (1975).

(Ngô 75b) C. Ngô, B. Tamain, J. Galin, M. Beiner, and R. J. Lombard, *Nucl. Phys.* **A240**:353 (1975).

(Ngô 76) C. Ngô, J. Péter, B. Tamain, M. Berlanger, and F. Hanappe, *Nucl. Phys.* **A267**:181 (1976).

(Ngô 77) C. Ngô and H. Hofmann, *Z. Phys.* **A282**:83 (1977).

(Ngô 77b) C. Ngô, J. Péter, B. Tamain, M. Berlanger, and F. Hanappe, *Z. Phys.* **A283**:161 (1977).

(Ngô 79) C. Ngô, *Proc. Int. Symp. on Continuum Spectra* held in San Antonio, Texas, December 1979, (T. Tamura, J. B. Natowitz, and D. H. Youngblood, eds.) Harwood Academic Publishers, CHUR, London, New York (1980), Vol. 2, p. 51.

(Ngô 80b) Ch. Ngô, Inv. Talk Int. Conf. on Extreme States in Nuclear Systems, Roumania, Céntres d'Etudes Nucléaires de Saclay (1980).

(Ngô 80c) Ch. Ngô, preprint, Inv. Lect. Predeal International Summer School, Roumania, Céntres d'Etudes Nucléaires de Saclay (1980).

(Ngô 81) C. Ngô, Invited Lectures, 14th Mazurian Summer School on Nuclear Physics, Mikolajki, Poland (1981).

(Nic 59) W. J. Nicholson and I. Halpern, *Phys. Rev.* **116**:175 (1959).

(Nix 64) J. R. Nix, LBL report UCRL-11338 (1964).

(Nix 65) J. R. Nix and W. J. Swiatecki, *Nucl. Phys.* **71**:1 (1965).

(Nix 69) J. R. Nix, *Nucl. Phys.* **A130**:241 (1969).

(Nix 74) J. R. Nix and A. J. Sierk, *Phys. Scr.* **10A**:94 (1974).

(Nix 77) J. R. Nix and A. J. Sierk, *Phys. Rev. C* **15**:2072 (1977).

(Nom 78) T. Nomura, H. Utsunomiya, T. Motobayashi, T. Inamura, and M. Yanokura, *Phys. Rev. Lett.* **40**:694 (1978).

(Nor 70) L. C. Northcliffe and R. F. Schilling, *Nucl. Data* **A7**:233 (1970).

(Nör 74) W. Nörenberg, *Phys. Lett.* **52B**:289 (1974).

(Nör 75) W. Nörenberg, *Z. Phys.* **A274**:241 (1975).

(Nör 76) W. Nörenberg, *J. Phys. (Paris) Colloq. C5* **37**:141 (1976).

(Nör 76b) W. Nörenberg and H. A. Weidenmüller, *Introduction to the Theory of Heavy-Ion Collisions* Springer-Verlag, Berlin (1976), Vol. 51.

(Nör 79) W. Nörenberg and C. Riedel, *Z. Phys.* **A290**:335 (1979).

(Nör 79b) W. Nörenberg and C. Riedel, Proceedings of Hirschegg Workshop VII (1979), p. 6.

(Nör 79c) W. Nörenberg, Lectures at Int. School in Physics, Predeal, Romania, 1978; report GSI-79-5 (1979).

(Nör 79d) W. Nörenberg, *Proc. Symp. on Deep-Inelastic and Fusion Reactions with Heavy Ions*, Springer-Verlag, Berlin (1979), Vol. 117, p. 50.

(Nör 79e) W. Nörenberg, *Proc. Int. Sym. on Continuum Spectra* held in San Antonio, Texas, December, 1979, (T. Tamura, J. B. Natowitz, and D. H. Youngblood, eds.), Harwood Academic Publishers, CHUR, London, New York (1980), Vol. 2, p. 383.

(Nör 80) W. Nörenberg and H. A. Weidenmüller, *Introduction to the Theory of Heavy Ion Collisions*, Springer-Verlag, Berlin (1980), Vol. 51.

(Nör 80c) W. Nörenberg, *J. Phys. (Paris) C* **10**:167 (1980).

(Nör 80d) W. Nörenberg and C. Riedel, *Proc. Int. Conf.*, Berkeley, LBL-11118 (1980), p. 437.

(Nör 81) W. Nörenberg, *Phys. Lett.* **104B**:107 (1981).

(Nör 83) W. Nörenberg, *Nucl. Phys.* **A409**:191C (1983).

(Nov 78) R. Novotny, G. Hammer, D. Pelte, H. Emling, and D. Schwalm, *Nucl. Phys.* **A294**:255 (1978).

(Nov 80) R. Novotny, U. Winkler, D. Pelte, H. Sann, and U. Lynen, *Nucl. Phys.* **A341**:301 (1980).

(Nov 80b) R. Novotny, D. Pelte, G. Strauch, and U. Winkler, *Proc. Int. Conf.*, Berkeley, LBL-11118 (1980), p. 486.

(Obe 78) F. E. Obenshain, R. L. Ferguson, M. L. Halbert, D. C. Hensley, H. Nakahara, F. Plasil, F. Pleasonton, A. H. Snell, and R. G. Stokstad, *Phys. Rev. C* **18**:264 (1978).

(Obe 80) F. E. Obenshain, R. L. Ferguson, A. I. Gavron, F. Plasil, R. L. Robinson, D. Shapira, A. H. Snell, and G. R. Young, *Proc. Int. Conf.*, Berkeley, LBL-11118 (1980), p. 608.

(Obl 78) P. Oblozinsky and I. Ribansky, *Phys. Lett.* **74B**:6 (1978).

(Oer 73) W. von Oertzen and W. Nörenberg, *Nucl. Phys.* **A207**:113 (1979).

(Oer 75) W. von Oertzen, In *Nuclear Spectroscopy* (J. Cerny, ed.), Academic Press, New York (1975).

(Oes 77) H. Oeschler, J. P. Coffin, P. Engelstein, A. Gallmann, K. S. Sim, and P. Wagner, *Phys. Lett.* **71B**:63 (1977).

(Oes 78) H. Oeschler, J. P. Coffin, P. Engelstein, A. Gallmann, K. S. Sim, and P. Wagner, *Phys. Rev. C* **18**:239 (1978).

(Oes 79b) H. Oeschler, P. Wagner, J. P. Coffin, P. Engelstein, and B. Heusch, *Phys. Lett.* **87B**:193 (1979).

(Oes 80) H. Oeschler, H. Freiesleben, K. D. Hildenbrand, P. Engelstein, J. P. Coffin, B. Heusch, and P. Wagner, *Phys. Rev. C* **22**:546 (1980).

(Oga 74) Yu. Ts. Oganesyan, Yu. E. Penionzhkevich, Nguyen Tac Anh, D. M. Nadkarnee, K. A. Gavrilov, Kim De En, and M. Yussonua, *Sov. J. Nucl. Phys.* **18**:377 (1974).

(Oga 74b) Yu. Ts. Oganesyan, D. M. Nadkarni, Yu. E. Penionzhkevich, B. I. Pustyl'nik, and Nguen Tak An, *Sov. J. Nucl. Phys.* **19**:244 (1974).

(Ohl 65) G. G. Ohlsen, *Nucl. Instrum. Methods* **37**:240 (1965).

(Olm 78b) A. Olmi, H. Sann, D. Pelte, Y. Eyal, A. Gobbi, W. Kohl, U. Lynen, G. Rudolf, H. Stelzer, and R. Bock, *Phys. Rev. Lett.* **41**:688 (1978).

(Olm 80) A. Olmi, U. Lynen, J. B. Natowitz, M. Dakowski, P. Doll, A. Gobbi, H. Sann, H. Stelzer, R. Bock, and D. Pelte, *Phys. Rev. Lett.* **44**:383 (1980).

(Ost 76) R. Ost, N. E. Sanderson, S. Mordechai, J. B. A. England, B. R. Fulton, J. M. Nelson, and G. C. Morrision, *Nucl. Phys.* **A265**:142 (1976).

(Ott 76) R. J. Otto, M. M. Fowler, D. Lee, and G. T. Seaborg, *Phys. Rev. Lett* **36**:135 (1976).

(Ott 78b) R. J. Otto, D. J. Morrissey, G. T. Seaborg, and W. D. Loveland, *Z. Phys.* **A287**:97 (1978).

(Oui 76) S. Ouichaoui, C. Ngô, J. Péter, F. Plasil, B. Tamain, Berlanger, and F. Hanappe, Caen, *Proc. Europ. Conf. Nucl. Phys. with Heavy Ions*, (1976), p. 112.

(Pai 81) G. Paic and B. Antolkovic, *Phys. Rev. C* **23**:1839 (1981).

(Pal 79) Z. Paltiel, and Z. Fraenkel, *Phys. Rev. C* **19**:2274 (1979).

(Pal 81) S. Pal and N. K. Ganguly, *Nucl. Phys.* **A370**:175 (1981).

(Pat 78) R. K. Pathria, *Statistical Mechanics*, Pergamon Press, New York (1981).

(Pau 28) W. Pauli, *Festschrift Zum 60. Geburtstage A. Sommerfelds* (P. Debye, ed.), Hirzel, Leipzig (1928), p. 30.

(Pel 81) D. Pelte, U. Winkler, R. Novotny, and H. Graf, *Nucl. Phys.* **371**:454 (1981).

(Pét 75) J. Péter, C. Ngô, and B. Tamain, *Nucl. Phys.* **A250**:351 (1975).

(Pét 75c) J. Péter, C. Ngô, and B. Tamain, *J. Phys. Lett.* **36**:L23 (1975).

(Pét 76) J. Péter, C. Ngô, F. Plasil, B. Tamain, F. Hanappe, and M. Berlanger, *Proceedings of Hirschegg Workshop IV*, (1976).

(Pét 77) J. Péter, M. Berlanger, C. Ngô, B. Tamain, B. Lucas, C. Mazur, M. Ribrag, and C. Signarbieux, *Z. Phys.* **A283**:413 (1977).

(Pét 77b) J. Péter, C. Ngô, F. Plasil, B. Tamain, M. Berlanger, and F. Hanappe, *Nucl. Phys.* **A279**:110 (1977).

(Pet 82) M. Petrovici, *Nucl. Phys.* **A387**:313C (1982).

(Pha 79) S. C. Phatak and B. Sinha, *Z. Phys.* **A290**:185 (1979).

(Pla 78) F. Plasil, R. L. Ferguson, H. C. Britt, B. H. Erkkila, P. D. Goldstone, R. H. Stokes, M. Blann, and H. H. Gutbrod, *Proceedings of Hirschegg Workshop VI*, (1978), p. 161.

(Pla 78c) F. Plasil, R. L. Ferguson, H. C. Britt, R. H. Stokes, B. H. Erkkila, P. D. Goldstone, M. Blann, and H. H. Gutbrod, *Phys. Rev. Lett.* **40**:1164 (1978).

(Pla 78d) F. Plasil, R. L. Ferguson, H. C. Britt, B. H. Erkkila, P. D. Goldstone, R. H. Stokes, and H. H. Gutbrod, *Phys. Rev. C* **18**:2603 (1978).

(Plo 82) M. Ploszajczak, M. E. Faber, *Phys. Rev. C* **25**:1538 (1982).

(Poi 79) J. Poitou, R. Lucas, J. V. Kratz, W. Bruchle, H. Gäggeler, M. Schadel, and G. Wirth, *Phys. Lett.* **88B**:69 (1979).

(Poi 81) J. Poitou, J. V. Kratz, G. Wirth, and R. Lucas, *Nucl. Instrum. Methods*, **180**:221 (1981).

(Pou 75) F. Pougheon, P. Roussel, C. Detraz, and G. Rotbard, *Phys. Rev. C* **11**:287 (1975).

(Pra 79) M. Praksah, V. S. Ramamurthy, S. K. Kataria, and S. S. Kapoor, *Phys. Lett.* **81B**:136 (1979).

(Pra 81) M. Prakash, S. Shlomo, V. S. Ramamurthy, S. S. Kapoor, and S. K. Kataria, *Phys. Lett.* **98B**:413 (1981).

(Puh 75) F. Pühlhofer, W. Pfeffer, B. Kohlmeyer, and W. F. W. Schneider, *Nucl. Phys.* **A244**:329 (1975).

(Puh 77b) F. Pühlhofer, W. F. W. Schneider, F. Busch, J. Barrette, P. Braun-Munzinger, C. K. Gelbke, and H. E. Wegner, *Phys. Rev. C* **16**:1010 (1977).

(Pui 79) R. J. Puigh, P. Dyer, R. Vandenbosch, T. D. Thomas, L. Nunnelley, and M. S. Zisman, *Phys. Lett.* **86B**:24 (1979).

(Pui 80) R. J. Puigh, H. Doubre, A. Lazzarini, A. Seamster, R. Vandenbosch, M. S. Zisman, and T. D. Thomas, *Nucl. Phys.* **A336**:279 (1980).

(Qui 60) A. R. Quinton, H. C. Britt, W. J. Knox, and C. E. Anderson, *Nucl. Phys.* **17**:74 (1960).

(Raj 79) M. Rajagopalan, L. Kowalski, D. Logan, M. Kaplan, J. M. Alexander, M. S. Zisman, and J. M. Miller, *Phys. Rev. C* **19**:54 (1979).

(Raj 82) M. Rajogopalan, D. Logan, J. W. Ball, M. Kaplan, H. Delagrange, M. F.

Rivet, J. M. Alexander, L. C. Vaz, and M. S. Zisman, *Phys. Rev. C* **25**:2417 (1982).

(Ram 80) V. S. Ramamurthy, M. Prakash, and S. S. Kapoor, *Phys. Rev. C* **21**:752 (1980).

(Ran 76) J. Randrup, *Nucl. Phys.* **A259**:253 (1976).

(Ran 77) J. Randrup, Proc. Fall Creek Falls, Tenn. Conf-770602 (1977), p. 218.

(Ran 77c) J. Randrup, *Nukleonika* **22**:1029 (1977).

(Ran 78) J. Randrup, *Ann. Phys. (N. Y.)* **112**:356 (1978).

(Ran 78c) J. Randrup, *Nucl. Phys.* **A307**:319 (1978).

(Ran 78d) J. Randrup, Nordita report 79 (1978).

(Ran 79c) J. Randrup, Proc. of Sym. on Heavy Ion Physics from 10 to 200 MeV/amu, report BNL-5115 (1979), Vol. 2, p. 713.

(Ran 79d) J. Randrup, *Nucl. Phys.* **A327**:490 (1979).

(Ran 80b) J. Randrup and W. J. Swiatecki, *Ann. Phys. (N. Y.)* **125**:193 (1980).

(Ran 81) J. Randrup, and S. E. Koonin, *Nucl. Phys.* **A356**:223 (1981).

(Ran 81b) J. Randrup, *Nucl. Phys.* **A383**:468 (1981).

(Ran 82) J. Randrup, *Phys. Lett.* **110B**:25 (1982).

(Raw 74) G. H. Rawitscher, *Phys. Rev. C* **9**:2210 (1974).

(Reg 78) R. Regimbart, A. N. Behkami, G. L. Wozniak, R. P. Schmitt, J. S. Sventek, and L. G. Moretto, *Phys. Rev. Lett.* **41**:1355 (1978).

(Reh 73) K. E. Rehm, W. Henning, R. Müller, M. Richter, H. P. Rother, and H. Schaller, *Phys. Lett.* **46B**:353 (1973).

(Reh 79) K. E. Rehm, H. Essel, K. Hartel, P. Kienle, H. J. Körner, R. E. Segel, P. Sperr, and W. Wagner, *Z. Phys.* **A293**:119 (1979).

(Reh 79b) K. E. Rehm, H. Essel, K. Hartel, P. Kienle, H. J. Körner, P. Sperr, and W. Wagner, *Phys. Lett.* **86B**:256 (1979).

(Reh 81) K. E. Rehm, H. Essel, P. Sperr, K. Hartel, P. Kienle, H. J. Körner, R. E. Segel, and W. W. Wagner, *Nucl. Phys.* **A366**:477 (1981).

(Rei 66) R. F. Reising, G. L. Bate, J. R. Huizenga, *Phys. Rev.* **141**:1161 (1966).

(Rei 76) W. Reisdorf, M. De Saint-Simon, L. Lessard, L. Remsberg, C. Thibault, E. Roeckl, R. Klapisch, I. V. Kuznetsov, Yu. Ts. Oganessian, and Yu. E. Penionshkevitch, *Phys. Lett.* **62B**:33 (1976).

(Reu 77) U. Reus, A. M. Habbestad Watzig, R. A. Esterlund, and P. Patzelt, *Phys. Rev. Lett.* **39**:171 (1977).

(Ric 79) A. Richter and C. Toepffer, *Heavy Ion Collisions VI* (R. Bock, ed.) North-Holland, Amsterdam (1979), p. 1.

(Rie 79) C. Riedel, G. Wolschin, and W. Nörenberg, *Z. Phys.* **A290**:47 (1979).

(Rie 79b) C. Riedel and W. Nörenberg, *Z. Phys.* **A290**:385 (1979).

(Rie 80) C. Riedel and G. Wolschin, *Z. Phys.* **A294**:17 (1980).

(Riv 77) M. F. Rivet, R. Bimbot, A. Fleury, D. Gardes, and Y. Llabador, *Nucl. Phys.* **A276**:157 (1977).

(Riv 80) M. F. Rivet, R. Bimbot, D. Gardes, G. Mouchaty, A. Fleury, F. Hubert, and Y. Llabador, *Z. Phys.* **A296**:201 (1980).

(Riv 82) M. F. Rivet, D. Logan, J. M. Alexander, D. Guerreau, E. Duek, M. S. Zisman, and M. Kaplan, *Phys. Rev. C* **25**:2430 (1982).

(Rob 79) M. C. Robel, Lawrence Berkeley Laboratory report LBL-8181 (1979).

(Rob 80) R. L. Robinson, R. L. Auble, J. B. Ball, F. E. Bertrand, R. L. Ferguson, C. B. Fulmer, I. Y. Lee, M. J. Martin, J. R. Wu, G. R. Young, J. C. Wells, and H. Yamada, *Proc. Int. Conf.*, Berkeley LBL-11118, (1980), p. 470.

(Ros 62) M. Rosen, *Nucl. Phys.* **31**:664 (1962).

(Ros 83) H. Rossner, D. Hilscher, E. Holub, G. Ingold, U. Jahnke, H. Orf, J. R. Huizenga, J. R. Birkelund, W. U. Schröder, and W. W. Wilcke, *Phys. Rev. C*

27:2666 (1983).

(Ros 84) H. H. Rossner, J. R. Huizenga, and W. U. Schröder, University of Rochester Report UR-NSRL-280 (1984).

(Roy 76) J. C. Roynette, H. Doubre, J. C. Jacmart, and N. Poffe, *European Conf. on Nucl. Phys. with Heavy Ions*, Communications, Caen (1976), p. 165.

(Roy 77) J. C. Roynette, H. Doubre, N. Frascaria, J. C. Jacmart, N. Poffe, and M. Riou, *Phys. Lett.* **67B**:395 (1977).

(Roy 81) J. C. Roynette, N. Frascaria, Y. Blumenfeld, J. C. Jacmart, E. Plagnol, J. P. Garron, A. Gamp, and H. Fuchs, *Z. Phys.* **A299**:73 (1981).

(Rud 79) G. Rudolf, A. Gobbi, H. Stelzer, U. Lynen, A. Olmi, H. Sann, R. G. Stokstad, and D. Pelte, *Nucl. Phys.* **A330**:243 (1979).

(Rud 79b) G. Rudolf, A. Gobbi, U. Lynen, A. Olmi, and H. Sann, Contributed Paper, Proc. Int. Symp., Texas (1979).

(Rud 81) G. Rudolf, U. Lynen, A. Gobbi, A. Olmi, and H. Sann, *Nucl. Phys.* **A367**:109 (1981).

(Rus 77) P. Russo, R. P. Schmitt, G. J. Wozniak, B. Cauvin, P. Glassel, R. C. Jared, and L. G. Moretto, *Phys. Lett.* **67B**:155 (1977).

(Rus 77b) P. Russo, R. P. Schmitt, G. J. Wozniak, R. C. Jared, P. Glassel, B. Cauvin, J. S. Sventek, and L. G. Moretto, *Nucl. Phys.* **A281**:509 (1977).

(Sai 78) M. De Saint-Simon, R. J. Otto, and G. T. Seaborg, *Phys. Rev. C* **18**:1651 (1978).

(Sal 80) D. Saloner and H. A. Weidenmüller, *Z. Phys.* **A294**:207 (1980).

(Sam 79b) S. K. Samaddar, M. I. Sobel, J. N. De, S. I. A. Garpman, D. Sperber, M. Zielinska-Pfabe, and S. Møller, *Nucl. Phys.* **A332**:210 (1979).

(Sam 81) S. K. Samaddar, D. Sperber, M. Zielinska-Pfabe, M. I. Sobel, and S. I. Garpman, *Phys. Rev. C* **23**:760 (1981).

(Sam 82) S. K. Samaddar, D. Sperber, M. Zielinska-Pfabe, and M. I. Sobel, *Phys. Scr.* **25**:517 (1982).

(San 75) H. Sann, D. Danjantschitsch, D. Hebbard, J. Junge, D. Pelte, B. Povh, D. Schwalm, and D. B. Tran Thoai, *Nucl. Instrum. Methods* **124**:509 (1975).

(San 76) A. Sandulescu, R. K. Gupta, W. Scheid, and W. Greiner, *Phys. Lett.* **60B**:225 (1976).

(San 77) H. Sann, A. Olmi, Y. Civelekoglu, U. Lynen, H. Stelzer, A. Gobbi, Y. Eyal, W. Kohl, R. Renfordt, I. Rode, G. Rudolf, D. Schwalm, and R. Bock, *Proc. Fall Creek Falls*, Tenn., Conf-770602 (1977), p. 281.

(San 78) A. Sandulescu, H. J. Lustig, J. Hahn, and W. Greiner, *J. Phys. G* **4**:L279 (1978).

(San 79b) M. Sandel, J. P. Vary, and S. I. A. Garpman, *Phys. Rev. C* **20**:744 (1979).

(San 80c) H. Sann, S. Bjørnholm, R. Bock, Y. T. Chu, A. Gobbi, E. Grosse, U. Lynen, E. Morenzoni, W. Müller, A. Olmi, D. Schwalm, and W. Wolfli, *Proc. Int. Conf.*, Berkeley, LBL-11118 (1980), p. 570.

(San 81) H. Sann, R. Bock, Y. T. Chu, A. Gobbi, A. Olmi, U. Lynen, W. Müller, S. Bjørnholm, and H. Esbensen, *Phys. Rev. Lett.* **47**:1248 (1981).

(Sar 76) D. G. Sarantites, J. H. Barker, M. L. Halbert, D. C. Hensley, R. A. Dayras, E. Eichler, N. R. Johnson, and S. A. Gronemeyer, *Phys. Rev. C* **14**:2138 (1976).

(Sar 78) D. G. Sarantites, L. Westerberg, R. A. Dayras, M. L. Halbert, D. C. Hensley, and J. H. Barker, *Phys. Rev. C* **17**:601 (1978).

(Sar 78b) D. G. Sarantites, L. Westerberg, M. L. Halbert, R. A. Dayras, D. C. Hensley, and J. H. Barker, *Phys. Rev. C* **18**:774 (1978).

(Sat 74) G. R. Satchler, *Phys. Rep.* **14**:97 (1974).

(Sat 75) G. R. Satchler, *Phys. Lett.* **58B**:408 (1975).

(Sat 75b) G. R. Satchler, *Phys. Lett.* **59B**:121 (1975).

(Sat 76b) G. R.Satchler and W. G. Love, *Phys. Lett.* **65B**:415 (1976).

(Sat 78) K. Sato, A. Iwamoto, K. Harada, S. Yamaji, and S. Yoshida, *Z. Phys.* **A288**:383 (1978).

(Sat 79c) G. R. Satchler and W. G. Love, *Phys. Rep.* **55**:183 (1979).

(Sau 76) G. Sauer, H. Chandra, and U. Mosel, *Nucl. Phys.* **A264**:221 (1976).

(Sch 58) H. Schopper, *Nucl. Instrum. Methods* **3**:158 (1958).

(Sch 65) H. W. Schmitt, W. E. Kiker, and C. W. Williams, *Phys. Rev.* **137**:B837 (1965).

(Sch 73) J. P. Schiffer, H. J. Körner, R. H. Siemssen, K. W. Jones, and A. Schwarzschild, *Phys. Lett.* **44B**:47 (1973).

(Sch 76) W. U. Schröder, J. R. Birkelund, J. R. Huizenga, K. L. Wolf, J. P. Unik, and V. E. Viola, Jr., *Phys. Rev. Lett.* **36**:514 (1976).

(Sch 76b) W. U. Schröder, J. R. Birkelund, J. R. Huizenga, K. L. Wolf, J. P. Unik, and V. E. Viola, Jr., Caen, *Proc. Europ. Conf. Nucl. Phys. with Heavy Ions*, (1976), p. 178.

(Sch 77) G. Schrieder, A. Müller-Arnke, and A. Richter, *Nucl. Phys.* **A279**:463 (1977).

(Sch 77c) W. U. Schröder, J. R. Birkelund, J. R. Huizenga, K. L. Wolf, and V. E. Viola, Jr., *Phys. Rev. C* **16**:623 (1977).

(Sch 77d) W. U. Schröder, J. R. Huizenga, J. R. Birkelund, K. L. Wolf, and V. E. Viola, Jr., *Phys. Lett.* **71B**:283 (1977).

(Sch 77e) W. U. Schröder and J. R. Huizenga, *Ann. Rev. Nucl. Sci.* **27**:465 (1977).

(Sch 77f) R. P. Schmitt, P. Russo, R. Babinet, R. Jared, and L. G. Moretto, *Nucl. Phys.* **A279**:141 (1977).

(Sch 78) W. U. Schröder, J. R. Huizenga, J. R. Birkelund, and W. Wilcke, *Proceedings of Hirschegg Workshop VI*, (1978), p. 85.

(Sch 78c) B. Schürmann, W. Nörenberg, and M. Simbel, *Z. Phy.* **A286**:263 (1978).

(Sch 78e) H. Schultheis and R. Schultheis, *Phys. Rev. C* **18**:1317 (1978).

(Sch 78f) M. Schädel, J. V. Kratz, H. Ahrens, W. Bruchle, G. Franz, H. Gäggeler, I. Warnecke, G. Wirth, G. Herrmann, N. Trautmann, and M. Weis, *Phys. Rev. Lett.* **41**:469 (1978).

(Sch 78g) W. U. Schröder, J. R. Birkelund, J. R. Huizenga, K. L. Wolf, and V. E. Viola, Jr., *Phys. Rep.* **45**:301 (1978).

(Sch 78h) R. P. Schmitt, G. Bizard, G. J. Wozniak, and L. G. Moretto, *Phys. Rev. Lett.* **41**:1152 (1978).

(Sch 78i) R. Schmidt, V. D. Toneev, and G. Wolschin, *Nucl. Phys.* **A311**:247 (1978).

(Sch 78m) R. Schaeffer, *Theoretical Methods in Medium-Energy and Heavy-Ion Physics* (K. W. Mcvoy and W. A. Friedman, eds.), Plenum Press, New York (1978).

(Sch 79) B. Schürmann, *Ann. Phys. (N. Y.)* **123**:102 (1979).

(Sch 79f) W. U. Schröder, *Proc. Int. Symp. on Continuum Spectra*, held in San Antonio, Texas, December, 1979 (T. Tamura, J. B. Natowitz, and D. H. Youngblood, eds.), Harwood Academic Publishers, CHUR, London, New York (1980), Vol. 2, p. 19.

(Sch 80) W. U. Schröder, J. R. Birkelund, J. R. Huizenga, W. W. Wilcke, and J. Randrup, *Phys. Rev. Lett.* **44**:308 (1980).

(Sch 80b) H. Schultheis and R. Schultheis, *Phys. Rev. C* **21**:755 (1980).

(Sch 81b) W. U. Schröder, J. R. Huizenga, and J. Randrup, *Phys. Lett.* **98B**:355 (1981).

(Sch 81c) R. P. Schmitt, G. J. Wozniak, G. U. Rattazzi, G. J. Mathews, R. Regimbart, and L. G. Moretto, *Phys. Rev. Lett.* **46**:522 (1981).

(Sch 81g) W. F. W. Schneider, F. Pühlhofer, R. P. Chestnut, C. Volant, H. Freiesleben, W. Pfeffer, B. Kohlmeyer, *Nucl. Phys.* **A371**:493 (1981).

(Sch 81h) D. Schull, W. C. Shen, H. Freiesleben, R. Bock, F. Busch, D. Bangert, W. Pfeffer, and F. Pühlhofer, *Phys. Lett.* **102B**:116 (1981).

(Sch 82b) R. Schmidt and R. Reif, Preprint, Central Institute for Nuclear Research,

Rossendorf 1982.
(Sch 82d) E. W. Schmid, S. Saito, and H. Fiedeldey, *Z. Phys.* **A306**:37 (1982).
(Sco 76) W. Scobel, H. H. Gutbrod, M. Blann, and A. Mignerey, *Phys. Rev. C* **14**:1808 (1976).
(Ser 47) R. Serber, *Phys. Rev.* **72**:1008 (1947).
(Sey 61) R. G. Seyler and C. H. Blanchard, *Phys. Rev.* **124**:227 (1961).
(Sey 63) R. G. Seyler and C. H. Blanchard, *Phys. Rev.* **131**:355 (1963).
(She 78) A. Sherman, D. Sperber, M. I. Sobel, and J. P. Bondorf, *Z. Phys.* **A286**:11 (1978).
(She 79) J. H. Shea, J. L. Durell, G. S. Foote, and I. S. Grant, *Nucl. Phys.* **A327**:207 (1979).
(Shl 79b) S. Shlomo, B. R. Barrett, and H. A. Weidenmüller, *Phys. Rev. C* **20**:1 (1979).
(Sie 71) P. J. Siemens, J. P. Bondorf, D. H. E. Gross, and F. Dickmann, *Phys. Lett.* **36B**:24 (1971).
(Sie 76) A. J. Sierk and J. R. Nix, *Proc. Symposium Macroscopic Features of Heavy-Ion Collisions*, Argone, 1976, Report ANL/PHY-76-2, Vol. I, p. 407.
(Sie 78) A. J. Sierk, S. E. Koonin, and J. R. Nix, *Phys. Rev. C* **17**:646 (1978).
(Sie 80) A. J. Sierk and J. R. Nix, *Phys. Rev. C* **21**:982 (1980).
(Sik 62) T. Sikkeland, E. L. Haines, and V. E. Viola, Jr., *Phys. Rev.* **125**:1350 (1962).
(Sik 63) T. Sikkeland and V. E. Viola, Jr., Third Conf. on Reactions Between Complex Nuclei Asilomar (1963), p. 232.
(Sik 64) T. Sikkeland, *Phys. Rev.* **135**:669 (1964).
(Sik 70b) T. Sikkeland, *Phys. Lett.* **31B**:451 (1970).
(Sik 71) T. Sikkeland, J. E. Clarkson, Naftali H. Steiger-Shafrir, and V. E. Viola, Jr., *Phys. Rev. C* **3**:329 (1971).
(Sik 80) B. Sikora, M. Blann, W. Scobel, J. Bisplinghoff, and M. Beckerman, *Phys. Rev. C* **21**:614 (1980).
(Sik 82) B. Sikora, J. Bisplinghoff, M. Blann, W. Scobel, M. Beckerman, F. Plasil, R. L. Ferguson, J. Birkelund, and W. Wilcke, *Phys. Rev. C* **25**:686 (1982).
(Sin 75) B. Sinha, *Phys. Rev. C* **20**:1 (1975).
(Sin 76b) B. C. Sinha and P. J. Siemens, *Phys. Lett.* **65B**:5 (1976).
(Sin 81) B. Sinha, *Phys. Lett.* **99B**:89 (1981).
(Siw 76) K. Siwek-Wilczyńska and J. Wilczyński, *Nucl. Phys.* **A264**:115 (1976).
(Siw 79) K. Siwek-Wilczyńska, E. H. Du Marchie Van Voorthuysen, J. Van Popta, R. H. Siemssen, and J. Wilczyński, *Proceedings of Hirschegg Workshop VI*, (1979), p. 111.
(Siw 79b) K. Siwek-Wilczyńska, E. H. Du Marchie Van Voorthuysen, J. Van Popta, R. H. Siemssen, and J. Wilczyński, *Phys. Rev. Lett.* **42**:1599 (1979).
(Siw 79c) K. Siwek-Wilczyńska, E. H. Du Marchie Van Voorthuysen, J. Van Popta, R. H. Siemssen, and J. Wilczyński, *Nucl. Phys.* **A330**:150 (1979).
(Sko 79) N. K. Skobelev, V. Z. Maidikov, G. S. Popeko, Yu. V. Gofman, O. F. Nemets, and Yu. Ts. Oganesyan, *Sov. J. Nucl. Phys.* **29**:316 (1979).
(Sob 81) L. G. Sobotka, C. C. Hsu, G. J. Wozniak, G. U. Rattazzi, R. J. Mcdonald, A. J. Pacheco, and L. G. Moretto, *Phys. Rev. Lett.* **46**:887 (1981).
(Sob 81b) L. G. Sobotka, C. C. Hsu, G. J. Wozniak, D. J. Morrissey, and L. G. Moretto, *Nucl. Phys.* **A371**:510 (1981).
(Sob 83) L. G. Sobotka, R. J. Mcdonald, G. J. Wozniak, D. J. Morrissey, A. J. Pacheco, and L. G. Moretto, *Phys. Rev. C* **28**:219 (1983).
(Spe 74) D. Sperber, *Phys. Scr.* **10A**:115 (1974).
(Spe 75) D. Sperber, *Nukleonika* **20**:755 (1975).
(Spe 79) P. Sperr, H. Essel, H. J. Körner, and K. E. Rehm, *Z. Phys.* **A291**:179 (1979).
(Spe 79b) H. J. Specht, *Proc. Int. Symp. on Continuum Spectra*, held in San Antonio,

Texas, December, 1979 (T. Tamura, J. B. Natowitz, and D. H. Youngblood, eds.), Harwood Academic Publishers, CHUR, London, New York (1980), Vol. 2, p. 341.

(Spe 80b) P. Sperr, H. J. Körner, W. A. Mayer, and W. Wagner, *Z. Phys.* **A297**:355 (1980).

(Spe 81) H. J. Specht, Invited Lecture, V General Conference of the EPS, Istanbul, Turkey 1981.

(Ste 72) E. P. Steinberg, S. B. Kaufman, B. D. Wilkins, C. E. Gross, and M. J. Fluss, *Nucl. Instrum. Methods* **99**:309 (1972).

(Ste 75) R. M. Steffen and K. Alder, *The Electromagnetic Interaction in Nuclear Spectroscopy* (W. D. Hamilton, ed.), North-Holland, Amsterdam (1975), p. 505.

(Ste 81b) H. Stelzer, *Nucl. Phys.* **A354**:433C (1981).

(Sto 73) W. Stocker and J. Burzlaff, *Nucl. Phys.* **A202**:265 (1973).

(Sto 77c) R. G. Stokstad, *Proc. Fall Creek Falls*, Tenn., Conf-770602 (1977), p. 22.

(Sto 79) H. Stocker, J. A. Maruhn, and W. Greiner, *Z. Phys.* **A293**:173 (1979).

(Str 56) V. M. Strutinsky, *Sov. Phys. JETP* **3**:638 (1956).

(Str 60) V. M. Strutinsky, *Sov. Phys. JETP* **10**:613 (1960).

(Str 63) V. M. Strutinskii, N. Ya. Lyashchenko, and N. A. Popov, *Sov. Phys. JETP* **16**:418 (1963).

(Str 64) V. M. Strutinskii, *Sov. Phys.* **19**:1401 (1964).

(Str 67) V. M. Strutinsky, *Nucl. Phys.* **A95**:420 (1967).

(Str 68) V. M. Strutinsky, *Nucl. Phys.* **A122**:1 (1968).

(Str 73) V. M. Strutinsky, *Phys. Lett.* **44B**:245 (1973).

(Str 78d) V. Strutinsky, *Z. Phys.* **A286**:77 (1978).

(Str 80b) V. M. Strutinsky, F. A. Ivanyuk, and V. V. Pashkevich, *Sov. J. Nucl. Phys.* **31**:46 (1980).

(Sug 77) K. Sugimoto, N. Takahashi, A. Mizobuchi, Y. Nojiri, T. Minamisono, M. Ishihara, K. Tanaka, and H. Kamitsubo, *Phys. Rev. Lett.* **39**:323 (1977).

(Sve 76) J. S. Sventek and L. G. Moretto, *Phys. Lett.* **65B**:326 (1976).

(Sve 78) J. S. Sventek and L. G. Moretto, *Phys. Rev. Lett.* **40**:697 (1978).

(Swi 50) W. J. Swiatecki, *Proc. Phys. Soc. (London)* **A63**:1208 (1950).

(Swi 51) W. J. Swiatecki, *Proc. Phys. Soc.* **A64**:226 (1951).

(Swi 72) W. J. Swiatecki, *J. Phys. (Paris) Colloq. C5* **37**:45 (1972).

(Swi 72b) W. J. Swiatecki and S. Bjørnholm, *Phys. Rep.* **4**:325 (1972).

(Swi 75) W. J. Swiatecki, Lawrence Berkeley Laboratory report LBL-4296 (1975), unpublished.

(Swi 76) Z. E. Switkowski, R. G. Stokstad, and R. M. Wieland, *Nucl. Phys.* **A274**:202 (1976).

(Swi 79) W. J. Swiatecki, Lawrence Berkeley Laboratory report LBL-8950 (1979), Unpublished (Lectures at Int. School of Nucl. Phys., Sicily, 1979).

(Swi 80) W. J. Swiatecki, *Prog. Part. Nucl. Phys.* **4**:383 (1980).

(Swi 81c) W. J. Swiatecki, *Phys. Scr.* **24**:113 (1981).

(Swi 82b) W. J. Swiatecki, *Nucl. Phys.* **A376**:275 (1982).

(Sym 80) T. J. M. Symons, P. Doll, M. Bini, D. L. Hendrie, J. Mahoney, G. Mantzouranis, D. K. Scott, K. Van Bibber, Y. P. Viyogi, H. H. Wieman, and C. K. Gelbke, *Phys. Lett.* **94B**:131 (1980).

(Tak 78) N. Takahashi, Y. Miake, Y. Nojiri, T. Minamisono, A. Mizobuchi, M. Ishihara, and K. Sugimoto, *Phys. Lett.* **78B**:397 (1978).

(Tak 79) N. Takigawa, *Nucl. Phys.* **A315**:186 (1979).

(Tak 80d) N. Takahashi, Y. Miake, K. H. Tanaka, Y. Nojiri, T. Minamisono and K.

Sugimoto, *Proc. Int. Sym. on Continuum Spectra*, held in San Antonio, Texas, December, 1979 (T. Tamura, J. B. Natowitz, and D. H. Youngblood, eds.), Harwood Academic Publishers, CHUR, London, New York (1980), Vol. 2, p. 109.

(Tak 81) E. Takada, T. Shimoda, N. Takahashi, T. Yamaya, K. Nagatani, T. Udagawa, and T. Tamura, *Phys. Rev. C* **23**:772 (1981).

(Tak 81b) N. Takigawa, K. Niita, Y. Okuhara, and S. Yoshida, *Nucl. Phys.* **A371**:130 (1981).

(Tam 75) B. Tamain, C. Ngô, J. Péter, and F. Hanappe, *Nucl. Phys.* **A252**:187 (1975).

(Tam 76) B. Tamain, F. Plasil, C. Ngô, J. Péter, M. Berlanger, and F. Hanappe, *Phys. Rev. Lett.* **36**:18 (1976).

(Tam 76) B. Tamain, F. Plasil, C. Ngô, J. Péter, M. Berlanger, and F. Hanappe, *Phys. Rev. Lett.* **36**:18 (1976).

(Tam 77) T. Tamura, T. Udagawa, D. H. Feng, and K.-K. Kan, *Phys. Lett.* **66B**:109 (1977).

(Tam 77b) T. Tamura and T. Udagawa, *Phys. Lett.* **71B**:273 (1977).

(Tam 79) B. Tamain and J. Péter, *Proc. Int. Symp. on Continuum Spectra*, held in San Antonio, Texas, December, 1979 (T. Tamura, J. B. Natowitz, and D. H. Youngblood, eds.), Harwood Academic Publishers, CHUR, London, New York (1980), Vol. 2, p. 281.

(Tam 79b) B. Tamain, R. Chechik, H. Fuchs, F. Hanappe, M. Morjean, C. Ngô, J. Péter, M. Dakowski, B. Lucas, C. Mazur, M. Ribrag, and C. Signarbieux, *Nucl. Phys.* **A330**:253 (1979).

(Tan 80c) T. Tanabe, R. Bock, M. Dakowski, A. Gobbi, H. Sann, H. Stelzer, U. Lynen, A. Olmi, and D. Pelte, *Nucl. Phys.* **A342**:194 (1980).

(Tas 76) L. Tassangot, N. Frascaria, J. P. Garron, J. C. Jacmart, N. Poffe, and C. Stephan, Caen, *Proc. Europ. Conf. Nucl. Phys. with Heavy Ions*, (1976), p. 174.

(Ter 59) J. Terrell, *Phys. Rev.* **113**:527 (1959).

(Tho 64) T. D. Thomas, *Nucl. Phys.* **53**:577 (1964).

(Tho 74) S. G. Thompson, L. G. Moretto, R. C. Jared, R. P. Babinet, J. Galin, M. M. Fowler, R. C. Gatti, and J. B. Hunter, *Phys. Scr.* **10A**:36 (1974).

(Tho 77) C. E. Thorn, M. J. Levine, J. J. Kolata, C. Flaum, P. D. Bond, and J. C. Sens, *Phys. Rev. Lett.* **38**:384 (1977).

(Tjo 74) P. O. Tjom, F. S. Stephens, R. M. Diamond, J. De Boer, and W. E. Meyerhof, *Phys. Rev. Lett.* **33**:593 (1974).

(Tra 75) W. Trautmann, D. Proetel, O. Häusser, W. Hering, and F. Riess, *Phys. Rev. Lett.* **35**:1694 (1975).

(Tra 77b) W. Trautmann, J. De Boer, W. Dünnweber, G. Graw, R. Kopp, C. Lauterbach, H. Puchta, and U. Lynen, *Phys. Rev. Lett.* **39**:1062 (1977).

(Tra 79) W. Trautmann, J. F. Sharpey-Schaefer, H. R. Andrews, B. Haas, O. Haüsser, P. S. Taras, and D. Ward, *Phys. Rev. Lett.* **43**:991 (1979).

(Tra 80) W. Trautmann, *Proc. Int. Symp. on Continuum Spectra*, held in San Antonio, Texas, December, 1979 (T. Tamura, J. B. Natowitz, and D. H. Youngblood, eds.), Harwood Academic Publishers, CHUR, London, New York (1980), Vol. 2, p. 115.

(Tra 81b) W. Trautmann, W. Dahme, W. Dünnweber, W. Hering, C. Lauterbach, H. Puchta, W. Kuhn, and J. P. Wurm, *Phys. Rev. Lett.* **46**:1188 (1981).

(Tra 81c) W. Trautmann, C. Lauterbach, J. Deboer, W. Dünnweber, G. Graw, W. Hamann, W. Hering, and H. Puchta, *Nucl. Instrum. Methods* **184**:449 (1981).

(Tra 81d) W. Trautmann, Habilitationsschrift, Universität München (1981).

(Tra 81e) W. Trautmann, W. Dünnweber, W. Hering, C. Lauterbach, H. Puchta, R. Ritzka, and W. Trombik, *Proc. Int. Winter Meeting on Nuclear Physics*, Bormio, 1981, p. 241.

(Tre 80) S. Trentalange, S. E. Koonin, and A. J. Sierk, *Phys. Rev. C* **22**:1159 (1980).

(Tri 81) H. Tricoire, C. Marty, and D. Vautherin, *Phys. Lett.* **100B**:109 (1981).

(Tsa 74) C. F. Tsang, *Phys. Scr.* **10A**:90 (1974).

(Tsa 81) M. B. Tsang, W. G. Lynch, R. J. Puigh, R. Vandenbosch, and A. G. Seamster, *Phys. Rev. C* **23**:1560 (1981).

(Tsa 83) M. B. Tsang, H. Utsunomiya, C. K. Gelbke, W. G. Lynch, B. B. Back, S. Saini, P. A. Baisden, and M. A. McMahan, *Phys. Lett.* **129**:18 (1983).

(Tsa 83b) M. B. Tsang, D. Ardouin, C. K. Gelbke, W. G. Lynch, Z. R. Xu, B. B. Back, R. Betts, S. Saini, P. A. Baisden, and M. A. McMahan, *Phys. Rev. C* **20**:747 (1983).

(Tse 81) I. Tserruya, A. Breskin, R. Chechik, Z. Fraenkel, S. Wald, N. Zwang, R. Bock, M. Dakowski, A. Gobbi, H. Sann, R. Bass, G. Kreyling, R. Renfordt, K. Stelzer, and U. Arlt, *Phys. Rev. Lett.* **47**:16 (1981).

(Uda 78) T. Udagawa and T. Tamura, *Phys. Rev. Lett.* **41**:1770 (1978).

(Uda 79) T. Udagawa and T. Tamura, *Phys. Rev. C* **20**:1949 (1979).

(Uda 79b) T. Udagawa and T. Tamura, *Proc. Int. Sym. on Continuum Spectra*, held in San Antonio, Texas, December, 1979, (T. Tamura, J. B. Natowitz, and D. H. Youngblood, eds.), Harwood Academic Publishers, CHUR, London, New York, (1980), Vol. 2, p. 155.

(Uda 80) T. Udagawa and T. Tamura, *Phys. Rev. Lett.* **45**:1311 (1980).

(Van 73) R. Vandenbosch and J. R. Huizenga, *Nuclear Fission*, Academic Press, New York (1973).

(Van 76) R. Vandenbosch, M. P. Webb, and T. D. Thomas, *Phys. Rev. Lett.* **36**:459 (1976).

(Van 76b) R. Vandenbosch, M. P. Webb, and T. D. Thomas, *Phys. Rev. C* **14**:743 (1976).

(Van 76c) R. Vandenbosch, M. P. Webb, T. D. Thomas, and M. S. Zisman, *Nucl. Phys.* **A269**:210 (1976).

(Van 76d) R. Vandenbosch, M. P. Webb, T. D. Thomas, S. W. Yates, and A. M. Friedman, *Phys. Rev. C* **13**:1893 (1976).

(Van 77) K. Van Bibber, R. Ledoux, S. G. Steadman, F. Videbaek, G. Young, and C. Flaum, *Phys. Rev. Lett.* **38**:334 (1977).

(Van 78) R. Vandenbosch, M. P. Webb, P. Dyer, R. J. Puigh, R. Weisfield, T. D. Thomas, and M. S. Zisman, *Phys. Rev. C* **17**:1672 (1978).

(Van 79b) R. Vandenbosch, *Phys. Rev. C* **20**:171 (1979).

(Van 80e) Mguyen Van Sen, J. C. Gondrand, F. Merchez, and R. Darves-Blanc, *Phys. Rev. C* **22**:2424 (1980).

(Var 73) J. P. Vary and C. B. Dover, *Phys. Rev. Lett.* **31**:1510 (1973).

(Vaz 78b) L. C. Vaz, J. M. Alexander, *Phys. Rev. C* **18**:833 (1978).

(Vaz 81) L. C. Vaz, J. M. Alexander, and G. R. Satchler, *Phys. Rep.* **69**:373 (1981).

(Ven 63) R. H. Venter, *Ann. Phys. (N. Y.)* **25**:405 (1963).

(Via 79) J. B. Viano, A. J. Cole, N. Longequeue, J. J. Lucas, J. Menet, J. C. Saulnier, and J. W. Sunier, *Phys. Rev. C* **20**:551 (1979).

(Vid 77) F. Videbaek, R. B. Goldstein, L. Grodzins, S. G. Steadman, T. A. Belote, and J. D. Garret, *Phys. Rev. C* **15**:954 (1977).

(Vig 76) S. E. Vigdor, *Proc. Symposium Macroscopic Features of Heavy-Ion Collisions*, Argonne, 1976, Report ANL/PHY-76-2, Vol. 1, p. 95.

(Vig 80) S. E. Vigdor, H. J. Karwowski, W. W. Jacobs, S. Kailas, P. P. Singh, F. Soga, and P. Yip, *Phys. Lett.* **90B**:384 (1980).

(Vio 62) V. E. Viola and T. Sikkeland, *Phys. Rev.* **128**:767 (1962).

(Vio 63) V. E. Viola, Jr., T. Darrah Thomas, and G. T. Seaborg, *Phys. Rev.* **129**:2710 (1963).

(Vio 63b) V. E. Viola, Jr., and T. Sikkeland, *Phys. Rev.* **130**:2044 (1963).

(Vio 66) V. E. Viola, Jr., *Nucl. Data* **A1**:391 (1966).

(Vio 76) V. E. Viola, Jr., R. G. Clark, W. G. Meyer, A. M. Zebelman, and R. G. Sextro, *Nucl. Phys.* **A261**:174 (1976).

(Vio 80) V. E. Viola, Jr., A. C. Mignerey, H. Breuer, K. L. Wolf, B. G. Glagola, W. W. Wilke, W. U. Schröder, J. R. Huizenga, and D. Hilscher, *Phys. Rev. C* **22**:122 (1980).

(Vio 80b) V. E. Viola, Jr., A. C. Mignerey, H. Breuer, K. L. Wolf, B. G. Glagola, W. W. Wilcke, W. U. Schröder, J. R. Huizenga, D. Hilscher, and J. R. Birkelund, *Phys. Rev. C* **22**:122 (1980).

(Vio 81) V. E. Viola, K. Kwiatkowski, H. Breuer, A. C. Mignerey, K. L. Wolf, B. G. Glagola, J. R. Birkelund, D. Hilscher, J. R. Huizenga, W. U. Schröder, and W. W. Wilcke, Preprint: Int. Workshop on Gross Properties of Nucl. and Nucl. Excitations Hirschegg (1981).

(Vol 74) V. V. Volkov, *Proc. Int. Conf. React. Complex Nuclei*, North-Holland, Amsterdam (1974), p. 363.

(Vol 74b) V. V. Volkov, *Proc. Symp. on Classical and Quantum Mechanical Aspects of Heavy-Ion Collisions*, Springer-Verlag, Berlin (1974), Vol. 33, p. 254.

(Vol 75) C. Volant, M. Conjeaud, S. Harar, S. M. Lee, A. Lepine, and E. F. Da Silveira, *Nucl. Phys.* **A238**:120 (1975).

(Vol 75b) V. V. Volkov, Workshop on Effects of Dynamics in Close Collisions Between Complex Nuclei, Gif-Sur-Yvette (1975).

(Vol 75c) V. V. Volkov, *Part. Nucl.* **6**:1040 (1975).

(Vol 76) V. V. Volkov, *Sov. J. Part. Nucl.* **6**:420 (1976).

(Vol 76b) V. V. Volkov, *Nukleonika* **21**:53 (1976).

(Vol 78) V. V. Volkov, *Phys. Rep.* **44**:93 (1978).

(Vol 80) V. V. Volkov, A. G. Artukh, G. F. Gridnev, M. Gruszecki, W. Karcz, A. N. Mezentsev, V. L. Mikheev, L. Polerski, A. Popescu, and D. G. Popescu, Proceedings of Hirschegg Workshop VIII (1980), p. 138.

(Was 79) P. Wastyn, H. Genz, M. Mutterer, A. Richter, G. Schrieder, J. C. Vanstaden, and J. P. Theobald, *Nucl. Phys.* **A320**:433 (1979).

(Was 79b) P. Wastyn, H. Feldmeier, F. Beck, M. Dworzecka, H. Genz, M. Mutterer, A. Richter, G. Schrieder, and J. P. Theobald, *Nucl. Phys.* **A332**:455 (1979).

(Web 76b) M. P. Webb, R. Vandenbosch, and T. D. Thomas, *Phys. Lett.* **62B**:407 (1976).

(Web 77) M. P. Webb, Univ. of Washington, Ph.D. thesis (1977), unpublished.

(Weg 74b) G. Wegmann, *Phys. Lett.* **50B**:327 (1974).

(Wei 35) C. F. von Weizsacker, *Z. Phys.* **96**:431 (1935).

(Wei 77) H. A. Weidenmüller, Proc. Fall Creek Falls, Tenn., Conf-770602 (1977).

(Wei 77b) H. A. Weidenmüller, Max Planck Institute, Heidelberg, Report MPI-H-1977-V30 (1977).

(Wei 77c) R. Weiner and M. Westrom, *Nucl. Phys.* **A286**:282 (1977).

(Wei 78) H. A. Weidenmüller, *Theoretical Methods in Medium Energy and Heavy Ion Physics* (K. W. McVoy and W. A. Friedman, eds.), Plenum Press, New York (1978).

(Wei 80) H. A. Weidenmüller, *Progress in Particle and Nuclear Physics*, Vol. 3 (D. H. Wilkinson, ed.), Pergamon, Oxford (1980).

(Wer 81) E. Werner, H. S. Wio, H. Hofmann, and K. Pomorski, *Z. Phys.* **A299**:231 (1981).

(Wes 78) G. D. Westfall, R. G. Sextro, A. M. Poskanzer, A. M. Zebelman, G. W. Butler, and E. K. Hyde, *Phys. Rev. C* **17**:1368 (1978).

(Wes 78d) L. Westerberg, D. G. Sarantites, D. C. Hensley, R. A. Dayras, M. L. Halbert, and J. H. Barker, *Phys. Rev. C* **18**:796 (1978).

(Wes 82) W. Westmeier, R. A. Esterlund, A. Rox, and P. Pazelt, *Phys. Lett.* **117B**:163 (1982).

(Wil 71) B. D. Wilkins, M. J. Fluss, S. B. Kaufman, C. E. Gross, and E. P. Steinberg, *Nucl. Instrum. Methods* **92**:381 (1971).

(Wil 73b) J. Wilczyński, *Phys. Lett.* **47B**:124 (1973).

(Wil 73c) J. Wilczyński, *Phys. Lett.* **47B**:484 (1973).

(Wil 75b) J. Wilczyński, K. Siwek-Wilczyńska, J. S. Larsen, J. C. Acquadro, and P. R. Christensen, *Nucl. Phys.* **A244**:147 (1975).

(Wil 76) J. Wilczyński and K. Siwek-Wilczyńska, *Nukleonika* **21**:517 (1976).

(Wil 76b) J. Wilczyński, *Proc. Symposium Macroscopic Features of Heavy-Ion Collisions*, Argonne, 1976, Report ANL/PHY-76-2, Vol. I, p. 211.

(Wil 79b) J. Wilczyński, *Proc. Sym. on Deep-Inelastic and Fusion Reactions with Heavy Ions*, Springer-Verlag, Berlin (1979), Vol. 117, p. 254.

(Wil 79d) J. Wilczyński, R. Kamermans, J. van Popta, R. H. Siemssen, K. Siwek-Wilczyńska, and S. Y. Van Der Werf, *Phys. Lett.* **88B**:65 (1979).

(Wil 80) W. W. Wilcke, J. R. Birkelund, A. D. Hoover, J. R. Huizenga, W. U. Schröder, V. E. Viola, Jr., K. L. Wolf, and A. C. Mignerey, *Phys. Rev. C* **22**:128 (1980).

(Wil 80b) J. Wilczyński, K. Siwek-Wilczyńska, J. Van Driel, S. Gonggrijp, D. C. J. M. Hageman, R. V. F. Janssens, J. Lukasiak, and R. H. Siemssen, *Phys. Rev. Lett.* **45**:606 (1980).

(Wil 80d) W. W. Wilcke, J. R. Birkelund, H. J. Wollersheim, A. D. Hoover, J. R. Huizenga, W. U. Schröder, and L. E. Tubbs, *Nucl. Data Tables* **25**:389 (1980).

(Wil 81) W. Willis, Report CERN-Ep/81-120 (1981).

(Wil 82) J. Wilczyński, K. Siwek-Wilczyńska, J. van Driel, S. Gonggrijp, D. C. J. M. Hageman, R. V. F. Janssens, J. Lukasiak, R. H. Siemssen, and S. Y. van Der Werf, *Nucl. Phys.* **A373**:109 (1982).

(Wil 83) W. W. Wilcke, J. P. Kosky, J. R. Birkelund, M. A. Butler, A. D. Dougan, J. R. Huizenga, W. U. Schröder, and H. J. Wollersheim, and D. Hilscher, *Phys. Rev. Lett.* **51**:99 (1983).

(Win 81) U. Winkler, R. Giraud, H. Graf, A. Karbach, R. Novotny, D. Pelte, and G. Strauch, *Nucl. Phys.* **A371**:477 (1981).

(Woj 77) H. Wojciechowski, L. R. Medsker, and R. H. Davis, *Phys. Rev C* **16**:1767 (1977).

(Wol 51) L. Wolfenstein, *Phys. Rev.* **82**:690 (1951).

(Wol 74) K. L. Wolf, J. P. Unik, J. R. Huizenga, J. R. Birkelund, H. Freiesleben, and V. E. Viola, *Phys. Rev. Lett.* **33**:1105 (1974).

(Wol 76) K. L. Wolf and C. T. Roche, *Proc. Symposium Macroscopic Features of Heavy-Ion Collisions*, Argonne, 1976, Report ANL/PHY-76, Vol. I, p. 295.

(Wol 76b) K. L. Wolf, C. T. Roche, R. L. Boudrie, J. R. Huizenga, J. R. Birkelund, W. U. Schröder, and V. E. Viola, Jr., Caen, Proc. Europ. Conf. Nucl. Phys. with Heavy Ions (1976), p. 176.

(Wol 77) G. Wolschin, *Nukleonika* **22**:1165 (1977).

(Wol 78) G. Wolschlin and W. Nörenberg, *Phys. Rev. Lett.* **41**:691 (1978).

(Wol 78b) G. Wolschin and W. Nörenberg, *Z. Phys.* **A284**:209 (1978).

(Wol 78c) G. Wolschin, *Nucl. Phys.* **A316**:146 (1979).

(Wol 79b) G. Wolschin, *Phys. Lett.* **88B**:35 (1979).

(Wol 79d) G. Wolschin, Presented at Int. School of Physics "Enrico Fermi," Vienna, Italy (1979).

(Wol 81) H. J. Wollersheim, W. W. Wilcke, J. R. Birkelund, J. R. Huizenga, W. U. Schröder, H. Freiesleben, and D. Hilscher, *Phys. Rev. C* **24**:2114 (1981).

(Wol 81b) G. Wolschin, *Proc. of the Nuclear Physics Workshop*, I. C. T. P., Trieste, Italy (October 1981), Nuclear Physics, (C. H. Dasso, ed.) North-Holland, Amsterdam (1982), p. 669.

(Wol 82) G. Wolschin, *Phys. Rev. Lett.* **48**:1004 (1982).

(Wol 82d) H. J. Wollersheim, W. W. Wilcke, J. R. Birkelund, and J. R. Huizenga, *Phys. Rev. C* **25**:338 (1982).

(Won 68) C. Y. Wong, *Phys. Lett.* **26B**:120 (1968).

(Won 78c) C. Y. Wong and H. H. K. Tang, *Phys. Rev. Lett.* **40**:1070 (1978).

(Won 79) C. Y. Wong and H. H. K. Tang, *Phys. Rev. C* **20**:1419 (1979).

(Won 79d) C. Y. Wong and H. H. K. Tang, Presented at the TDHF Workshop, Paris (1979).

(Won 79f) C. Y. Wong and C. Toepffer, Proc. Conf. Large Ampl. Coll. Nucl. Motion Hungary, 1979.

(Woz 78) G. J. Wozniak, R. P. Schmitt, P. Glässel, and L. G. Moretto, *Nucl. Phys.* **A298**:169 (1978).

(Woz 78b) G. J. Wozniak, R. P. Schmitt, P. Glässel, R. C. Jared, G. Bizard, and L. G. Moretto, *Phys. Rev. Lett.* **40**:1436 (1978).

(Woz 80) G. J. Wozniak, R. J. McDonald, A. J. Pacheco, C. C. Hsu, D. J. Morrissey, L. G. Sobotka, L. G. Moretto, S. Shih, C. Schuck, R. M. Diamond, H. Kluge, and F. S. Stephens, *Phys. Rev. Lett.* **45**:1081 (1980).

(Wu 78) J. R. Wu, C. C. Chang, and H. D. Holmgren, *Phys. Rev. Lett.* **40**:1013 (1978).

(Wu 79) J. R. Wu, C. C. Chang, H. D. Holmgren, and R. W. Koontz, *Phys. Rev. C* **20**:1284 (1979).

(Wu 80c) Wu Zhong Li, Zhu Yong Tai, Xia Guo Zhong, Liu Bu Sheng, Li Fa Wei, and Fan Guo Ying, Proc. Int. Conf., Berkeley, LBL-11118 (1980), p. 512.

(Wu 81) E. C. Wu, K. D. Hildenbrand, H. Freiesleben, A. Gobbi, A. Olmi, H. Sann, and U. Lynen, GSI Darmstadt, preprint GSI-81-30 (1981).

(Wur 80) J. P. Wurm, *Proc. Int. Symp. on Continuum Spectra*, held in San Antonio, Texas, December, 1979, (T. Tamura, J. B. Natowitz, and D. H. Youngblood, eds.) Harwood Academic Publishers, CHUR, London, New York (1980), Vol. 2, p. 227.

(Yam 76) S. Yamaji, W. Scheid, H. J. Fink, and W. Greiner, *Z. Phys.* **A278**:69 (1976).

(Yam 76b) S. Yamaji, W. Scheid, H. J. Fink, and W. Greiner, *J. Phys. G* **2**:L189 (1976).

(Yam 77) S. Yamaji, K.-H. Ziegenhain, H. J. Fink, W. Greiner, and W. Scheid, *J. Phys. G* **3**:1283 (1977).

(Yam 79) H. Yamada, D. R. Zolnowski, S. E. Cala, A. C. Kahler, J. Pierce, and T. T. Sugihara, *Phys. Rev. Lett.* **43**:605 (1979).

(Yam 80) S. Yamaji, A. Iwamoto, K. Harada, and S. Yoshida, *Proc. Int. Conf.*, Berkeley, LBL-11118 (1980) p. 396.

(Yas 76) K. Yasue, *Phys. Lett.* **64B**:239 (1976).

(You 76) G. Young, E. Blum, A. J. Lazzarini, M. A. Neuhausen, S. G. Steadman, and N. Tsoupas, *Proc. Symposium Macroscopic Features of Heavy-Ion Collisions*, Argonne, 1976, Report ANL/PHY-76-2, Vol. I, p. 849.

(You 80) G. R. Young, R. L. Ferguson, A. Gavron, D. C. Hensley, F. E. Obenshain, F. Plasil, A. H. Snell, M. P. Webb, C. F. Maquire, and G. A. Petitt, *Phys. Rev. Lett.* **45**:1389 (1980).

(Zab 78) J. G. Zabolitzky and W. Ey, *Phys. Lett.* **76B**:527 (1978).

(Zeb 74) A. M. Zebelman, L. Kowalski, J. Miller, K. Beg, Y. Eyal, G. Jaffe, A. Kandil, and D. Logan, *Phys. Rev. C* **10**:200 (1974).

(Zie 74) M. Zielinska-Pfabe and K. Dietrich, *Phys. Lett.* **49B**:123 (1974).

(Zie 79) R. Zierl, H. Essel, P. Kienle, H. J. Korner, K. E. Rehm, P. Sperr, and W. Wagner, *Z. Phys.* **A291**:91 (1979).

(Zin 75) P. G. Zint and U. Mosel, *Phys. Lett.* **56B**:424 (1975).
(Zin 75b) P. G. Zint and U. Mosel, *Phys. Lett.* **58B**:269 (1975).
(Zin 76) P. G. Zint and U. Mosel, *Phys. Rev. C* **14**:1488 (1976).
(Zin 77) P. G. Zint, *Z. Phys.* **A281**:373 (1977).
(Zoh 75) O. Zohni, J. Maruhn, W. Scheid, and W. Greiner, *Z. Phys.* **A275**:235 (1975).
(Zoh 76) O. Zohni, *Nukleonika* **21**:801 (1976).
(Zoh 78) O. Zohni and M. Blann, *Nucl. Phys.* **A297**:163 (1978).
(Zoh 78b) O. Zohni, C. Cerruti, M. Namboodiri, O. Hartin, and J. B. Natowitz, preprint Texas A & M (1978).
(Zoh 79) O. Zohni, Seminar Report, USDOE-Ames Lab., Iowa (1979).
(Zoh 81) O. Zohni, *Phys. Rev. A* **23**:2297 (1981).
(Zol 78) D. R. Zolnowski, H. Yamada, S. E. Cal, A. C. Kahler, and T. T. Sugihara, *Phys. Rev. Lett.* **41**:92 (1978).
(Zwa 61) R. Zwanzig, *Lectures in Theoretical Physics*, (W. E. Britten, W. B. Downs, and J. Downs, eds.) Wiley and Sons, Interscience, New York and London, Vol. III: 106 (1961).

Index